THE x86 PC

Assembly Language, Design, and Interfacing

Fifth Edition

Muhammad Ali Mazidi

Janice Gillispie Mazidi

Danny Causey

Prentice Hall

Boston Columbus Indianapolis New York San Francisco Upper Saddle River

Amsterdam Cape Town Dubai London Madrid Milan Munich Paris Montreal Toronto

Delhi Mexico City Sao Paulo Sydney Hong Kong Seoul Singapore Taipei Tokyo

Editor in Chief: Vernon Anthony
Acquisitions Editor: Wyatt Morris
Editorial Assistant: Chris Reed
Director of Marketing: David Gesell
Marketing Manager: Kara Clark
Marketing Assistant: Les Roberts
Senior Managing Editor: JoEllen Gohr
Project Manager: Rex Davidson
Senior Operations Supervisor: Pat Tonneman
Operations Specialist: Laura Weaver
Art Director: Candace Rowley
Cover Designer: Rachel Hirschi
Cover Art: Shutterstock
Printer/Binder: Edwards Brothers
Cover Printer: Lehigh-Phoenix Color
Text Font: Times Roman

Library of Congress Cataloging in Publication Data

Mazidi, Muhammad Ali.
The x86 PC : Assembly language, design, and interfacing / Muhammad Ali Mazidi, Janice Gillispie Mazidi, Danny Causey. -- 5th ed.
p. cm.
Includes bibliographical references and index.
1. Intel 80x86 series microprocessors--Programming. 2. Assembler language (Computer program language) 3. IBM microcomputers--Programming. I. Mazidi, Janice Gillispie. II. Causey, Danny. III. Title. IV. Title: Assembly language, design, and interfacing.
QA76.8.I292M37 2010
005.265--dc22

2009016076

Prentice Hall
is an imprint of
PEARSON

www.pearsonhighered.com

10 9 8 7 6 5 4 3 2 1
ISBN-10: 0-13-502648-2
ISBN-13: 978-0-13-502648-9

Trademark Information and Acknowledgments

All the figures, tables, and instructions related to the x86 family of microprocessors used in this textbook belong to Intel. Copyright of Intel Corporation.

MS-DOS and Windows are trademarks of Microsoft Corp.

PC AT and PS/2 are registered trademarks of IBM Corporation.

Regard man as a mine rich in gems of inestimable value. Education can, alone, cause it to reveal its treasures, and enable mankind to benefit therefrom.

Baha'u'llah

BRIEF CONTENTS

CHAPTERS

APPENDICES

CONTENTS

DEDICATIONS

This book is dedicated to the memory of Muhammad Ali's parents, who raised 10 children and persevered through more than 50 years of hardship together with dignity and faith.

We feel especially blessed to have the support, love, and encouragement of Janice's parents, whose kindness, wisdom, and sense of humor have been the bond that has welded us into a family.

In addition, we must also mention our two most important collaborations: our sons Robert Nabil and Michael Jamal, who have taught us the meaning of love and patience.

We would also like to honor the memory of a dear friend, Kamran Lotfi.

– Muhammad Ali Mazidi
and Janice Mazidi

PREFACE

Purpose

This book is intended for use in college-level courses in which both Assembly language programming and x86 PC interfacing are discussed. It not only builds the foundation of Assembly language programming, but also provides a comprehensive treatment of x86 PC design and interfacing for students in engineering and computer science disciplines. This volume is intended for those who wish to gain an in-depth understanding of the internal working of the x86 PC. It builds a foundation for the design and interfacing of microprocessor-based systems using the real-world example of the x86 PC. In addition, it can also be used by practicing technicians, hardware engineers, computer scientists, and hobbyists who want to do PC interfacing and data acquisition.

Prerequisites

Readers should have taken an introductory digital course. Knowledge of other programming languages would be helpful, but is not necessary.

Although a vast majority of current PCs use x86 such as Pentium microprocessors, their design is based on the IBM PC/AT, an 80286 microprocessor system introduced in 1984. A good portion of the features of the PC/AT, hence its limitations, are based on the original IBM PC, an 8088 microprocessor system, introduced in 1981. In other words, one cannot expect to understand fully the architectural philosophy of the x86 PC and its internal architecture unless the 80286 PC/AT and its subset, the IBM PC/XT, are first understood. For this reason, we describe the 8088 and 80286 microprocessors in Chapter 9.

Contents

A systematic, step-by-step approach has been used in covering various aspects of Assembly language programming. Many examples and sample programs are given to clarify concepts and provide students an opportunity to learn by doing. Review questions are provided at the end of each section to reinforce the main points of the section. We feel that one of the functions of a textbook is to familiarize the student with terminology used in technical literature and in industry, so we have followed that guideline in this text.

Chapter 0 covers concepts in number systems (binary, decimal, and hex) and computer architecture. Most students will have learned these concepts in previous courses, but Chapter 0 provides a quick overview for those students who have not learned these concepts, or who may need to refresh their memory.

Chapter 1 provides a brief history of the evolution of x86 microprocessors and an overview of the internal workings of the 8086 as a basis of all x86 processors. Chapter 1 should be used in conjunction with Appendix A (a tutorial introduction to DEBUG) so that the student can experiment with concepts being learned on the x86 PC. The order of topics in Appendix A has been designed to correspond to the order of topics presented in Chapter 1. Thus, the student can begin programming with DEBUG without having to learn how to use an assembler.

Chapter 2 explains the use of assemblers to create programs. Although the programs in the book were developed and tested with Microsoft's MASM assembler, any Intel-compatible assembler such as Borland's TASM may be used.

Chapter 3 introduces the bulk of the logic and arithmetic instructions for unsigned numbers, plus bitwise operations in C.

Chapter 4 introduces DOS and BIOS interrupts. Programs in Assembly allow the student to get input from the keyboard and send output to the monitor. In addition, macro programming in assembly is described.

Chapter 5 describes how to program the keyboard and mouse.

Chapter 6 covers arithmetic and logic instructions for signed numbers as well as string processing instructions.

Chapter 7 discusses modular programming and how to develop larger Assembly language programs by breaking them into smaller modules to be coded and tested separately. In addition, in-line Assembly language within C programs is explained.

Chapter 8 introduces some 32-bit concepts of x86 programming. Although this book emphasizes 16-bit programming, the 386 is introduced to help the student appreciate the power of 32-bit CPUs.

Chapter 9 describes the 8088 and 286 microprocessors and supporting chips in detail and shows how they are used in the original IBM PC/XT/AT. In addition, the origin and function of the address, data, and control signals of the ISA expansion slot are described.

Chapter 10 provides an introduction to various types of RAM and ROM memories, their interfacing to the microprocessor, the memory map of the x86 PC, the timing issue in interfacing memory, and the checksum byte and parity bit techniques of ensuring data integrity in RAM and ROM.

Chapter 11 is dedicated to the interfacing of I/O ports, the use of IN and OUT instructions in the x86, and interfacing and programming of the 8255 programmable peripheral chip. We describe I/O programming in several languages, as well.

Chapter 12 covers the interfacing of PCs to devices for data acquisition such as LCDs, stepper motors, ADC, DAC, and sensors.

Chapter 13 discusses the use of the 8253/54 timer chip in the x86 PC, as well as how to generate music and time delays.

Chapter 14 is dedicated to the explanation of hardware and software interrupts, the use of the 8259 interrupt controller, the origin and assignment of IRQ signals on the expansion slots of the ISA bus, and exception interrupts in 80x86 microprocessors.

Chapter 15 is dedicated to direct memory access (DMA) concepts, the use of the 8237 DMA chip in the x86 PC, and DMA channels and associated signals on the ISA bus.

Chapter 16 covers the basics of video monitors and various video modes and adapters of the PC, in addition to the memory requirements of various video boards in graphics mode.

Chapter 17 discusses serial communication principles and programming of the PC COM port in Assembly and C.

Chapter 18 covers the interfacing and programming of the keyboard in the x86 PC, in addition to printer port interfacing and programming. In addition, a discussion of various types of parallel ports such as EPP and ECP is included.

Chapter 19 discusses hard disk storage organization and terminology.

Chapter 20 examines the x87 math coprocessor, its programming, and IEEE single- and double-precision floating point data types.

Chapter 21 explores the programming and hardware of the 386 microprocessor, contrasts and explains real and protected modes, and discusses the implementation of virtual memory.

Chapter 22 is dedicated to the interfacing of high-speed memories and describes various types of DRAM, including EDO and SDRAM, and examines cache memory and various cache organizations and terminology in detail.

In Chapter 23 we describe the main features of the 486, Pentium, and Pentium Pro and compare these microprocessors with the RISC processors. Chapter 23 also provides a discussion of MMX technology and how to write programs to detect which CPU a PC has.

Chapter 24 examines the new generation of 64-bit microprocessors from Intel and AMD.

Chapter 25 provides an overview of the IC technology and failure analysis, describes IC interfacing and system design issues, and covers error detection and correction.

Chapter 26 is dedicated to the discussion of the various types of PC buses, such as ISA and PC104, their performance comparisons, and features of the PCI bus.

Chapter 27 describes the USB port in detail and shows how to use the C programming language to access USB devices connected to the USB port of x86 PCs.

Appendices

The appendices have been designed to provide all reference material required for the topics covered in this combined volume so that no additional references should be necessary.

Appendix A provides a tutorial introduction to DEBUG. Appendix B provides a listing of Intel's 8086 instructions. Appendix C describes assembler directives with examples of their use. Appendix D lists commonly used interrupt fuction calls and legacy software. Appendix E lists the I/O maps of x86 PCs. Appendix F provides a table of ASCII codes.

Lab Manual

The lab manual contains some very basic labs and can be found at the **www.MicroDigitalEd.com** website. The more advanced and rigorous lab assignments are left up to the instructor depending on the course objectives, class level, and whether the course is graduate or undergraduate. The support materials for this and other books by the authors can be found on this website, too.

Solutions Manual/PowerPoint® Slides

The end-of-chapter problems cover some very basic concepts. The more challenging and rigorous homework assignments are left up to the instructor depending on the course objectives, class level, and whether the course is graduate or undergraduate. The solutions manual was produced with the help of Mr. Sepehr Naimi. The solutions manual and PowerPoint® slides for the drawings are available online for instructors only.

Online Instructor Resources

To access supplementary materials online, instructors need to request an instructor access code. Go to **www.prenhall.com**, click the **Instructor Resource Center** link, and then click **Register Today** for an instructor access code. Within 48 hours after registering you will receive a confirming e-mail including an instructor access code. Once you have received your code, go to the site and log on for full instructions on downloading the materials you wish to use.

Acknowledgments

This book is the result of the dedication, work, and love of many individuals. Thanks must go to the many students and professors around the world whose comments have helped to shape this book since its original publication in 1992. Special thanks to Dimitri Moonen for his detailed reading of the book and pointing out errors. Thanks to Pedran Mazidi for redoing the tables and figures for this new edition.

We would like to thank the people at Prentice Hall, in particular our editor Wyatt Morris, who continues to support and encourage our writing, and our project manager Rex Davidson, who made the book a reality. We were lucky to get the best copy editor in the world, Bret Workman. Thank you for your fantastic job, as usual.

Finally, we would like to sincerely thank Dr. Roger S. Walker of the Computer Science Engineering Department, University of Texas at Arlington for his constant encouragement. We enjoyed writing this book, and hope you enjoy reading it and using it for your courses and projects. Please let us know if you have any suggestions or find any errors.

ABOUT THE AUTHORS

Muhammad Ali Mazidi went to Tabriz University and holds Master's degrees from both Southern Methodist University and the University of Texas at Dallas. He is currently a.b.d. on his Ph.D. in the Electrical Engineering Department of Southern Methodist University. He is co-author of some widely used textbooks, including *The 8051 Microcontroller and Embedded Systems, The PIC Microcontroller and Embedded Systems,* and *The HCS12 Microcontroller and Embedded Systems,* also available from Prentice Hall. He teaches microprocessor-based system design at DeVry University in Dallas, Texas. He is the founder of MicroDigitalEd.com, which provides support for this book.

Janice Gillispie Mazidi has a Master of Science degree in Computer Science from the University of North Texas. She has several years experience as a software engineer in Dallas. She is currently teaching computer science and math courses in a local high school.

Danny Causey is a graduate of DeVry University and holds a Bachelor's of Science in Computer Engineering Technology. He is co-author of *The PIC Microcontroller and Embedded Systems* and *The HCS12 Microcontroller and Embedded Systems.* He is a partner in MicroDigitalEd.com. He works in the software development field, providing tools to accelerate daily business.

The authors can be contacted at the following email addresses if you have any comments or suggestions, or if you find any errors.

mdebooks@yahoo.com
mmazidi@microdigitaled.com
dcausey@microdigitaled.com

CHAPTER 0

INTRODUCTION TO COMPUTING

OBJECTIVES

Upon completion of this chapter, you will be able to:

- **>> Convert any number from base 2, base 10, or base 16 to either of the other two bases**
- **>> Add and subtract hex numbers**
- **>> Add binary numbers**
- **>> Represent any binary number in 2's complement**
- **>> Represent an alphanumeric string in ASCII code**
- **>> Describe logical operations AND, OR, NOT, XOR, NAND, NOR**
- **>> Use logic gates to diagram simple circuits**
- **>> Explain the difference between a bit, a nibble, a byte, and a word**
- **>> Give precise mathematical definitions of the terms *kilobyte*, *megabyte*, *gigabyte*, and *terabyte***
- **>> Explain the difference between RAM and ROM and describe their use**
- **>> Describe the purpose of the major components of a computer system**
- **>> List the three types of buses found in computers and describe the purpose of each type of bus**
- **>> Describe the role of the CPU in computer systems**
- **>> List the major components of the CPU and describe the purpose of each**

To understand the software and hardware of a microcontroller-based system, one must first master some very basic concepts underlying computer design. In this chapter (which in the tradition of digital computers is called Chapter 0), the fundamentals of numbering and coding systems are presented. After an introduction to logic gates, an overview of the workings inside the computer is given. Finally, in the last section we give a brief history of CPU architecture. Although some readers may have an adequate background in many of the topics of this chapter, it is recommended that the material be scanned, however briefly.

SECTION 0.1: NUMBERING AND CODING SYSTEMS

Whereas human beings use base 10 (*decimal*) arithmetic, computers use the base 2 (*binary*) system. In this section we explain how to convert from the decimal system to the binary system, and vice versa. The convenient representation of binary numbers in base 16, called *hexadecimal,* also is covered. Finally, the binary format of the alphanumeric code, called *ASCII*, is explored.

Decimal and binary number systems

Although there has been speculation that the origin of the base 10 system is the fact that human beings have 10 fingers, there is absolutely no speculation about the reason behind the use of the binary system in computers. The binary system is used in computers because 1 and 0 represent the two voltage levels of on and off. Whereas in base 10 there are 10 distinct symbols, 0, 1, 2, ..., 9, in base 2 there are only two, 0 and 1, with which to generate numbers. Base 10 contains digits 0 through 9; binary contains digits 0 and 1 only. These two binary digits, 0 and 1, are commonly referred to as *bits*.

Converting from decimal to binary

One method of converting from decimal to binary is to divide the decimal number by 2 repeatedly, keeping track of the remainders. This process continues until the quotient becomes zero. The remainders are then written in reverse order to obtain the binary number. This is demonstrated in Example 0-1.

Example 0-1

Convert 25_{10} to binary.

Solution:

```
          Quotient  Remainder
25/2 =    12        1      LSB (least significant bit)
12/2 =    6         0
6/2  =    3         0
3/2  =    1         1
1/2  =    0         1      MSB (most significant bit)
```

Therefore, $25_{10} = 11001_2$.

Converting from binary to decimal

To convert from binary to decimal, it is important to understand the concept of weight associated with each digit position. First, as an analogy, recall the weight of numbers in the base 10 system, as shown in the diagram. By the same token, each digit position in a number in base 2 has a weight associated with it:

740683_{10}	=	
3×10^0	=	3
8×10^1	=	80
6×10^2	=	600
0×10^3	=	0000
4×10^4	=	40000
7×10^5	=	700000
		740683

110101_2 =				**Decimal**	**Binary**
1×2^0	=	1×1	=	1	1
0×2^1	=	0×2	=	0	00
1×2^2	=	1×4	=	4	100
0×2^3	=	0×8	=	0	0000
1×2^4	=	1×16	=	16	10000
1×2^5	=	1×32	=	32	100000
				53	110101

Knowing the weight of each bit in a binary number makes it simple to add them together to get its decimal equivalent, as shown in Example 0-2.

Example 0-2

Convert 11001_2 to decimal.

Solution:

Weight:	16	8	4	2	1
Digits:	1	1	0	0	1
Sum:	16 +	8 +	0 +	0 +	1 = 25_{10}

Knowing the weight associated with each binary bit position allows one to convert a decimal number to binary directly instead of going through the process of repeated division. This is shown in Example 0-3.

Example 0-3

Use the concept of weight to convert 39_{10} to binary.

Solution:

Weight:	32	16	8	4	2	1
	1	0	0	1	1	1
	32 +	0 +	0 +	4 +	2 +	1 = 39

Therefore, $39_{10} = 100111_2$.

Hexadecimal system

Base 16, or the *hexadecimal* system as it is called in computer literature, is used as a convenient representation of binary numbers. For example, it is much easier for a human being to represent a string of 0s and 1s such as 100010010110 as its hexadecimal equivalent of 896H. The binary system has 2 digits, 0 and 1. The base 10 system has 10 digits, 0 through 9. The hexadecimal (base 16) system has 16 digits. In base 16, the first 10 digits, 0 to 9, are the same as in decimal, and for the remaining six digits, the letters A, B, C, D, E, and F are used. Table 0-1 shows the equivalent binary, decimal, and hexadecimal representations for 0 to 15.

Table 0-1: Base 16 Number Systems

Decimal	Binary	Hex
0	0000	0
1	0001	1
2	0010	2
3	0011	3
4	0100	4
5	0101	5
6	0110	6
7	0111	7
8	1000	8
9	1001	9
10	1010	A
11	1011	B
12	1100	C
13	1101	D
14	1110	E
15	1111	F

Converting between binary and hex

To represent a binary number as its equivalent hexadecimal number, start from the right and group 4 bits at a time, replacing each 4-bit binary number with its hex equivalent shown in Table 0-1. To convert from hex to binary, each hex digit is replaced with its 4-bit binary equivalent. See Examples 0-4 and 0-5.

Example 0-4

Represent binary 100111110101 in hex.

Solution:

First the number is grouped into sets of 4 bits: 1001 1111 0101.
Then each group of 4 bits is replaced with its hex equivalent:

1001	1111	0101
9	F	5

Therefore, 100111110101_2 = 9F5 hexadecimal.

Example 0-5

Convert hex 29B to binary.

Solution:

	2	9	B
=	0010	1001	1011

Dropping the leading zeros gives 1010011011.

Converting from decimal to hex

Converting from decimal to hex could be approached in two ways:

1. Convert to binary first and then convert to hex. Example 0-6 shows this method of converting decimal to hex.
2. Convert directly from decimal to hex by repeated division, keeping track of the remainders. Experimenting with this method is left to the reader.

Example 0-6

(a) Convert 45_{10} to hex.

32	16	8	4	2	1
1	0	1	1	0	1

First, convert to binary.
32 + 8 + 4 + 1 = 45

$45_{10} = 0010\ 1101_2$ = 2D hex

(b) Convert 629_{10} to hex.

512	256	128	64	32	16	8	4	2	1
1	0	0	1	1	1	0	1	0	1

$629_{10} = (512 + 64 + 32 + 16 + 4 + 1) = 0010\ 0111\ 0101_2$ = 275 hex

(c) Convert 1714_{10} to hex.

1024	512	256	128	64	32	16	8	4	2	1
1	1	0	1	0	1	1	0	0	1	0

$1714_{10} = (1024 + 512 + 128 + 32 + 16 + 2) = 0110\ 1011\ 0010_2$ = 6B2 hex

Converting from hex to decimal

Conversion from hex to decimal can also be approached in two ways:

1. Convert from hex to binary and then to decimal. Example 0-7 demonstrates this method of converting from hex to decimal.
2. Convert directly from hex to decimal by summing the weight of all digits.

Example 0-7

Convert the following hexadecimal numbers to decimal.

(a) $6B2_{16} = 0110\ 1011\ 0010_2$

1024	512	256	128	64	32	16	8	4	2	1
1	1	0	1	0	1	1	0	0	1	0

$1024 + 512 + 128 + 32 + 16 + 2 = 1714_{10}$

(b) $9F2D_{16} = 1001\ 1111\ 0010\ 1101_2$

32768	16384	8192	4096	2048	1024	512	256	128	64	32	16	8	4	2	1
1	0	0	1	1	1	1	1	0	0	1	0	1	1	0	1

$32768 + 4096 + 2048 + 1024 + 512 + 256 + 32 + 8 + 4 + 1 = 40{,}749_{10}$

Table 0-2: Counting in Bases

Decimal	Binary	Hex
0	00000	0
1	00001	1
2	00010	2
3	00011	3
4	00100	4
5	00101	5
6	00110	6
7	00111	7
8	01000	8
9	01001	9
10	01010	A
11	01011	B
12	01100	C
13	01101	D
14	01110	E
15	01111	F
16	10000	10
17	10001	11
18	10010	12
19	10011	13
20	10100	14
21	10101	15
22	10110	16
23	10111	17
24	11000	18
25	11001	19
26	11010	1A
27	11011	1B
28	11100	1C
29	11101	1D
30	11110	1E
31	11111	1F

Counting in bases 10, 2, and 16

To show the relationship between all three bases, in Table 0-2 we show the sequence of numbers from 0 to 31 in decimal, along with the equivalent binary and hex numbers. Notice in each base that when one more is added to the highest digit, that digit becomes zero and a 1 is carried to the next-highest digit position. For example, in decimal, 9 + 1 = 0 with a carry to the next-highest position. In binary, 1 + 1 = 0 with a carry; similarly, in hex, F + 1 = 0 with a carry.

Table 0-3: Binary Addition

A + B	Carry	Sum
0 + 0	0	0
0 + 1	0	1
1 + 0	0	1
1 + 1	1	0

Addition of binary and hex numbers

The addition of binary numbers is a very straightforward process. Table 0-3 shows the addition of two bits. The discussion of subtraction of binary numbers is bypassed since all computers use the addition process to implement subtraction. Although computers have adder circuitry, there is no separate circuitry for subtractors. Instead, adders are used in conjunction with *2's complement* circuitry to perform subtraction. In other words, to implement "*x* – *y*", the computer takes the 2's complement of y and adds it to x. The concept of 2's complement is reviewed next. Example 0-8 shows the addition of binary numbers.

Example 0-8

Add the following binary numbers. Check against their decimal equivalents.

Solution:

	Binary	*Decimal*
	1101	13
+	1001	9
	10110	22

2's complement

To get the 2's complement of a binary number, invert all the bits and then add 1 to the result. Inverting the bits is simply a matter of changing all 0s to 1s and 1s to 0s. This is called the *1's complement*. See Example 0-9.

Example 0-9

Take the 2's complement of 10011101.

Solution:

	10011101	binary number
	01100010	1's complement
+	1	
	01100011	2's complement

Addition and subtraction of hex numbers

In studying issues related to software and hardware of computers, it is often necessary to add or subtract hex numbers. Mastery of these techniques is essential. Hex addition and subtraction are discussed separately below.

Addition of hex numbers

This section describes the process of adding hex numbers. Starting with the least significant digits, the digits are added together. If the result is less than 16, write that digit as the sum for that position. If it is greater than 16, subtract 16 from it to get the digit and carry 1 to the next digit. The best way to explain this is by example, as shown in Example 0-10.

Example 0-10

Perform hex addition: 23D9 + 94BE.

Solution:

	23D9	LSD:	9 − 14 = 23	23 – 16 = 7 with a carry
+	94BE		1 − 13 + 11 = 25	25 – 16 = 9 with a carry
	B897		1 − 3 + 4 = 8	
		MSD:	2 − 9 = B	

Subtraction of hex numbers

In subtracting two hex numbers, if the second digit is greater than the first, borrow 16 from the preceding digit. See Example 0-11.

Example 0-11

Perform hex subtraction: 59F – 2B8.

Solution:

	59F	LSD:	8 from 15 = 7
–	2B8		11 from 25 (9 + 16) = 14 (E)
	2E7		2 from 4 (5 – 1) = 2

ASCII code

The discussion so far has revolved around the representation of number systems. Since all information in the computer must be represented by 0s and 1s, binary patterns must be assigned to letters and other characters. In the 1960s a standard representation called *ASCII* (American Standard Code for Information Interchange) was established. The ASCII (pronounced "ask-E") code assigns binary patterns for numbers 0 to 9, all the letters of the English alphabet, both uppercase (capital) and lowercase, and many control codes and punctuation marks. The great advantage of this system is that it is used by most computers, so that information can be shared among computers. The ASCII system uses a total of 7 bits to represent each code. For example, 100 0001 is assigned to the uppercase letter "A" and 110 0001 is for the lowercase "a". Often, a zero is placed in the most significant bit position to make it an 8-bit code. Figure 0-1 shows ASCII codes. A complete list of extended ASCII codes is given in Appendix F. The use of ASCII is not only standard for keyboards used in the United States and many other countries but also provides a standard for printing and displaying characters by output devices such as printers and monitors.

Notice that the pattern of ASCII codes was designed to allow for easy manipulation of ASCII data. For example, digits 0 through 9 are represented by ASCII codes 30 through 39. This enables a program to easily convert ASCII to decimal by masking off the "3" in the upper nibble. Also notice that there is a relationship between the uppercase and lowercase letters. The uppercase letters are represented by ASCII codes 41 through 5A while lowercase letters are represented by codes 61 through 7A. Looking at the binary code, the only bit that is different between the uppercase "A" and lowercase "a" is bit 5. Therefore, conversion between uppercase and lowercase is as simple as changing bit 5 of the ASCII code.

Hex	Ch	Hex	Ch	Hex	Ch	Hex	Ch	Hex	Ch	Hex	Ch	Hex	Ch	Hex	Ch
00		10	►	20		30	0	40	@	50	P	60	`	70	p
01	☺	11	◄	21	!	31	1	41	A	51	Q	61	a	71	q
02	☻	12	↕	22	"	32	2	42	B	52	R	62	b	72	r
03	♥	13	‼	23	#	33	3	43	C	53	S	63	c	73	s
04	♦	14	¶	24	$	34	4	44	D	54	T	64	d	74	t
05	♣	15	§	25	%	35	5	45	E	55	U	65	e	75	u
06	♠	16	▬	26	&	36	6	46	F	56	V	66	f	76	v
07	•	17	↨	27	'	37	7	47	G	57	W	67	g	77	w
08	◘	18	↑	28	(	38	8	48	H	58	X	68	h	78	x
09	○	19	↓	29	)	39	9	49	I	59	Y	69	i	79	y
0A	◙	1A	→	2A	*	3A	:	4A	J	5A	Z	6A	j	7A	z
0B	♂	1B	←	2B	+	3B	;	4B	K	5B	[	6B	k	7B	{
0C	♀	1C	∟	2C	,	3C	<	4C	L	5C	\	6C	l	7C	¦
0D	♪	1D	↔	2D	-	3D	=	4D	M	5D	]	6D	m	7D	}
0E	♫	1E	▲	2E	.	3E	>	4E	N	5E	^	6E	n	7E	~
0F	☼	1F	▼	2F	/	3F	?	4F	O	5F	_	6F	o	7F	⌂

Figure 0-1. Selected ASCII Codes

Review Questions

1. Why do computers use the binary number system instead of the decimal system?
2. Convert 34_{10} to binary and hex.
3. Convert 110101_2 to hex and decimal.

4. Perform binary addition: 101100 + 101.
5. Convert 101100_2 to its 2's complement representation.

SECTION 0.2: DIGITAL PRIMER

This section gives an overview of digital logic and design. First, we cover binary logic operations, then we show gates that perform these functions. Next, logic gates are put together to form simple digital circuits. Finally, we cover some logic devices commonly found in microcontroller interfacing.

Binary logic

As mentioned earlier, computers use the binary number system because the two voltage levels can be represented as the two digits 0 and 1. Signals in digital electronics have two distinct voltage levels. For example, a system may define 0 V as logic 0 and +5 V as logic 1. Figure 0-2 shows this system with the built-in tolerances for variations in the voltage. A valid digital signal in this example should be within either of the two shaded areas.

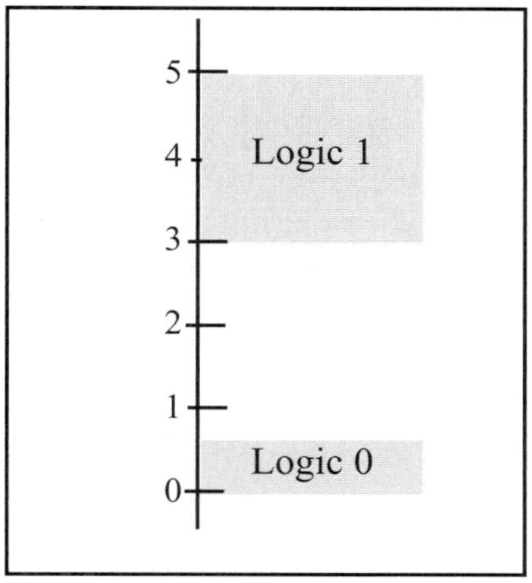

Figure 0-2. Binary Signals

Logic gates

Binary logic gates are simple circuits that take one or more input signals and send out one output signal. Several of these gates are defined below.

AND gate

The AND gate takes two or more inputs and performs a logic AND on them. See the truth table and diagram of the AND gate. Notice that if both inputs to the AND gate are 1, the output will be 1. Any other combination of inputs will give a 0 output. The example shows two inputs, x and y. Multiple outputs are also possible for logic gates. In the case of AND, if all inputs are 1, the output is 1. If any input is 0, the output is zero.

Logical AND Function

Inputs		Output
X	**Y**	**X AND Y**
0	0	0
0	1	0
1	0	0
1	1	1

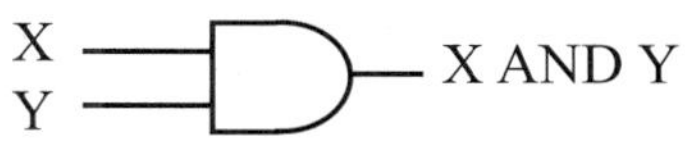

OR gate

The OR logic function will output a 1 if one or more inputs is 1. If all inputs are 0, then and only then will the output be 0.

Logical OR Function

Inputs		Output
X	**Y**	**X OR Y**
0	0	0
0	1	1
1	0	1
1	1	1

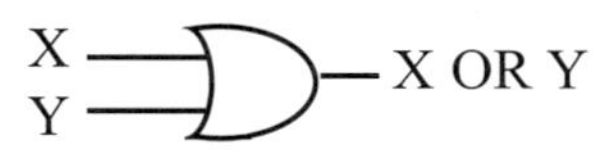

Tri-state buffer

A buffer gate does not change the logic level of the input. It is used to isolate or amplify the signal.

Buffer

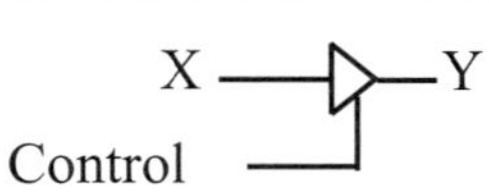

Inverter

The inverter, also called NOT, outputs the value opposite to that input to the gate. That is, a 1 input will give a 0 output, while a 0 input will give a 1 output.

XOR gate

The XOR gate performs an exclusive-OR operation on the inputs. Exclusive-OR produces a 1 output if one (but only one) input is 1. If both operands are 0, the output is zero. Likewise, if both operands are 1, the output is also zero. Notice from the XOR truth table, that whenever the two inputs are the same, the output is zero. This function can be used to compare two bits to see if they are the same.

NAND and NOR gates

The NAND gate functions like an AND gate with an inverter on the output. It produces a zero output when all inputs are 1; otherwise, it produces a 1 output. The NOR gate functions like an OR gate with an inverter on the output. It produces a 1 if all inputs are 0; otherwise, it produces a 0. NAND and NOR gates are used extensively in digital design because they are easy and inexpensive to fabricate. Any circuit that can be designed with AND, OR, XOR, and INVERTER gates can be implemented using only NAND and NOR gates. A simple example of this is given below. Notice in NAND, that if any input is zero, the output is one. Notice in NOR, that if any input is one, the output is zero.

Logic design using gates

Next we will show a simple logic design to add two binary digits. If we add two binary digits there are four possible outcomes:

	Carry	*Sum*
0 + 0 =	0	0
0 + 1 =	0	1
1 + 0 =	0	1
1 + 1 =	1	0

Logical Inverter

Input	**Output**
X	**NOT X**
0	1
1	0

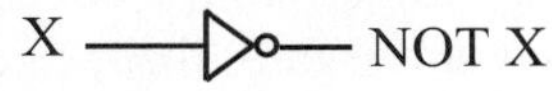

Logical XOR Function

Inputs	**Output**
X Y	**X XOR Y**
0 0	0
0 1	1
1 0	1
1 1	0

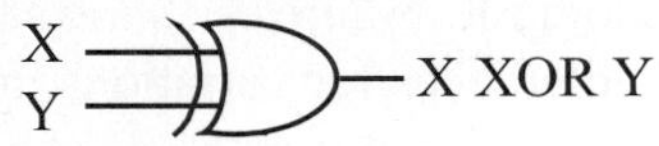

Logical NAND Function

Inputs	**Output**
X Y	**X NAND Y**
0 0	1
0 1	1
1 0	1
1 1	0

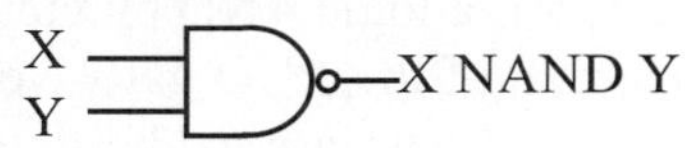

Logical NOR Function

Inputs	**Output**
X Y	**X NOR Y**
0 0	1
0 1	0
1 0	0
1 1	0

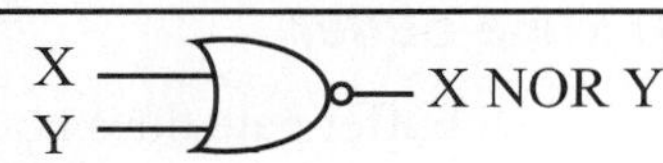

Notice that when we add 1 + 1 we get 0 with a carry to the next higher place. We will need to determine the sum and the carry for this design. Notice that the sum column above matches the output for the XOR function, and that the carry column matches the output for the AND function. Figure 0-3 (a) shows a simple adder implemented with XOR and AND gates. Figure 0-3 (b) shows the same logic circuit implemented with AND and OR gates.

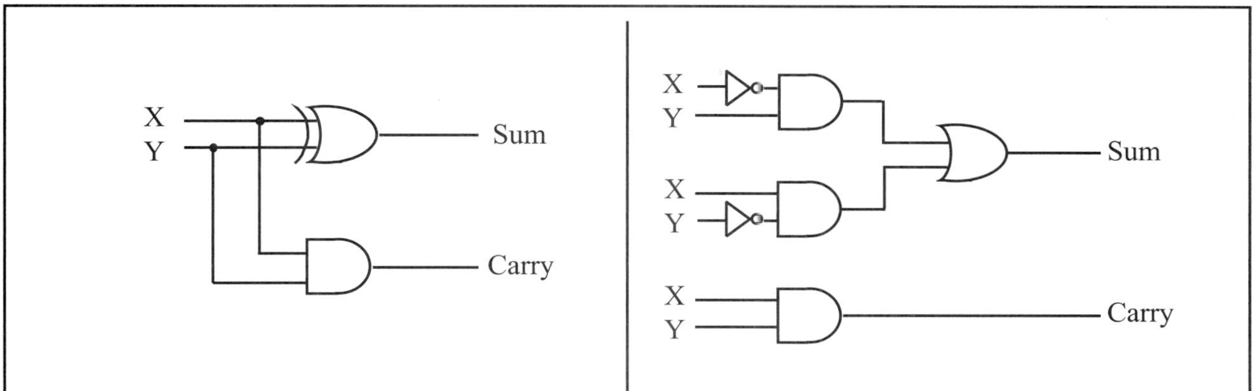

(a) Half-Adder Using XOR and AND

(a) Half-Adder Using AND, OR, Inverters

Figure 0-3. Two Implementations of a Half-Adder

Figure 0-4 shows a block diagram of a half-adder. Two half-adders can be combined to form an adder that can add three input digits. This is called a full-adder. Figure 0-5 shows the logic diagram of a full-adder, along with a block diagram that masks the details of the circuit. Figure 0-6 shows a 3-bit adder using three full-adders.

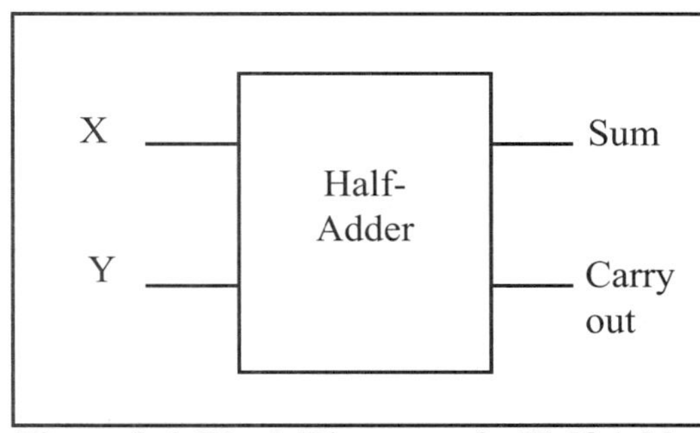

Figure 0-4. Block Diagram of a Half-Adder

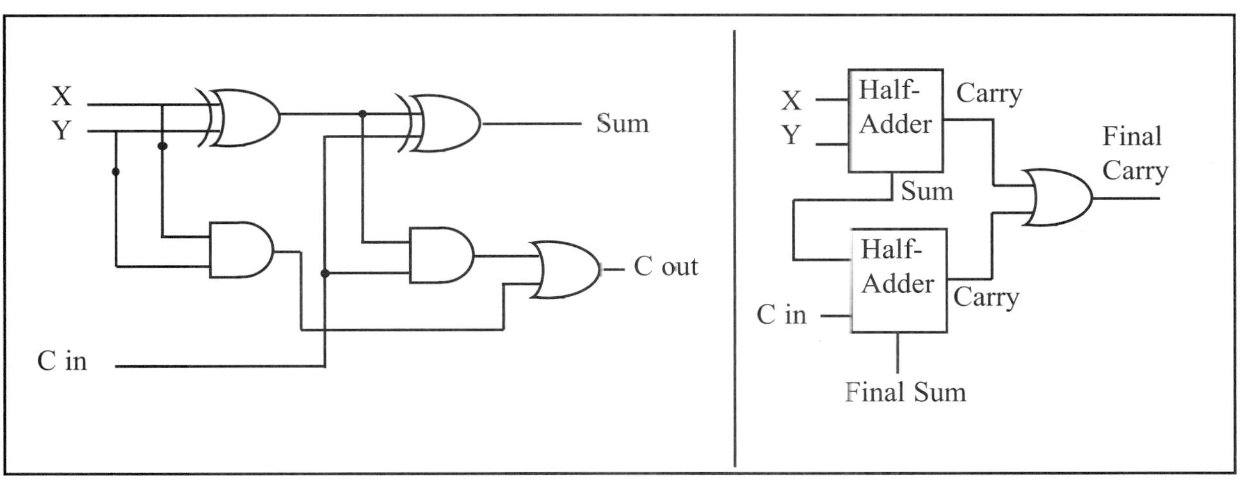

Figure 0-5. Full-Adder Built from a Half-Adder

Decoders

Another example of the application of logic gates is the decoder. Decoders are widely used for address decoding in computer design. Figure 0-7 shows decoders for 9 (1001 binary) and 5 (0101) using inverters and AND gates.

Flip-flops

A widely used component in digital systems is the flip-flop. Frequently, flip-flops are used to store data. Figure 0-8 shows the logic diagram, block diagram, and truth table for a flip-flop.

The D flip-flop (D-FF) is widely used to latch data. Notice from the truth table that a D-FF grabs the data at the input as the clock is activated. A D-FF holds the data as long as the power is on.

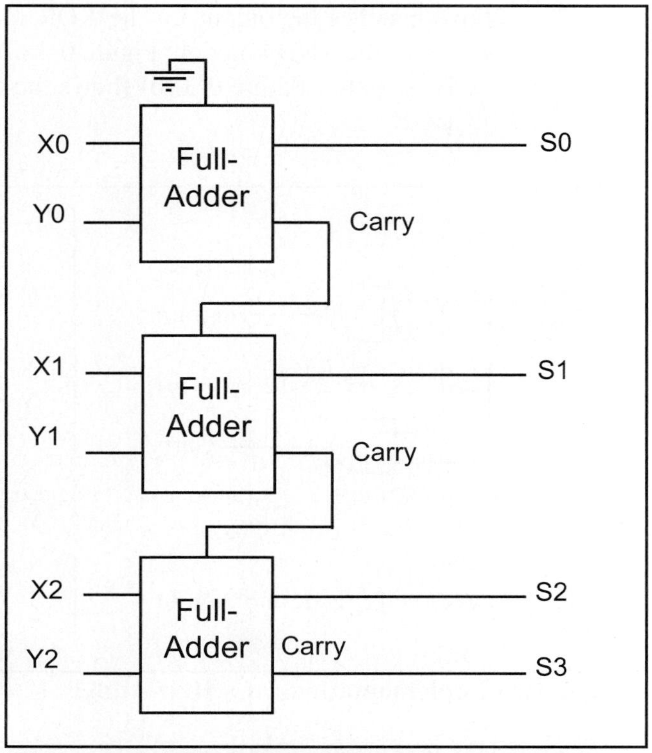

Figure 0-6. 3-Bit Adder Using Three Full-Adders

(a) Address decoder for 9 (binary 1001) The output of the AND gate will be 1 if and only if the input is binary 1001.

(b) Address decoder for 5 (binary 0101) The output of the AND gate will be 1 if and only if the input is binary 0101.

Figure 0-7. Address Decoders

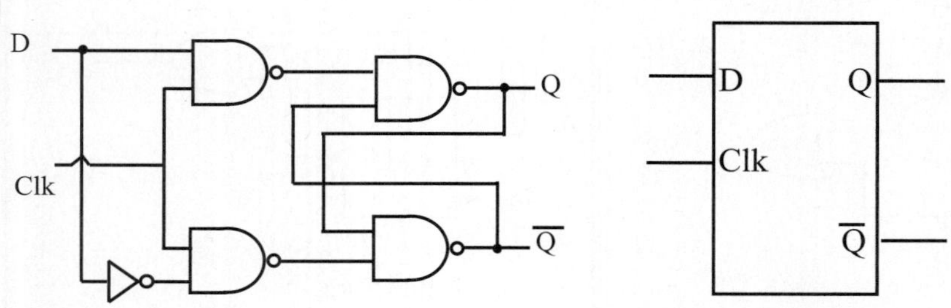

(a) Circuit diagram

(b) Block diagram

Clk	D	Q
No	x	no change
↓	0	0
↓	1	1

x = don't care

(c) Truth table

Figure 0-8. D Flip-Flops

Review Questions

1. The logical operation ______ gives a 1 output when all inputs are 1.
2. The logical operation ______ gives a 1 output when 1 or more of its inputs is 1.
3. The logical operation ______ is often used to compare if two inputs have the same value.
4. A ______ gate does not change the logic level of the input.
5. Name a common use for flip-flops.
6. An address _______ is used to identify a predetermined binary address.

SECTION 0.3: INSIDE THE COMPUTER

In this section we provide an introduction to the organization and internal working of computers. The model used is generic, but the concepts discussed are applicable to all computers, including the IBM PC, PS/2, and compatibles. Before embarking on this subject, it will be helpful to review definitions of some of the most widely used terminology in computer literature, such as *K*, *mega*, *giga*, *byte*, *ROM*, *RAM*, and so on.

Some important terminology

One of the most important features of a computer is how much memory it has. Here we review terms used to describe amounts of memory in x86 PCs and compatibles. Recall from the discussion above that a *bit* is a binary digit that can have the value 0 or 1. A *byte* is defined as 8 bits. A *nibble* is half a byte, or 4 bits. A *word* is two bytes, or 16 bits. In a 32-bit computer a word is 32 bits. The display is intended to show the relative size of these units. Of course, they could all be composed of any combination of zeros and ones.

```
Bit                                                        0
Nibble                                                  0000
Byte                                               0000 0000
Word                                     0000 0000 0000 0000
32-bit Word  0000 0000 0000 0000 0000 0000 0000 0000
```

A *kilobyte* is 2^{10} bytes, which is 1024 bytes. The abbreviation *K* is often used. A *megabyte*, or meg as some call it, is 2^{20} bytes. That is a little over 1 million bytes; it is exactly 1,048,576 bytes. Moving rapidly up the scale in size, a *gigabyte* is 2^{30} bytes (over 1 billion), and a *terabyte* is 2^{40} bytes (over 1 trillion). As an example of how some of these terms are used, suppose that a given computer has 16 megabytes of memory. That would be 16×2^{20}, or $2^4 \times 2^{20}$, which is 2^{24}. Therefore, 16 megabytes is 2^{24} bytes. See Table 0-4.

Table 0-4: Power of 2

2^{10} = 1024 = 1K
2^{20} = 1024K = 1M
2^{30} = 1024M = 1G
2^{40} = 1024G = 1T

Two types of memory commonly used in microcomputers are *RAM*, which stands for "random access memory" (sometimes called *read/write memory*), and *ROM*, which stands for "read-only memory." RAM is used by the computer for temporary storage of programs that it is running. That data is lost when the computer is turned off. For this reason, RAM is sometimes called *volatile memory*. ROM contains programs and information essential to operation of the computer. The information in ROM is permanent, cannot be changed by the user, and is not lost when the power is turned off. Therefore, it is called *nonvolatile memory*.

Internal organization of computers

The internal working of every computer can be broken down into three parts: CPU (central processing unit), memory, and I/O (input/output) devices (see Figure 0-9). The function of the CPU is to execute (process) information stored in memory. The function of I/O devices such as the keyboard and video monitor is to provide a means of communicating with the CPU. The CPU is connected to memory and I/O through strips of wire called a *bus*. The bus carries information from place to place inside a computer just as a street bus carries people from place to place. In every computer there are three types of buses: address bus, data bus, and control bus.

For a device (memory or I/O) to be recognized by the CPU, it must be assigned an address. The address assigned to a given device must be unique; no two devices are allowed to have the same address. The CPU puts the address (in binary, of course) on the address bus, and the decoding circuitry finds the device. Then the CPU uses the data bus either to get data from that device or to send data to it. The control buses are used to provide read or write signals to the device to indicate if the CPU is asking for information or sending it information. Of the three buses, the address bus and data bus determine the capability of a given CPU.

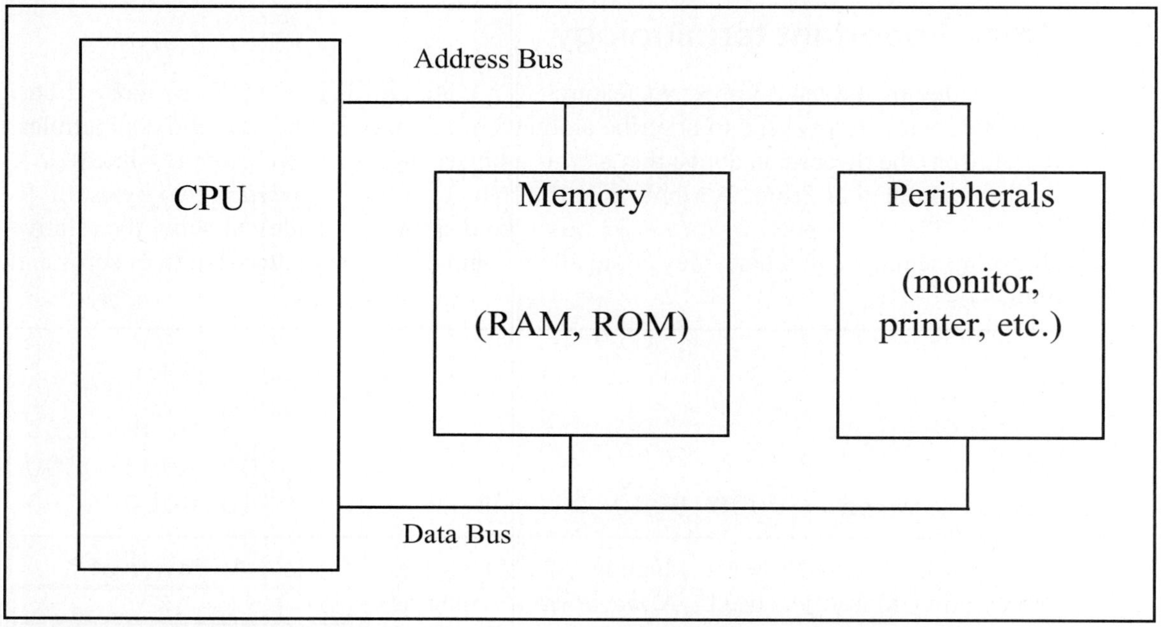

Figure 0-9. Inside the Computer

More about the data bus

Since data lines are used to carry information in and out of a CPU, the more data lines available, the better the CPU. If one thinks of data lines as highway lanes, it is clear that more lanes provide a better pathway between the CPU and its external devices (such as printers, RAM, ROM, etc.; see Figure 0-10). By the same token, that increase in the number of lanes increases the cost of construction. More data buses mean a more expensive CPU and computer. The grouping of data lines is called *data bus*. The average size of data buses in CPUs varies between 8 and 64. Early computers such as Apple 2 used an 8-bit data bus, while supercomputers such as Cray use a 64-bit data bus. Data buses are bidirectional, since the CPU must use them either to receive or to send data. The processing power of a computer is related to the size of its buses, since an 8-bit bus can send out 1 byte a time, but a 16-bit bus can send out 2 bytes at a time, which is twice as fast.

More about the address bus

Since the address bus is used to identify the devices and memory connected to the CPU, the more address buses available, the larger the number of devices that can be addressed. In other words, the number of address buses for a CPU determines the number of locations with which it can communicate. The number of locations is always equal to 2^x, where x is the number of address lines, regardless of the size of the data bus. For example, a CPU with 16 address lines can provide a total of 65,536 (2^{16}) or 64K bytes of addressable memory. Each location can have a maximum of 1 byte of data. This is due to the fact that all general-purpose microprocessor CPUs are what is called *byte addressable*. As another example, the IBM PC AT used a CPU with 24 address lines and 16 data lines. In this case the total accessible memory is 16 megabytes (2^{24} = 16 megabytes). In this example there would be 2^{24} locations, and since each location is one byte, there would be 16 megabytes of memory. The address bus is a *unidirectional* bus, which means that the CPU uses the address bus only to send out addresses. To summarize: The total number of memory locations addressable by a given CPU is always equal to 2^x where x is the number of address bits, regardless of the size of the data bus.

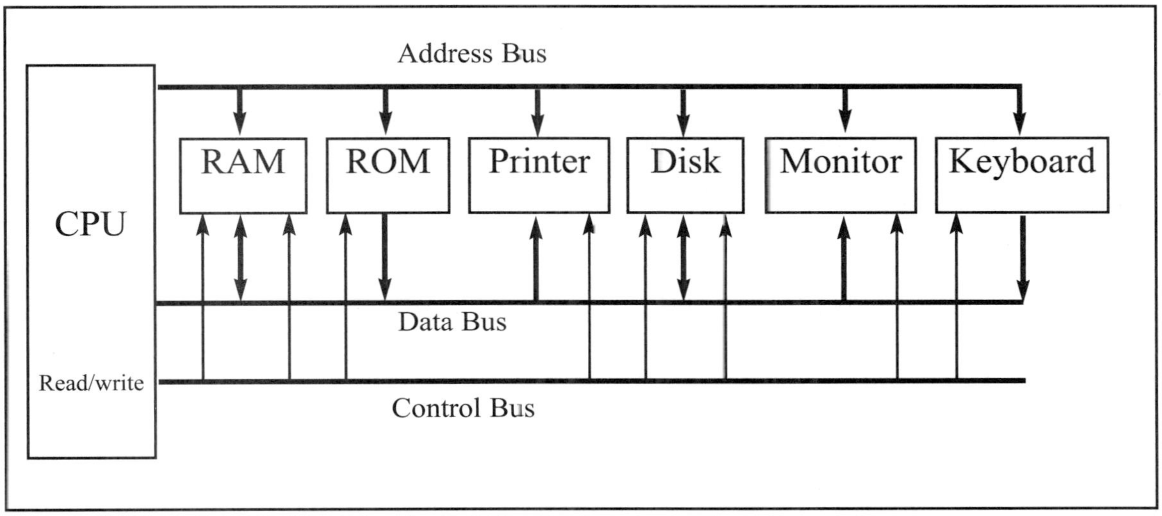

Figure 0-10. Internal Organization of Computers

CPU and its relation to RAM and ROM

For the CPU to process information, the data must be stored in RAM or ROM. The function of ROM in computers is to provide information that is fixed and permanent. This is information such as tables for character patterns to be displayed on the video monitor, or programs that are essential to the working of the computer, such as programs for testing and finding the total amount of RAM installed on the system, or programs to display information on the video monitor. In contrast, RAM is used to store information that is not permanent and can change with time, such as various versions of the operating system and application packages such as word processing or tax calculation packages. These programs are loaded into RAM to be processed by the CPU. The CPU cannot get the information directly from the disk since the disk is too slow. In other words, the CPU first seeks the information to be processed from RAM (or ROM). Only if it is not there does the CPU seek it from a mass storage device such as a disk, and then it transfers the information to RAM. For this reason, RAM and ROM are sometimes referred to as *primary memory* and disks are called *secondary memory*. Figure 0-11 shows a block diagram of the internal organization of the PC.

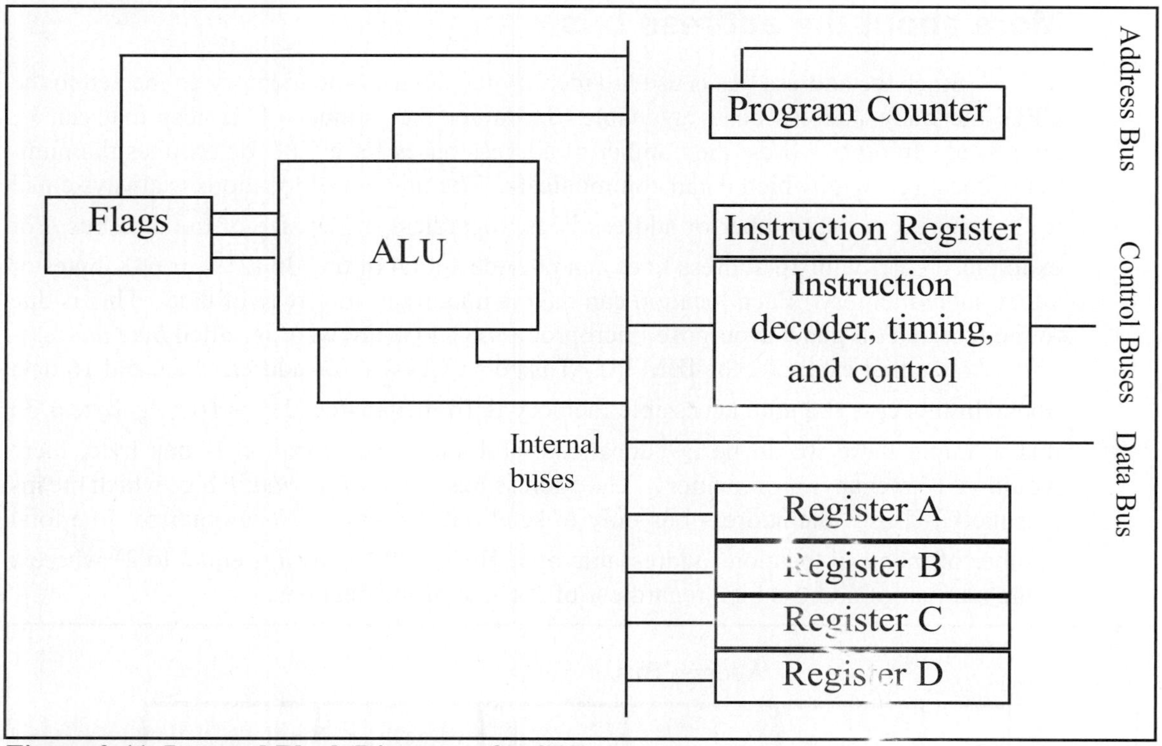

Figure 0-11. Internal Block Diagram of a CPU

Inside CPUs

A program stored in memory provides instructions to the CPU to perform an action. The action can simply be adding data such as payroll data or controlling a machine such as a robot. It is the function of the CPU to fetch these instructions from memory and execute them. To perform the actions of fetch and execute, all CPUs are equipped with resources such as the following:

1. Foremost among the resources at the disposal of the CPU are a number of *registers*. The CPU uses registers to store information temporarily. The information could be two values to be processed, or the address of the value needed to be fetched from memory. Registers inside the CPU can be 8-bit, 16-bit, 32-bit, or even 64-bit registers, depending on the CPU. In general, the more and bigger the registers, the better the CPU. The disadvantage of more and bigger registers is the increased cost of such a CPU.
2. The CPU also has what is called the *ALU* (arithmetic/logic unit). The ALU section of the CPU is responsible for performing arithmetic functions such as add, subtract, multiply, and divide, and logic functions such as AND, OR, and NOT.
3. Every CPU has what is called a *program counter*. The function of the program counter is to point to the address of the next instruction to be executed. As each instruction is executed, the program counter is incremented to point to the address of the next instruction to be executed. The contents of the program counter are placed on the address bus to find and fetch the desired instruction. In the IBM PC, the program counter is a register called IP, or the instruction pointer.
4. The function of the *instruction decoder* is to interpret the instruction fetched into the CPU. One can think of the instruction decoder as a kind of dictionary, storing the meaning of each instruction and what steps the CPU should take upon receiving a given instruction. Just as a dictionary requires more pages the more words it defines,

a CPU capable of understanding more instructions requires more transistors to design.

Internal working of computers

To demonstrate some of the concepts discussed above, a step-by-step analysis of the process a CPU would go through to add three numbers is given next. Assume that an imaginary CPU has registers called A, B, C, and D. It has an 8-bit data bus and a 16-bit address bus. Therefore, the CPU can access memory from addresses 0000 to FFFFH (for a total of 10000H locations). The action to be performed by the CPU is to put hexadecimal value 21 into register A, and then add to register A values 42H and 12H. Assume that the code for the CPU to move a value to register A is 1011 0000 (B0H) and the code for adding a value to register A is 0000 0100 (04H). The necessary steps and code to perform them are as follows.

Action	*Code*	*Data*
Move value 21H into register A	B0H	21H
Add value 42H to register A	04H	42H
Add value 12H to register A	04H	12H

If the program to perform the actions listed above is stored in memory locations starting at 1400H, the following would represent the contents for each memory address location:

Memory address	*Contents of memory address*
1400	(B0)code for moving a value to register A
1401	(21)value to be moved
1402	(04)code for adding a value to register A
1403	(42)value to be added
1404	(04)code for adding a value to register A
1405	(12)value to be added
1406	(F4)code for halt

The actions performed by the CPU to run the program above would be as follows:

1. The CPU's program counter can have a value between 0000 and FFFFH. The program counter must be set to the value 1400H, indicating the address of the first instruction code to be executed. After the program counter has been loaded with the address of the first instruction, the CPU is ready to execute.
2. The CPU puts 1400H on the address bus and sends it out. The memory circuitry finds the location while the CPU activates the READ signal, indicating to memory that it wants the byte at location 1400H. This causes the contents of memory location 1400H, which is B0, to be put on the data bus and brought into the CPU.
3. The CPU decodes the instruction B0 with the help of its instruction decoder dictionary. When it finds the definition for that instruction it knows it must bring into register A of the CPU the byte in the next memory location. Therefore, it commands its controller circuitry to do exactly that. When it brings in value 21H from memory location 1401, it makes sure that the doors of all registers are closed except register A. Therefore, when value 21H comes into the CPU it will go directly into register A. After completing one instruction, the program counter points to the address of the next instruction to be executed, which in this case is 1402H. Address 1402 is sent out on the address bus to fetch the next instruction.

4. From memory location 1402H it fetches code 04H. After decoding, the CPU knows that it must add to the contents of register A the byte sitting at the next address (1403). After it brings the value (in this case 42H) into the CPU, it provides the contents of register A along with this value to the ALU to perform the addition. It then takes the result of the addition from the ALU's output and puts it in register A. Meanwhile the program counter becomes 1404, the address of the next instruction.
5. Address 1404H is put on the address bus and the code is fetched into the CPU, decoded, and executed. This code is again adding a value to register A. The program counter is updated to 1406H.
6. Finally, the contents of address 1406 are fetched in and executed. This HALT instruction tells the CPU to stop incrementing the program counter and asking for the next instruction. In the absence of the HALT, the CPU would continue updating the program counter and fetching instructions.

Now suppose that address 1403H contained value 04 instead of 42H. How would the CPU distinguish between data 04 to be added and code 04? Remember that code 04 for this CPU means move the next value into register A. Therefore, the CPU will not try to decode the next value. It simply moves the contents of the following memory location into register A, regardless of its value.

Review Questions

1. How many bytes is 24 kilobytes?
2. What does "RAM" stand for? How is it used in computer systems?
3. What does "ROM" stand for? How is it used in computer systems?
4. Why is RAM called volatile memory?
5. List the three major components of a computer system.
6. What does "CPU" stand for? Explain its function in a computer.
7. List the three types of buses found in computer systems and state briefly the purpose of each type of bus.
8. State which of the following is unidirectional and which is bidirectional.
   (a) data bus (b) address bus
9. If an address bus for a given computer has 16 lines, what is the maximum amount of memory it can access?
10. What does "ALU" stand for? What is its purpose?
11. How are registers used in computer systems?
12. What is the purpose of the program counter?
13. What is the purpose of the instruction decoder?

PROBLEMS

SECTION 0.1: NUMBERING AND CODING SYSTEMS

1. Convert the following decimal numbers to binary.
   (a) 12 (b) 123 (c) 63 (d) 128 (e) 1000
2. Convert the following binary numbers to decimal.
   (a) 100100 (b) 1000001 (c) 11101 (d) 1010 (e) 00100010
3. Convert the values in Problem 2 to hexadecimal.
4. Convert the following hex numbers to binary and decimal.
   (a) 2B9H (b) F44H (c) 912H (d) 2BH (e) FFFFH
5. Convert the values in Problem 1 to hex.

6. Find the 2's complement of the following binary numbers.
   (a) 1001010 (b) 111001 (c) 10000010 (d) 111110001
7. Add the following hex values.
   (a) 2CH + 3FH (b) F34H + 5D6H (c) 20000H + 12FFH (d) FFFFH + 2222H
8. Perform hex subtraction for the following.
   (a) 24FH – 129H (b) FE9H – 5CCH (c) 2FFFFH – FFFFFH (d) 9FF25H – 4DD99H
9. Show the ASCII codes for numbers 0, 1, 2, 3, ..., 9 in both hex and binary.
10. Show the ASCII code (in hex) for the following string:

    "U.S.A. is a country" CR, LF
    "in North America" CR, LF
    CR is carriage return
    LF is line feed

SECTION 0.2: DIGITAL PRIMER

11. Draw a 3-input OR gate using a 2-input OR gate.
12. Show the truth table for a 3-input OR gate.
13. Draw a 3-input AND gate using a 2-input AND gate.
14. Show the truth table for a 3-input AND gate.
15. Design a 3-input XOR gate with a 2-input XOR gate. Show the truth table for a 3-input XOR.
16. List the truth table for a 3-input NAND.
17. List the truth table for a 3-input NOR.
18. Show the decoder for binary 1100.
19. Show the decoder for binary 11011.
20. List the truth table for a D-FF.

SECTION 0.3: INSIDE THE COMPUTER

21. Answer the following:
    (a) How many nibbles are 16 bits?
    (b) How many bytes are 32 bits?
    (c) If a word is defined as 16 bits, how many words is a 64-bit data item?
    (d) What is the exact value (in decimal) of 1 meg?
    (e) How many K is 1 meg?
    (f) What is the exact value (in decimal) of 1 giga?
    (g) How many K is 1 giga?
    (h) How many meg is 1 giga?
    (i) If a given computer has a total of 8 megabytes of memory, how many bytes (in decimal) is this? How many kilobytes is this?
22. A given mass storage device such as a hard disk can store 2 gigabytes of information. Assuming that each page of text has 25 rows and each row has 80 columns of ASCII characters (each character = 1 byte), approximately how many pages of information can this disk store?
23. In a given byte-addressable computer, memory locations 10000H to 9FFFFH are available for user programs. The first location is 10000H and the last location is 9FFFFH. Calculate the following:
    (a) The total number of bytes available (in decimal)
    (b) The total number of kilobytes (in decimal)
24. A given computer has a 32-bit data bus. What is the largest number that can be carried into the CPU at a time?

25. Below are listed several computers with their data bus widths. For each computer, list the maximum value that can be brought into the CPU at a time (in both hex and decimal).
    (a) Apple 2 with an 8-bit data bus
    (b) IBM PC with a 16-bit data bus
    (c) IBM PC with a 32-bit data bus
    (d) Cray supercomputer with a 64-bit data bus
26. Find the total amount of memory, in the units requested, for each of the following CPUs, given the size of the address buses.
    (a) 16-bit address bus (in K)
    (b) 24-bit address bus (in megabytes)
    (c) 32-bit address bus (in megabytes and gigabytes)
    (d) 48-bit address bus (in megabytes, gigabytes, and terabytes)
27. Regarding the data bus and address bus, which is unidirectional and which is bidirectional?
28. Which register of the CPU holds the address of the instruction to be fetched?
29. Which section of the CPU is responsible for performing addition?
30. List the three bus types present in every CPU.

ANSWERS TO REVIEW QUESTIONS

SECTION 0.1: NUMBERING AND CODING SYSTEMS

1. Computers use the binary system because each bit can have one of two voltage levels: on and off.
2. $34_{10} = 100010_2 = 22_{16}$
3. $110101_2 = 35_{16} = 53_{10}$
4. 1110001
5. 010100

SECTION 0.2: DIGITAL PRIMER

1. AND
2. OR
3. XOR
4. Buffer
5. Storing data
6. Decoder

SECTION 0.3: INSIDE THE COMPUTER

1. 24,576
2. Random access memory; it is used for temporary storage of programs that the CPU is running, such as the operating system, word processing programs, etc.
3. Read-only memory; it is used for permanent programs such as those that control the keyboard, etc.
4. The contents of RAM are lost when the computer is powered off.
5. The CPU, memory, and I/O devices
6. Central processing unit; it can be considered the "brain" of the computer; it executes

the programs and controls all other devices in the computer.

7. The address bus carries the location (address) needed by the CPU; the data bus carries information in and out of the CPU; the control bus is used by the CPU to send signals controlling I/O devices.
8. (a) bidirectional (b) unidirectional
9. 64K, or 65,536 bytes
10. Arithmetic/logic unit; it performs all arithmetic and logic operations.
11. It is for temporary storage of information.
12. It holds the address of the next instruction to be executed.
13. It tells the CPU what steps to perform for each instruction.

CHAPTER 1

THE x86 MICROPROCESSOR

OBJECTIVES

Upon completion of this chapter, you will be able to:

- >> Describe the Intel family of microprocessors from the 8085 to the Pentium in terms of bus size, physical memory, and special features
- >> Explain the function of the EU (execution unit) and BIU (bus interface unit)
- >> Describe pipelining and how it enables the CPU to work faster
- >> List the registers of the 8086
- >> Code simple MOV and ADD instructions and describe the effect of these instructions on their operands
- >> State the purpose of the code segment, data segment, stack segment, and extra segment
- >> Explain the difference between a logical address and a physical address
- >> Describe the "little endian" storage convention of x86 microprocessors
- >> State the purpose of the stack
- >> Explain the function of PUSH and POP instructions
- >> List the bits of the flag register and briefly state the purpose of each bit
- >> Demonstrate the effect of ADD instructions on the flag register
- >> List the addressing modes of the 8086 and recognize examples of each mode
- >> Know how to use flowcharts and pseudocode in program development

This chapter examines the architecture of the 8086 with some examples of Assembly language programming. Section 1.1 gives a history of the evolution of Intel's family of x86 microprocessors. An overview of the internal workings of 8086 microprocessors is given in Section 1.2. An introduction to 8086 Assembly language programming is covered in Section 1.3. Sections 1.4 and 1.5 cover code and stack segments respectively and show how physical addresses are generated. Section 1.6 explores the flag register and its use in Assembly language programming. Finally, Section 1.7 describes in detail the addressing modes of the 8086.

SECTION 1.1: BRIEF HISTORY OF THE x86 FAMILY

In this section we trace the evolution of Intel's family of microprocessors from the late 1970s, when the personal computer had not yet found widespread acceptance, to the powerful microcomputers widely in use today.

Evolution from 8080/8085 to 8086

In 1978, Intel Corporation introduced a 16-bit microprocessor called the 8086. This processor was a major improvement over the previous generation 8080/8085 series Intel microprocessors in several ways. First, the 8086's capacity of 1 megabyte of memory exceeded the 8080/8085's capability of handling a maximum of 64K bytes of memory. Second, the 8080/8085 was an 8-bit system, meaning that the microprocessor could work on only 8 bits of data at a time. Data larger than 8 bits had to be broken into 8-bit pieces to be processed by the CPU. In contrast, the 8086 is a 16-bit microprocessor. Third, the 8086 was a pipelined processor, as opposed to the nonpipelined 8080/8085. In a system with pipelining, the data and address buses are busy transferring data while the CPU is processing information, thereby increasing the effective processing power of the microprocessor. Although pipelining was a common feature of mini- and mainframe computers, Intel was a pioneer in putting pipelining on a single-chip microprocessor. Section 1.2 discusses pipelining. Table 1-1 shows the evolution of Intel microprocessors up to the 8088.

Table 1-1: Evolution of Intel's Microprocessors (from the 8008 to the 8088)

Product	8008	8080	8085	8086	8088
Year introduced	1972	1974	1976	1978	1979
Technology	PMOS	NMOS	NMOS	NMOS	NMOS
Number of pins	18	40	40	40	40
Number of transistors	3000	4500	6500	29,000	29,000
Number of instructions	66	111	113	133	133
Physical memory	16K	64K	64K	1M	1M
Virtual memory	None	None	None	None	None
Internal data bus	8	8	8	16	16
External data bus	8	8	8	16	8
Address bus	8	16	16	20	20
Data types	8	8	8	8/16	8/16

Evolution from 8086 to 8088

The 8086 is a microprocessor with a 16-bit data bus internally and externally, meaning that all registers are 16 bits wide and there is a 16-bit data bus to transfer data in and out of the CPU. Although the introduction of the 8086 marked a great advancement over the previous generation of microprocessors, there was still some resistance in using the 16-bit external data bus since at that time all peripherals were designed around an 8-bit microprocessor. In addition, a printed circuit board with a 16-bit data bus was much more expensive. Therefore, Intel came out with the 8088 version. It is identical to the 8086 as far as programming is concerned, but externally it has an 8-bit data bus instead of a 16-bit bus. It has the same memory capacity, 1 megabyte.

Success of the 8088

In 1981, Intel's fortunes changed forever when IBM picked up the 8088 as their microprocessor of choice in designing the IBM PC. The 8088-based IBM PC was an enormous success, largely because IBM and Microsoft (the developer of the MS-DOS/Windows operating system) made it an open system, meaning that all documentation and specifications of the hardware and software of the PC were made public. This made it possible for many other vendors to clone the hardware successfully and thus spawned a major growth in both hardware and software designs based on the IBM PC. This is in contrast with the Apple computer, which was a closed system, blocking any attempt at cloning by other manufacturers, both domestically and overseas.

Other microprocessors: the 80286, 80386, and 80486

With a major victory behind Intel and a need from PC users for a more powerful microprocessor, Intel introduced the 80286 in 1982. Its features included 16-bit internal and external data buses; 24 address lines, which give 16 megabytes of memory (2^{24} = 16 megabytes); and most significantly, virtual memory. The 80286 can operate in one of two modes: real mode or protected mode. Real mode is simply a faster 8088/8086 with the same maximum of 1 megabyte of memory. Protected mode allows for 16M of memory but is also capable of protecting the operating system and programs from accidental or deliberate destruction by a user, a feature that is absent in the single-user 8088/8086. Virtual memory is a way of fooling the microprocessor into thinking that it has access to an almost unlimited amount of memory by swapping data between disk storage and RAM. IBM picked up the 80286 for the design of the IBM PC AT, and the clone makers followed IBM's lead.

With users demanding even more powerful systems, in 1985 Intel introduced the 80386 (sometimes called 80386DX), internally and externally a 32-bit microprocessor with a 32-bit address bus. It is capable of handling physical memory of up to 4 gigabytes (2^{32}). Virtual memory was increased to 64 terabytes (2^{46}). All microprocessors discussed so far were general-purpose microprocessors and could not handle mathematical calculations rapidly. For this reason, Intel introduced numeric data processing chips, called math coprocessors, such as the 8087, 80287, and 80387. Later Intel introduced the 386SX, which is internally identical to the 80386 but has a 16-bit external data bus and a 24-bit address bus, which gives a capacity of 16 megabytes (2^{24}) of memory. This makes the 386SX system much cheaper. With the introduction of the 80486 in 1989, Intel put a greatly enhanced version of the 80386 and the math coprocessor on a single chip plus additional features such as cache memory. Cache memory is static RAM with a very fast access time. Table 1-2 summarizes the evolution of Intel's microprocessors from the 8086 to the Pentium Pro. It must be noted that all programs written for the 8088/86 will run on 286, 386, and 486 computers.

Table 1-2: Evolution of Intel's Microprocessors (from the 8086 to the Pentium Pro)

Product	8086	80286	80386	80486	Pentium	Pentium Pro
Year Introduced	1978	1982	1985	1989	1993	1995
Technology	NMOS	NMOS	CMOS	CMOS	BICMOS	BICMOS
Clock rate (MHz)	3–10	10–16	16–33	25–33	60, 66	150
Number of pins	40	68	132	168	273	387
Number of transistors	29,000	134,000	275,000	1.2 mill.	3.1 mill.	5.5 mill.
Physical memory	1M	16M	4G	4G	4G	64G
Virtual memory	None	1G	64T	64T	64T	64T
Internal data bus	16	16	32	32	32	32
External data bus	16	16	32	32	64	64
Address bus	20	24	32	32	32	36
Data types	8/16	8/16	8/16/32	8/16/32	8/16/32	8/16/32

e-source

See Intel's Microprocessor Quick Reference Guide at:

http://www.intel.com/pressroom/kits/quickreffam.htm

Table 1-3: Evolution of Intel x86 Microprocessors: From Pentium II to Itanium

Product	Pentium II	Pentium III	Pentium 4	Itanium II
Year introduced	1997	1999	2000	2002
Technology	BICMOS	BICMOS	BICMOS	BICMOS
Number of transistors	7.5 mill.	9.5 mill.	42 mill.	220 mill.
Cache size	512K	512K	512K	3MB
Physical memory	64G	64G	64G	64G
Virtual memory	64T	64T	64T	64T
Internal data bus	32	32	32	64
External data bus	64	64	64	64
Address bus	36	36	36	64
Data types	8/16/32	8/16/32	8/16/32	8/16/32/64

Pentium and Pentium Pro

In 1992, Intel announced release of the newest x86 microprocessor, the Intel Pentium. It was given the name Pentium instead of the expected name 80586 because numbers cannot be copyrighted, whereas a name such as Pentium can be copyrighted. By using submicron fabrication technology, Intel designers were able to utilize more than 3 million transistors on the Pentium chip. Upon its release, the Pentium had speeds of 60 and 66 MHz, but new design features made its processing speed twice that of the 66-MHz 80486 and over 300 times faster than that of the original 8088. The Pentium processor is fully compatible with previous x86 processors but includes several new features, including separate 8K cache memory for code and data, a 64-bit bus, and a vastly improved floating-point processor. The Pentium is packaged in a 273-pin PGA chip. It uses BICMOS technology, which combines the speed of bipolar transistors with the power efficiency of CMOS technology. Although it has a 64-bit data bus, its registers are 32-bit and it has a 32-bit address bus capable of addressing 4 gigabytes of memory. In 1995 Intel introduced the Pentium Pro, the sixth generation of the x86 family. It is an enhanced version of the Pentium that uses 5.5 million transistors. It was designed to be used primarily for 32-bit servers and workstations. Table 1-3 shows the evolution of Intel's microprocessors from the Pentium II to the Itanium II.

Pentium II

In 1997 Intel introduced its Pentium II processor. This 7.5-million-transistor processor featured MMX (MultiMedia extention) technology incorporated into the CPU. MMX allows for fast graphics and audio processing. In 1998 the Pentium II Xeon processor was released. Its primary market is for servers and workstations. In 1999 the Celeron was released. Its lower cost and good performance make it ideal for PCs used to meet educational and home business needs.

Pentium III

In 1999 Intel released the Pentium III. This 9.5-million-transistor processor includes 70 new instructions called SIMD that enhance video and audio performance in such areas as 3-D imaging, and streaming audio that have become common features of online computing. In 1999 Intel also introduced the Pentium III Xeon processor, designed more for servers and business workstations with multiprocessor configurations.

Pentium 4

The Pentium 4, which debuted late in 1999, boasts the speeds of 1.4 to 1.5 GHz. The Pentium 4 represents the first completely new architecture since the development of the Pentium Pro. The new 32-bit architecture, called NetBurst, is designed for heavy multimedia processing such as video, music, and graphic file manipulation on the Internet. The system bus operates at 400 MHz. In addition, new cache and pipelining technology and an expansion of the multimedia instruction set are designed to make the P4 a high-end media processing microprocessor.

Intel 64 Architecture

Intel has selected Itanium as the new brand name for the first product in its 64-bit family of processors, formerly called Merced. The evolution of microprocessors is increasingly influenced by the evolution of the Internet. The Itanium architecture is designed to meet Internet-driven needs for powerful servers and high-performance workstations. The Itanium will have the ability to execute many instructions simultaneously plus extremely large memory capabilities. See Chapter 24 for more.

Review Questions

1. Name three features of the 8086 that were improvements over the 8080/8085.
2. What is the major difference between 8088 and 8086 microprocessors?
3. Give the size of the address bus and physical memory capacity of the following:
   (a) 8086 (b) 80286 (c) 80386
4. The 80286 is a _______-bit microprocessor, whereas the 80386 is a _______-bit microprocessor.
5. State the major difference between the 80386 and the 80386SX.
6. List additional features introduced with the 80286 that were not present in the 8086.
7. List additional features of the 80486 that were not present in the 80386.
8. List additional features of the Pentium that were not present in the 80486.
9. How many transistors did the Pentium II use?
10. Which microprocessor was the first to incorporate MMX technology on-chip?
11. Give the additional features of the Pentium II that were not present in the Pentium.
12. Give all the data types supported by the Pentium 4.
13. Give all the data types supported by the Itanium.
14. True or false. Itanium has a 64-bit architecture.

SECTION 1.2: INSIDE THE 8088/86

In this section we explore concepts important to the internal operation of the 8088/86, such as pipelining and registers. See the block diagram in Figure 1-1.

Pipelining

There are two ways to make the CPU process information faster: increase the working frequency or change the internal architecture of the CPU. The first option is technology dependent, meaning that the designer must use whatever technology is available at the time, with consideration for cost. The technology and materials used in making ICs (integrated circuits) determine the working frequency, power consumption, and the number of transistors packed into a single-chip microprocessor. More discussion of IC technology is given in Chapter 25. It is sufficient for the purpose at hand to say that designers can make the CPU work faster by increasing the frequency under which it runs if technology and cost allow. The second option for improving the processing power of the CPU has to do with the internal working of the CPU. In the 8085 microprocessor, the CPU could either fetch or execute at a given time. In other words, the CPU had to fetch an instruction from memory, then execute it and then fetch again, execute it, and so on.

The idea of pipelining in its simplest form is to allow the CPU to fetch and execute at the same time as shown in Figure 1-2.

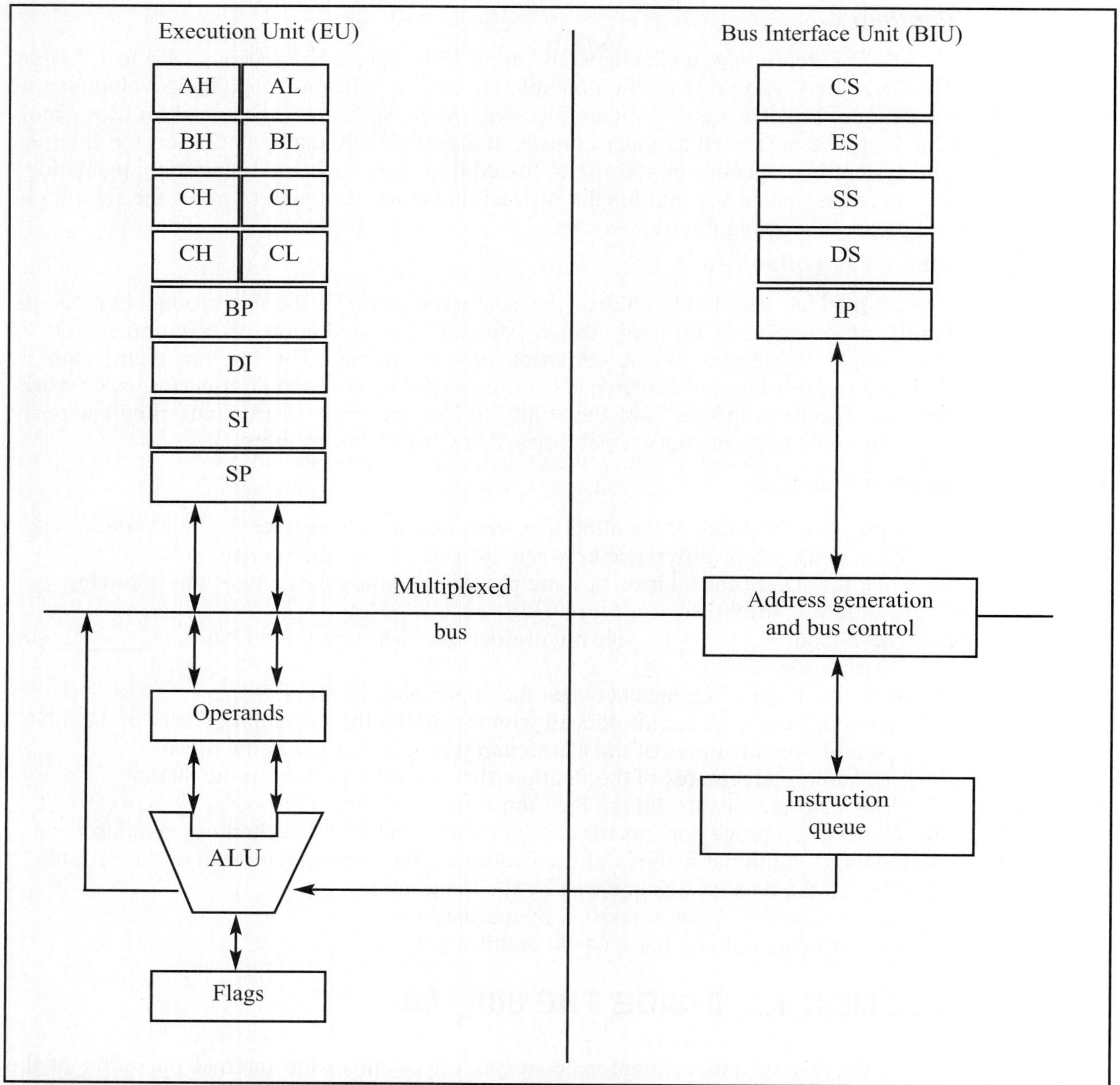

Figure 1-1. Internal Block Diagram of the 8088/86 CPU
(Reprinted by permission of Intel Corporation, Copyright Intel Corp. 1989)

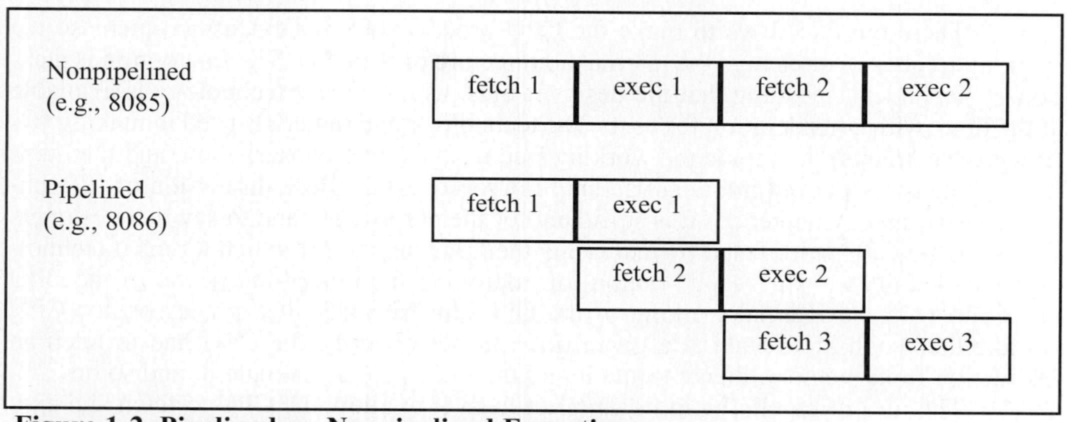

Figure 1-2. Pipelined vs. Nonpipelined Execution

Intel implemented the concept of pipelining in the 8088/86 by splitting the internal structure of the microprocessor into two sections: the execution unit (EU) and the bus interface unit (BIU). These two sections work simultaneously. The BIU accesses memory and peripherals while the EU executes instructions previously fetched. This works only if the BIU keeps ahead of the EU; thus the BIU of the 8088/86 has a buffer, or queue (see Figure 1-1). The buffer is 4 bytes long in the 8088 and 6 bytes in the 8086. If any instruction takes too long to execute, the queue is filled to its maximum capacity and the buses will sit idle. The BIU fetches a new instruction whenever the queue has room for 2 bytes in the 6-byte 8086 queue, and for 1 byte in the 4-byte 8088 queue. In some circumstances, the microprocessor must flush out the queue. For example, when a jump instruction is executed, the BIU starts to fetch information from the new location in memory and information in the queue that was fetched previously is discarded. In this situation the EU must wait until the BIU fetches the new instruction. This is referred to in computer science terminology as a branch penalty. In a pipelined CPU, this means that too much jumping around reduces the efficiency of a program. Pipelining in the 8088/86 has two stages, fetch and execute, but in more powerful computers pipelining can have many stages. The concept of pipelining combined with an increased number of data bus pins has, in recent years, led to the design of very powerful microprocessors.

Registers

AX 16-bit register	
AH 8-bit register	AL 8-bit register

In the CPU, registers are used to store information temporarily. That information could be one or two bytes of data to be processed or the address of data. The registers of the 8088/86 fall into the six categories outlined in Table 1-4. The general-purpose registers in 8088/86 microprocessors can be accessed as either 16-bit or 8-bit registers. All other registers can be accessed only as the full 16 bits. In the 8088/86, data types are either 8 or 16 bits. To access 12-bit data, for example, a 16-bit register must be used with the highest 4 bits set to 0. The bits of a register are numbered in descending order, as shown below.

8-bit register:

D7	D6	D5	D4	D3	D2	D1	D0

16-bit register:

D15	D14	D13	D12	D11	D10	D9	D8	D7	D6	D5	D4	D3	D2	D1	D0

Different registers in the 8088/86 are used for different functions, and since some instructions use only specific registers to perform their tasks, the use of registers will be described in the context of instructions and their application in a given program. The first letter of each general register indicates its use. AX is used for the accumulator, BX as a base addressing register, CX as a counter in loop operations, and DX to point to data in I/O operations. Table 1-4 lists the registers of the 8088/86/286.

Table 1-4: Registers of the 8088/86/286 by Category

Category	Bits	Register Names
General	16	AX, BX, CX, DX
	8	AH, AL, BH, BL, CH, CL, DH, DL
Pointer	16	SP (stack pointer), BP (base pointer)
Index	16	SI (source index), DI (destination index)
Segment	16	CS (code segment), DS (data segment), SS (stack segment), ES (extra segment)
Instruction	16	IP (instruction pointer)
Flag	16	FR (flag register)

Note: The general registers can be accessed as the full 16 bits (such as AX), or as the high byte only (AH) or low byte only (AL).

Review Questions

1. Explain the functions of the EU and the BIU.
2. What is pipelining, and how does it make the CPU execute faster?
3. Registers of the 8086 are either _____ bits or _____ bits in length.
4. List the 16-bit registers of the 8086.

SECTION 1.3: INTRODUCTION TO ASSEMBLY PROGRAMMING

While the CPU can work only in binary, it can do so at very high speeds. However, it is quite tedious and slow for humans to deal with 0s and 1s in order to program the computer. A program that consists of 0s and 1s is called *machine language*, and in the early days of the computer, programmers actually coded programs in machine language. Although the hexadecimal system was used as a more efficient way to represent binary numbers, the process of working in machine code was still cumbersome for humans. Eventually, *Assembly languages* were developed, which provided mnemonics for the machine code instructions, plus other features that made programming faster and less prone to error. The term *mnemonic* is typically used in computer science and engineering literature to refer to codes and abbreviations that are relatively easy to remember. Assembly language programs must be translated into machine code by a program called an *assembler*. Assembly language is referred to as a *low-level language* because it deals directly with the internal structure of the CPU. To program in Assembly language, the programmer must know the number of registers and their size, as well as other details of the CPU.

Today, one can use many different programming languages, such as C/C++, BASIC, C#, and numerous others. These languages are called *high-level languages* because the programmer does not have to be concerned with the internal details of the CPU. Whereas an assembler is used to translate an Assembly language program into machine code (sometimes called *object code*), high-level languages are translated into machine code by a program called a *compiler*. For instance, to write a program in C, one must use a C compiler to translate the program into machine language.

There are numerous assemblers available for translating x86 Assembly language programs into machine code. One of the most commonly used assemblers, MASM by Microsoft, is introduced in Chapter 2. The present chapter is designed to correspond to Appendix A: DEBUG Programming. The program in this chapter can be entered and run with the use of the DEBUG program. If you are not familiar with DEBUG, refer to Appendix A for a tutorial introduction. The DEBUG utility is provided with the Microsoft Windows operating system and therefore is widely accessible.

Assembly language programming

An Assembly language program consists of, among other things, a series of lines of Assembly language instructions. An Assembly language instruction consists of a mnemonic, optionally followed by one or two operands. The operands are the data items being manipulated, and the mnemonics are the commands to the CPU, telling it what to do with those items. We introduce Assembly language programming with two widely used instructions: the move and add instructions.

MOV instruction

Simply stated, the MOV instruction copies data from one location to another. It has the following format:

```
MOV  destination,source ;copy source operand to destination
```

This instruction tells the CPU to move (in reality, copy) the source operand to the destination operand. For example, the instruction "`MOV DX,CX`" copies the contents of register CX to register DX. After this instruction is executed, register DX will have the same value as register CX. The MOV instruction does not affect the source operand. The following program first loads CL with value 55H, then moves this value around to various registers inside the CPU.

```
MOV   CL,55H ;move 55H into register CL
MOV   DL,CL ;copy the contents of CL into DL (now DL=CL=55H)
MOV   AH,DL ;copy the contents of DL into AH (now AH=DL=55H)
MOV   AL,AH ;copy the contents of AH into AL (now AL=AH=55H)
MOV   BH,CL ;copy the contents of CL into BH (now BH=CL=55H)
MOV   CH,BH ;copy the contents of BH into CH (now CH=BH=55H)
```

The use of 16-bit registers is demonstrated below.

```
MOV   CX,468FH   ;move 468FH into CX (now CH=46,CL=8F)
MOV   AX,CX    ;copy contents of CX to AX (now AX=CX=468FH)
MOV   DX,AX    ;copy contents of AX to DX (now DX=AX=468FH)
MOV   BX,DX    ;copy contents of DX to BX (now BX=DX=468FH)
MOV   DI,BX    ;now DI=BX=468FH
MOV   SI,DI    ;now SI=DI=468FH
MOV   DS,SI    ;now DS=SI=468FH
MOV   BP,DI    ;now BP=DI=468FH
```

In the 8086 CPU, data can be moved among all the registers shown in Table 1-4 (except the flag register) as long as the source and destination registers match in size. Code such as "MOV AL,DX" will cause an error, since one cannot move the contents of a 16-bit register into an 8-bit register. There is no such instruction as "MOV FR,AX". Loading the flag register is done through other means, discussed in later chapters.

If data can be moved among all registers including the segment registers, can data be moved directly into all registers? The answer is no. Data can be moved directly into nonsegment registers only, using the MOV instruction. For example, look at the following instructions to see which are legal and which are illegal.

```
MOV   AX,58FCH   ;move 58FCH into AX   (LEGAL)
MOV   DX,6678H   ;move 6678H into DX   (LEGAL)
MOV   SI,924BH   ;move 924B  into SI   (LEGAL)
MOV   BP,2459H   ;move 2459H into BP   (LEGAL)
MOV   DS,2341H   ;move 2341H into DS   (ILLEGAL)
MOV   CX,8876H   ;move 8876H into CX   (LEGAL)
MOV   CS,3F47H   ;move 3F47H into CS   (ILLEGAL)
MOV   BH,99H     ;move 99H into BH     (LEGAL)
```

From the discussion above, note the following three points:

1. Values cannot be loaded directly into any segment register (CS, DS, ES, or SS). To load a value into a segment register, first load it to a nonsegment register and then move it to the segment register, as shown next.

```
MOV   AX,2345H   ;load 2345H into AX
MOV   DS,AX    ;then load the value of AX into DS

MOV   DI,1400H   ;load 1400H into DI
MOV   ES,DI    ;then move it into ES, now ES=DI=1400
```

2. If a value less than FFH is moved into a 16-bit register, the rest of the bits are assumed to be all zeros. For example, in "MOV BX,5" the result will be BX = 0005; that is, BH = 00 and BL = 05.
3. Moving a value that is too large into a register will cause an error.

```
MOV   BL,7F2H      ;ILLEGAL: 7F2H is larger than 8 bits
MOV   AX,2FE456H   ;ILLEGAL: the value is larger than AX
```

ADD instruction

The ADD instruction has the following format:

```
ADD   destination,source    ;ADD the source operand to the destination
```

The ADD instruction tells the CPU to add the source and the destination operands and put the result in the destination. To add two numbers such as 25H and 34H, each can be moved to a register and then added together:

```
MOV   AL,25H    ;move 25 into AL
MOV   BL,34H    ;move 34 into BL
ADD   AL,BL     ;AL = AL + BL
```

Executing the program above results in AL = 59H (25H + 34H = 59H) and BL = 34H. Notice that the contents of BL do not change. The program above can be written in many ways, depending on the registers used. Another way might be:

```
MOV   DH,25H    ;move 25 into DH
MOV   CL,34H    ;move 34 into CL
ADD DH,CL     ;add CL to DH: DH = DH + CL
```

The program above results in DH = 59H and CL = 34H. There are always many ways to write the same program. One question that might come to mind after looking at the program above is whether it is necessary to move both data items into registers before adding them together. No, it is not necessary. Look at the following variation:

```
MOV DH,25H    ;load one operand into DH
ADD DH,34H    ;add the second operand to DH
```

In the case above, while one register contained one value, the second value followed the instruction as an operand. This is called an *immediate operand*. The examples shown so far for the ADD and MOV instructions show that the source operand can be either a register or immediate data. In the examples above, the destination operand has always been a register. The format for Assembly language instructions, descriptions of their use, and a listing of legal operand types are provided in Appendix B.

The largest number that an 8-bit register can hold is FFH. To use numbers larger than FFH (255 decimal), 16-bit registers such as AX, BX, CX, or DX must be used. For example, to add 34EH and 6A5H, the following program can be used:

```
MOV AX,34EH    ;move 34EH into AX
MOV DX,6A5H    ;move 6A5H into DX
ADD   DX,AX     ;add AX to DX: DX = DX + AX
```

Running the program above gives DX = 9F3H (34E + 6A5 = 9F3) and AX = 34E. Again, any 16-bit nonsegment registers could have been used to perform the action above:

```
MOV   CX,34EH   ;load 34EH into CX
ADD CX,6A5H ;add 6A5H to CX (now CX=9F3H)
```

The general-purpose registers are typically used in arithmetic operations. Register AX is sometimes referred to as the *accumulator*.

Review Questions

1. Write the Assembly language instruction to move value 1234H into register BX.
2. Write the Assembly language instructions to add the values 16H and ABH. Place the result in register AX.
3. No value can be moved directly into which registers?
4. What is the largest hex value that can be moved into a 16-bit register? Into an 8-bit register? What are the decimal equivalents of these hex values?

SECTION 1.4: INTRODUCTION TO PROGRAM SEGMENTS

A typical Assembly language program consists of at least three segments: a code segment, a data segment, and a stack segment. The *code segment* contains the Assembly language instructions that perform the tasks that the program was designed to accomplish. The *data segment* is used to store information (data) that needs to be processed by the instructions in the code segment. The *stack* is used by the CPU to store information temporarily. In this section we describe the code and data segments of a program in the context of some examples and discuss the way data is stored in memory. The stack segment is covered in Section 1.5.

Origin and definition of the segment

A segment is an area of memory that includes up to 64K bytes and begins on an address evenly divisible by 16 (such an address ends in 0H). The segment size of 64K bytes came about because the 8085 microprocessor could address a maximum of 64K bytes of physical memory since it had only 16 pins for the address lines (2^{16} = 64K). This limitation was carried into the design of the 8088/86 to ensure compatibility. Whereas in the 8085 there was only 64K bytes of memory for all code, data, and stack information, in the 8088/86 there can be up to 64K bytes of memory assigned to each category. Within an Assembly language program, these categories are called the *code segment*, *data segment*, and *stack segment*. For this reason, the 8088/86 can only handle a maximum of 64K bytes of code, 64K bytes of data, and 64K bytes of stack at any given time, although it has a range of 1 megabyte of memory because of its 20 address pins (2^{20} = 1 megabyte). How to move this window of 64K bytes to cover all 1 megabyte of memory is discussed below, after we discuss logical address and physical address.

Logical address and physical address

In Intel literature concerning the 8086, there are three types of addresses mentioned frequently: the physical address, the offset address, and the logical address. The *physical address* is the 20-bit address that is actually put on the address pins of the 8086 microprocessor and decoded by the memory interfacing circuitry. This address can have a range of 00000H to FFFFFH for the 8086 and real-mode 286, 386, and 486 CPUs. This is an actual physical location in RAM or ROM within the 1 megabyte memory range. The *offset address* is a location within a 64K-byte segment range. Therefore, an offset address can range from 0000H to FFFFH. The *logical address* consists of a segment value and an offset address. The differences among these addresses and the process of converting from one to another is best understood in the context of some examples, as shown next.

Code segment

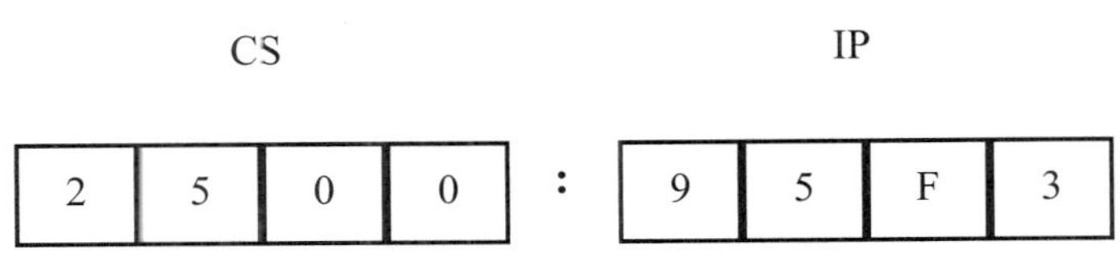

To execute a program, the 8086 fetches the instructions (opcodes and operands) from the code segment. The logical address of an instruction always consists of a CS (code segment) and an IP (instruction pointer), shown in CS:IP format. The physical address for the location of the instruction is generated by shifting the CS left one hex digit and then adding it to the IP. IP contains the offset address. The resulting 20-bit address is called the physical address since it is put on the external physical address bus pins to be decoded by the memory decoding circuitry. To clarify this important concept, assume values in CS and IP as shown in the diagram. The offset address is contained in IP; in this case it is 95F3H. The logical address is CS:IP, or 2500:95F3H. The physical address will be 25000 + 95F3 = 2E5F3H. The physical address of an instruction can be calculated as follows:

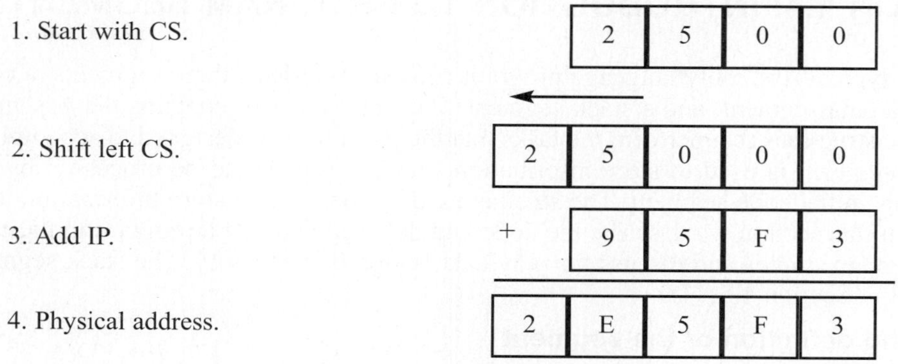

The microprocessor will retrieve the instruction from memory locations starting at 2E5F3. Since IP can have a minimum value of 0000H and a maximum of FFFFH, the logical address range in this example is 2500:0000 to 2500:FFFF. This means that the lowest memory location of the code segment above will be 25000H (25000 + 0000) and the highest memory location will be 34FFFH (25000 + FFFF). What happens if the desired instructions are located beyond these two limits? The answer is that the value of CS must be changed to access those instructions. See Example 1-1.

Example 1-1

If CS = 24F6H and IP = 634AH, show (a) the logical address, and (b) the offset address. Calculate (c) the physical address, (d) the lower range, and (e) the upper range of the code segment.

Solution:

(a) 24F6:634A (b) 634A (c) 2B2AA (24F60 + 634A)
(d) 24F60 (24F60 + 0000) (e) 34F5F (24F60 + FFFF)

Logical address vs. physical address in the code segment

In the code segment, CS and IP hold the logical address of the instructions to be executed. The following Assembly language instructions have been assembled (translated into machine code) and stored in memory. The three columns show the logical address of CS:IP, the machine code stored at that address, and the corresponding Assembly language code.

```
LOGICAL ADDRESS     MACHINE LANGUAGE       ASSEMBLY LANGUAGE
CS:IP               OPCODE AND OPERAND     MNEMONICS AND OPERAND
1132:0100           B057                   MOV   AL,57
1132:0102           B686                   MOV   DH,86
1132:0104           B272                   MOV   DL,72
1132:0106           89D1                   MOV   CX,DX
1132:0108           88C7                   MOV   BH,AL
1132:010A           B39F                   MOV   BL,9F
1132:010C           B420                   MOV   AH,20
1132:010E           01D0                   ADD   AX,DX
1132:0110           01D9                   ADD   CX,BX
1132:0112           05351F                 ADD   AX,1F35
```

The program above shows that the byte at address 1132:0100 contains B0, which is the opcode for moving a value into register AL, and address 1132:0101 contains the operand (in this case 57) to be moved to AL. Therefore, the instruction "`MOV AL,57`" has a machine code of B057, where B0 is the opcode and 57 is the operand. Similarly, the machine code B686 is located in memory locations 1132:0102 and 1132:0103 and represents the opcode and the operand for the instruction "`MOV DH,86`". The physical address is an actual location within RAM (or even ROM). The following are the physical addresses and the contents of each location for the program above. Remember that it is the physical address that is put on the address bus by the 8086 CPU to be decoded by the memory circuitry:

LOGICAL ADDRESS	PHYSICAL ADDRESS	MACHINE CODE CONTENTS
1132:0100	11420	B0
1132:0101	11421	57
1132:0102	11422	B6
1132:0103	11423	86
1132:0104	11424	B2
1132:0105	11425	72
1132:0106	11426	89
1132:0107	11427	D1
1132:0108	11428	88
1132:0109	11429	C7
1132:010A	1142A	B3
1132:010B	1142B	9F
1132:010C	1142C	B4
1132:010D	1142D	20
1132:010E	1142E	01
1132:010F	1142F	D0
1132:0110	11430	01
1132:0111	11431	D9
1132:0112	11432	05
1132:0113	11433	35
1132:0114	11434	1F

Data segment

Assume that a program is being written to add 5 bytes of data, such as 25H, 12H, 15H, 1FH, and 2BH, where each byte represents a person's daily overtime pay. One way to add them is as follows:

```
MOV   AL,00H   ;initialize AL
ADD   AL,25H   ;add 25H to AL
ADD   AL,12H   ;add 12H to AL
ADD   AL,15H   ;add 15H to AL
ADD   AL,1FH   ;add 1FH to AL
ADD   AL,2BH   ;add 2BH to AL
```

In the program above, the data and code are mixed together in the instructions. The problem with writing the program this way is that if the data changes, the code must be searched for every place the data is included, and the data retyped. For this reason, the idea arose to set aside an area of memory strictly for data. In x86 microprocessors, the area of memory set aside for data is called the *data segment*. Just as the code segment is associated with CS and IP as its segment register and offset, the data segment uses register DS and an offset value.

The following demonstrates how data can be stored in the data segment and the program rewritten so that it can be used for any set of data. Assume that the offset for the data segment begins at 200H. The data is placed in memory locations:

```
DS:0200 = 25
DS:0201 = 12
DS:0202 = 15
DS:0203 = 1F
DS:0204 = 2B
```

and the program can be rewritten as follows:

```
MOV  AL,0       ;clear AL
ADD  AL,[0200]  ;add the contents of DS:200 to AL
ADD  AL,[0201]  ;add the contents of DS:201 to AL
ADD  AL,[0202]  ;add the contents of DS:202 to AL
ADD  AL,[0203]  ;add the contents of DS:203 to AL
ADD  AL,[0204]  ;add the contents of DS:204 to AL
```

Notice that the offset address is enclosed in brackets. The brackets indicate that the operand represents the address of the data and not the data itself. If the brackets were not included, as in "`MOV AL,0200`", the CPU would attempt to move 200 into AL instead of the contents of offset address 200. Keep in mind that there is one important difference in the format of code for MASM and DEBUG in that DEBUG assumes that all numbers are in hex (no "H" suffix is required), whereas MASM assumes that they are in decimal and the "H" must be included for hex data.

This program will run with any set of data. Changing the data has no effect on the code. Although this program is an improvement over the preceding one, it can be improved even further. If the data had to be stored at a different offset address, say 450H, the program would have to be rewritten. One way to solve this problem would be to use a register to hold the offset address, and before each ADD, to increment the register to access the next byte. Next a decision must be made as to which register to use. The 8088/86 allows only the use of registers BX, SI, and DI as offset registers for the data segment. In other words, while CS uses only the IP register as an offset, DS uses only BX, DI, and SI to hold the offset address of the data. The term *pointer* is often used for a register holding an offset address. In the following example, BX is used as a pointer:

```
MOV  AL,0        ;initialize AL
MOV  BX,0200H    ;BX points to offset addr of first byte
ADD  AL,[BX]     ;add the first byte to AL
INC  BX          ;increment BX to point to the next byte
ADD  AL,[BX]     ;add the next byte to AL
INC  BX          ;increment the pointer
ADD  AL,[BX]     ;add the next byte to AL
INC  BX          ;increment the pointer
ADD  AL,[BX]     ;add the last byte to AL
```

The INC instruction adds 1 to (increments) its operand. "`INC BX`" achieves the same result as "`ADD BX,1`". For the program above, if the offset address where data is located is changed, only one instruction will need to be modified and the rest of the program will be unaffected. Examining the program above shows that there is a pattern of two instructions being repeated. This leads to the idea of using a loop to repeat certain instructions. Implementing a loop requires familiarity with the flag register, discussed later in this chapter.

Logical address and physical address in the data segment

The physical address for data is calculated using the same rules as for the code segment. That is, the physical address of data is calculated by shifting DS left one hex digit and adding the offset value, as shown in Examples 1-2, 1-3, and 1-4.

Little endian convention

Previous examples used 8-bit or 1-byte data. In this case the bytes are stored one after another in memory. What happens when 16-bit data is used? For example:

```
MOV  AX,35F3H ;load 35F3H into AX
MOV  [1500],AX  ;copy the contents of AX to offset 1500H
```

In cases like this, the low byte goes to the low memory location and the high byte goes to the high memory address. In the example above, memory location DS:1500 contains F3H and memory location DS:1501 contains 35H (DS:1500 = F3 DS:1501 = 35).

This convention is called little endian versus big endian. The origin of the terms *big endian* and *little endian* is from a *Gulliver's Travels* story about how an egg should be opened: from the little end or the big end. In the big endian method, the high byte goes to the low address, whereas in the little endian method, the high byte goes to the high address and the low byte to the low address. See Example 1-5. All Intel microprocessors and many microcontrollers use the little endian convention. Freescale (formerly Motorola) microprocessors, along with some other microcontrollers, use big endian. This difference might seem as trivial as whether to break an egg from the big end or little end, but it is a nuisance in converting software from one camp to be run on a computer of the other camp.

Example 1-2

Assume that DS is 5000 and the offset is 1950. Calculate the physical address.

Solution:

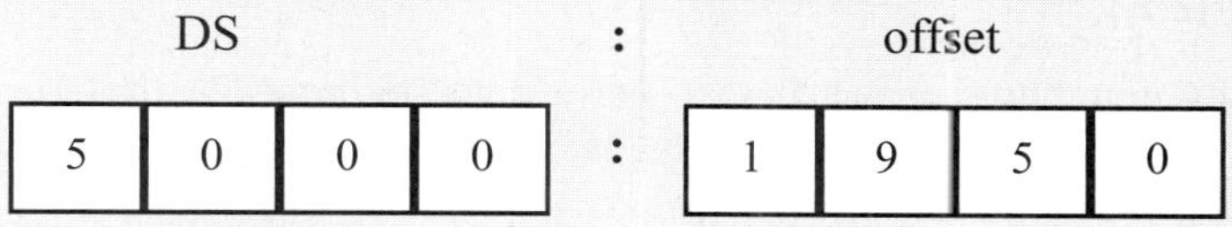

The physical address will be 50000 + 1950 = 51950.

1. Start with DS.

2. Shift DS left.

3. Add the offset.

4. Physical address.

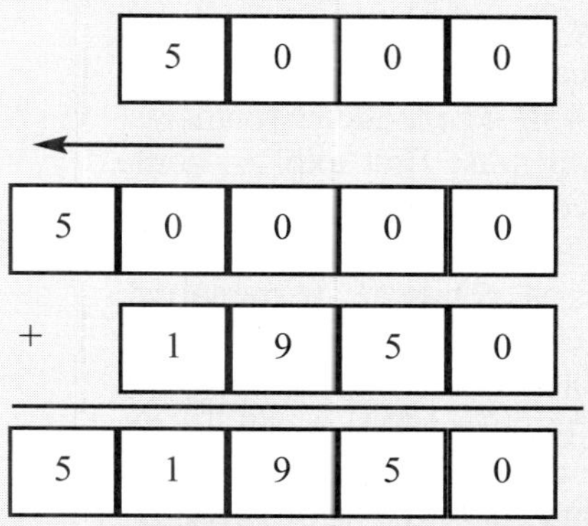

Example 1-3

If DS = 7FA2H and the offset is 438EH, calculate (a) the physical address, (b) the lower range, and (c) the upper range of the data segment. Show (d) the logical address.

Solution:

(a) 83DAE (7FA20 + 438E)   (b) 7FA20 (7FA20 + 0000)

(c) 8FA1F (7FA20 + FFFF)   (d) 7FA2:438E

Example 1-4

Assume that the DS register is 578C. To access a given byte of data at physical memory location 67F66, does the data segment cover the range where the data resides? If not, what changes need to be made?

Solution:

No, since the range is 578C0 to 678BF, location 67F66 is not included in this range. To access that byte, DS must be changed so that its range will include that byte.

Example 1-5

Assume memory locations with the following contents: DS:6826 = 48 and DS:6827 = 22. Show the contents of register BX in the instruction "MOV BX,[6826]".

Solution:

According to the little endian convention used in all x86 microprocessors, register BL should contain the value from the low offset address 6826 and register BH the value from the offset address 6827, giving BL = 48H and BH = 22H.

DS:6826 = 48
DS:6827 = 22

BH	BL
22	48

Extra segment (ES)

ES is a segment register used as an extra data segment. Although in many normal programs this segment is not used, its use is absolutely essential for string operations and is discussed in detail in Chapter 6.

Memory map of the IBM PC

For a program to be executed on the PC, Windows must first load it into RAM. Where in RAM will it be loaded? To answer that question, we must first explain some very important concepts concerning memory in the PC. The 20-bit address of the 8088/86 allows a total of 1 megabyte (1024K bytes) of memory space with the address range 00000–FFFFF. During the design phase of the first IBM PC, engineers had to decide on the allocation of the 1-megabyte memory space to various sections of the PC. This memory allocation is called a *memory map*. The memory map of the IBM PC is shown in Figure 1-3. Of this 1 megabyte, 640K bytes from addresses 00000–9FFFFH were set aside for RAM. The 128K bytes from A0000H to BFFFFH were allocated for video memory. The remaining 256K bytes from C0000H to FFFFFH were set aside for ROM.

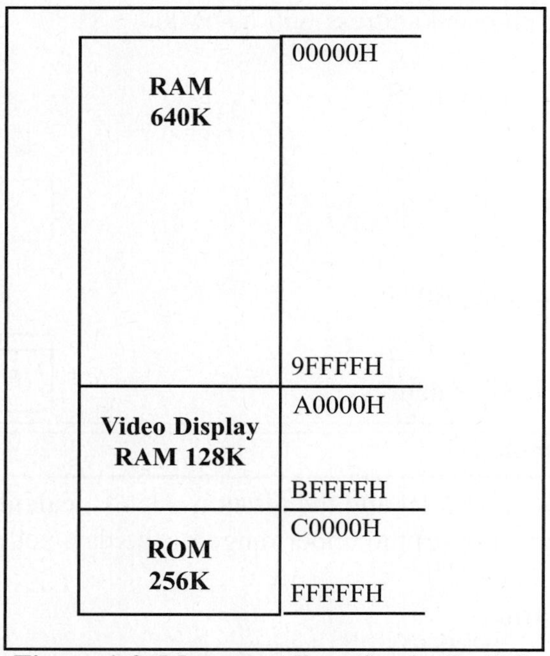

Figure 1-3. Memory Allocation in the PC

More about RAM

In the early 1980s, most PCs came with only 64K to 256K bytes of RAM memory, which was considered more than adequate at the time. Users had to buy memory expansion boards to expand memory up to 640K if they needed additional memory. The need for expansion depends on the Windows version being used and the memory needs of the application software being run. The Windows operating system first allocates the available RAM on the PC for its own use and then lets the rest be used for applications such as word processors. The complicated task of managing RAM memory is left to Windows since the amount of memory used by Windows varies among its various versions and since different computers have different amounts of RAM, plus the fact that the memory needs of application packages vary. For this reason we do not assign any values for the CS, DS, and SS registers since such an assignment means specifying an exact physical address in the range 00000–9FFFFH, and this is beyond the knowledge of the user. Another reason is that assigning a physical address might work on a given PC but it might not work on a PC with a different OS version and RAM size. In other words, the program would not be portable to another PC. Therefore, memory management is one of the most important functions of the operating system and should be left to Windows. This is very important to remember because in many examples in this book we have values for the segment registers CS, DS, and SS that will be different from the values that readers will get on their PCs. Do not try to assign the value to the segment registers to comply with the values in this book.

Video RAM

From A0000H to BFFFFH is set aside for video. The amount used and the location vary depending on the video board installed on the PC. Table E-2 of Appendix E lists the starting addresses for video boards.

More about ROM

From C0000H to FFFFFH is set aside for ROM. Not all the memory space in this range is used by the PC's ROM. Of this 256K bytes, only the 64K bytes from location F0000H–FFFFFH are used by BIOS (basic input/output system) ROM. Some of the remaining space is used by various adapter cards (such as the network card), and the rest is free. In recent years, newer versions of Windows have gained some very powerful memory management capabilities and can put to good use all the unused memory space beyond 640. The 640K-byte memory space from 00000 to 9FFFFH is referred to as *conventional memory*, while the 384K bytes from A0000H to FFFFFH are called the UMB (*upper memory block*) in Microsoft literature.

Function of BIOS ROM

Since the CPU can only execute programs that are stored in memory, there must be some permanent (nonvolatile) memory to hold the programs telling the CPU what to do when the power is turned on. This collection of programs held by ROM is referred to as BIOS in the PC literature. BIOS, which stands for *basic input-output system*, contains programs to test RAM and other components connected to the CPU. It also contains programs that allow Windows to communicate with peripheral devices such as the keyboard, video, printer, and disk. It is the function of BIOS to test all the devices connected to the PC when the computer is turned on and to report any errors. For example, if the keyboard is disconnected from the PC before the computer is turned on, BIOS will report an error on the screen, indicating that condition. It is only after testing and setting up the peripherals that BIOS will load Windows from disk into RAM and hand over control of the PC to Windows. Although there are occasions when either Windows or applications programs need to use programs in BIOS ROM, Windows always controls the PC once it is loaded.

Review Questions

1. A segment is an area of memory that includes up to ____ bytes.
2. How large is a segment in the 8086? Can the physical address 346E0 be the starting address for a segment? Why or why not?
3. State the difference between the physical and logical addresses.
4. A physical address is a ____-bit address; an offset address is a ____-bit address.
5. Which register is used as the offset register with segment register CS?
6. If BX = 1234H and the instruction "MOV [2400],BX" were executed, what would be the contents of memory locations at offsets 2400 and 2401?

SECTION 1.5: THE STACK

In this section we examine the concept of the stack, its use in x86 microprocessors, and its implementation in the stack segment. Then more advanced concepts relating to segments are discussed, such as overlapping segments.

What is a stack, and why is it needed?

The *stack* is a section of read/write memory (RAM) used by the CPU to store information temporarily. The CPU needs this storage area since there are only a limited number of registers. There must be some place for the CPU to store information safely and temporarily. Now one might ask, why not design a CPU with more registers? The reason is that in the design of the CPU, every transistor is precious and not enough of them are available to build hundreds of registers. In addition, how many registers should a CPU have to satisfy every possible program and application? All applications and programming techniques are not the same. In a similar manner, it would be too costly in terms of real estate and construction costs to build a 50-room house to hold everything one might possibly buy throughout his or her lifetime. Instead, one builds or rents a shed for storage.

Having looked at the advantages of having a stack, what are the disadvantages? The main disadvantage of the stack is its access time. Since the stack is in RAM, it takes much longer to access compared to the access time of registers. After all, the registers are inside the CPU and RAM is outside. This is the reason that some very powerful (and consequently, expensive) computers do not have a stack; the CPU has a large number of registers to work with.

How stacks are accessed

If the stack is a section of RAM, there must be registers inside the CPU to point to it. The two main registers used to access the stack are the SS (stack segment) register and the SP (stack pointer) register. These registers must be loaded before any instructions accessing the stack are used. Every register inside the x86 (except segment registers and SP) can be stored in the stack and brought back into the CPU from the stack memory. The storing of a CPU register in the stack is called a *push*, and loading the contents of the stack into the CPU register is called a *pop*. In other words, a register is pushed onto the stack to store it and popped off the stack to retrieve it. The job of the SP is very critical when push and pop are performed. In the x86, the stack pointer register (SP) points at the current memory location used for the top of the stack and as data is pushed onto the stack it is decremented. It is incremented as data is popped off the stack into the CPU. When an instruction pushes or pops a general-purpose register, it must be the entire 16-bit register. In other words, one must code "`PUSH AX`"; there are no instructions such as "`PUSH AL`" or "`PUSH AH`". The reason that the SP is decremented after the push is to make sure that the stack is growing downward from upper addresses to lower addresses. This is the opposite of the IP (instruction pointer). As was seen in the preceding section, the IP points to the next instruction to be executed and is incremented as each instruction is executed. To ensure that the code section and stack section of the program never write over each other, they are located at opposite ends of the RAM memory set aside for the program and they grow toward each other but must not meet. If they meet, the program will crash. To see how the stack grows, look at the following examples.

Pushing onto the stack

Notice in Example 1-6 that as each PUSH is executed, the contents of the register are saved on the stack and SP is decremented by 2. For every byte of data saved on the stack, SP is decremented once, and since push is saving the contents of a 16-bit register, it is decremented twice. Notice also how the data is stored on the stack. In the x86, the lower byte is always stored in the memory location with the lower address. That is the reason that 24H, the contents of AH, is saved in the memory location with the address 1235 and AL in location 1234.

Popping the stack

Popping the contents of the stack back into the x86 CPU is the opposite process of pushing. With every pop, the top 2 bytes of the stack are copied to the register specified by the instruction and the stack pointer is incremented twice. Although the data actually remains in memory, it is not accessible since the stack pointer is beyond that point. Example 1-7 demonstrates the POP instruction.

Example 1-6

Assuming that SP = 1236, AX = 24B6, DI = 85C2, and DX = 5F93, show the contents of the stack as each of the following instructions is executed.

```
PUSH  AX
PUSH  DI
PUSH  DX
```

Solution:

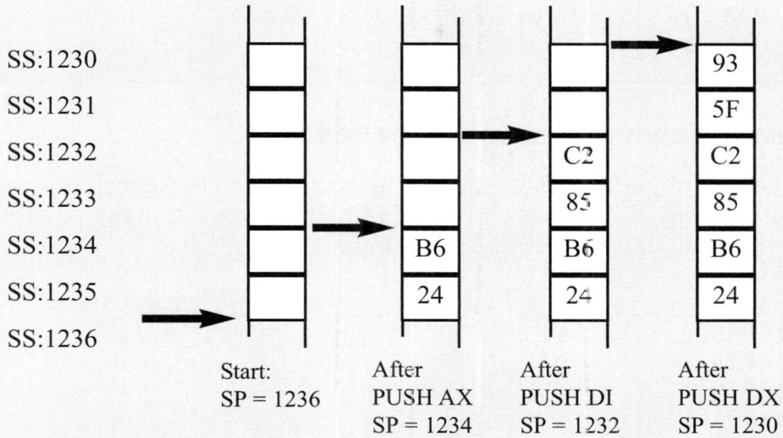

Example 1-7

Assuming that the stack is as shown below, and SP = 18FA, show the contents of the stack and registers as each of the following instructions is executed:

```
POP   CX
POP   DX
POP   BX
```

Solution:

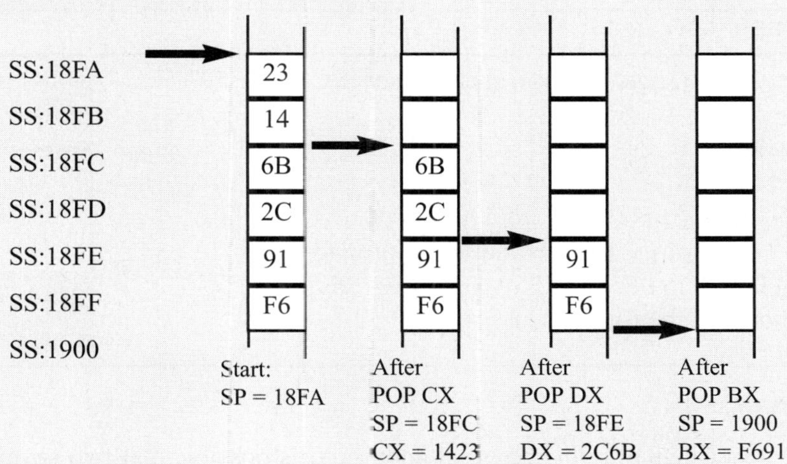

Logical address vs. physical address for the stack

What is the exact physical location of the stack? That depends on the value of the stack segment (SS) register and SP, the stack pointer. To compute physical addresses for the stack, the same principle is applied as was used for the code and data segments. We shift left SS and then add offset SP, the stack pointer register. See Example 1-8.

What values are assigned to the SP and SS, and who assigns them? It is the job of the Windows operating system to assign the values for the SP and SS since memory management is the responsibility of the operating system. Before leaving the discussion of the stack, two points must be made. First, in the x86 literature, the top of the stack is the last stack location occupied. This is different from other CPUs. Second, BP is another register that can be used as an offset into the stack, but it has very special applications and is widely used to access parameters passed between Assembly language programs and high-level language programs such as C.

Example 1-8

If SS = 3500H and the SP is FFFEH,
(a) Calculate the physical address of the stack. (b) Calculate the lower range.
(c) Calculate the upper range of the stack segment. (d) Show the stack's logical address.

Solution:

(a) 44FFE (35000 + FFFE) (b) 35000 (35000 + 0000)
(c) 44FFF (35000 + FFFF) (d) 3500:FFFE

A few more words about segments in the x86

Can a single physical address belong to many different logical addresses? Yes, look at the case of a physical address value of 15020H. There are many possible logical addresses that represent this single physical address:

Logical address (hex)	Physical address (hex)
1000:5020	15020
1500:0020	15020
1502:0000	15020
1400:1020	15020
1302:2000	15020

This shows the dynamic behavior of the segment and offset concept in the 8086 CPU. One last point that must be clarified is the case when adding the offset to the shifted segment register results in an address beyond the maximum allowed range of FFFFFH. In that situation, wrap-around will occur. This is shown in Example 1-9.

Example 1-9

What is the range of physical addresses if CS = FF59?

Solution:

The low range is FF590 (FF590 + 0000).
The range goes to FFFFF and wraps around,
from 00000 to 0F58F (FF590 + FFFF = 0F58F),
as shown in the illustration.

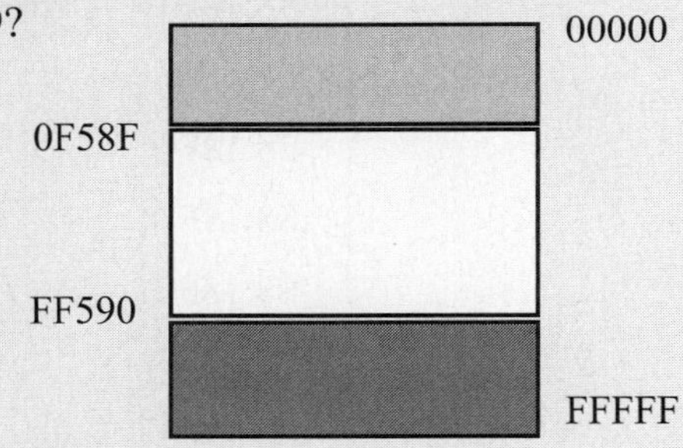

Overlapping

In calculating the physical address, it is possible that two segments can overlap, which is desirable in some circumstances. Figure 1-4 illustrates overlapping and nonoverlapping segments.

Review Questions

1. Which registers are used to access the stack?
2. With each PUSH instruction, the stack pointer register SP is (circle one) incremented/decremented by 2.
3. With each POP instruction, SP is (circle one) incremented/decremented by 2.
4. List three possible logical addresses corresponding to physical address 143F0.

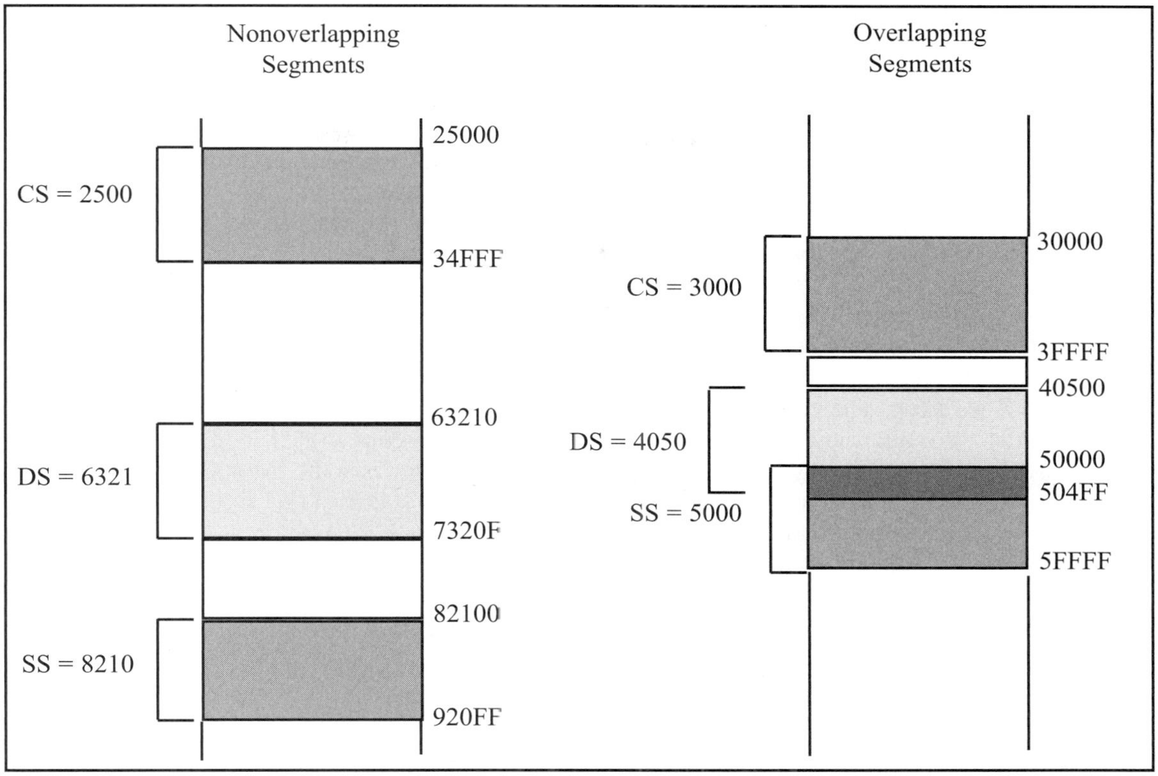

Figure 1-4. Nonoverlapping vs. Overlapping Segments

SECTION 1.6: FLAG REGISTER

In this section we describe the flag register. Many Assembly language instructions alter bits of the flag register and some instructions will function differently based on the information in the flag register. After describing the bits of the flag register and their function, programming examples are given to demonstrate the use of the flag register.

The flag register is a 16-bit register sometimes referred to as the *status register*. Although the register is 16 bits wide, only some of the bits are used. The rest are either undefined or reserved by Intel. Six of the flags are called *conditional flags*, meaning that they indicate some condition that resulted after an instruction was executed. These six are CF, PF, AF, ZF, SF, and OF. The three remaining flags are sometimes called *control flags* since they are used to control the operation of instructions before they are executed. A diagram of the flag register is shown in Figure 1-5.

15	14	13	12	11	10	9	8	7	6	5	4	3	2	1	0
R	R	R	R	OF	DF	IF	F	SF	ZF	U	AF	U	PF	U	CF

R = reserved
U = undefined
OF = overflow flag
DF = direction flag
IF = interrupt flag
TF = trap flag

SF = sign flag
ZF = zero flag
AF = auxiliary carry flag
PF = parity flag
CF = carry flag

Figure 1-5. Flag Register

Bits of the flag register

Below are listed the bits of the flag register that are used in x86 Assembly language programming. A brief explanation of each bit is given. How these flag bits are used will be seen in programming examples throughout the textbook.

CF, the Carry Flag. This flag is set whenever there is a carry out, either from d7 after an 8-bit operation, or from d15 after a 16-bit data operation.

PF, the Parity Flag. After certain operations, the parity of the result's low-order byte is checked. If the byte has an even number of 1s, the parity flag is set to 1; otherwise, it is cleared.

AF, Auxiliary Carry Flag. If there is a carry from d3 to d4 of an operation, this bit is set; otherwise, it is cleared (set equal to zero). This flag is used by the instructions that perform BCD (binary coded decimal) arithmetic.

ZF, the Zero Flag. The zero flag is set to 1 if the result of an arithmetic or logical operation is zero; otherwise, it is cleared.

SF, the Sign Flag. Binary representation of signed numbers uses the most significant bit as the sign bit. After arithmetic or logic operations, the status of this sign bit is copied into the SF, thereby indicating the sign of the result.

TF, the Trap Flag. When this flag is set it allows the program to single-step, meaning to execute one instruction at a time. Single-stepping is used for debugging purposes.

IF, Interrupt Enable Flag. This bit is set or cleared to enable or disable only the external maskable interrupt requests.

DF, the Direction Flag. This bit is used to control the direction of string operations, which are described in Chapter 6.

OF, the Overflow Flag. This flag is set whenever the result of a signed number operation is too large, causing the high-order bit to overflow into the sign bit. In general, the carry flag is used to detect errors in unsigned arithmetic operations. The overflow flag is only used to detect errors in signed arithmetic operations.

Example 1-10

Show how the flag register is affected by the addition of 38H and 2FH.

Solution:

```
        MOV  BH,38H          ;BH= 38H
        ADD  BH,2FH          ;add 2F to BH, now BH=67H

    38      0011  1000
+   2F      0010  1111
    67      0110  0111
```

CF = 0 since there is no carry beyond d7

AF = 1 since there is a carry from d3 to d4

PF = 0 since there is an odd number of 1s in the result

ZF = 0 since the result is not zero

SF = 0 since d7 of the result is zero

Flag register and ADD instruction

In this section we examine the impact of the ADD instruction on the flag register as an example of the use of the flag bits. The flag bits affected by the ADD instruction are CF (carry flag), PF (parity flag), AF (auxiliary carry flag), ZF (zero flag), SF (sign flag), and OF (overflow flag). The overflow flag will be covered in Chapter 6, since it relates only to signed number arithmetic. To understand how each of these flag bits is affected, look at Examples 1-10 and 1-11.

The same concepts apply for 16-bit addition, as shown in Examples 1-12 and 1-13. It is important to notice the differences between 8-bit and 16-bit operations in terms of their impact on the flag bits. The parity bit only counts the lower 8 bits of the result and is set accordingly. Also notice the CF bit. The carry flag is set if there is a carry beyond bit d15 instead of bit d7.

Notice the zero flag (ZF) status after the execution of the ADD instruction. Since the result of the entire 16-bit operation is zero (meaning the contents of BX), ZF is set to

Example 1-11

Show how the flag register is affected by

```
        MOV   AL,9CH      ;AL=9CH
        MOV   DH,64H      ;DH=64H
        ADD   AL,DH       ;now AL=0
```

Solution:

```
        9C          1001  1100
   –    64          0110  0100
        00          0000  0000
```

CF = 1 since there is a carry beyond d7

AF = 1 since there is a carry from d3 to d4

PF = 1 since there is an even number of 1s in the result

ZF = 1 since the result is zero

SF = 0 since d7 of the result is zero

Example 1-12

Show how the flag register is affected by

```
        MOV   AX,34F5H    ;AX= 34F5H
        ADD   AX,95EBH    ;now AX= CAE0H
```

Solution:

```
        34F5        0011  0100  1111  0101
   +    95EB        1001  0101  1110  1011
        CAE0        1100  1010  1110  0000
```

CF = 0 since there is no carry beyond d15

AF = 1 since there is a carry from d3 to d4

PF = 0 since there is an odd number of 1s in the lower byte

ZF = 0 since the result is not zero

SF = 1 since d15 of the result is one

Example 1-13

Show how the flag register is affected by

```
        MOV   BX,AAAAH    ;BX= AAAAH
        ADD   BX,5556H    ;now BX= 0000H
```

Solution:

```
        AAAA        1010  1010  1010  1010
   +    5556        0101  0101  0101  0110
        0000        0000  0000  0000  0000
```

CF = 1 since there is a carry beyond d15

AF = 1 since there is a carry from d3 to d4

PF = 1 since there is an even number of 1s in the lower byte

ZF = 1 since the result is zero

SF = 0 since d15 of the result is zero

Example 1-14

Show how the flag register is affected by

```
        MOV   AX,94C2H    ;AX=94C2H
        MOV   BX,323EH    ;BX=323EH
        ADD   AX,BX       ;now AX=C700H
        MOV   DX,AX       ;now DX=C700H
        MOV   CX,DX       ;now CX=C700H
```

Solution:

```
        94C2        1001  0100  1100  0010
   +    323E        0011  0010  0011  1110
        C700        1100  0111  0000  0000
```

After the ADD operation, the following are the flag bits:

CF = 0 since there is no carry beyond d15

AF = 1 since there is a carry from d3 to d4

PF = 1 since there is an even number of 1s in the lower byte

ZF = 0 since the result is not zero

SF = 1 since d15 of the result is 1

high. Do all instructions affect the flag bits? The answer is no; some instructions such as data transfers (MOV) affect no flags. As an exercise, run these examples on DEBUG to see the effect of various instructions on the flag register.

Running the instructions in Example 1-14 in DEBUG will verify that MOV instructions have no effect on the flag. How these flag bits are used in programming is discussed in future chapters in the context of many applications. In Appendix B we give additional information about the effect of various instructions on the flags.

Use of the zero flag for looping

One of the most widely used applications of the flag register is the use of the zero flag to implement program loops. The term *loop* refers to a set of instructions that is repeated a number of times. For example, to add 5 bytes of data, a counter can be used to keep track of how many times the loop needs to be repeated. Each time the addition is performed the counter is decremented and the zero flag is checked. When the counter becomes zero, the zero flag is set (ZF = 1) and the loop is stopped. The following shows the implementation of the looping concept in the program, which adds 5 bytes of data. Register CX is used to hold the counter and BX is the offset pointer (SI or DI could have been used instead). AL is initialized before the start of the loop. In each iteration, ZF is checked by the JNZ instruction. JNZ stands for "Jump Not Zero" meaning that if ZF = 0, jump to a new address. If ZF = 1, the jump is not performed and the instruction below the jump will be executed. Notice that the JNZ instruction must come immediately after the instruction that decrements CX since JNZ needs to check the effect of "DEC CX" on ZF. If any instruction were placed between them, that instruction might affect the zero flag.

```
        MOV  CX,05     ;CX holds the loop count
        MOV  BX,0200H  ;BX holds the offset data address
        MOV  AL,00     ;initialize AL
ADD_LP: ADD  AL,[BX]   ;add the next byte to AL
        INC  BX        ;increment the data pointer
        DEC  CX        ;decrement the loop counter
        JNZ  ADD_LP    ;jump to next iteration if counter not zero
```

Review Questions

1. The ADD instruction can affect which bits of the flag register?
2. The carry flag will be set to 1 in an 8-bit ADD if there is a carry out from bit ___.
3. CF will be set to 1 in a 16-bit ADD if there is a carry out from bit ____.

SECTION 1.7: x86 ADDRESSING MODES

The CPU can access operands (data) in various ways, called addressing modes. The number of addressing modes is determined when the microprocessor is designed and cannot be changed. The x86 provides a total of seven distinct addressing modes:

1. register
2. immediate
3. direct
4. register indirect
5. based relative
6. indexed relative
7. based indexed relative

Each addressing mode is explained below, and application examples are given in later chapters. ADD and MOV instructions are used below to explain addressing modes.

Register addressing mode

The register addressing mode involves the use of registers to hold the data to be manipulated. Memory is not accessed when this addressing mode is executed; therefore, it is relatively fast. Examples of register addressing mode follow:

```
MOV  BX,DX  ;copy the contents of DX into BX
MOV  ES,AX  ;copy the contents of AX into ES
```

```
ADD   AL,BH   ;add the contents of BH to contents of AL
```

It should be noted that the source and destination registers must match in size. In other words, coding "MOV CL,AX" will give an error, since the source is a 16-bit register and the destination is an 8-bit register.

Immediate addressing mode

In the immediate addressing mode, the source operand is a constant. In immediate addressing mode, as the name implies, when the instruction is assembled, the operand comes immediately after the opcode. For this reason, this addressing mode executes quickly. However, in programming it has limited use. Immediate addressing mode can be used to load information into any of the registers except the segment registers and flag registers. Examples:

```
MOV   AX,2550H        ;move 2550H into AX
MOV CX,625            ;load the decimal value 625 into CX
MOV BL,40H            ;load 40H into BL
```

To move information to the segment registers, the data must first be moved to a general-purpose register and then to the segment register. Example:

```
MOV   AX,2550H
MOV   DS,AX
MOV   DS,0123H ;illegal! cannot move data into segment reg.
```

In the first two addressing modes, the operands are either inside the microprocessor or tagged along with the instruction. In most programs, the data to be processed is often in some memory location outside the CPU. There are many ways of accessing the data in the data segment. The following describes those different methods.

Direct addressing mode

In the direct addressing mode the data is in some memory location(s) and the address of the data in memory comes immediately after the instruction. Note that in immediate addressing, the operand itself is provided with the instruction, whereas in direct addressing mode, the address of the operand is provided with the instruction. This address is the offset address and one can calculate the physical address by shifting left the DS register and adding it to the offset as follows:

```
MOV DL,[2400]    ;move contents of DS:2400H into DL
```

In this case the physical address is calculated by combining the contents of offset location 2400 with DS, the data segment register. Notice the bracket around the address. In the absence of this bracket executing the command will give an error since it is interpreted to move the value 2400 (16-bit data) into register DL, an 8-bit register. See Example 1-15.

Example 1-15

Find the physical address of the memory location and its contents after the execution of the following, assuming that DS = 1512H.

```
MOV    AL,99H
MOV    [3518],AL
```

Solution:

First AL is initialized to 99H, then in line two, the contents of AL are moved to logical address DS:3518, which is 1512:3518. Shifting DS left and adding it to the offset gives the physical address of 18638H (15120H + 3518H = 18638H). That means after the execution of the second instruction, the memory location with address 18638H will contain the value 99H.

Register indirect addressing mode

In the register indirect addressing mode, the address of the memory location where the operand resides is held by a register. The registers used for this purpose are SI, DI, and BX. If these three registers are used as pointers, that is, if they hold the offset of the memory location, they must be combined with DS in order to generate the 20-bit physical address. For example:

```
MOV AL,[BX]  ;moves into AL the contents of the memory
             ;location pointed to by DS:BX.
```

Notice that BX is in brackets. In the absence of brackets, the code is interpreted as an instruction moving the contents of register BX to AL (which gives an error because source and destination do not match) instead of the contents of the memory location whose offset address is in BX. The physical address is calculated by shifting DS left one hex position and adding BX to it. The same rules apply when using register SI or DI.

```
MOV   CL,[SI]       ;move contents of DS:SI into CL
MOV   [DI],AH       ;move contents of AH into DS:DI
```

The examples above moved byte-sized data. Example 1-16 shows 16-bit data.

Example 1-16

Assume that DS = 1120, SI = 2498, and AX = 17FE. Show the contents of memory locations after the execution of "`MOV [SI],AX`".

Solution:

The contents of AX are moved into memory locations with logical address DS:SI and DS:SI + 1; therefore, the physical address starts at DS (shifted left) + SI = 13698. According to the little endian convention, low address 13698H contains FE, the low byte, and high address 13699H will contain 17, the high byte.

Based relative addressing mode

In the based relative addressing mode, base registers BX and BP, as well as a displacement value, are used to calculate what is called the *effective address*. The default segments used for the calculation of the physical address (PA) are DS for BX and SS for BP. For example:

```
MOV CX,[BX]+10      ;move DS:BX+10 and DS:BX+10+1 into CX
                    ;PA = DS (shifted left) + BX + 10
```

Alternative codings are "`MOV  CX,[BX+10]`" or "`MOV CX,10[BX]`". Again the low address contents will go into CL and the high address contents into CH. In the case of the BP register,

```
MOV   AL,[BP]+5     ;PA = SS (shifted left) + BP + 5
```

Again, alternative codings are "`MOV AL,[BP+5]`" or "`MOV AL,5[BP]`". A brief mention should be made of the terminology *effective address* used in Intel literature. In "`MOV  AL,[BP]+5`", BP+5 is called the effective address since the fifth byte from the beginning of the offset BP is moved to register AL. Similarly in "`MOV  CX,[BX]+10`", BX+10 is called the effective address.

Indexed relative addressing mode

The indexed relative addressing mode works the same as the based relative addressing mode, except that registers DI and SI hold the offset address. Examples:

```
MOV   DX,[SI]+5     ;PA = DS (shifted left) + SI + 5
MOV   CL,[DI]+20    ;PA = DS (shifted left) + DI + 20
```

Example 1-17 gives further examples of indexed relative addressing mode.

Example 1-17

Assume that DS = 4500, SS = 2000, BX = 2100, SI = 1486, DI = 8500, BP = 7814, and AX = 2512. All values are in hex. Show the exact physical memory location where AX is stored in each of the following. All values are in hex.
(a) `MOV[BX]+20,AX` (b) `MOV[SI]+10,AX`
(c) `MOV[DI]+4,AX` (d) `MOV[BP]+12,AX`

Solution:
In each case PA = segment register (shifted left) + offset register + displacement.
(a) DS:BX+20 location 47120 = (12) and 47121 = (25)
(b) DS:SI+10 location 46496 = (12) and 46497 = (25)
(c) DS:DI+4 location 4D504 = (12) and 4D505 = (25)
(d) SS:BP+12 location 27826 = (12) and 27827 = (25)

Based indexed addressing mode

By combining based and indexed addressing modes, a new addressing mode is derived called the *based indexed addressing mode*. In this mode, one base register and one index register are used. Examples:

```
MOV   CL,[BX][DI]+8    ;PA = DS (shifted left) + BX + DI + 8
MOV   CH,[BX][SI]+20   ;PA = DS (shifted left) + BX + SI + 20
MOV   AH,[BP][DI]+12   ;PA = SS (shifted left) + BP + DI + 12
MOV   AH,[BP][SI]+29   ;PA = SS (shifted left) + BP + SI + 29
```

The coding of the instructions above can vary; for example, the last example could have been written in either of the following two ways:

```
MOV   AH,[BP+SI+29]
MOV   AH,[SI+BP+29]    ;the register order does not matter
```

Note that "`MOV AX,[SI][DI]+displacement`" is illegal.

Table 1-5: Offset Registers for Various Segments

Segment register:	CS	DS	ES	SS
Offset register(s):	IP	SI, DI, BX	SI, DI, BX	SP, BP

In many of the examples above, the MOV instruction was used for the sake of clarity, even though one can use any instruction as long as that instruction supports the addressing mode. For example, the instruction "`ADD DL,[BX]`" would add the contents of the memory location pointed at by DS:BX to the contents of register DL.

Segment overrides

Table 1-5 summarizes the offset registers that can be used with the four segment registers. The x86 CPU allows the program to override the default segment and use any segment register. To do that, specify the segment in the code. For example, in "`MOV AL,[BX]`", the physical address of the operand to be moved into AL is DS:BX, as was shown earlier since DS is the default segment for pointer BX. To override that default, specify the desired segment in the instruction as "`MOV AL,ES:[BX]`". Now the address of the operand being moved to AL is ES:BX instead of DS:BX. Extensive use of all these addressing modes is shown in future chapters in the context of program examples.

Table 1-6 shows more examples of segment overrides shown next to the default address in the absence of the override. Table 1-7 summarizes addressing modes of the 8088/86.

Table 1-6: Sample Segment Overrides

Instruction	Segment Used	Default Segment
MOV AX, CS:[BP]	CS:BP	SS:BP
MOV DX,SS:[SI]	SS:SI	DS:SI
MOV AX,DS:[BP]	DS:BP	SS:BP
MOV CX,ES:[BX]+12	ES:BX+12	DS:BX+12
MOV SS:[BX][DI]+32,AX	SS:BX+DI+32	DS:BX+DI+32

Table 1-7: Summary of the x86 Addressing Modes

Addressing Mode	Operand	Default Segment
Register	reg	none
Immediate	data	none
Direct	[offset]	DS
Register indirect	[BX]	DS
	[SI]	DS
	[DI]	DS
Based relative	[BX]+disp	DS
	[BP]+disp	SS
Indexed relative	[DI]+disp	DS
	[SI]+disp	SS
Based indexed relative	[BX][SI]+disp	DS
	[BX][DI]+disp	DS
	[BP][SI]+disp	SS
	[BP][DI]+disp	SS

Review Questions

1. Can the x86 programmer make up new addressing modes?
2. Is the IP (instruction pointer) register also available in low-byte and high-byte formats?
3. Is the CS (code segment) register also available in low-byte and high-byte formats?
4. Which segment is used for the direct addressing mode?
5. Which registers can be used for the register indirect addressing mode?

PROBLEMS

SECTION 1.1: BRIEF HISTORY OF THE x86 FAMILY

1. Which microprocessor, the 8088 or the 8086, was released first?
2. If the 80286 and 80386SX both have 16-bit external data buses, what is the difference between them?
3. What does "16-bit" or "32-bit" microprocessor mean? Does it refer to the internal or external data path?
4. Do programs written for the 8088/86 run on 80286-, 80386-, and 80486-based CPUs?
5. What does the term *upward compatibility* mean?

6. Name a major difference between the 8088 and the 8086.
7. Which has the larger queue, the 8088 or the 8086?

SECTION 1.2: INSIDE THE 8088/86

8. State another way to increase the processing power of the CPU other than increasing the frequency.
9. What do "BIU" and "EU" stand for, and what are their functions?
10. Name the general-purpose registers of the 8088/86.
    (a) 8-bit (b) 16-bit
11. Which of the following registers cannot be split into high and low bytes?
    (a) CS (b) AX (c) DS
    (d) SS (e) BX (f) DX
    (g) CX (h) SI (i) DI

SECTION 1.3: INTRODUCTION TO ASSEMBLY PROGRAMMING

12. Which of the following instructions cannot be coded in 8088/86 Assembly language? Give the reason why not, if any. To verify your answer, code each in DEBUG. Assume that all numbers are in hex.
    (a) MOV AX,27 (b) MOV AL,97F (c) MOV DS,9BF2
    (d) MOV CX,397 (e) MOV SI,9516 (f) MOV CS,3490
    (g) MOV DS,BX (h) MOV BX,CS (i) MOV CH,AX
    (j) MOV AX,23FB9 (k) MOV CS,BH (l) MOV AX,DL

SECTION 1.4: INTRODUCTION TO PROGRAM SEGMENTS

13. Name the segment registers and their functions in the 8088/86.
14. If CS = 3499H and IP = 2500H, find:
    (a) The logical address (b) The physical address
    (c) The lower and upper ranges of the code segment
15. Repeat Problem 14 with CS = 1296H and IP = 100H.
16. If DS = 3499H and the offset = 3FB9H, find:
    (a) The physical address (b) The logical address of the data being fetched
    (c) The lower and upper range addresses of the data segment
17. Repeat Problem 16 using DS = 1298H and the offset = 7CC8H.
18. Assume that the physical address for a location is 0046CH. Suggest a possible logical address.
19. If an instruction that needs to be fetched is in physical memory location 389F2 and CS = 2700, does the code segment range include it or not? If not, what value should be assigned to CS if the IP must equal 1282?
20. Using DEBUG, assemble and unassemble the following program and provide the logical address, physical address, and the content of each address location. The CS value is decided by Windows, but use IP = 170H.

```
MOV AL,76H
MOV BH,8FH
ADD BH,AL
ADD BH,7BH
MOV BL,BH
ADD BL,AL
```

21. Repeat Problem 20 for the following program from page 36.

```
MOV AL,0        ;clear AL
ADD AL,[0200]   ;add the contents of DS:200 to AL
ADD AL,[0201]   ;add the contents of DS:201 to AL
ADD AL,[0202]   ;add the contents of DS:202 to AL
ADD AL,[0203]   ;add the contents of DS:203 to AL
ADD AL,[0204]   ;add the contents of DS:204 to AL
```

SECTION 1.5: THE STACK

22. The stack is:
    (a) A section of ROM
    (b) A section of RAM used for temporary storage
    (c) A 16-bit register inside the CPU
    (d) Some memory inside the CPU
23. In problem 22, choose the correct answer for the stack pointer.
24. When data is pushed onto the stack, the stack pointer is ______________, but when data is popped off the stack, the stack pointer is ___________.
25. Choose the correct statement:
    (a) The stack segment and code segment start at the same point of read/write memory and grow upward.
    (b) The stack segment and code segment start at opposite points of read/write memory and grow toward each other.
    (c) There will be no problem if the stack and code segments meet each other.
26. What is the main disadvantage of the stack as temporary storage compared to having a large number of registers inside the CPU?
27. If SS = 2000 and SP = 4578, find:
    (a) The physical address
    (b) The logical address
    (c) The lower range of the stack segment
    (d) The upper range of the stack segment
28. If SP = 24FC, what is the offset address of the first location of the stack that is available to push data into?
29. Assume that SP = FF2EH, AX = 3291H, BX = F43CH, and CX = 09. Find the contents of the stack and SP after the execution of each of the following instructions.
```
PUSH   AX
PUSH   BX
PUSH   CX
```
30. Show the sequence of instructions needed to restore all the registers to their original values in Problem 29. Show the content of SP at each point.
31. The following registers are used as offsets. Assuming that the default segment is used to get the logical address, give the segment register associated with each offset.
    (a) BP (b) DI (c) IP
    (d) SI (e) SP (f) BX
32. Show the override segment register and the default segment register used (if there were no override) in each of the following cases.
    (a) MOV SS:[BX],AX (b) MOV SS:[DI],BX
    (c) MOV DX,DS:[BP+6]

SECTION 1.6: FLAG REGISTER

33. Find the status of the CF, PF, AF, ZF, and SF for the following operations.
    (a)MOV BL,9FH (b) MOV AL,23H (c) MOV DX,10FFH
    ADD BL,61H ADD AL,97H ADD DX,1

SECTION 1.7: x86 ADDRESSING MODES

34. Assume that the registers have the following values (all in hex) and that CS = 1000, DS = 2000, SS = 3000, SI = 4000, DI = 5000, BX = 6080, BP = 7000, AX = 25FF, CX = 8791, and DX = 1299. Calculate the physical address of the memory where the operand is stored and the contents of the memory locations in each of the following addressing examples.
    (a) MOV [SI],AL (b) MOV [SI+BX+8],AH
    (c) MOV [BX],AX (d) MOV [DI+6],BX
    (e) MOV [DI][BX]+28,CX (f) MOV [BP][SI]+10,DX

(g) MOV [3600],AX
(h) MOV [BX]+30,DX
(i) MOV [BP]+200,AX
(j) MOV [BP+SI+100],BX
(k) MOV [SI]+50,AH
(l) MOV [DI+BP+100],AX.

35. Give the addressing mode for each of the following:
(a) MOV AX,DS
(b) MOV BX,5678H
(c) MOV CX,[3000]
(d) MOV AL,CH
(e) MOV [DI],BX
(f) MOV AL,[BX]
(g) MOV DX,[BP+DI+4]
(h) MOV CX,DS
(i) MOV [BP+6],AL
(j) MOV AH,[BX+SI+50]
(k) MOV BL,[SI]+10
(l) MOV [BP][SI]+12,AX

36. Show the contents of the memory locations after the execution of each instruction.

(a)
```
MOV BX,129FH
MOV [1450],BX
DS:1450 ....
DS:1451 ....
```
(b)
```
MOV DX,8C63H
MOV [2348],DX
DS:2348 ....
DS:2349 ....
```

ANSWERS TO REVIEW QUESTIONS

SECTION 1.1: BRIEF HISTORY OF THE x86 FAMILY

1. (1) increased memory capacity from 64K to 1 megabyte;
   (2) the 8086 is a 16-bit microprocessor instead of an 8-bit microprocessor;
   (3) the 8086 was a pipelined processor
2. The 8088 has an 8-bit external data bus whereas the 8086 has a 16-bit data bus.
3. (a) 20-bit, 1 megabyte (b) 24-bit, 16 megabytes (c) 32-bit, 4 gigabytes
4. 16, 32
5. The 80386 has 32-bit address and data buses, whereas the 80386SX has a 24-bit address bus and a 16-bit external data bus.
6. Virtual memory, protected mode
7. Math coprocessor on the CPU chip, cache memory and controller
8. A 64-bit bus, faster floating point, and separate cache memory for code and data.
9. 7.5 million
10. Pentium II
11. MMX
12. 8-bit, 16-bit, and 32-bit
13. 8-bit, 16-bit, 32-bit, and 64-bit
14. True

SECTION 1.2: INSIDE THE 8088/86

1. The execution unit executes instructions; the bus interface unit fetches instructions.
2. Pipelining divides the microprocessor into two sections: the execution unit and the bus interface unit; this allows the CPU to perform these two functions simultaneously; that is, the BIU can fetch instructions while the EU executes the instructions previously fetched.
3. 8, 16
4. AX, BX, CX, DX, SP, BP, SI, DI, CS, DS, SS, ES, IP, FR

SECTION 1.3: INTRODUCTION TO ASSEMBLY PROGRAMMING

1. MOV BX,1234H
2. MOV AX,16H
   ADD AX,ABH
3. The segment registers CS, DS, ES, and SS
4. FFFFH = 65535_{10}, FFH = 255_{10}

SECTION 1.4: INTRODUCTION TO PROGRAM SEGMENTS

1. 64K
2. A segment contains 64K bytes; yes because 346E0H is evenly divisible by 16.
3. The physical address is the 20-bit address that is put on the address bus to locate a byte; the logical address is the address in the form xxxx:yyyy, where xxxx is the segment address and yyyy is the offset into the segment.
4. 20, 16
5. IP
6. 2400 would contain 34 and 2401 would contain 12.

SECTION 1.5: THE STACK

1. SS is the segment register; SP and BP are used as pointers into the stack.
2. Decremented
3. Incremented
4. 143F:0000, 1000:43F0, 1410:02F0

SECTION 1.6: FLAG REGISTER

1. CF, PF, AF, ZF, SF, and OF
2. 7
3. 15

SECTION 1.7: x86 ADDRESSING MODES

1. No
2. No
3. No
4. DS and ES
5. BX, SI, and DI

CHAPTER 2

ASSEMBLY LANGUAGE PROGRAMMING

OBJECTIVES

Upon completion of this chapter, you will be able to:

- **>> Explain the difference between Assembly language instructions and pseudo-instructions**
- **>> Identify the segments of an Assembly language program**
- **>> Code simple Assembly language instructions**
- **>> Assemble, link, and run a simple Assembly language program**
- **>> Code control transfer instructions such as conditional and unconditional jumps and call instructions**
- **>> Code Assembly language data directives for binary, hex, decimal, or ASCII data**
- **>> Write an Assembly language program using either the simplified segment definition or the full segment definition**
- **>> Explore the use of the MASM and emu8086 assemblers**

This chapter is an introduction to Assembly language programming for the x86 microprocessors. In Section 2.1, the basic form of a program is explained. In Section 2.2 the steps required to edit, assemble, link, and run a program are explored. Further examples of x86 Assembly language programming are given in Section 2.3. Next, control transfer instructions such as jump and call are discussed in Section 2.4. The data types and data directives in x86 Assembly language are explained in Section 2.5. Then the full segment definition is discussed in Section 2.6 followed by the discussion of a popular assembler called emu8086. Finally, flowchart and pseudocode are explained in Section 2.7.

SECTION 2.1: DIRECTIVES AND A SAMPLE PROGRAM

In this section we explain the components of a simple Assembly language program to be assembled by the assembler. A given Assembly language program (see Figure 2-1) is a series of statements, or lines, which are either Assembly language instructions such as ADD and MOV, or statements called directives. *Directives* (also called *pseudo-instructions*) give directions to the assembler about how it should translate the Assembly language instructions into machine code. An Assembly language instruction consists of four fields:

[label:] mnemonic [operands] [;comment]

Brackets indicate that the field is optional. Do not type in the brackets.

1. The label field allows the program to refer to a line of code by name. The label field cannot exceed 31 characters. Labels for directives do not need to end with a colon. A label must end with a colon when it refers to an opcode generating instruction; the colon indicates to the assembler that this refers to code within this code segment. Appendix C, Section 2 gives more information about labels.

2,3. The Assembly language mnemonic (instruction) and operand(s) fields together perform the real work of the program and accomplish the tasks for which the program was written. In Assembly language statements such as

```
ADD   AL,BL
MOV   AX,6764
```

ADD and MOV are the mnemonic opcodes, and "AL,BL" and "AX,6764" are the operands. Instead of a mnemonic and operand, these two fields could contain assembler pseudo-instructions, or directives. They are used by the assembler to organize the program as well as other output files. Directives do not generate any machine code and are used only by the assembler as opposed to instructions, which are translated into machine code for the CPU to execute. In Figure 2-1 the commands DB, END, and ENDP are examples of directives.

4. The comment field begins with a ";". Comments may be at the end of a line or on a line by themselves. The assembler ignores comments, but they are indispensable to programmers. Comments are optional, but are highly recommended to make it easier for someone to read and understand the program.

Model definition

The first statement in Figure 2-1 after the first two comments is the MODEL directive. This directive selects the size of the memory model. Among the options for the memory model are SMALL, MEDIUM, COMPACT, and LARGE.

```
.MODEL SMALL       ;this directive defines the model as small
```

SMALL is one of the most widely used memory models for Assembly language programs and is sufficient for the programs in this book. The small model uses a maximum of 64K bytes of memory for code and another 64K bytes for data. The other models are defined as follows:

```
.MODEL MEDIUM      ;the data must fit into 64K bytes
                   ;but the code can exceed 64K bytes of memory
.MODEL COMPACT     ;the data can exceed 64K bytes
                   ;but the code cannot exceed 64K bytes
.MODEL LARGE       ;both data and code can exceed 64K
                   ;but no single set of data should exceed 64K
.MODEL HUGE        ;both code and data can exceed 64K
                   ;data items (such as arrays) can exceed 64K
.MODEL TINY        ;used with COM files in which data and code
                   ;must fit into 64K bytes
```

Notice in the above list that MEDIUM and COMPACT are opposites. Also note that the TINY model cannot be used with the simplified segment definition described in this section.

Segment definition

As mentioned in Chapter 1, the x86 CPU has four segment registers: CS (code segment), DS (data segment), SS (stack segment), and ES (extra segment). Every line of an Assembly language program must correspond to one of these segments. The simplified segment definition format uses three simple directives: "`.CODE`", "`.DATA`", and "`.STACK`", which correspond to the CS, DS, and SS registers, respectively. There is another segment definition style called the *full segment definition*, described in Section 2.6.

Segments of a program

Although one can write an Assembly language program that uses only one segment, normally a program consists of at least three segments: the stack segment, the data segment, and the code segment.

```
.STACK       ;marks the beginning of the stack segment
.DATA        ;marks the beginning of the data segment
.CODE        ;marks the beginning of the code segment
```

Assembly language statements are grouped into segments in order to be recognized by the assembler and consequently by the CPU. The stack segment defines storage for the stack, the data segment defines the data that the program will use, and the code segment contains the Assembly language instructions. In Chapter 1 we gave an overview of how these segments were stored in memory. In the following pages we describe the stack, data, and code segments as they are defined in Assembly language programming.

Stack segment

The following directive reserves 64 bytes of memory for the stack:

```
.STACK  64
```

Data segment

The data segment in the program of Figure 2-1 defines three data items: DATA1, DATA2, and SUM. Each is defined as DB (define byte). The DB directive is used by the assembler to allocate memory in byte-sized chunks. Memory can be allocated in different sizes, such as 2 bytes, which has the directive DW (define word). More of these pseudo-instructions are discussed in detail in Section 2.5. The data items defined in the data segment will be accessed in the code segment by their labels. DATA1 and DATA2 are given initial values in the data section. SUM is not given an initial value, but storage is set aside for it.

```
;THE FORM OF AN ASSEMBLY LANGUAGE PROGRAM
;NOTE: USING SIMPLIFIED SEGMENT DEFINITION
            .MODEL SMALL
            .STACK 64
            .DATA
DATA1       DB    52H
DATA2       DB    29H
SUM         DB    ?
            .CODE
MAIN        PROC  FAR          ;this is the program entry point
            MOV   AX,@DATA     ;load the data segment address
            MOV   DS,AX        ;assign value to DS
            MOV   AL,DATA1     ;get the first operand
            MOV   BL,DATA2     ;get the second operand
            ADD   AL,BL        ;add the operands
            MOV   SUM,AL       ;store the result in location SUM
            MOV   AH,4CH       ;set up to return to OS
            INT   21H          ;
MAIN        ENDP
            END   MAIN         ;this is the program exit point
```

Figure 2-1. Simple Assembly Language Program

Code segment definition

The last segment of the program in Figure 2-1 is the code segment. The first line of the segment after the .CODE directive is the PROC directive. A *procedure* is a group of instructions designed to accomplish a specific function. A code segment may consist of only one procedure, but usually is organized into several small procedures in order to make the program more structured. Every procedure must have a name defined by the PROC directive, followed by the assembly language instructions and closed by the ENDP directive. The PROC and ENDP statements must have the same label. The PROC directive may have the option FAR or NEAR. The operating system that controls the computer must be directed to the beginning of the program in order to execute it. The OS requires that the entry point to the user program be a FAR procedure. From then on, either FAR or NEAR can be used. The differences between a FAR and a NEAR procedure, as well as where and why each is used, are explained later in this chapter. For now, just remember that in order to run a program, FAR must be used at the program entry point.

A good question to ask at this point is: What value is actually assigned to the CS, DS, and SS registers for execution of the program? The operating system must pass control to the program so that it may execute, but before it does that it assigns values for the segment registers. The operating system must do this because it knows how much memory is installed in the computer, how much of it is used by the system, and how much is available. In the IBM PC, the operating system first finds out how many kilobytes of RAM memory are installed, allocates some for its own use, and then allows the user program to use the portions that it needs. Various OS versions require different amounts of memory, and since the user program must be able to run across different versions, one cannot tell the OS to give the program a specific area of memory, say from 25FFF to 289E2. Therefore, it is the job of the OS to assign exact values for the segment registers. When the program begins executing, of the three segment registers, only CS and SS have the proper values. The DS value (and ES, if used) must be initialized by the program. This is done as follows:

```
MOV AX,@DATA ;DATA refers to the start of the data segment
MOV DS,AX
```

Remember from Chapter 1 that no segment register can be loaded directly. That is the reason the two lines of code above are needed. You cannot code "`MOV DS,@DATA`".

After these housekeeping chores, the Assembly language program instructions can be coded to perform the desired tasks. The program in Figure 2-1 loads AL and BL with DATA1 and DATA2, respectively, ADDs them together, and stores the result in SUM.

```
MOV   AL,DATA1
MOV   BL,DATA2
ADD   AL,BL
MOV   SUM,AL
```

The two last instructions in the shell are "`MOV AH,4CH`" and "`INT 21H`." Their purpose is to return control to the operating system. The last two lines end the procedure and the program, respectively. Note that the label for ENDP (MAIN) matches the label for PROC. The END pseudo-instruction ends the entire program by indicating to OS that the entry point MAIN has ended. For this reason the labels for the entry point and END must match.

Figure 2-2 shows a sample shell of an Assembly language program. When writing your first few programs, it is handy to keep a copy of this shell on your disk and simply fill it in with the instructions and data for your program.

```
;THE FORM OF AN ASSEMBLY LANGUAGE PROGRAM
; USING SIMPLIFIED SEGMENT DEFINITION
            .MODEL SMALL
            .STACK 64
            .DATA
            ;
            ;place data definitions here
            ;
            .CODE
MAIN        PROC   FAR            ;this is the program entry point
            MOV    AX,@DATA       ;load the data segment address
            MOV    DS,AX          ;assign value to DS
            ;
            ;place code here
            ;
            MOV    AH,4CH         ;set up to
            INT    21H            ;return to OS
MAIN        ENDP
            END    MAIN           ;this is the program exit point
```

Figure 2-2. Shell of an Assembly Language Program

Review Questions

1. What is the purpose of pseudo-instructions?
2. ____________ are translated by the assembler into machine code, whereas ________________ are not.
3. Write an Assembly language program with the following characteristics:
   (a) A data item named HIGH_DAT, which contains 95
   (b) Instructions that move HIGH_DAT to registers AH, BH, and DL
   (c) A program entry point named START
4. Find the errors in the following:

```
      .MODEL ENORMOUS
      .STACK
      .CODE
      .DATA
MAIN  PROC  FAR
      MOV   AX,DATA
      MOV   DS,@DATA
      MOV   AL,34H
      ADD   AL,4FH
      MOV   DATA1,AL
START ENDP
      END
```

SECTION 2.2: ASSEMBLE, LINK, AND RUN A PROGRAM

Now that the basic form of an Assembly language program has been given, the next question is: How is it created and assembled? The three steps to create an executable Assembly language program are outlined as follows:

Step	Input	Program	Output
1. Edit the program	keyboard	editor	myfile.asm
2. Assemble the program	myfile.asm	MASM or TASM	myfile.obj
3. Link the program	myfile.obj	LINK or TLINK	myfile.exe

The MASM and LINK programs are the assembler and linker programs for Microsoft's MASM assembler. If you are using another assembler, such as Borland's TASM, consult the manual for the procedure to assemble and link a program. Many excellent editors or word processors are available that can be used to create and/or edit the program. The editor must be able to produce an ASCII file. Although filenames follow the usual OS conventions, the source file must end in ".asm" for the assembler used in this book. This ".asm" source file is assembled by an assembler, such as Microsoft's MASM or Borland's TASM. The assembler will produce an object file and a list file, along with other files that may be useful to the programmer. The extension for the object file must be ".obj". This object file is input to the LINK program, which produces the executable program that ends in ".exe". The ".exe" file can be executed by the microprocessor. Before feeding the ".obj" file into LINK, all syntax errors produced by the assembler must be corrected. Of course, fixing these errors will not guarantee that the program will work as intended since the program may contain conceptual errors. Figure 2-3 shows the steps in producing an executable file.

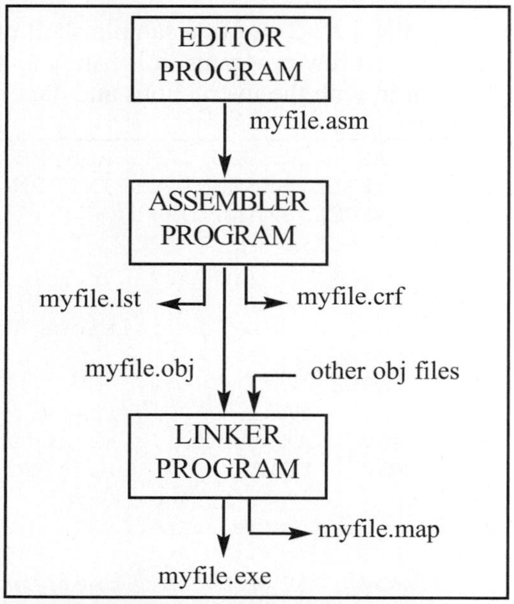

Figure 2-3. Steps to Create a Program

Figure 2-4 shows how an executable program is created by following the steps outlined above, and then run under DEBUG. The portions in bold indicate what the user would type in to perform these steps. Figure 2-4 assumes that the MASM, LINK, and DEBUG programs are on drive C and the Assembly language program is on drive A. The drives used will vary depending on how the system is set up.

.asm and .obj files

The ".asm" file (the source file) is the file created with a word processor or line editor. The MASM (or other) assembler converts the .asm file's Assembly language instructions into machine language (the ".obj" object file). In addition to creating the object program, MASM also creates the ".lst" list file.

.lst file

The ".lst" file, which is optional, is very useful to the programmer because it lists all the opcodes and offset addresses as well as errors that MASM detected. MASM assumes that the list file is not wanted (NUL.LST indicates no list). To get a list file, type in a filename after the prompt. This file can be displayed on the monitor or sent to the printer. The programmer uses it to help debug the program. It is only after fixing all the errors indicated in the ".lst" file that the ".obj" file can be input to the LINK program to create the executable program. One way to look at the list file is to use the following command at the OS level. This command will print myfile.lst to the monitor, one screen at a time.

C>type myfile.lst | more

Another way to look at the list file is to bring it into a word processor. Then you can read it or print it. There are two assembler directives that can be used to make the ".lst" file more readable: PAGE and TITLE.

```
C>MASM C:MYFILE.ASM <enter>

Microsoft (R) Macro Assembler  Version  5.10
Copyright (C) Microsoft Corp 1981, 1988.  All rights reserved.

Object filename [ C:MYFILE.OBJ] : C: <enter>
Source listing  [ NUL.LST] :C:MYFILE.LST  <enter>
Cross-reference [ NUL.CRF] : <enter>

        47962 + 413345 Bytes symbol space free

      0 Warning Errors
      0 Severe  Errors


C>LINK C:MYFILE.OBJ <enter>

Microsoft (R) Overlay Linker  Version 3.64
Copyright (C) Microsoft Corp 1983-1988.  All rights reserved.

Run File [ C:MYFILE.EXE] :C:<enter>
List File [ NUL.MAP] : <enter>
Libraries [ .LIB] :<enter>
LINK : warning L4021: no stack segment

C>DEBUG C:MYFILE.EXE <enter>
-U CS:0 1 <enter>
1064:0000 B86610            MOV       AX,1066
-D 1066:0 F <enter>
1066:0000 52 29 00 00 00 00 00 00-C0 00 00 00 00 00 00 00  R).............
-G <enter>
Program terminated normally
-D 1066:0 F <enter>
1066:0000 52 29 7B 00 00 00 00 00-C0 00 00 00 00 00 00 00  R){............
-Q <enter>
C>
```

Figure 2-4. Creating and Running the .exe File
Note: The parts you type in are printed in **bold**.

PAGE and TITLE directives

The format of the PAGE directive is

```
PAGE [ lines] ,[ columns]
```

and its function is to tell the printer how the list should be printed. In the default mode, meaning that the PAGE directive is coded with no numbers coming after it, the output will have 66 lines per page with a maximum of 80 characters per line. In this book, programs will change the default settings to 60 and 132 as follows:

```
PAGE 60,132
```

The range for number of lines is 10 to 255 and for columns is 60 to 132. When the list is printed and it is more than one page, the assembler can be instructed to print the title of the program on top of each page. What comes after the TITLE pseudo-instruction is up to the programmer, but it is common practice to put the name of the program as stored on the disk immediately after the TITLE pseudo-instruction and then a brief description of the function of the program. The text after the TITLE pseudo-instruction cannot be more than 60 ASCII characters.

.crf file

MASM produces another optional file, the cross-reference, which has the extension ".crf". It provides an alphabetical list of all symbols and labels used in the program as well as the program line numbers in which they are referenced. This can be a great help in large programs with many data segments and code segments.

LINKing the program

The assembler (MASM) creates the opcodes, operands, and offset addresses under the ".obj" file. It is the LINK program that produces the ready-to-run version of a program that has the ".exe" (EXEcutable) extension. The LINK program sets up the file so that it can be loaded by the OS and executed.

In Figure 2-4 we used DEBUG to execute the program in Figure 2-1 and analyze the result. In the program in Figure 2-1, three data items are defined in the data segment. Before running the program, one could look at the data in the data segment by dumping the contents of DS:offset as shown in Figure 2-4. Now what is the value for the DS register? This can vary from PC to PC and from OS to OS. For this reason it is important to look at the value in "MOV AX,xxxx" as was shown and use that number. The result of the program can be verified after it is run as shown in Figure 2-4. When the program is working successfully, it can be run at the OS level. To execute myfile.exe, simply type in

```
C>myfile
```

However, since this program produces no output, there would be no way to verify the results. When the program name is typed in at the OS level, as shown above, the OS loads the program in memory. This is sometimes referred to as *mapping*, which means that the program is mapped into the physical memory of the PC.

.map file

When there are many segments for code or data, there is a need to see where each is located and how many bytes are used by each. This is provided by the map file. This file, which is optional, gives the name of each segment, where it starts, where it stops, and its size in bytes.

Download Microsoft Assembler (MASM) and a Tutorial on how to use it from the following website:

http://www.MicroDigitalEd.com

Review Questions

1. (a) The input file to the MASM assembler program has the extension _____.
   (b) The input file to the LINK program has the extension _____.
2. Select all the file types from the second column that are the output of the program in the first column.

______ Editor	(a) .obj	(b) .asm
______ Assembler	(c) .exe	(d) .lst
______ Linker	(e) .crf	(f) .map

SECTION 2.3: MORE SAMPLE PROGRAMS

Now that some familiarity with Assembly language programming in the IBM PC has been achieved, in this section we look at more example programs in order to allow the reader to master the basic features of Assembly programming. The following pages show Program 2-1 and the list file generated when the program was assembled. After the program was assembled and linked, DEBUG was used to dump the code segment to see what value is assigned to the DS register. Precisely where the OS loads a program into RAM depends on many factors, including the amount of RAM on the system and the version of OS used. Therefore, remember that the value you get could be different for "MOV AX,xxxx" as well as for CS in the program examples. Do not attempt to modify the segment register contents to conform to those in the examples, or your system may crash!

Write, run, and analyze a program that adds 5 bytes of data and saves the result. The data should be the following hex numbers: 25, 12, 15, 1F, and 2B.

```
PAGE       60,132
TITLE      PROG2-1   (EXE)    PURPOSE: ADDS 5 BYTES OF DATA
           .MODEL SMALL
           .STACK 64
;---------------
           .DATA
DATA_IN    DB              25H,12H,15H,1FH,2BH
SUM        DB              ?
;---------------
           .CODE
MAIN       PROC    FAR
           MOV     AX,@DATA
           MOV     DS,AX
           MOV     CX,05                ;set up loop counter CX=5
           MOV     BX,OFFSET DATA_IN    ;set up data pointer BX
           MOV     AL,0                 ;initialize AL
AGAIN:     ADD     AL,[BX]              ;add next data item to AL
           INC     BX                   ;make BX point to next data item
           DEC     CX                   ;decrement loop counter
           JNZ     AGAIN                ;jump if loop counter not zero
           MOV     SUM,AL               ;load result into sum
           MOV     AH,4CH               ;set up return
           INT     21H                  ;return to OS
MAIN       ENDP
           END     MAIN
```

After the program was assembled and linked, it was run using DEBUG:

```
C>debug prog2-1.exe
-u cs:0 19
1067:0000 B86610  MOV   AX,1066
1067:0003 8ED8    MOV   DS,AX
1067:0005 B90500  MOV   CX,0005
1067:0008 BB0000  MOV   BX,0000
1067:000D 0207    ADD   AL,[BX]
1067:000F 43            INC   BX
1067:0010 49            DEC   CX
1067:0013 A20500  MOV   [0005],AL
1067:0016 B44C    MOV   AH,4C
1067:0018 CD21    INT   21
-d 1066:0 f
1066:0000  25 12 15 1F 2B 00 00 00-00 00 00 00 00 00 00 00 %...+..........
-g
Program terminated normally
-d 1066:0 f
1066:0000  25 12 15 1F 2B 96 00 00-00 00 00 00 00 00 00 00 %...+..........
-q
C>
```

Program 2-1

Analysis of Program 2-1

The DEBUG program is explained thoroughly in Appendix A. The commands used in running Program 2-1 were (1) u, to unassemble the code from cs:0 for 19 bytes; (2) d, to dump the contents of memory from 1066:0 for the next F bytes; and (3) g, to go, that is, run the program.

Notice in Program 2-1 that when the program was run in DEBUG, the contents of the data segment memory were dumped before and after execution of the program to verify that the program worked as planned. Normally, it is not necessary to unassemble this much code, but it was done here because in later sections of the chapter we examine the jump instruction in this program. Also notice that the first 5 bytes dumped above are the data items defined in the data segment of the program and the sixth item is the sum of those five items, so it appears that the program worked correctly (25H + 12H + 15H + 1FH + 2BH = 96H). Program 2-1 is explained below, instruction by instruction.

"`MOV CX,05`" will load the value 05 into the CX register. This register is used by the program as a counter for iteration (looping).

"`MOV BX,OFFSET DATA_IN`" will load into BX the offset address assigned to DATA. The assembler starts at offset 0000 and uses memory for the data and then assigns the next available offset memory for SUM (in this case, 0005).

"`ADD AL,[ BX]` " adds the contents of the memory location pointed at by the register BX to AL. Note that [BX] is a pointer to a memory location.

"`INC BX`" simply increments the pointer by adding 1 to register BX. This will cause BX to point to the next data item, that is, the next byte.

"`DEC CX`" will decrement (subtract 1 from) the CX counter and will set the zero flag high if CX becomes zero.

"`JNZ AGAIN`" will jump back to the label AGAIN as long as the zero flag is indicating that CX is not zero. "JNZ AGAIN" will not jump (that is, execution will resume with the next instruction after the JNZ instruction) only after the zero flag has been set high by the "`DEC CX`" instruction (that is, CX becomes zero). When CX becomes zero, this means that the loop is completed and all five numbers have been added to AL.

Various approaches to Program 2-1

There are many ways in which any program may be written. The method shown for Program 2-1 defined one field of data and used pointer [BX] to access data elements. In the method used below, a name is assigned to each data item that will be accessed in the program. Variations of Program 2-1 are shown below to clarify the use of addressing modes in the context of a real program and also to show that the x86 can use any general-purpose register to do arithmetic and logic operations. In earlier-generation CPUs, the accumulator had to be the destination of all arithmetic and logic operations, but in the x86 this is not the case. Since the purpose of these examples is to show different ways of accessing operands, it is left to the reader to run and analyze the programs.

```
;from the data segment:
DATA1   DB 25H
DATA2   DB 12H
DATA3   DB 15H
DATA4   DB 1FH
DATA5   DB 2BH
SUM   DB  ?
;from the code segment:
MOV  AL,DATA1        ;MOVE DATA1 INTO AL
ADD  AL,DATA2        ;ADD DATA2 TO AL
ADD  AL,DATA3
ADD  AL,DATA4
ADD  AL,DATA5
MOV  SUM,AL          ;SAVE AL IN SUM
```

There is quite a difference between these two methods of writing the same program. While in the first one the register indirect addressing mode was used to access the data, in the second method the direct addressing mode was used.

```
Microsoft (R) Macro Assembler Version 5.10              2/13/7
PROG2-1  (EXE)   PURPOSE: ADDS 5 BYTES OF DATA              Page     1-1


                              PAGE    60,132
                              TITLE   PROG2-1  (EXE)   PURPOSE: ADDS 5 BYTES OF DATA
                                      .MODEL SMALL
                                      .STACK 64
                              ;----------------------------------
                                      .DATA
 0000  25 12 15 1F 2B         DATA_IN         DB      25H,12H,15H,1FH,2BH
 0005  00                     SUM             DB      ?
                              ;----------------------------------
                                      .CODE
 0000                         MAIN            PROC    FAR
 0000  B8 ---- R                      MOV     AX,@DATA
 0003  8E D8                          MOV     DS,AX
 0005  B9 0005                        MOV     CX,05              ;set up loop counter CX=5
 0008  BB 0000 R              MOV     BX,OFFSET DATA_IN          ;set up data pointer BX
 000B  B0 00                          MOV     AL,0               ;initialize AL
 000D  02 07                  AGAIN:          ADD AL,[EX]        ;add next data item to AL
 000F  43                     INC     BX              ;make BX point to next data item
 0010  49                             DEC     CX                 ;decrement loop counter
 0011  75 FA                          JNZ     AGAIN              ;jump if loop counter not zero
 0013  A2 0005 R              MOV     SUM,AL          ;load result into sum
 0016  B4 4C                          MOV     AH,4CH             ;set up return
 0018  CD 21                          INT     21H                ;return to OS
 001A                         MAIN            ENDP
                                      END     MAIN


Microsoft (R) Macro Assembler Version 5.10              2/13/7
PROG2-1  (EXE)   PURPOSE: ADDS 5 BYTES OF DATA              Symbols-1

Segments and Groups:
            N a m e        Length   Align    Combine Class

DGROUP . . . . . . . . . . . . .    GROUP
  _DATA . . . . . . . . . . . .  0006   WORD  PUBLIC'DATA'
  STACK . . . . . . . . . . . .         0040  PARA   STACK 'STACK'
_TEXT . . . . . . . . . . . . .  001A   WORD  PUBLIC'CODE'

Symbols:
            N a m e        Type      Value    Attr

AGAIN . . . . . . . . . . . . .      L NEAR000D    _TEXT

DATA_IN . . . . . . . . . . . .      L BYTE 0000   _DATA

MAIN . . . . . . . . . . . . .   F PROC 0000   _TEXT  Length = 001A

SUM . . . . . . . . . . . . . .  L BYTE 0005   _DATA

@CODE . . . . . . . . . . . . .      TEXT  _TEXT
@CODESIZE . . . . . . . . . . .      TEXT  0
@CPU . . . . . . . . . . . . .   TEXT  0101h
@DATASIZE . . . . . . . . . . .      TEXT  0
@FILENAME . . . . . . . . . . .      TEXT  prog2_1
@VERSION . . . . . . . . . . .       TEXT  510

   25 Source Lines
   25 Total Lines
   25 Symbols
 45756 + 410160 Bytes symbol space free 0 Warning Errors 0 Severe Errors
```

MASM List File for Program 2-1

Write and run a program that adds four words of data and saves the result. The values will be 234DH, 1DE6H, 3BC7H, and 566AH. Use DEBUG to verify the sum is D364.

```
TITLE         PROG2-2   (EXE)   PURPOSE: ADDS 4 WORDS OF DATA
PAGE   60,132
              .MODEL SMALL
              .STACK 64
;-------------
              .DATA
DATA_IN       DW            234DH,1DE6H,3BC7H,566AH
              ORG    10H
SUM           DW            ?
;-------------
              .CODE
MAIN          PROC          FAR
              MOV    AX,@DATA
              MOV    DS,AX
              MOV    CX,04                   ;set up loop counter CX=4
              MOV    DI,OFFSET DATA_IN       ;set up data pointer DI
              MOV    BX,00                   ;initialize BX
ADD_LP:       ADD    BX,[DI]      ;add contents pointed at by [DI] to BX
              INC    DI                      ;increment DI twice
              INC    DI                      ;to point to next word
              DEC    CX                      ;decrement loop counter
              JNZ    ADD_LP                  ;jump if loop counter not zero
              MOV    SI,OFFSET SUM     ;load pointer for sum
              MOV    [SI],BX                 ;store in data segment
              MOV    AH,4CH                  ;set up return
              INT    21H                     ;return to OS
MAIN          ENDP
              END    MAIN

After the program was assembled and linked, it was run using DEBUG:
C>debug c:prog2-2.exe
1068:0000   B86610        MOV    AX,1066
-D 1066:0 1F
1066:0000 4D 23 E6 1D C7 3B 6A 56-00 00 00 00 00 00 00 00 M#f.G;jV........
1066:0010 00 00 00 00 00 00 00 00-00 00 00 00 00 00 00 00 ................
-G
Program terminated normally
-D 1066:0 1F
1066:0000 4D 23 E6 1D C7 3B 6A 56-00 00 00 00 00 00 00 00 M#f.G;jV........
1066:0010 64 D3 00 00 00 00 00 00-00 00 00 00 00 00 00 00 dS..............
-Q
C>
```

Program 2-2

Analysis of Program 2-2

First notice that the 16-bit data (a word) is stored with the low-order byte first. For example, "234D" as defined in the data segment is stored as "4D23", meaning that the lower address, 0000, has the least significant byte, 4D, and the higher address, 0001, has the most significant byte, 23. This is shown in the DEBUG display of the data segment. Similarly, the sum, D364, is stored as 64D3. As discussed in Chapter 1, this method of low byte to low address and high byte to high address operand assignment is referred to in computer literature as "little endian."

Second, note that the address pointer is incremented twice, since the operand being accessed is a word (two bytes). The program could have used "`ADD DI,2`" instead of using "`INC DI`" twice. When storing the result of word addition, "`MOV SI,OFFSET SUM`" was used to load the pointer (in this case 0010, as defined by ORG 0010H) for the memory allocated for the label SUM, and then "`MOV [SI],BX`" was used to move the contents of register BX to memory locations with offsets 0010 and 0011. As was done previously, it could have been coded simply as "`MOV SUM,BX`", using direct addressing mode.

Program 2-2 uses the ORG directive. In previous programs where ORG was not used, the assembler would start at offset 0000 and use memory for each data item. The ORG directive can be used to set the offset addresses for data items. Although the programmer cannot assign exact physical addresses, one is allowed to assign offset addresses. The ORG directive in Program 2-2 caused SUM to be stored at DS:0010, as can be seen by looking at the DEBUG display of the data segment.

Write and run a program that transfers 6 bytes of data from memory locations with offset of 0010H to memory locations with offset of 0028H.

```
TITLE       PROG2-3  (EXE)    PURPOSE: TRANSFERS 6 BYTES OF DATA
PAGE   60,132
            .MODEL SMALL
            .STACK 64
            .DATA
            ORG    10H
DATA_IN      DB          25H,4FH,85H,1FH,2BH,0C4H
            ORG    28H
COPY         DB          6 DUP(?)
;------------------
            .CODE
MAIN        PROC         FAR
            MOV    AX,@DATA
            MOV    DS,AX
            MOV    SI,OFFSET DATA_IN ;SI points to data to be copied
            MOV    DI,OFFSET COPY    ;DI points to copy of data
            MOV    CX,06H            ;loop counter = 6
MOV_LOOP:    MOV    AL,[SI]          ;move the next byte from DATA area to AL
            MOV    [DI],AL      ;move the next byte to COPY area
            INC    SI                        ;increment DATA pointer
            INC    DI                        ;increment COPY pointer
            DEC    CX                        ;decrement LOOP counter
            JNZ    MOV_LOOP          ;jump if loop counter not zero
            MOV    AH,4CH                    ;set up to return
            INT    21H                       ;return to OS
MAIN        ENDP
            END    MAIN
```

After the program was assembled and linked, it was run using DEBUG:

```
C>debug prog2-3.exe
-u cs:0 1
1069:0000   B86610       MOV    AX,1066
-d 1066:0 2f
1066:0000 00 00 00 00 00 00 00 00-00 00 00 00 00 00 00 00 ................
1066:0010 25 4F 85 1F 2B C4 00 00-00 00 00 00 00 00 00 00 %O..+D..........
1066:0020 00 00 00 00 00 00 00 00-00 00 00 00 00 00 00 00 ................
-g

Program terminated normally
-d 1066:0 2f
1066:0000 00 00 00 00 00 00 00 00-00 00 00 00 00 00 00 00 ................
1066:0010 25 4F 85 1F 2B C4 00 00-00 00 00 00 00 00 00 00 %O..+D..........
1066:0020 00 00 00 00 00 00 00 00-25 4F 85 1F 2B C4 00 00 %O..+D..........
-q
C>
```

Program 2-3

Analysis of Program 2-3

Program 2-3 shows the data segment being dumped before and after the program was run to verify that the data was copied successfully. Notice that C4 was coded in the data segments as 0C4. This is required by the assembler to indicate that C is a hex number and not a letter. This is required if the first digit is a hex digit A through F.

This program uses two registers, SI and DI, as pointers to the data items being manipulated. The first is used as a pointer to the data item to be copied and the second as a pointer to the location the data item is to be copied to. With each iteration of the loop, both data pointers are incremented to point to the next byte.

Stack segment definition revisited

One of the primary functions of the operating system is to determine the total amount of RAM installed on the PC and then manage it properly. The OS uses the portion it needs for the operating system and allocates the rest. Since memory requirements vary for different OS versions, a program cannot dictate the exact physical memory location for the stack or any segment. Since memory management is the responsibility of the OS, it will map Assembly programs into the memory of the PC with the help of LINK.

Although in the OS environment a program can have multiple code segments and data segments, it is strongly recommended that it have only one stack segment, to prevent RAM fragmentation by the stack. It is the function of LINK to combine all different code and data segments to create a single executable program with a single stack, which is the stack of the system. Various options for segment definition are discussed in Chapter 7 and many of these concepts are explained there.

Review Questions

1. What is the purpose of the INC instruction?
2. What is the purpose of the DEC instruction?
3. In Program 2-1, why does the label AGAIN have a colon after it, whereas the label MAIN does not?
4. State the difference between the following two instructions:
```
        MOV    BX,DATA1
        MOV    BX,OFFSET DATA1
```
5. State the difference between the following two instructions:
```
        ADD    AX,BX
        ADD    AX,[ BX]
```

SECTION 2.4: CONTROL TRANSFER INSTRUCTIONS

In the sequence of instructions to be executed, it is often necessary to transfer program control to a different location. There are many instructions in the x86 to achieve this. This section covers the control transfer instructions available in the 8086 Assembly language. Before that, however, it is necessary to explain the concept of FAR and NEAR as it applies to jump and call instructions.

FAR and NEAR

If control is transferred to a memory location within the current code segment, it is NEAR. This is sometimes called *intrasegment* (within segment). If control is transferred outside the current code segment, it is a FAR or intersegment (between segments) jump. Since the CS:IP registers always point to the address of the next instruction to be executed, they must be updated when a control transfer instruction is executed. In a NEAR jump, the IP is updated and CS remains the same, since control is still inside the current code segment. In a FAR jump, because control is passing outside the current code segment, both CS and IP have to be updated to the new values. In other words, in any control transfer instruction such as jump or call, the IP must be changed, but only in the FAR case is the CS changed, too.

Conditional jumps

Conditional jumps, summarized in Table 2-1, have mnemonics such as JNZ (jump not zero) and JC (jump if carry). In the conditional jump, control is transferred to a new location if a certain condition is met. The flag register is the one that indicates the current condition. For example, with "JNZ label", the processor looks at the zero flag to see if it is raised. If not, the CPU starts to fetch and execute instructions from the address of the label. If ZF = 1, it will not jump but will execute the next instruction below the JNZ.

Table 2-1: 8086 Conditional Jump Instructions

Mnemonic	Condition Tested	"Jump IF ..."
JA/JNBE	(CF = 0) and (ZF = 0)	above/not below nor zero
JAE/JNB	CF = 0	above or equal/not below
JB/JNAE	CF = 1	below/not above nor equal
JBE/JNA	(CF or ZF) = 1	below or equal/not above
JC	CF = 1	carry
JE/JZ	ZF = 1	equal/zero
JG/JNLE	((SF xor OF) or ZF) = 0	greater/not less nor equal
JGE/JNL	(SF xor OF) = 0	greater or equal/not less
JL/JNGE	(SF xor OR) = 1	less/not greater nor equal
JLE/JNG	((SF xor OF) or ZF) = 1	less or equal/not greater
JNC	CF = 0	not carry
JNE/JNZ	ZF = 0	not equal/not zero
JNO	OF = 0	not overflow
JNP/JPO	PF = 0	not parity/parity odd
JNS	SF = 0	not sign
JO	OF = 1	overflow
JP/JPE	PF = 1	parity/parity equal
JS	SF = 1	sign

Note:
"Above" and "below" refer to the relationship of two unsigned values; "greater" and "less" refer to the relationship of two signed values.
(Reprinted by permission of Intel Corporation, Copyright Intel Corp. 1989)

Short jumps

All conditional jumps are short jumps. In a short jump, the address of the target must be within −128 to +127 bytes of the IP. In other words, the conditional jump is a two-byte instruction: One byte is the opcode of the J condition and the second byte is a value between 00 and FF. An offset range of 00 to FF gives 256 possible addresses; these are split between backward jumps (to −128) and forward jumps (to +127).

In a jump backward, the second byte is the 2's complement of the displacement value. To calculate the target address, the second byte is added to the IP of the instruction after the jump. To understand this, look at the unassembled code of Program 2-1 for the instruction JNZ AGAIN, repeated below.

```
1067:0000 B86610        MOV   AX,1066
1067:0003 8ED8     MOV  DS,AX
1067:0005 B90500        MOV   CX,0005
1067:0008 BB0000        MOV   BX,0000
1067:000D 0207     ADD  AL,[BX]
1067:000F 43            INC   BX
1067:0010 49            DEC   CX
1067:0011 75FA     JNZ  000D
1067:0013 A20500        MOV   [0005],AL
1067:0016 B44C     MOV  AH,4C
1067:0018 CD21     INT  21
```

The instruction "JNZ AGAIN" was assembled as "`JNZ 000D`", and 000D is the address of the instruction with the label AGAIN. The instruction "`JNZ 000D`" has the opcode 75 and the target address FA, which is located at offset addresses 0011 and 0012. This is followed by "`MOV SUM,AL`", which is located beginning at offset address 0013. The IP value of MOV, 0013, is added to FA to calculate the address of label AGAIN (0013

+ FA = 000D) and the carry is dropped. In reality, FA is the 2's complement of −6, meaning that the address of the target is −6 bytes from the IP of the next instruction.

Similarly, the target address for a forward jump is calculated by adding the IP of the following instruction to the operand. In that case the displacement value is positive, as shown next. Below is a portion of a list file showing the opcodes for several conditional jumps.

```
0005        8A 47 02 AGAIN:     MOV    AL,[ BX] +2
0008        3C 61               CMP    AL,61H
000A        72 06               JB     NEXT
000C        3C 7A               CMP    AL,7AH
000E        77 02               JA     NEXT
0010        24 DF               AND    AL,ODFH
0012        88 04     NEXT:     MOV    [ SI] ,AL
```

In the program above, "`JB NEXT`" has the opcode 72 and the target address 06 and is located at IP = 000A and 000B. The jump will be 6 bytes from the next instruction, which is IP = 000C. Adding gives us 000CH + 0006H = 0012H, which is the exact address of the NEXT label. Look also at "`JA NEXT`", which has 77 and 02 for the opcode and displacement, respectively. The IP of the following instruction, 0010, is added to 02 to get 0012, the address of the target location.

It must be emphasized that regardless of whether the jump is forward or backward, for conditional jumps the address of the target address can never be more than −128 to +127 bytes away from the IP associated with the instruction following the jump (− for the backward jump and + for the forward jump). If any attempt is made to violate this rule, the assembler will generate a "relative jump out of range" message. These conditional jumps are sometimes referred to as SHORT jumps.

Unconditional jumps

"`JMP label`" is an unconditional jump in which control is transferred unconditionally to the target location label. The unconditional jump can take the following forms:

1. SHORT JUMP, which is specified by the format "`JMP SHORT label`". This is a jump in which the address of the target location is within −128 to +127 bytes of memory relative to the address of the current IP. In this case, the opcode is EB and the operand is 1 byte in the range 00 to FF. The operand byte is added to the current IP to calculate the target address. If the jump is backward, the operand is in 2's complement. This is exactly like the J condition case. Coding the directive "short" makes the jump more efficient in that it will be assembled into a 2-byte instruction instead of a 3-byte instruction.
2. NEAR JUMP, which is the default, has the format "`JMP label`". This is a near jump (within the current code segment) and has the opcode E9. The target address can be any of the addressing modes of direct, register, register indirect, or memory indirect:

(a) Direct JUMP is exactly like the short jump explained earlier, except that the target address can be anywhere in the segment within the range +32767 to −32768 of the current IP.

(b) Register indirect JUMP; the target address is in a register. For example, in "`JMP BX`", IP takes the value BX.

(c) Memory indirect JMP; the target address is the contents of two memory locations pointed at by the register. Example: "`JMP [ DI]` " will replace the IP with the contents of memory locations pointed at by DI and DI+1.

3. FAR JUMP, which has the format "`JMP FAR PTR label`". This is a jump out of the current code segment, meaning that not only the IP but also the CS is replaced with new values.

CALL statements

Another control transfer instruction is the CALL instruction, which is used to call a procedure. CALLs to procedures are used to perform tasks that need to be performed frequently. This makes a program more structured. The target address could be in the current segment, in which case it will be a NEAR call or outside the current CS segment, which is a FAR call. To make sure that after execution of the called subroutine the microprocessor knows where to come back, the microprocessor automatically saves the address of the instruction following the call on the stack. It must be noted that in the NEAR call only the IP is saved on the stack, and in a FAR call both CS and IP are saved. When a subroutine is called, control is transferred to that subroutine and the processor saves the IP (and CS in the case of a FAR call) and begins to fetch instructions from the new location. After finishing execution of the subroutine, for control to be transferred back to the caller, the last instruction in the called subroutine must be RET (return). In the same way that the assembler generates different opcode for FAR and NEAR calls, the opcode for the RET instruction in the case of NEAR and FAR is different, as well. For NEAR calls, the IP is restored; for FAR calls, both CS and IP are restored. This will ensure that control is given back to the caller. As an example, assume that SP = FFFEH and the following code is a portion of the program unassembled in DEBUG:

```
12B0:0200   BB1295   MOV BX,9512
12B0:0203   E8FA00   CALL 0300
12B0:0206   B82F14   MOV AX,142F
```

Since the CALL instruction is a NEAR call, meaning that it is in the same code segment (different IP, same CS), only IP is saved on the stack. In this case, the IP address of the instruction after the call is saved on the stack as shown in Figure 2-5. That IP will be 0206, which belongs to the "`MOV AX,142F`" instruction.

The last instruction of the called subroutine must be a RET instruction that directs the CPU to POP the top 2 bytes of the stack into the IP and resume executing at offset address 0206. For this reason, the number of PUSH and POP instructions (which alter the SP) must match. In other words, for every PUSH there must be a POP.

```
12B0:0300   53   PUSH BX
12B0:0301   ...   ...... ..
.........   ...   ...... ..
12B0:0309   5B   POP BX
12B0:030A   C3   RET
```

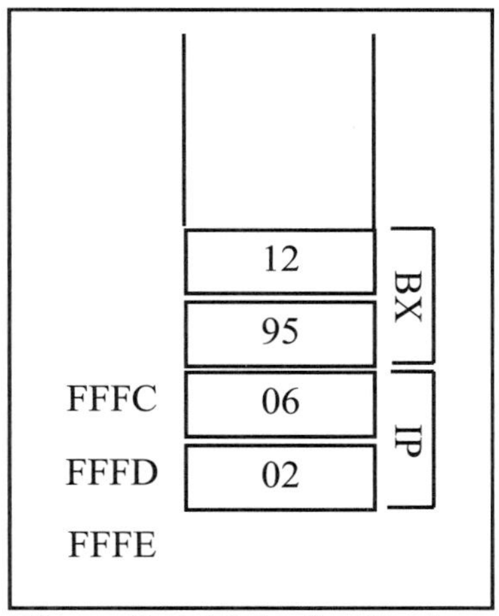

Figure 2-5. IP in the Stack

Assembly language subroutines

In Assembly language programming it is common to have one main program and many subroutines to be called from the main program. This allows you to make each subroutine into a separate module. Each module can be tested separately and then brought together, as will be shown in Chapter 7. The main program is the entry point from the OS and is FAR, as explained earlier, but the subroutines called within the main program can be FAR or NEAR. Remember that NEAR routines are in the same code segment, while FAR routines are outside the current code segment. If there is no specific mention of FAR after the directive PROC, it defaults to NEAR, as shown in Figure 2-6. From now on, all code segments will be written in that format

Rules for names in Assembly language

By choosing label names that are meaningful, a programmer can make a program much easier to read and maintain. There are several rules that names must follow. First, each label name must be unique. The names used for labels in Assembly language programming consist of alphabetic letters in both upper- and lowercase, the digits 0 through 9, and the special characters question mark (?), period (.), at (@), underline (_), and dollar sign ($). The first character of the name must be an alphabetic character or special character. It cannot be a digit. The period can only be used as the first character, but this is not recommended since later versions of MASM have several reserved words that begin with a period. Names may be up to 31 characters long. A list of reserved words is given at the end of Appendix C.

```
        .CODE
MAIN    PROC  FAR               ;THIS IS THE ENTRY POINT FOR OS
        MOV   AX,@DATA
        MOV   DS,AX
        CALL  SUBR1
        CALL  SUBR2
        CALL  SUBR3
        MOV   AH,4CH
        INT   21H
MAIN    ENDP
;-------------------------------
SUBR1   PROC
        ...
        ...
        RET
SUBR1   ENDP
;-------------------------------
SUBR2   PROC
        ...
        ...
        RET
SUBR2   ENDP
;-------------------------------
SUBR3   PROC
        ...
        ...
        RET
SUBR3   ENDP
;-------------------------------
        END         MAIN        ;THIS IS THE EXIT POINT
```

Figure 2-6. Shell of Assembly Language Subroutines

Review Questions

1. If control is transferred outside the current code segment, is it NEAR or FAR?
2. If a conditional jump is not taken, what is the next instruction to be executed?
3. In calculating the target address to jump to, a displacement is added to the contents of register _______.
4. What is the advantage in coding the operator "SHORT" in an unconditional jump?
5. A(n) _______ jump is within −128 to +127 bytes of the current IP. A(n) ________ jump is within the current code segment, whereas a(n) __________ jump is outside the current code segment.
6. How does the CPU know where to return to after executing a RET?
7. Describe briefly the function of the RET instruction.
8. State why the following label names are invalid.
   (a) GET.DATA (b) 1_NUM (c) TEST-DATA (d) RET

SECTION 2.5: DATA TYPES AND DATA DEFINITION

The assembler supports all the various data types of the x86 microprocessor by providing data directives that define the data types and set aside memory for them. In this section we study these directives and how they are used to represent different data types of the x86. The application of these directives becomes clearer in the context of examples in subsequent chapters.

x86 data types

The 8088/86 microprocessor supports many data types, but none are longer than 16 bits wide since the size of the registers is 16 bits. It is the job of the programmer to break down data larger than 16 bits (0000 to FFFFH, or 0 to 65535 in decimal) to be processed by the CPU. Many of these programs are shown in Chapter 3. The data types used by the 8088/86 can be 8-bit or 16-bit, positive or negative. If a number is less than 8 bits wide, it still must be coded as an 8-bit register with the higher digits as zero. Similarly, if the number is less than 16 bits wide it must use all 16 bits, with the rest being 0s. For example, the number 5 is only 3 bits wide (101) in binary, but the 8088/86 will accept it as 05 or "0000 0101" in binary. The number 514 is "10 0000 0010" in binary, but the 8088/86 will accept it as "0000 0010 0000 0010" in binary. The discussion of signed numbers is postponed until later chapters since their representation and application are unique.

Assembler data directives

All the assemblers designed for the x86 microprocessors have standardized the directives for data representation. The following are some of the data directives used by the x86 microprocessor and supported by all software vendors.

ORG (origin)

ORG is used to indicate the beginning of the offset address. The number that comes after ORG can be either in hex or in decimal. If the number is not followed by H, it is decimal and the assembler will convert it to hex. Although the ORG directive is used extensively in this book in the data segment to separate fields of data to make it more readable for the student, it can also be used for the offset of the code segment (IP).

DB (define byte)

The DB directive is one of the most widely used data directives in the assembler. It allows allocation of memory in byte-sized chunks. This is indeed the smallest allocation unit permitted. DB can be used to define numbers in decimal, binary, hex, and ASCII. For decimal, the D after the decimal number is optional, but using B (binary) and H (hexadecimal) for the others is required. Regardless of which one is used, the assembler will convert numbers into hex. To indicate ASCII, simply place the string in single quotation marks ('like this'). The assembler will assign the ASCII code for the numbers or characters automatically. DB is the only directive that can be used to define ASCII strings larger than two characters; therefore, it should be used for all ASCII data definitions.

```
0000 19                     DATA1  DB   25                ;DECIMAL
0001 89                     DATA2  DB   10001001B         ;BINARY
0002 12                     DATA3  DB   12H               ;HEX
0010                               ORG  0010H
0010 32 35 39 31            DATA4  DB   '2591'            ;ASCII NUMBERS
0018                               ORG  0018H
0018 00                     DATA5  DB   ?                 ;SET ASIDE A BYTE
0020                               ORG  0020H
0020 4D 79 20 6E 61 6D      DATA6  DB 'My name is Joe'    ;ASCII CHARACTERS
     65 20 69 73 20 4A
     6F 65
```

List File for DB Examples

Following are some DB examples:

```
DATA1       DB      25                          ;DECIMAL
DATA2       DB      10001001B                   ;BINARY
DATA3       DB      12H                         ;HEX
                   ORG    0010H
DATA4       DB      '2591'                      ;ASCII NUMBERS
```

```
                    ORG       0018H
DATA5         DB    ?                                ;SET ASIDE A BYTE
                    ORG     0020H
DATA6         DB    My name is Joe'                  ;ASCII CHARACTERS
```

Either single or double quotes can be used around ASCII strings. This can be useful for strings, which should contain a single quote such as "O'Leary".

DUP (duplicate)

DUP is used to duplicate a given number of characters. This can avoid a lot of typing. For example, contrast the following two methods of filling six memory locations with FFH:

```
       ORG 0030H
DATA7  DB 0FFH,0FFH,0FFH,0FFH,0FFH,0FFH   ;FILL 6 BYTES WITH FF
       ORG 38H
DATA8  DB 6 DUP(0FFH)      ;FILL 6 BYTES WITH FF
; the following reserves 32 bytes of memory with no initial value given
       ORG 40H
DATA9  DB 32 DUP (?)       ;SET ASIDE 32 BYTES
;DUP can be used inside another DUP
; the following fills 10 bytes with 99
DATA10 DB 5 DUP (2 DUP (99)) ;FILL 10 BYTES WITH 99
```

```
0030                              ORG    0030H
0030  FF FF FF FF FF FF    DATA7  DB     0FFH,0FFH,0FFH,0FFH,0FFH,0FFH  ; 6 FF
0038                              ORG    38H
0038  0006[                DATA8  DB     6 DUP(0FFH)      ;FILL 6 BYTES WITH FF
             FF
                     ]
0040                              ORG    40H
0040 0020 [                DATA9  DB     32 DUP (?)       ;SET ASIDE 32 BYTES
        ??
                     ]
0060                              ORG    60H
0060 0005[                 DATA10 DB     5 DUP (2 DUP (99))    ;FILL 10 BYTES WITH 99
             0002[
                     63
                     ]     ]
```

List File for DUP Examples

DW (define word)

DW is used to allocate memory 2 bytes (one word) at a time. The following are some examples of DW:

```
              ORG    70H
DATA11        DW      954                           ;DECIMAL
DATA12        DW      100101010100B                 ;BINARY
DATA13        DW      253FH                         ;HEX
                     ORG    78H
DATA14        DW     9,2,7,0CH,00100000B,5,'HI'    ;MISC. DATA
DATA15        DW      8 DUP (?)                   ;SET ASIDE 8 WORDS
```

```
0070                                 ORG     70H
0070  03BA                    DATA11  DW    954                  ;DECIMAL
0072  0954                    DATA12  DW    100101010100B        ;BINARY
0074  253F                    DATA13  DW    253FH                ;HEX
0078                                 ORG     78H
0078 0009 0002 0007 000C      DATA14  DW    9,2,7,0CH,00100000B,5,'HI'    ;MISC. DATA
      0020 0005 4849
0086 0008[                    DATA15  DW    8 DUP (?)            ;SET ASIDE 8 WORDS
         ????          ]
```

List File for DW Examples

EQU (equate)

This is used to define a constant without occupying a memory location. EQU does not set aside storage for a data item but associates a constant value with a data label so that when the label appears in the program, its constant value will be substituted for the label. EQU can also be used outside the data segment, even in the middle of a code segment. Using EQU for the counter constant in the immediate addressing mode:

```
COUNT   EQU 25
```

When executing the instructions "MOV CX,COUNT", the register CX will be loaded with the value 25. This is in contrast to using DB:

```
COUNT   DB 25
```

When executing the same instruction "MOV CX,COUNT" it will be in the direct addressing mode. Now what is the real advantage of EQU? First, note that EQU can also be used in the data segment:

```
COUNT               EQU         25
COUNTER1      DB              COUNT
COUNTER2      DB              COUNT
```

Assume that there is a constant (a fixed value) used in many different places in the data and code segments. By the use of EQU, one can change it once and the assembler will change all of them, rather than making the programmer try to find every location and correct it.

DD (define doubleword)

The DD directive is used to allocate memory locations that are 4 bytes (two words) in size. Again, the data can be in decimal, binary, or hex. In any case the data is converted to hex and placed in memory locations according to the rule of low byte to low address and high byte to high address. DD examples are:

```
               ORG   00A0H
DATA16         DD      1023                                 ;DECIMAL
DATA17         DD      10001001011001011100B         ;BINARY
DATA18         DD      5C2A57F2H                     ;HEX
DATA19         DD      23H,34789H,65533
```

```
00A0                                  ORG 00A0H
00A0 000003FF                 DATA16   DD   1023                    ;DECIMAL
00A4 0008965C                 DATA17   DD   10001001011001011100B   ;BINARY
00A8 5C2A57F2                 DATA18   DD   5C2A57F2H               ;HEX
00AC 00000023 00034789        DATA19   DD   23H,34789H,65533
     0000FFFD
```

List File for DD Examples

DQ (define quadword)

DQ is used to allocate memory 8 bytes (four words) in size. This can be used to represent any variable up to 64 bits wide:

```
               ORG   00C0H
DATA20         DQ     4523C2H              ;HEX
DATA21         DQ     'HI'                         ;ASCII CHARACTERS
DATA22         DQ     ?                            ;NOTHING
```

```
00C0                                ORG 00C0H
00C0 C223450000000000        DATA20   DQ   4523C2H     ;HEX
00C8 4948000000000000        DATA21   DQ   'HI'        ;ASCII CHARACTERS
00D0 0000000000000000        DATA22   DQ   ?           ;NOTHING
```

List File for DQ Examples

DT (define ten bytes)

DT is used for memory allocation of packed BCD numbers. The application of DT will be seen in the multibyte addition of BCD numbers in Chapter 3. For now, observe how they are located in memory. Notice that the "H" after the data is not needed. This directive allocates 10 bytes, but a maximum of 18 digits can be entered.

```
                ORG   00E0H
DATA23          DT      867943569829     ;BCD
DATA24          DT      ?                ;NOTHING
```

```
00E0                            ORG 00E0H
00E0 299856437986000000         DATA23 DT   867943569829     ;BCD
     00
00EA 000000000000000000         DATA24  DT   ?               ;NOTHING
     00
```

List File for DT Examples

DT can also be used to allocate 10-byte integers by using the "D" option:

```
DEC    DT    65535d       ;the assembler will convert the decimal
                          ;number to hex and store it
```

Figure 2-7 shows the memory dump of the data section, including all the examples in this section. It is essential to understand the way operands are stored in memory. Looking at the memory dump shows that all of the data directives use the little endian format for storing data, meaning that the least significant byte is located in the memory location of the lower address and the most significant byte resides in the memory location of the higher address. For example, look at the case of "DATA20 DQ 4523C2", residing in memory starting at offset 00C0H. C2, the least significant byte, is in location 00C0, with 23 in 00C1, and 45, the most significant byte, in 00C2. It must also be noted that for ASCII data, only the DB directive can be used to define data of any length, and the use of DD, DQ, or DT directives for ASCII strings of more than 2 bytes gives an assembly error. When DB is used for ASCII numbers, notice how it places them backwards in memory. For example, see "DATA4 DB '2591'" at origin 10H: 32, ASCII for 2, is in memory location 10H; 35, ASCII for 5, is in 11H; and so on.

```
-D 1066:0 100
1066:0000 19 89 12 00 00 00 00 00-00 00 00 00 00 00 00 00 ................
1066:0010 32 35 39 31 00 00 00 00-00 00 00 00 00 00 00 00 2591............
1066:0020 4D 79 20 6E 61 6D 65 20-69 73 20 4A 6F 65 00 00 My name is Joe..
1066:0030 FF FF FF FF FF FF 00 00-FF FF FF FF FF 00 00 00 ................
1066:0040 00 00 00 00 00 00 00 00-00 00 00 00 00 00 00 00 ................
1066:0060 63 63 63 63 63 63 63 63-63 63 00 00 00 00 00 00 cccccccccc......
1066:0070 BA 03 54 09 3F 25 00 00-09 00 02 00 07 00 0C 00 :.T.?%..........
1066:0080 20 00 05 00 4F 48 00 00-00 00 00 00 00 00 00 00 ...OH...........
1066:0090 00 00 00 00 00 00 00 00-00 00 00 00 00 00 00 00 ................
1066:00A0 FF 03 00 00 5C 96 08 00-F2 57 2A 5C 23 00 00 00 ....\...rW*\#...
1066:00B0 89 47 03 00 FD FF 00 00-00 00 00 00 00 00 00 00 B#E......IH.....
1066:00C0 C2 23 45 00 00 00 00 00-49 48 00 00 00 00 00 00 ................
1066:00D0 00 00 00 00 00 00 00 00-00 00 00 00 00 00 00 00 ................
1066:00E0 29 98 56 43 79 86 00 00-00 00 00 00 00 00 00 00 9.VCy6..........
```

Figure 2-7. DEBUG Dump of Data Segment

Review Questions

1. The ________ directive is always used for ASCII strings longer than 2 bytes.
2. How many bytes are defined by the following?
   DATA_1 DB 6 DUP (4 DUP (0FFH))
3. Do the following two data segment definitions result in the same storage in bytes at offset 10H and 11H? If not, explain why.

	ORG	10H			ORG	10H
DATA_1	DB	72		DATA_1	DW	7204H
DATA_2	DB	04H				

4. The DD directive is used to allocate memory locations that are ____ bytes in length. The DQ directive is used to allocate memory locations that are ________ bytes in length.
5. State briefly the purpose of the ORG directive.
6. What is the advantage in using the EQU directive to define a constant value?
7. How many bytes are set aside by each of the following directives?
   (a) ASC_DATA DB '1234' (b) HEX_DATA DW 1234H
8. Does the little endian storage convention apply to the storage of ASCII data?

SECTION 2.6: FULL SEGMENT DEFINITION

The way that segments have been defined in the programs above is a newer definition referred to as *simple segment definition*. It is supported by Microsoft's MASM 5.0 and higher, Borland's TASM version 1 and higher, and many other compatible assemblers. The older, more traditional definition is called the *full segment definition*. Although the simplified segment definition is much easier to understand and use, especially for beginners, it is good to become familiar with full segment definition since many older programs use it.

Segment definition

The SEGMENT and ENDS directives indicate to the assembler the beginning and ending of a segment and have the following format:

```
label SEGMENT     [ options]
      ;place the statements belonging to this segment here
label ENDS
```

The label, or name, must follow naming conventions (see the end of Section 2.4) and must be unique. The [options] field gives important information to the assembler for organizing the segment, but is not required. The ENDS label must be the same label as in the SEGMENT directive. In the full segment definition, the "`.MODEL`" directive is not used. Further, the directives "`.STACK`", "`.DATA`", and "`.CODE`" are replaced by SEGMENT and ENDS directives that surround each segment. Figure 2-8 shows the full segment definition and simplified format, side by side. This is followed by Programs 2-2 and 2-3, rewritten using the full segment definition.

Stack segment definition

The stack segment shown below contains the line: "`DB 64 DUP (?)`" to reserve 64 bytes of memory for the stack. The following three lines in full segment definition are comparable to "`.STACK 64`" in simple definition:

```
STSEG  SEGMENT              ;the "SEGMENT" directive begins the segment
       DB 64 DUP (?)        ;this segment contains only one line
STSEG  ENDS                 ;the "ENDS" segment ends the segment
```

```
;FULL SEGMENT DEFINITION                    ;SIMPLIFIED FORMAT
      ;—-- stack segment —--                .MODEL  SMALL
      name1 SEGMENT                         .STACK   64
            DB    64 DUP (?)                ;
      name1 ENDS                            ;
      ;—-- data segment —--                 ;——————————————————
      name2 SEGMENT                         . DATA
      ;place data definitions here          ;place data definitions here
      name2 ENDS                            ;
      ;—-- code segment —--                 ;————————————————————
name3 SEGMENT                               .CODE
      MAIN  PROC  FAR                       MAIN  PROC   FAR
            ASSUME ...                            MOV    AX,@DATA
            MOV   AX,name2                        MOV    DS,AX
            MOV   DS,AX                           ...
            ...                                   ...
      MAIN  ENDP                            MAIN  ENDP
      name3 ENDS                                  END    MAIN
            END   MAIN
```

Figure 2-8. Full versus Simplified Segment Definition

```
TITLE         PURPOSE: ADDS 4 WORDS OF DATA
PAGE  60,132
STSEG         SEGMENT
              DB    32 DUP (?)
STSEG         ENDS
DTSEG         SEGMENT
DATA_IN       DW          234DH,1DE6H,3BC7H,566AH
              ORG   10H
SUM           DW          ?
DTSEG         ENDS
;——————————————
CDSEG         SEGMENT
MAIN          PROC        FAR
              ASSUME CS:CDSEG,DS:DTSEG,SS:STSEG
              MOV   AX,DTSEG
              MOV   DS,AX
              MOV   CX,04                   ;set up loop counter CX=4
              MOV   DI,OFFSET DATA_IN       ;set up data pointer DI
              MOV   BX,00                   ;initialize BX
ADD_LP:       ADD   BX,[DI]       ;add contents pointed at by [DI]  to BX
              INC   DI                      ;increment DI twice
              INC   DI                      ;to point to next word
              DEC   CX                      ;decrement loop counter
              JNZ   ADD_LP                  ;jump if loop counter not zero
              MOV   SI,OFFSET SUM           ;load pointer for sum
              MOV   [SI],BX                 ;store in data segment
              MOV   AH,4CH                  ;set up return
              INT   21H                     ;return to OS
MAIN          ENDP
CDSEG         ENDS
              END   MAIN
```

Program 2-2, rewritten with full segment definition

```
TITLE        PURPOSE: TRANSFERS 6 BYTES OF DATA
PAGE         60,132
STSEG        SEGMENT
             DB          32 DUP (?)
STSEG        ENDS
;----------------
DTSEG        SEGMENT
             ORG   10H
DATA_IN      DB          25H,4FH,85H,1FH,2BH,0C4H
             ORG   28H
COPY         DB          6 DUP(?)
DTSEG        ENDS
;----------------
CDSEG        SEGMENT
MAIN         PROC        FAR
             ASSUME CS:CDSEG,DS:DTSEG,SS:STSEG
             MOV   AX,DTSEG
             MOV   DS,AX
             MOV   SI,OFFSET DATA_IN ;SI points to data to be copied
             MOV   DI,OFFSET COPY    ;DI points to copy of data
             MOV   CX,06H            ;loop counter = 6
MOV_LOOP:    MOV   AL,[SI]           ;move next byte from DATA area to AL
             MOV   [DI],AL           ;move the next byte to COPY area
             INC   SI                ;increment DATA pointer
             INC   DI                ;increment COPY pointer
             DEC   CX                ;decrement LOOP counter
             JNZ   MOV_LOOP          ;jump if loop counter not zero
             MOV   AH,4CH            ;set up to return
             INT   21H               ;return to OS
MAIN         ENDP
CDSEG        ENDS
             END   MAIN
```

Program 2-3, rewritten with full segment definition

Data segment definition

In full segment definition, the SEGMENT directive names the data segment and must appear before the data. The ENDS segment marks the end of the data segment:

```
DTSEG        SEGMENT      ;the SEGMENT directive begins the segment
             ;define your data here
DTSEG        ENDS         ;the ENDS segment ends the segment
```

Code segment definition

The code segment also begins and ends with SEGMENT and ENDS directives:

```
CDSSEG       SEGMENT      ;the SEGMENT directive begins the segment
             ;your code is here
CDSEG        ENDS         ;the ENDS segment ends the segment
```

In full segment definition, immediately after the PROC directive is the ASSUME directive, which associates segment registers with specific segments by assuming that the segment register is equal to the segment labels used in the program. If an extra segment had been used, ES would also be included in the ASSUME statement. The ASSUME statement is needed because a given Assembly language program can have several code segments, one or two or three or more data segments and more than one stack segment, but only one of each can be addressed by the CPU at a given time since there is only one of each of the segment registers available inside the CPU. Therefore, ASSUME tells the assembler which of the segments defined by the SEGMENT directives should be used. It also helps the assembler to calculate the offset addresses from the beginning of that segment. For example, in "`MOV AL,[ BX]` " the BX register is the offset of the data segment.

Upon transfer of control from OS to the program, of the three segment registers, only CS and SS have the proper values. The DS value (and ES, if used) must be initialized by the program. This is done as follows in full segment definition:

```
MOV AX,DTSEG        ;DTSEG is the label for the data segment
MOV DS,AX
```

Using the emu8086 assembler

There is a simple and popular assembler called emu8086 that one can use for assembling the 8086 Assembly language programs. It is available from the www.emu8086.com website. Examine Figures 2-9 to 2-13 for screenshots using emu8086.

Download the emu8086 assembler from the following website:
http://www.emu8086.com

See a tutorial on emu8086 on the following website:
http://www.MicroDigitalEd.com

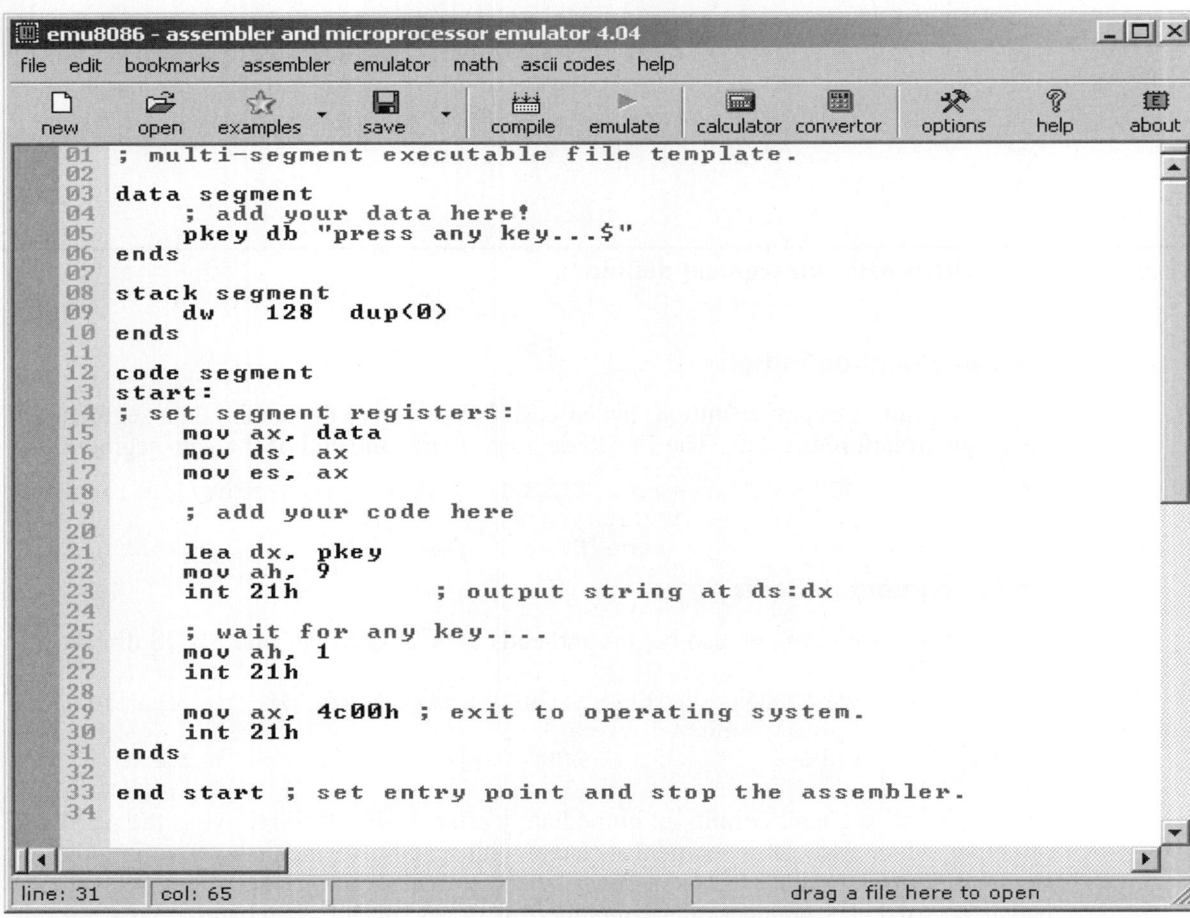

Figure 2-9. emu8086 Screen Shot

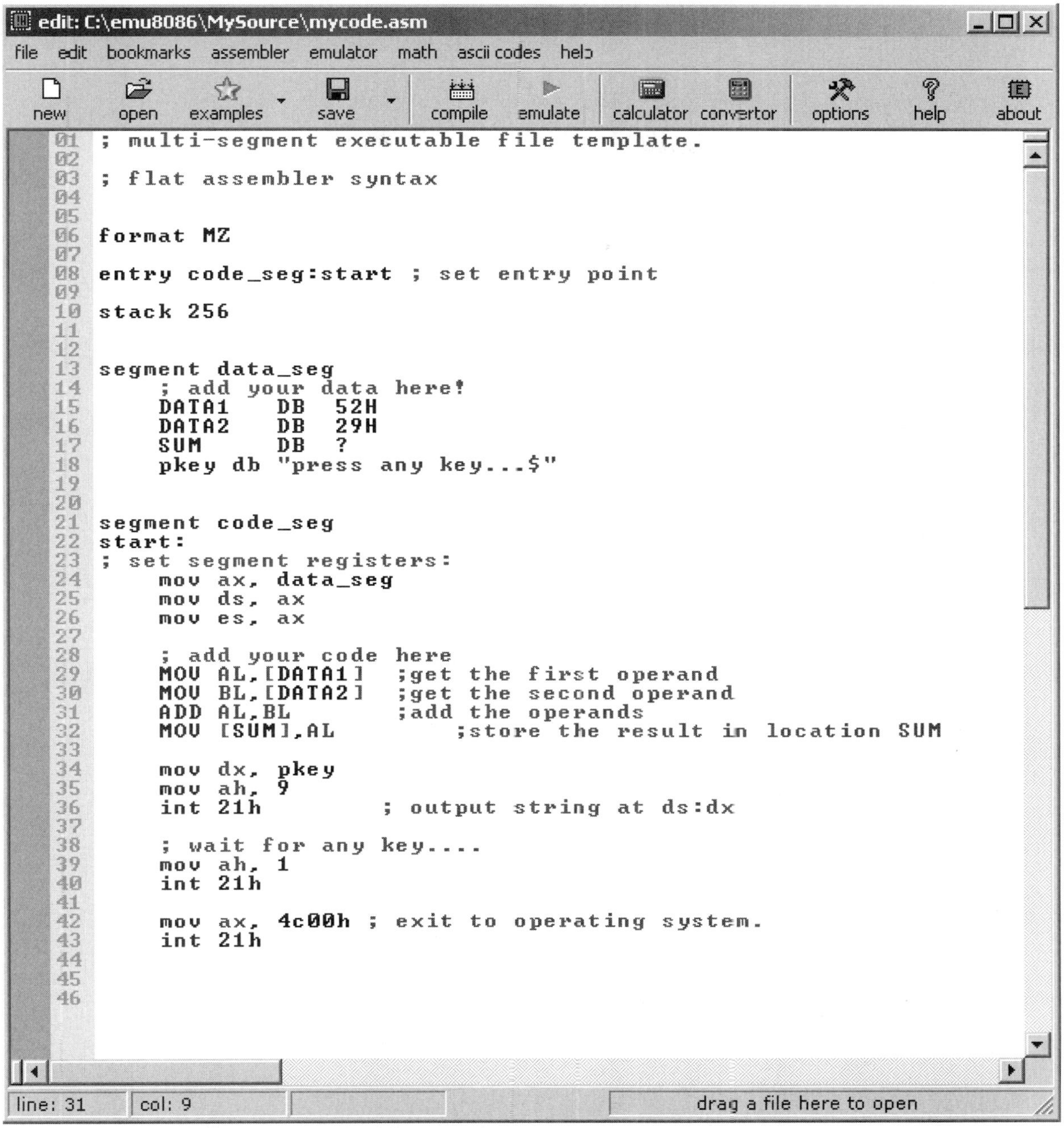

Figure 2-10. emu8086 with Figure 2-1 Inserted

Note: emu8086 requires putting brackets around variables, unlike MASM.

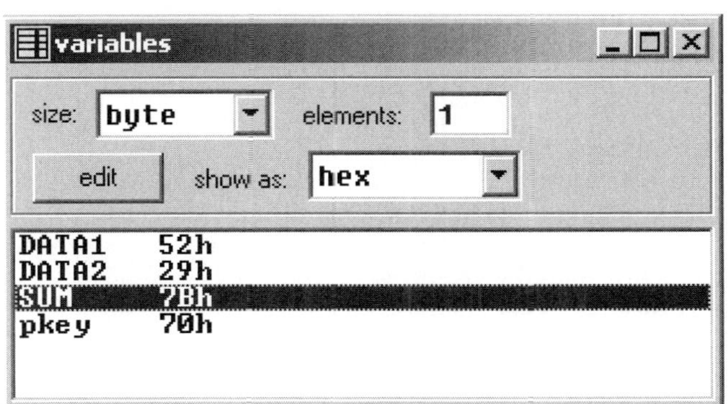

Figure 2-11. emu8086 Screenshot of Variables for Program in Figure 2-10

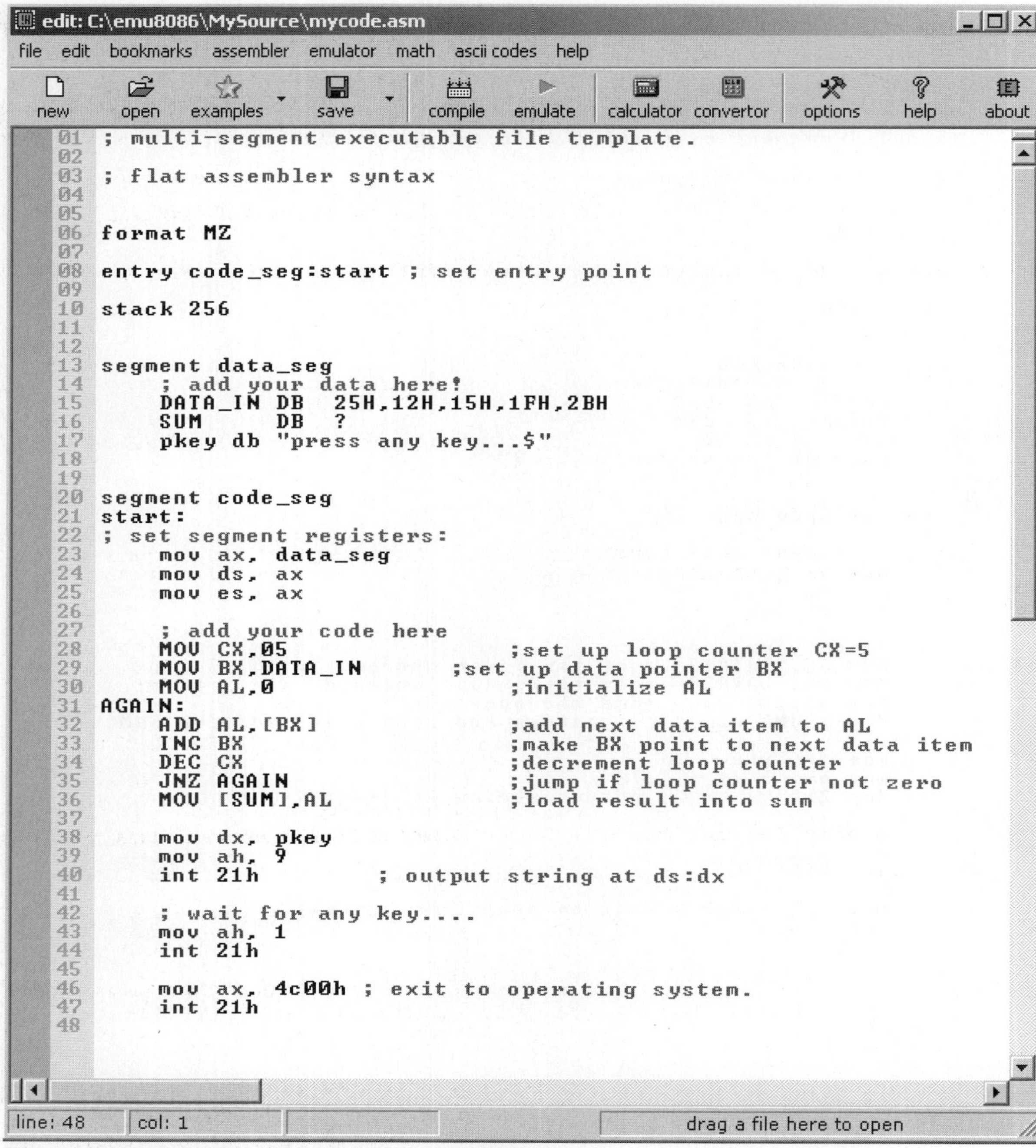

Figure 2-12. emu8086 with Program 2-1 Inserted

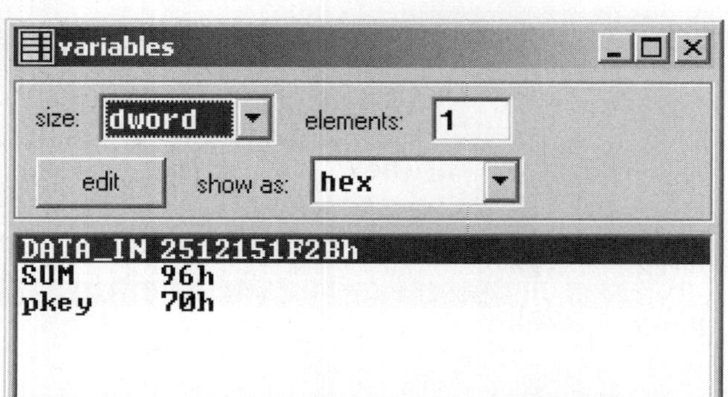

Figure 2-13. emu8086 Screenshot of Variables for Program in Figure 2-12

EXE vs. COM Files

All program examples so far were designed to be assembled and linked into EXE files. This section looks at the COM file, which like the EXE file contains the executable machine code and can be run at the OS level.

Why COM files?

There are occasions when, due to a limited amount of memory, one needs to have very compact code. This is the time when the COM file is useful. The fact that the EXE file can be of any size is one of the main reasons that EXE files are used so widely. On the other hand, COM files are used because of their compactness since they cannot be greater than 64K bytes. The reason for the 64K-byte limit is that the COM file must fit into a single segment, and since in the x86 the size of a segment is 64K bytes, the COM file cannot be larger than 64K. To limit the size of the file to 64K bytes requires defining the data inside the code segment and also using an area (the end area) of the code segment for the stack. One of the distinguishing features of the COM file program is the fact that in contrast to the EXE file, it has no separate data segment definition. One can summarize the differences between COM and EXE files as shown in Table 2-2.

Table 2-2: EXE vs. COM File Format

EXE File	COM File
unlimited size	maximum size 64K bytes
stack segment is defined	no stack segment definition
data segment is defined	data segment defined in code segment
code, data defined at any offset address	code and data begin at offset 0100H
larger file (takes more memory)	smaller file (takes less memory)

Another reason for the difference in the size of the EXE and COM files is the fact that the COM file does not have a header block. The header block, which occupies 512 bytes of memory, precedes every EXE file and contains information such as size, address location in memory, and stack address of the EXE module.

Review Questions

1. The ______ (full, simple) segment definition is the newer definition style.
2. True or false. The ASSUME directive is used in simple segment definition.
3. In full segment definition, each segment begins with the ________ directive and ends with a matching ________ directive.

SECTION 2.7: FLOWCHARTS AND PSEUDOCODE

Structured programming is a term used to denote programming techniques that can make a program easier to code, debug, and maintain over time. There are certain principles that every structured program should follow. Some of these are as follows:

1. The program should be designed before it is coded. By using techniques of flowcharting or pseudocode, described below, the design of the program is clear to the person coding it as well as those who will maintain the program later.
2. Using comments within the program and documentation accompanying the program also will help someone else figure out what the program does and how it does it. It may even help the programmer who wrote the program remember how it worked years later!
3. The main routine should consist primarily of calls to subroutines that perform the work of the program. This is sometimes called *top-down programming*. Use subroutines to accomplish tasks that are repeated. This saves time in coding and also makes the program easier to read. Also use subroutines to perform related tasks, such as "housekeeping" chores to keep the main program clear of clutter.
4. Data control is very important. It can be very frustrating and time consuming to track through a long program to find where a variable was changed. First of all, the pro-

grammer should document the purpose of each variable, and which subroutines might alter its value. Further, each subroutine should document its input and output variables, and which input variables might be altered within it.

Flowcharts

If you have taken any previous programming courses, you are probably familiar with flowcharting. Flowcharts use graphic symbols to represent different types of program operations. These symbols are connected together into a flowchart to show the flow of execution of the program. Figure 2-14 shows some of the more commonly used symbols. Flowchart templates are available to help you draw the symbols quickly and neatly.

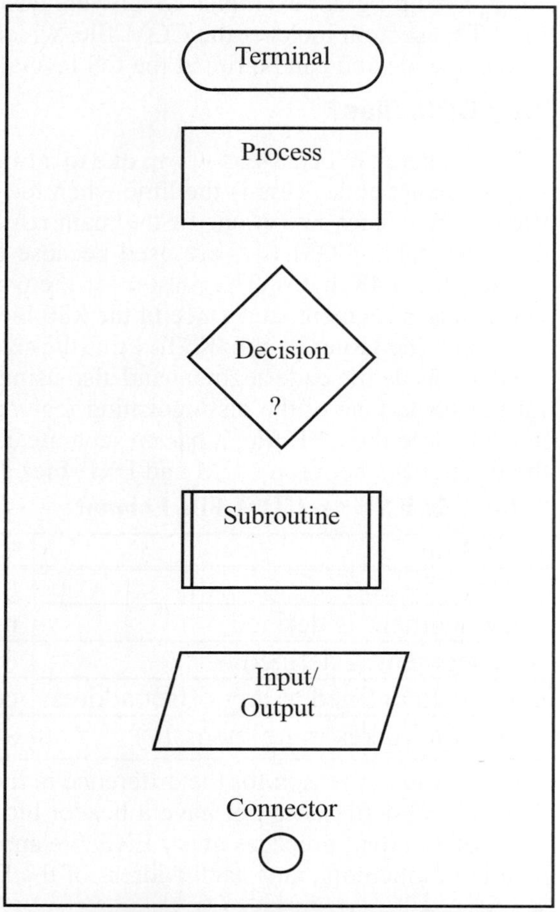

Figure 2-14. Commonly Used Flowchart Symbols

Pseudocode

Flowcharting has been standard practice in industry for decades. However, some programs find limitations in using flowcharts, such as the fact that you can't write much in the little boxes, and it is hard to get the "big picture" of what the program does without getting bogged down in the details. An alternative to using flowcharts is pseudocode, which involves writing brief descriptions of the flow of the code. Figures 2-15 through 2-19 show flowcharts and pseudocode for commonly used control structures. Then we show an example of flowcharts and pseudocode for selected programs in this chapter.

Control structures

Structured programming uses three basic types of program control structures: sequence, control, and iteration. Sequence is simply executing instructions one after the other. Figure 2-15 shows how sequence can be represented in pseudocode and flowcharts. Figures 2-16 and 2-17 show two control programming structures: IF-THEN and IF-THEN-ELSE in both pseudocode and flowcharts. Note in all of these figures that "statement" can indicate one statement or a group of statements. For example, a statement might be "Compute net pay." Obviously it would require many lines of code to complete this "statement."

Figures 2-18 and 2-19 show two iteration control structures: REPEAT-UNTIL and WHILE-DO. Both structures execute a statement or group of statements repeatedly. The difference between them is that the REPEAT-UNTIL structure always executes the statement(s) at least once, and checks the condition after each iteration, whereas the WHILE-DO may not execute the statement(s) at all since the condition is checked at the beginning of each iteration.

Figure 2-20 shows a flowchart versus pseudocode for Program 2-1. In this example, more program details are given than one usually finds. For example, this flowchart shows steps for initializing and decrementing counters. Another programmer may not include these steps in the flowchart or pseudocode. Notice that housekeeping chores such as initializing the data segment register in the MAIN procedure are not included in the flowchart or pseudocode.

Code specific to a certain language or operating platform is not described in the pseudocode or flowchart. Ideally, one could take a flowchart or pseudocode for a given program and then code the program in any language. For example, if a company needed to have both a C and BASIC language version of the same program, these two programs would have the same flowchart or pseudocode, even though the actual program code would look quite different.

It is important to remember that the purpose of flowcharts or pseudocode is to show the flow of the program and what the program does, not the specific Assembly language instructions that

```
Statement 1
Statement 2
```

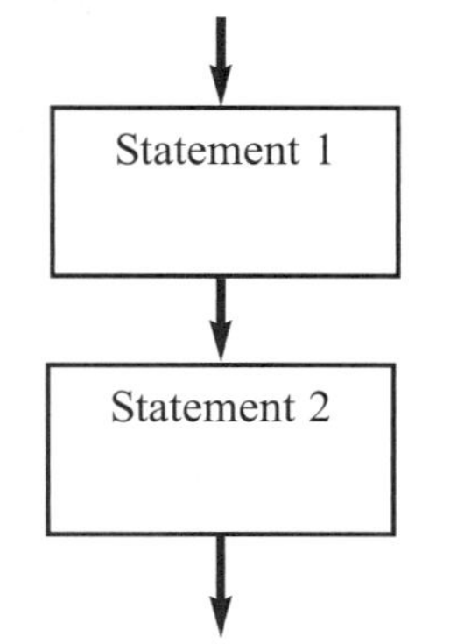

Figure 2-15. SEQUENCE Pseudocode vs. Flowchart

```
IF (condition) THEN
    Statement 1
ELSE
    Statement 2
```

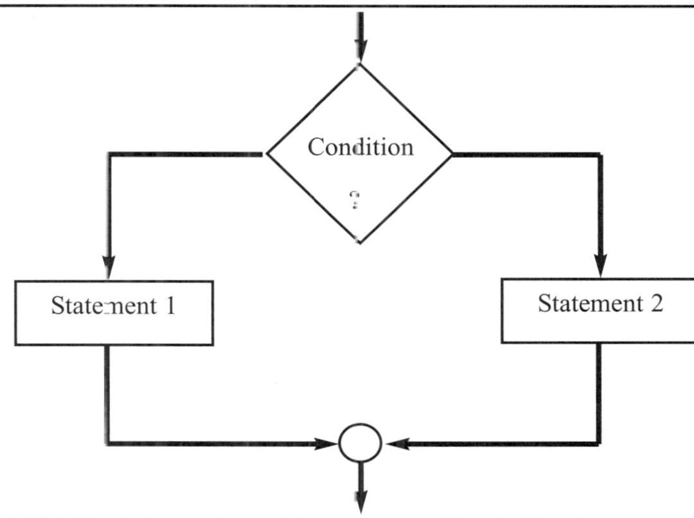

Figure 2-16. IF-THEN-ELSE Pseudocode vs. Flowchart

```
IF (condition) THEN
    Statement 1
```

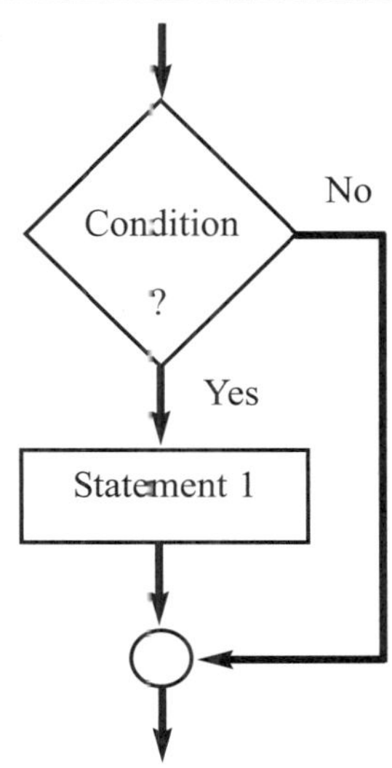

Figure 2-17. IF-THEN Pseudocode vs. Flowchart

accomplish the program's objectives.

Notice also that pseudocode gives the same information in a much more compact form compared to the flowchart. Often pseudocode is written in layers, in a top-down manner, so that the top level shows the flow of the program and subsequent levels show more details of how the program accomplishes its assigned tasks.

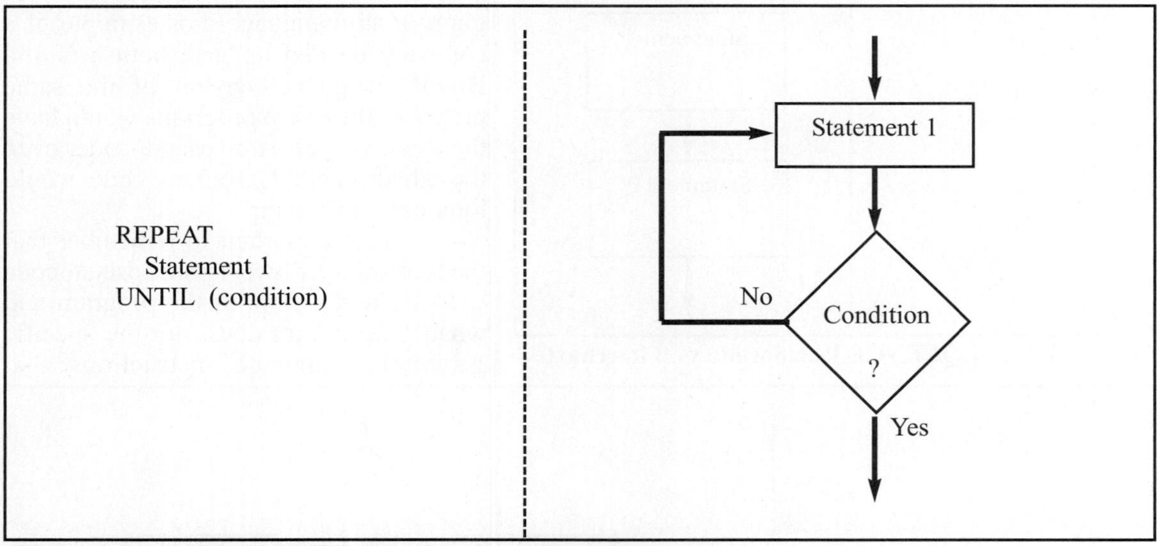

Figure 2-18. REPEAT-UNTIL Pseudocode vs. Flowchart

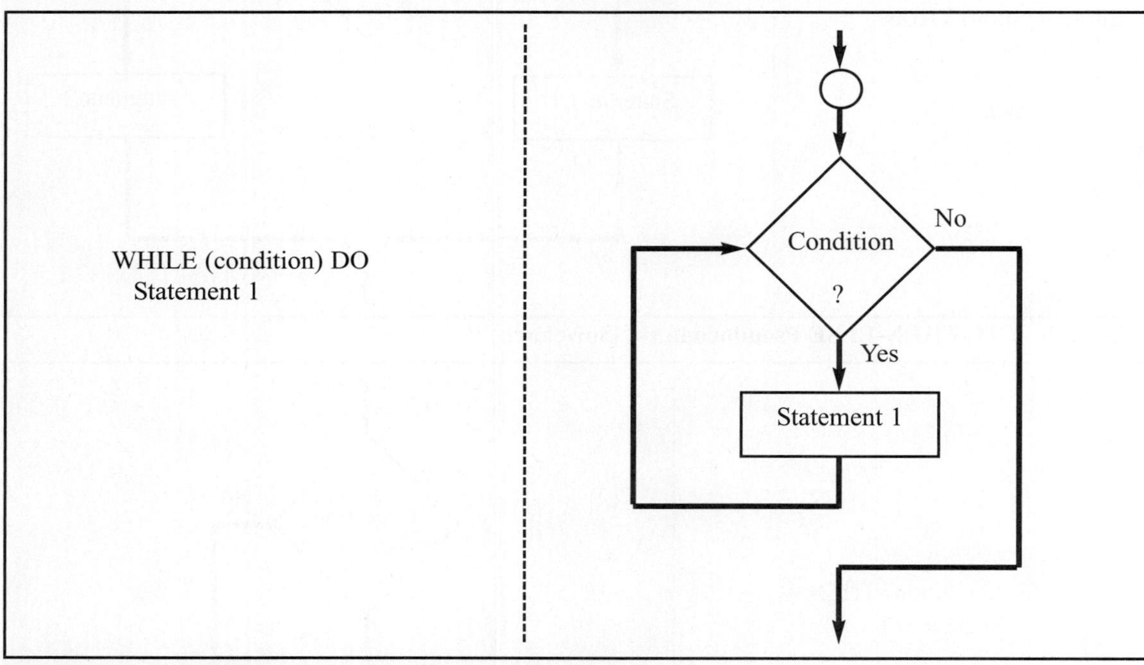

Figure 2-19. WHILE-DO Pseudocode vs. Flowchart

Review Questions

1. List four principles of structure programming.
2. ________ (flowcharts, pseudocode) use(s) graphic symbols to illustrate program flow.
3. Name two control programming structures.
4. True or false. The WHILE-DO control structure always executes the statement(s) at least once.

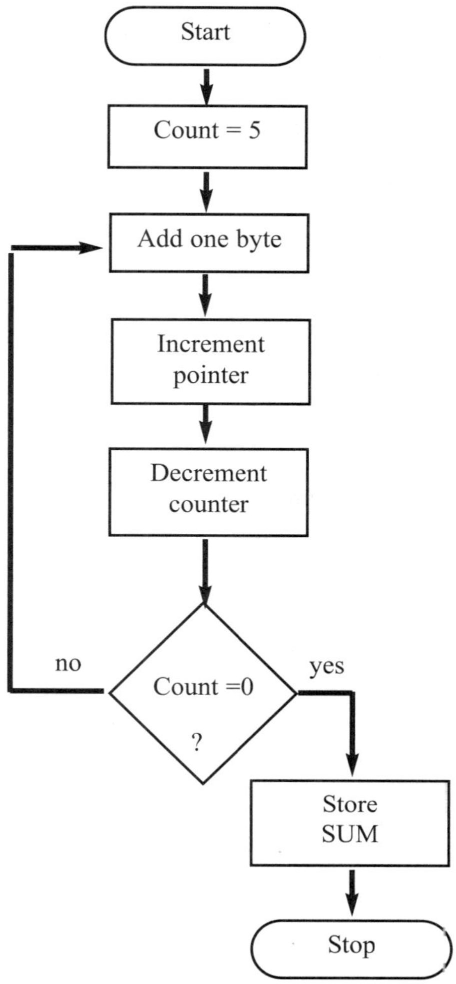


```
Count = 5

Repeat
        Add next byte
        Increment pointer
        Decrement count
Until Count = 0

Store SUM
```

Figure 2-20. Flowchart vs. Pseudocode for Program 2-1

PROBLEMS

SECTION 2.1: DIRECTIVES AND A SAMPLE PROGRAM
SECTION 2.2: ASSEMBLE, LINK, AND RUN A PROGRAM
SECTION 2.3: MORE SAMPLE PROGRAMS

1. Rewrite Program 2-3 to transfer one word at a time instead of one byte.
2. List the steps in getting a ready-to-run program.
3. Which program produces the ".exe" file?
4. Which program produces the ".obj" file?
5. True or false: The ".lst" file is produced by the assembler regardless of whether or not the programmer wants it.
6. The source program file must have the ".asm" extension in some assemblers such as MASM. Is this true for the assembler you are using?
7. Circle one: The linking process comes (after, before) assembling.

SECTION 2.4: CONTROL TRANSFER INSTRUCTIONS

8. In some applications it is common practice to save all registers at the beginning of a subroutine. Assume that SP = 1288H before a subroutine CALL. Show the contents of the stack pointer and the exact memory contents of the stack after PUSHF for the following:

```
1132:0450 CALL PROC1
```

```
1132:0453    INC BX
             PROC1 PROC
             PUSH AX
             PUSH BX
             PUSH CX
             PUSH DX
             PUSH  SI
             PUSH DI
             PUSHF
             ....
             PROC1 ENDP
```

9. To restore the original information inside the CPU at the end of a CALL to a subroutine, the sequence of POP instructions must follow a certain order. Write the sequence of POP instructions that will restore the information in Problem 8. At each point, show the contents of the SP.
10. When a CALL is executed, how does the CPU know where to return?
11. In a FAR CALL, ___________ and ___________ are saved on the stack, whereas in a NEAR CALL, ________ is saved on the stack.
12. Compare the number of bytes of stack taken due to NEAR and FAR CALLs.
13. Find the contents of the stack and stack pointer after execution of the CALL instruction shown below. SUM is a near procedure. Assume the value SS:1296 right before the execution of CALL.

    CS : IP
    2450:673A CALL SUM
    2450:673D DEC AH
14. The following is a section of BIOS of the IBM PC which is described in detail in Chapter 3. All the jumps below are short jumps, meaning that the labels are in the range –128 to +127.

```
IP                    Code
E06C 733F             JNC   ERROR1
...            ...
E072 7139             JNO   ERROR1
...   ...
E08C 8ED8      C8:    MOV   DS,AX
 ...           ...
E0A7 EBE3             JMP   C8
...            ...
E0AD F4               ERROR1: HLT
```

    Verify the address calculations of:
    (a) JNC ERROR1 (b) JNO ERROR1 (c) JMP C8

SECTION 2.5: DATA TYPES AND DATA DEFINITION

15. Find the precise offset location of each ASCII character or data in the following:

```
      ORG   20H
DATA1 DB    '1-800-555-1234'
      ORG   40H
DATA2 DB    'Name: John Jones'
      ORG   60H
DATA3 DB    '5956342'
      ORG   70H
DATA4 DW    2560H,1000000000110B
DATA5 DW    49
      ORG   80H
DATA6 DD    25697F6EH
DATA7 DQ    9E7BA21C99F2H
      ORG   90H
DATA8 DT    439997924999828
DATA9 DB    6 DUP (0EEH)
```

16. The following program contains some errors. Fix the errors and make the program run correctly. Verify it through the DEBUG program. This program adds four words

and saves the result.

```
        TITLE PROBLEM   (EXE)    PROBLEM 16 PROGRAM
        PAGE  60,132
        .MODEL SMALL
        .STACK 32
        ;----------------------------
        .DATA
        DATA          DW        234DH,DE6H,3BC7H,566AH
                      ORG      10H
        SUM           DW        ?
        ;----------------------------
        .CODE
START:  PROC  FAR
        MOV    AX,DATA
        MOV    DS,AX
        MOV    CX,04                ;SET UP LOOP COUNTER CX=4
        MOV    BX,0                 ;INITIALIZE BX TO ZERO
        MOV    DI,OFFSET DATA       ;SET UP DATA POINTER BX
LOOP1:ADD      BX,[DI]    ;ADD CONTENTS POINTED AT BY [DI]  TO BX
        INC    DI                   ;INCREMENT DI
        JNZ    LOOP1                ;JUMP IF COUNTER NOT ZERO
        MOV    SI,OFFSET RESULT     ;LOAD POINTER FOR  RESULT
        MOV    [SI],BX              ;STORE THE SUM
        MOV    AH,4CH
        INT    21H
START ENDP
        END    STRT
```

ANSWERS TO REVIEW QUESTIONS

SECTION 2.1: DIRECTIVES AND A SAMPLE PROGRAM

1. Pseudo-instructions direct the assembler as to how to assemble the program.
2. Instructions, pseudo-instructions or directives
3.

```
                    .MODEL SMALL
                    .STACK 64
                    .DATA
     HIGH_DAT       DB 95
                    .CODE
     START          PROC FAR
                    MOV AX,@DATA
                    MOV DS,AX
                    MOV AH,HIGH_DAT
                    MOV BH,AH
                    MOV DL,BH
                    MOV AH,4CH
                    INT 21H
     START          ENDP
                    END START
```

4. (1) There is no ENORMOUS model.
   (2) ENDP label does not match label for PROC directive.
   (3) .CODE and .DATA directives need to be switched.
   (4) "MOV AX,DATA" should be "MOV AX,@DATA".
   (5) "MOV DS,@DATA" should be "MOV DS,AX".
   (6) END must have the entry point label "MAIN".

SECTION 2.2: ASSEMBLE, LINK, AND RUN A PROGRAM

1. (a) MASM must have the ".asm" file as input.
   (b) LINK must have the ".obj" file as input.
2. Editor outputs : (b) .asm
   Assembler outputs: (a) .obj, (d) .lst, and (e) .crf files
   Linker outputs: (c) .exe and (f) .map files

SECTION 2.3: MORE SAMPLE PROGRAMS

1. Increments the operand; that is, it causes 1 to be added to the operand.

2. Decrements the operand; that is, it causes 1 to be subtracted from the operand.
3. A colon is required after labels referring to instructions; colons are not placed after labels for directives.
4. The first moves the contents of the word beginning at offset DATA1, and the second moves the offset address of DATA1.
5. The first adds the contents of BX to AX, and the second adds the contents of the memory location at offset BX.

SECTION 2.4: CONTROL TRANSFER INSTRUCTIONS

1. FAR
2. The instruction right below the jump
3. IP
4. The machine code for the instruction will take up 1 less byte.
5. SHORT, NEAR, FAR
6. The contents of CS and IP were stored on the stack when the call executed.
7. It restores the contents of CS:IP and returns control to the instruction immediately following the CALL.
8. (a) GET.DATA, invalid because "." is only allowed as the first character
   (b) 1_NUM, because the first character cannot be a number
   (c) TEST-DATA, because "-" is not allowed
   (d) RET, is a reserved word

SECTION 2.5: DATA TYPES AND DATA DEFINITION

1. DB
2. 24
3. No because of the little endian storage conventions, which will cause the word "7204H" to be stored with the lower byte (04) at offset 10H and the upper byte at offset 11H; DB allocates each byte as it is defined.
4. 4, 8
5. It is used to assign the offset address.
6. If the value is to be changed later, it can be changed in one place instead of at every occurrence.
7. (a) 4 (b) 2
8. No

SECTION 2.6: FULL SEGMENT DEFINITION

1. Simple
2. False
3. SEGMENT, ENDS

SECTION 2.7: FLOWCHARTS AND PSEUDOCODE

1. Design before coding, use top-down programming, documentation, data control
2. Flowcharts
3. IF-THEN and IF-THEN-ELSE
4. False

CHAPTER 3

ARITHMETIC AND LOGIC INSTRUCTIONS AND PROGRAMS

OBJECTIVES

Upon completion of this chapter, you will be able to:

>> Demonstrate how 8-bit and 16-bit unsigned numbers are added in the x86
>> Convert data to any of the forms: ASCII, packed BCD, unpacked BCD
>> Explain the effect of unsigned arithmetic instructions on the flags
>> Code the following Assembly language unsigned arithmetic instructions:
>> Addition instructions ADD and ADC
>> Subtraction instructions SUB and SBB
>> Multiplication and division instructions MUL and DIV
>> Code BCD arithmetic instructions:
>> DAA, DAS
>> Code the following Assembly language logic instructions:
>> AND, OR, and XOR
>> Logical shift instructions SHR and SHL
>> The compare instruction CMP
>> Code bitwise rotation instructions ROR, ROL, RCR, and RCL
>> Demonstrate an ability to use all of the above instructions in Assembly language programs
>> Perform bitwise manipulation using the C language

In this chapter, most of the arithmetic and logic instructions are discussed and program examples are given to illustrate the application of these instructions. Unsigned numbers are used in this discussion of arithmetic and logic instructions. Signed numbers are discussed separately in Chapter 6. In Section 3.1 we examine the addition and subtraction of unsigned numbers. The multiplication and division of unsigned numbers are discussed in Section 3.2. The logic instructions and programs are covered in Section 3.3. Section 3.4 is dedicated to BCD and ASCII data conversion, while the instructions related to rotate and shift operations are examined in Section 3.5. The last section of the chapter describes bitwise operations in the C language.

SECTION 3.1: UNSIGNED ADDITION AND SUBTRACTION

Unsigned numbers are defined as data in which all the bits are used to represent data and no bits are set aside for the positive or negative sign. This means that the operand can be between 00 and FFH (0 to 255 decimal) for 8-bit data and between 0000 and FFFFH (0 to 65535 decimal) for 16-bit data. This section covers the ADD and SUB instructions.

Addition of unsigned numbers

The form of the ADD instruction is

```
ADD destination,source ;destination = destination + source
```

The instructions ADD and ADC are used to add two operands. The destination operand can be a register or in memory. The source operand can be a register, in memory, or immediate. Remember that memory-to-memory operations are never allowed in x86 Assembly language. The instruction could change any of the ZF, SF, AF, CF, or PF bits of the flag register, depending on the operands involved. The effect of the ADD instruction on the overflow flag is discussed in Chapter 6 since it is used in signed number operations. Look at Example 3-1.

Example 3-1

Show how the flag register is affected by

```
MOV   AL,0F5H
ADD   AL,0BH
```

Solution:

```
     F5H          1111 0101
 +   0BH      +   0000 1011
    100H          0000 0000
```

After the addition, the AL register (destination) contains 00 and the flags are as follows:
CF = 1, since there is a carry out from D7
SF = 0, the status of D7 of the result
PF = 1, the number of 1s is zero (zero is an even number)
AF = 1, there is a carry from D3 to D4
ZF = 1, the result of the action is zero (for the 8 bits)

In discussing addition, the following two cases will be examined:

1. Addition of individual byte and word data
2. Addition of multibyte data

Write a program to calculate the total sum of 5 bytes of data. Each byte represents the daily wages of a worker. This person does not make more than $255 (FFH) a day. The decimal data is as follows: 125, 235, 197, 91, and 48.

```
TITLE       PROG3-1A (EXE) ADDING 5 BYTES
PAGE        60,132
.MODEL SMALL
.STACK 64
;--------------------------------
            .DATA
COUNT       EQU   05
DATA        DB          125,235,197,91,48
            ORG   0008H
SUM         DW    ?
;--------------------------------
     .CODE
MAIN PROC   FAR
     MOV    AX,@DATA
     MOV    DS,AX
     MOV    CX,COUNT            ;CX is the loop counter
     MOV    SI,OFFSET DATA      ;SI is the data pointer
     MOV    AX,00               ;AX will hold the sum
BACK:ADD    AL,[ SI]            ;add the next byte to AL
     JNC    OVER          ;if no carry, continue
     INC    AH            ;else accumulate carry in AH
OVER:INC    SI            ;increment data pointer
     DEC    CX            ;decrement loop counter
     JNZ    BACK          ;if not finished, go add next byte
     MOV    SUM,AX        ;store sum
     MOV    AH,4CH
     INT    21H           ;go back to OS
MAIN ENDP
     END    MAIN
```

Program 3-1a

CASE 1: Addition of individual byte and word data

In Chapter 2 there was a program that added 5 bytes of data. The total sum was purposely kept less than FFH, the maximum value an 8-bit register can hold. To calculate the total sum of any number of operands, the carry flag should be checked after the addition of each operand. Program 3-1a uses AH to accumulate carries as the operands are added to AL.

Analysis of Program 3-1a

These numbers are converted to hex by the assembler as follows: 125 = 7DH, 235 = 0EBH, 197 = 0C5H, 91 = 5BH, 48 = 30H. Three iterations of the loop are shown below. The tracing of the program is left to the reader as an exercise.

1. In the first iteration of the loop, 7DH is added to AL with CF = 0 and AH = 00. CX = 04 and ZF = 0.
2. In the second iteration of the loop, EBH is added to AL, which results in AL = 68H and CF = 1. Since a carry occurred, AH is incremented. CX = 03 and ZF = 0.
3. In the third iteration, C5H is added to AL, which makes AL = 2DH. Again a carry occurred, so AH is incremented again. CX = 02 and ZF = 0.

This process continues until CX = 00 and the zero flag becomes 1, which will cause JNZ to fall through. Then the result will be saved in the word-sized memory set aside in the data segment. Although this program works correctly, due to pipelining it is strongly recommended that the following lines of the program be replaced:

Replace these lines	With these lines
`BACK: ADD   AL,[SI]`	`BACK: ADD   AL,[SI]`
`      JNC   OVER`	`      ADC   AH,00 ;add 1 to AH if CF=1`
`      INC   AH`	`      INC   SI`
`OVER: INC   SI`	

The "ADC AH,00" instruction in reality means add 00 + AH + CF and place the result in AH. This is much more efficient since the instruction "JNC OVER" has to empty the queue of pipelined instructions and fetch the instructions from the OVER target every time the carry is zero (CF = 0).

The addition of many word operands works the same way. Register AX (or CX, DX, or BX) could be used as the accumulator and BX (or any general-purpose 16-bit register) for keeping the carries. Program 3-1b is the same as Program 3-1a, rewritten for word addition.

Write a program to calculate the total sum of five words of data. Each data value represents the yearly wages of a worker. This person does not make more than $65,555 (FFFFH) a year. The decimal data is as follows: 27345, 28521, 29533, 30105, and 32375.

```
TITLE      PROG3-1B (EXE) ADDING 5 WORDS
PAGE       60,132
.MODEL SMALL
.STACK 64
;-----------------
       .DATA
COUNT  EQU        05
DATA   DW         27345,28521,29533,30105,32375
       ORG        0010H
SUM    DW         2 DUP(?)
;-----------------
     .CODE
MAIN PROC   FAR
     MOV    AX,@DATA
     MOV    DS,AX
     MOV    CX,COUNT            ;CX is the loop counter
     MOV    SI,OFFSET DATA      ;SI is the data pointer
     MOV    AX,00               ;AX will hold the sum
     MOV    BX,AX               ;BX will hold the carries
BACK:ADD    AX,[SI]             ;add the next word to AX
     ADC    BX,0          ;add carry to BX
     INC    SI            ;increment data pointer twice
     INC    SI            ;to point to next word
     DEC    CX            ;decrement loop counter
     JNZ    BACK          ;if not finished, continue adding
     MOV    SUM,AX              ;store the sum
     MOV    SUM+2,BX      ;store the carries
     MOV    AH,4CH
     INT    21H           ;go back to OS
MAIN ENDP
     END    MAIN
```

Program 3-1b

```
TITLE       PROG3-2 (EXE) MULTIWORD ADDITION
PAGE        60,132
.MODEL SMALL
.STACK 64
;-----------------------------------
      .DATA
DATA1  DQ    548FB9963CE7H
       ORG   0010H
DATA2  DQ    3FCD4FA23B8DH
       ORG   0020H
DATA3  DQ    ?
;-----------------------------------
      .CODE
MAIN  PROC   FAR
      MOV    AX,@DATA
      MOV    DS,AX
      CLC                       ;clear carry before first addition
      MOV    SI,OFFSET DATA1    ;SI is pointer for operand1
      MOV    DI,OFFSET DATA2    ;DI is pointer for operand2
      MOV    BX,OFFSET DATA3    ;BX is pointer for the sum
      MOV    CX,04              ;CX is the loop counter
BACK:MOV     AX,[SI]                  ;move the first operand to AX
      ADC    AX,[DI]                  ;add the second operand to AX
      MOV    [BX],AX                  ;store the sum
      INC    SI                       ;point to next word of operand1
      INC    SI
      INC    DI                 ;point to next word of operand2
      INC    DI
      INC    BX                 ;point to next word of sum
      INC    BX
      LOOP   BACK               ;if not finished, continue adding
      MOV    AH,4CH
      INT    21H                ;go back to OS
MAIN  ENDP
      END    MAIN
```

Program 3-2

CASE 2: Addition of multiword numbers

Assume a program is needed that will add the total U.S. budget for the last 100 years or the mass of all the planets in the solar system. In cases like this, the numbers being added could be up to 8 bytes wide or more. Since registers are only 16 bits wide (2 bytes), it is the job of the programmer to write the code to break down these large numbers into smaller chunks to be processed by the CPU. If a 16-bit register is used and the operand is 8 bytes wide, that would take a total of four iterations. However, if an 8-bit register is used, the same operands would require eight iterations. This obviously takes more time for the CPU. This is one reason to have wide registers in the design of the CPU. Powerful new CPUs such as the Itanium have registers of 64 bits wide and larger.

Analysis of Program 3-2

In writing this program, the first thing to be decided was the directive used for coding the data in the data segment. DQ was chosen since it can represent data as large as 8 bytes wide. The question is: Which add instruction should be used? In the addition of multibyte (or multiword) numbers, the ADC instruction is always used since the carry must be added to the next-higher byte (or word) in the next iteration. Before executing ADC, the carry flag must be cleared (CF = 0) so that in the first iteration, the carry would not be added. Clearing the carry flag is achieved by the CLC (clear carry) instruction. Three pointers have been used: SI for DATA1, DI for DATA2, and BX for DATA3 where the result is saved. There is a new instruction in that program, "`LOOP    XXXX`", which replaces the often used "`DEC CX`" and "`JNZ XXXX`". In other words:

```
LOOP   xxxx   ;is equivalent to the following two instructions

DEC    CX
JNZ    xxxx
```

When "`LOOP xxxx`" is executed, CX is decremented automatically, and if CX is not 0, the microprocessor will jump to target address xxxx. If CX is 0, the next instruction (the one below "`LOOP xxxx`") is executed.

Subtraction of unsigned numbers

```
SUB  dest,source;dest = dest - source
```

In subtraction, the x86 microprocessors (indeed, almost all modern CPUs) use the 2's complement method. Although every CPU contains adder circuitry, it would be too cumbersome (and take too many transistors) to design separate subtractor circuitry. For this reason, the x86 uses internal adder circuitry to perform the subtraction command. Assuming that the x86 is executing simple subtract instructions, one can summarize the steps of the hardware of the CPU in executing the SUB instruction for unsigned numbers as follows.

1. Take the 2's complement of the subtrahend (source operand).
2. Add it to the minuend (destination operand).
3. Invert the carry.

These three steps are performed for every SUB instruction by the internal hardware of the x86 CPU regardless of the source and destination of the operands as long as the addressing mode is supported. It is after these three steps that the result is obtained and the flags are set. Example 3-2 illustrates the three steps.

Example 3-2

Show the steps involved in the following:

```
MOV    AL,3FH        ;load AL=3FH
MOV    BH,23H        ;load BH=23H
SUB    AL,BH         ;subtract BH from AL. Place result in AL.
```

Solution:

```
 AL    3F          0011 1111          0011 1111
-BH   -23        - 0010 0011        + 1101 1101 (2's complement)
       1C                           1 0001 1100 CF=0 (step 3)
```

The flags would be set as follows: CF = 0, ZF = 0, AF = 0, PF = 0, and SF = 0.
The programmer must look at the carry flag (not the sign flag) to determine if the result is positive or negative.

After the execution of SUB, if CF = 0, the result is positive; if CF = 1, the result is negative and the destination has the 2's complement of the result. Normally, the result is left in 2's complement, but the NOT and INC instructions can be used to change it. The NOT instruction performs the 1's complement of the operand; then the operand is incremented to get the 2's complement. See Example 3-3.

SBB (subtract with borrow)

This instruction is used for multibyte (multiword) numbers and will take care of the borrow of the lower operand. If the carry flag is 0, SBB works like SUB. If the carry flag is 1, SBB subtracts 1 from the result. Notice the "`PTR`" operand in Example 3-4. The PTR (pointer) data specifier directive is widely used to specify the size of the operand when it differs from the defined size. In Example 3-4, "`WORD PTR`" tells the assembler to use a word operand, even though the data is defined as a doubleword.

Example 3-3

Analyze the following program:

```
;from the data segment:
DATA1        DB      4CH
DATA2        DB      6EH
DATA3        DB      ?
;from the code segment:
             MOV     DH,DATA1        ;load DH with DATA1 value (4CH)
             SUB     DH,DATA2        ;subtract DATA2 (6E) from DH (4CH)
             JNC     NEXT            ;if CF=0 jump to NEXT target
             NOT     DH              ;if CF=1 then take 1's complement
             INC     DH              ;and increment to get 2's complement
NEXT:        MOV     DATA3,DH        ;save DH in DATA3
```

Solution:

Following the three steps for "SUB DH,DATA2":

```
      4C     0100 1100           0100 1100
     -6E     0110 1110      +    1001 0010    (2's complement)
     -22                        01101 1110 CF=1  (step 3)result is negative
```

Example 3-4

Analyze the following program:

```
DATA_A       DD      62562FAH
DATA_B       DD      412963BH
RESULT       DD      ?
...          ...     ...
             MOV     AX,WORD PTR DATA_A           ;AX=62FA
             SUB     AX,WORD PTR DATA_B           ;SUB 963B from AX
             MOV     WORD PTR RESULT,AX           ;save the result
             MOV     AX,WORD PTR DATA_A +2        ;AX=0625
             SBB     AX,WORD PTR DATA_B +2        ;SUB 0412 with borrow
             MOV     WORD PTR RESULT+2,AX         ;save the result
```

Solution:

After the SUB, AX = 62FA – 963B = CCBF and the carry flag is set. Since CF = 1, when SBB is executed, AX = 625 – 412 – 1 = 212. Therefore, the value stored in RESULT is 0212CCBF.

Review Questions

1. The ADD instruction that has the syntax 'ADD destination, source" replaces the __________ operand with the sum of the two operands.
2. Why is the instruction "`ADD DATA_1,DATA_2`" illegal?
3. Rewrite the instruction above in a correct form.
4. The ADC instruction that has the syntax 'ADC destination, source" replaces the _________ operand with the sum of _____________.
5. The execution of part (a) below results in ZF = 1, whereas the execution of part (b) results in ZF = 0. Explain why.
   (a) MOV BL,04FH   (b) MOV BX,04FH
       ADD BL,0B1H   ADD BX,0B1H
6. The instruction "`LOOP ADD_LOOP`" is equivalent to what two instructions?
7. Show how the CPU would subtract 05H from 43H.
8. If CF = 1, AL = 95, and BL = 4F prior to the execution of "`SBB AL,BL`", what will be the contents of AL after the subtraction?

SECTION 3.2: UNSIGNED MULTIPLICATION AND DIVISION

One of the major changes from the 8080/85 microprocessor to the 8086 was inclusion of instructions for multiplication and division. In this section we cover each one with examples. This is multiplication and division of unsigned numbers. Signed numbers are treated in Chapter 6.

In multiplying or dividing two numbers in the x86 microprocessor, the use of registers AX, AL, AH, and DX is necessary since these functions assume the use of those registers.

Multiplication of unsigned numbers

In discussing multiplication, the following cases will be examined: (1) byte times byte, (2) word times word, and (3) byte times word.

byte × byte: In byte-by-byte multiplication, one of the operands must be in the AL register and the second operand can be either in a register or in memory as addressed by one of the addressing modes discussed in Chapter 1. After the multiplication, the result is in AX. See the following example:

```
RESULT   DW    ?            ;result is defined in the data segment
         ...
         MOV   AL,25H       ;a byte is moved to AL
         MOV   BL,65H       ;immediate data must be in a register
         MUL   BL           ;AL = 25 x 65H
         MOV   RESULT,AX    ;the result is saved
```

In the program above, 25H is multiplied by 65H and the result is saved in word-sized memory named RESULT. In that example, the register addressing mode was used. The next three examples show the register, direct, and register indirect addressing modes.

```
;from the data segment:
DATA1        DB       25H
DATA2        DB       65H
RESULT       DW       ?
;from the code segment:
      MOV   AL,DATA1
      MOV   BL,DATA2
      MUL   BL                     ;register addressing mode
      MOV   RESULT,AX
```

or

```
      MOV   AL,DATA1
      MUL   DATA2                  ;direct addressing mode
      MOV   RESULT,AX
```

or

```
      MOV   AL,DATA1
      MOV   SI,OFFSET DATA2
      MUL   BYTE PTR [ SI]         ;register indirect addressing mode
      MOV   RESULT,AX
```

In the register addressing mode example, any 8-bit register could have been used in place of BL. Similarly, in the register indirect example, BX or DI could have been used as pointers. If the register indirect addressing mode is used, the operand size must be specified with the help of the PTR pseudo-instruction. In the absence of the "BYTE PTR" directive in the example above, the assembler could not figure out if it should use a byte or word operand pointed at by SI. This confusion would cause an error.

word × word: In word-by-word multiplication, one operand must be in AX and the second operand can be in a register or memory. After the multiplication, registers AX and DX will contain the result. Since word-by-word multiplication can produce a 32-bit result, AX will hold the lower word and DX the higher word. Example:

```
DATA3       DW     2378H
DATA4       DW     2F79H
RESULT1     DW     2 DUP(?)
...         ....
            MOV   AX,DATA3       ;load first operand into AX
            MUL   DATA4          ;multiply it by the second operand
            MOV   RESULT1,AX     ;store the lower word result
            MOV   RESULT1+2,DX   ;store the higher word result
```

word × byte: This is similar to word-by-word multiplication except that AL contains the byte operand and AH must be set to zero. Example:

```
;from the data segment:
DATA5       DB     6BH
DATA6       DW     12C3H
RESULT3     DW     2 DUP(?)
;from the code segment:
            MOV   AL,DATA5       ;AL holds byte operand
            SUB   AH,AH          ;AH must be cleared
            MUL   DATA6          ;byte in AL mult. by word operand
            MOV   BX,OFFSET RESULT3  ;BX points to product
            MOV   [BX],AX            ;AX holds lower word
            MOV   [BX]+2,DX          ;DX holds higher word
```

Table 3-1: Unsigned Multiplication Summary

Multiplication	Operand 1	Operand 2	Result
byte × byte	AL	register or memory	AX
word × word	AX	register or memory	DX AX
word × byte	AL = byte, AH = 0	register or memory	DX AX

Table 3-1 gives a summary of multiplication of unsigned numbers.

Division of unsigned numbers

In the division of unsigned numbers, the following cases are discussed:

1. Byte over byte
2. Word over word
3. Word over byte
4. Doubleword over word

In divide, there could be cases where the CPU cannot perform the division. In these cases an interrupt is activated. This is referred to as an *exception*. In what situation can the microprocessor not handle the division and must call an interrupt? They are

1. if the denominator is zero (dividing any number by 00), and
2. if the quotient is too large for the assigned register.

In the IBM PC and compatibles, if either of these cases happens, the PC will display the "divide error" message.

byte/byte: In dividing a byte by a byte, the numerator must be in the AL register and AH must be set to zero. The denominator cannot be immediate but can be in a register or memory as supported by the addressing modes. After the DIV instruction is performed, the quotient is in AL and the remainder is in AH. The following shows the various addressing modes that the denominator can take.

```
DATA7       DB     95
DATA8       DB     10
```

Table 3-2: Unsigned Division Summary

Division	Numerator	Denominator	Quotient	Rem.
byte/byte	AL = byte, AH = 0	register or memory	AL[1]	AH
word/word	AX = word, DX = 0	register or memory	AX[2]	DX
word/byte	AX = word	register or memory	AL[1]	AH
doubleword/word	DXAX = doubleword	register or memory	AX[2]	DX

Notes:
1. Divide error interrupt if AL > FFH.
2. Divide error interrupt if AX > FFFFH.

```
QOUT1         DB     ?
REMAIN1       DB     ?

;using immediate addressing mode will give an error
              MOV   AL,DATA7               ;move data into AL
              SUB   AH,AH                  ;clear AH
              DIV   10                     ;immed. mode not allowed!!
;allowable modes include:
;using direct mode
              MOV   AL,DATA7               ;AL holds numerator
              SUB   AH,AH                  ;AH must be cleared
              DIV   DATA8                  ;divide AX by DATA8
              MOV   QOUT1,AL               ;quotient = AL = 09
              MOV   REMAIN1,AH             ;remainder = AH = 05
;using register addressing mode
              MOV   AL,DATA7               ;AL holds numerator
              SUB   AH,AH                  ;AH must be cleared
              MOV   BH,DATA8               ;move denom. to register
              DIV   BH                     ;divide AX by BH
              MOV   QOUT1,AL               ;quotient = AL = 09
              MOV   REMAIN1,AH             ;remainder = AH = 05
;using register indirect addressing mode
              MOV   AL,DATA7               ;AL holds numerator
              SUB   AH,AH                  ;AH must be cleared
              MOV   BX,OFFSET DATA8        ;BX holds offset of DATA8
              DIV   BYTE PTR [ BX]         ;divide AX by DATA8
              MOV   QOUT2,AX
              MOV   REMAIND2,DX
```

word/word: In this case the numerator is in AX and DX must be cleared. The denominator can be in a register or memory. After the DIV, AX will have the quotient and the remainder will be in DX.

```
              MOV   AX,10050       ;AX holds numerator
              SUB   DX,DX          ;DX must be cleared
              MOV   BX,100         ;BX used for denominator
              DIV   BX
              MOV   QOUT2,AX       ;quotient = AX = 64H = 100
              MOV   REMAIND2,DX    ;remainder = DX = 32H = 50
```

word/byte: Again, the numerator is in AX and the denominator can be in a register or memory. After the DIV instruction, AL will contain the quotient, and AH will contain the remainder. The maximum quotient is FFH. The following program divides AX = 2055 by CL = 100. Then AL = 14H (20 decimal) is the quotient and AH = 37H (55 decimal) is the remainder.

```
              MOV   AX,2055        ;AX holds numerator
              MOV   CL,100         ;CL used for denominator
              DIV   CL
              MOV   QUO,AL         ;AL holds quotient
              MOV   REMI,AH        ;AH holds remainder
```

doubleword/word: The numerator is in AX and DX, with the most significant word in DX and the least significant word in AX. The denominator can be in a register or in memory. After the DIV instruction, the quotient will be in AX, and the remainder in DX. The maximum quotient is FFFFH.

```
;from the data segment:
DATA1         DD     105432
DATA2         DW     10000
QUOT          DW     ?
REMAIN        DW     ?
;from the code segment:
              MOV   AX,WORD PTR DATA1         ;AX holds lower word
              MOV   DX,WORD PTR DATA1+2;DX higher word of numerator
              DIV   DATA2
              MOV   QUOT,AX                   ;AX holds quotient
              MOV   REMAIN,DX                 ;DX holds remainder
```

In the program above, the contents of DX:AX are divided by a word-sized data value, 10000. Now one might ask: How does the CPU know that it must use the doubleword in DX:AX for the numerator? The 8088/86 automatically uses DX:AX as the numerator anytime the denominator is a word in size, as was seen earlier in the case of a word divided by a word. This explains why DX had to be cleared in that case. Notice in the example above that DATA1 is defined as DD but fetched into a word-size register with the help of WORD PTR. In the absence of WORD PTR, the assembler will generate an error. A summary of the results of division of unsigned numbers is given in Table 3-2.

Review Questions

1. In unsigned multiplication of a byte in DATA1 with a byte in AL, the product will be placed in register(s) __________ .
2. In unsigned multiplication of AX with BX, the product is placed in register(s) ______________.
3. In unsigned multiplication of CX with a byte in AL, the product is placed in register(s) ________________.
4. In unsigned division of a byte in AL by a byte in DH, the quotient will be placed in _________ and the remainder in __________.
5. In unsigned division of a word in AX by a word in DATA1, the quotient will be placed in _________ and the remainder in __________.
6. In unsigned division of a word in AX by a byte in DATA2, the quotient will be placed in _________ and the remainder in ____________.
7. In unsigned division of a doubleword in DXAX by a word in CX, the quotient will be placed in _________ and the remainder in ___________.

SECTION 3.3: LOGIC INSTRUCTIONS

In this section we discuss the logic instructions AND, OR, XOR, SHIFT, and COMPARE in the context of many examples.

AND

```
AND destination, source
```

This instruction will perform a logical AND on the operands and place the result in the destination. The destination operand can be a register or in memory. The source operand can be a register, in memory, or immediate.

AND will automatically change the CF and OF to zero, and PF, ZF, and SF are set according to the result. The rest of the flags are either undecided or unaffected. As seen in Example 3-5, AND can be used to mask certain bits of the operand. It can also be used to test for a zero operand:

Logical AND Function

Inputs		Output
X	**Y**	**X AND Y**
0	0	0
0	1	0
1	0	0
1	1	1

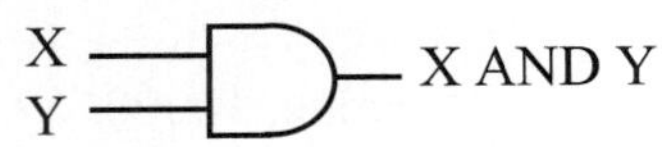


```
      AND   DH,DH
      JZ    XXXX
      ...
XXXX: ...
```

The above code will AND DH with itself and set ZF = 1 if the result is zero, making the CPU fetch from the target address XXXX. Otherwise, the instruction below JZ is executed. AND can thus be used to test if a register contains zero.

Example 3-5

Show the results of the following:

```
      MOV   BL,35H
      AND   BL,0FH         ;AND BL with 0FH. Place the result in BL.
```

Solution:

```
35H   0 0 1 1 0 1 0 1
0FH   0 0 0 0 1 1 1 1
05H   0 0 0 0 0 1 0 1
```

Flag settings will be: SF = 0, ZF = 0, PF = 1, CF = OF = 0.

OR

```
OR    destination,source
```

The destination and source operands are ORed and the result is placed in the destination. OR can be used to set certain bits of an operand to 1. The destination operand can be a register or in memory. The source operand can be a register, in memory, or immediate.

The flags will be set the same as for the AND instruction. CF and OF will be reset to zero, and SF, ZF, and PF will be set according to the result. All other flags are not affected. See Example 3-6.

The OR instruction can also be used to test for a zero operand. For example, "`OR BL,0`" will OR the register BL with 0 and make ZF = 1 if BL is zero. "`OR BL,BL`" will achieve the same result.

Logical OR Function

Inputs		Output
X	**Y**	**X OR Y**
0	0	0
0	1	1
1	0	1
1	1	1

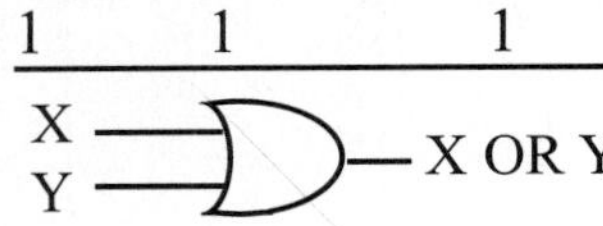

Example 3-6

Show the results of the following:

```
MOV AX,0504        ;AX = 0504
OR  AX,0DA68H      ;AX = DF6C
```

Solution:

```
0504H   0000 0101 0000 0100
DA68H   1101 1010 0110 1000   Flags will be: SF = 1 , ZF = 0, PF = 1, CF = OF = 0.
DF6C    1101 1111 0110 1100   Notice that parity is checked for the lower 8 bits only.
```

XOR

```
XOR  dest,src
```

The XOR instruction will eXclusive-OR the operands and place the result in the destination. XOR sets the result bits to 1 if they are not equal; otherwise, they are reset to 0. The flags are set the same as for the AND instruction. CF = 0 and OF = 0 are set internally and the rest are changed according to the result of the operation. The rules for the operands are the same as in the AND and OR instructions. See Examples 3-7 and 3-8.

XOR can also be used to see if two registers have the same value. "XOR BX,CX" will make ZF = 1 if both registers have the same value, and if they do, the result (0000) is saved in BX, the destination.

Logical XOR Function

Inputs		Output
A	**B**	**A XOR B**
0	0	0
0	1	1
1	0	1
1	1	0

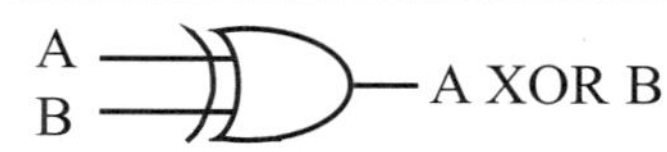

Another widely used application of XOR is to toggle bits of an operand. For example, to toggle bit 2 of register AL:

```
XOR  AL,04H        ;XOR  AL with 0000 0100
```

This would cause bit 2 of AL to change to the opposite value; all other bits would remain unchanged.

Example 3-7

Show the results of the following:

```
MOV    DH,54H
XOR    DH,78H
```

Solution:

```
54H    0 1 0 1 0 1 0 0
78H    0 1 1 1 1 0 0 0
2C     0 0 1 0 1 1 0 0   Flag settings will be: SF = 0, ZF = 0, PF = 0, CF = OF = 0.
```

Example 3-8

The XOR instruction can be used to clear the contents of a register by XORing it with itself. Show how "XOR AH,AH" clears AH, assuming that AH = 45H.

Solution:

```
45H    01000101
45H    01000101
00     00000000     Flag settings will be: SF = 0, ZF = 1, PF =1 , CF = OF = 0.
```

SHIFT

There are two kinds of shifts: logical and arithmetic. The logical shift is for unsigned operands, and the arithmetic shift is for signed operands. Logical shift will be discussed in this section and the discussion of arithmetic shift is postponed to Chapter 6. Using shift instructions shifts the contents of a register or memory location right or left. The number of times (or bits) that the operand is shifted can be specified directly if it is once only, or through the CL register if it is more than once.

SHR: This is the logical shift right. The operand is shifted right bit by bit, and for every shift the LSB (least significant bit) will go to the carry flag (CF) and the MSB (most significant bit) is filled with 0. Examples 3-9 and 3-10 should help to clarify SHR.

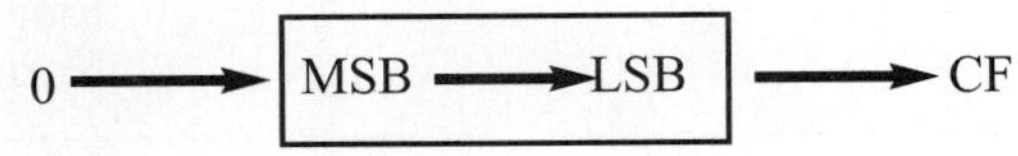

Example 3-9

Show the result of SHR in the following:

```
MOV   AL,9AH
MOV   CL,3     ;set number of times to shift
SHR   AL,CL
```

Solution:

```
9AH =     10011010
          01001101     CF = 0 (shifted once)
          00100110     CF = 1 (shifted twice)
          00010011     CF = 0 (shifted three times)
```

After shifting right three times, AL = 13H and CF = 0.

If the operand is to be shifted once only, this is specified in the SHR instruction itself rather than placing 1 in the CL. This saves coding of one instruction:

```
MOV   BX,0FFFFH    ;BX=FFFFH
SHR   BX,1         ;shift right BX once only
```

After the shift above, BX = 7FFFH and CF = 1. Although SHR does affect the OF, SF, PF, and ZF flags, they are not important in this case. The operand to be shifted can be in a register or in memory, but immediate addressing mode is not allowed for shift instructions. For example, "`SHR 25,CL`" will cause the assembler to give an error.

Example 3-10

Show the results of SHR in the following:

```
;from the data segment:
DATA1       DW     7777H
;from the code segment:
TIMES       EQU    4
            MOV    CL,TIMES      ;CL=04
            SHR    DATA1,CL      ;shift DATA1 CL times
```

Solution:

After the four shifts, the word at memory location DATA1 will contain 0777. The four LSBs are lost through the carry, one by one, and 0s fill the four MSBs.

SHL: Shift left is also a logical shift. It is the reverse of SHR. After every shift, the LSB is filled with 0 and the MSB goes to CF. All the rules are the same as for SHR.

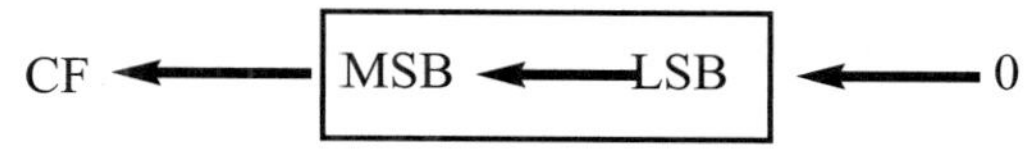

Example 3-11

Show the effects of SHL in the following:

```
MOV    DH,6
MOV    CL,4
SHL    DH,CL
```

Solution:

```
                   00000110
CF=0               00001100    (shifted left once)
CF=0               00011000
CF=0               00110000
CF=0               01100000    (shifted four times)
```

After the four shifts left, the DH register has 60H and CF = 0.

Example 3-11 could have been coded as

```
MOV    DH,6
SHL    DH,1
SHL    DH,1
SHL    DH,1
SHL    DH,1
```

COMPARE of unsigned numbers

```
CMP    destination,source    ;compare dest and src
```

The CMP instruction compares two operands and changes the flags according to the result of the comparison. The operands themselves remain unchanged. The destination operand can be in a register or in memory and the source operand can be in a register, in memory, or immediate. Although all the CF, AF, SF, PF, ZF, and OF flags reflect the result of the comparison, only CF and ZF are used, as outlined in Table 3-3.

Table 3-3: Flag Settings for Compare Instruction

Compare operands	CF	ZF
destination > source	0	0
destination = source	0	1
destination < source	1	0

The following demonstrates how the CMP instruction is used:

```
DATA1 DW     235FH
      ...
      MOV    AX,0CCCCH
      CMP    AX,DATA1     ;compare CCCC with 235F
      JNC    OVER         ;jump if CF=0
      SUB    AX,AX
OVER: INC    DATA1
```

In the program above, AX is greater than the contents of memory location DATA1 (0CCCCH > 235FH); therefore, CF = 0 and JNC (jump no carry) will go to target OVER. In contrast, look at the following:

```
       MOV    BX,7888H
       MOV    CX,9FFFH
       CMP    BX,CX           ;compare 7888 with 9FFF
       JNC    NEXT
       ADD    BX,4000H
NEXT:  ADD    CX,250H
```

In the above, BX is smaller than CX (7888H < 9FFFH), which sets CF = 1, making "`JNC NEXT`" fall through so that "`ADD BX,4000H`" is executed. In the example above, CX and BX still have their original values (CX = 9FFFH and BX = 7888H) after the execution of "`CMP BX,CX`". Notice that CF is always checked for cases of greater or smaller than, but for equal, ZF must be used. The next program sample has a variable named TEMP, which is being checked to see if it has reached 99:

```
TEMP   DB     ?
...
       MOV    AL,TEMP         ;move the TEMP variable into AL
       CMP    AL,99           ;compare AL with 99
       JZ     HOT_HOT         ;if ZF=1 (TEMP = 99) jump to HOT_HOT
       INC    BX              ;otherwise (ZF=0) increment BX
...
HOT_HOT: HLT                  ;halt the system
```

The compare instruction is really a SUBtraction except that the values of the operands do not change. The flags are changed according to the execution of SUB. Although all the flags are affected, the only ones of interest are ZF and CF. It must be emphasized that in CMP instructions, the operands are unaffected regardless of the result of the comparison. Only the flags are affected. This is despite the fact that CMP uses the SUB operation to set or reset the flags. Program 3-3 uses the CMP instruction to search for the highest byte in a series of 5 bytes defined in the data segment. The instruction "`CMP AL,[ BX]` " works as follows, where [BX] is the contents of the memory location pointed at by register BX.

If AL < [BX], then CF = 1 and [BX] becomes the basis of the new comparison.
If AL > [BX], then CF = 0 and AL is the larger of the two values and remains the basis of comparison.

Although JC (jump carry) and JNC (jump no carry) check the carry flag and can be used after a compare instruction, it is recommended that JA (jump above) and JB (jump below) be used for two reasons. One reason is that assemblers will unassemble JC as JB, and JNC as JA, which may be confusing to beginning programmers. Another reason is that "jump above" and "jump below" are easier to understand than "jump carry" and "jump no carry," since it is more immediately apparent that one number is larger than another, than whether a carry would be generated if the two numbers were subtracted.

Program 3-3 searches through five data items to find the highest grade. The program has a variable called “Highest” that holds the highest grade found so far. One by one, the grades are compared to Highest. If any of them is higher, that value is placed in Highest. This continues until all data items are checked. A REPEAT-UNTIL structure was chosen in the program design. Figure 3-1 shows the flowchart for Program 3-3. This design could be used to code the program in many different languages.

Program 3-3 as coded in Assembly language, uses the CMP instruction to search through 5 bytes of data to find the highest grade. The program uses register AL to hold the highest grade found so far. AL is given the initial value of 0. A loop is used to compare each of the 5 bytes with the value in AL. If AL contains a higher value, the loop continues to check the next byte. If AL is smaller than the byte being checked, the contents of

Assume that there is a class of five people with the following grades: 69, 87, 96, 45, and 75. Find the highest grade.

```
TITLE       PROG3-3 (EXE) CMP EXAMPLE
PAGE        60,132
.MODEL SMALL
.STACK 64
;------------------
            .DATA
GRADES      DB     69,87,96,45,75
            ORG    0008
HIGHEST     DB     ?
;------------------
            .CODE
MAIN        PROC   FAR
            MOV    AX,@DATA
            MOV    DS,AX
            MOV    CX,5                ;set up loop counter
            MOV    BX,OFFSET GRADES    ;BX points to GRADE data
            SUB    AL,AL         ;AL holds highest grade found so far
AGAIN:      CMP    AL,[BX]             ;compare next grade to highest
            JA     NEXT                ;jump if AL still highest
            MOV    AL,[BX]             ;else AL holds new highest
NEXT:       INC    BX                  ;point to next grade
            LOOP   AGAIN               ;continue search
            MOV    HIGHEST,AL          ;store highest grade
            MOV    AH,4CH
            INT    21H                 ;go back to OS
MAIN        ENDP
            END    MAIN
```

Program 3-3

AL are replaced by that byte and the loop continues.

Program 3-4 uses the CMP instruction to determine if an ASCII character is uppercase or lowercase. Note that small and capital letters in ASCII have the following values:

Letter	Hex	Binary	Letter	Hex	Binary
A	41	0100 0001	a	61	0110 0001
B	42	0100 0010	b	62	0110 0010
C	43	0100 0011	c	63	0110 0011
...	...	...	...	...	...
Y	59	0101 1001	y	79	0111 1001
Z	5A	0101 1010	z	7A	0111 1010

As can be seen, there is a relationship between the pattern of lowercase and uppercase letters, as shown below for A and a:

```
A      0100 0001    41H
a      0110 0001    61H
```

The only bit that changes is d5. To change from lowercase to uppercase, d5 must be masked. Program 3-4 first detects if the letter is in lowercase, and if it is, it is ANDed with 1101 1111B = DFH. Otherwise, it is simply left alone. To determine if it is a lowercase letter, it is compared with 61H and 7AH to see if it is in the range a to z. Anything above or below this range should be left alone.

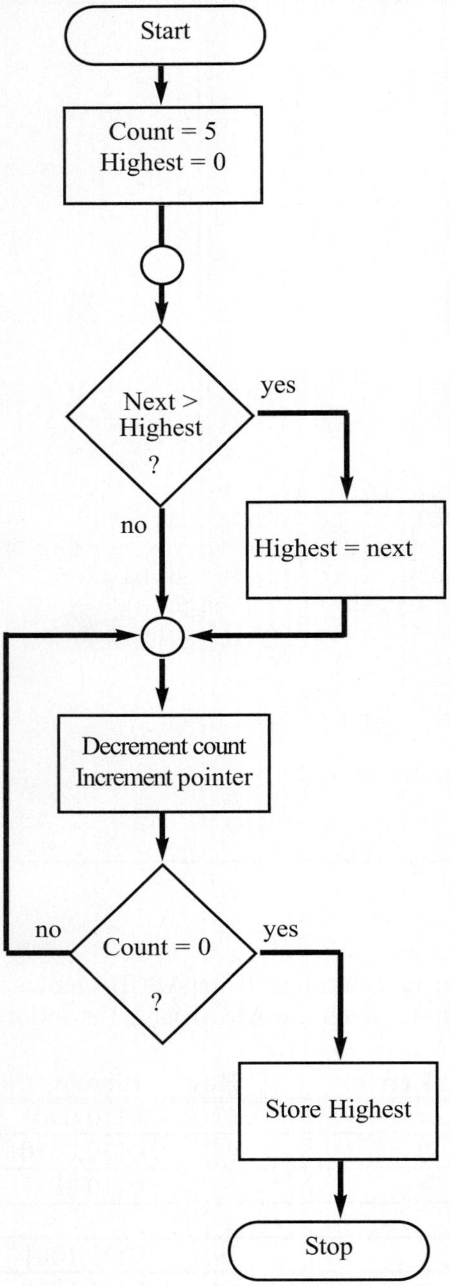


```
Count = 5
Highest = 0

REPEAT
    IF (Next > Highest)
    THEN
          Highest = Next
    ENDIF
    Decrement Count
UNTIL Count = 0

Store Highest
```

Figure 3-1. Flowchart and Pseudocode for Program 3-3

In Program 3-4, 20H could have been subtracted from the lowercase letters instead of ANDing with 1101 1111B. That is what IBM did in their BIOS, as shown next.

```
IBM BIOS method of converting from lowercase to uppercase
;----- CONVERT ANY LOWERCASE TO UPPERCASE

EBFB           K60:                       ;LOWER TO UPPER
EBFB 3C61          CMPAL,'a'          ;FIND OUT IF ALPHABETIC
EBFD 7206          JBK61              ;NOT_CAPS_STATE
EBFF 3C7A          CMPAL,'z'
EC01 7702          JAK61              ;NOT_CAPS_STATE
EC03 2C20          SUBAL,'a'-'A'      ;CONVERT TO UPPERCASE
EC05           K61: ....
```

```
TITLE       PROG3-4 (EXE) LOWERCASE TO UPPERCASE CONVERSION
 PAGE       60,132
 .MODEL SMALL
 .STACK 64
 ;----------------
            .DATA
 DATA1      DB     `mY NAME is jOe'
            ORG    0020H
 DATA2      DB     14 DUP(?)
 ;----------------
      .CODE
 MAIN PROC  FAR
      MOV   AX,@DATA
      MOV   DS,AX
      MOV   SI,OFFSET DATA1   ;SI points to original data
      MOV   BX,OFFSET DATA2   ;BX points to uppercase data
      MOV   CX,14             ;CX is loop counter
 BACK:MOV   AL,[SI]           ;get next character
      CMP   AL,61H            ;if less than `a'
      JB    OVER              ;then no need to convert
      CMP   AL,7AH            ;if greater than `z'
      JA    OVER              ;then no need to convert
      AND   AL,11011111B      ;mask d5 to convert to uppercase
 OVER:MOV   [BX],AL           ;store uppercase character
      INC   SI                ;increment pointer to original
      INC   BX                ;increment pointer to uppercase data
      LOOP  BACK              ;continue looping if CX > 0
      MOV   AH,4CH
      INT   21H               ;go back to OS
 MAIN ENDP
      END   MAIN
```

Program 3-4

Review Questions

1. Use operands 4FCAH and C237H to perform:
   (a) AND (b) OR (c) XOR
2. ANDing a word operand with FFFFH will result in what value for the word operand? To set all bits of an operand to 0, it should be ANDed with ______.
3. To set all bits of an operand to 1, it could be ORed with _______.
4. XORing an operand with itself results in what value for the operand?
5. Show the steps if value A0F2H were shifted left three times. Then show the steps if A0F2H were shifted right three times.
6. The CMP instruction works by performing a(n) _________ operation on the operands and setting the flags accordingly.
7. True or false. The CMP instruction alters the contents of its operands.

BIOS examples of logic instructions

Next we examine some real-life examples from the original IBM PC BIOS programs. The purpose is to see the instructions discussed so far in the context of real-life applications.

When the computer is turned on, the CPU starts to execute the programs stored in BIOS in order to set the computer up for the OS. If anything has happened to the BIOS programs, the computer can do nothing. The first subroutine of BIOS is to test the CPU. This involves checking the flag register bit by bit as well as checking all other registers. The BIOS program for testing the flags and registers is given on the next page followed by their explanation:

```
              306    ASSUME CS:CODE,DS:NOTHING,ES:NOTHING,SS:NOTHING
E05B          307    ORG 0E05BH
E05B          ......
E05B          309 START:
E05B FA       310    CLI            ; DISABLE INTERRUPTS
E05C B4D5     311    MOV AH,0D5H; SET SF, CF, ZF, AND AF FLAGS ON
E05E 9E       312    SAHF
E05F 734C     313    JNC ERR01      ; GO TO ERR ROUTINE IF CF NOT SET
E061 754A     314    JNZ ERR01      ; GO TO ERR ROUTINE IF ZF NOT SET
E063 7B48     315    JNP ERR01      ; GO TO ERR ROUTINE IF PF NOT SET
E065 7946     316    JNS ERR01      ; GO TO ERR ROUTINE IF SF NOT SET
E067 9F       317    LAHF           ; LOAD FLAG IMAGE TO AH
E068 B105     318    MOV CL,5       ; LOAD CNT REG WITH SHIFT CNT
E06A D2EC     319    SHR AH,CL      ; SHIFT AF INTO CARRY BIT POS
E06C 733F     320    JNC ERR01      ; GO TO ERR ROUTINE IF AF NOT SET
E06E B040     321    MOV AL,40H     ; SET THE OF FLAG ON
E070 D0E0     322    SHL AL,1       ; SETUP FOR TESTING
E072 7139     323    JNO ERR01      ; GO TO ERR ROUTINE IF OF NOT SET
E074 32E4     324    XOR AH,AH      ; SET AH = 0
E076 9E       325    SAHF           ; CLEAR SF, CF, ZF, AND PF
E077 7634     326    JBE ERR01      ; GO TO ERR ROUTINE IF CF ON
              327    ; OR GO TO ERR ROUTINE IF ZF ON
E079 7832     328    JS ERR01       ; GO TO ERR ROUTINE IF SF ON
E07B 7A30     329    JP ERR01       ; GO TO ERR ROUTINE IF PF ON
E07D 9F       330    LAHF           ; LOAD FLAG IMAGE TO AH
E07E B105     331    MOV CL,5       ; LOAD CNT REG WITH SHIFT CNT
E080 D2EC     332    SHR AH,CL      ; SHIFT 'AF' INTO CARRY BIT POS
E082 7229     333    JC ERR01       ; GO TO ERR ROUTINE IF ON
E084 D0E4     334    SHL AH,1       ; CHECK THAT 'OF' IS CLEAR
E086 7025     335    JO ERR01       ; GO TO ERR ROUTINE IF ON
              336
              337;--- READ/WRITE THE 888 GENERAL AND SEGMENTATION REGISTERS
              338;    WITH ALL ONES AND ZEROES.
              339
E088 B8FFFF   340    MOV AX,0FFFFH       ; SET UP ONE'S PATTERN IN AX
E08B F9       341    STC
E08C          342C8:
E08C 8ED8     343    MOV DS,AX      ; WRITE PATTERN TO ALL REGS
E08E 8CDB     344    MOV BX,DS
E090 8EC3     345    MOV ES,BX
E092 8CC1     346    MOV CX,ES
E094 8ED1     347    MOV SS,CX
E096 8CD2     348    MOV DX,SS
E098 8BE2     349    MOV SP,DX
E09A 8BEC     350    MOV BP,SP
E09C 8BF5     351    MOV SI,BP
E09E 8BFE     352    MOV DI,SI
E0A0 7307     353    JNC C9         ; TST1A
E0A2 33C7     354    XOR AX,DI      ; PATTERN MAKE IT THRU ALL REGS
E0A4 7507     355    JNZ ERR01      ; NO - GO TO ERR ROUTINE
E0A6 F8       356    CLC
E0A7 EBE3     357    JMP C8
E0A9          358C9:
E0A9 0BC7     359    OR AX,DI       ; ZERO PATTERN MAKE IT THRU?
E0AB 7401     360    JZ C10         ; YES - GO TO NEXT TEST
E0AD F4       361ERR01: HLT         ; HALT SYSTEM
```


Line-by-line explanation:

Line	Explanation
310	CLI ensures that no interrupt will occur while the test is being conducted.
311	MOVAH,0D5H:

Line	Explanation
	flag S Z - AC - P - C D5H 1 1 0 1 0 1 0 1

312	SAHF (store AH into lower byte of the flag register) is one way to move data to flags. Another is to use the stack:
	`MOVAX,00D5H`
	`PUSHAX`
	`POPF`
	However, there is no RAM available yet to use for the stack because the CPU is tested before memory is tested.

313–316	Will make the CPU jump to HLT if any flag does not work.
317	LAHF (load AH with the lower byte of flag register) is the opposite of SAHF.
318	Loads CL for five shifts.
319	"SHR AH,CL". By shifting AH five times, AF (auxiliary carry) will be in the CF position.
320	If no AF, there is an error. Lines 317 to 320 are needed because there is no jump condition instruction for AF.
321–323	Checks the OF flag. This is discussed in Chapter 6 when signed numbers are discussed.
324–335	Checks the same flags for zero. Remember that JNZ is the same as JBE.
340	Loads AX with FFFFH.
341	STC (set the carry) makes CF = 1.
343–352	Moves the AX value (FFFFH) into every register and ends up with DI = FFFFH if the registers are good.
353	Since CF = 1 (remember STC) it falls through.
354	Exclusive-ORing AX and DI with both having the same FFFFH value makes AX = 0000 and ZF = 1 if the registers are good (see lines 343–352). If ZF = 0, one of the registers must have corrupted the data FFFF, therefore the CPU is bad.
355	If ZF = 0, there is an error.
356	CLC clears the carry flag. This is the opposite of STC.
357	Jumps to C8 and repeats the same process, this time with value 0000. The contents of AX are moved around to every register until DI = 0000, and at 353 the JNC C9 will jump since CF = 0 by the CLC instruction before it went to the loop.
359	At C9, AX and DI are ORed. If 0000, the contents of AX are copied successfully to all registers, and DI will be 0000; therefore, ORing will raise the ZF, making ZF = 1.
360	If ZF = 1, the CPU is good and the system can perform the next test. Otherwise, ZF = 0, meaning that the CPU is bad and the system should be halted.

SECTION 3.4: BCD AND ASCII CONVERSION

This section covers BCD and ASCII conversions with some examples.

BCD number system

BCD stands for binary coded decimal. BCD is needed because we use the digits 0 to 9 for numbers in everyday life. Binary representation of 0 to 9 is called BCD (see Figure 3-2). In computer literature one encounters two terms for BCD numbers: (1) unpacked BCD, and (2) packed BCD.

Digit	BCD
0	0000
1	0001
2	0010
3	0011
4	0100
5	0101
6	0110
7	0111
8	1000
9	1001

Figure 3-2. BCD Code

Unpacked BCD

In unpacked BCD, the lower 4 bits of the number represent the BCD number and the rest of the bits are 0. Example: "0000 1001" and "0000 0101" are unpacked BCD for 9 and 5, respectively. In the case of unpacked BCD it takes 1 byte of memory location or a register of 8 bits to contain the number.

Packed BCD

In the case of packed BCD, a single byte has two BCD numbers in it, one in the lower 4 bits and one in the upper 4 bits. For example, "0101 1001" is packed BCD for 59. It takes only 1 byte of memory to store the packed BCD operands. This is one reason to use packed BCD since it is twice as efficient in storing data.

ASCII numbers

In ASCII keyboards, when key "0" is activated, for example, "011 0000" (30H) is provided to the computer. In the same way, 31H (011 0001) is provided for key "1", and so on, as shown in the following list:

Key	ASCII (hex)	Binary	BCD (unpacked)
0	30	011 0000	0000 0000
1	31	011 0001	0000 0001
2	32	011 0010	0000 0010
3	33	011 0011	0000 0011
4	34	011 0100	0000 0100
5	35	011 0101	0000 0101
6	36	011 0110	0000 0110
7	37	011 0111	0000 0111
8	38	011 1000	0000 1000
9	39	011 1001	0000 1001

It must be noted that although ASCII is standard in the United States (and many other countries), BCD numbers have universal application. Now since the keyboard, printers, and monitors are all in ASCII, how does data get converted from ASCII to BCD, and vice versa? These are the subjects covered next.

ASCII to BCD conversion

To process data in BCD, first the ASCII data provided by the keyboard must be converted to BCD. Whether it should be converted to packed or unpacked BCD depends on the instructions to be used. There are instructions that require that data be in unpacked BCD and there are others that must have packed BCD data to work properly. Each is covered separately.

ASCII to unpacked BCD conversion

To convert ASCII data to BCD, the programmer must get rid of the tagged "011" in the higher 4 bits of the ASCII. To do that, each ASCII number is ANDed with "0000 1111" (0FH), as shown in the next example. This example is written in three different ways using different addressing modes. Programs 3-5a, 3-5b, and 3-5c show three different methods for converting the 10 ASCII digits to unpacked BCD. All use the same data segment:

```
ASC         DB          '9562481273'
            ORG         0010H
UNPACK      DB          10 DUP(?)
```

In Program 3-5a, notice that although the data was defined as DB, a byte definition directive, it was accessed in word-sized chunks. This is a workable approach; however, using the PTR directive as shown in Program 3-5b makes the code more readable for programmers.

In both of the solutions so far, registers BX and DI were used as pointers for an array of data. An array is simply a set of data located in consecutive memory locations. Now one might ask: What happens if there are four, five, or six arrays? How can they all be accessed with only three registers as pointers: BX, DI, and SI? Program 3-5c shows how this can be done with a single register used as a pointer to access two arrays. However, to do that, the arrays must be of the same size and defined similarly.

```
         MOV    CX,5
         MOV    BX,OFFSET ASC      ;BX points to ASCII data
         MOV    DI,OFFSET UNPACK   ;DI points to unpacked BCD data
AGAIN:   MOV    AX,[BX]            ;move next 2 ASCII numbers to AX
         AND    AX,0F0FH           ;remove ASCII 3s
         MOV    [DI],AX            ;store unpacked BCD
         ADD    DI,2               ;point to next unpacked BCD data
         ADD    BX,2               ;point to next ASCII data
         LOOP   AGAIN
```

Program 3-5a

```
         MOV    CX,5               ;CX is loop counter
         MOV    BX,OFFSET ASC      ;BX points to ASCII data
         MOV    DI,OFFSET UNPACK   ;DI points to unpacked BCD data
AGAIN:   MOV    AX,WORD PTR [BX]   ;move next 2 ASCII numbers to AX
         AND    AX,0F0FH           ;remove ASCII 3s
         MOV    WORD PTR [DI],AX   ;store unpacked BCD
         ADD    DI,2               ;point to next unpacked BCD data
         ADD    BX,2               ;point to next ASCII data
         LOOP   AGAIN
```

Program 3-5b

Program 3-5c uses the based addressing mode since BX+ASC is used as a pointer. ASC is the displacement added to BX. Either DI or SI could have been used for this purpose. For word-sized operands, "`WORD PTR`" would be used since the data is defined as DB. This is shown below.

```
MOV    AX,WORD PTR ASC[BX]
AND    AX,0F0FH
MOV    WORD PTR UNPACKED[BX],AX
```

```
         MOV    CX,10              ;load the ccunter
         SUB    BX,BX              ;clear BX
AGAIN:   MOV    AL,ASC[BX]         ;move to AL content of mem [BX+ASC]
         AND    AL,0FH             ;mask the upper nibble
         MOV    UNPACK[BX],AL      ;move to mem [BX+UNPACK] the AL
         INC    BX                 ;point to next byte
         LOOP   AGAIN              ;loop until it is finished
```

Program 3-5c

ASCII to packed BCD conversion

To convert ASCII to packed BCD, it is first converted to unpacked BCD (to get rid of the 3) and then combined to make packed BCD. For example, for 9 and 5 the keyboard gives 39 and 35, respectively. The goal is to produce 95H or "1001 0101", which is called packed BCD, as discussed earlier. This process is illustrated in detail below.

Key	ASCII	Unpacked BCD	Packed BCD
4	34	00000100	
7	37	00000111	01000111 or 47H

```
            ORG    0010H
VAL_ASC     DB     '47'
VAL_BCD     DB     ?
;reminder:  DB will put 34 in 0010H location and 37 in 0011H
            MOV    AX,WORD PTR VAL_ASC      ;AH=37,AL=34
            AND    AX,0F0FH               ;mask 3 to get unpacked BCD
            XCHG   AH,AL                  ;swap AH and AL.
            MOV    CL,4                   ;CL=04 to shift 4 times
            SHL    AH,CL                  ;shift left AH to get AH=40H
            OR     AL,AH                  ;OR them to get packed BCD
            MOV    VAL_BCD,AL             ;save the result
```

After this conversion, the packed BCD numbers are processed and the result will be in packed BCD format. As will be seen later in this section, there are special instructions, such as DAA and DAS, which require that the data be in packed BCD form and give the result in packed BCD. For the result to be displayed on the monitor or be printed by the printer, it must be in ASCII format. Conversion from packed BCD to ASCII is discussed next.

Packed BCD to ASCII conversion

To convert packed BCD to ASCII, it must first be converted to unpacked and then the unpacked BCD is tagged with 011 0000 (30H). The following shows the process of converting from packed BCD to ASCII.

Packed BCD	Unpacked BCD	ASCII
29H	02H & 09H	32H & 39H
0010 1001	0000 0010 & 0000 1001	011 0010 & 011 1001

```
VAL1_BCD    DB     29H
VAL3_ASC    DW     ?
            ...
            MOV    AL,VAL1_BCD
            MOV    AH,AL         ;copy AL to AH. now AH=29,AL=29H
            AND    AX,0F00FH     ;mask 9 from AH and 2 from AL
            MOV    CL,4          ;CL=04 for shift
            SHR    AH,CL         ;shift right AH to get unpacked BCD
            OR     AX,3030H      ;combine with 30 to get ASCII
            XCHG   AH,AL         ;swap for ASCII storage convention
            MOV    VAL3_ASC,AX   ;store the ASCII
```

BCD addition and subtraction

After learning how to convert ASCII to BCD, the application of BCD numbers is the next step. There are two instructions that deal specifically with BCD numbers: DAA and DAS. Each is discussed separately.

BCD addition and correction

There is a problem with adding BCD numbers, which must be corrected. The problem is that after adding packed BCD numbers, the result is no longer BCD. Look at this example:

```
MOV AL,17H
ADD AL,28H
```

Adding them gives 0011 1111B (3FH), which is not BCD! A BCD number can only have digits from 0000 to 1001 (or 0 to 9). In other words, adding two BCD numbers must give a BCD result. The result above should have been 17 + 28 = 45 (0100 0101). To correct this problem, the programmer must add 6 (0110) to the low digit: 3F + 06 = 45H. The same problem could have happened in the upper digit (for example, in 52H + 87H = D9H). Again to solve this problem, 6 must be added to the upper digit (D9H + 60H = 139H), to ensure that the result is BCD (52 + 87 = 139). This problem is so pervasive that the vast majority of microprocessors have an instruction to deal with it.

DAA

The DAA (decimal adjust for addition) instruction in x86 microprocessors is provided exactly for the purpose of correcting the problem associated with BCD addition. DAA will add 6 to the lower nibble or higher nibble if needed; otherwise, it will leave the result alone. The following example will clarify these points:

```
DATA1    DB  47H
DATA2    DB  25H
DATA3    DB?
      MOV AL,DATA1        ;AL holds first BCD operand
      MOV BL,DATA2        ;BL holds second BCD operand
      ADD AL,BL           ;BCD addition
      DAA                 ;adjust for BCD addition
      MOV DATA3,AL        ;store result in correct BCD form
```

After the program is executed, the DATA3 field will contain 72H (47 + 25 = 72). Note that DAA works only on AL. In other words, while the source can be an operand of any addressing mode, the destination must be AL in order for DAA to work. It needs to be emphasized that DAA must be used after the addition of BCD operands and that BCD operands can never have any digit greater than 9. In other words, no A–F digit is allowed. It is also important to note that DAA works only after an ADD instruction; it will not work after the INC instruction.

Summary of DAA action

1. If after an ADD or ADC instruction the lower nibble (4 bits) is greater than 9, or if AF = 1, add 0110 to the lower 4 bits.
2. If the upper nibble is greater than 9, or if CF = 1, add 0110 to the upper nibble.

In reality there is no other use for the AF (auxiliary flag) except for BCD addition and correction. For example, adding 29H and 18H will result in 41H, which is incorrect as far as BCD is concerned. See the following code:

```
     Hex      BCD
     29       0010 1001
   + 18     + 0001 1000
     41       0100 0001     Because AF = 1,
   +  6     +      0110     DAA adds 6 to lower nibble.
     47       0100 0111     The final result is BCD.
```

Program 3-6 demonstrates the use of DAA after addition of multibyte packed BCD numbers.

Two sets of ASCII data have come in from the keyboard. Write and run a program to:
1. Convert from ASCII to packed BCD.
2. Add the multibyte packed BCD and save it.
3. Convert the packed BCD result to ASCII.

```
TITLE      PROG3-6 (EXE) ASCII TO BCD CONVERSION AND ADDITION
PAGE       60,132
.MODE SMALL
.STACK 64
;---------------------
           .DATA
DATA1_ASC  DB     `0649147816'
           ORG    0010H
DATA2_ASC  DB     `0072687188'
           ORG    0020H
DATA3_BCD  DB     5 DUP (?)
           ORG    0028H
DATA4_BCD  DB     5 DUP (?)
           ORG    0030H
DATA5_ADD  DB     5 DUP (?)
           ORG    0040H
DATA6_ASC  DB     10 DUP (?)
;-----------------------
           .CODE
MAIN PROC  FAR
     MOV   AX,@DATA
     MOV   DS,AX
     MOV   BX,OFFSET DATA1_ASC    ;BX points to first ASCII data
     MOV   DI,OFFSET DATA3_BCD    ;DI points to first BCD data
     MOV   CX,10                  ;CX holds number bytes to convert
     CALL  CONV_BCD               ;convert ASCII to BCD
     MOV   BX,OFFSET DATA2_ASC    ;BX points to second ASCII data
     MOV   DI,OFFSET DATA4_BCD    ;DI points to second BCD data
     MOV   CX,10                  ;CX holds number bytes to convert
     CALL  CONV_BCD               ;convert ASCII to BCD
     CALL  BCD_ADD                ;add the BCD operands
     MOV   SI,OFFSET DATA5_ADD    ;SI points to BCD result
     MOV   DI,OFFSET DATA6_ASC    ;DI points to ASCII result
     MOV   CX,05                  ;CX holds count for convert
     CALL  CONV_ASC               ;convert result to ASCII
     MOV   AH,4CH
     INT   21H                    ;go back to OS
MAIN ENDP
;-----------------------
;THIS SUBROUTINE CONVERTS ASCII TO PACKED BCD
CONV_BCD PROC
AGAIN:     MOV    AX,[ BX]      ;BX=pointer for ASCII data
     XCHG  AH,AL
     AND   AX,0F0FH     ;mask ASCII 3s
     PUSH  CX           ;save the counter
     MOV   CL,4         ;shift AH left 4 bits
     SHL   AH,CL        ;to get ready for packing
     OR    AL,AH        ;combine to make packed BCD
     MOV   [ DI] ,AL          ;DI=pointer for BCD data
     ADD   BX,2         ;point to next 2 ASCII bytes
     INC   DI           ;point to next BCD data
     POP   CX           ;restore loop counter
     LOOP  AGAIN
     RET
CONV_BCD ENDP
;--------------
```

Program 3-6 *(continued on the following page)*

```
;THIS SUBROUTINE ADDS TWO MULTIBYTE PACKED BCD OPERANDS
BCD_ADD PROC
       MOV  BX,OFFSET DATA3_BCD      ;BX=pointer for operand 1
       MOV  DI,OFFSET DATA4_BCD      ;DI=pointer for operand 2
       MOV  SI,OFFSET DATA5_ADD      ;SI=pointer for sum
       MOV  CX,05
       CLC
BACK:  MOV  AL,[BX]+4      ;get next byte of operand 1
       ADC  AL,[DI]+4      ;add next byte of operand 2
       DAA                 ;correct for BCD addition
       MOV  [SI]  +4,AL    ;save sum
       DEC  BX             ;point to next byte of operand 1
       DEC  DI             ;point to next byte of operand 2
       DEC  SI             ;point to next byte of sum
       LOOP      BACK
       RET
BCD_ADD ENDP
;-----------------
;THIS SUBROUTINE CONVERTS FROM PACKED BCD TO ASCII
CONV_ASC PROC
AGAIN2: MOV AL,[SI]        ;SI=pointer for BCD data
      MOV   AH,AL          ;duplicate to unpack
      AND   AX,0F00FH      ;unpack
      PUSH  CX             ;save counter
      MOV   CL,04          ;shift right 4 bits to unpack
      SHR   AH,CL          ;the upper nibble
      OR    AX,3030H       ;make it ASCII
      XCHG  AH,AL          ;swap for ASCII storage convention
      MOV   [DI],AX              ;store ASCII data
      INC   SI             ;point to next BCD data
      ADD   DI,2           ;point to next ASCII data
      POP   CX             ;restore loop counter
      LOOP  AGAIN2
      RET
CONV_ASC ENDP
      END   MAIN
```

Program 3-6 *(continued from the preceding page)*

The following shows that 6 is added to the upper nibble due to the fact it is greater than 9:

```
    Hex     BCD
    53      0010 0011
  + 75   +  0111 0101
    D8      1101 1000     Because the upper nibble is greater than 9,
  + 6    +  0110          DAA adds 6 to upper nibble.
   128      0010 1000     The final result is BCD.
```

See Appendix B for more examples of DAA.

BCD subtraction and correction

The problem associated with the addition of packed BCD numbers also shows up in subtraction. Again, there is an instruction (DAS) specifically designed to solve the problem. Therefore, when subtracting packed BCD (single-byte or multibyte) operands, the DAS instruction is put after the SUB or SBB instruction. AL must be used as the destination register to make DAS work.

Summary of DAS action

1 If after a SUB or SBB instruction the lower nibble is greater than 9, or if AF = 1, subtract 0110 from the lower 4 bits.
2. If the upper nibble is greater than 9, or CF = 1, subtract 0110 from the upper nibble.

Due to the widespread use of BCD numbers, a specific data directive, DT, has been created. DT can be used to represent BCD numbers from 0 to $10^{20} - 1$ (that is, twenty 9s). Assume that the following operands represent the budget, the expenses, and the balance, which is the budget minus the expenses.

```
BUDGET        DT    87965141012
EXPENSES      DT    31610640392
BALANCE       DT    ?              ;balance = budget - expenses

      MOV    CX,10                 ;counter=10
      MOV    BX,00                 ;pointer=0
      CLC                          ;clear carry for the 1st iteration
      BACK: MOVAL,BYTE PTR BUDGET[ BX]  ;get a byte of the BUDGET
      SBB    AL,BYTE PTR EXPENSES[ BX]   ;subtract a byte from it
      DAS                                ;correct the result for BCD
      MOV    BYTE PTR BALANCE[ BX] ,AL  ;save it in BALANCE
      INC    BX                          ;increment for the next byte
      LOOP   BACK                        ;continue until CX=0
```

Notice in the code section above that (1) no H (hex) indicator is needed for BCD numbers when using the DT directive, and (2) the use of the based relative addressing mode (BX + displacement) allows access to all three arrays with a single register BX.

Review Questions

1. For the following decimal numbers, give the packed BCD and unpacked BCD representations.
   (a) 15 (b) 99
2. True or false. The DAA instruction must be used after the ADD instruction.
3. True or false. The DAS instruction must be used after the SUB instruction.
4. Find the value of AL after the following code is executed.
   MOV AL,29H
   ADD AL,18H
   DAA

SECTION 3.5: ROTATE INSTRUCTIONS

In many applications there is a need to perform a bitwise rotation of an operand. The rotation instructions ROR, ROL and RCR, RCL are designed specifically for that purpose. They allow a program to rotate an operand right or left. In this section we explore the rotate instructions, which frequently have highly specialized applications. In rotate instructions, the operand can be in a register or memory. If the number of times an operand is to be rotated is more than 1, this is indicated by CL. This is similar to the shift instructions. There are two types of rotations. One is a simple rotation of the bits of the operand, and the other is a rotation through the carry. Each is explained below.

Rotating the bits of an operand right and left

ROR rotate right

In rotate right, as bits are shifted from left to right they exit from the right end (LSB) and enter the left end (MSB). In addition, as each bit exits the LSB, a copy of it is given to the carry flag. In other words, in ROR the LSB is moved to the MSB and is also copied to CF, as shown in the diagram. If the operand is to be rotated once, the 1 is coded, but if it is to be rotated more than once, register CL is used to hold the number of times it is to be rotated.

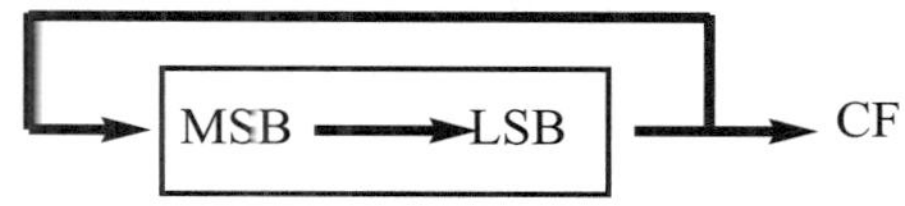


```
        MOV    AL,36H         ;AL=0011 0110
        ROR    AL,1           ;AL=0001 1011   CF=0
        ROR    AL,1           ;AL=1000 1101   CF=1
        ROR    AL,1           ;AL=1100 0110   CF=1
;or:
        MOV    AL,36H         ;AL=0011 0110
        MOV    CL,3           ;CL=3 number of times to rotate
        ROR    AL,CL          ;AL=1100 0110 CF=1
;the operand can be a word:
        MOV    BX,0C7E5H      ;BX=1100 0111 1110 0101
        MOV    CL,6           ;CL=6 number of times to rotate
        ROR    BX,CL          ;BX=1001 0111 0001 1111 CF=1
```

ROL rotate left

In rotate left, as bits are shifted from right to left they exit the left end (MSB) and enter the right end (LSB). In addition, every bit that leaves the MSB is copied to the carry flag. In other words, in ROL the MSB is moved to the LSB and is also copied to CF, as shown in the diagram. If the operand is to be rotated once, the 1 is coded. Otherwise, the number of times it is to be rotated is in CL.

```
        MOV    BH,72H         ;BH=0111 0010
        ROL    BH,1           ;BH=1110 0100   CF=0
        ROL    BH,1           ;BH=1100 1001   CF=1
        ROL    BH,1           ;BH=1001 0011   CF=1
        ROL    BH,1           ;BH=0010 0111   CF=1
;or:
        MOV    BH,72H         ;BH=0111 0010
        MOV    CL,4           ;CL=4 number of times to rotate
        ROL    BH,CL          ;BH=0010 0111   CF=1

;The operand can be a word:
        MOV    DX,672AH       ;DX=0110 0111 0010 1010
        MOV    CL,3           ;CL=3 number of times to rotate
        ROL    DX,CL ;DX=0011 1001 0101 0011 CF=1
```

Program 3-7 shows an application of the rotation instruction. The maximum count in Program 3-7 will be 8 since the program is counting the number of 1s in a byte of data. If the operand is a 16-bit word, the number of 1s can go as high as 16. Program 3-8 is Program 3-7, rewritten for a word-sized operand. It also provides the count in BCD format instead of hex. Reminder: AL is used to make a BCD counter because the DAA instruction works only on AL.

RCR rotate right through carry

In RCR, as bits are shifted from left to right, they exit the right end (LSB) to the carry flag, and the carry flag enters the left end (MSB). In other words, in RCR the LSB is moved to CF and CF is moved to the MSB. In reality, CF acts as if it is part of the operand. This is shown in the diagram. If the operand is to be rotated once, the 1 is

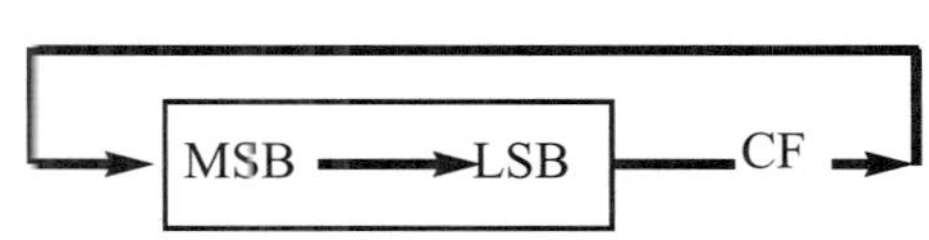

Write a program that finds the number of 1s in a byte.

```
;From the data segment:
DATA1       DB          97H
COUNT       DB          ?
;From the code segment:
            SUB   BL,BL         ;clear BL to keep the number of 1s
            MOV   DL,8          ;rotate total of 8 times
            MOV   AL,DATA1
AGAIN:      ROL   AL,1          ;rotate it once
            JNC   NEXT          ;check for 1
            INC   BL            ;if CF=1 then add one to count
NEXT:       DEC   DL            ;go through this 8 times
            JNZ   AGAIN         ;if not finished go back
            MOV   COUNT,BL      ;save the number of 1s
```

Program 3-7

Write a program to count the number of 1s in a word. Provide the count in BCD.

```
DATAW1      DW          97F4H
COUNT2      DB          ?
            ...
            SUB   AL,AL         ;clear AL to keep the number of 1s in BCD
            MOV   DL,16         ;rotate total of 16 times
            MOV   BX,DATAW1     ;move the operand to BX
AGAIN:      ROL   BX,1          ;rotate it once
            JNC   NEXT          ;check for 1. If CF=0 then jump
            ADD   AL,1          ;if CF=1 then add one to count
            DAA                 ;adjust the count for BCD
NEXT:       DEC   DL            ;go through this 16 times
            JNZ   AGAIN         ;if not finished go back
            MOV   COUNT2,AL     ;save the number of 1s in COUNT2
```

Program 3-8

coded, but if it is to be rotated more than once, the register CL holds the number of times.

```
      CLC                       ;make CF=0
      MOV   AL,26H              ;AL=0010 0110
      RCR   AL,1                ;AL=0001 0011 CF=0
      RCR   AL,1                ;AL=0000 1001 CF=1
      RCR   AL,1                ;AL=1000 0100 CF=1
or:
      CLC                       ;make CF=0
      MOV   AL,26H              ;AL=0010 0110
      MOV   CL,3                ;CL=3 number of times to rotate
      RCR   AL,CL               ;AL=1000 0100 CF=1

;the operand can be a word
      STC                       ;make CF=1
      MOV   BX,37F1H            ;BX=0011 0111 1111 0001
      MOV   CL,5                ;CL=5 number of times to rotate
      RCR   BX,CL               ;BX=0001 1001 1011 1111 CF=0
```

RCL rotate left through carry

In RCL, as bits are shifted from right to left they exit the left end (MSB) and enter the carry flag, and the carry flag enters the right end (LSB). In other words, in RCL the MSB is moved to CF and CF is moved to the LSB. In reality, CF acts as if it is part of the operand. This is shown in the diagram. If the operand is to be rotated once, the 1 is coded, but if it is to be rotated more than once, register CL holds the number of times.

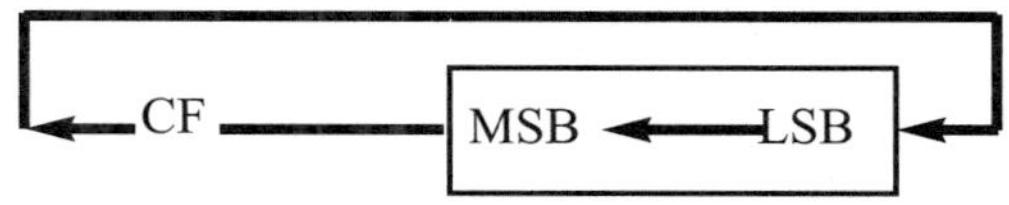


```
       STC                   ;make CF=1
       MOV    BL,15H         ;BL=0001 0101
       RCL    BL,1           ;0010 1011 CF=0
       RCL    BL,1           ;0101 0110 CF=0
or:
       STC                   ;make CF=1
       MOV    BL,15H         ;BL=0001 0101
       MOV    CL,2           ;CL=2 number of times for rotation
       RCL    BL,CL          ;BL=0101 0110 CF=0

;the operand can be a word:
       CLC                   ;make CF=0
       MOV    AX,191CH       ;AX=0001 1001 0001 1100
       MOV    CL,5           ;CL=5 number of times to rotate
       RCL    AX,CL          ;AX=0010 0011 1000 0001 CF=1
```

Review Questions

1. What is the value of BL after the following?
```
       MOV    BL,25H
       MOV    CL,4
       ROR    BL,CL
```
2. What are the values of DX and CF after the following?
```
       MOV    DX,3FA2H
       MOV    CL,7
       ROL    DX,CL
```
3. What is the value of BH after the following?
```
       SUB    BH,BH
       STC
       RCR    BH,1
       STC
       RCR    BH,1
```
4. What is the value of BX after the following?
```
       MOV    BX,FFFFH
       MOV    CL,5
       CLC
       RCL    BX,CL
```
5. Why does "`ROR  BX,4`" give an error in the 8086? How would you change the code to make it work?

SECTION 3.6: BITWISE OPERATORS IN THE C LANGUAGE

One of the most important and powerful features of the C language is its ability to perform bit manipulation. Due to the fact that many books on C do not cover this important topic, it is appropriate to discuss it in this section. This section describes the action of operators and provides examples.

Bitwise operators in C

While every C programmer is familiar with the logical operators AND (&&), OR (||), and NOT (!), many C programmers are less familiar with the bitwise operators AND (&), OR (|), EX-OR (^), inverter (~), Shift Right (>>), and Shift Left (<<). These bitwise operators are widely used in software engineering and control; consequently, their understanding and mastery are critical in system design and interfacing. See Table 3-4.

Table 3-4: Bitwise Logic Operators for C

		AND	OR	EX-OR	Inverter
A	B	A&B	A|B	A^B	Y=~B
0	0	0	0	0	1
0	1	0	1	1	0
1	0	0	1	1	
1	1	1	1	0	

The following code shows Examples 3-5 through 3-7 using the C logical operators. Recall that "0x" in the C language indicates that the data is in hex format.

```
0x35 & 0x0F = 0x05              /* ANDing: see Example 3-5 */
0x0504 | 0xDA68 = 0xDF6C        /* ORing:  see Example 3-6 */
0x54 ^ 0x78 = 0x2C              /* XORing: see Example 3-7 */
~0x37 = 0xC8                    /* inverting 37H */
```

The last one is like the NOT instruction in x86 microprocessors:

```
MOV AL,37H          ;AL=37H
NOT AL              ;AFTER INVERTING 37, AL=C8H
```

Bitwise shift operators in C

There are two bitwise shift operators in C: Shift Right (>>) and Shift Left (<<). They perform exactly the same operation as SHR and SHL in Assembly language, as discussed in Section 3.3. Their format in C is as follows:

```
data >> number of bits to be shifted/* shifting right */
data << number of bits to be shifted/* shifting left */
```

The following example shows the use of shift operators in C:

```
0x9A >> 3= 0x13          //shifting right 3 times: see Example 3-9
0x7777 >> 4 = 0x0777    //shifting right 4 times: see Example 3-10
0x6 << 4 = 0x60         // shifting left 4 times: see Example 3-11
```

Program 3-9 demonstrates the syntax of bitwise operators in C. Next we show some real-world examples of their usage.

Packed BCD to ASCII conversion in C

Section 3.4 showed one way to convert a BCD number to ASCII. This conversion is widely used when dealing with a real-time clock chip. Many of the real-time clock chips provide very accurate time and date for up to ten years without the need for external power. There is a real-time clock in every x86 IBM PC or compatible computer. However, these chips provide the time and date in packed BCD. In order to display the data, it needs to be converted to ASCII. Program 3-10 is a C version of the packed BCD-to-ASCII conversion example discussed in Section 3.4. Program 3-10 converts a byte of packed BCD data into two ASCII characters and displays them using the C bitwise operators.

Notice in Program 3-10 that if the packed BCD data is displayed without conversion to ASCII, we get the parenthesis ")". See Appendix F.

```
/* Program 3-9 Repeats Examples 3-5 through 3-11 in C */
#include    <stdio.h>
main()
{
 // Notice the way data is defined in C for hex format using 0x
unsigned char data_1 = 0x35;
unsigned int data_2 = 0x504;
unsigned int data_3 = 0xDA66;
unsigned char data_4= 0x54;
unsigned char data_5=0x78;
unsigned char data_6=0x37;
unsigned char data_7=0x09A;
unsigned char temp;
unsigned int temp_2;

temp=data_1&0x0F;        //ANDing
printf("\nMasking the upper four bits of %X (hex) we get %X
(hex)\n",data_1,temp);

temp_2=data_2|data_3;    //ORing
printf("The result of %X hex ORed with %X hex is %X
hex\n",data_2,data_3,temp_2);

temp= data_4^data_5;     //EX-ORing
printf("The results of %X hex EX-ORed with %X hex is %X
hex\n",data_4,data_5,temp);

temp=~data_6;            //INVERTING
printf("The result of %X hex inverted is %X hex\n",data_6,temp);

temp=data_7>>3;          //SHIFTING Right
printf("When %X hex is shifted right three times we get %X
hex\n",data_7,temp);

printf("When %X hex is shifted right four times we get %X
hex\n",0x7777,0x7777>>4);

temp=(0x6<<4);           //SHIFTING Left
printf("When %X hex is shifted left %d times we get %X
hex\n",0x6,4,temp);
}
```

Program 3-9

```
/* Program 3-10 shows packed BCD-to-ASCII conversion using logical
bitwise operators in C */
#include    <stdio.h>
main()
{
unsigned char mybcd=0x29;     /* declare a BCD number in hex */
unsigned char asci_1;
unsigned char asci_2;
asci_1=mybcd&0x0f;            /* mask the upper four bits */
asci_1=asci_1|0x30;           /* make it an ASCII character */
asci_2=mybcd&0xf0;            /* mask the lower four bits */
asci_2=asci_2>>4;             /* shift it right 4 times */
asci_2=asci_2|0x30;           /* make it an ASCII character */
printf("BCD data %X is %c , %c in ASCII\n",mybcd,asci_1,asci_2);
printf("My BCD data is %c if not converted to ASCII\n",mybcd);
}
```

Program 3-10

Testing bits in C

In many cases of system programming and hardware interfacing, it is necessary to test a given bit to see if it is high. For example, many devices send a high signal to state that they are ready for an action or to indicate that they have data. How can the bit (or bits) be tested? In such cases, the unused bits are masked and then the remaining data is tested. Program 3-11 asks the user for a byte and tests to see whether or not D0 of that byte is high.

```
/* Program 3-11 shows how to test bit D0 to see if it is high */
#include    <stdio.h>
main()
{
unsigned char status;
unsigned char temp;
printf("\nType in a Hex value\n");
scanf("%X",&status);            //get the data
temp=status&0x01;               //mask all bits except D0
if (temp==0x01)                 //is it high?
 printf("D0 is high");          //if yes, say so
else printf("D0 is low");       //if no, say no
}
```

Program 3-11

The Assembly language version of Program 3-11 is as follows:

```
;assume AL=value (in hex)
      AND AL,01                 ;MASK ALL BITS EXCEPT D0
      CMP AL,01                 ;IS D0 HIGH
      JNE BELOW                 ;MAKE A DECISION
      ....                      ;YES D0 IS HIGH
BELOW:....                      ;D0 IS LOW
```

Review Questions

1. What is the result of 0x2F&0x27 ?
2. What is the result of 0x2F|0x27 ?
3. What is the result of 0x2F^0x27 ?
4. What is the result of ~0x2F ?
5. What is the result of 0x2F >>>> 3 ?
6. What is the result of 0x27 <<<< 4 ?
7. In Program 3-10 if mybcd = 0x32, what is displayed if it is not converted to BCD?
8. Modify Program 3-11 to test D3.

PROBLEMS

SECTION 3.1: UNSIGNED ADDITION AND SUBTRACTION

1. Find CF, ZF, and AF for each of the following. Also indicate the result of the addition and where the result is saved.

(a)	(b)	(c)

MOV BH,3FH
ADD BH,45H
ADD CX,DX
	MOV DX,4599H
	MOV CX,3458H
	ADC AX,00
		MOV AX,255
		STC

(d)	(e)	(f)

MOV BX,0FF01H
ADD BL,BH
ADC CX,00
	MOV CX,0FFFFH
	STC
	ADC AH,00
		MOV AH,0FEH
		STC

2. Write, run, and analyze a program that calculates the total sum paid to a salesperson for eight months. The following are the monthly paychecks for those months: $2300, $4300, $1200, $3700, $1298, $4323, $5673, $986.
3. Rewrite Program 3-2 (in Section 3.1) using byte addition.
4. Write a program that subtracts two multibytes and saves the result. Subtraction should be done a byte at a time. Use the data in Program 3-2.
5. State the three steps involved in a SUB and show the steps for the following data.
   (a) 23H – 12H (b) 43H – 51H (c) 99 – 99

SECTION 3.2: UNSIGNED MULTIPLICATION AND DIVISION

6. Write, run, and analyze the result of a program that performs the following:
   (1) (a) byte1 × byte2 (b) byte1 × word1 (c) word1 × word2
   (2) (a) byte1 / byte2 (b) word1 / word2 (c) doubleword / byte1
   Assume byte1 = 230, byte2 = 100, word1 = 9998, word2 = 300, and doubleword = 100000.

SECTION 3.3: LOGIC INSTRUCTIONS

7. Assume that the following registers contain these hex contents: AX = F000, BX = 3456, and DX = E390. Perform the following operations. Indicate the result and the register where it is stored. Give also ZF and CF in each case.
   Note: the operations are independent of each other.

```
(a) AND DX,AX          (b) OR DH,BL
(c) XOR AL,76H         (d) AND DX,DX
(e) XOR AX,AX          (f) OR BX,DX
(g) AND AH,OFF         (h) OR AX,9999H
(i) XOR DX,0EEEEH      (j) XOR BX,BX
(k) MOV CL,04          (l) SHR DX,1
    SHL AL,CL
(m) MOV CL,3           (n) MOV CL,5
    SHR DL,CL              SHL BX,CL
(o) MOV CL,6
    SHL DX,CL
```

8. Indicate the status of ZF and CF after CMP is executed in each of the following cases.

```
(a) MOV BX,2500        (b)  MOV AL,0FFH       (c)  MOV DL,34
    CMP BX,1400             CMP AL,6FH             CMP DL,88
(d) SUB AX,AX          (e)  XOR DX,DX         (f)  SUB CX,CX
    CMP AX,0000             CMP DX,0FFFFH          DEC CX
    CMP CX,0FFFFH
(g) MOV BX,2378H       (h)  MOV AL,0AAH
    MOV DX,4000H            AND AL,55H
    CMP DX,BX               CMP AL,00
```

9. Indicate whether or not the jump happens in each case.

```
(a)  MOV CL,5          (b)  MOV BH,65H        (c)  MOV AH,55H
     SUB AL,AL              MOV AL,48H             SUB DL,DL
     SHL AL,CL              OR  AL,BH              OR  DL,AH
     JNC TARGET             SHL AL,1               MOV CL,AH
     JC  TARGET             AND CL,0FH
     SHR DL,CL
     JNC TARGET
```

10. Rewrite Program 3-3 to find the lowest grade in that class.
11. Rewrite Program 3-4 to convert all uppercase letters to lowercase.
12. In the IBM BIOS program for testing flags and registers, verify every jump (conditional and unconditional) address calculation. Reminder: As mentioned in Chapter 2, in forward jumps the target address is calculated by adding the displacement value to

IP of the instruction after the jump and by subtracting in backward jumps.

SECTION 3.4: BCD AND ASCII CONVERSION

13. In Program 3-6 rewrite BCD_ADD to do subtraction of the multibyte BCD.
14. Using the DT directive, write a program to subtract two 10-byte BCD numbers.
15. Using the DT directive, write a program to add two 10-byte BCD numbers.
16. We would like to make a counter that counts up from 0 to 99 in BCD. What instruction would you place in the dotted area?

```
        SUB  AL,AL
        ADD  AL,1
        ....  ....
```

17. Write Problem 16 to count down (from 99 to 0).
18. An instructor has the following grading policy: "Curving of grades is achieved by adding to every grade the difference between 99 and the highest grade in the class." If the following are the grades of the class, write a program to calculate the grades after they have been curved: 81, 65, 77, 82, 73, 55, 88, 78, 51, 91, 86, 76. Your program should work for any set of grades.
19. If we try to divide 1,000,000 by 2:
(a) What kind of problem is associated with this operation in the 8086 CPU?
(b) How does the CPU let us know that there is a problem?
20. Which of the following groups of code perform the same operation as LOOP XXX?

```
(a) DEC CL       (b) DEC CH      (c) DEC BX     (d) DEC CX
    JNZ XXX          JNZ XXX         JNZ            JNZ XXX
```

SECTION 3.5: ROTATE INSTRUCTIONS

21. Write a program that finds the number of zeros in a 16-bit word.
22. Explain the difference between RCL and ROL instructions.

SECTION 3.6: BITWISE OPERATORS IN THE C LANGUAGE

23. Write a C program with the following components:
(a) have two hex values: data1 = 55H and data2 = AAH, both defined as unsigned char,
(b) mask the upper 4 bits of data1 and display it in hex,
(c) perform AND, OR, and EX-OR operations between the two data items and then display each result,
(d) invert one and display it,
(e) shift left data1 four times and shift right data2 two times, then display each result.
24. Repeat the above problem with two values input from the user. Use the scanf("%X") function to get the hex data.
25. In the same way that the real-time clock chip provides data in BCD, it also expects data in BCD when it is being initialized. However, data coming from the keyboard is in ASCII. Write a C program to convert two ASCII bytes of data to packed BCD.
26. Write a C program in which the user is prompted for a hex value. Then the data is tested to see if the two least significant bits are high. If so, a message states "D0 and D1 are both high"; otherwise, it states which bit is not high.
27. Repeat the above problem for bits D0 and D7.

ANSWERS TO REVIEW QUESTIONS

SECTION 3.1: UNSIGNED ADDITION AND SUBTRACTION

1. Destination
2. In x86 Assembly language, there are no memory-to-memory operations.
3. MOV AX,DATA_2
   ADD DATA_1,AX
4. Destination, source + destination + CF
5. In (a), the byte addition results in a carry to CF; in (b), the word addition results in a carry to the high byte BH.
6. DEC CX
   JNZ ADD_LOOP
7.
```
 43H 0100 0011       0100 0011
–05H 0000 0101  2's complement = +1111 1011
 3EH                 0011 1110
```
   CF = 0; therefore, the result is positive
8. AL = 95 – 4F – 1 = 45

SECTION 3.2: UNSIGNED MULTIPLICATION AND DIVISION

1. AX 2. DX and AX 3. AX 4. AL, AH
5. AX, DX 6. AL, AH 7. AX, DX

SECTION 3.3: LOGIC INSTRUCTIONS AND SAMPLE PROGRAMS

1. (a) 4202 (b) CFFF (c) 8DFD
2. The operand will remain unchanged; all zeros
3. All ones
4. All zeros
5.
```
A0F2 =        1010 0000 1111 0010
shift left:   0100 0001 1110 0100  CF = 1
shift again:  1000 0011 1100 1000  CF = 0
shift again:  0000 0111 1001 0000  CF = 1
A0F2 shifted left three times = 0790
A0F2 =        1010 0000 1111 0010
shift right:  0101 0000 0111 1001 CF = 0
shift again:  0010 1000 0011 1100 CF = 1
shift again:  0001 0100 0001 1110 CF = 0
A0F2 shifted right three times = 141E
```
6. SUB
7. False

SECTION 3.4: BCD AND ASCII CONVERSION

1. (a) 15 = 0001 0101 packed BCD = 0000 0001 0000 0101 unpacked BCD
   (b) 99 = 1001 1001 packed BCD = 0000 1001 0000 1001 unpacked BCD
2. True
3. True
4. AL = 47H

SECTION 3.5: ROTATE INSTRUCTIONS

1. BL = 52H, CF = 0
2. DX = D11FH, CF = 1
3. BH = C0H
4. BX = FFEFH
5. The source operand cannot be immediate; to fix it:
   MOV CL,4
   ROR BX,CL

SECTION 3.6: BITWISE OPERATION IN THE C LANGUAGE

1. 0x27
2. 0x2F
3. 0x08
4. 0xD0
5. 0x05
6. 0x70
7. 2
8.
```
/* This program shows how to test Bit D3 to see if it is high
*/
#include     <<stdio.h>>
main()
{
unsigned char status;
unsigned char temp;
printf("\nType in a Hex value\n");
scanf("%X",&status);
temp=status&0x08;
if (temp==0x08)
    printf("D3 is high");
else printf("D3 is low");
}
```

CHAPTER 4

INT 21H AND INT 10H PROGRAMMING AND MACROS

OBJECTIVES

Upon completion of this chapter, you will be able to:

- **>>** **Use INT 10H function calls to:**
- **>>** **Clear the screen**
- **>>** **Set the cursor position**
- **>>** **Write characters to the screen in text mode**
- **>>** **Draw lines on the screen in graphics mode**
- **>>** **Change the video mode**
- **>>** **Use INT 21H function calls to:**
- **>>** **Input characters from the keyboard**
- **>>** **Output characters to the screen**
- **>>** **Input or output strings**
- **>>** **Use the LABEL directive to set up structured data items**
- **>>** **Code Assembly language instructions to define and invoke macros**
- **>>** **Explain how macros are expanded by the assembler**
- **>>** **Use the LOCAL directive to define local variables within macros**
- **>>** **Use the INCLUDE directive to retrieve macros from other files**

There are some extremely useful subroutines within BIOS and the OS that are available to the user through the INT (interrupt) instruction. In this chapter, some of them are studied to see how they are used in the context of applications. First, a few words about the interrupt itself. The INT instruction is somewhat like a FAR call. When it is invoked, it saves CS:IP and the flags on the stack and goes to the subroutine associated with that interrupt. The INT instruction has the following format:

```
INT   xx;the interrupt number xx can be 00 - FFH
```

Since interrupts are numbered 00 to FF, this gives a total of 256 interrupts in x86 microprocessors. Of these 256 interrupts, two of them are the most widely used: INT 10H and INT 21H. Each one can perform many functions. INT 10H is examined in Section 4.1, while INT 21H is covered in Section 4.2. Section 4.3 discusses the concept of the macro and how it is used. Various functions of INT 21H and INT 10H are selected by the value put into the AH register, as shown in Appendices D and E. Interrupt instructions are discussed in detail in Appendix B.

SECTION 4.1: BIOS INT 10H PROGRAMMING

INT 10H subroutines are burned into the ROM BIOS of the x86-based IBM PC and compatibles and are used to communicate with the computer's screen video. The manipulation of screen text or graphics can be done through INT 10H. There are many functions associated with INT 10H. Among them are changing the color of characters or the background color, clearing the screen, and changing the location of the cursor. These options are chosen by putting a specific value in register AH. In this section we show how to use INT 10H to clear the screen, change the cursor position, change the screen color, and draw lines on the screen.

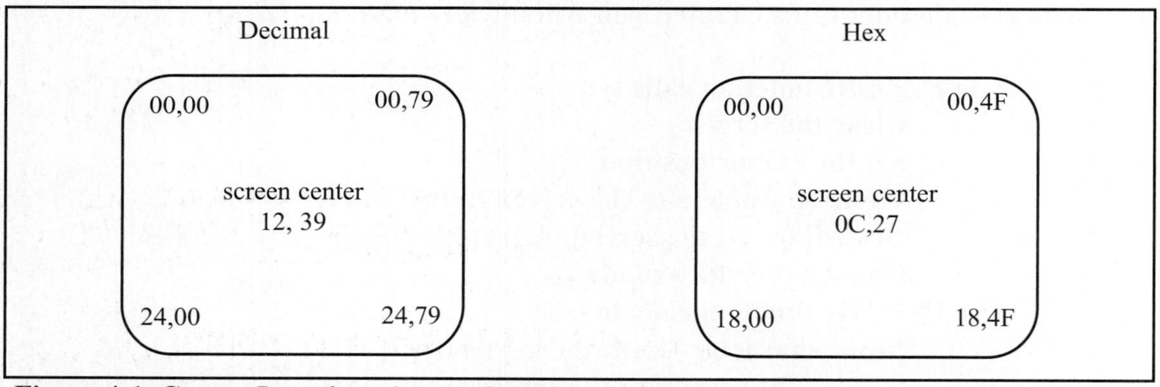

Figure 4-1. Cursor Locations (row, column)

Monitor screen in text mode

The monitor screen in the x86 PC is divided into 80 columns and 25 rows in normal text mode (see Figure 4-1). In other words, the text screen is 80 characters wide by 25 characters long. Since both a row and a column number are associated with each location on the screen, one can move the cursor to any location on the screen simply by changing the row and column values. The 80 columns are numbered from 0 to 79 and the 25 rows are numbered 0 to 24. The top left corner has been assigned 00,00 (row = 00, column = 00). Therefore, the top right corner will be 00,79 (row = 00, column = 79). Similarly, the bottom left corner is 24,00 (row = 24, column = 00) and the bottom right corner of the monitor is 24,79 (row = 24, column = 79). Figure 4-1 shows each location of the screen in both decimal and hex.

Clearing the screen using INT 10H function 06H

It is often desirable to clear the screen before displaying data. To use INT 10H to clear the screen, the following registers must contain certain values before INT 10H is called: AH = 06, AL = 00, BH = 07, CX = 0000, DH = 24, and DL = 79. The code will look like this:

```
MOV   AH,06        ;AH=06 to select scroll function
MOV   AL,00        ;AL=00 the entire page
MOV   BH,07        ;BH=07 for normal attribute
MOV   CH,00        ;CH=00 row value of start point
MOV   CL,00        ;CL=00 column value of start point
MOV   DH,24        ;DH=24 row value of ending point
MOV   DL,79        ;DL=79 column value of ending point
INT   10H          ;invoke the interrupt
```

Remember that DEBUG assumes immediate operands to be in hex; therefore, DX would be entered as 184F. However, MASM assumes immediate operands to be in decimal. In that case DH = 24 and DL = 79.

In the program above, one of many options of INT 10H was chosen by putting 06 into AH. Option AH = 06, called the scroll function, will cause the screen to scroll upward. The CH and CL registers hold the starting row and column, respectively, and DH and DL hold the ending row and column. To clear the entire screen, one must use the top left cursor position of 00,00 for the start point and the bottom right position of 24,79 for the end point.

Option AH = 06 of INT 10H is in reality the "scroll window up" function; therefore, one could use that to make a window of any size by choosing appropriate values for the start and end rows and columns. However, to clear the screen, the top left and bottom right values are used for start and stop points in order to scroll up the entire screen. It is more efficient coding to clear the screen by combining some of the lines above as follows:

```
MOV   AX,0600H     ;scroll entire screen
MOV   BH,07        ;normal attribute
MOV   CX,0000      ;start at 00,00
MOV   DX,184FH     ;end at 24,79 (hex = 18,4F)
INT   10H          ;invoke the interrupt
```

INT 10H function 02: setting the cursor to a specific location

INT 10H function AH = 02 will change the position of the cursor to any location. The desired position of the cursor is identified by the row and column values in DX, where DH = row and DL = column. Video RAM can have multiple pages of text, but only one of them can be viewed at a time. When AH = 02, to set the cursor position, page zero is chosen by making BH = 00.

It must be pointed out that after INT 10H (or INT 21H) has executed, the registers that have not been used by the interrupt remain unchanged. In other words, these registers have the same values after execution of the interrupt as before the interrupt was invoked. Examples 4-1 and 4-2 demonstrate setting the cursor to a specific location.

Example 4-1

Write the code to set the cursor position to row = 15 = 0FH and column = 25 = 19H.

Solution:

```
MOV   AH,02        ;set cursor option
MOV   BH,00        ;page 0
MOV   DL,25        ;column position
MOV   DH,15        ;row position
INT   10H          ;invoke interrupt 10H
```

Example 4-2

Write a program that (1) clears the screen and (2) sets the cursor at the center of the screen.

Solution:

The center of the screen is the point at which the middle row and middle column meet. Row 12 is at the middle of rows 0 to 24 and column 39 (or 40) is at the middle of columns 0 to 79. By setting row = DH = 12 and column = DL = 39, the cursor is set to the screen center.

```
;clearing the screen
       MOV    AX,0600H       ;scroll the entire page
       MOV    BH,07          ;normal attribute
       MOV    CX,0000        ;row and column of top left
       MOV    DX,184FH       ;row and column of bottom right
       INT    10H            ;invoke the video BIOS service

;setting the cursor to the center of screen
       MOV    AH,02          ;set cursor option
       MOV    BH,00          ;page 0
       MOV    DL,39          ;center column position
       MOV    DH,12          ;center row position
       INT    10H            ;invoke interrupt 10H
```

INT 10H function 03: get current cursor position

In text mode, one is able to determine where the cursor is located at any time by executing the following:

```
MOV    AH,03          ;option 03 of BIOS INT 10H
MOV    BH,00          ;page 00
INT    10H            ;interrupt 10H routine
```

After execution of the program above, registers DH and DL will have the current row and column positions, and CX provides information about the shape of the cursor. The reason that page 00 was chosen is that the video memory could contain more than one page of data, depending on the video board installed on the PC. In text mode, page 00 is chosen for the currently viewed page.

Changing the video mode

To change the video mode, one can use INT 10H with AH = 00 and AL = video mode. A list of video modes is given in Appendix D, Table D-2.

Attribute byte in monochrome monitors

There is an attribute associated with each character on the screen. The attribute provides information to the video circuitry, such as color and intensity of the character (foreground) and the background. The attribute byte for each character on the monochrome monitor is limited. Figure 4-2 shows bit definitions of the monochrome attribute byte.

Foreground refers to the actual character displayed. Normal, highlighted intensity and blinking are for the foreground only. The following are some possible variations of the attributes in Figure 4-2.

Binary	Hex	Result
0000 0000	00	white on white (no display)
0000 0111	07	white on black normal
0000 1111	0F	white on black highlight
1000 0111	87	white on black blinking
0111 0111	77	black on black (no display)
0111 0000	70	black on white
1111 0000	F0	black on white blinking

For example, "00000111" would give the normal screen mode where the background is black and the foreground is normal intensity, nonblinking. "00001111" would give the same mode with the foreground highlighted. "01110000" would give a reverse video screen mode with the foreground black and the background normal intensity. See Example 4-3.

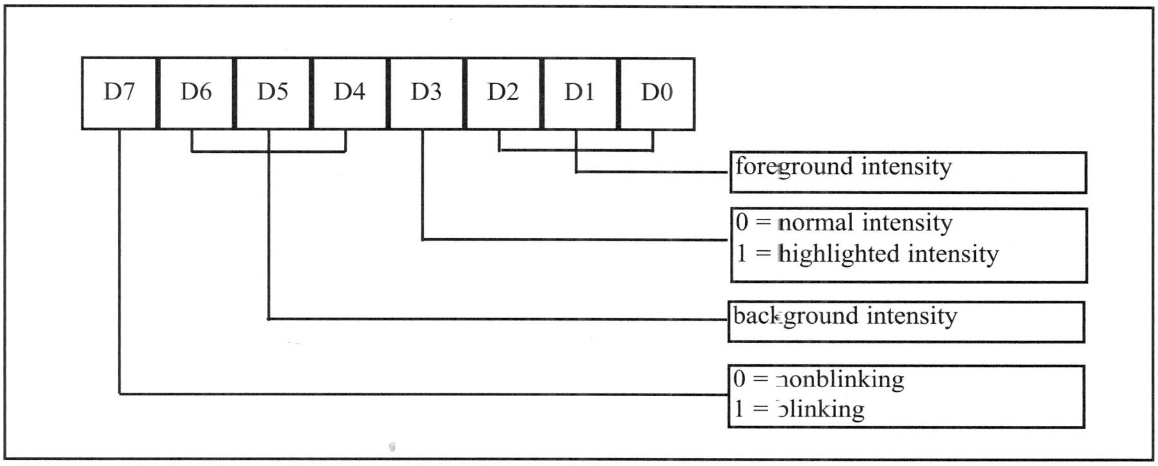

Figure 4-2. Attribute Byte for Monochrome Monitors

Attribute byte in CGA text mode

Since all color monitors and their video circuitry are upwardly compatible, in examples concerning color, in this chapter we use CGA mode, the common denominator for all color monitors. The bit definition of the attribute byte in CGA text mode is as shown in Figure 4-3. From the bit definition it can be seen that the background can take eight different colors by combining the prime colors red, blue, and green. The foreground can be any of 16 different colors by combining red, blue, green, and intensity. Example 4-4 shows the use of the attribute byte in CGA mode. Table 4-1 lists the possible colors. As examples of some possible variations look at the following cases:

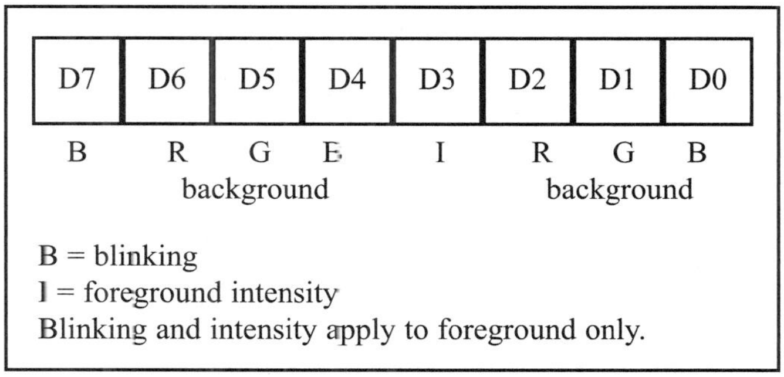

Figure 4-3. CGA Attribute Byte

Binary	Hex	Color effect
0000 0000	00	Black on black
0000 0001	01	Blue on black
0001 0010	12	Green on blue
0001 0100	14	Red on blue
0001 1111	1F	High-intensity white on blue

Example 4-3

Write a program using INT 10H to:
(a) Change the video mode.
(b) Display the letter "D" in 200H locations with attributes black on white blinking (blinking letters "D" are black and the screen background is white).
(c) Then use DEBUG to run and verify the program.

Solution:

(a) INT 10H function AH = 00 is used with AL = video mode to change the video mode. Use AL = 03.

```
        MOV    AH,00         ;SET MODE OPTION
        MOV    AL,03         ;CHANGE THE VIDEO MODE
        INT    10H           ;MODE OF 80X25 FOR ANY COLOR MONITOR
```

(b) With INT 10H function AH = 09, one can display a character a certain number of times with specific attributes.

```
        MOV    AH,09         ;DISPLAY OPTION
        MOV    BH,00         ;PAGE 0
        MOV    AL,44H        ;THE ASCII FOR LETTER "D"
        MOV    CX,200H       ;REPEAT IT 200H TIMES
        MOV    BL,0F0H       ;BLACK ON WHITE BLINKING
        INT    10H
```

(c) Reminder: DEBUG assumes that all the numbers are in hex.

```
C>debug
-A
1131:0100 MOV AH,00
1131:0102 MOV AL,03     ;CHANGE THE VIDEO MODE
1131:0104 INT 10
1131:0106 MOV AH,09
1131:0108 MOV BH,00
1131:010A MOV AL,44
1131:010C MOV CX,200
1131:010F MOV BL,F0
1131:0111 INT 10
1131:0113 INT 3
1131:0114
-
```

Now see the result by typing in the command -G. Make sure that IP = 100 before running it. As an exercise, change the BL register to other attribute values given earlier. For example, BL = 07 white on black, or BL = 87H white on black blinking.

Graphics: pixel resolution and color

In the text mode, the screen is viewed as a matrix of rows and columns of characters. In graphics mode, the screen is viewed as a matrix of horizontal and vertical pixels. The number of pixels varies among monitors and depends on monitor resolution and the video board. In this section we show how to access and program pixels on the screen. Before embarking on pixel programming, the relationship between pixel resolution, the number of colors available, and the amount of video memory in a given video board must be clarified. There are two facts associated with every pixel on the screen: (1) the location of the pixel, and (2) its attributes, color and intensity. These two facts must be stored in the video RAM. Therefore, the higher the number of pixels and colors, the larger the amount of memory that is needed to store them. In other words, the memory requirement goes up as the resolution and the number of colors on the monitor go up. The CGA mode can have a maximum of 16K bytes of video memory due to its inherent design structure. The 16K bytes of memory can be used in three different ways.

Table 4-1: The 16 Possible Colors

I	R	G	B	Color
0	0	0	0	black
0	0	0	1	blue
0	0	1	0	green
0	0	1	1	cyan
0	1	0	0	red
0	1	0	1	magenta
0	1	1	0	brown
0	1	1	1	white
1	0	0	0	gray
1	0	0	1	light blue
1	0	1	0	light green
1	0	1	1	light cyan
1	1	0	0	light red
1	1	0	1	light magenta
1	1	1	0	yellow
1	1	1	1	high intensity white

1. Text mode of 80 × 25 characters; This takes a total of 2K bytes (80 × 25 = 2000) for the characters plus 2K bytes of memory for their attributes, since each character has one attribute byte. That means that each screen (frame) takes 4K bytes, and that results in CGA supporting a total of four pages of data, where each page represents one full screen. In this mode, 16 colors are supported. To select this mode, use AL = 03 for mode selection in INT 10H option AH = 00.
2. Graphics mode of 320 × 200 (medium resolution); In this mode there are a total of 64,000 pixels (320 columns × 200 rows = 64,000). Dividing the total video RAM memory of 128K bits (16K × 8 bits = 128K bits) by the 64,000 pixels gives 2 bits for the color of each pixel. These 2 bits give four possibilities. Therefore, the 320 × 200 resolution CGA can support no more than 4 colors. To select this mode, use AL = 04.
3. Graphics resolution of 640 × 200 (high resolution); In this mode there are a total of 128,000 pixels (200 × 640 = 128,000). Dividing the 16K bytes of memory by this gives 1 bit (128,000/128,000 = 1) for color. The bit can be on (white) or off (black). Therefore, the 640 × 200 high-resolution CGA can support only black and white. To select this mode, use AL = 06.

Example 4-4

Write a program that puts 20H (ASCII space) on the entire screen. Use high-intensity white on a blue background attribute for any characters to be displayed.

Solution:

```
MOV   AH,00      ;SET MODE OPTION
MOV   AL,03      ;CGA COLOR TEXT MODE OF 80 × 25
INT   10H
MOV   AH,09      ;DISPLAY OPTION
MOV   BH,00      ;PAGE 0
MOV   AL,20H     ;ASCII FOR SPACE
MOV   CX,800H    ;REPEAT IT 800H TIMES
MOV   BL,1FH     ;HIGH-INTENSITY WHITE ON BLUE
INT   10H
```

From the discussion above one can conclude that with a fixed amount of video RAM, the number of supported colors decreases as the resolution increases. That is the reason that to create more colors in video boards there must be memory available to store the extra colors.

INT 10H and pixel programming

To address a single pixel on the screen, use INT 10H with AH = 0CH. The X and Y coordinates of the pixel must be known. The values for X (column) and Y (row) vary, depending on the resolution of the monitor. The registers holding these values are CX = the column point (the X coordinate) and DX = the row point (Y coordinate). If the display mode supports more than one page, BH = page number; otherwise, it is ignored. To turn the pixel on or off, AL = 1 or AL = 0 for black and white. The value of AL can be modified for various colors.

Drawing horizontal or vertical lines in graphics mode

To draw a horizontal line, choose values for the row and column to point to the beginning of the line and then continue to increment the column until it reaches the end of the line, as shown in Example 4-5.

Example 4-5

Write a program to: (a) clear the screen, (b) set the mode to CGA of 640 × 200 resolution, and (c) draw a horizontal line starting at column = 100, row = 50, and ending at column 200, row 50.

Solution:

```
      MOV   AX,0600H      ;SCROLL THE SCREEN
      MOV   BH,07         ;NORMAL ATTRIBUTE
      MOV   CX,0000       ;FROM ROW=00,COLUMN=00
      MOV   DX,184FH      ;TO ROW=18H,COLUMN=4FH
      INT   10H           ;INVOKE INTERRUPT TO CLEAR SCREEN
      MOV   AH,00         ;SET MODE
      MOV   AL,06         ;MODE = 06 (CGA HIGH RESOLUTION)
      INT   10H           ;INVOKE INTERRUPT TO CHANGE MODE
      MOV   CX,100        ;START LINE AT COLUMN =100 AND
      MOV   DX,50         ;ROW = 50
BACK: MOV   AH,0CH        ;AH=0CH TO DRAW A LINE
      MOV   AL,01         ;PIXELS = WHITE
      INT   10H           ;INVOKE INTERRUPT TO DRAW LINE
      INC   CX            ;INCREMENT HORIZONTAL POSITION
      CMP   CX,200        ;DRAW LINE UNTIL COLUMN = 200
      JNZ   BACK
```

As an exercise, put INT 3 at the end of the program above and run it in DEBUG to get a feeling of the concept. To draw a vertical line, simply increment the vertical value held by the DX register and keep CX constant. The linear equation $y = mx + b$ can be used to draw any line.

Review Questions

1. Interrupt 10H function calls perform what services?
2. The monitor in text mode has _____ columns and _____ rows. The top left position is (__,__) and the bottom right position is (__,__).
3. Fill in the blanks in the following program, which clears the screen. Write comments on each line stating the purpose of each line of code.

```
MOV AH,___
MOV AL,___
MOV BH,___
MOV CH,___
MOV CL,___
MOV DH,___
MOV DL,___
```

```
        INT 10H
```

4. INT 10 function AH = 03 was used. Afterward, DH = 05 and DL = 34. What does this indicate?
5. What is the purpose of the attribute byte for monochrome monitors?
6. In text mode, there is one attribute byte associated with each _____ on the screen.
7. Write the attribute byte to display background green, foreground white blinking.
8. State the purpose of the following program, which is for a monochrome monitor.

```
        MOV  AH,02
        MOV  BH,00
        MOV  DX,0000
        INT  10H
        MOV  AH,09
        MOV  BH,00
        MOV  AL,2AH
        MOV  CX,80
        MOV  BL,0F0H
        INT  10H
```

SECTION 4.2: DOS INTERRUPT 21H

INT 21H is provided by DOS in contrast to INT 10H, which is BIOS-ROM based. When the OS is loaded into the computer, INT 21H can be invoked to perform some extremely useful functions. These functions are commonly referred to as DOS INT 21H function calls. A partial list of these options is provided in Appendix D. In this section we use only the options dealing with inputting information from the keyboard and displaying it on the screen. In previous chapters, a fixed set of data was defined in the data segment and the results were viewed in a memory dump. Starting with this chapter, data will come from the keyboard and after it is processed, the results will be displayed on the screen. This is a much more dynamic way of processing information and is the main reason for placing this chapter at this point of the book. Although data is input and output through the keyboard and monitor, there is still a need to dump memory to verify the data when troubleshooting programs.

INT 21H option 09: outputting a string of data to the monitor

INT 21H can be used to send a set of ASCII data to the monitor. To do that, the following registers must be set: AH = 09 and DX = the offset address of the ASCII data to be displayed. Then INT 21H is invoked. The address in the DX register is an offset address and DS is assumed to be the data segment. INT 21H option 09 will display the ASCII data string pointed at by DX until it encounters the dollar sign "$". In the absence of encountering a dollar sign, DOS function call 09 will continue to display any garbage that it can find in subsequent memory locations until it finds "$". For example, to display the message "The earth is but one country", the following is from the data segment and code segment.

```
DATA_ASC     DB    'The earth is but one country','$'

        MOV    AH,09          ;option 09 to display string of data
        MOV    DX,OFFSET DATA_ASC     ;DX= offset address of data
        INT    21H                    ;invoke the interrupt
```

INT 21H option 02: outputting a single character to the monitor

There are occasions when it is necessary to output to the monitor only a single character. To do that, 02 is put in AH, DL is loaded with the character to be displayed, and then INT 21H is invoked. The following displays the letter "J".

```
        MOV    AH,02        ;option 02 displays one character
        MOV    DL,'J'       ;DL holds the character to be displayed
        INT    21H          ;invoke the interrupt
```

This option can also be used to display '$' on the monitor since the string display option (option 09) will not display '$'.

INT 21H option 01: inputting a single character, with echo

This functions waits until a character is input from the keyboard, then echoes it to the monitor. After the interrupt, the input character will be in AL.

```
        MOV     AH,01 ;option 01 inputs one character
        INT     21H   ;after the interrupt, AL = input character (ASCII)
```

Program 4-1 combines INT 10H and INT 21H. The program does the following: (1) clears the screen, (2) sets the cursor to the center of the screen, and (3) starting at that point of the screen, displays the message "This is a test of the display routine".

INT 21H option 0AH: inputting a string of data from the keyboard

Option 0AH of INT 21H provides a means by which one can get data from the keyboard and store it in a predefined area of memory in the data segment. To do that, registers are set as follows: AH = 0AH and DX = offset address at which the string of data is stored. This is commonly referred to as a *buffer area*. DOS requires that a buffer area be defined in the data segment and the first byte specifies the size of the buffer. DOS will put the number of characters that came in through the keyboard in the second byte and the keyed-in data is placed in the buffer starting at the third byte. For example, the following program will accept up to six characters from the keyboard, including the return (carriage return) key. Six locations were reserved for the buffer and filled with FFH. The following shows portions of the data segment and code segment.

```
ORG     0010H
DATA1 DB      6,?,6 DUP (FF);0010H=06, 0012H to 0017H = FF

        MOV     AH,0AH              ;string input option of INT 21H
        MOV     DX,OFFSET DATA1     ;load the offset address of buffer
        INT     21H                 ;invoke interrupt 21H
```

The following shows the memory contents of offset 0010H:

```
0010    0011    0012    0013    0014    0015    0016    0017
06      00      FF      FF      FF      FF      FF      FF
```

When this program is executed, the computer waits for the information to come in from the keyboard. When the data comes in, the IBM PC will not exit the INT 21H routine until it encounters the return key. Assuming the data that was entered through the keyboard was "USA" <RETURN>, the contents of memory locations starting at offset 0010H would look like this:

```
0010    0011    0012    0013    0014    0015    0016    0017
06      03      55      53      41      0D      FF      FF
USACR
```

The following is a step-by-step analysis:

0010H = 06	DOS requires the size of the buffer in the first location.
0011H = 03	The keyboard was activated three times (excluding the RETURN key) to key in the letters U, S, and A.
0012H = 55H	This is the ASCII hex value for letter U.
0013H = 53H	This is the ASCII hex value for letter S.
0014H = 41H	This is the ASCII hex value for letter A.
0015H = 0DH	This is the ASCII hex value for CR (carriage return).

```
TITLE       PROG4-1 SIMPLE DISPLAY PROGRAM
PAGE        60,132
            .MODEL SMALL
            .STACK       64
;------------------------------
            .DATA
MESSAGE     DB    'This is a test of the display routine','$'
;------------------------------
      .CODE
MAIN PROC   FAR
     MOV    AX,@DATA
     MOV    DS,AX
     CALL   CLEAR                ;CLEAR THE SCREEN
     CALL   CURSOR               ;SET CURSOR POSITION
     CALL   DISPLAY              ;DISPLAY MESSAGE
     MOV    AH,4CH
     INT    21H                  ;GO BACK TO DOS
MAIN ENDP
;-------------------------------
;THIS SUBROUTINE CLEARS THE SCREEN
CLEAR PROC
     MOV    AX,0600H             ;SCROLL SCREEN FUNCTION
     MOV    BH,07                ;NORMAL ATTRIBUTE
     MOV    CX,0000              ;SCROLL FROM ROW=00,COL=00
     MOV    DX,184FH             ;TO ROW=18H,COL=4FH
     INT    10H                  ;INVOKE INTERRUPT TO CLEAR SCREEN
     RET
CLEAR ENDP
;-------------------------------
;THIS SUBROUTINE SETS THE CURSOR AT THE CENTER OF THE SCREEN
CURSOR PROC
     MOV    AH,02                ;SET CURSOR FUNCTION
     MOV    BH,00                ;PAGE 00
     MOV    DH,12                ;CENTER ROW
     MOV    DL,39                ;CENTER COLUMN
     INT    10H                  ;INVOKE INTERRUPT TO SET CURSOR POSITION
     RET
CURSOR ENDP
;-------------------------------
;THIS SUBROUTINE DISPLAYS A STRING ON THE SCREEN
DISPLAY  PROC
     MOV    AH,09                ;DISPLAY FUNCTION
     MOV    DX,OFFSET MESSAGE    ;DX POINTS TO OUTPUT BUFFER
     INT    21H                  ;INVOKE INTERRUPT TO DISPLAY STRING
     RET
DISPLAY  ENDP
     END    MAIN
```

Program 4-1

One might ask where the value 03 in 0011H came from. DOS puts that value there to indicate that three characters were entered. How can this character count byte be accessed? See the following:

```
MOV   AH,0AH
MOV   DX,OFFSET DATA1
INT   21H
```

```
;After data has been keyed in, next fetch the count value
        MOV    BX,OFFSET DATA1
        SUB    CH,CH                 ;CH=00
        MOV    CL,[ BX] +1           ;move count to CL
```

To locate the CR value 0DH in the string and replace it, say, with 00, simply code the following line next:

```
        MOV    SI,CX
        MOV    BYTE PTR[ BX+SI] +2,00
```

The actual keyed-in data is located beginning at location [BX]+2.

Inputting more than the buffer size

Now what happens if more than six characters (five, the maximum length + the CR = 6) are keyed in? Entering a message like "USA a country in North America" <RETURN> will cause the computer to sound the speaker and the contents of the buffer will look like this:

0010	0011	0012	0013	0014	0015	0016	0017
06	05	55	53	41	20	61	0D
		U	S	A	SP	a	CR

Location 0015 has ASCII 20H for space, and 0016 has ASCII 61H for "a", and finally, the 0D for RETURN key is at 0017. The actual length is 05 at memory offset 0011H. Another question is: What happens if only the CR key is activated and no other character is entered? For example, in the following,

```
        ORG    20H
DATA4 DB       10,?,10 DUP (FF)
```

which puts 0AH in memory 0020H, the 0021H is for the count and the 0022H is the first location that will have the data that was entered. So if only the return key is activated, 0022H has 0DH, the hex code for CR.

0020	0021	0022	0023	0024	0025	0026	0027	0028	0029	002A	002B	002C
0A	00	0D	FF	FF	FF	FF	FF	FF	FF	FF	FF	FF

The actual number of characters entered is 0 at location 0021. Remember that CR is not included in the count. It must be noted that as data is entered it is displayed on the screen. This is called an *echo*. So the 0AH option of INT 21H accepts the string of data from the keyboard and echoes (displays) it on the screen as it is keyed in.

Use of carriage return and line feed

In Program 4-2, the EQU statement is used to equate CR (carriage return) with its ASCII value of 0DH, and LF (line feed) with its ASCII value of 0AH. This makes the program much more readable. Since the result of the conversion was to be displayed in the next line, the string was preceded by CR and LF. In the absence of CR the string would be displayed wherever the cursor happened to be. In the case of CR and no LF, the string would be displayed on the same line after it had been returned to the beginning of the line by the CR and, consequently, would write over some of the characters on that line.

Program 4-3 prompts the user to type in a name. The name can have a maximum of eight letters. After the name is typed in, the program gets the length of the name and prints it to the screen.

Program 4-4 demonstrates many of the functions described in this chapter.

```
;Program 4-2 performs the following: (1) clears the screen, (2) sets
;the cursor at the beginning of the third line from the top of the
;screen, (3) accepts the message "IBM perSonal COmputer" from the
;keyboard, (4) converts lowercase letters of the message to uppercase,
;(5) displays the converted results on the next line.

TITLE       PROG4-2
PAGE        60,132
            .MODEL SMALL
            .STACK      64

;-----------------------------
            .DATA
BUFFER      DB    22,?,22 DUP (?)         ;BUFFER FOR KEYED-IN DATA
            ORG   18H
DATAREA     DB    CR,LF,22 DUP (?),'$'    ;DATA HERE AFTER CONVERSION
;DTSEG      ENDS
CR    EQU 0DH
LF    EQU 0AH

;-----------------------------
      .CODE
MAIN PROC  FAR
      MOV   AX,@DATA
      MOV   DS,AX
      CALL  CLEAR               ;CLEAR THE SCREEN
      CALL  CURSOR              ;SET CURSOR POSITION
      CALL  GETDATA             ;INPUT A STRING INTO BUFFER
      CALL  CONVERT             ;CONVERT STRING TO UPPERCASE
      CALL  DISPLAY             ;DISPLAY STRING DATAREA
      MOV   AH,4CH
      INT   21H         ;GO BACK TO DOS
MAIN ENDP

;-----------------------------
;THIS SUBROUTINE CLEARS THE SCREEN
CLEAR PROC
      MOV   AX,0600H            ;SCROLL SCREEN FUNCTION
      MOV   BH,07               ;NORMAL ATTRIBUTE
      MOV   CX,0000                   ;SCROLL FROM ROW=00,COL=00
      MOV   DX,184FH            ;TO ROW=18H,4FH
      INT   10H                 ;INVOKE INTERRUPT TO CLEAR SCREEN
      RET
CLEARENDP
;THIS SUBROUTINE SETS THE CURSOR TO THE BEGINNING OF THE 3RD LINE
CURSOR PROC
      MOV   AH,02               ;SET CURSOR FUNCTION
      MOV   BH,00               ;PAGE 0
      MOV   DL,01               ;COLUMN 1
      MOV   DH,03               ;ROW 3
      INT   10H                 ;INVOKE INTERRUPT TO SET CURSOR
      RET
CURSOR ENDP

;-----------------------------
;THIS SUBROUTINE DISPLAYS A STRING ON THE SCREEN
DISPLAY PROC
      MOV   AH,09               ;DISPLAY STRING FUNCTION
      MOV   DX,OFFSET DATAREA ;DX POINTS TO BUFFER
      INT   21H                 ;INVOKE INTERRUPT TO DISPLAY STRING
      RET
DISPLAY     ENDP
```

Program 4-2 *(continued on the next page)*

```
;------------------------------
;THIS SUBROUTINE PUTS DATA FROM THE KEYBOARD INTO A BUFFER
GETDATA PROC
     MOV   AH,0AH             ;INPUT STRING FUNCTION
     MOV   DX,OFFSET BUFFER   ;DX POINTS TO BUFFER
     INT   21H                ;INVOKE INTERRUPT TO INPUT STRING
     RET
GETDATA ENDP
;------------------------------
;THIS SUBROUTINE CONVERTS ANY SMALL LETTER TO ITS CAPITAL
CONVERT PROC
     MOV   BX,OFFSET BUFFER
     MOV   CL,[BX]+1                ;GET THE CHAR COUNT
     SUB   CH,CH                    ;CX = TOTAL CHARACTER COUNT
     MOV   DI,CX                    ;INDEXING INTO BUFFER
     MOV   BYTE PTR[BX+DI]+2,20H    ;REPLACE CR WITH SPACE
     MOV   SI,OFFSET DATAREA+2      ;STRING ADDRESS
AGAIN: MOV AL,[BX]+2                ;GET THE KEYED-IN DATA
     CMP   AL,61H                   ;CHECK FOR 'a'
     JB    NEXT                     ;IF BELOW, GO TO NEXT
     CMP   AL,7AH                   ;CHECK FOR 'z'
     JA    NEXT                     ;IF ABOVE GO TO NEXT
     AND   AL,11011111B             ;CONVERT TO CAPITAL
NEXT: MOV  [SI],AL                  ;PLACE IN DATA AREA
     INC   SI                       ;INCREMENT POINTERS
     INC   BX
     LOOP  AGAIN                    ;LOOP IF COUNTER NOT ZERO
     RET
CONVERT ENDP
     END   MAIN
```

Program 4-2 *(continued from the preceding page)*

INT 21H option 07: keyboard input without echo

Option 07 of INT 21H requires the user to enter a single character but that character is not displayed (or echoed) on the screen. After execution of the interrupt, the PC waits until a single character is entered and provides the character in AL.

```
     MOV   AH,07 ;keyboard input without echo
     INT   21H
```

Using the LABEL directive to define a string buffer

A more systematic way of defining the buffer area for the string input is to use the LABEL directive. The LABEL directive can be used in the data segment to assign multiple names to data. When used in the data segment it looks like this:

```
name  LABEL attribute
```

The attribute can be either BYTE, WORD, DWORD, FWORD, QWORD, or TBYTE. Simply put, the LABEL directive is used to assign the same offset address to two names. For example, in the following,

```
JOE   LABEL BYTE
TOM   DB    20 DUP(0)
```

the offset address assigned to JOE is the same offset address for TOM since the LABEL directive does not occupy any memory space (see Appendix C for many examples of the use of the LABEL directive).

```
TITLE       PROG4-3     READS IN LAST NAME AND DISPLAYS LENGTH
PAGE        60,132
            .MODEL SMALL
            .STACK 64 (?)
;----------------------------
            .DATA
MESSAGE1    DB     'What is your last name?','$'
            ORG    20H
BUFFER1     DB     9,?,9 DUP (0)
            ORG    30H
MESSAGE2    DB     CR,LF,'The number of letters in your name is: ','$'
ROW         EQU 08
COLUMN      EQU 05
CR          EQU 0DH      ;EQUATE CR WITH ASCII CODE FOR CARRIAGE RETURN
LF          EQU 0AH      ;EQUATE LF WITH ASCII CODE FOR LINE FEED
;----------------------------
            .CODE
MAIN        PROC  FAR
     MOV    AX,@DATA
     MOV    DS,AX
     CALL   CLEAR
     CALL   CURSOR
     MOV    AH,09              ;DISPLAY THE PROMPT
     MOV    DX,OFFSET MESSAGE1
     INT    21H
     MOV    AH,0AH             ;GET LAST NAME FROM KEYBOARD
     MOV    DX,OFFSET BUFFER1
     INT    21H
     MOV    BX,OFFSET BUFFER1 ;FIND OUT NUMBER OF LETTERS IN NAME
     MOV    CL,[ BX+1]         ;GET NUMBER OF LETTERS
     OR     CL,30H             ;MAKE IT ASCII
     MOV    MESSAGE2+40,CL     ;PLACE AT END OF STRING
     MOV    AH,09              ;DISPLAY SECOND MESSAGE
     MOV    DX,OFFSET MESSAGE2
     INT    21H
     MOV    AH,4CH
     INT    21H          ;GO BACK TO DOS
MAIN ENDP
;-----------------------------
CLEAR PROC                     ;CLEAR THE SCREEN
     MOV    AX,0600H
     MOV    BH,07
     MOV    CX,0000
     MOV    DX,184FH
     INT    10H
     RET
CLEAR ENDP
;-----------------------------
CURSOR PROC                          ;SET CURSOR POSITION
     MOV    AH,02
     MOV    BH,00
     MOV    DL,COLUMN
     MOV    DH,ROW
     INT    10H
     RET
CURSOR ENDP
     END    MAIN
```

Program 4-3

Write a program to perform the following: (1) clear the screen, (2) set the cursor at row 5 and column 1 of the screen, (3) prompt "There is a message for you from Mr. Jones. To read it enter Y ". If the user enters 'Y' or 'y' then the message "Hi! I must leave town tomorrow, therefore I will not be able to see you" will appear on the screen. If the user enters any other key, then the prompt "No more messages for you" should appear on the next line.

```
TITLE PROGRAM 4-4
PAGE 60,132
      .MODEL SMALL
      .STACK 64
;----------------------------
            .DATA
PROMPT1     DB    'There is a message for you from Mr. Jones. '
            DB    'To read it enter Y','$'
MESSAGE     DB    CR,LF,'Hi! I must leave town tomorrow, '
            DB    'therefore I will not be able to see you','$'
PROMPT2     DB    CR,LF,'No more messages for you','$'
;DTSEG      ENDS
CR          EQU 0DH
LF          EQU 0AH
;----------------------------
      .CODE
MAIN PROC   FAR
     MOV    AX,@DATA
     MOV    DS,AX
     CALL   CLEAR        ;CLEAR THE SCREEN
     CALL   CURSOR       ;SET CURSOR POSITION
     MOV    AH,09        ;DISPLAY THE PROMPT
     MOV    DX,OFFSET PROMPT1
     INT    21H
     MOV    AH,07        ;GET ONE CHAR, NO ECHO
     INT    21H
     CMP    AL,'Y'       ;IF 'Y', CONTINUE
     JZ     OVER
     CMP    AL,'y'
     JZ     OVER
     MOV    AH,09        ;DISPLAY SECOND PROMPT IF NOT Y
     MOV    DX,OFFSET PROMPT2
     INT    21H
     JMP    EXIT
OVER:MOV    AH,09        ;DISPLAY THE MESSAGE
     MOV    DX,OFFSET MESSAGE
     INT    21H
EXIT:MOV    AH,4CH
     INT    21H          ;GO BACK TO DOS
MAIN        ENDP
;-----------------------------
CLEAR PROC               ;CLEARS THE SCREEN
     MOV    AX,0600H
     MOV    BH,07
     MOV    CX,0000
     MOV    DX,184FH
     INT    10H
     RET
CLEAR       ENDP
;-----------------------------
CURSOR PROC                    ;SET CURSOR POSITION
     MOV    AH,02
     MOV    BH,00
     MOV    DL,05        ;COLUMN 5
     MOV    DH,08        ;ROW 8
     INT    10H
     RET
CURSOR ENDP
     END    MAIN
```

Program 4-4

```
TITLE PROGRAM 4-4 REWRITTEN WITH FULL SEGMENT DEFINITION
PAGE 60,132
STSEG SEGMENT
      DB 64 DUP (?)
STSEG ENDS
;----------------------------
DTSEG SEGMENT
PROMPT1     DB    'There is a message for you from Mr. Jones. '
            DB    'To read it enter Y','$'
MESSAGE     DB    CR,LF,'Hi! I must leave town tomorrow, '
            DB    'therefore I will not be able to see you','$'
PROMPT2     DB    CR,LF,'No more messages for you','$'
DTSEG  ENDS
CR          EQU 0DH
LF          EQU 0AH
;----------------------------
CDSEG SEGMENT
MAIN PROC  FAR
     ASSUME CS:CDSEG,DS:DTSEG,SS:STSEG
     MOV   AX,DTSEG
     MOV   DS,AX
     CALL  CLEAR        ;CLEAR THE SCREEN
     CALL  CURSOR       ;SET CURSOR POSITION
     MOV   AH,09        ;DISPLAY THE PROMPT
     MOV   DX,OFFSET PROMPT1
     INT   21H
     MOV   AH,07        ;GET ONE CHAR, NO ECHO
     INT   21H
     CMP   AL,'Y'       ;IF 'Y', CONTINUE
     JZ    OVER
     CMP   AL,'y'
     JZ    OVER
     MOV   AH,09        ;DISPLAY SECOND PROMPT IF NOT Y
     MOV   DX,OFFSET PROMPT2
     INT   21H
     JMP   EXIT
OVER: MOV  AH,09        ;DISPLAY THE MESSAGE
     MOV   DX,OFFSET MESSAGE
     INT   21H
EXIT: MOV  AH,4CH
     INT   21H          ;GO BACK TO DOS
MAIN       ENDP
;-----------------------------
CLEAR PROC              ;CLEARS THE SCREEN
     MOV   AX,0600H
     MOV   BH,07
     MOV   CX,0000
     MOV   DX,184FH
     INT   10H
     RET
CLEAR      ENDP
;-----------------------------
CURSOR PROC                   ;SET CURSOR POSITION
     MOV   AH,02
     MOV   BH,00
     MOV   DL,05        ;COLUMN 5
     MOV   DH,08        ;ROW 8
     INT   10H
     RET
CURSOR     ENDP
CDSEG      ENDS
     END   MAIN
```

Program 4-4 (rewritten with full segment definition)

Next we show how to use this directive to define a buffer area for the string keyboard input:

```
DATA_BUF      LABEL BYTE
MAX_SIZE      DB    10
BUF_COUNT     DB    ?
BUF_AREA      DB    10 DUP(20H)
```

Now in the code segment the data can be accessed by name as follows:

```
      MOV    AH,0AH         ;load string into buffer
      MOV    DX,OFFSET DATA_BUF
      INT    21H
      MOV    CL,BUF_COUNT;load the actual length of string
      MOV    SI,OFFSET BUF_AREA;SI=address of first byte of string
```

This is much more structured and easier to follow. By using this method, it is easy to refer to any parameter by its name. For example, using the LABEL directive, one can rewrite the CONVERT subroutine in Program 4-2 as follows:

```
      ;In that data segment the BUFFER is redefined as
BUFFER        LABEL BYTE
BUFSIZE       DB    22
BUFCOUNT      DB    ?
REALDATA      DB    22 DUP(' ')
      ;and in the code segment, in place of the CONVERT procedure:
CONVERT PROC
      MOV CL,BUFCOUNT;load the counter
      SUB CH,CH;CX=counter
      MOV DI,CX;index into data field
      MOV BX,OFFSET REALDATA  ;actual data address in buffer
      MOV BYTE PTR[ BX+DI] ,20H ;replace the CR with space
      MOV SI,OFFSET DATAREA   ;SI=address of converted data
AGAIN:MOV AL,[ BX]            ;move the char into AL
      CMP AL,61H              ;check if is below 'a'
      JB NEXT                 ;if yes then go to next
      CMP AL,7AH              ;check for above 'z'
      JA NEXT                 ;if yes then go to next
      AND AL,11011111B        ;if not then mask it to capital
NEXT: MOV [ SI] ,AL           ;move the character
      INC SI                  ;increment the pointer
      INC BX                  ;increment the pointer
      LOOP AGAIN              ;repeat if CX not zero yet
      RET                     ;return to main procedure
CONVERT ENDP
```

Review Questions

1. INT ____ function calls reside in ROM BIOS, whereas INT ___ function calls are provided by DOS.
2. What is the difference between the following two programs?

```
      MOV AH,09                   MOV AH,0AH
      MOV DX,OFFSET BUFFER        MOV DX,OFFSET BUFFER
      INT 21H                     INT 21H
```

3. INT 21H function 09 will display a string of data beginning at the location specified in register DX. How does the system know where the end of the string is?
4. Fill in the blanks to display the following string using INT 21H.

```
      MESSAGE1     DB         'What is your last name?$'
            MOV    AH,_____
            MOV    DX,_____
            INT       21H
```

5. The following prompt needs to be displayed. What will happen if this string is output using INT 21H function 09?
```
PROMPT1      DB 'Enter (to nearest $) your annual salary'
```
6. Use the EQU directive to equate the name "BELL" with the ASCII code for sounding the bell.
7. Write a program to sound the bell.
8. Code the data definition directives for a buffer area where INT 21H Option 0AH will input a social security number.

SECTION 4.3: WHAT IS A MACRO AND HOW IS IT USED?

There are applications in Assembly language programming where a group of instructions performs a task that is used repeatedly. For example, INT 21H function 09H for displaying a string of data and function 0AH for keying in data are used repeatedly in the same program. So it does not make sense to rewrite them every time they are needed. Therefore, to reduce the time that it takes to write these codes and reduce the possibility of errors, the concept of macros was born. Macros allow the programmer to write the task (set of codes to perform a specific job) once only and to invoke it whenever it is needed, wherever it is needed.

MACRO definition

Every macro definition must have three parts, as follows:

```
name   MACRO  dummy1,dummy2,...,dummyN
...    ...
...    ...
       ENDM
```

The MACRO directive indicates the beginning of the macro definition and the ENDM directive signals the end. What goes in between the MACRO and ENDM directives is called the body of the macro. The name must be unique and must follow Assembly language naming conventions. The dummies are names, or parameters, or even registers that are mentioned in the body of the macro. After the macro has been written, it can be invoked (or called) by its name, and appropriate values are substituted for dummy parameters. Displaying a string of data using function 09 of INT 21H is a widely used service. The following is a macro for that service:

```
STRING         MACRO DATA1
               MOV   AH,09
               MOV   DX,OFFSET DATA1
               INT   21H
               ENDM
```

The above code is the macro definition. Note that dummy argument DATA1 is mentioned in the body of the macro. In the following example, assume that a prompt has already been defined in the data segment as shown below In the code segment, the macro can be invoked by its name with the user's actual data:

```
MESSAGE1      DB           'What is your name?','$'
              ...
              STRING       MESSAGE1
```

The instruction "`STRING MESSAGE1`" invokes the macro.

The assembler expands the macro by providing the following code in the .LST file:

```
1 MOV   AH,09
1 MOV   DX,OFFSET MESSAGE1
1 INT   21H
```

The (1) indicates that the code is from the macro. In earlier versions of MASM, a plus sign (+) indicated lines from macros.

Example 4-6 demonstrates the macro definition.

Example 4-6

Write macro definitions for setting the cursor position, displaying a string, and clearing the screen.

Solution:

```
CURSOR MACRO ROW,COLUMN
 ;THIS MACRO SETS THE CURSOR LOCATION TO ROW,COLUMN
 ;;USING BIOS INT 10H FUNCTION 02
 MOV   AH,02             ;SET CURSOR FUNCTION
 MOV BH,00          ;PAGE 00
 MOV DH,ROW ;ROW POSITION
 MOV DL,COLUMN      ;COLUMN POSITION
 INT 10H            ;INVOKE THE INTERRUPT
 ENDM

 DISPLAY MACRO STRING
 ;THIS MACRO DISPLAYS A STRING OF DATA
 ;;DX = ADDRESS OF STRING. USES FUNCTION 09 INT 21H.
 MOV AH,09                ;DISPLAY STRING FUNCTION
 MOV DX,OFFSET STRING ;DX = OFFSET ADDRESS OF DATA
 INT 21H                  ;INVOKE THE INTERRUPT
 ENDM

 CLEARSCR MACRO
 ;THIS MACRO CLEARS THE SCREEN
 ;;USING OPTION 06 OF INT 10H
 MOV AX,0600H       ;SCROLL SCREEN FUNCTION
 MOV BH,07          ;NORMAL ATTRIBUTE
 MOV CX,0           ;FROM ROW=00,COLUMN=00
 MOV DX,184FH       ;TO ROW=18H,COLUMN=4FH
 INT 10H            ;INVOKE THE INTERRUPT
 ENDM
```

Remember that the comments marked with ";;" will not be listed in the list file as seen in the list file for Program 4-5.

Comments in a macro

Now the question is: Can macros contain comments? The answer is yes, but there is a way to suppress comments and make the assembler show only the lines that generate opcodes. There are basically two types of comments in the macro: listable and non-listable. If comments are preceded by a single semicolon (;) as is done in Assembly language programming, they will show up in the ".lst" file, but if comments are preceded by a double semicolon (;;) they will not appear in the ".lst" file when the program is assembled. There are also three directives designed to make programs that use macros more readable, meaning that they only affect the ".lst" file and have no effect on the ".obj" or

Using the macro definition in Example 4-6, write a program that clears the screen and then at each of the following screen locations displays the indicated message:

at row 2 and column 4 "My name" at row 12 and column 44 "What is"
at row 7 and column 24 "is Joe" at row 19 and column 64 "your name?"

```
TITLE      PROG4-5
PAGE       60,132
;-------------------
CLEARSCR   MACRO
;THIS MACRO CLEARS THE SCREEN
;;USING OPTION 06 OF INT 10H
     MOV   AX,0600H    ;SCROLL SCREEN FUNCTION
     MOV   BH,07       ;NORMAL ATTRIBUTE
     MOV   CX,0        ;FROM ROW=00,COLUMN=00
     MOV   DX,184FH    ;TO ROW=18H,COLUMN=4FH
     INT   10H         ;INVOKE THE INTERRUPT
     ENDM
;---------------------
DISPLAY    MACRO STRING
;THIS MACRO DISPLAYS A STRING OF DATA
;;DX = ADDRESS OF STRING. USES FUNCTION 09 INT 21H.
     MOV   AH,09              ;DISPLAY STRING FUNCTION
     MOV   DX,OFFSET STRING   ;DX = OFFSET ADDRESS OF DATA
     INT   21H                ;INVOKE THE INTERRUPT
     ENDM
;----------------------
CURSOR     MACRO ROW,COLUMN
;THIS MACRO SETS THE CURSOR LOCATION TO ROW,COLUMN
;;USING BIOS INT 10H FUNCTION 02
     MOV   AH,02       ;SET CURSOR FUNCTION
     MOV   BH,00       ;PAGE 00
     MOV   DH,ROW      ;ROW POSITION
     MOV   DL,COLUMN   ;COLUMN POSITION
     INT   10H         ;INVOKE THE INTERRUPT
     ENDM
;---------------------
     .MODEL SMALL
     .STACK 64
     .DATA
MESSAGE1 DB `My name `,'$'
MESSAGE2 DB `is Joe','$'
MESSAGE3 DB `What is `,'$'
MESSAGE4 DB `your name?','$'
     .CODE
MAIN PROC  FAR
     MOV   AX,@DATA
     MOV   DS,AX
     .LALL                 ;LIST ALL
     CLEARSCR              ;INVOKE CLEAR SCREEN MACRO
     CURSOR 2,4            ;SET CURSOR TO ROW2,COL 2
     DISPLAY MESSAGE1      ;INVOKE DISPLAY MACRO
     .XALL                 ;LIST ALL EXECUTABLE
     CURSOR 7,24           ;SET CURSOR TO ROW 7,COL 24
     DISPLAY MESSAGE2      ;INVOKE DISPLAY MACRO
     .SALL                 ;SUPPRESS ALL
     CURSOR 12,44          ;SET CURSOR TO ROW 12,COL 44
     DISPLAY MESSAGE3      ;INVOKE DISPLAY MACRO
     CURSOR 19,64          ;SET CURSOR TO ROW 19,COL 64
     DISPLAY MESSAGE4      ;INVOKE DISPLAY MACRO
     MOV  AH,4CH
     INT  21H              ;GO BACK TO DOS
MAIN ENDP
     END    MAIN
```

Program 4-5

".exe" files. They are as follows:

.LALL (List ALL) will list all the instructions and comments that are preceded by a single semicolon in the ".lst" file. The comments preceded by a double semicolon cannot be listed in the ".lst" file in any way.

.SALL (Suppress ALL) is used to make the list file shorter and easier to read. It suppresses the listing of the macro body and the comments. This is especially useful if the macro is invoked many times within the same program and there is no need to see it listed every time. It must be emphasized that the use of .SALL will not eliminate any opcode from the object file. It only affects the listing in the ".lst" file.

.XALL (eXecutable ALL), which is the default listing directive, is used to list only the part of the macro that generates opcodes.

Analysis of Program 4-5

Compare the ".asm" and ".lst" files to see the use of .LALL, .XALL, and .SALL. The .LALL directive was used for each macro and then .XALL was used for two of them. From then on, all were suppressed.

```
Microsoft (R) Macro Assembler Version 5.10 1/13/92 00:17:15
PROG4-5                                                    Page 1-1
            TITLE PROG4-5
            PAGE  60,132
            ;----------------------
            CLEARSCR   MACRO
            ;THIS MACRO CLEARS THE SCREEN
            ;;USING OPTION 06 OF INT 10H
                   MOV  AX,0600H     ;SCROLL SCREEN FUNCTION
                   MOV  BH,07        ;NORMAL ATTRIBUTE
                   MOV  CX,0         ;FROM ROW=00,COLUMN=00
                   MOV  DX,184FH     ;TO ROW=18H,COLUMN=4FH
                   INT  10H          ;INVOKE THE INTERRUPT
                   ENDM
            ;------------------------
            DISPLAY MACRO STRING
            ;THIS MACRO DISPLAYS A STRING OF DATA
            ;;DX = ADDRESS OF STRING. USES FUNCTION 09 INT 21H.
                   MOV  AH,09         ;DISPLAY STRING FUNCTION
                   MOV  DX,OFFSET STRING ;DX = OFFSET ADDRESS OF DATA
                   INT  21H           ;INVOKE THE INTERRUPT
                   ENDM
            ;---------------------------
            CURSOR MACRO ROW,COLUMN
            ;THIS MACRO SETS THE CURSOR LOCATION TO ROW,COLUMN
            ;;USING BIOS INT 10H FUNCTION 02
                   MOV  AH,02        ;SET CURSOR FUNCTION
                   MOV  BH,00        ;PAGE 00
                   MOV  DH,ROW       ;ROW POSITION
                   MOV  DL,COLUMN    ;COLUMN POSITION
                   INT  10H          ;INVOKE THE INTERRUPT
                   ENDM
```

List File for Program 4-5 *(continued on the next page)*

LOCAL directive and its use in macros

In the discussion of macros so far, examples have been chosen that do not have a label or name in the body of the macro. This is because if a macro is expanded more than once in a program and there is a label in the label field of the body of the macro, these labels must be declared as LOCAL. Otherwise, an assembler error would be generated when the same label was encountered in two or more places.

```
Microsoft (R) Macro Assembler Version 5.10 1/13/92 00:17:15
PROG4-5                                                    Page 1-2

 0000                              .MODEL SMALL
 0000                              .STACK 64
                                   ;---------------------
 0000                              .DATA
 0000 4D 79 20 6E 61 6D            MESSAGE1 DB `My name `,'$'
      65 20 24
 0009 69 73 20 4A 6F 65            MESSAGE2 DB `is Joe','$'
 0010 57 68 61 74 20 69            MESSAGE3 DB `What is `,'$'
      73 20 24
 0019 79 6F 75 72 20 6E            MESSAGE4 DB `your name?','$'
      61 6D 65 3F 24
                                   ;-----------------------
 0024                              .CODE
 0000                              MAIN   PROC   FAR
 0000 B8 0000s                            MOV    AX,@DATA
 0003 8E D8                               MOV    DS,AX
                                          .LALL         ;LIST ALL
                                          CLEARSCR      ;INVOKE CLEAR SCREEN MACRO

                              1 ;THIS MACRO CLEARS THE SCREEN
                              1 ;
 0005 B8 0600                 1       MOV AX,0600H      ;SCROLL SCREEN FUNCTION
 0008 B7 07                   1       MOV BH,07         ;NORMAL ATTRIBUTE
 000A B9 0000                 1       MOV CX,0          ;FROM ROW=00,COLUMN=00
 000D BA 184F                 1       MOV DX,184FH      ;TO ROW=18H,COLUMN=4FH
 0010 CD 10                   1       INT 10H           ;INVOKE THE INTERRUPT
                                      CURSOR 2,4 ;CURSOR MACRO WILL SET CURSOR
                              1 ;THIS MACRO SETS THE CURSOR LOCATION
                              1 ;
 0012 B4 02                   1       MOV AH,02         ;SET CURSOR FUNCTION
 0014 B7 00                   1       MOV BH,00         ;PAGE 00
 0016 B6 02                   1       MOV DH,2          ;ROW POSITION
 0018 B2 04                   1       MOV DL,4          ;COLUMN POSITION
 001A CD 10                   1       INT 10H           ;INVOKE THE INTERRUPT
                                      DISPLAY MESSAGE1 ;INVOKE DISPLAY MACRO
                              1 ;THIS MACRO DISPLAYS A STRING OF DATA
                              1 ;
 001C B4 09                   1       MOV AH,09  ;DISPLAY STRING FUNCTION
 001E BA 0000 R               1       MOV DX,OFFSET MESSAGE1 ;DX = OFFSET ADDR
 0021 CD 21                   1       INT 21H           ;INVOKE THE INTERRUPT
                                      .XALL             ;LIST ALL EXECUTABLE
                                      CURSOR 7,24       ;SET CURSOR TO ROW=7,COL=
 0023 B4 02                   1       MOV AH,02         ;SET CURSOR FUNCTION
 0025 B7 00                   1       MOV BH,00         ;PAGE 00
 0027 B6 07                   1       MOV DH,7          ;ROW POSITION
 0029 B2 18                   1       MOV DL,24         ;COLUMN POSITION
 002B CD 10                   1       INT 10H           ;INVOKE THE INTERRUPT
                                      DISPLAY MESSAGE2 ;INVOKE DISPLAY MACRO
 002D B4 09                   1       MOV AH,09         ;DISPLAY STRING FUNCTION
 002F BA 0009 R               1       MOV DX,OFFSET MESSAGE2 ;DX = OFFSET ADDR
 0032 CD 21                   1       INT 21H           ;INVOKE THE INTERRUPT
                                      .SALL             ;SUPPRESS ALL
                                      CURSOR 12,44      ;SET CURSOR TO
                                      DISPLAY MESSAGE3 ;INVOKE DISPLAY MACRO
                                      CURSOR 19,64      ;SET CURSOR TO
                                      DISPLAY MESSAGE4 ;INVOKE DISPLAY MACRO
 0056 B4 4C                           MOV AH,4CH
 0058 CD 21                           INT 21H    ;GO BACK TO DOS
 005A                         MAIN    ENDP
                                      END MAIN
```

List File for Program 4-5 *(continued from the preceding page)*

The following rules must be observed in the body of the macro:

1. All labels in the label field must be declared LOCAL.
2. The LOCAL directive must be right after the MACRO directive. In other words, it must be placed even before comments and the body of the macro; otherwise, the assembler gives an error.
3. The LOCAL directive can be used to declare all names and labels at once as follows:

```
LOCAL name1,name2,name3
```

or one at a time as:

```
LOCAL name1
LOCAL name2
LOCAL name3
```

To clarify these points, look at Example 4-7.

Example 4-7

Write a macro that multiplies two words by repeated addition, then saves the result.

Solution:

The following macro can be expanded as often as desired in the same program since the label "back" has been declared as LOCAL.

```
MULTIPLY MACRO   VALUE1, VALUE2, RESULT
      LOCAL BACK
;     THIS MACRO COMPUTES RESULT = VALUE1 X VALUE2
;;    BY REPEATED ADDITION
;;VALUE1 AND VALUE2 ARE WORD OPERANDS; RESULT IS A DOUBLEWORD
            MOV    BX,VALUE1     ;BX=MULTIPLIER
            MOV    CX,VALUE2     ;CX=MULTIPLICAND
            SUB    AX,AX         ;CLEAR AX
            MOV    DX,AX         ;CLEAR DX
BACK:       ADD    AX,BX         ;ADD BX TO AX
            ADC    DX,00         ;ADD CARRIES IF THERE IS ONE
            LOOP   BACK          ;CONTINUE UNTIL CX=0
            MOV    RESULT,AX     ;SAVE THE LOW WORD
            MOV    RESULT+2,DX   ;SAVE THE HIGH WORD
            ENDM
```

Notice in Example 4-7 that the "BACK" label is defined as LOCAL right after the MACRO directive. Defining this anywhere else causes an error. The use of a LOCAL directive allows the assembler to define the labels separately each time it encounters them. The list file below shows that when the macro is expanded for the first time, the list file has "??0000". For the second time it is "??0001", and for the third time it is "??0002" in place of the "BACK" label, indicating that the label "BACK" is local. To clarify this concept, try Example 4-7 without the LOCAL directive to see how the assembler will give an error.

INCLUDE directive

Assume that there are several macros that are used in every program. Must they be rewritten every time? The answer is no if the concept of the INCLUDE directive is known. The INCLUDE directive allows a programmer to write macros and save them in a file, and later bring them into any file. For example, assume that the following widely used macros were written and then saved under the filename "MYMACRO1.MAC".

Use the macro definition in Example 4-7 to write a program that multiplies the following:
(1) 2000 × 500 (2) 2500 × 500 (3) 300 × 400

```
TITLE PROG4-6
PAGE 60,132
;--------------------------------------
MULTIPLY MACRO VALUE1, VALUE2, RESULT
LOCAL BACK
;THIS MACRO COMPUTES RESULT = VALUE1 X VALUE2
;;BY REPEATED ADDITION
;;VALUE1 AND VALUE2 ARE WORD OPERANDS; RESULT IS A DOUBLEWORD
     MOV   BX,VALUE1    ;BX=MULTIPLIER
     MOV   CX,VALUE2    ;CX=MULTIPLICAND
     SUB   AX,AX        ;CLEAR AX
     MOV   DX,AX        ;CLEAR DX
BACK: ADD  AX,BX        ;ADD BX TO AX
     ADC   DX,00        ;ADD CARRIES IF THERE IS ONE
     LOOP  BACK         ;CONTINUE UNTIL CX=0
     MOV   RESULT,AX    ;SAVE THE LOW WORD
     MOV   RESULT+2,DX  ;SAVE THE HIGH WORD
     ENDM
;-------------------
           .MODEL SMALL
           .STACK 64
;--------------------
           .DATA
RESULT1    DW    2 DUP (0)
RESULT2    DW    2 DUP (0)
RESULT3    DW    2 DUP (0)
;--------------------
     .CODE
MAIN PROC  FAR
     MOV   AX,@DATA
     MOV   DS,AX
     MULTIPLY 2000,500,RESULT1
     MULTIPLY 2500,500,RESULT2
     MULTIPLY 300,400,RESULT3
     MOV   AH,4CH
     INT   21H          ;GO BACK TO DOS
MAIN ENDP
     END   MAIN
```

Program 4-6

```
Microsoft (R) Macro Assembler Version 5.10 1/13/92 00:33:14
PROG4-6                                                     Page 1-1

                    TITLE PROG4-6
                    PAGE 60,132
                    ;--------------------
                          MULTIPLY MACRO VALUE1, VALUE2, RESULT
                          LOCAL BACK
                    ;THIS MACRO COMPUTES RESULT = VALUE1 X VALUE2
                    ;;BY REPEATED ADDITION
                    ;;VALUE1 AND VALUE2 ARE WORD OPERANDS; RESULT IS
                          MOV BX,VALUE1      ;BX=MULTIPLIER
                          MOV CX,VALUE2      ;CX=MULTIPLICAND
                          SUB AX,AX          ;CLEAR AX
                          MOV DX,AX          ;CLEAR DX
                    BACK: ADD AX,BX          ;ADD BX TO AX
                          ADC DX,00    ;ADD CARRIES IF THERE IS ONE
                          LOOP BACK          ;CONTINUE UNTIL CX=0
                          MOV RESULT,AX      ;SAVE THE LOW WORD
                          MOV RESULT+2,DX    ;SAVE THE HIGH WORD
                          ENDM
```

List File for Program 4-6 *(continued on the next page)*

```
Microsoft (R) Macro Assembler Version 5.10 1/13/92 00:33:14
PROG4-6                                            Page 1-2

                     ;-----------------
 0000                        .MODEL SMALL
 0000                        .STACK 64
 0040                        STSEG       ENDS
                     ;-----------------
 0000                        .DATA
 0000 0002 [                 RESULT1 DW 2 DUP (0)
 0000
 ]

 0004 0002 [                 RESULT2 DW 2 DUP (0)
 0000
 ]

 0008 0002 [                 RESULT3 DW 2 DUP (0)
                             ;-------------------
 000C                        .CODE
 0000                MAIN PROC   FAR
                             ASSUME CS:CDSEG,DS:DTSEG,SS:STSEG
 0000 B8 0000s               MOV   AX,@DATA
 0003 8E D8                  MOV   DS,AX
                             MULTIPLY 2000,500,RESULT1
 0005 BB 07D0  1             MOV   BX,2000          ;BX=MULTIPLIER
 0008 B9 01F4  1             MOV   CX,500           ;CX=MULTIPLICAND
 000B 2B C0    1             SUB   AX,AX       ;CLEAR AX
 000D 8B D0    1             MOV   DX,AX       ;CLEAR DX
 000F 03 C3    1 ??0000:     ADD   AX,BX       ;ADD BX TO AX
 0011 83 D2 00 1             ADC   DX,00       ;ADD CARRIES IF THERE
 0014 E2 F9    1             LOOP  ??0000      ;CONTINUE UNTIL CX=0
 0016 A3 0000 R 1            MOV   RESULT1,AX  ;SAVE THE LOW WORD
 0019 89 16 0002 R 1         MOV   RESULT1+2,DX;SAVE THE HIGH WORD
                             MULTIPLY 2500,500,RESULT2
 001D BB 09C4  1             MOV   BX,2500          ;BX=MULTIPLIER
 0020 B9 01F4  1             MOV   CX,500           ;CX=MULTIPLICAND
 0023 2B C0    1             SUB   AX,AX       ;CLEAR AX
 0027 03 C3    1??0001:      ADD   AX,BX       ;ADD BX TO AX
 0029 83 D2 00 1             ADC   DX,00       ;ADD CARRIES IF THERE
 002C E2 F9    1             LOOP  ??0001      ;CONTINUE UNTIL CX=0
 002E A3 0004 R 1            MOV   RESULT2,AX  ;SAVE THE LOW WORD
 0031 89 16 0006 R 1         MOV   RESULT2+2,DX ;SAVE THE HIGH WORD
                             MULTIPLY 300,400,RESULT3
 0035 BB 012C  1             MOV   BX,300           ;BX=MULTIPLIER
 0038 B9 0190  1             MOV   CX,400           ;CS=MULTIPLICAND
 003B 2B C0    1             SUB   AX,AX       ;CLEAR AX
 003D 8B D0    1             MOV   DX,AX       ;CLEAR DX
 003F 03 C3    1??0002:      ADD   AX,BX       ;ADD BX TO AX
 0041 83 D2 00 1             ADC   DX,00       ;ADD CARRIES IF THERE
 0044 E2 F9    1             LOOP  ??0002           ;CONTINUE UNTIL
 0046 A3 0008 R 1            MOV   RESULT3,AX  ;SAVE THE LOW WORD
 0049 89 16 000A R 1         MOV   RESULT3+2,DX ;SAVE THE HIGH WORD
 004D B4 4C                  MOV   AH,4CH
 004F CD 21                  INT   21H   ;GO BACK TO DOS
 0051                MAIN    ENDP
                             END   MAIN
```

List File for Program 4-6 *(continued from the preceding page)*

```
CLEARSCR    MACRO              ;the clear screen macro
            MOV   AX,0600H
            MOV   BH,07
            MOV   CX,0000
            MOV   DX,184FH
            INT   10H
            ENDM
DISPLAY     MACRO  STRING      ;the string display macro
            MOV   AH,09
            MOV   DX,OFFSET STRING
            INT  21H
            ENDM
REGSAVE     MACRO         ;this macro saves all the registers
            PUSH  AX
            PUSH  BX
            PUSH  CX
            PUSH  DX
            PUSH  DI
            PUSH  SI
            PUSH  BP
            PUSHF
            ENDM
REGRESTO    MACRO         ;this macro restores all the registers
            POPF
            POP BP
            POPSI
            POPDI
            POPDX
            POPCX
            POPBX
            POPAX
            ENDM
```

Assuming that these macros are saved on a disk under the filename "MYMACRO1.MAC", the INCLUDE directive can be used to bring this file into any ".asm" file and then the program can call upon any of the macros as many times as needed. When a file includes all macros, the macros are listed at the beginning of the ".lst" file and as they are expanded, they will be part of the program. To understand this, see Program 4 -7. Program 4-7 includes macros to clear the screen, set the cursor, and display strings. These macros are all saved under the "MYMACRO2.MAC" filename. The ".asm" and ".lst" versions of the program that use the clear screen and display string macros only to display "This is a test of macro concepts" are shown on the following pages.

```
TITLE       PROG4-7
PAGE        60,132
            INCLUDE MYMACRO2.MAC
            .MODEL SMALL
            .STACK 64
;--------------------------
            .DATA
MESSAGE1 DB `This is a test of macro concepts'.'$'
            .CODE
MAIN        PROC  FAR
            MOV   AX,@DATA
            MOV   DS,AX
            CLEARSCR          ;INVOKE CLEAR SCREEN MACRO
            DISPLAY MESSAGE1  ;INVOKE DISPLAY MACRO
            MOV   AH,4CH
            INT   21H         ;GO BACK TO DOS
MAIN        ENDP
            END   MAIN
```

Program 4-7

```
Microsoft (R) Macro Assembler Version 5.10 1/13/92 00:41:49
PROG4-7                                                  Page 1-1

                TITLE PROG4-7
                PAGE 60,132
                ;
                INCLUDE MYMACRO2.MAC
            C ; MYMACRO2 (MAC) FOR PROGRAM5-3
            C ;-------------------------
            C CURSOR MACRO ROW,COLUMN
            C ;THIS MACRO SETS THE CURSOR LOCATION AT ROW,COLUMN
            C ;;USING BIOS INT 10H FUNCTION 02
            C     MOV AH,02          ;SET CURSOR FUNCTION
            C     MOV BH,00          ;PAGE 00
            C     MOV DH,ROW         ;ROW POSITION
            C     MOV DL,COLUMN      ;COLUMN POSITION
            C     INT 10H            ;INVOKE THE INTERRUPT
            C     ENDM
            C ;-----------------------------
            C DISPLAY MACRO STRING
            C ;THIS MACRO DISPLAYS A STRING OF DATA
            C ;;DX = ADDRESS OF STRING. USES FUNCTION 09 INT 21H.
            C     MOV AH,09                ;DISPLAY STRING FUNCTION
            C     MOV DX,OFFSET STRING     ;DX = OFFSET ADDRESS OF DATA
            C     INT 21H                  ;INVOKE THE INTERRUPT
            C     ENDM
            C ;----------------------------
            C CLEARSCR MACRO
            C ;THIS MACRO CLEARS THE SCREEN
            C ;;USING OPTION 06 OF INT 10H
            C     MOV AX,0600H       ;SCROLL SCREEN FUNCTION
            C     MOV BH,07          ;NORMAL ATTRIBUTE
                        C      MOV CX,0        ;FROM ROW=00,COLUMN=00
            C     MOV DX,184FH             ;TO ROW=18H,COLUMN=4FH
            C     INT 10H            ;INVOKE THE INTERRUPT
            C     ENDM
            C ;-------------------------
 0000                   .MODEL SMALL
 0000                   .STACK 64
                        ;-----------------
Microsoft (R) Macro Assembler Version 5.10 1/13/92 00:41:49
PROG4-7                                                  Page 1-2

 0000                   .DATA
 0000 54 68 69 73 20 69 MESSAGE1 DB `This is a test of macro concepts','$'
 73 20 61 20 74 65
 73 74 20 6F 66 20
 6D 61 63 72 6F 20
 63 6F 6E 63 65 70
 74 73 24
                         ;---------------------
 0000                    .CODE
 0000                   MAIN  PROC FAR
 0000 B8 0000s                MOV   AX,@DATA
 0003 8E D8                   MOV   DS,AX
                              CLEARSCR     ;INVOKE CLEAR SCREEN MACRO
 0005 B8 0600                 MOV   AX,0600H
```

List File for Program 4-7 *(continued on the next page)*

```
0008 B7 07          1          MOV BH,07           ;NORMAL ATTRIBUTE
000A B9 0000        1          MOV CX,0            ;FROM ROW=00,COLUMN=00
000D BA 184F        1          MOV DX,184FH        ;TO ROW=18H,COLUMN=4FH
0010 CD 10          1          INT 10H             ;INVOKE THE INTERRUPT
                               DISPLAY MESSAGE1    ;INVOKE DISPLAY MACRO
0012 B4 09          1          MOV AH,09           ;DISPLAY STRING
0014 BA 0000 R      1          MOV DX,OFFSET MESSAGE1 ;DX =OFFSET ADDR
0017 CD 21          1          INT 21H             ;INVOKE THE INTERRUPT
0019 B4 4C                     MOV AH,4CH
001B CD 21                     INT 21H             ;GO BACK TO DOS
001D                      MAIN ENDP
                               END MAIN
```

List File for Program 4-7 *(continued from the preceding page)*

Notice that in the list file of Program 4-7, the letter "C" in front of the lines indicates that they are copied from another file and included in the present file.

Review Questions

1. Discuss the benefits of macro programming.
2. List the three parts of a macro.
3. Explain and contrast macro definition, invoking the macro, and expanding the macro.
4. True or false. A label defined within a macro is automatically understood by the assembler to be local.
5. True or false. In the list file for Program 4-7, the "C" at the beginning of a line indicates that it is a comment.

PROBLEMS

SECTION 4.1: BIOS INT 10H PROGRAMMING

1. Write a program that:
   (a) Clears the screen, and (b) sets the cursor position at row = 5 and column = 12.
2. What is the function of the following program?
```
MOV   AH,02
MOV   BH,00
MOV   DL,20
MOV   DH,10
INT      10H
```
3. The following program is meant to set the cursor at position row = 14 and column = 20. Fix the error and run the program to verify your solution.
```
MOV   AH,02
MOV   BH,00
MOV   DH,14H
MOV   DL,20H
INT      10H
```
4. Write a program that sets the cursor at row = 12, column = 15, then use the code below to get the current cursor position in register DX with DH = row and DL = column. Is the cursor position in DH and DL in decimal or hex? Verify your answer.
```
MOV   AH,03
MOV   BH,00
INT   10H
```
5. In clearing the screen, does the sequence of code prior to INT 10H matter? In setting a cursor position? Verify by rearranging and executing the instructions.

6. You want to clear the screen using the following program, but there are some errors. Fix the errors and run the program to verify it.

```
MOV   AX,0600H
MOV   BH,07
MOV   CX,0000
MOV   DX,184F
INT   10H
```

7. Write a program that:
(a) Clears the screen
(b) Sets the cursor at row = 8 and column = 14
(c) Displays the string "IBM Personal Computer"

SECTION 4.2: DOS INTERRUPT 21H

8. Run the following program and dump the memory to verify the contents of memory locations 0220H to 022FH if "IBM PC with 8088 CPU" is keyed in.

```
ORG             220H
BUFFER          DB        15, 16 DUP (0FFH)
```

and for the code:

```
MOV   AH,0AH
MOV   DX,OFFSET BUFFER
INT   21H
```

9. Write a program that:
(a) Clears the screen.
(b) Puts the cursor on position row = 15 and column = 20.
(c) Displays the prompt "What is your name?"
(d) Gets a response from the keyboard and displays it at row = 17 and column = 20.
10. Write a program that sets the mode to medium resolution, draws a vertical line in the middle of the screen, then draws a horizontal line across the middle of the screen.
11. Write a program to input a social security number in the form 123-45-6789 and transfer it to another area with the hyphens removed, as in 123456789. Use the following data definition.

```
SS_AREA       LABEL   BYTE
SS_SIZE       DB      12
SS_ACTUAL     DB      ?
SS_DASHED     DB      12 DUP (?)
SS_NUM        DB      9 DUP (?)
```

12. Write a program (use the simplified segment definition) to input two seven-digit numbers in response to the prompts "Enter the first number" and "Enter the second number". Add them together and display the sum with the message "The total sum is: ".

SECTION 4.3: WHAT IS A MACRO AND HOW IS IT USED?

13. Every macro must start with directive ________ and end with directive ________.
14. Identify the name, body, and dummy argument in the following macro:

```
WORK_HOUR     MACRO     OVRTME_HR
              MOV       AL,40;WEEKLY HRS
              ADD       AL,OVRTME_HR          ;TOTAL HRS WORKED
              ENDM
```

15. Explain the difference between the .SALL, .LALL, and .XALL directives.
16. What is the total value in registers DX and AX after invoking the following macro? The macro is invoked as: `WAGES 60000,25000,3000`

```
WAGES MACRO SALARY,OVERTIME,BONUSES
;TOTAL WAGES=SALARY + OVERTIME + BONUSES
        SUB   AX,AX          ;CLEAR
```

```
MOV   DX,AX         ;AX AND DX
ADD   AX,SALARY
ADD   AX,OVERTIME
ADC   DX,0          ;TAKE CARE OF CARRY
ADD   AX,BONUSES
ADC   DX,0
ENDM
```

17. In Problem 16, in the body of the macro, dummies were used as they are listed from left to right. Can they be used in any order? Rewrite the body (leave the dummies alone) by adding OVERTIME first.
18. In Problem 16, state the comments that are listed if the macro is expanded as:

```
.LALL
WAGES X,Y,Z
```

19. Macros can use registers as dummies. Show the ".lst" file and explain what the macro in Problem 16 does if it is invoked as follows:

```
WAGES BX,CX,SI
```

20. Fill in the blanks for the following macro to add an array of bytes. Some blanks might not need to be filled.

```
SUMMING      MACROCOUNT,VALUES
LOCAL        ......
;;this macro adds an array of byte size elements.
;;ax will hold the total sum
MOV     CX,....                 ;size of array
MOV     SI,OFFSET .... .;load offset address of array
SUB     AX,AX                  ;clear AX
AGAIN: ADD   AL,[ SI]
ADC     AH,0               ;add bytes and takes care of carries
INC         SI                  ;point to next byte
LOOP    AGAIN                  ;continue until finished
ENDM   .....
```

21. Invoke and run the macro above for the following data.

```
;In the data segment
   DATA1 DB    89,75,98,91,65
   SUM1  DW  ?
   DATA2 DB    86,69,99,14,93,99,54,39
   SUM2  DW  ?
   DATA3 DB    10 DUP (99)
   SUM3  DW  ?
```

(Hint: For the format, see Problem 20.)

22. Insert the listing directives in Problem 21 as follows and analyze the ".lst" file. From the code segment:

```
.LALL
SUMMING      5,DATA1      ;adding and saving data1 MOVSUM1,AX
.XALL
SUMMING      ........     ;adding and saving data2
.....
.SALL                     ;adding and saving data3
.....
```

23. Rewrite Problem 20 to have a third dummy argument for SUM. Then rework Problem 19.
24. Rewrite Program 4-6 using the DD directive for RESULT1, RESULT2, and RESULT3.

ANSWERS TO REVIEW QUESTIONS

SECTION 4.1: BIOS INT 10H PROGRAMMING

1. Perform screen I/O
2. 80, 25; 00,00 and 24,79
3.
```
MOV AH,06          ;SELECT CLEAR SCREEN FUNCTION
MOV AL,00          ;AH=0 TO SCROLL ENTIRE PAGE
MOV BH,07          ;BH=07 FOR NORMAL ATTRIBUTE
MOV CH,00          ;START AT ROW 00
MOV CL,00          ;START AT COLUMN 00
MOV DH,24          ;END AT ROW 24
MOV DL,79          ;END AT ROW 79
INT 10H            ;INVOKE THE INTERRUPT
```
4. Indicates that the cursor is at row 5, column 34
5. It provides information about the foreground and background intensity, whether the foreground is blinking and/or highlighted.
6. Character
7. 10100111
8. The first time INT 10H is invoked, it sets the cursor to position 00,00; the second time it is invoked, it displays the character '*' 80 times with attributes of white on black, blinking.

SECTION 4.2: DOS INTERRUPT 21H

1. 10H, 21H
2. The rightmost code inputs a string from the keyboard into a buffer; the code on the left outputs a string from a buffer to the monitor.
3. The end of the string is the dollar sign '$'.
4. 0AH, OFFSET MESSAGE1
5. When the '$' within the string is encountered, the computer will stop displaying the string.
6. BELL EQU 07H
7. Using the EQU in Answer 6, the code segment would include the following:
```
MOV AH,02
MOV DL,BELL
INT 21H
```
8.
```
SS_AREA        LABEL BYTE
SS_SIZE        DB    12
SS_ACTUAL      DB    ?
SS_NUM         DB    12 DUP (?)
```

SECTION 4.3: WHAT IS A MACRO AND HOW IS IT USED?

1. Macro programming can save the programmer time by allowing a set of frequently repeated instructions to be invoked within the program by a single line. This can also make the code easier to read.
2. The three parts of a macro are the MACRO directive, the body, and the ENDM directive.
3. The macro definition is the list of statements the macro will perform. It begins with the MACRO directive and ends with the ENDM directive. Invoking the macro is when the macro is called from within an assembly language program. Expanding the macro is when the assembly replaces the line invoking the macro with the Assembly language code in the body of the macro.
4. False. A label that is to be local to a macro must be declared local with the LOCAL directive.
5. False. The "C" at the beginning of a line indicates that this line of code was brought in from another file by the INCLUDE directive.

CHAPTER 5

KEYBOARD AND MOUSE PROGRAMMING

OBJECTIVES

Upon completion of this chapter, you will be able to:

- **>> Code Assembly language instructions using INT 16H to get and check the keyboard input buffer and status bytes**
- **>> Code Assembly language instructions for key press and detection**
- **>> Use INT 33H to control mouse functions in text and graphics modes**
- **>> Code Assembly language instructions to initialize the mouse**
- **>> Code Assembly language instructions to set or get the mouse cursor position**
- **>> Use INT 33H functions to retrieve mouse button press or release information**
- **>> Limit mouse cursor postions by setting boundaries or defining exclusion areas**

This chapter explores keyboard and mouse programming in x86 PC. In Section 5.1, we utilize INT 16H function calls to access keyboard input. Then in Section 5.2, we demonstrate the use of INT 33H to control mouse functions.

SECTION 5.1: INT 16H KEYBOARD PROGRAMMING

In this section we first look at the keyboard used in IBM PC and compatible computers. Next we examine the scan codes that uniquely identify each key on the keyboard so that the computer can determine which key was pressed. Then the use of INT 16H is described for checking if a key has been pressed, identifying which key is pressed, and other functions.

IBM PC/XT keyboard

The original IBM PC keyboard had 83 keys, arranged in three major groupings:

1. The standard typewriter keys
2. Ten function keys, F1 to F10
3. 15-key keypad

These 83 keys are shown in Table 5-1. In later years, *enhanced keyboards* have become popular which feature 101 keys.

IBM PC scan codes

In IBM PC and compatible computers, each key on the keyboard is associated with a scan code. Tables 5-1, 5-2, and 5-3 provide the scan codes for both the original PC and enhanced keyboards.

Table 5-1: PC Scan Codes for 83 PC Keys

Hex	Key	Hex	Key	Hex	Key	Hex	Key
01	Esc	17	I and i	2D	X and x	43	F9
02	! and 1	18	O and o	2E	C and c	44	F10
03	@ and 2	19	P and p	2F	V and v	45	Num Lock
04	# and 3	1A	{ and [	30	B and b	46	Scroll Lock
05	$ and 4	1B	} and]	31	N and n	47	7 and Home
06	% and 5	1C	enter	32	M and m	48	8 and ↑
07	^ and 6	1D	ctrl	33	< and ,	49	9 and PgUp
08	& and 7	1E	A and a	34	> and .	4A	- (gray)
09	* and 8	1F	S and s	35	? and /	4B	4 and ←
0A	(and 9	20	D and d	36	Right Shift	4C	5 (keypad)
0B	) and 0	21	F and f	37	PrtSc and *	4D	6 and →
0C	_ and -	22	G and g	38	Alt	4E	+ (gray)
0D	+ and =	23	H and h	39	space bar	4F	1 and End
0E	backspace	24	J and j	3A	Caps Lock	50	2 and ↓
0F	tab	25	K and k	3B	F1	51	3 and PgDn
10	Q and q	26	L and l	3C	F2	52	0 and Ins
11	W and w	27	: and ;	3D	F3	53	. and Del
12	E and e	28	" and '	3E	F4		
13	R and r	29	~ and `	3F	F5		
14	T and t	2A	Left Shift	40	F6		
15	Y and y	2B	| and \	41	F7		
16	U and u	2C	Z and z	42	F8		

(Reprinted by permission from "IBM BIOS Technical Reference" c. 1987 by International Business Machines Corporation)

In Table 5-1, notice that the same scan code is used for a given lowercase letter and its capital. The same is true for all the keys with dual labels. If the scan code is the same for both of them, how does the system distinguish between them? This is taken care of by the keyboard shift status byte. Some INT 16H function calls provide the status byte in AL, as we will see in later examples. The meaning of each bit of the keyboard status byte is given in Table 5-4. Notice that some of the bits are used for the 101-key enhanced keyboards.

Table 5-2: Combination Key Scan Codes

Hex	Key	Hex	Key	Hex	Key	Hex	Key
54	Shift F1	60	Ctrl F3	6C	Alt F5	78	Alt 1
55	Shift F2	61	Ctrl F4	6D	Alt F6	79	Alt 2
56	Shift F3	62	Ctrl F5	6E	Alt F7	7A	Alt 3
57	Shift F4	63	Ctrl F6	6F	Alt F8	7B	Alt 4
58	Shift F5	64	Ctrl F7	70	Alt F9	7C	Alt 5
59	Shift F6	65	Ctrl F8	71	Alt F10	7D	Alt 6
5A	Shift F7	66	Ctrl F9	72	Ctrl PrtSc	7E	Alt 7
5B	Shift F8	67	Ctrl F10	73	Ctrl ←	7F	Alt 8
5C	Shift F9	68	Alt F1	74	Ctrl →	80	Alt 9
5D	Shift F10	69	Alt F2	75	Ctrl End	81	Alt 10
5E	Ctrl F1	6A	Alt F3	76	Ctrl PgDn		
5F	Ctrl F2	6B	Alt F4	77	Ctrl Home		

Table 5-3: Extended Keyboard Scan Codes

Hex	Key	Hex	Key	Hex	Key	Hex	Key
85	F11	8E	Ctrl -	97	Alt Home	A0	Alt ↓
86	F12	8F	Ctrl 5	98	Alt ↑	A1	Alt PgDn
87	Shift F11	90	Ctrl +	99	Alt PgUp	A2	Alt Ins
88	Shift F12	91	Ctrl ↓	9A		A3	Alt Del
89	Ctrl F11	92	Ctrl Ins	9B	Alt ←	A4	Alt /
8A	Ctrl F12	93	Ctrl Del	9C		A5	Alt Tab
8B	Alt F11	94	Ctrl Tab	9D	Alt →	A6	Alt Enter
8C	Alt F12	95	Ctrl /	9E			
8D	Ctrl ↑	96	Ctrl *	9F	Alt End		

When a key is pressed, the operating system stores its scan code in memory locations called a *keyboard buffer*, located in the BIOS data area. To relieve programmers from the details of keyboard interaction with the motherboard, IBM has provided INT 16H. Next we look at the services provided by the BIOS INT 16H.

Table 5-4: Keyboard Status Byte

Bit	If = 1	Mask Code (OR)
0	Right shift pressed	FEH
1	Left shift pressed	FDH
2	Ctrl pressed	FBH
3	Alt pressed	F7H
4	Scroll Lock toggled	EFH
5	NumLock toggled	DFH
6	CapsLock toggled	BFH
7	Ins toggled	7FH

Checking a key press

Chapter 4 demonstrated the use of INT 21H function AH = 07, which waits for the user to input a character. What if a program must run a certain task continuously while checking for a keypress? Such cases require the use of INT 16H, a BIOS interrupt used exclusively for the keyboard. To check a key press we use INT 16H function AH = 01.

```
MOV AH,01          ;check for key press
INT 16H            ;using INT 16H
```

Upon return, ZF = 0 if there is a key press; ZF = 1 if there is no key press. Notice that this function does not wait for the user to press a key. It simply checks to see if there is a key press. If a character is available, it returns its scan code in AH and its ASCII code in AL. The use of this function is best understood in the context of examples. Program 5-1 sends the ASCII bell character, 07 hex (see Appendix F), to the screen continuously. To stop the bell sound, the user must press any key.

```
TITLE PROGRAM 5-1:KEYBOARD HIT USING INT 16H
;THIS PROGRAM SOUNDS THE BELL CONTINUOUSLY UNLESS ANY KEY IS PRESSED
      .MODEL SMALL
      .STACK
      .DATA
MESSAGE  DB 'TO STOP THE BELL SOUND PRESS ANY KEY$'
      .CODE
MAIN  PROC
      MOV   AX,@DATA
      MOV   DS,AX
      MOV   AH,09
      MOV   DX,OFFSET MESSAGE ;DISPLAY THE MESSAGE
      INT   21H
AGAIN: MOV  AH,02          ;SENDING TO MONITOR A SINGLE CHAR
      MOV   DL,07          ;SEND OUT THE BELL CHAR
      INT   21H
      MOV   AH,01          ;CHECK THE KEY PRESS
      INT   16H            ;USING INT 16H
      JZ    AGAIN          ;IF NO KEY PRESS STAY IN THE LOOP
      MOV   AH,4CH         ;IF ANY KEY PRESSED GO BACK TO DOS
      INT   21H
MAIN  ENDP
      END
```

Program 5-1

Which key is pressed?

There are times when the program needs to know not only if a key has been pressed but also which key was pressed. To do that, INT 16H function AH = 0 can be used immediately after the call to INT 16H function AH = 01.

```
MOV AH,0           ;get key pressed
INT 16H            ;using INT 16H
```

Upon return, AL contains the ASCII character of the pressed key; its scan key is in AH. Notice that this function must be used immediately after calling INT 16H function AH = 01. Program 5-2 demonstrates how it works.

INT 16H function AH = 0 can also be used by itself to get a character from the keyboard. The difference between these two functions is that AH = 1 comes back whether or not a key has been pressed whereas AH = 0 doesn't return until a key is pressed. For characters such as F1–F10 for which there is no ASCII code, it simply provides the scan code in AH and AL = 0. Therefore, if AL = 0, a special function key was pressed. This option simply provides the code for the character and does not display it.

```
TITLE PROGRAM 5-2: MODIFIED VERSION OF PROGRAM 5-1
;THIS PROGRAM SOUNDS THE BELL CONTINUOUSLY UNTIL 'Q' OR 'q' IS PRESSED
      .MODEL SMALL
      .STACK
      .DATA
MESSAGE DB 'TO STOP THE BELL SOUND PRESS Q (or q) KEY$'
      .CODE
MAIN  PROC
      MOV   AX,@DATA
      MOV   DS,AX
      MOV   AH,09
      MOV   DX,OFFSET MESSAGE ;DISPLAY THE MESSAGE
      INT   21H
AGAIN:MOV   AH,02
      MOV   DL,07         ;SOUND THE BELL BY SENDING OUT BELL CHAR
      INT   21H
      MOV   AH,01         ;CHECK FOR KEY PRESS
      INT   16H           ;USING INT 16H
      JZ    AGAIN         ;IF NO KEY PRESS KEEP SOUNDING THE BELL
      MOV   AH,0          ;TO GET THE CHARACTER
      INT   16H           ;WE MUST USE INT 16H ONE MORE TIME
      CMP   AL,'Q'        ;IS IT 'Q'?
      JE    EXIT          ;IF YES EXIT
      CMP   AL,'q'        ;IS IT 'q'
      JE    EXIT          ;IF YES EXIT
      JMP   AGAIN         ;NO. KEEP SOUNDING THE BELL
EXIT: MOV   AH,4CH        ;GO BACK TO DOS
      INT   21H
MAIN  ENDP
      END
```

Program 5-2

Other INT 16H functions

Due to additional keys on the IBM extended keyboard, BIOS added the following additional services to INT 16H.

INT 16H, AH = 10H (read a character)

This is the same as AH = 0 except that it also accepts the additional keys on the IBM extended (enhanced) keyboard.

INT 16H, AH = 11H (find if a character is available)

This is the same as AH = 1 except that it also accepts the additional keys on the IBM extended (enhanced) keyboard.

Review Questions

1. Which function of INT 16H is used for key press detection?
2. In the above question, how do you know if a key is pressed?
3. In the above question, how can the ASCII value for the pressed key be obtained?
4. Indicate the main difference between INT 21H function AH = 07, and INT 16H function AH = 01.
5. Write a simple program to sound the bell unless the letter 'X' is pressed. If 'X' is pressed, the program should exit.

SECTION 5.2: MOUSE PROGRAMMING WITH INT 33H

Next to the keyboard, the mouse is one of the most widely used input devices. This section describes how to use INT 33H to add mouse capabilities to programs.

INT 33H

The original IBM PC and DOS did not provide support for the mouse. For this reason, mouse interrupt INT 33H is not part of BIOS or DOS. This is in contrast to INT 21H and INT 10H, which are the DOS and BIOS interrupts, respectively. Now, INT 33H is part of the mouse driver software that is installed when the PC is booted.

Detecting the presence of a mouse

While new PCs come with a mouse and driver already installed by the PC manufacturer, many older-generation PCs in use do not have a mouse. Therefore, the first task of any INT 33H program should be to verify the presence of a mouse and the number of buttons it supports. This is the purpose of INT 33H function AX = 0. Upon return from INT 33H, if AX = 0 then no mouse is supported. If AX = FFFFH, the mouse is supported and the number of mouse buttons will be contained in register BX. Although most mice have two buttons, right and left, there are some with middle buttons as well. See the following code.

```
        MOV    AX,0          ;mouse initialization option
        INT    33H
        CMP    AX,0          ;check AX contents after INT 33H
        JE     EXIT          ;exit if AX=0 since no mouse available
        MOV    M_BUTTON,BX ;mouse is there, save number of buttons
        ...
EXIT:
```

Notice the following points about the way INT 33H is called.

1. In INT 21H and INT 10H, the AH register is used to select the functions. This is not the case in INT 33H. In INT 33H the register AL is used to select various functions and AH is set to 0. That is the reason behind the instruction "`MOV AX,0`".
2. Do not forget the "H" indicating hex in coding INT 33H. In the absence of the "H", the compiler assumes it is decimal and will execute DOS INT 21H since 33 decimal is equal to 21H.

Some mouse terminology

Before further discussion of INT 33H, some terminology concerning the mouse needs to be clarified. The *mouse pointer* (or *cursor*) is the pointer on the screen indicating where the mouse is pointing at a given time. In graphics mode, the mouse pointer (cursor) is an arrow; in text mode, the mouse pointer is a flashing block. In either mode, as the mouse is moved, the mouse cursor is also moved. While the movement of the mouse is measured in inches (or centimeters), the movement of the mouse cursor (arrowhead) on the screen is measured in units called *mickeys*. Mickey units indicate *mouse sensitivity*. For example, a mouse that can move the cursor 200 units for every inch of mouse movement has a sensitivity of 200 mickeys. In this case, one mickey represents 1/200 of an inch on the screen. Some mice have a sensitivity of 400 mickeys in contrast to the commonly used 200 mickeys. In that case, for every inch of mouse movement, the mouse cursor moves 400 mickeys.

Displaying and hiding the mouse cursor

The AX = 01 function of INT 33H is used to display the mouse cursor.

```
        MOV    AX,01
        INT    33H
```

After executing the above code, the mouse pointer is displayed. If the video mode is graphics, the mouse arrow becomes visible. If the video mode is text, a rectangular block representing the mouse cursor becomes visible. In text mode, the color of the mouse cursor block is the opposite of the background color in order to be visible. It is best to hide the mouse cursor after making it visible by executing option AX = 02 of INT 33H. This is shown in Example 5-1. Try Example 5-1 in DEBUG (remember to omit the "H" and place INT 3 as the last instruction when in DEBUG). Then try it with mode AH = 03 for INT 10H to see the mouse cursor in text mode.

Video resolution vs. mouse resolution in text mode

As discussed in Chapter 4, the video screen is divided into 640 × 200 pixels in text mode. This means that in text mode of 80 × 25 characters, each character will use 8 × 8 pixels (80 × 8 = 640 and 25 × 8 = 200). When the video mode is set to text mode (AH = 03 of INT 10H), the mouse will automatically adopt the same resolution of 640 × 200 for its horizontal and vertical coordinates. Therefore, in text mode when a program gets the mouse cursor position, the values are provided in pixels and must be divided by 8 to get the mouse cursor position in terms of character locations 0 to 79 (horizontal) and 0 to 24 (vertical) on the screen.

Video resolution vs. mouse resolution in graphics mode

In graphics, resolution is not only 640 × 200 but also 640 × 350 and 640 × 480. When the video resolution is changed to these video modes, the mouse also adopts the graphics resolutions. See Table 5-5.

Table 5-5: Video and Mouse Resolution for Some Video Modes

Video Mode	Video Resolution	Type	Mouse Resolution	Characters per Screen
AL = 03	640 × 200	Text	640 × 200	80 × 25
AL = 0EH	640 × 200	Graphics	640 × 200	80 × 25
AL = 0FH	640 × 350	Graphics	640 × 350	80 × 44
AL = 10H	640 × 350	Graphics	640 × 350	80 × 44
AL = 11H	640 × 480	Graphics	640 × 480	80 × 60
AL = 12H	640 × 480	Graphics	640 × 480	80 × 60

Example 5-1

(a) Use INT 10H option 0F to get the current video mode and save it in BL;
(b) set the video mode to VGA graphics using option AH = 10H of INT 10H;
(c) initialize the mouse with AX = 0, INT 33H; (d) make the mouse visible;
(e) use INT 21H option AH = 01 to wait for key press;
(f) if any key is pressed restore the original video mode.

Solution:

```
MOV   AH,0FH       ;get the current video mode
INT   10H
MOV   BL,AL        ;and save it
MOV   AH,0         ;set the video mode
MOV   AL,10H       ;to VGA graphics
INT   10H
MOV   AX,0         ;initialize the mouse
INT   33H
MOV   AX,01        ;make the mouse cursor visible
INT   33H
MOV   AH,01        ;wait for key press
INT   21H
MOV   AX,2         ;when any key is pressed
INT   33H          ;make mouse invisible
MOV   AH,0
MOV   AL,BL        ;and restore original video mode
INT   10H
```

Getting the current mouse cursor position (AX = 03)

Option AX = 03 of INT 33H gets the current position of the mouse cursor. Upon return, the X and Y coordinates are in registers CX (horizontal) and DX (vertical). BX contains the button status as follows: D0 = left button status, D1 = right button status, D2 = center button status. The status is 1 if down, 0 if up. Notice that the cursor position provided by this function is given in pixels. For example, the position returned will be in the range of 0–639 (horizontal) and 0–199 (vertical) for a 640 × 200 screen in most text and graphics video modes. However, the mouse cursor position is often needed in terms of character positions such as 80 × 25 and not in terms of pixels. To get the mouse cursor character position, divide both the horizontal and vertical values of CX and DX by 8. See Programs 5-3 and 5-4.

```
TITLE PROGRAM 5-3: DISPLAYING MOUSE POSITION
;Performs the following tasks: (a) gets the current video mode and
;saves it, (b) sets the mode to a new video mode, (c) gets the mouse
;pointer position, converts it to character position and displays it
;continuously unless a key is pressed, (d) upon pressing any key, it
;restores the original video mode and exits to DOS.
CURSOR MACRO ROW,COLUMN
      MOV   AH,02H
      MOV   BH,00
      MOV   DH,ROW
      MOV   DL,COLUMN
      INT   10H
      ENDM
DISPLAY MACRO STRING
      MOV   AH,09H
      MOV   DX,OFFSET STRING  ;load string address
      INT   21H
      ENDM
            .MODEL SMALL
            .STACK
            .DATA
MESSAGE_1   DB 'PRESS ANY KEY TO GET OUT','$'
MESSAGE_2   DB 'THE MOUSE CURSOR IS LOCATED AT ','$'
POS_HO      DB ?,?, ' AND $'
POS_VE            DB ?,?,'$'
OLDVIDEO    DB ?              ;current video mode
NEWVIDEO    DB 0EH            ;new video mode
            .CODE
MAIN        PROC
            MOV   AX,@DATA
            MOV   DS,AX
            MOV   AH,0FH            ;get current video mode
            INT   10H
            MOV   OLDVIDEO,AL             ;save it
            MOV   AX,0600H          ;clear screen
            MOV   BH,07
            MOV   CX,0
            MOV   DX,184FH
            INT   10H
            MOV   AH,00H            ;set new video mode
            MOV   AL,NEWVIDEO
            INT   10H
            MOV   AX,0              ; initialize mouse
            INT   33H
      MOV   AX,01             ;show mouse cursor
      INT   33H
```

Program 5-3 *(continued on the next page)*

```
        CURSOR 20,20
        DISPLAY MESSAGE_1
AGAIN:MOV    AX,03H              ;get mouse location
        INT    33H
        MOV    AX,CX               ;get the hor. pixel position
        CALL   CONVERT             ;convert to displayable data
        MOV    POS_HO,AL           ;save the LSD
        MOV    POS_HO+1,AH         ;save the MSD
        MOV    AX,DX               ;get the vert. pixel position
        CALL   CONVERT             ;convert
        MOV    POS_VE,AL           ;save
        MOV    POS_VE+1,AH
        CURSOR 5,20                ;
        DISPLAY MESSAGE_2 ;
        DISPLAY POS_HO
        DISPLAY POS_VE
        MOV    AH,01               ;check for key press
        INT    16H
        JZ     AGAIN ;if no key press, keep monitoring mouse position
        MOV    AH,02               ;hide mouse
        INT    33H
        MOV    AH,0         ;restore original video mode
        MOV    AL,OLDVIDEO                ;load original video mode
        INT    10H
        MOV    AH,4CH              ;go back to DOS
        INT    21H
MAIN    ENDP
        END
```

Program 5-3 *(continued from the previous page)*

Setting the mouse pointer position (AX = 04)

This function allows a program to set the mouse pointer to a new location anywhere on the screen. Before calling this function, the coordinates for the new location must be placed in registers CX for the horizontal (x coordinate) and DX for the vertical (y coordinate). These values must be in pixels in the range of 0–639 and 0–199 for 640 × 200 resolution. Coordinate (0,0) is the upper left corner of the screen. For example, to set the mouse cursor at location 9 × 5 (on a 80 × 25 screen), simply multiply both by 8 to get the pixel location. Therefore, a character coordinate of 9 × 5 becomes 72 × 40 in pixel coordinates.

Getting mouse button press information (AX = 05)

This function is used to get information about specific button presses since the last call to this function. It is set up as follows.

AX = 05
BX = 0 for left button; 1 for right button; 2 for center button
Upon return:
AX = button status where
D0 = Left button, if 1 it is down and if 0 it is up
D1 = Right button, if 1 it is down and if 0 it is up
D2 = Center button, if 1 it is down and if 0 it is up
BX = button press count
CX = x-coordinate at the last button press in pixels (horizontal)
DX = y-coordinate at the last button press in pixels (vertical)

Notice in function AX = 05 that upon returning from INT 33H, register AX has the button status (up or down), while register BX has the number of times the specific button is pressed since the last call to this function. Program 5-4 shows one way to use this function.

```
TITLE PROGRAM 5-4:MOUSE BOX PROGRAM
;Performs the following:
;(a) gets the current video mode and saves it,
;(b) sets the video mode to a new one and clears screen,
;(c) draws a colored box and gets the mouse position,
;(d) displays different messages depending on whether the mouse is
; clicked inside or outside the box  Pressing any key will return to DOS.
;Thanks to Travis Erck and Gary Hudson for their input on this program

CURSOR       MACRO ROW,COLUMN
             MOV   AH,02H
             MOV   BH,00
             MOV   DH,ROW
             MOV   DL,COLUMN
             INT   10H
             ENDM
DISPLAY      MACRO STRING
             MOV   AH,09H
             MOV   DX,OFFSET STRING  ;load string address
             INT   21H
             ENDM
FILL         MACRO ROW_START,COL_START,ROW_END,COL_END,COLOR
             LOCAL START,AGAIN
             MOV   DX,ROW_START
START:       MOV   CX,COL_START
AGAIN:       MOV   AH,0CH
             MOV   AL,COLOR
             INT   10H
             INC   CX
             CMP   CX,COL_END
             JNE   AGAIN
             INC   DX
             CMP   DX,ROW_END
             JNE   START
             ENDM

                   .MODEL SMALL
                   .STACK
                   .DATA
MESSAGE_1       DB 'AN EXAMPLE OF HOW TO USE INTERRUPT 33H FOR MOUSE.','$'
MESSAGE_2       DB 'IT WORKS!','$'
MESSAGE_3       DB 'CLICK IN THE BOX TO SEE WHAT HAPPENS!','$'
MESSAGE_4       DB 'No, NO, NO I SAID IN THE BOX!','$'
MESSAGE_5       DB 'NOW PRESS ANY KEY TO GET BACK TO DOS. $'
OLDVIDEO        DB ?
NEWVIDEO        DB 12H

                   .CODE
MAIN         PROC
             MOV   AX,@DATA
             MOV   DS,AX
             MOV   AH,0FH                    ;get the current video mode
             INT   10H
             MOV   OLDVIDEO,AL               ;save it
```

Program 5-4 *(continue on the next page)*

```
            MOV    AX,0600H                  ;clear screen
            MOV    BH,07
            MOV    CX,0
            MOV    DX,184FH
            INT    10H
            MOV    AH,00H                    ;set new video mode
            MOV    AL,NEWVIDEO
            INT    10H
            CURSOR 0,0
            FILL 150,250,250,350,4           ;draw red box
            CURSOR 1,1
            DISPLAY MESSAGE_1;
            CURSOR 5,22
            DISPLAY MESSAGE_3;
            MOV    AX,0000H                  ;initialize mouse
            INT    33H
            MOV    AX,01H
            INT    33H                       ;show mouse cursor
BACK:       MOV    AX,03H                    ;check for mouse button press
            INT    33H                       ;now CX =COL and DX=ROW location
            CMP    BX,0001H     ;check to see if left button is pressed
            JNE    BACK                      ;If not keep checking
            CMP    CX,250                    ;see if on right side of box
            JB     NOT_INSIDE           ;if less it must be outside box
            CMP    CX,350                    ;see if on left side of box
            JA     NOT_INSIDE           ;if not then it is outside the box
            CMP    DX,150                    ;check for the top of the box
            JB     NOT_INSIDE           ;if not then outside the box
            CMP    DX,250                    ;see if bottom of the box
            JA     NOT_INSIDE
            CURSOR 18,18                ;then it must be inside box
            DISPLAY MESSAGE_2           ;indicate mouse is inside the box
            JMP    EXIT                      ;go prepare to exit to DOS
NOT_INSIDE:
            CURSOR 20,18                ;indicate mouse is not inside box
            DISPLAY MESSAGE_4
EXIT:       MOV    AH,02H               ;hide mouse before exiting to DOS
            INT    33H
            CURSOR 22,18
            DISPLAY MESSAGE_5
            MOV    AH,07                     ;wait for a key press
            INT    21H
            MOV    AH,0                      ;restore original video mode
            MOV    AL,OLDVIDEO
            INT    10H
            MOV    AH,4CH                    ;exit
            INT    21H                       ;to DOS
MAIN        ENDP
            END
```

Program 5-4 *(continued from the previous page)*

Monitoring and displaying the button press count program

Program 5-5 uses the AX = 05 function to monitor the number of times the left button is pressed and then displays the count. It prompts the user to press the left button a number of times. When the user is ready to see how many times the button was pressed, any key can be pressed.

```
TITLE    PROGRAM 5-5: DISPLAY MOUSE PRESS COUNT
;THIS PROGRAM WAITS FOR THE MOUSE PRESS COUNT AND DISPLAYS IT WHEN
;ANY KEY IS PRESSED.
;PRESS ANY KEY TO GO BACK TO DOS
PAGE     60,132

CURSOR        MACRO ROW,COLUMN
              MOV   AH,02H
              MOV   BH,00
              MOV   DH,ROW
              MOV   DL,COLUMN
              INT         10H
              ENDM

DISPLAY       MACRO STRING
              MOV   AH,09H
              MOV   DX,OFFSET STRING ;LOAD STRING ADDRESS
              INT         21H
              ENDM

              .MODEL SMALL
              .STACK
              .DATA
MESSAGE_1     DB 'PRESS LEFT BUTTON A NUMBER OF TIMES:LESS THAN 99.','$'
MESSAGE_2     DB  'TO FIND OUT HOW MANY TIMES, PRESS ANY KEY','$'
MESSAGE_3     DB 'YOU PRESSED IT ','$'
P_COUNT       DB ?,?, ' TIMES $'
MESSAGE_4     DB 'NOW PRESS ANY KEY TO GO BACK TO DOS','$'
OLDVIDEO      DB ?              ;current video mode
NEWVIDEO      DB 12H            ;new video mode
                    .CODE
MAIN          PROC
              MOV   AX,@DATA
              MOV   DS,AX
              MOV   AH,0FH             ;get current video mode
              INT   10H
              MOV   OLDVIDEO,AL        ;save it
              MOV   AX,0600H           ;clear screen
              MOV   BH,07
              MOV   CX,0
              MOV   DX,184FH
              INT   10H
              MOV   AH,00H             ;set new video mode
              MOV   AL,NEWVIDEO
              INT   10H
              MOV   AX,0               ;initialize mouse
              INT   33H
```

Program 5-5 *(continued on the next page)*

```
            MOV    AX,01               ;show mouse cursor
            INT    33H
            CURSOR 2,1
            DISPLAY MESSAGE_1
            CURSOR 4,1
            DISPLAY MESSAGE_2
            MOV    AH,07               ;wait for key press
            INT    21H
            MOV    AX,05H              ;get mouse press count
            MOV    BX,0                ;check press count for left button
            INT    33H
            MOV    AX,BX               ;BX=button press count
            MOV    BL,10
            DIV    BL
            OR     AX,3030H            ;convert it to ASCII
            MOV    P_COUNT,AL          ;save the number
            MOV    P_COUNT+1,AH
            CURSOR 10,2
            DISPLAY MESSAGE_3
            DISPLAY P_COUNT
            CURSOR 20,2
            DISPLAY MESSAGE_4
            MOV    AH,07               ;wait for a key press to get out
            INT    21H
            MOV    AH,02               ;hide mouse
            INT    33H
            MOV    AH,0                ;restore original video mode
            MOV    AL,OLDVIDEO         ;load original vide mode
            INT    10H
            MOV    AH,4CH              ;go back to DOS
            INT    21H
MAIN        ENDP
            END    MAIN
```

Program 5-5 *(continued from the previous page)*

Review Questions

1. Which function of INT 33H is used to detect the presence of a mouse in a PC? In which register do we expect to get that information? In which register do we find the number of buttons in the mouse?
2. The following code is an attempt to call INT 33H function 2. Is it correct? If not, correct it.
   ```
   MOV AH,02
   INT 33H
   ```
3. Why do we need to save the original video mode before changing it for the mouse?
4. In INT 33H function AX = 03, how can a left button press be detected?

PROBLEMS

SECTION 5.1: INT 16H KEYBOARD PROGRAMMING

1. Which function of INT 16H is used to find out which key has been pressed?
2. In the above question, how can the ASCII value for the pressed key be obtained?

SECTION 5.2: MOUSE PROGRAMMING WITH INT 33H

3. Using INT 33H, write and test an Assembly language program to check the presence of a mouse in a PC. If a mouse driver is installed, it should state the number of buttons it supports. If no mouse driver is installed, it should state this.

ANSWERS TO REVIEW QUESTIONS

SECTION 5.1: INT 16H KEYBOARD PROGRAMMING

1. INT 16H function AH = 01
2. After return from INT 16H function AH = 01, if ZF = 1 there is no key press; if ZF = 0 then a key has been pressed.
3. If ZF = 0, then we use INT 16H function AH = 0 to get the ASCII character for the pressed key.
4. INT 21H waits for the user to press the key; INT 16H scans the keyboard, allowing the program to continue executing other tasks while scanning for the key press.
5.

```
AGAIN:    MOV AH,02    ;USE FUNCTION AH=02 OF INT 21H
          MOV DL,07    ;SOUND THE BELL BY SENDING OUT BELL CHAR
          INT 21H
          MOV AH,01    ;CHECK FOR KEY PRESS
          INT 16H      ;USING INT 16H
          JZ  AGAIN    ;IF NO KEY PRESS KEEP SOUNDING THE BELL
          MOV AH,0     ;TO GET THE CHARACTER
          INT 16H      ;WE MUST USE INT 16H ONE MORE TIME
          CMP AL,'X'   ;IS IT 'X'?.
          JE  EXIT     ;IF YES EXIT
          JMP AGAIN    ;NO. KEEP SOUNDING THE BELL
EXIT:     MOV AH,4CH   ;GO BACK TO DOS
          INT 21H
```

SECTION 5.2: MOUSE PROGRAMMING WITH INT 33H

1. AX = 0; register AX; number of buttons in register BX
2. It is wrong. Register AL = 02 and AH = 0.

```
MOV AX,02
INT 33H
```

3. Because if we don't do that, when we go back to DOS we lose our cursor if the mouse program has changed the video to graphics.
4. We check the contents of register BX for value 01.

CHAPTER 6

SIGNED NUMBERS, STRINGS, AND TABLES

OBJECTIVES

Upon completion of this chapter, you will be able to:

- **>> Represent 8- or 16-bit signed numbers as used in computers**
- **>> Convert a number to its 2's complement**
- **>> Code signed arithmetic instructions ADD, SUB, IMUL, and IDIV**
- **>> Demonstrate how arithmetic instructions affect the sign flag**
- **>> Explain the difference between a carry and an overflow**
- **>> Prevent overflow errors by sign-extending data**
- **>> Code signed shift instructions SAL and SAR**
- **>> Code logic instruction CMP for signed numbers and explain its effect on the flag register**
- **>> Code conditional jump instructions after CMP of signed data**
- **>> Explain the function of registers SI and DI in string instructions**
- **>> Describe the operation of the direction flag in string instructions**
- **>> Code instructions CLD and STD to control the direction flag**
- **>> Describe the operation of the REP prefix**
- **>> Code string instructions:**
  - **>> MOVSB and MOVSW for data transfer**
  - **>> STOS, LODS to store and load the contents of AX**
  - **>> CMPS to compare two strings of data**
  - **>> SCAS to scan a string for data matching that in AX**
  - **>> XLAT for table processing**

This chapter deals with signed numbers and tables. In Section 6.1, we focus on the concept of signed numbers in software engineering. Signed number operations are explained along with examples. We discuss string operations and table processing in Section 6.2.

SECTION 6.1: SIGNED NUMBER ARITHMETIC OPERATIONS

All data items used so far have been unsigned numbers, meaning that the entire 8-bit or 16-bit operand was used for the magnitude. Many applications require signed data. In this section the concept of signed numbers is discussed along with related instructions.

Concept of signed numbers in computers

In everyday life, numbers are used that could be positive or negative. For example, a temperature of 5 degrees below zero can be represented as –5, and 20 degrees above zero as +20. Computers must be able to accommodate such numbers. To do that, computer scientists have devised the following arrangement for the representation of signed positive and negative numbers: The most significant bit (MSB) is set aside for the sign (+ or –) and the rest of the bits are used for the magnitude. The sign is represented by 0 for positive (+) numbers and 1 for negative (–) numbers. Signed byte and word representations are discussed below.

Signed byte operands

In signed byte operands, D7 (MSB) is the sign and D0 to D6 are set aside for the magnitude of the number. If D7 = 0, the operand is positive, and if D7 = 1, it is negative.

D7	D6	D5	D4	D3	D2	D1	D0
sign	magnitude						

Positive numbers

The range of positive numbers that can be represented by the format above is 0 to +127.

```
0       0000 0000
+1      0000 0001
+5      0000 0101
...     ...  ....
+127    0111 1111
```

If a positive number is larger than +127, a word-sized operand must be used. Word operands are discussed later.

Negative numbers

For negative numbers D7 is 1, but the magnitude is represented in 2's complement. Although the assembler does the conversion, it is still important to understand how the conversion works. To convert to negative number representation (2's complement), follow these steps:

1. Write the magnitude of the number in 8-bit binary (no sign).
2. Invert each bit.
3. Add 1 to it.

Examples 6-1, 6-2, and 6-3 demonstrate these three steps. From the examples above it is clear that the range of byte-sized negative numbers is –1 to –128. The following lists byte-sized signed number ranges:

```
Decimal        Binary       Hex
-128           1000 0000    80
-127           1000 0001    81
-126           1000 0010    82
...            .... ...     ..
-2             1111 1110    FE
-1             1111 1111    FF
 0             0000 0000    00
+1             0000 0001    01
+2             0000 0010    02
...            ... ...      ...
+127           0111 1111    7F
```

Example 6-1

Show how the computer would represent –5.

Solution:

```
1. 0000 0101     5 in 8-bit binary
2. 1111 1010     invert each bit
3. 1111 1011     add 1 (hex = FBH)
```

This is the signed number representation in 2's complement for –5.

Example 6-2

Show –34H as it is represented internally.

Solution:

```
1. 0011 0100
2. 1100 1011
3. 1100 1100      (which is CCH)
```

Example 6-3

Show the representation for -128_{10}.

Solution:

```
1.   1000 0000
2.   0111 1111
3.   1000 0000
```
Notice that this is not negative zero (–0).

Word-sized signed numbers

In x86 computers a word is 16 bits in length. Setting aside the MSB (D15) for the sign leaves a total of 15 bits (D14–D0) for the magnitude. This gives a range of –32768 to +32767. If a number is larger than this, it must be treated as a multiword operand and be processed chunk by chunk the same way as unsigned numbers (as discussed in Chapter 3). The following shows the range of signed word operands. To convert a negative number to its word operand representation, the three steps discussed in negative byte operands are used.

D15	D14	D13	D12	D11	D10	D9	D8	D7	D6	D5	D4	D3	D2	D1	D0
sign	magnitude														

Decimal	Binary	Hex
−32,768	1000 0000 0000 0000	8000
−32,767	1000 0000 0000 0001	8001
−32,766	1000 0000 0000 0010	8002
...	...	...
...	...	...
−2	1111 1111 1111 1110	FFFE
−1	1111 1111 1111 1111	FFFF
0	0000 0000 0000 0000	0000
+1	0000 0000 0000 0001	0001
+2	0000 0000 0000 0010	0002
...	...	...
...	...	...
+32,766	0111 1111 1111 1110	7FFE
+32,767	0111 1111 1111 1111	7FFF

Overflow problem in signed number operations

When using signed numbers, a serious problem arises that must be dealt with. This is the overflow problem. The CPU indicates the existence of the problem by raising the OF (overflow) flag, but it is up to the programmer to take care of it. The CPU understands only 0s and 1s and ignores the human convention of positive and negative numbers. Now what is an overflow? If the result of an operation on signed numbers is too large for the register, an overflow occurs and the programmer must be notified. Look at Example 6-4.

Example 6-4

Look at the following code and data segments:

```
DATA1       DB     +96
DATA2       DB     +70
            ...    ....
            MOV    AL,DATA1      ;AL=0110 0000 (AL=60H)
            MOV    BL,DATA2      ;BL=0100 0110 (BL=46H)
            ADD    AL,BL         ;AL=1010 0110 (AL=A6H= 90 invalid!)

+ 96 0110 0000
+ 70 0100 0110
+166 1010 0110
```

According to the CPU, this is 90, which is wrong. (OF = 1, SF = 1, CF = 0)

In the example above, +96 is added to +70 and the result according to the CPU is −90. Why? The reason is that the result was more than what AL could handle. Like all other 8-bit registers, AL could only contain up to +127. The designers of the CPU created the overflow flag specifically for the purpose of informing the programmer that the result of the signed number operation is erroneous.

When the overflow flag is set in 8-bit operations

In 8-bit signed number operations, OF is set to 1 if either of the following two conditions occurs:

1. There is a carry from D6 to D7 but no carry out of D7 (CF = 0).
2. There is a carry from D7 out (CF = 1) but no carry from D6 to D7.

In other words, the overflow flag is set to 1 if there is a carry from D6 to D7 or from D7 out, but not both. This means that if there is a carry both from D6 to D7 and from D7 out, OF = 0. In Example 6-4, since there is only a carry from D6 to D7 and no carry from D7 out, OF = 1. Examples 6-5, 6-6, and 6-7 give further illustrations of the overflow flag in signed arithmetic.

Example 6-5

Observe the results of the following:

```
MOV   DL,-128     ;DL=1000 0000 (DL=80H)
MOV   CH,-2       ;CH=1111 1110 (CH=FEH)
ADD   DL,CH       ;DL=0111 1110 (DL=7EH=+126 invalid!)

   -128   1000 0000
+    -2   1111 1110
   -130   0111 1110 OF=1, SF=0 (positive), CF=1
```

According to the CPU, the result is +126, which is wrong. The error is indicated by the fact that OF = 1.

Example 6-6

Observe the results of the following:

```
MOV   AL,-2       ;AL=1111 1110 (AL=FEH)
MOV   CL,-5       ;CL=1111 1011 (CL=FBH)
ADD   CL,AL       ;CL=1111 1001 (CL=F9H=7 which is correct)

  -2 1111 1110
+ -5 1111 1011
  -7 1111 1001
```

OF = 0, CF = 0 , and SF = 1 (negative); the result is correct since OF = 0.

Example 6-7

Observe the results of the following:

```
MOV   DH,+7       ;DH=0000 0111     (DH=07H)
MOV   BH,+18      ;BH=0001 0010     (BH=12H)
ADD   BH,DH       ;BH=0001 1001     (BH=19H=+25, correct)

   +7     0000 0111
+ +18     0001 0010
  +25     0001 1001
```

OF = 0, CF = 0, and SF = 0 (positive).

Overflow flag in 16-bit operations

In a 16-bit operation, OF is set to 1 in either of two cases:

1. There is a carry from D14 to D15 but no carry out of D15 (CF = 0).
2. There is a carry from D15 out (CF = 1) but no carry from D14 to D15.

Again the overflow flag is low (not set) if there is a carry from both D14 to D15 and from D15 out. The OF is set to 1 only when there is a carry from D14 to D15 or from D15 out, but not from both. See Examples 6-8 and 6-9.

Avoiding erroneous results in signed number operations

To avoid the problems associated with signed number operations, one can sign-extend the operand. Sign extension copies the sign bit (D7) of the lower byte of a register into the upper bits of the register, or copies the sign bit of a 16-bit register into another register. CBW (convert signed byte to signed word) and CWD (convert signed word to signed double word) are used to perform sign extension. They work as follows:

Example 6-8

Observe the results in the following:

```
      MOV   AX,6E2FH     ; 28,207
      MOV   CX,13D4H     ;+ 5,076
      ADD   AX,CX        ;= 33,283 is the expected answer

 6E2F             0110 1110 0010 1111
+13D4             0001 0011 1101 0100
 8203             1000 0010 0000 0011 = - 32,253 incorrect!
                  OF = 1, CF = 0, SF = 1
```

Example 6-9

Observe the results in the following:

```
      MOV   DX,542FH     ; 21,551
      MOV   BX,12E0H     ; +4,832
      ADD   DX,BX        ;=26,383

 543F        0101 0100 0010 1111
+12E0        0001 0010 1110 0000
 670F        0110 0111 0000 1111 = 26,383 (correct answer); OF = 0, CF = 0, SF = 0
```

CBW will copy D7 (the sign flag) to all bits of AH. This is demonstrated below. Notice that the operand is assumed to be AL and the previous contents of AH are destroyed.

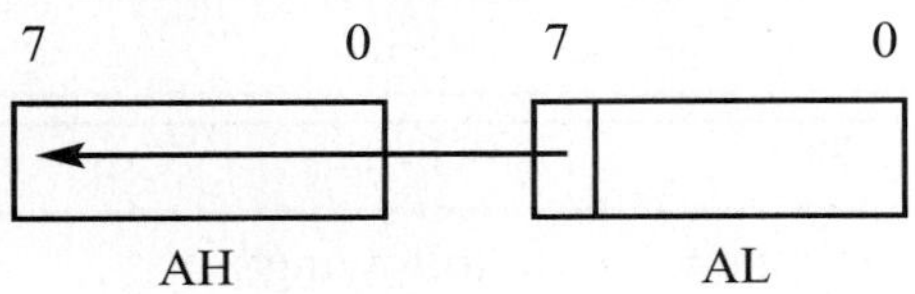


```
MOV   AL,+96        ;AL=0110 0000
CBW                 ;now AH=0000 0000 and AL=0110 0000
```

or:

```
MOV   AL,-2         ;AL=1111 1110
CBW                 ;AH=1111 1111 and AL=1111 1110
```

CWD sign-extends AX. It copies D15 of AX to all bits of the DX register. This is used for signed word operands. This is illustrated below.

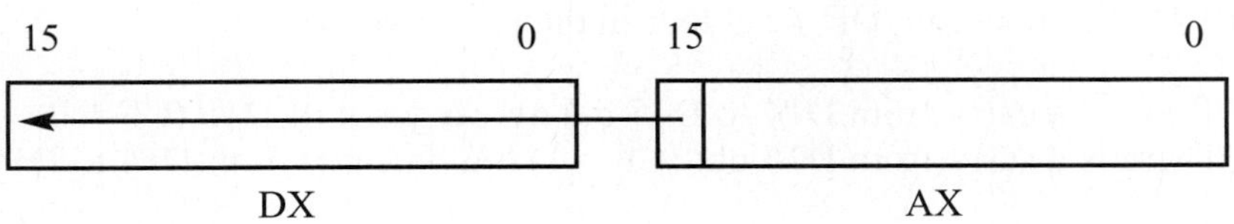

Look at the following example:

```
MOV   AX,+260       ;AX=0000 0001 0000 0100 or AX=0104H
CWD                 ;DX=0000H and AX=0104H
```

Another example:

```
MOV   AX,-32766     ;AX=1000 0000 0000 0010B or AX=8002H
CWD                 ;DX=FFFF and AX=8002
```

As can be seen in the examples above, CWD does not alter AX. The sign of AX is copied to the DX register. How can these instructions help correct the overflow error? To answer that question, Example 6-10 shows Example 6-4 rewritten to correct the overflow problem.

In Example 6-10, if the overflow flag is not raised (OF = 0), the result of the signed number is correct and JNO (jump if no overflow) will jump to OVER. However, if OF = 1, which means that the result is erroneous, each operand must be sign-extended and then added. That is the function of the code below the JNO instruction. The program in Example 6-10 works for addition of any two signed bytes.

Example 6-10

Rewrite Example 6-4 to provide for handling the overflow problem.

Solution:

```
DATA1       DB     +96
DATA2       DB     +70
RESULT      DW     ?
            ......
            SUB    AH,AH         ;AH=0
            MOV    AL,DATA1      ;GET OPERAND 1
            MOV    BL,DATA2      ;GET OPERAND 2
            ADD    AL,BL         ;ADD THEM
            JNO    OVER          ;IF OF=0 THEN GO TO OVER
            MOV    AL,DATA2      ;OTHERWISE GET OPERAND 2 TO
            CBW                  ;SIGN EXTEND IT
            MOV    BX,AX         ;SAVE IT IN BX
            MOV    AL,DATA1      ;GET BACK OPERAND 1 TO
            CBW                  ;SIGN EXTEND IT
            ADD    AX,BX         ;ADD THEM AND
OVER:       MOV    RESULT,AX     ;SAVE IT
```

The following is an analysis of the values in Example 6-10. Each is sign-extended and then added as follows:

```
S     AH          AL
0     000 0000    0110 0000    +96    after sign extension
0     000 0000    0100 0110    +70    after sign extension
0     000 0000    1010 0110    +166
```

As a rule, if the possibility of overflow exists, all byte-sized signed numbers should be sign-extended into a word, and similarly, all word-sized signed operands should be sign-extended before they are processed. This will be shown shortly in Program 6-1. Before discussing that, it is important to understand the division and multiplication of signed operands.

IDIV (Signed number division)

The Intel manual says that IDIV means "integer division"; it is used for signed number division. In actuality, all arithmetic instructions of the 8088/86 are for integer numbers regardless of whether the operands are signed or unsigned. To perform operations on real numbers, the 8087 coprocessor is used. Remember that real numbers are the ones with decimal points such as "3.56". Division of signed numbers is very similar to the division of unsigned numbers discussed in Chapter 3. Table 6-1 summarizes signed number division.

Table 6-1: Signed Division Summary

Division	Numerator	Denominator	Quotient	Rem.
byte/byte	AL = byte CBW	register or memory	AL	AH
word/word	AX = word CWD	register or memory	AX	DX
word/byte	AX = word	register or memory	AL[1]	AH
doubleword/word	DXAX = doubleword	register or memory	AX[2]	DX

Notes:
1. Divide error interrupt if –127 > AL > +127.
2. Divide error interrupt if –32,767 > AL > +32,767.

Table 6-2: Signed Multiplication Summary

Multiplication	Operand 1	Operand 2	Result
byte × byte	AL	register or memory	AX[1]
word × word	AX	register or memory	DX AX[2]
word × byte	AL = byte CBW	register or memory	DX AX[2]

Notes:
1. CF = 1 and OF = 1 if AH has part of the result, but if the result is not large enough to need the AH, the sign bit is copied to the unused bits and the CPU makes CF = 0 and OF = 0 to indicate that.
2. CF = 1 and OF = 1 if DX has part of the result, but if the result is not large enough to need the DX, the sign bit is copied to the unused bits and the CPU makes CF = 0 and OF = 0 to indicate that. One can use the J condition to find out which of the conditions above has occurred. The rest of the flags are undefined.

```
TITLE       PROG 6-1            FIND THE AVERAGE TEMPERATURE
PAGE        60,132
            .MODEL STMALL
            .STACK 64
;-----------------------
            .DATA
SIGN_DAT    DB +13,-10,+19,+14,-18,-9,+12,-19,+16
            ORG 0010H
AVERAGE     DW ?
REMAINDER   DW ?
;-----------------------
     .CODE
MAIN PROC   FAR
     MOV    AX,@DATA
     MOV    DS,AX
     MOV    CX,9                    ;LOAD COUNTER
     SUB    BX,BX                   ;CLEAR BX, USED AS ACCUMULATOR
     MOV    SI,OFFSET SIGN_DAT      ;SET UP POINTER
BACK:MOV    AL,[ SI]                ;MOVE BYTE INTO AL
     CBW                            ;SIGN EXTEND INTO AX
     ADD    BX,AX                   ;ADD TO BX
     INC    SI                      ;INCREMENT POINTER
     LOOP   BACK                    ;LOOP IF NOT FINISHED
     MOV    AL,9                    ;MOVE COUNT TO AL
     CBW                            ;SIGN EXTEND INTO AX
     MOV    CX,AX                   ;SAVE DENOMINATOR IN CX
     MOV    AX,BX                   ;MOVE SUM TO AX
     CWD                            ;SIGN EXTEND THE SUM
     IDIV   CX                      ;FIND THE AVERAGE
     MOV    AVERAGE,AX              ;STORE THE AVERAGE (QUOTIENT)
     MOV    REMAINDER,DX            ;STORE THE REMAINDER
     MOV    AH,4CH
     INT    21H                     ;GO BACK TO DOS
MAIN ENDP
     END    MAIN
```

Program 6-1

IMUL (Signed number multiplication)

Signed number multiplication is similar in its operation to the unsigned multiplication described in Chapter 3. The only difference between them is that the operands in signed number operations can be positive or negative; therefore, the result must indicate the sign. Table 6-2 summarizes signed number multiplication; it is similar to Table 3-1.

An application of signed number arithmetic is given in Program 6-1. It computes the average of the following Celsius temperatures: +13, −10, +19, +14, −18, −9, +12, −19, and +16.

The program is written in such a way as to handle any overflow that may occur. In Program 6-1, each byte of data was sign-extended and added to BX, computing the total sum, which is a signed word. Then the sum and the count were sign-extended, and by dividing the total sum by the count (number of bytes, which in this case is 9), the average was calculated.

The following is the ".lst" file of Program 6-1. Notice the signed number format provided by the assembler.

```
Microsoft (R) Macro Assembler Version 5.10       12/30/91 06:14:2
PROG6-1                                                   Page 1-1


                          TITLE PROG6-1 FIND THE AVERAGE TEMPERATURE
                          PAGE 60,132
 0000                     .MODEL SMALL
 0000                     .STACK 64
                          ;------------------
 0000                     .DATA
 0000 0D F6 13 0E EE F7   SIGN_DAT   DB +13,-10,+19,+14,-18,-9,+12,-19,+16
 0C F7 10
 0010                         ORG        0010H
 0010 0000                    AVERAGE    DW ?
 0012 0000                    REMAINDER  DW ?
                              ;------------------
 0014                   .CODE
 0000                   MAIN  PROC  FAR
 0000 B8 0000s                MOV   AX,@DATA
 0003 8E D8                   MOV   DS,AX
 0005 B9 0009                 MOV   CX,9  ;LOAD COUNTER
 0008 2B DB                   SUB   BX,BX       ;CLEAR BX, USED AS ACC
 000A BE 0000 R               MOV   SI,OFFSET SIGN_DAT ;SET UP POINTER
 000D 8A 04             BACK: MOV   AL,[SI]     ;MOVE BYTE INTO AL
 000F 98                      CBW         ;SIGN EXTEND INTO AX
 0010 03 D8                   ADD   BX,AX       ;ADD TO BX
 0012 46                      INC   SI    ;INCREMENT POINTER
 0013 E2 F8                   LOOP  BACK   ;LOOP IF NOT FINISHED
 0015 B0 09                   MOV   AL,9  ;MOVE COUNT TO AL
 0017 98                      CBW         ;SIGN EXTEND INTO AX
 0018 8B C8                   MOV   CX,AX       ;SAVE DENOMINATOR IN CX
 001A 8B C3                   MOV   AX,BX       ;MOVE SUM TO AX
 001C 99                      CWD         ;SIGN EXTEND THE SUM
 001D F7 F9                   IDIV  CX    ;FIND THE AVERAGE
 001F A3 0010 R               MOV   AVERAGE,AX ;STORE THE AVERAGE (QU
 0022 89 16 0012 R            MOV   REMAINDER,DX ;STORE THE REMAINDER
 0026 B4 4C                   MOV   AH,4CH
 0028 CD 21                   INT   21H   ;GO BACK TO DOS
 002A                   MAIN  ENDP
                              END   MAIN
```

List File for Program 6-1

Arithmetic shift

As discussed in Chapter 3, there are two types of shifts: logical and arithmetic. Logical shift, which is used for unsigned numbers, was discussed previously. The arithmetic shift is used for signed numbers. It is basically the same as the logical shift, except that the sign bit is copied to the shifted bits. SAR (shift arithmetic right) and SAL (shift arithmetic left) are two instructions for the arithmetic shift.

SAR (shift arithmetic right)

```
SAR    destination,count
```

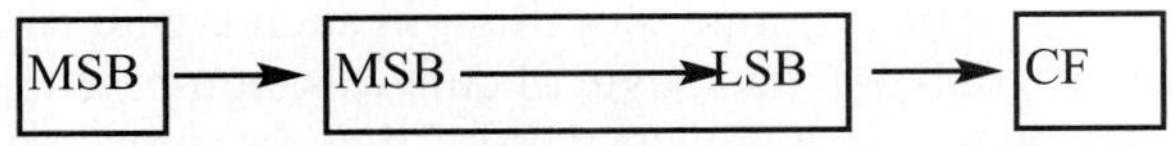

As the bits of the destination are shifted to the right into CF, the empty bits are filled with the sign bit. One can use the SAR instruction to divide a signed number by 2, as shown next:

```
MOV    AL,-10          ;AL=-10=F6H=1111 0110
SAR    AL,1            ;AL is arithmetic shifted right once
                       ;AL=1111 1011=FDH=-5
```

Example 6-11 demonstrates the use of the SAR instruction.

Example 6-11

Using DEBUG, evaluate the results of the following:

```
MOV    AX,-9
MOV    BL,2
IDIV   BL              ;divide -9 by 2 results in FCH
MOV    AX,-9
SAR    AX,1            ;divide -9 by 2 with arithmetic shift
                       ;results in FBH
```

Solution:

The DEBUG trace demonstrates that an IDIV of –9 by 2 gives FCH (– 4), whereas SAR –9 gives FBH (–5). This is because SAR rounds negative numbers down but IDIV rounds up.

SAL (shift arithmetic left) and SHL (shift left)

These two instructions do exactly the same thing. It is basically the same instruction with two mnemonics. As far as signed numbers are concerned, there is no need for SAL. For a discussion of SHL (SAL), see Chapter 3.

Signed number comparison

```
CMP    dest,source
```

Although the CMP (compare) instruction is the same for both signed and unsigned numbers, the J condition instruction used to make a decision for the signed numbers is different from that used for the unsigned numbers. While in unsigned number comparisons CF and ZF are checked for conditions of larger, equal, and smaller (see Chapter 3), in signed number comparison, OF, ZF, and SF are checked:

```
destination > source    OF=SF or ZF=0
destination = source    ZF=1
destination < source    OF=negation of SF
```

The mnemonics used to detect the conditions above are as follows:

```
JG     Jump Greater              jump if OF=SF or ZF=0
JGE    Jump Greater or Equal     jump if OF=SF
JL     Jump Less                 jump if OF=inverse of SF
JLE    Jump Less or Equal        jump if OF=inverse of SF or ZF=1
JE     Jump if Equal             jump of ZF = 1
```

Example 6-12 should help clarify how the condition flags are affected by the compare instruction. Program 6-2 is an example of the application of the signed number comparison. It uses the data in Program 6-1 and finds the lowest temperature.

Example 6-12

Show the DEBUG trace of the following instructions comparing several signed numbers.

```
        MOV   AL,-5
        CMP   AL,-9
        CMP   AL,-2
        CMP   AL,-5
        CMP   AL,+7
```

Solution:

```
C>debug
-a 100
103D:0100 mov al,-5
103D:0102 cmp al,-9
103D:0104 cmp al,-2
103D:0106 cmp al,-5
103D:0108 cmp al,7
103D:010A int 3
103D:010B
-t=100,5

AX=00FB BX=0000 CX=0000 DX=0000 SP=CFDE BP=0000 SI=0000 DI=0000
DS=103D ES=103D SS=103D CS=103D IP=0102 NV UP DI PL NZ NA PO NC
103D:0102 3CF7 CMP AL,F7

AX=00FB BX=0000 CX=0000 DX=0000 SP=CFDE BP=0000 SI=0000 DI=0000
DS=103D ES=103D SS=103D CS=103D IP=0104 NV UP DI PL NZ NA PO NC
103D:0104 3CFE CMP AL,FE

AX=00FB BX=0000 CX=0000 DX=0000 SP=CFDE BP=0000 SI=0000 DI=0000
DS=103D ES=103D SS=103D CS=103D IP=0106 NV UP DI NG NZ AC PO CY
103D:0106 3CFB CMP AL,FB

AX=00FB BX=0000 CX=0000 DX=0000 SP=CFDE BP=0000 SI=0000 DI=0000
DS=103D ES=103D SS=103D CS=103D IP=0108 NV UP DI PL ZR NA PE NC
103D:0108 3C07 CMP AL,07

AX=00FB BX=0000 CX=0000 DX=0000 SP=CFDE BP=0000 SI=0000 DI=0000
DS=103D ES=103D SS=103D CS=103D IP=010A NV UP DI NG NZ NA PO NC
103D:010A CC INT 3

-q
```

```
TITLE        PROG6-2      ;FIND THE LOWEST TEMPERATURE
PAGE         60,132
;-----------------
             .MODEL SMALL
             .STACK 64
;-----------------
             .DATA
SIGN_DAT     DB     +13,-10,+19,+14,-18,-9,+12,-19,+16
             ORG    0010H
LOWEST       DB     ?
;-----------------
        .CODE
MAIN  PROC   FAR
      MOV    AX,@DATA
      MOV    DS,AX
      MOV    CX,8                       ;LOAD COUNTER (NUMBER ITEMS - 1)
      MOV    SI,OFFSET SIGN_DAT         ;SET UP POINTER
      MOV    AL,[SI]              ;AL HOLDS LOWEST VALUE FOUND SO FAR
BACK:INC     SI                   ;INCREMENT POINTER
      CMP    AL,[SI]              ;COMPARE NEXT BYTE TO LOWEST
      JLE    SEARCH               ;IF AL IS LOWEST, CONTINUE SEARCH
      MOV    AL,[SI]              ;OTHERWISE SAVE NEW LOWEST
SEARCH:LOOP BACK                  ;LOOP IF NOT FINISHED
      MOV    LOWEST,AL            ;SAVE LOWEST TEMPERATURE
      MOV    AH,4CH
      INT    21H                  ;GO BACK TO DOS
MAIN ENDP
      END    MAIN
```

Program 6-2

Review Questions

1. In an 8-bit operand, bit _____ is used for the sign bit, whereas in a 16-bit operand, bit _____ is used for the sign bit.
2. Convert 16H to its 2's complement representation.
3. The range of byte-sized signed operands is –______ to +_____. The range of word-sized signed operands is –__________ to +__________.
4. Explain the difference between an overflow and a carry .
5. Explain the purpose of the CBW and CWD instructions. Demonstrate the effect of CBW on AL = F6H. Demonstrate the effect of CWD on AX = 124CH.
6. The instruction for signed multiplication is ________. The instruction for signed division is _______.
7. Explain the difference between the SHR (discussed in Chapter 3) and SAR instructions.
8. For each of the following instructions, indicate the flag condition necessary for each jump to occur:
   (a) JLE   (b) JG

SECTION 6.2: STRING AND TABLE OPERATIONS

There is a group of instructions referred to as *string instructions* in the x86 family of microprocessors. They are capable of performing operations on a series of operands located in consecutive memory locations. For example, while the CMP instruction can compare only 2 bytes (or words) of data, the CMPS (compare string) instruction is capable of comparing two arrays of data located in memory locations pointed at by the SI and DI registers. These instructions are very powerful and can be used in many applications, as will be shown shortly.

Use of SI and DI, DS and ES in string instructions

For string operations to work, designers of CPUs must set aside certain registers for specific functions. These registers must permanently provide the source and destination operands. This is exactly what the designers of the x86 have done. In 8088/86 microprocessors, the SI and DI registers always point to the source and destination operands, respectively. Now the question is: Which segments are they combined with to generate the 20-bit physical address? To generate the physical address, the 8088/86 always uses SI as the offset of the DS (data segment) register and DI as the offset of ES (extra segment). This is the default mode. It must be noted that the ES register must be initialized for the string operation to work.

Byte and word operands in string instructions

In each of the string instructions, the operand can be a byte or a word. Operands are distinguished by the letters B (byte) and W (word) in the instruction mnemonic. Table 6-3 provides a summary of all the string instructions. Each one will be discussed separately in the context of examples.

Table 6-3: String Operation Summary

Instruction	Mnemonic	Destination	Source	Prefix
move string byte	MOVSB	ES:DI	DS:SI	REP
move string word	MOVSW	ES:DI	DS:SI	REP
store string byte	STOSB	ES:DI	AL	
store string word	STOSW	ES:DI	AX	REP
load string byte	LODSB	AL	DS:SI	none
load string word	LODSW	AX	DS:SI	none
compare string byte	CMPSB	ES:DI	DS:SI	REPE/REPNE
compare string word	CMPSW	ES:DI	DS:SI	REPE/REPNE
scan string byte	SCASB	ES:DI	AL	REPE/REPNE
scan string byte	SCASW	ES:DI	AX	REPE/REPNE

DF, the direction flag

To process operands located in consecutive memory locations requires that the pointer be incremented or decremented. In string operations this is achieved by the direction flag. Of the 16 bits of the flag register (D0–D15), bit 11 (D10) is set aside for the direction flag (DF). It is the job of the string instruction to increment or decrement the SI and DI pointers, but it is the job of the programmer to specify the choice of increment or decrement by setting the direction flag to high or low. The instructions CLD (clear direction flag) and STD (set direction flag) are specifically designed for that.

CLD (clear direction flag) will reset (put to zero) DF, indicating that the string instruction should increment the pointers automatically. This automatic incrementation sometimes is referred to as *autoincrement*.

STD (set the direction flag) performs the opposite function of the CLD instruction. It sets DF to 1, indicating to the string instruction that the pointers SI and DI should be decremented automatically.

REP prefix

The REP (repeat) prefix allows a string instruction to perform the operation repeatedly. Now the question is: How many times is it repeated? REP assumes that CX holds the number of times that the instruction should be repeated. In other words, the REP prefix tells the CPU to perform the string operation and then decrements the CX register automatically. This process is repeated until CX becomes zero. To understand some of the concepts discussed so far, look at Example 6-13.

Example 6-13

Using string instructions, write a program that transfers a block of 20 bytes of data.

Solution:

```
;in the data segment:
DATA1 DB          'ABCDEFGHIJKLMNOPQRST'
      ORG  30H
DATA2 DB          20 DUP (?)

;in the code segment:
     MOV   AX,@DATA
     MOV   DS,AX               ;INITIALIZE THE DATA SEGMENT
     MOV   ES,AX               ;INITIALIZE THE EXTRA SEGMENT
     CLD                       ;CLEAR DIRECTION FLAG FOR AUTOINCREMENT
     MOV   SI,OFFSET DATA1     ;LOAD THE SOURCE POINTER
     MOV   DI,OFFSET DATA2     ;LOAD THE DESTINATION POINTER
     MOV   CX,20               ;LOAD THE COUNTER
     REP   MOVSB               ;REPEAT UNTIL CX BECOMES ZERO
```

In Example 6-13, after the transfer of every byte by the MOVSB instruction, both the SI and DI registers are incremented automatically once only (notice CLD). The REP prefix causes the CX counter to be decremented and MOVSB is repeated until CX becomes zero. Notice in Example 6-13 that both DS and ES are set to the same value.

An alternative solution for Example 6-13 would change only two lines of code:

```
MOV   CX,10
REP   MOVSW
```

In this case the MOVSW will transfer a word (2 bytes) at a time and increment the SI and DI registers each twice. REP will repeat that process until CX becomes zero. Notice that the CX has the value of 10 in it since 10 words is equal to 20 bytes.

STOS and LODS instructions

The STOSB instruction stores the byte in the AL register into memory locations pointed at by ES:DI and increments (if DF = 0) DI once. If DF = 1, then DI is decremented. The STOSW instruction stores the contents of AX in memory locations ES:DI and ES:DI+1 (AL into ES:DI and AH into ES:DI+1), then increments DI twice (if DF = 0). If DF = 1, DI is decremented twice.

The LODSB instruction loads the contents of memory locations pointed at by DS:SI into AL and increments (or decrements) SI once if DF = 0 (or DF = 1). LODSW loads the contents of memory locations pointed at by DS:SI into AL and DS:SI+1 into AH. The SI is incremented twice if DF = 0. Otherwise, it is decremented twice. LODS is never used with a REP prefix.

Testing memory using STOSB and LODSB

Example 6-14 uses string instructions STOSB and LODSB to test an area of RAM memory. In the program in Example 6-14, first AAH is written into 100 locations by using word-sized operand AAAAH and a count of 50. In the test part, LODSB brings in the contents of memory locations into AL one by one, and each time it is eXclusive-ORed with AAH (the AH register has the hex value of AA). If they are the same, ZF = 1 and the process is continued. Otherwise, the pattern written there by the previous routine is not there and the program will exit. This, in concept, is somewhat similar to the routine used in the IBM PC's BIOS except that the BIOS routine is much more involved and uses several different patterns of data for the test and it can be used to test any part of RAM, either the main RAM or the video RAM.

Example 6-14

Write a program that:
(1) Uses STOSB to store byte AAH in 100 memory locations.
(2) Uses LODS to test the contents of each location to see if AAH is there. If the test fails, the system should display the message "bad memory".

Solution:

Assuming that ES and DS have been assigned in the ASSUME directive, the following is from the code segment:

```
            ;PUT PATTERN AAAAH IN TO 50 WORD LOCATIONS
            MOV    AX,DTSEG              ;INITIALIZE
            MOV    DS,AX                 ;DS REG
            MOV    ES,AX                 ;AND ES REG
            CLD                          ;CLEAR DF FOR INCREMENT
            MOV    CX,50                 ;LOAD THE COUNTER (50 WORDS)
            MOV    DI,OFFSET MEM_AREA ;LOAD THE POINTER FOR DESTINATION
            MOV    AX,0AAAAH             ;LOAD THE PATTERN
            REP    STOSW                 ;REPEAT UNTIL CX=0
            ;BRING IN THE PATTERN AND TEST IT ONE BY ONE
            MOV    SI,OFFSET MEM_AREA ;LOAD THE POINTER FOR SOURCE
            MOV    CX,100                ;LOAD THE COUNT (COUNT 100 BYTES)
AGAIN:      LODSB                        ;LOAD INTO AL FROM DS:SI
            XOR    AL,AH                 ;IS PATTERN THE SAME?
            JNZ    OVER                  ;IF NOT THE SAME THEN EXIT
            LOOP   AGAIN                 ;CONTINUE UNTIL CX=0
            JMP    EXIT                  ;EXIT PROGRAM
OVER:       MOV    AH,09                 ;{ DISPLAY
            MOV    DX, OFFSET MESSAGE ;{ THE MESSAGE
            INT    21H                   ;{ ROUTINE
EXIT: ..
```

The REPZ and REPNZ prefixes

These prefixes can be used with the CMPS and SCAS instructions for testing purposes. They are explained below.

REPZ (repeat zero), which is the same as REPE (repeat equal), will repeat the string operation as long as the source and destination operands are equal (ZF = 1) or until CX becomes zero.

REPNZ (repeat not zero), which is the same as REPNE (repeat not equal), will repeat the string operation as long as the source and destination operands are not equal (ZF = 0) or until CX becomes zero. These two prefixes will be used in the context of applications after the explanation of the CMPS and SCANS instructions.

CMPS (compare string) allows the comparison of two arrays of data pointed at by the SI and DI registers. One can test for the equality or inequality of data by use of the REPE or REPNE prefixes, respectively. The comparison can be performed a byte at a time or a word at time by using CMPSB or CMPSW.

For example, if comparing "Euorop" and "Europe" for equality, the comparison will continue using the REPE CMPS as long as the two arrays are the same.

```
;from the data segment:

DATA1       DB'Europe'
DATA2       DB'Euorop'

     ;from the code segment:
     CLD                     ;DF=0 for increment
     MOV   SI,OFFSET DATA1   ;SI=DATA1 offset
     MOV   DI,OFFSET DATA2   ;DI=DATA2 offset
     MOV   CX,06             ;load the counter
     REPE  CMPSB             ;repeat until not equal or CX=0
```

Example 6-15

Assuming that there is a spelling of "Europe" in an electronic dictionary and a user types in "Euorope", write a program that compares these two and displays the following message, depending on the result:
1. If they are equal, display "The spelling is correct".
2. If they are not equal, display "Wrong spelling".

Solution:

```
 DAT_DICT    DB     'Europe'
 DAT_TYPED   DB     'Euorope'
 MESSAGE1    DB     'The spelling is correct','$'
 MESSAGE2    DB     'Wrong spelling','$'

;from the code segment:
             CLD                              ;DF=0 FOR INCREMENT
             MOV    SI,OFFSET DAT_DICT        ;SI=DATA1 OFFSET
             MOV    DI,OFFSET DAT_TYPED       ;DI=DATA2 OFFSET
             MOV    CX,06                     ;LOAD THE COUNTER
             REPE   CMPSB          ;REPEAT AS LONG AS EQUAL OR UNTIL CX=0
             JE     OVER                  ;IF ZF=1 THEN DISPLAY MESSAGE1
             MOV    DX,OFFSET MESSAGE2    ;IF ZF=0 THEN DISPLAY MESSAGE2
             JMP    DISPLAY
 OVER:       MOV    DX,OFFSET MESSAGE1
 DISPLAY:    MOV    AH,09
             INT    21H
```

In the case above, the two arrays are to be compared letter by letter. The first characters pointed at by SI and DI are compared. In this case they are the same ("E"), so the zero flag is set to 1 and both SI and DI are incremented. Since ZF = 1, the REPE prefix repeats the comparison. This process is repeated until the third letter is reached. The third letters "o" and "r" are not the same; therefore, ZF is reset to zero and the comparison will stop. ZF can be used to make the decision as shown in Example 6-15.

One could juggle the code in Example 6-15 to make it more efficient and use fewer jumps, but for the sake of clarity it is presented in this manner.

CMPS can be used to test inequality of two arrays using "REPNE CMPSB". For example, when comparing the following social security numbers, the comparison will continue to the last digit since no two digits in the same position are the same.

```
231-24-7659              564-77-1338
```

SCASB (scan string)

The SCASB string instruction compares each byte of the array pointed at by ES:DI with the contents of the AL register, and depending on which prefix, REPE or REPNE, is used, a decision is made for equality or inequality. For example, in the array "Mr. Gones", one can scan for the letter "G" by loading the AL register with the character "G" and then using the "`REPNE SCASB`" operation to look for that letter.

```
;in the data segment:
;DATA1        DB    'Mr. Gones'
;and in the code segment:
       CLD                          ;DF=0 FOR INCREMENT
       MOV    DI,OFFSET DATA1       ;DI=ARRAY OFFSET
       MOV    CX,09                 ;LENGTH OF ARRAY
       MOV    AL,'G'                ;SCANNING FOR THE LETTER 'G'
       REPNE  SCASB                 ;REPEAT THE SCANNING IF NOT EQUAL
                                    ;OR UNTIL THE CX IS ZERO
```

Example 6-16

Write a program that scans the name "Mr. Gones" and replaces the "G" with the letter "J", then displays the corrected name.

Solution:

```
 ;in the data segment:
 DATA1       DB     'Mr. Gones','$'

 ;and in the code segment:

             MOV    AX,@DATA
             MOV    DS,AX
             MOV    ES,AX
             CLD                         ;DF=0 FOR INCREMENT
             MOV    DI,OFFSET DATA1      ;ES:DI=ARRAY OFFSET
             MOV    CX,09                ;LENGTH OF ARRAY
             MOV    AL,'G'               ;SCANNING FOR THE LETTER 'G'
             REPNE  SCASB                ;REPEAT THE SCANNING IF NOT EQUAL
;or
             JNE    OVER                 ;UNTIL CX IS ZERO. JUMP IF Z=0
             DEC    DI                   ;DECREMENT TO POINT AT 'G'
             MOV    BYTE PTR [ DI] ,'J'  ;REPLACE 'G' WITH 'J'
 OVER:       MOV    AH,09                ;DISPLAY
             MOV    DX,OFFSET DATA1      ;THE
             INT    21H                  ;CORRECTED NAME
```

In Example 6-16, the letter "G" is compared with "M". Since they are not equal, DI is incremented and CX is decremented, and the scanning is repeated until the letter "G" is found or the CX register is zero. In this example, since 'G" is found, ZF is set to 1 (ZF = 1), indicating that there is a letter 'G" in the array.

Replacing the scanned character

SCASB can be used to search for a character in an array, and if it is found, it will be replaced with the desired character. See Example 6-16.

In string operations the pointer is incremented after each execution (that is, if DF = 0). Therefore, in the example above, DI must be decremented, causing the pointer to point to the scanned character and then replace it.

XLAT instruction and look-up tables

There is often a need in computer applications for a table that holds some important information. To access the elements of the table, 8088/86 microprocessors provide the XLAT (translate) instruction. To understand the XLAT instruction, one must first understand tables. The table is commonly referred to as a look-up table. Assume that one needs a table for the values of x^2, where x is between 0 and 9. First the table is generated and stored in memory:

```
SQUR_TABLE   DB     0,1,4,9,16,25,36,49,64,81
```

Now one can access the square of any number from 0 to 9 by the use of XLAT. To do that, the register BX must have the offset address of the look-up table, and the number whose square is sought must be in the AL register. Then after the execution of XLAT, the AL register will have the square of the number. The following shows how to get the square of 5 from the table:

```
        MOV    BX,OFFSET SQUR_TABLE ;load the offset address of table
        MOV    AL,05           ;AL=05 will retrieve 6th element
        XLAT                   ;pull the element out of table
                               ;and put in AL
```

After execution of this program, the AL register will have 25 (19H), the square of 5. It must be noted that for XLAT to work, the entries of the look-up table must be in sequential order and must have a one-to-one relation with the element itself. This is because of the way XLAT works. In actuality, XLAT is one instruction, which is equivalent to the following code:

```
        SUB    AH,AH           ;AH=0
        MOV    SI,AX           ;SI=000X
        MOV    AL,[ BX+SI]     ;GET THE SIth ENTRY FROM BEGINNING
                               ;OF THE TABLE POINTED AT BY BX
```

In other words, if there was no XLAT instruction, the code above would do the same thing, and this is the way many RISC processors perform this operation. Now why would one want to use XLAT to get the square of a number from a look-up table when there is the MUL instruction? The answer is that MUL takes longer.

Code conversion using XLAT

In many microprocessor-based systems, the keyboard is not an ASCII type of keyboard. One can use XLAT to translate the hex keys of such keyboards to ASCII. Assuming that the keys are 0–F, the following is the program to convert the hex digits of 0–F to their ASCII equivalents.

```
;data segment:
ASC_TABL     DB     '0','1','2','3','4','5','6','7','8'
             DB     '9','A','B','C','D','E','F'
HEX_VALU     DB     ?
ASC_VALU     DB     ?
;code segment:
             MOV    BX,OFFSET ASC_TABL       ;BX= TABLE OFFSET
             MOV    AL,HEX_VALU              ;AL=THE HEX DATA
             XLAT                       ;GET THE ASCII EQUIVALENT
             MOV    ASC_VALU,AL              ;MOVE IT TO MEMORY
```

Review Questions

1. In string operations, register _____ is used to point to the source operand and register _____ is used to point to the destination operand.
2. SI is used as an offset into the _____ segment, and DI is used as an offset into the

_____ segment.

3. The _____ flag, bit ____ of the flag register, is used to tell the CPU whether to increment or decrement pointers in repeated string operations.
4. State the purpose of instructions CLD and STD.
5. If a string instruction is repeatedly executed because of a REP prefix, how does the CPU know when to stop repeating it?
6. In the following program segment, what condition will cause the REPNZ to fail?

```
MOV    SI, OFFSET DATA1
MOV    DI, OFFSET DATA2
MOV    CX,LENGTH
REPNZ  CMPSB
```

PROBLEMS

SECTION 6.1: SIGNED NUMBER ARITHMETIC OPERATIONS

1. Show how the x86 computer would represent the following numbers and verify each with DEBUG.
   (a) –23 (b) +12 (c) –28H (d) +6FH
   (e) –128 (f) +127 (g) +365 (h) –32,767
2. Find the overflow flag for each case and verify the result using DEBUG.
   (a) (+15) + (–12) (b) (–123) + (–127) (c) (+25H) + (+34)
   (d) (–127) + (+127) (e) (+1000) + (–100)
3. Sign-extend the following and write simple programs in DEBUG to verify them.
   (a) –122 (b) –999H (c) +17H
   (d) +127 (e) –129
4. Modify Program 6-2 to find the highest temperature. Verify your program.

SECTION 6.2: STRING AND TABLE OPERATIONS

5. Which instructions are used to set and reset the direction flag? State the purpose of the direction flag.
6. The REP instruction can be used with which of the following instructions?
   (a) MOVSB (b) MOVSW (c) CMPSB
   (d) LODSB (e) STOSW (f) SCASW
7. In Problem 6, state the source and destination operand for each instruction.
8. Write and verify a program that transfers a block of 200 words of data.
9. Use instructions LODSx and STOSx to mask the 3 from a set of 50 ASCII digits and transfer the result to a different memory location. This involves converting from ASCII to unpacked BCD, then storing it at a different location; for example,

	source	destination
ASCII for '5'	0011 0101	0000 0101

10. Which prefix is used for the inequality case for CMPS and SCAS instructions?
11. Write a program that scans the initials "IbM" and replaces the lowercase "b" with uppercase "B".
12. Using the timing chart in Appendix B.2, compare the clock count of the instruction XLAT and its equivalent to see which is more efficient.
13. Write a program using a look-up table and XLAT to retrieve the y value in the equation $y = x^2 + 2x + 5$ for x values of 0 to 9.

ANSWERS TO REVIEW QUESTIONS

SECTION 6.1: SIGNED NUMBER ARITHMETIC OPERATIONS

1. D7, D15
2. 16H = 0001 0110_2 in 2's complement: 1110 1010_2
3. –128 to +127; –32,768 to +32,767 (decimal)
4. An overflow is a carry into the sign bit; a carry is a carry out of the register.
5. The CBW instruction sign extends the sign bit of a byte into a word; the CWD instruction sign extends the sign bit of a word into a doubleword.
   F6H sign-extended into AX = FFF6H
   124C sign-extended into DX AX would be DX = 0000 and AX = 124CH.
6. IMUL, IDIV
7. SHR shifts each bit right one position and fills the MSB with zero.
   SAR shifts each bit right one position and fills the MSB with the sign bit in each; the LSB is shifted into the carry flag.
8. (a) JLE will jump if OF is the inverse of SF, or if ZF = 1.
   (b) JG will jump if OF equals SF, or if ZF = 0.

SECTION 6.2: STRING AND TABLE OPERATIONS

1. SI, DI
2. Data, extra
3. Direction, 11 or D10
4. CLD clears DF to 0; STD sets DF to 1
5. When CX = 0
6. If CX = 0 or the point at which DATA1 and DATA2 are not equal

CHAPTER 7

MODULES AND MODULAR PROGRAMMING

OBJECTIVES

Upon completion of this chapter, you will be able to:

- **>> Discuss the advantages of modular programming**
- **>> Break large programs into modules, code modules and calling programs**
- **>> Declare names that are defined externally via the EXTRN directive**
- **>> Link subprograms together into one executable program**
- **>> Code segment directives to link data, code, or stack segments from different modules into one segment**
- **>> Code programs using the full segment definitions**
- **>> List the various methods of passing parameters to modules and discuss the advantages and disadvantages of each**
- **>> Code programs passing the parameters via registers, memory, or stack**

In this chapter the concept of modules is presented. In Section 7.1, modules are discussed along with rules for writing modules and linking them together. In Sections 7.2 and 7.3, some very useful modules are given, along with the methods of passing parameters among various modules.

SECTION 7.1: WRITING AND LINKING MODULES

Why modules?

It is common practice in writing software packages to break down the project into small modules and distribute the task of writing those modules among several programmers. This not only makes the project more manageable but also has other advantages, such as:

1. Each module can be written, debugged, and tested individually.
2. The failure of one module does not stop the entire project.
3. The task of locating and isolating any problem is easier and less time consuming.
4. One can use the modules to link with high-level languages such as C/C++, C#, or Visual Basic.
5. Parallel development shortens considerably the time required to complete a project.

In this section we explain how to write and link modules to create a single executable program.

Writing modules

In previous chapters, a main procedure was written that called many other subroutines. In those examples, if one subroutine did not work properly, the entire program would have to be rewritten and reassembled. A more efficient way to develop software is to treat each subroutine as a separate program (or module) with a separate filename. Then each one can be assembled and tested. After testing each program and making sure that each works, they can all be brought together (linked) to make a single program. To enable these modules to be linked together, certain Assembly language directives must be used. Among these directives, the two most widely used are EXTRN (external) and PUBLIC. Each is discussed below.

EXTRN directive

The EXTRN directive is used to notify the assembler and linker that certain names and variables which are not defined in the present module are defined externally somewhere else. In the absence of the EXTRN directive, the assembler would show an error since it cannot find where the names are defined. The EXTRN directive has the following format:

```
EXTRN name1:type              ;each name in a separate EXTRN
EXTRN name2:type              ;or
EXTRN name1:type,name2:type   ;many listed in the same EXTRN
```

External procedure names can be NEAR, FAR, or PROC (which will be NEAR for small models or FAR for larger models). The following are the types for data names, with the number of bytes indicated in parentheses: BYTE (1), WORD (2), DWORD (4), FWORD (6), QWORD (8), or TBYTE (10).

PUBLIC directive

Those names or parameters defined as EXTRN (indicating that they are defined outside the present module) must be defined as PUBLIC in the module where they are defined. Defining a name as PUBLIC allows the assembler and linker to match it with its

EXTRN counterpart(s). The following is the format for the PUBLIC directive:

```
PUBLIC name1         ;each name can be in a separate directive
PUBLIC name2

PUBLIC name1, name2 ;or many can be listed in the same PUBLIC
```

Example 7-1 should help to clarify these concepts. It demonstrates that for every EXTRN definition there is a PUBLIC directive defined in another module. In Example 7-1 the EXTRN and PUBLIC directives are related to the name of a FAR procedure.

Example 7-1

Assume there is a program that constitutes the main routine, and two smaller subroutines named SUBPROG1 and SUBPROG2. The subprograms are called from the main routine. The following shows the use of the EXTRN and PUBLIC directives:

Solution:

```
;-------------------------------------------------------------------
;one file will contain the main module:
            EXTRN   SUBPROG1:FAR
            EXTRN   SUBPROG2:FAR
            .MODEL SMALL
            .CODE
MAIN        PROC  FAR
            ...
            CALL    SUBPROG1
            CALL    SUBPROG2
            ...
            MOV     AH,4CH
            INT     21H
MAIN        ENDP
            END     MAIN

;----------------- and in a separate file: ----------------------
             PUBLIC SUBPROG1
             .MODEL SMALL
             .CODE
SUBPROG1     PROC FAR
             ...
             RET
SUBPROG1     ENDP
             END

;--------------- and in another file: ----------------------------
             PUBLIC SUBPROG2
             .MODEL SMALL
             .CODE
SUBPROG2     PROC FAR
             ...
             RET
SUBPROG2     ENDP
             END
```

END directive in modules

In Example 7-1, notice the entry and exit points of the program. The entry point is MAIN and the exit point is "END MAIN". Modules that are called by the main module have the END directive with no label or name after it. Notice that SUBPROG1 and SUBPROG2 each have the END directive with no labels after them.

Linking modules together into one executable unit

Assuming that each program module in Example 7-1 is assembled separately and saved under the filenames EXAMPLE1.OBJ, PROC1.OBJ, and PROC2.OBJ, the following shows how to link them together in MASM in order to generate a single executable file:

```
C>LINK EXAMPLE1.OBJ + PROC1.OBJ + PROC2.OBJ
```

Program 7-1 shows how the EXTRN and PUBLIC directives can also be applied to data variables. In Program 7-1, the main module contains a data segment and a stack segment, but the subroutine modules do not. Each module can have its own data and stack segment. While it is entirely permissible and possible that the modules have their own data segments if they need them, generally there is only one stack that is defined in the main program and it must be defined so that it is combined with the system stack. Later in this chapter we show how to combine many segments of different modules to generate one uniform segment for each segment of code, data, and stack.

Use the program shells in Example 7-1 to:
1. Add two words.
2. Multiply two words.
Each one should be performed by a separate module. The data is defined in the main module, and the add and multiply modules have no data segment of their own.

```
TITLE PROG7-1MM DEMONSTRATES MODULAR PROGRAMMING
PAGE 60,132
      EXTRN SUBPROG1:FAR
      EXTRN SUBPROG2:FAR
      PUBLIC VALUE1, VALUE2, SUM, PRODUCT
            .MODEL SMALL
            .STACK 64
            .DATA
VALUE1      DW 2050
VALUE2      DW 500
SUM         DW 2 DUP (?)
PRODUCT     DW 2 DUP (?)
;-----------------------------
            .CODE
MAIN  PROC  FAR
      MOV   AX,@DATA
      MOV   DS,AX
      CALL  SUBPROG1     ;CALL SUBPROG TO ADD VALUE1 + VALUE2
      CALL  SUBPROG2     ;CALL SUBPROG TO MUL VALUE1 * VALUE2
      MOV   AH,4CH
      INT   21H          ;GO BACK TO OS
MAIN  ENDP
      END   MAIN
```

Program 7-1: Main Module

Analysis of Program 7-1

Notice in the main module that each of the two subroutines was declared with the EXTRN directive, indicating that these procedures would be defined in another file. The external subroutines were defined as FAR in this case. In the files where each subroutine is defined, it is declared as PUBLIC, so that other programs can call it. In the main module, the names VALUE1, VALUE2, SUM, and PRODUCT were defined as PUBLIC, so that other programs could access these data items. In the subprograms, these data items were declared as EXTRN. These three programs would be linked together as follows:

```
C>LINK PROG7-1MM.OBJ + PROG7-1M2 + PROG7-1M3
```

The linker program resolves external references by matching PUBLIC and EXTRN names. The linker program will search through the files specified in the LINK command for the external subroutines. Notice that the filenames are unrelated to the procedure names. "MAIN" is contained in file "PROG7-1MM.OBJ".

```
;THIS PROGRAM FINDS THE SUM OF TWO EXTERNALLY DEFINED WORDS
;AND STORES THE SUM IN A LOCATION DEFINED BY THE CALLING MODULE
;
TITLE PROG7-1M2 PROGRAM TO ADD TWO WORDS
PAGE   60,132
       EXTRN VALUE1:WORD
       EXTRN VALUE2:WORD
       EXTRN SUM:WORD
       PUBLIC SUBPROG1
             .MODEL SMALL
             .CODE
SUBPROG1     PROC  FAR
             SUB   BX,BX        ;INITIALIZE CARRY COUNT
             MOV   AX,VALUE1
             MOV   DX,VALUE2
             ADD   AX,DX        ;ADD VALUE1 + VALUE2
             ADC   BX,00        ;ACCUMULATE CARRY
             MOV   SUM,AX       ;STORE SUM
             MOV   SUM+2,BX     ;STORE CARRY
             RET
SUBPROG1     ENDP
             END
```

Program 7-1: Module 2

```
;THIS PROGRAM FINDS THE PRODUCT OF TWO EXTERNALLY DEFINED WORDS
;AND STORES THE PRODUCT IN A LOCATION DEFINED BY THE CALLING MODULE
;
TITLE        PROG7-1M3 PROGRAM TO MULTIPLY TWO WORDS
PAGE         60,132
             EXTRN VALUE1:WORD
             EXTRN VALUE2:WORD
             EXTRN PRODUCT:WORD
             PUBLIC SUBPROG2
             .MODEL       SMALL
             .CODE
SUBPROG2     PROC  FAR
             MOV   AX,VALUE1
             MOV   CX,VALUE2
             MUL   CX           ;MULTIPLY VALUE1 * VALUE2
             MOV   PRODUCT,AX   ;STORE PRODUCT
             MOV   PRODUCT+2,DX       ;STORE PRODUCT HIGH WORD
             RET
SUBPROG2     ENDP
             END
```

Program 7-1: Module 3

Example 7-2 shows the shell of modular programs using the simplified segment definition. Modular programming with full segment definition is defined later in this section. Notice that in the main module of Example 7-2, the name MAIN has a colon after it and is used for the first executable instruction. This is the entry point of the program. The exit point of the program is indicated by the same label, which must be named in the END directive. No program can have more than one entry and one exit point. The label MAIN was chosen in this instance, but of course any name could have been chosen. Remember that the END directives in other modules do not have a label after the word "END".

Example 7-2

Create a shell for modular programming using the simplified segment definition.

Solution:

Modular program shells for the simplified segment directives are as follows.
The main file will contain:

```
            .MODEL SMALL
            .STACK 64
            .DATA
            ....
            .CODE
            EXTRN  SUBPROG1:NEAR
            EXTRN  SUBPROG2:NEAR
MAIN:       MOV    AX,@DATA      ;this is the program entry point
            MOV    DS,AX
            CALL   SUBPROG1
            CALL   SUBPROG2
            MOV    AH,4CH
            INT    21H
            END    MAIN          ;this is the program exit point

;------------- and in a separate file: -----------------------
            .MODEL SMALL
            .CODE
            PUBLIC SUBPROG1
SUBPROG1    PROC
            ...
            RET
SUBPROG1    ENDP
            END

;--------------- and in another file: -------------------------
            .MODEL SMALL
            .CODE
            PUBLIC SUBPROG2
SUBPROG2    PROC
            ...
            RET
SUBPROG2    ENDP
            END
```

Program 7-2 is the same as Program 7-1, rewritten for the full segment definition. Compare the two programs to see the ease of the simplified segment definition. When using the simplified segment definition shown in Example 7-2, procedures will default to NEAR for small or compact models and to FAR for medium, large, or huge models.

Modular programming and full segment definition

Program 7-2 uses full segment definition to redefine all the segments of Program 7-1. An analysis of how the segments are combined as shown in the link map follows the program. The code segments were not made PUBLIC in this example. Notice that in order to combine various segments from different modules into one segment, the segment names must be the same.

```
TITLE PROG7-2MM   PROG7-1 REWRITTEN WITH FULL SEGMENT DEFINITION
PAGE  60,132
      EXTRN SUBPROG1:FAR
      EXTRN SUBPROG2:FAR
      PUBLIC VALUE1, VALUE2, SUM, PRODUCT
;----------------------------
STSEG SEGMENT PARA STACK 'STACK'
      DB    100 DUP(?)
STSEG ENDS
;----------------------------
DTSEG        SEGMENT PARA 'DATA'
VALUE1       DW 2050
VALUE2       DW 500
SUM          DW 2 DUP (?)
PRODUCT      DW 2 DUP (?)
DTSEG        ENDS
;-----------------------------
CODSG_A      SEGMENT PARA 'CODE'
MAIN  PROC   FAR
      ASSUME CS:CODSG_A,DS:DTSEG,SS:STSEG
      MOV    AX,DTSEG
      MOV    DS,AX
      CALL   SUBPROG1     ;CALL SUBPROG TO ADD VALUE1 + VALUE2
      CALL   SUBPROG2     ;CALL SUBPROG TO MUL VALUE1 * VALUE2
      MOV    AH,4CH
      INT    21H          ;GO BACK TO OS
MAIN  ENDP
CODSG_A      ENDS
      END    MAIN
```

Program 7-2: Main Module

```
;THIS PROGRAM FINDS THE SUM OF TWO EXTERNALLY DEFINED WORDS
;AND STORES THE SUM IN A LOCATION DEFINED BY THE CALLING MODULE
TITLE PROG7-2M2 PROGRAM TO ADD TWO WORDS
PAGE  60,132
      EXTRN VALUE1:WORD
      EXTRN VALUE2:WORD
      EXTRN SUM:WORD
      PUBLIC SUBPROG1
CODSG_B SEGMENT PARA 'CODE'
SUBPROG1     PROC  FAR
             ASSUME CS:CODSG_B
             SUB   BX,BX          ;INITIALIZE CARRY COUNT
             MOV   AX,VALUE1
             MOV   DX,VALUE2
             ADD   AX,DX          ;ADD VALUE1 + VALUE2
             ADC   BX,00          ;ACCUMULATE CARRY
             MOV   SUM,AX         ;STORE SUM
             MOV   SUM+2,BX       ;STORE CARRY
             RET
SUBPROG1     ENDP
CODSG_B      ENDS
             END
```

Program 7-2: Module 2

```
;THIS PROGRAM FINDS THE PRODUCT OF TWO EXTERNALLY DEFINED WORDS
;AND STORES THE PRODUCT IN A LOCATION DEFINED BY THE CALLING MODULE
TITLE PROG7-2M3 PROGRAM TO MULTIPLY TWO WORDS
PAGE  60,132
      EXTRN VALUE1:WORD
      EXTRN VALUE2:WORD
      EXTRN PRODUCT:WORD
      PUBLIC SUBPROG2
CODSG_C     SEGMENT PARA 'CODE'
SUBPROG2    PROC  FAR
            ASSUME CS:CODSG_C
            MOV   AX,VALUE1
            MOV   CX,VALUE2
            MUL   CX                  ;MUL VALUE1 * VALUE2
            MOV   PRODUCT,AX          ;STORE PRODUCT
            MOV   PRODUCT+2,DX        ;STORE PRODUCT HIGH WORD
            RET
SUBPROG2    ENDP
CODSG_C     ENDS
      END
 Start Stop Length Name Class
 00000H 00063H 00064H STSEG STACK
 00070H 0007BH 0000CH DTSEG DATA
 00080H 00092H 00013H CODSG_A CODE
 000A0H 000B5H 00016H CODSG_B CODE
 000C0H 000D0H 00011H CODSG_C CODE
```

Program 7-2: Module 3 and the Link Map

Analysis of Program 7-2 link map

The link map shows the start and end of each segment. Notice that each segment starts at a 16-byte boundary: 00070H, 00080H, etc. The code segment for the main module has the name "CODSG_A", starts at 00080H, and ends at 00092H, taking a total of 00013H bytes. It was classified as 'CODE'. The next code segment is defined under the name "CODSG_B". Notice that it starts at the 16-byte boundary 000A0H since it was defined as PARA. This means that from 00093H to 0009FH is unused. Similarly, the third module starts at 000C0H. Notice that each code segment is separate. They can all be merged together into one segment by using the PUBLIC option. This is shown in Example 7-3. To merge the code segments together, each code segment must have the same name and be declared PUBLIC.

Example 7-3

Show the link map for Program 7-2 rewritten to combine code segments (use PARA boundaries) using directive: `CDSEG SEGMENT PARA PUBLIC 'CODE'`

Solution:

Start	Stop	Length	Name	Class
00000H	00063H	00064H	STSEG	STACK
00070H	0007BH	0000CH	DTSEG	DATA
00080H	000D0H	00051H	CDSEG	CODE

The following are the SEGMENT directives using word boundaries:

```
STSEG SEGMENT WORD STACK 'STACK'
DTSEG SEGMENT WORD 'DATA'
CDSEG SEGMENT WORD PUBLIC 'CODE'
```

The following is the link map when the program used WORD boundaries:

Start	Stop	Length	Name	Class
00000H	00063H	00064H	STSEG	STACK
00064H	0006FH	0000CH	DTSEG	DATA
00070H	000AAH	0003BH	CDSEG	CODE

SEGMENT directive

In previous chapters, when a segment was defined using full segment definition, no other attributes were mentioned after it. It was simply written

```
name   SEGMENT
```

This kind of definition of segments was acceptable since there was only one of each segment of code, data, and stack. However, when there are many modules to be linked together, the segment definition must be adjusted. The complete segment definition used widely in modular programming is as follows:

```
name   SEGMENT alignment combine type class name
```

Appendix C (see SEGMENT) gives a complete description of the fields of the SEGMENT directive. A brief explanation of each field is given below.

The *alignment* field indicates whether a segment should start on a byte, word, paragraph, or page boundary. For example, if WORD is given in the alignment field, the segment will start at the next available word. When the WORD boundary is used, if a previous segment ended at offset 0048H, the next segment will start at 004AH. The default alignment is PARA, meaning that each segment will start on a paragraph boundary. A paragraph in OS is defined as 16 bytes; therefore, each segment will start on a 16-byte boundary. When PARA is used, if the previous segment ended at 0048H, the next segment would begin at the next paragraph boundary, which is 0050H. Paragraph boundaries end in 0; they are evenly divisible by 16 (10H).

The *combine type* field indicates to the linker whether segments of the same type should be linked together. Typical options for combine type are STACK or PUBLIC. An example below shows how to use this field in the stack segment definition to combine the stack segment of a program with the system stack to eliminate the "Warning: no stack segment" message generated by the linker. If the combine type is PUBLIC, the linker will combine that segment with other segments of the same type in other modules. This can be used to combine code segments with various names under a single name.

The *class name* field has four options: 'CODE', 'STACK', 'DATA', and 'EXTRA'. It must be enclosed in single quotes. It is used in combining segments of the same type from various modules.

Complete stack segment definition

The following stack segment definition in the main module will eliminate the "Warning: no stack segment" message generated by the linker:

```
name   SEGMENT      PARA STACK 'STACK'
```

Complete data and code segment definitions

The following is a data segment definition that can be used if no other module has defined any data segment:

```
name   SEGMENT      PARA 'DATA'
```

If any other module has defined a data segment then PUBLIC should be placed between PARA and 'DATA'. The following are the code and data segment definitions to combine segments from different modules:

```
name   SEGMENT      PARA PUBLIC 'CODE'
name   SEGMENT      PARA PUBLIC 'DATA'
```

Example 7-4 rewrites Example 7-2 to define segments using the complete segment definition.

Example 7-4

Create a shell for modular programming using the complete segment definition.

Solution:

The main file will contain:

```
TITLE        PROG  PROGRAM SHELL WITH COMPLETE SEGMENT DEFINITION
PAGE  60,132
      EXTRN SUBPROG1:FAR
      EXTRN SUBPROG2:FAR
      PUBLIC ...              ;declare data here to be shared
;-----------------------------
STSEG        SEGMENT PARA STACK 'STACK'
             DB          100 DUP(?)
STSEG        ENDS
;-----------------------------
DTSEG        SEGMENT PARA 'DATA'
             ;define data here
DTSEG        ENDS
;------------------------------
CODSG_A      SEGMENT PARA 'CODE'
MAIN  PROC   FAR
             ASSUME CS:CODSG_A,DS:DTSEG,SS:STSEG
             MOV    AX,DTSEG
             MOV    DS,AX
             CALL   SUBPROG1     ;CALL SUBPROG
             CALL   SUBPROG2     ;CALL SUBPROG
             ...
             MOV    AH,4CH
             INT    21H   ;GO BACK TO OS
MAIN         ENDP
CODSG_A      ENDS
             END    MAIN

;-----------------and in another file: ----------------------------
TITLE        SUBPROG1 PROGRAM ...
PAGE   60,132
      EXTRN ...             ;declare data that is defined externally
      PUBLIC SUBPROG1       ;declare procedures that are called externally
CODSG_B      SEGMENT PARA 'CODE'
SUBPROG1     PROC  FAR
             ASSUME CS:CODSG_B
 ; the instructions that perform the work of the subroutine go here
             RET
SUBPROG1     ENDP
CODSG_B      ENDS
             END

;-----------------and in another file: ----------------------------
TITLE        SUBPROG2 PROGRAM TO ...
PAGE   60,132
      EXTRN ...             ;declare data that is defined externally
      PUBLIC SUBROG2        ;declare procedures that are called externally
CODSG_C      SEGMENT PARA 'CODE'
SUBPROG2     PROC  FAR
             ASSUME CS:CODSG_C
; the instructions that perform the work of the subroutine go here
             RET
SUBPROG2     ENDP
CODSG_C      ENDS
             END
```

Review Questions

1. List three advantages of modular programming.
2. The ______ directive is used within a module to indicate that the named variable can be used by another module.
3. The ______ directive is used within a module to indicate that the named variable was defined in another module.
4. How does the system determine the entry and exit points of a program consisting of more than one module?
5. What is a paragraph?
6. Write the directive used in complete segment definition that will define the stack segment so that it will be combined with the system stack.
7. If a word-sized data item named TOTAL was defined in module 1, code the directive to define TOTAL in module 2.
8. If PARA were used for the alignment type of a code segment that ended at 56H, where would the next code segment begin?
9. Write the code segment directives for a calling program and a module so that they will be combined into one code segment.

SECTION 7.2: SOME VERY USEFUL MODULES

This section shows the development of two very useful programs that convert from hex to decimal, and vice versa. Then they are rewritten as modules that can be called from any program. Finally, the calling program is written.

Binary (hex)-to-ASCII (decimal) conversion

The result of arithmetic operations is, of course, in binary. To display the result in decimal, the number is first converted to decimal, and then each digit is tagged with 30H to put it in ASCII form so that it can be displayed or printed. The first step is to convert the binary number to decimal. Look at the following example, which converts 34DH to decimal.

```
34DH = (3 × 16²)  + (4 × 16¹) + (D = 13 × 16⁰)
= (3 × 256)       + (4 × 16)  + (13 × 1)
= 768             + 64        + 13
= 845
```

Another method to convert a hex number to decimal is to divide it repeatedly by 10 (0AH), storing each remainder, until the quotient is less than 10. The following steps would be performed:

```
34DH / A = 84 remainder 5
84H  / A = 8  remainder 4
8 (< A, so the process stops)
Taking the remainders in reverse order gives: 845 decimal
```

Program 7-3 shows the conversion process for a word-sized (16-bit) number using the method of repeated division demonstrated above. Since a word-sized hex number is between 0 and FFFFH, the result in decimal can be as high as 65,535. Therefore, a string length of 5 should be sufficient to hold the result. The binary number to be converted is in data item BINNUM. Notice in Program 7-3 that as each decimal digit (the remainder) is placed in DL, it is tagged with 30H to convert it to ASCII. It is then placed in a memory area called ASCNUM. The ASCII digits are placed in memory so that the lowest digit is in high memory, as is the convention of ASCII storage in Windows OS.

```
TITLE          PROG7-3 CONVERT BINARY TO ASCII
;USING SIMPLIFIED SEGMENT DEFINITION
PAGE   60,132
               .MODEL       SMALL
;-----------------------------
               .STACK       64
               .DATA
BINNUM         DW 246DH
               ORG    10H
ASCNUM         DB     5 DUP ('0')
;------------------------------
               .CODE
B2ASC_CON      PROC   FAR
               MOV    AX,@DATA
               MOV    DS,AX
               MOV    BX,10          ;BX=10 THE DIVISOR
               MOV    SI,OFFSET ASCNUM  ;SI = BEGINNING OF ASCII STRING
               ADD    SI,5           ;ADD LENGTH OF STRING
               DEC    SI             ;SI POINTS TO LAST ASCII DIGIT
               MOV    AX,BINNUM      ;LOAD BINARY (HEX) NUMBER
BACK:          SUB    DX,DX          ;DX MUST BE 0 IN WORD DIVISION
               DIV    BX             ;DIVIDE HEX NUMBER BY 10 (BX=10)
               OR     DL,30H         ;TAG '3' TO MAKE IT ASCII
               MOV    [ SI] ,DL            ;MOVE THE ASCII DIGIT
               DEC    SI             ;DECREMENT POINTER
               CMP    AX,0           ;CONTINUE LOOPING WHILE AX >0
               JA     BACK
               MOV    AH,4CH
               INT    21H            ;GO BACK TO OS
B2ASC_CON      ENDP
               END    B2ASC_CON
```

Program 7-3

ASCII (decimal)-to-binary (hex) conversion

When a user keys in digits 0 to 9, the keyboard provides the ASCII version of the digits to the computer. For example, when the key marked 9 is pressed, in reality the keyboard provides its ASCII version 00111001 (39H) to the system. In Chapter 3 we showed how in some cases, such as addition, the numbers can be processed in ASCII and there is no need to convert them to hex (binary). However, in the majority of cases the number needs to be converted to hex in order to be processed by the CPU. Look at the example of converting decimal 482 to hex. The following shows the steps to convert the number to hex:

$$482 / 16^2 = 482 / 256 = 1$$
$$482 - (1 \times 256) = 226 \qquad 226 / 16^1 = 226 / 16 = 14 = E$$
$$226 - (14 \times 16) = 2$$
$$482 \text{ decimal} = 1E2 \text{ hexadecimal}$$

However, a computer would use a different method since it works in binary arithmetic, not decimal. First the 30H would be masked off each ASCII digit. Then each digit is multiplied by a weight (a power of 10) such as 1, 10, 100, or 1000 and they are then added together to get the final hex (binary) result. Converting decimal 482 to hex involves the following steps. First a user types in '482' through the PC ASCII keyboard, yielding 343832, the ASCII version of 482. Then the following steps are performed:

```
2 × 1          =       =      2
8 × 10         =   80  =    50H
4 × 100        =  400  =   190H
                           1E2 hexadecimal
```

```
TITLE        PROG7-4 CONVERT ASCII TO BINARY
PAGE         60,132
             .MODEL      SMALL
             .STACK 64
;----------------------------
             .DATA
TEN          DW     10
ASCNUM       DB     '09325'
STRLEN       DB     5
             ORG    10H
BINNUM       DW     0
;----------------------------
             .CODE
ASC2B_CON PROC FAR
             MOV    AX,@DATA
             MOV    DS,AX
             SUB    DI,DI                ;CLEAR DI FOR BINARY(HEX) RESULT
             MOV    SI,OFFSET ASCNUM     ;SI = BEGINNING OF ASCII STRING
             MOV    BL,STRLEN            ;BL = LENGTH OF ASCII STRING
             SUB    BH,BH                ;BH=0 USE BX IN BASED INDEX MODE
             DEC    BX                   ;BX IS OFFSET TO LAST DIGIT
             MOV    CX,1                 ;CX = WEIGHT FACTOR
AGAIN:       MOV    AL,[ SI+BX]          ;GET THE ASCII DIGIT
             AND    AL,0FH               ;STRIP OFF '3'
             SUB    AH,AH                ;CLEAR AH FOR WORD MULTIPLICATION
             MUL    CX                   ;MULTIPLY BY THE WEIGHT
             ADD    DI,AX                ;ADD IT TO BINARY (HEX )RESULT
             MOV    AX,CX                ;MULTIPLY THE WEIGHT FACTOR
             MUL    TEN                  ; BY TEN
             MOV    CX,AX                ; FOR NEXT ITERATION
             DEC    BX                   ;DECREMENT DIGIT POINTER
             JNS    AGAIN                ;JUMP IF COUNTER >= 0
             MOV    BINNUM,DI            ;SAVE THE BINARY(HEX)RESULT
             MOV    AH,4CH
             INT    21H                  ;GO BACK TO OS
ASC2B_CON    ENDP
             END    ASC2B_CON
```

Program 7-4

Program 7-4 converts an ASCII number to binary. It assumes the maximum size of the decimal number to be 65535. Therefore, the maximum hex result is FFFFH, a 16-bit word. It begins with the least significant digit, masks off the 3, and multiplies it by its weight factor. Register CX holds the weight, which is 1 for the least significant digit. For the next digit CX becomes 10 (0AH), for the next it becomes 100 (64H), and so on. The program assumes that the least significant ASCII digit is in the highest memory location of the data. This is consistent with the conventions of storing ASCII numbers with the most significant digit in the lower memory address and the least significant digit in the highest memory address. For example, placing '749' at memory offset 200 gives offset 200 = (37), 201 = (34), and 202 = (39). OS 21H function call 0A also places ASCII numbers this way.

Programs 7-3 and 7-4 have been written and tested with sample data, and now can be changed from programs into modules that can be called by any program.

Binary-to-ASCII module

Program 7-6 is the modularized Program 7-3. The procedure is declared as public, so it can be called by another program. All values used are declared external since the data will be provided by the calling program. Therefore, this module does not need its own data segment. Notice the following points about the module:

```
TITLE       PROG7-5 BINARY TO DECIMAL CONVERSION MODULE
PAGE  60,132
;this module converts a binary (hex) number up to FFFFH to decimal
; then makes it displayable (ASCII)
;CALLING PROGRAM SETS
; AX = BINARY VALUE TO BE CONVERTED TO ASCII
; SI = OFFSET ADDRESS WHERE ASCII VALUE IS TO BE STORED
      .MODEL SMALL
      PUBLIC B2ASC_CON
      .CODE
B2ASC_CON PROC FAR
      PUSHF             ;STORE REGS CHANGED BY THIS MODULE
      PUSH BX
      PUSH DX
      MOV  BX,10        ;BX=10 THE DIVISOR
      ADD  SI,4         ;SI POINTS TO LAST ASCII DIGIT
B2A_LOOP: SUB DX,DX     ;DX MUST BE 0 IN WORD DIVISION
      DIV  BX           ;DIVIDE HEX NUMBER BY 10 (BX=10)
      OR   DL,30H       ;TAG '3' TO REMAINDER TO MAKE IT ASCII
      MOV  [SI],DL      ;MOVE THE ASCII DIGIT
      DEC  SI           ;DECREMENT POINTER
      CMP  AX,0         ;CONTINUE LOOPING WHILE AX >0
      JA   B2A_LOOP
      POP  DX           ;RESTORE REGISTERS
      POP  BX
      POPF
      RET
B2ASC_CON ENDP
      END
```

Program 7-5

```
TITLE       PROG7-6 ASCII TO BINARY CONVERSION MODULE
PAGE  60,132
;this module converts any ASCII number between 0 to 65535 to binary
;CALLING PROGRAM SETS
; SI = OFFSET OF ASCII STRING
; BX = STRING LENGTH - 1 (USED AS INDEX INTO ASCII NUMBER)
;THIS MODULE SETS
; AX = BINARY NUMBER
            .MODEL SMALL
            EXTRN TEN:WORD
            PUBLIC ASC2B_CON
            .CODE
ASC2B_CON   PROC FAR
            PUSHF               ;STORE REGS CHANGED IN THIS MODULE
            PUSH  DI
            PUSH  CX
            SUB   DI,DI         ;CLEAR DI FOR THE BINARY (HEX) RESULT
            MOV   CX,1          ;CX = WEIGHT FACTOR
A2B_LOOP:   MOV   AL,[SI+BX]    ;GET THE ASCII DIGIT
            AND   AL,0FH        ;STRIP OFF '3'
            SUB   AH,AH         ;CLEAR AH FOR WORD MULTIPLICATION
            MUL   CX            ;MULTIPLY BY THE WEIGHT
            ADD   DI,AX         ;ADD IT TO BINARY (HEX) RESULT
            MOV   AX,CX         ;MULTIPLY THE WEIGHT FACTOR
            MUL   TEN           ; BY TEN
            MOV   CX,AX         ; FOR NEXT ITERATION
            DEC   BX            ;DECREMENT DIGIT POINTER
            JNS   A2B_LOOP      ;JUMP IF OFFSET >=0
            MOV   AX,DI         ;STORE BINARY NUMBER IN AX
            POP   CX            ;RESTORE FLAGS
            POP   DI
            POPF
            RET
ASC2B_CON   ENDP
            END
```

Program 7-6

```
TITLE        PROG7-7 CALLING PROGRAM TO CONVERT ASCII TO BINARY
PAGE    60,132
        PUBLIC TEN
        .MODEL SMALL
        .STACK      64
        .DATA
ASC_AREA     LABEL BYTE
MAX_LEN      DB 6
ACT_LEN      DB ?
ASC_NUM      DB 6 DUP (?)
             ORG 10H
BINNUM       DW 0
PROMPT1      DB 'PLEASE ENTER A 5 DIGIT NUMBER','$'
TEN          DW 10
        .CODE
        EXTRN ASC2B_CON:FAR
MAIN    PROC   FAR
        MOV    AX,@DATA
        MOV    DS,AX
        MOV    AH,09                    ;DISPLAY THE PROMPT
        MOV    DX,OFFSET PROMPT1
        INT    21H
        MOV    AH,0AH                   ;INPUT STRING
        MOV    DX,OFFSET ASC_AREA
        INT    21H
        MOV    SI,OFFSET ASC_NUM
        MOV    BH,00
        MOV    BL,ACT_LEN
        DEC    BX
        CALL   ASC2B_CON
        MOV    BINNUM,AX    ;SAVE THE BINARY (HEX) RESULT
        MOV    AH,4CH
        INT    21H    ;GO BACK TO OS
MAIN    ENDP
        END    MAIN
```

Program 7-7

1. Since this module will be called by another module, no entry point and exit point were given. Therefore, the END directive does not have the label B2ASC_CON.
2. The module must return to the caller and not OS as was the case in Program 7-4.
3. This module does not need its own data or stack segments.

ASCII-to-binary module

Program 7-6 is the modularized version of Program 7-4. Notice the following points about the module:

1. TEN is defined in the calling program.
2. This module must return to the caller and not OS.

Calling module

Program 7-7 shows the calling program for the module that converts ASCII to binary. This program sets up the data segment, inputs the ASCII data from the keyboard, places it in memory, then calls the routine to convert the number to binary. Finally, the hex result is stored in memory. The flowchart and pseudocode for the main module and the subroutine are given next.

Review Questions

1. Show a step-by-step analysis of Program 7-3 with data F624H. Show the sequence of instructions and the data values.
2. Show a step-by-step analysis of Program 7-4 with data '1456'. Show the sequence of instructions and the data values.

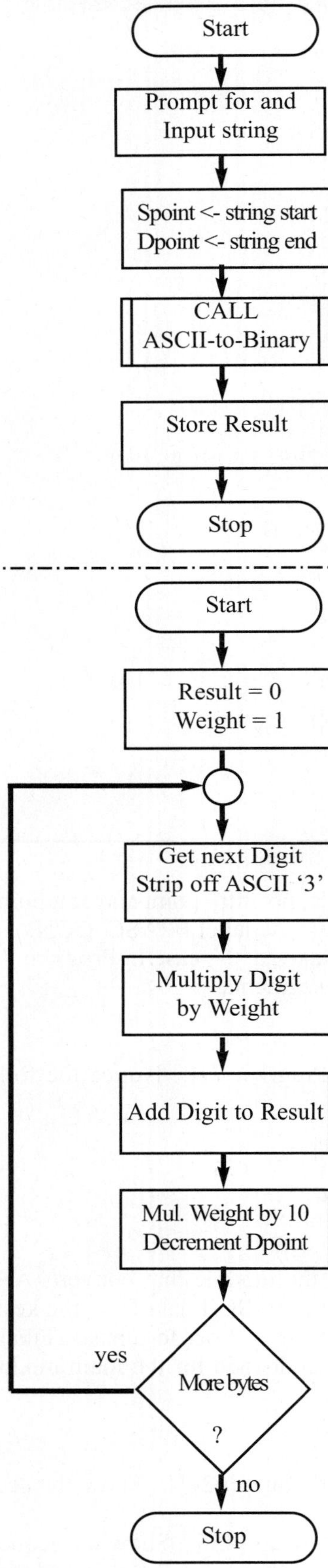


```
                Main Module

Prompt for string
Input string
Spoint <- points to start of string
Dpoint <- points to end of string
CALL precedure to convert ASCII to binary
        Send: Spoint, Dpoint
        Return: binary result
Store binary result
```

```
                Procedure
        to convert ASCII to binary

Receive:
  Spoint (points to start of string)
  Dpoint (points to end of string)

Result = 0
Weight = 1
REPEAT
  Get Digit pointed to by Dpoint
  Strip off ASCII '3'
  Multiply Digit by Weight
  Add Digit to Result
  Multiply Weight by 10
  Decrement Dpoint
UNTIL (all bytes converted)

Return:
  Binary result
```

Flowchart and Pseudocode for Program 7-7

SECTION 7.3: PASSING PARAMETERS AMONG MODULES

Occasionally, there is a need to pass parameters among different Assembly language modules or between Assembly language and C/C++, C#, and Visual Basic language programs. The parameter could be fixed values, variables, arrays of data, or even pointers to memory. Parameters can be passed from one module to another through registers, memory, or the stack. In this section we explore passing parameters between Assembly language modules.

Passing parameters via registers

When there is a need to pass parameters among various modules, one could use the CPU's registers. For example, if a main routine is calling a subroutine, the values are placed in the registers in the main routine and then the subroutine is called upon to process the data. In such cases the programmer must clearly document the registers used for the incoming data and the registers that are expected to have the result after the execution of the subroutine. In Chapter 4 this concept was demonstrated with INT 21H and INT 10H. Program 7-7 demonstrated this method. In that program, registers BX and SI were set to point to certain data items before the module was called, and the called module placed its result in register AX prior to returning to the calling routine.

Passing parameters via memory

Although parameter passing via registers is widely used in many of the OS and BIOS interrupt function calls, the limited number of registers inside the CPU is a major limitation associated with this method of parameter passing. This makes register management a cumbersome task. One alternative is to pass parameters via memory by defining an area of RAM and passing parameters to these RAM locations. OS and IBM BIOS use this method frequently. The problem with passing parameters to a fixed area of memory is that there must be a universal agreement to the address of the memory area in order to make sure that modules can be run on the hardware and software of various companies. This kind of standardization is hard to come by. The only reason that BIOS and OS use an area of memory for passing parameters is because IBM and Microsoft worked closely together to decide on the memory addresses. Another option, and indeed the most widely used method of passing parameters, is via the stack, as discussed next. Passing parameters via the stack makes the parameters both register and memory independent.

Passing parameters via the stack

The stack is a very critical part of every program and playing with it can be risky. When a module is called, it is the stack that holds the address where the program must return after execution. Therefore, if the contents of the stack are altered, the program can crash. This is the reason that working with the stack and passing parameters through it must be understood very thoroughly before one embarks on it.

Program 7-8 demonstrates this method of parameter passing and is written with the following requirements. The main module gets three word-sized operands from the data segment, stores them on the stack, and then calls the subroutine. The subroutine gets the operands from the stack, adds them together, holds the result in a register, and then returns control to the main module. The main module stores the result of the addition. Following the program is a detailed stack contents analysis that will show how the parameters are stored on the stack by the main routine and retrieved from the stack by the called routine.

Stack contents analysis for Program 7-8

To clarify the concept of parameter passing through the stack, the following is a step-by-step analysis of the stack pointer and stack contents. Assume that the stack pointer has the value SP = 17FEH before the "PUSH VALUE3" instruction in the main module is executed.

```
TITLE PROG7-8 PASSING PARAMETERS VIA THE STACK
PAGE  60,132
      .MODEL        SMALL
      EXTRN SUBPROG6:FAR
      .STACK       64
      .DATA
VALUE1       DW 3F62H
VALUE2       DW 1979H
VALUE3       DW 25F1H
RESULT       DW 2 DUP (?)
      .CODE
MAIN  PROC  FAR
      MOV   AX,@DATA
      MOV   DS,AX
      PUSH  VALUE3      ;SAVE VALUE3 ON STACK
      PUSH  VALUE2      ;SAVE VALUE2 ON STACK
      PUSH  VALUE1      ;SAVE VALUE1 ON STACK
      CALL  SUBPROG6    ;CALL THE ADD ROUTINE
      MOV   RESULT,AX   ;STORE
      MOV   RESULT+2,BX ; THE RESULT
      MOV   AH,4CH
      INT   21H
MAIN  ENDP
      END   MAIN
```

Program 7-8: Main Module

```
;------------------ in a separate file: ---------------------------
TITLE       SUBPROG6 MODULE TO ADD THREE WORDS BROUGHT IN FROM THE
STACK
PAGE  60,132
            .MODEL SMALL
      PUBLIC SUBPROG6
            .CODE
SUBPROG6    PROC  FAR
      SUB   BX,BX        ;CLEAR BX FOR CARRIES
      PUSH  BP     ;SAVE BP
      MOV   BP,SP        ;SET BP FOR INDEXING
      MOV   AX,[BP]+6    ;MOV VALUE1 TO AX
      MOV   CX,[BP]+8    ;MOV VALUE2 TO CX
      MOV   DX,[BP]+10   ;MOV VALUE3 TO DX
      ADD   AX,CX        ;ADD VALUE2 TO VALUE1
      ADC   BX,00        ;KEEP THE CARRY IN BX
      ADD   AX,DX        ;ADD VALUE3
      ADC   BX,00        ;KEEP THE CARRY IN BX
      POP   BP     ;RESTORE BP BEFORE RETURNING
      RET   6      ;RETURN AND ADD 6 TO SP TO BYPASS DATA
SUBPROG6    ENDP
      END
```

Program 7-8: Module 2

1. VALUE3 = 25F1H is pushed and SP = 17FC (remember little endian: low byte to low address and high byte to high address).
2. VALUE2 = 1979H is pushed and then SP = 17FA.
3. VALUE1 = 3F62H is pushed and then SP = 17F8.
4. CALL SUBPROG6 is a FAR call; therefore, both CS and IP are pushed onto the stack, making SP = 17F4. If it had been a near call, only IP would have been saved.
5. In the subprogram module, register BP is saved by PUSHing BP onto the stack, which makes SP = 17F2. In the subprogram, BP is used to access values in the stack. First SP is copied to BP since only BP can be used in indexing mode with the stack segment (SS) register. In other words, "`MOV AX,[SP+4]`" will cause an error. "`MOV`

AX,[BP] +6" loads VALUE1 into AX. [BP] + 6 = 17F2 + 6 = 17F8, which is exactly where VALUE1 is located. Similarly, BP + 8 = 17F2 + 8 = 17FA is the place where VALUE2 is located, and BP + 10 = 17F2H + 10 = 17FCH is the location of VALUE3.

6. After all the parameters are brought into the CPU by the present module and are processed (in this case added), the module restores the original BP contents by POPping BP from stack. Then SP = 17F4.
7. RET 6: This is a new instruction. The RETurns shown previously did not have numbers right after them. The "RET n" instruction means first to POP CS:IP (IP only if the CALL was NEAR) off the top of the stack and then add n to the SP. As can be seen from the Program 7-8 diagram, after popping CS and IP off the stack, the stack pointer is incremented four times, making SP = 17F8. Then adding 6 to it to bypass the six locations of the stack where the parameters are stored makes the SP = 17FEH, its original value. Now what would happen if the program had a RET instruction instead of the "RET 6"? The problem is that every time this subprogram is executed it will cause the stack to lose six locations. If that had been done in the example above, when the same routine is called again the stack starts at 17F8 instead of 17FE. If this practice of losing some area of the stack continues, eventually the stack could be reduced to a point where the program would run out of stack and crash.

Address	Contents
17F0 17F1	
17F2 17F3	BP
17F4 17F5	IP
17F6 17F7	CS
17F8 17F9	62 VALUE1 3F
17FA 17FB	79 VALUE2 19
17FC 17FD	F1 VALUE3 25
17FE	

Program 7-8: Stack Contents Diagram

Review Questions

1. List one advantage and one disadvantage of each method of parameter passing.
   (a) via register (b) via stack (c) via memory
2. Assume that we would like to access some parameters from the stack. Which of the following are correct ways of accessing the stack?
   (a) MOV AX,[BP]+20 (b) MOV AX,[SP]+20
   (c) MOV AX,[BP+DI] (d) MOV AX,[SP+SI]

PROBLEMS

SECTION 7.1: WRITING AND LINKING MODULES

1. Fill in the blanks in the following program. The main program defines the data and calls another module to add 5 bytes of data, then saves the result. Note: Some blanks may not need anything.

```
.MODEL SMALL
.STACK 100H
.DATA
PUBLIC          _______,________
DATA1       DB     25,12,34,56,98
RESULT      DW     ?
.CODE
EXTRN            _______:FAR
HERE: MOV      AX,@DATA
      MOV      DS,AX
      CALL     SUM
      MOV      AH,4CH
      INT      21H
      END      ______
```

In another file there is the module for summing 5 bytes of data:

```
        .MODEL   SMALL
________DATA1:BYTE
________RESULT:WORD
        COUNT EQU     5
.CODE
_________SUM
SUMPROC        ________
          MOV    BX,OFFSET DATA1
          SUB    AX,AX
          MOV    CX,COUNT
_______:  ADD    AL, BYTE PTR [ BX]
          ADC    AH,0
          INC    BX
          LOOP   AGAIN
          MOV    RESULT,AX
          RET
_____     ENDP
          END        _______
```

2. If a label or parameter is not defined in a module, it must be declared as _____.
3. If a label or parameter is used by other modules, it must be declared as ________ in the present module.
4. List the options for the EXTRN directive when it is referring to a procedure.
5. List the options for the EXTRN directive when it is referring to a data item.
6. List the options for the PUBLIC directive when it is referring to a procedure.
7. List the options for the PUBLIC directive when it is referring to a data item.
8. Convert Program 4-1 to the modular format, making each of the INT subroutines a separate module. Each module should be NEAR. Assemble and test the program.

SECTION 7.2: SOME VERY USEFUL MODULES

9. Write a program that accepts two unsigned numbers (each less than 999) from the keyboard, converts them to hex, takes the average, and displays the result on the monitor. Use the hex-to-decimal and ASCII-to-hex conversion modules in the text.

SECTION 7.3: PASSING PARAMETERS AMONG MODULES

10. Write a program (similar to Program 7-1) with the following components.
    (a) In the main program, two values are defined: 1228 and 52400.
    (b) The main program calls two separate modules, passing the values by stack.
    (c) In the first module, the two numbers are multiplied and the result is passed back to the main module.
    (d) The second module performs division of the two numbers (52400 /122) and passes both the quotient and remainder back to the main program to be stored.
    (e) Analyze the stack and its contents for each module if SP = FFF8H immediately before the first CALL instruction in the main module.

ANSWERS TO REVIEW QUESTIONS

SECTION 7.1: WRITING AND LINKING MODULES

1. (1) Each module can be developed individually, allowing parallel development of modules, which shortens development time; (2) easier to locate source of bugs; (3) these modules can be linked with high-level languages such as C.
2. PUBLIC
3. EXTRN
4. The module that is the entry and exit point will have a label after the END statement.
5. A paragraph consists of 16 bytes and begins on an address ending in 0H.
6. name SEGMENT PARA STACK 'STACK'
7. EXTRN TOTAL:WORD
8. 60H
9. name SEGMENT PARA PUBLIC 'CODE'

SECTION 7.2: SOME VERY USEFUL MODULES

1. 1st iteration: AX = F624 F624/A = 189D remainder DL = 2
2nd iteration: AX = 189D 189D/A = 0276 remainder DL = 1
3rd iteration: AX = 0276 0276/A = 003F remainder DL = 0
4th iteration: AX = 003F 003F/A = 6 remainder DL = 3
5th iteration: AX = 0006 0006/A = 0 remainder DL = 6
AX is now zero, so the conversion is complete: F624H = 63,012
2. 1st iteration: AL = 36 06 × 1 = 6 DI = 6
2nd iteration: AL = 35 05 × A = 32 DI = 6 + 32 = 38
3rd iteration: AL = 34 04 × 64 = 190 DI = 38 + 1C8
4th iteration: AL = 31 01 × 03E8 = 03E8 DI = 1C8 + 03E8 = 05B0
5th iteration: AL = 30 00 × 2710 = 0 DI = 05B0
BX has been decremented from 4 to 0, and is now –1, so the conversion is complete: 01456 = 05B0H

SECTION 7.3: PASSING PARAMETERS AMONG MODULES

1. (a) By register; one advantage is the execution speed of registers; one disadvantage is that there is a limited number of registers available so that not many values can be passed.
(b) By stack; one advantage is it does not use up available registers; one disadvantage is that errors in processing the stack can cause the system to crash.
(c) By memory; one advantage is a large area available to store data; one disadvantage is that the program would not be portable to other computers.
2. (a) and (c) are correct; (b) and (d) are not correct because SP cannot be used in indexing mode with SS.

CHAPTER 8

32-BIT PROGRAMMING FOR x86

OBJECTIVES

Upon completion of this chapter, you will be able to:

- >> **Discuss the major differences between the 16-bit and 32-bit CPUs**
- >> **List the 32-bit registers of the x86 CPU**
- >> **Diagram the register sizes available in the 32-bit CPUs**
- >> **Explain the difference in register usage between the 16-bit and the 32-bit systems**
- >> **Discuss how the increased register size of the 32-bit systems relates to an increased memory range**
- >> **Diagram how the "little endian" storage convention of x86 machines stores doubleword-sized operands**
- >> **Code programs for the 32-bit CPU using extended registers and new directives**
- >> **Code arithmetic statements using the extended registers of the 32-bit CPUs**
- >> **Code Assembly language within C programs by using in-line coding**

All programs discussed so far used 16-bit registers in x86 computers. In this chapter we discuss the 32-bit registers of x86 microprocessors in addition to combining C and Assembly languages. In Section 8.1, we discuss the 32-bit registers, and some program examples will be given that use the 32-bit capability of x86 machines. We also show how to combine Assembly and C in Visual Studio .NET.

SECTION 8.1: 32-BIT PROGRAMMING IN x86

In this section we concentrate on some of the most important differences between the 16-bit and 32-bit modes. While in the 8086/286 microprocessors the register size was 16 bits wide, in the 386 and higher CPUs the size of registers was extended to 32 bits. All register names have been changed to reflect this extension. Therefore, AX has become EAX, BX is now EBX, and so on, as illustrated below and outlined in Table 8-1. For example, the 386 and higher CPUs contain registers AL, AH, AX, and EAX with 8, 8, 16, and 32 bits, respectively. In the 16-bit, register AX is accessible either as AL or AH or AX, while in the 32-bit, register EAX can be accessed only as AL or AH or AX or EAX. In other words, the upper 16 bits of EAX are not accessible as a separate register. The same rule applies to EBX, ECX, and EDX. See Figure 8-1. Registers DI, SI, BP, and SP have become EDI, ESI, EBP, and ESP, respectively. In addition to the CS, DS, SS, and ES segment registers, there are also two new segment registers: FS and GS. There are also several control registers: CR0, CR1, CR2, and CR3. See Table 8-1.

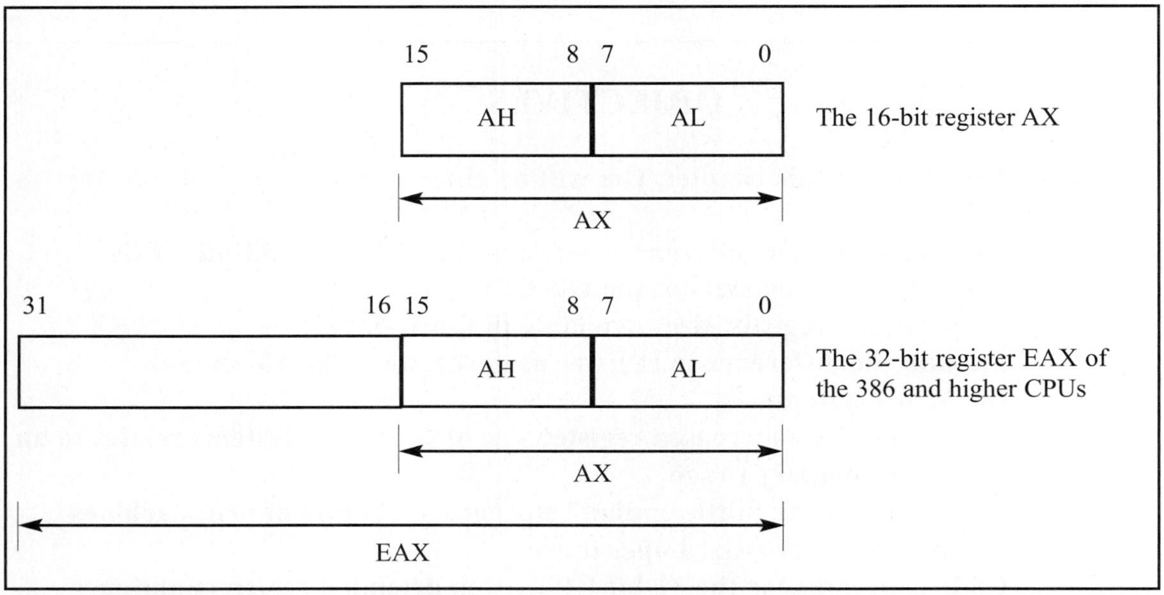

Figure 8-1: 16-bit vs. 32-bit Mode

32-bit general registers as pointers

Another major change from 16-bit to 32-bit is the ability of general registers such as EAX, ECX, and EDX to be used as pointers. As shown in previous chapters, AX, CX, and DX could not be used as pointers. For example, an instruction such as "MOV CL,[AX]" would have caused an error in the 16-bit since only BX, SI, DI, and BP were allowed to be used as pointers to memory. This has changed. Starting with the 386 CPUs, the following instructions are perfectly legal:

```
MOV    AX,[ ECX]
ADD    SI,[ EDX]
OR     EBX,[ EAX] +20
```

Table 8-1: Registers of the 32-bit x86 by Category

Category	Bits	Register Names
General	32	EAX, EBX, ECX, EDX
	16	AX, BX, CX, DX
	8	AH, AL, BH, BL, CH, CL, DH, DL
Pointer	32	ESP (extended SP), EBP (extended BP)
	16	SP (stack pointer), BP (base pointer)
Index	32	ESI (extended SI), EDI (extended DI)
	16	SI (source index), DI (destination index)
Segment	16	CS (code segment), DS (data segment)
		SS (stack segment), ES (extra segment)
		FS (extra segment), GS (extra segment)
Instruction	32	EIP (extended instruction pointer)
Flag	32	EFT (extended flag register)
Control	32	CR0, CR1, CR2, CR3

Note: Only bit 0 of CR0 is available in real mode. All other control registers are available in protected mode only.

It must be noted that when EAX, ECX, or EDX are used as offset addresses, DS is the default segment register. That means that SS is the default segment register for ESP and EBP, CS for EIP, and DS for all other registers. Table 8-2 summarizes addressing modes for 32-bit programming of x86 processors.

Table 8-2: Addressing Modes for 32-bit Programming

Addressing Mode	Operand	Default Segment
Register	register	none
Immediate	data	none
Direct	[offset]	DS
Register indirect	[BX]	DS
	[SI]	DS
	[DI]	DS
	[EAX]	DS
	[EBX]	DS
	[ECX]	DS
	[EDX]	DS
	[ESI]	DS
	[EDI]	DS
Based relative	[BX]+disp	DS
	[BX]+disp	SS
	[EAX]+disp	DS
	[EBX]+disp	DS
	[ECX]+disp	DS
	[EDX]+disp	DS
	[EBP]+disp	SS
Indexed relative	[DI]+disp	DS
	[SI]+disp	DS
	[EDI]+disp	DS
	[ESI]+disp	DS
Based indexed relative	[R1][R2]+disp where RI and R2 are any of the above	If BP is used, the segment is SS; otherwise; DS is the segment

Note: In based indexed relative addressing, disp is optional.

Accessing 32-bit registers with commonly used assemblers

In Assembly language the directive ".386" is used to access the 32-bit registers of 386 and higher CPUs and to employ the new instructions of the 386 microprocessor. Every new generation of x86 has some new instructions that do not execute on lower processors, meaning that they are upwardly compatible. In other words, using the ".386" directive in a program means that the program must be run only on 386 and higher CPUs and cannot be run on 8086/286. In contrast, all programs in previous chapters were written to be run on any x86 computer. The following are additional assembler directives, which indicate the type of microprocessor supported by Microsoft's assembler (MASM).

MASM	Meaning
.86	will run on any x86 CPU (default)
.386	will run on any 386 and higher CPU; also allows use of new 386 instructions

Write a program using the 32-bit registers of the 386 to add the values 100000, 200000, and 40000. Then subtract from the total sum the values 80000, 35000, and 250. Place the result in memory locations allocated using the DD, doubleword directive, used for 32-bit numbers.

```
TITLE       ADD AND SUBTRACT USING 32-BIT REGISTERS IN 386 MACHINES
 PAGE 60,132
            .MODEL SMALL
            .386
            .STACK      200H
            .DATA
 RESULT     DD     ?
            .CODE
 BEGIN:     MOV    AX,@DATA
            MOV    DS,AX
            SUB    EAX,EAX
            ADD    EAX,100000   ;EAX = 186A0H
            ADD    EAX,200000   ;EAX = 186A0H + 30D40 H = 493E4H
            ADD    EAX,40000    ;EAX = 493E4H + 9C40H = 53020H
            SUB    EAX,80000    ;EAX = 53020H - 13880H = 3F7A0H
            SUB    EAX,35000    ;EAX = 3F7A0H - 88B8H = 36EE8H
            SUB    EAX,250      ;EAX = 36338H - FAH = 36DEEH (224750)
            MOV    RESULT,EAX
            MOV    AH,4CH
            INT    21H
            END    BEGIN
```

Program 8-1

Program 8-1 demonstrates the use of the ".386" directive and the 80386 32-bit instructions. The simplified segment definition was used. The program used the 32-bit register EAX to add and subtract values of various size to demonstrate 32-bit programming of the x86. Now the question is how to run this program and see the register contents in x86 machines. Unfortunately, the DEBUG utility used in earlier chapters cannot be used since it shows only the 16-bit registers. In many assemblers, including MASM, there are advanced debugging tools that one can use to see the execution of the 32-bit programs. In the case of MASM, the CodeView utility is the tool that allows one to monitor the execution of 16-bit in addition to 32-bit programs.

Following Program 8-1 is shown a trace of the program, using Microsoft's CodeView program. To examine the 32-bit registers in CodeView, press F2 to display registers; then select the Options menu from the top of the screen, and a drop-down menu appears. Select "386" from the drop-down menu to display the registers in 32-bit format.

```
File View Search Run Watch Options Language Calls Help | F8=Trace F5=Go

4833:0000 B83648 MOV AX,4836 EAX=00036DEE
4833:0003 8ED8 MOV DS,AX EBX=00000000
4833:0005 662BC0 SUB EAX,EAX ECX=00000000
4833:0008 6605A0860100 ADD EAX,000186A0 EDX=00000000
4833:000E 6605400D0300 ADD EAX,00030D40 ESP=00000200
4833:0014 6605409C0000 ADD EAX,00009C40 EBP=00000000
4833:001A 662D80380100 SUB EAX,00013880 ESI=00000000
4833:0020 662DB8880000 SUB EAX,000088B8 EDI=00000000
4833:0026 662DFA000000 SUB EAX,000000FA DS=...4836
 ES=....4823
File View Search Run Watch Options Language Calls Help | F8=Trace F5=Go
4833:0030 B44C MOV AH,4C FS=....0000
4833:0032 CD21 INT 21 GS=....0000
4833:0034 0000 ADD Byte Ptr [BX+SI],AL SS=....4837
4833:0036 0000 ADD Byte Ptr [BX+SI],AL CS=....4833
4833:0038 0000 ADD Byte Ptr [BX+SI],AL IP=0000002C
4833:003A 0000 ADD Byte Ptr [BX+SI],AL
4833:003C 0000 ADD Byte Ptr [BX+SI],AL NV UP
4833:003E 0000 ADD Byte Ptr [BX+SI],AL EI NG
 NZ AC
>t PE NC
>t
>t DS:0004
> 00000000                NEEDS WORK!
```

Figure 8-2: CodeView Screen for Execution of Program 8-1

Little endian revisited

In analyzing how the x86 stores 32-bit data in memory or loads a 32-bit operand into a register, recall the little endian convention: The low byte goes to the low address and the high byte to the high address. For example, an instruction such as "MOV RESULT,EAX" in Program 8-1 will store the data in this way:

OFFSET	CONTENTS
RESULT	d0-d7
RESULT+1	d8-d15
RESULT+2	d16-d23
RESULT+3	d24-d31

Example 8-1

Assuming that SI = 1298 and EAX = 41992F56H, show the contents of memory locations after the instruction "`MOV [SI],EAX`".

Solution (in hex):

DS:1298 = (56)
DS:1299 = (2F)
DS:129A = (99)
DS:129B = (41)

Some more examples of 32-bit programming

One way to increase the processing power of the microprocessor is to widen the register size. This allows processing large numbers as a whole rather than breaking them into smaller chunks to fit into small registers. The 32-bit registers have become standard in all recent microprocessors. Powerful supercomputers use 64-bit registers. In this section we show revisions of some earlier programs using the 32-bit capability of the x86 machines to see the impact of the wider registers in programming. By comparing 32-bit versions of these programs with the 16-bit versions, one can see the increased efficiency of 32-bit coding. The impact on speed is discussed in the final section.

Adding 16-bit words using 32-bit registers

Program 3-1b used 16-bit registers for adding several words of data. The sum was accumulated in one register and another register was used to add up the carries. This is not necessary when using 32-bit registers. First, refresh your memory by looking at Program 3-1b and then examine Program 8-2, a 32-bit version of the same program, written for 386 and higher CPUs.

```
TITLE        REVISION OF PROGRAM 3-1B USING 32-BIT REGISTERS
PAGE   60,132
             .MODEL SMALL
             .386
             .STACK 200H
             .DATA
DATA1 DD     27345,28521,29533,30105,32375
SUM   DD     ?
COUNT EQU    5
             .CODE
BEGIN:       MOV    AX,@DATA
             MOV    DS,AX
             MOV    CX,COUNT              ;CX is loop counter
             MOV    SI,OFFSET DATA1       ;SI is data pointer
             SUB    EAX,EAX               ;EAX will hold sum
BACK:        ADD    EAX,DWORD PTR[SI]     ;add next word to EAX
             ADD    SI,4                  ;SI points to next dword
             DEC    CX                    ;decrement loop counter
             JNZ    BACK                  ;continue adding
             MOV    SUM,EAX               ;store sum
             MOV    AH,4CH
             INT    21H
             END    BEGIN
```

Program 8-2

Adding multiword data in 32-bit

In Program 3-2, two multiword numbers were added using 16-bit registers. Each number could be as large as 8 bytes wide. That program required a total of four iterations. Using the 32-bit registers of the 386/46 requires only two iterations, as shown in Program 8-3a. This loop version of the multiword addition program is very long and inefficient. It can be made more efficient by saving the flag register that holds the carry bit of the first 32-bit addition on the stack and then adding 4 to each pointer instead of incrementing the pointers four times. The loop is shown below in Program 8-3b.

Due to the high penalty associated with branch instructions such as the LOOP and Jcondition instructions in the 386 and higher CPUs, it is better to use the nonloop version of this program, shown in Program 8-4.

First notice that the data is stored exactly the same way as in the loop version of the program. Data directive DQ is used to set up storage for the 8-byte numbers. First, the lower dword (4 bytes) of DATA1 is moved into EAX, and the lower dword of DATA2 is

Rewrite Program 3-2 in Chapter 3 to add two 8-byte operands using 32-bit registers.

```
TITLE       ADD TWO 8-BYTE NUMBER USING 32-BIT REGISTERS IN THE 386
PAGE        60,132
     .MODEL SMALL
     .386
     .STACK 200H
     .DATA
DATA1       DQ     548FB9963CE7H
            ORG    0010H
DATA2       DQ     3FCD4FA23B8DH
            ORG    0020H
DATA3       DQ     ?
            .CODE
BEGIN:MOV   AX,@DATA
     MOV    DS,AX
     CLC                       ;clear carry before first addition
     MOV    SI,OFFSET DATA1    ;SI is pointer for operand1
     MOV    DI,OFFSET DATA2    ;DI is pointer for operand2
     MOV    BX,OFFSET DATA3    ;BX is pointer for the sum
     MOV    CX,02              ;CX is the loop counter
BACK:MOV    EAX,DWORD PTR [ SI]      ;move the operand to EAX
     ADC    EAX,DWORD PTR [ DI]      ;add the operand to EAX
     MOV    DWORD PTR [ BX] ,EAX     ;store the sum
     INC    SI                 ;point to next dword of operand1
     INC    SI
     INC    SI
     INC    SI
     INC    DI                 ;point to next dword of operand2
     INC    DI
     INC    DI
     INC    DI
     INC    BX
     INC    BX                 ;point to next dword of sum
     INC    BX
     INC    BX
     LOOP   BACK               ;if not finished, continue adding
     MOV    AH,4CH
     INT    21H                ;go back to DOS
     END    BEGIN
```

Program 8-3a

```
;this revision of Program 3-1a shows how to save the flags
;     before updating the pointers
BACK:MOV    EAX,DWORD PTR [ SI]      ;move the operand to EAX
     ADC    EAX,DWORD PTR [ DI]      ;add the operand to EAX
     MOV    DWORD PTR [ BX] ,EAX     ;store the sum
     PUSHF                           ;save the flags
     ADD    SI,4         ;point to next dword of operand1
     ADD    DI,4         ;point to next dword of operand2
     ADD    BX,4         ;point to next dword of sum
     POPF                ;restore the flags
     LOOP   BACK         ;if not finished, continue adding
```

Program 8-3b

added to EAX. Then the upper dword of DATA1 is moved into EBX, and the upper dword of DATA2 is added to EBX, with any carry that may have been generated in the addition of the lower dwords. EAX now holds the lower 4 bytes of the result, and EBX holds the upper 4 bytes of the result.

Program 8-4 is much more efficient than using the loop concept. To see why and for a discussion of the impact of branching instructions on the performance of programs in the 386 and higher CPUs, see Chapter 23.

```
TITLE       ADD TWO 8-BYTE NUMBERS USING 32-BIT REGISTERS IN THE 386
PAGE 60,132              ;(NO-LOOP VERSION)
            .MODEL SMALL
            .386
            .STACK 200H
            .DATA
DATA1       DQ    548FB9963CE7H
            ORG   0010H
DATA2       DQ 3FCD4FA23B8DH
            ORG   0020H
DATA3       DQ    ?
            .CODE
BEGIN:MOV   AX,@DATA
      MOV   DS,AX
      MOV   EAX,DWORD PTR DATA1     ;move lower dword of DATA1 into EAX
      ADD   EAX,DWORD PTR DATA2     ;add lower dword of DATA2 to EAX
      MOV   EBX,DWORD PTR DATA1+4   ;move upper dword of DATA1 into EBX
      ADC   EBX,DWORD PTR DATA2+4   ;add upper dword of DATA2 to EBX
      MOV   DWORD PTR DATA3,EAX     ;store lower dword of result
      MOV   DWORD PTR DATA3+4,EBX   ;store upper dword of result
      MOV   AH,4CH
      INT   21H
      END   BEGIN
```

Program 8-4

Combining C with Assembly

Although Assembly language is the fastest language available for a given CPU, it cannot be run on different CPUs. For example, Intel's x86 Assembly programs cannot be run on IBM's PowerPC RISC processor based computers since the opcode, mnemonics, register names, and size are totally different. Therefore, a portable language is needed.

Why C?

Although the dream of a universal language among the peoples of the world is still unrealized, C language is becoming the universal language among all the various CPUs. Today, a large portion of programs written for all computers are in the C/C++ language. C/C++ is such a universal programming language that it can be run on any CPU architecture with little or no modification. It is simply recompiled for that CPU. The fact that C/C++ is such a portable language is making it the dominant language of programmers. However, C/C++ is not as fast as Assembly language. Combining C and Assembly language takes advantage of C/C++'s portability and Assembly's speed. Today it is very common to see a software project written for embedded systems using 90 to 95% C and the rest Assembly language. There are two ways to mix C/C++ and Assembly. One is simply to insert the Assembly code in C programs, which is commonly referred to as in-line assembly. The second method is to make the C/C++ language call an external Assembly language procedure. In this section we discuss how to do in-line assembly coding and leave the other method to the reader to explore. This section covers Microsoft's Visual C++ and C#.

Inserting x86 assembly code into Visual C++ programs

In this section we discuss in-lining with Microsoft C++. For other C compilers, consult their C manual. The following code demonstrates how to change the cursor position to row = 10 and column = 20 in a C program. Assembly instructions are prefaced with "asm", which is a reserved word. Microsoft uses the keyword "_asm". Note that in Microsoft, not all interrupts may be supported in the latest versions of Visual C++. The following shows two variations of Microsoft format for in-line assembly.

```
/* version 1: using keyword asm before each line of in-line code */
/* Microsoft uses keyword "_asm"  */
/* compiled in Visual C++ 2005 Express Edition - a free download
-- from Microsoft website*/


#include <iostream>
#include <windows.h>
#include <tchar.h>
using namespace std;

int _tmain(int argc, _TCHAR* argv[])
{
  int data1=0xFFFFFF;
  int data2=0xFFFFFF;
  int sum;
  _asm{
    mov eax,data1
    mov ebx,data2
    add eax,ebx
    mov sum,eax
  }
  cout<<sum;
  return 0;
}
```

As shown above, each line of in-line code is prefaced by the keyword "asm", or a block of in-line code is prefaced by "asm". Each line must end in a semicolon or new-line, and any comments must be in the correct form for C.

Review Questions

1 In the 32-bit, the bits of register EDX can be accessed either as DL, bits __ to __; or DH, bits __ to __; or DX, bits __ to __; or EAX, bits __ to __.
2. True or false: The instruction "MOV DX,[AX]" is illegal in the 8086 but "MOV DX,[EAX]" is legal in the 32-bit x86.
3. What is the default segment register when EAX is used as a pointer?
4. What is the purpose of the ".386" directive?
5. Compare the number of iterations for adding two 8-byte numbers for the following CPUs.
   (a) 8085 (an 8-bit) CPU
   (b) 8086/286
   (c) x86 and higher CPUs (32-bit mode)
   (d) Itanium and 64-bit x86 supercomputer (64-bit system)

6. What data directives are used to define 32-bit and 64-bit operands?
7. What directive is used in MASM to inform the assembler that the program is using 32-bit instructions?

PROBLEMS

SECTION 8.1: 32-BIT PROGRAMMING IN x86

1. In an x86 program, show the content of each register indicated in parentheses after execution of the instruction.
   (a) MOV EAX,9823F4B6H (AL, AH, AX, and EAX)
   (b) MOV EBX,985C2H (BL, BH, BX, and EBX)

(c) MOV EDX,2000000 (DL, DH, DX, and EDX)
(d) MOV ESI,120000H (SI, ESI)

2. Show the destination and its contents in each of the following cases.

```
(a) MOV   EAX,299FF94H
    ADD   EAX,34FFFFH
(b) MOV   EBX,500 000
    ADD   EBX,700 000
(c) MOV   EDX,40 000 000
    SUB   EDX,1 500 000
(d) MOV  EAX,39393834H
    AND  EAX,0F0F0F0FH
(e) MOV  EBX,9FE35DH
    XOR  EBX,0F0F0F0H
```

3. Using the little endian convention, show the contents of the destination in each case.

```
(a) MOV  [SI],EAX ;ASSUME SI = 2000H AND EAX = 9823F456H
(b) MOV  [BX],ECX ;ASSUME BX, 348CH AND ECX = 1F23491H
(c) MOV  EBX,[DI] ;ASSUME DI = 4044H WITH THE
         ;FOLLOWING DATA.   ALL IN HEX.
         DS:4044 = (92)
         DS:4045 = (6D)
         DS:4046 = (A2)
         DS:4047 = (4C)
```

ANSWERS TO REVIEW QUESTIONS

SECTION 8.1: 32-BIT PROGRAMMING IN x86

1. 0 to 7, 8 to 15, 0 to 15, 0 to 31
2. True
3. DS
4. Allows use of 386 instructions
5. (a) 8; (b) 4; (c) 2; (d) 1
6. DD, DQ
7. .386

CHAPTER 9

8088, 80286 MICROPROCESSORS AND ISA BUS

OBJECTIVES

Upon completion of this chapter, you will be able to:

- >> **State the function of the pins of the 8088**
- >> **List the functions of the 8088 data, address, and control buses**
- >> **State the differences in the 8088 microprocessor in maximum mode versus minimum mode**
- >> **Describe the function of the pins of the 8284 clock generator chip**
- >> **Describe the function of the pins of the 8288 bus controller chip**
- >> **Explain the role of the 8088, 8284A, and 8288 in the PC**
- >> **Explain how bus arbitration between the CPU and DMA is accomplished**
- >> **State the function of the pins of the 80286**
- >> **Describe the differences between real and protected modes**
- >> **Describe the operation of the 80286 data, address, and control buses**
- >> **Describe the purpose of the expansion slots of the IBM PC AT (ISA) bus**
- >> **Describe the ISA bus system**

Since the original IBM PCs used 8088 and 80286 microprocessors, this chapter is a detailed hardware study of these two microprocessors, as well as the major signals of the ISA bus. In Section 9.1, a detailed look at the 8088 CPU, including pin descriptions, is provided. Two IC chips that support the 8088, the 8284 clock generator and the 8288 chip, are discussed in Section 9.2. Next, the IBM PC address, data, and control buses are covered in Section 9.3. In Section 9.4, the 80286 microprocessor is discussed, including pin descriptions. Finally, in Section 9.5, the PC's ISA buses are covered.

SECTION 9.1: 8088 MICROPROCESSOR

The first IBM PC used the 8088 microprocessor, and modern PCs still carry that legacy. In this section, the function of each pin of the 8088 CPU is described, as well as how the microprocessor chip is connected with some simple logic gates to create the address, data, and control signals. The 8088 is a 40-pin microprocessor chip that can work in two modes: minimum mode and maximum mode. Maximum mode is used when we need to connect the 8088 to an 8087 math coprocessor. If we do not need a math coprocessor, the 8088 is used in minimum mode. First we look at the 8088 in minimum mode since it is much simpler and easy to understand. Maximum mode and supporting chips are discussed in Section 9.2.

In 1978 Intel introduced the 16-bit microprocessor called the 8086. It was 16-bit both internally and externally. A year later Intel introduced the 8088 to allow the use of 8-bit peripheral chips and to make system boards cheaper. The 8088 is internally identical to the 8086, but has only an 8-bit external data bus. Since the original IBM PC introduced in 1981 used the 8088, we explore the 8088 instead of the 8086.

Microprocessor buses

Every microprocessor-based system must have three sets of separate buses: the address bus, the data bus, and the control bus. The address bus provides the path for the address to locate the targeted device, while the data bus is used to transfer data between the CPU and the targeted device. The control bus provides the signals to indicate the type of operation being executed, such as read or write. Next we discuss how these signals are provided by the 8088 microprocessor.

Data bus in 8088

Figure 9-1 shows the 8088/86 in minimum mode. Pins 9-16 (AD0–AD7) are used for both data and addresses in the 8088. At the time of the design of this microprocessor in the late 1970s, due to IC chip packaging limitations, there was a great effort to use the minimum number of pins for external connections. Therefore, designers multiplexed the address and data buses, meaning that Intel used the same pins to carry two sets of information: address and data. Notice that the name of the pins reflects this dual function. In the 8088, the address/data bus pins are named AD0–AD7, "AD" standing for "address/data." The ALE (address latch enable) pin signals whether the information on pins AD0–AD7 is address or data. Every time the microprocessor sends out an address, it activates (sets high) the ALE to indicate that the information on pins AD0–AD7 is the address (A0–A7). This information must be latched, then pins AD0–AD7 are used to carry data. When data is to be sent out or in, ALE is low, which indicates that AD0–AD7 will be used as data buses (D0–D7). This process of separating address and data from pins AD0–AD7 is called *demultiplexing*.

Address bus in 8088

The 8088 has 20 address pins (A0–A19), allowing it to address a maximum of one megabyte of memory (2^{20} = 1M). Pins AD0–AD7 provide the A0–A7 addresses with the assistance of a latch. To demultiplex the address signals from the address/data pins, a latch

and I/O bus activities. For example, in the read timing, ALE latches the address in the first clock cycle. In the second and third clock cycles, the read signal is provided. Finally, by the end of the fourth clock cycle the data must be at the pins of the CPU to be fetched in. Notice that the entire read or write cycle time is only 4 clock cycles. If the task of reading or writing takes more than 4 clocks due to the slowness of memory or I/O devices, wait states (WS) can be requested from the CPU. This will be demonstrated in Chapter 10.

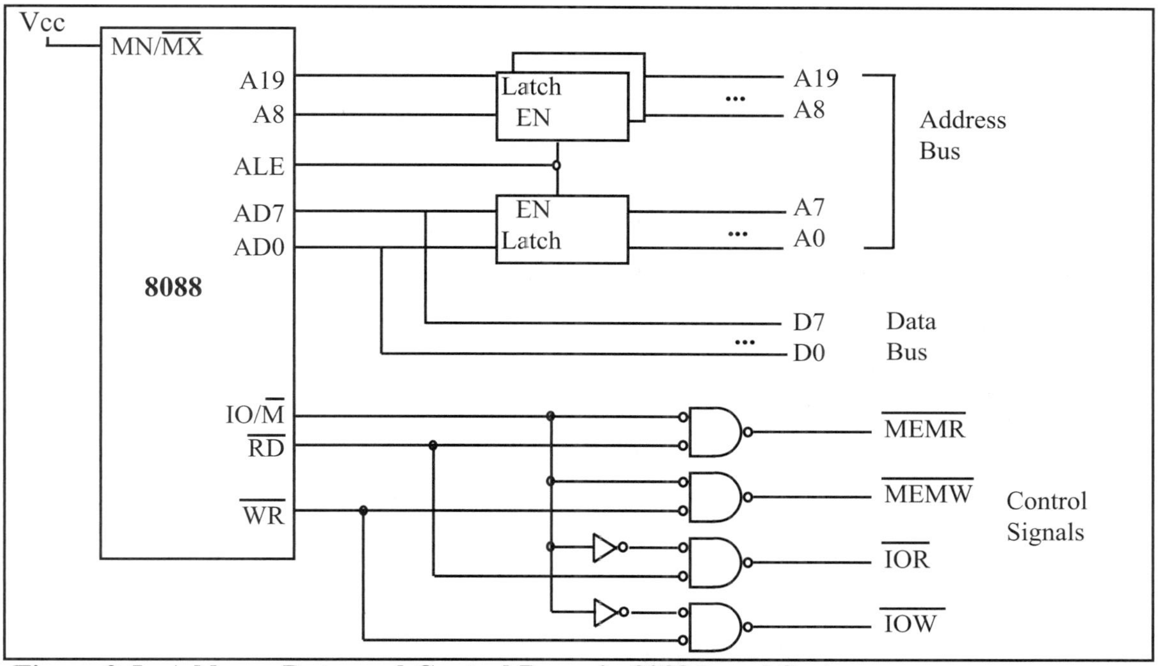

Figure 9-5. Address, Data, and Control Buses in 8088-based System

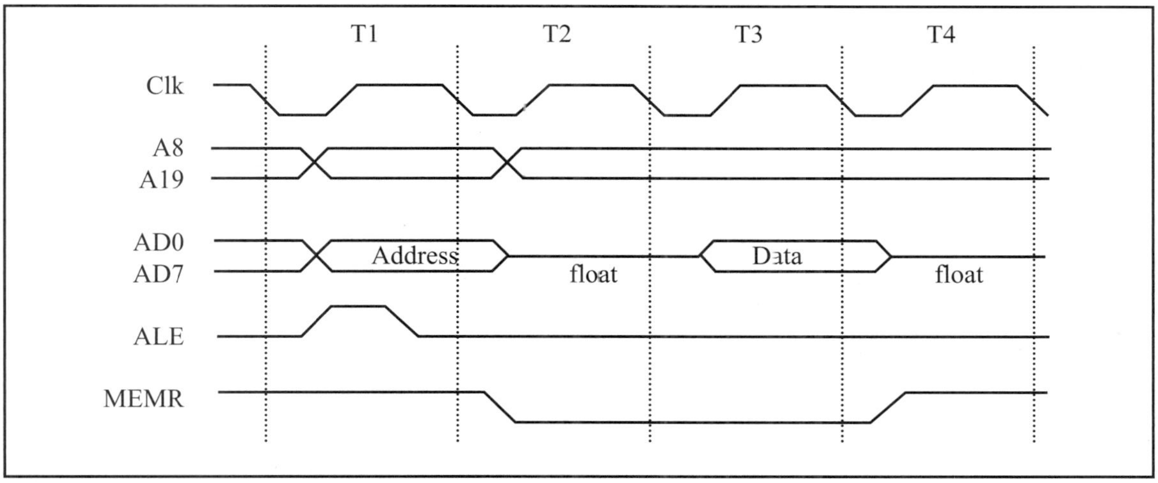

Figure 9-6. ALE Timing

Other 8088 pins

Pins 24–32 of the 8088 have different functions depending on whether the 8088 is used in minimum mode or maximum mode. As stated earlier, maximum mode is used only when we want to connect the 8088 to an 8087 math coprocessor. In maximum mode, the 8088 needs supporting chips to generate the control signals, as described in the next section. Table 9-2 lists the functions of pins 24–32 of the 8088 in minimum mode.

Other pins of the 8088 are described below.

MN/MX (minimum/maximum)

Minimum mode is selected by connecting MN/MX (pin number 33) directly to +5 V; maximum mode is selected by grounding this pin.

NMI (nonmaskable interrupt)

This is an edge-triggered (going from low to high) input signal to the processor that will make the microprocessor jump to the interrupt vector table after it finishes the current instruction. This interrupt cannot be masked by software, as we will see in Chapter 14.

INTR (interrupt request)

INTR is an active-high level-triggered input signal that is continuously monitored by the microprocessor for an external interrupt. This pin and INTA are connected to the 8259 interrupt controller chip, as we will see in Chapter 14.

CLOCK

Microprocessors require a very accurate clock for synchronization of events and driving the CPU. For this reason, Intel has designed the 8284 clock generator to be used with the 8088 processor. CLOCK is an input and is connected to the 8284 clock generator. It acts as the heartbeat of the CPU. Any irregularity causes the CPU to malfunction. The 8284 chip is used whether the 8088 is connected in minimum mode or in maximum mode. The details of the 8284 chip are covered in the next section.

READY

READY is an input signal used to insert a wait state for slower memories and I/O.

Table 9-2: Pins 24–32 in Minimum Mode

Pin	Name and Function
24	INTA (interrupt acknowledge) Active-low output signal. Informs interrupt controller that an INTR has occurred and that the vector number is available on the lower 8 lines of the data bus.
25	ALE (address latch enable) Active-high output signal. Indicates that a valid address is available on the external address bus.
26	DEN (data enable) Active-low output signal. Enables the 74LS245. This allows isolation of the CPU from the system bus.
27	DT/R (data transmit/receive) Active-low output signal used to control the direction of data flow through the 74LS245 transceiver.
28	IO/M (input-output or memory) Indicates whether the address bus is accessing memory or an I/O device. In the 8088, it is low when accessing memory and high when accessing I/O. This pin is used along with RD and WR pins to generate the four control signals MEMR, MEMW, IOR, and IOW.
29	WR (write) Active-low output signal. Indicates that the data on the data bus is being written to memory or an I/O device. Used along with signal IO/M (pin 28) to generate the MEMW and IOW control signals for write operations.
30	HLDA (hold acknowledge) Active-high output signal. After input on HOLD, the CPU responds with HLDA to signal that the DMA controller can use the buses.
31	HOLD (hold) Active-high input from the DMA controller that indicates that the device is requesting access to memory and I/O space and that the CPU should release control of the local buses.
32	RD (Read) Active-low output signal. Indicates that the data is being read (brought in) from memory or I/O to the CPU. Used along with signal IO/M (pin 28) to generate MEMR and IOR control signals for read operations.

It inserts wait states when it is low. The READY signal is needed to interface the CPU to low-speed memories and I/O devices.

TEST

In minimum mode this is not used. In maximum mode, however, this is an input from the 8087 math coprocessor to coordinate communications between the two processors.

RESET

To terminate the present activities of the microprocessor, a high is applied to the RESET input pin. A presence of high will force the microprocessor to stop all activity and set the major registers to the values shown in Table 9-3. The data in Table 9-3 has certain implications in the allocation of memory space to RAM and ROM that we will clarify next.

Table 9-3: IP and Segment Register Contents after Reset

Register	Contents
CS	FFFF
IP	0000
DS	0000
SS	0000
ES	0000

(Reprinted by permission of Intel Corporation, Copyright Intel Corp. 1983)

At what address does the 8088 wake up?

According to Table 9-3, when power is applied to the 8088, it wakes up at physical address FFFF0H, since a CS:IP address of FFFF:0000 leads to a physical address of FFFF0H. Therefore, we must have a nonvolatile memory such as ROM at the FFFF0H address. This is discussed further in Chapter 10.

Review Questions

1. Describe the differences between the external data bus of the 8086 and 8088.
2. In the 8088, pins AD0–AD7 are used for both data and addresses. How does the CPU indicate whether the information on these pins is data or an address?
3. The 8088 memory or I/O read cycles take ____ clock pulses to complete.
4. If we do not need an 8087 math coprocessor, the 8088 is connected in _______ mode.
5. Indicate whether each of the following pins are input pins, output pins, or both.
   (a) AD0–AD7 (b) ALE (c) A8–A15
6. Give the status of the IO/M and RD pins when MEMR is active.
7. Give the status of the IO/M and WR pins when MEMW is active.

SECTION 9.2: 8284 AND 8288 SUPPORTING CHIPS

The original IBM PC introduced in 1981 used the 8088 in maximum mode with a socket for the 8087 math coprocessor. In maximum mode, the 8088 requires the use of the 8288 to generate some of the control signals. In this section we cover the 8088's supporting chips, the 8284 and 8288, and their use in maximum mode. Modern microprocessors such as the Pentium have all these chips incorporated into a single chip. Therefore, this section can be skipped unless you are interested in the design of the original PC.

Figure 9-7 shows the 8086/88 in maximum mode. Comparing Figure 9-7 with Figure 9-1, we see that pins 24–32 have different functions. To use the 8088 in maximum mode we must use the 8288 supporting chip. We describe the 8288 next and how it is used with the 8088 in maximum mode.

8288 bus controller

As shown in Figure 9-8, the 8288 is a 20-pin chip specially designed to provide all the control signals when the 8088 is in maximum mode. The input and output signals are described below.

Input signals

$\overline{S0}$, $\overline{S1}$, $\overline{S2}$ (status input)

Input to these pins comes from the 8088. Depending upon the input from the CPU, the 8288 will provide one of the commands or control signals shown in Table 9-4.

CLK (clock)

This is input from the 8284 clock generator, providing the clock pulse to the 8288 to synchronize all command and control signals with the CPU. The 8284 chip is discussed later in this section.

$\overline{AEN}$ (address enable)

AEN, an active-low signal, activates the 8288 command output at least 115 ns after its activation. In the IBM PC it is connected to the AEN generation circuitry.

CEN (command enable)

An active-high signal is used to activate/enable the command signals and DEN. In the IBM PC it is connected to the AEN generating circuitry.

IOB (input/output bus mode)

An active-high signal makes the 8288 operate in input/output bus mode rather than in system bus mode. Since the IBM PC is designed with system buses, it is connected to low.

Output signals

The following are the output signals of the 8288 bus controller chip.

$\overline{MRDC}$ (memory read command)

This is active low and provides the $\overline{\text{MEMR}}$ (memory read) control signal. It acti-

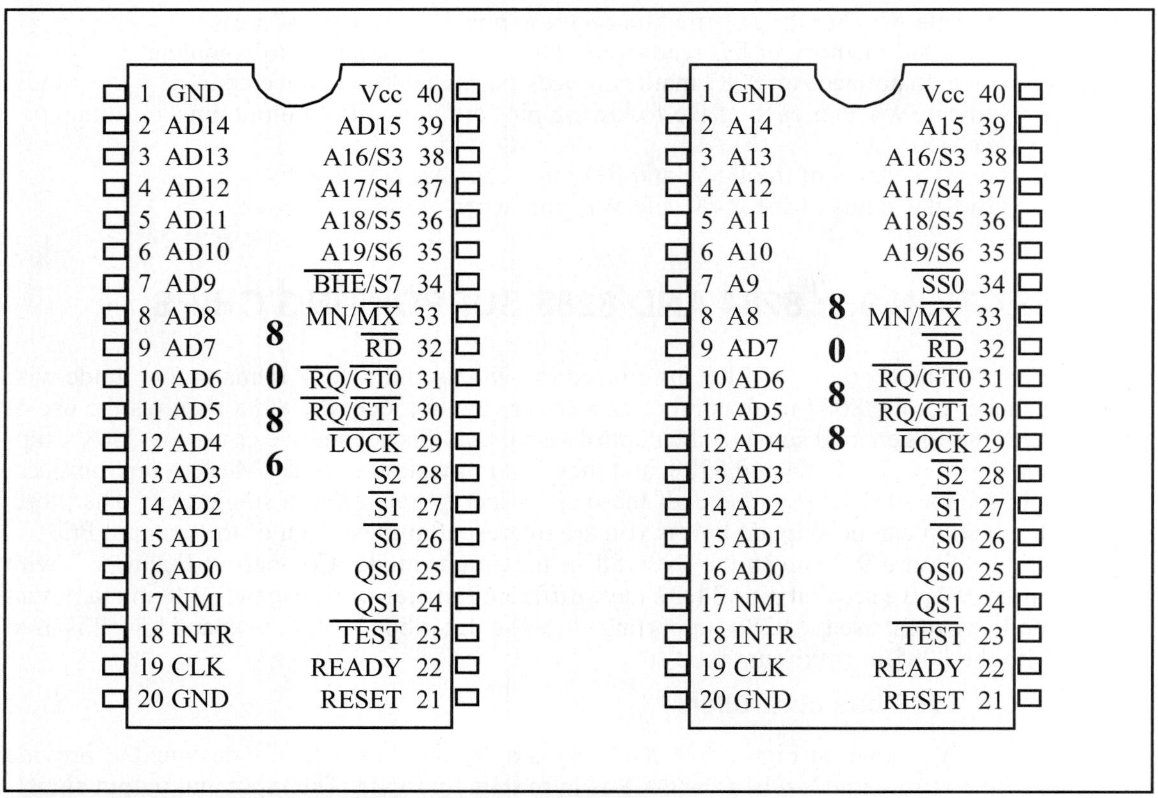

Figure 9-7. The 8086 and 8088 in Maximum Mode
(Reprinted by permission of Intel Corporation, Copyright Intel, 1989)

vates the selected device or memory to release its data to the data bus.

$\overline{\text{MWTC}}$ (memory write command), $\overline{\text{AMWC}}$ (advanced memory write)

These two active-low signals are used to tell memory to record the data present on the data bus. These two are the same as the $\overline{\text{MEMW}}$ (memory write) signal, the only difference being that AMWC is activated slightly earlier in order to give extra time to slow devices.

$\overline{\text{IORC}}$ (I/O read command)

$\overline{\text{IORC}}$ is an active-low signal that tells the I/O device to release its data to the data bus. In the PC it is called the $\overline{\text{IOR}}$ (I/O read) control signal.

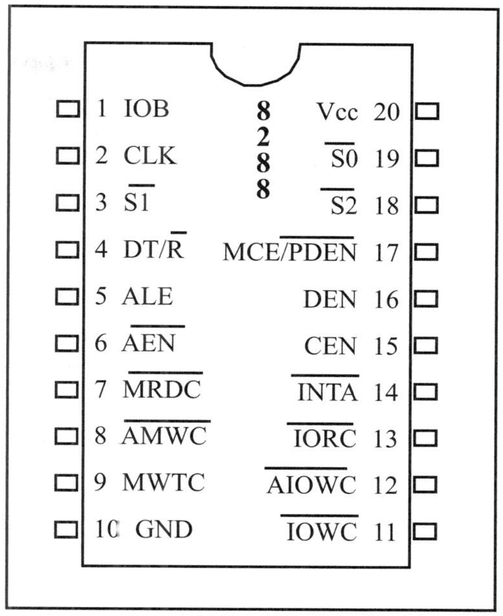

Figure 9-8. 8288 Bus Controller
(Reprinted by permission of Intel Corporation, Copyright Intel, 1983)

$\overline{\text{IOWC}}$ (I/O write command), $\overline{\text{AIOWC}}$ (advanced I/O write command)

Both are active-low signals used to tell the I/O device to pick up the data on the data bus. $\overline{\text{AIOWC}}$ is available a little bit early to give sufficient time to slow devices. It is unused in the IBM PC. In the PC, $\overline{\text{IOWC}}$ is labeled as $\overline{\text{ICW}}$.

Table 9-4: Status Pins of the 8288 and Their Meaning

S2	S1	S0	Processor State	8288 Command
0	0	0	Interrupt acknowledge	$\overline{\text{INTA}}$
0	0	1	Read input/output port	$\overline{\text{IORC}}$
0	1	0	Write input/output port	$\overline{\text{IOWC}}$, $\overline{\text{AIOWC}}$
0	1	1	Halt	None
1	0	0	Code access	$\overline{\text{MRDC}}$
1	0	1	Read memory	$\overline{\text{MRDC}}$
1	1	0	Write memory	$\overline{\text{MWTC}}$, $\overline{\text{AMWC}}$
1	1	1	Passive	None

(Reprinted by permission of Intel Corporation, Copyright Intel Corp. 1989)

$\overline{\text{INTA}}$ (interrupt acknowledge)

An active-low signal will inform the interrupting device that its interrupt has been acknowledged and will provide the vector address to the data bus. In the IBM PC this is connected to INTA of the 8259 interrupt controller chip.

DT/$\overline{\text{R}}$ (data transmit/receive)

DT/$\overline{\text{R}}$ is used to control the direction of data in and out of the 8088. In the IBM PC it is connected to DIR of the 74LS245. When the 8088 is writing data, this signal is high and will allow data to go from the A side to the B side of the 74LS245, so that data is released to the system bus. Conversely, when the CPU is reading data, this signal is low, which allows data to come in from the B to the A side of the 74LS245 data transceiver chip so that it can be received by the CPU.

DEN (data enable)

An active-high signal will make the data bus either a local data bus or the system data bus. In the IBM PC it is used along with a signal from the 8259 interrupt controller to activate G of the 74LS245 transceiver.

MCE/PDEN (master cascade enable/peripheral data enable)

This is used along with the 8259 interrupt controller in master configuration. In the IBM PC the 8259 is used as a slave; therefore, this pin is ignored.

ALE (address latch enable)

ALE is an active-high signal used to activate address latches. The 8088 multiplexes address and data through AD0–AD7 in order to save pins. In the IBM PC, ALE is connected to the G input of the 74LS373, making demultiplexing of the addresses possible.

8284 clock generator

The 8284 is used in both minimum and maximum modes since it provides the clock and timing for the 8088-based system. Figure 9-9 shows the 8284A, an 18-pin chip especially designed for use with the 8088/86 microprocessor. It provides not only the clock and synchronization for the microprocessor, but also the READY signal for the insertion of wait states into the CPU bus cycle. A description of each pin and how it is connected in the IBM PC follows.

Input pins

$\overline{\text{RES}}$ (reset in)

This is an input active-low signal to generate RESET. In the IBM PC, it is connected to the power-good signal from the power supply. When the power switch in the IBM PC is turned on, assuming that the power supply is good, a low signal is provided to this pin and the 8284 in turn will activate the RESET pin, forcing the 8088 to reset; then the microprocessor takes over. This is called a *cold boot*.

X1 and X2 (crystal in)

X1 and X2 are the pins to which a crystal is attached. The crystal frequency must be 3 times the desired frequency for the microprocessor. The maximum crystal for the 8284A is 24 MHz. The IBM PC is connected to a crystal of 14.31818 MHz.

F/$\overline{\text{C}}$ (frequency/clock)

This pin provides an option for the way the clock is generated. If connected to low, the clock is generated by the 8284 with the help of a crystal oscillator. If it is connected to high, it expects to receive clocks at the EFI pin. Since the IBM PC uses a crystal, this pin is connected to low.

EFI (external frequency in)

External frequency is connected to this pin if F/$\overline{\text{C}}$ has been connected to high. In the IBM PC this is not connected since a crystal is used instead of an external frequency generator.

CSYNC (clock synchronization)

This active-high signal is used to allow several 8284 chips to be connected together and synchronized. The IBM PC uses only one 8284; therefore, this pin is connected to low.

RDY1 and $\overline{\text{AEN1}}$

RDY1 is active high and $\overline{\text{AEN1}}$ (address enable) is active low. They are used together to provide a ready signal to the microprocessor, which will insert a WAIT state to the CPU read/write cycle. In the PC, RDY1 is connected to DMAWAIT, and $\overline{\text{AEN1}}$ is connected to RDY/WAIT. This allows a wait state to be inserted by either the CPU or DMA.

RDY2 and $\overline{AEN2}$

These function exactly like RDY1 and $\overline{AEN1}$ but are designed to allow for a multiprocessing system. In the IBM PC, RDY2 is connected to low, and $\overline{AEN2}$ is connected to high, which permanently disables this function since there is only one 8088 microprocessor in the system.

ASYNC

This is called *ready synchronization* select. An active low is used for devices that are not able to adhere to the very strict RDY setup time requirement. In the IBM PC this is connected to low, making the timing design of the system easier with slower logic gates.

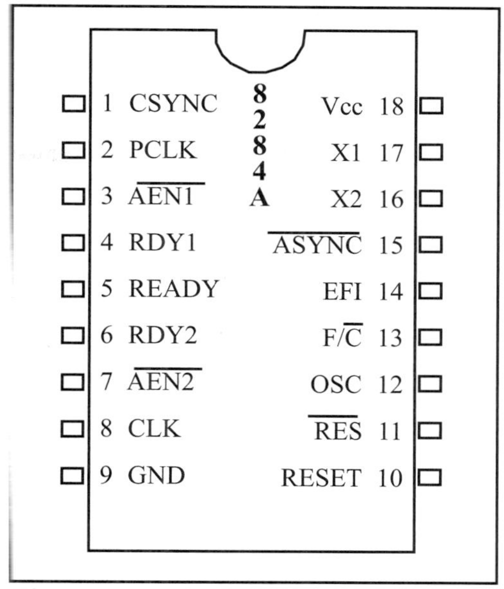

Figure 9-9. 8284A Chip
(Reprinted by permission of Intel Corporation, Copyright Intel, 1983)

Output signals

RESET

This is an active-high signal that provides a RESET signal to the 8088. It is activated by the RES input signal discussed earlier.

OSC (oscillator)

This provides a clock frequency equal to the crystal oscillator and is TTL compatible. Since the IBM crystal oscillator is 14.31818 MHz, OSC will provide this frequency to the expansion slot of the IBM PC.

CLK (clock)

This is an output clock frequency equal to one-third of the crystal oscillator, or EFI input frequency, with a duty cycle of 33%. This is connected to the clock input of the 8088 and all other devices that must be synchronized with the CPU. In the IBM PC it is connected to pin 19 of the 8088 microprocessor and other circuitry under the CLK88 label. This frequency, 4.772776 MHz (14.31818 divided by 3), is the processor frequency on which all of the timing calculations of the memory and I/O cycle are based.

PCLK (peripheral clock)

PCLK is one-half of CLK (or one-sixth of the crystal) with a duty cycle of 50% and is TTL compatible. In the IBM PC this 2.386383 MHz is provided to the 8253 timer to be used to generate speaker tones, and for other functions.

READY

This signal is connected to READY of the CPU. In the IBM PC it is used to signal the 8088 that the CPU needs to insert a wait state due to the slowness of the devices that the CPU is trying to contact.

Review Questions

1. Pin RESET is an _________ (input, output) for the 8284 and an _________ (input, output) for the 8088.
2. True or false. Regardless of whether the 8088 is in minimum or maximum mode, the 8284 clock generator is needed to provide a reliable clock.
3. True or false. The 8288 is used to provide control signals for the 8088 when it is in minimum mode.
4. The 8288 output pin _______ controls the direction data flows in and out of the CPU.

SECTION 9.3: 8-BIT SECTION OF ISA BUS

Previous sections have explained the 8088 CPU and supporting chips. This section will explain how they are all connected in the original IBM PC to produce the required buses to communicate with memory, input/output peripherals, and the 8-bit section of the ISA bus. The study of the 8-bit section of ISA is the main topic of this section.

A bit of bus history

The original IBM PC introduced in 1981 used an 8088 microprocessor, whose 8-bit data bus gave birth to the 8-bit section of the ISA bus. In 1984 when IBM introduced the IBM PC/AT using the 80286 microprocessor, the data bus was expanded to 16 bits. The 8-bit data bus can be seen as a subsection of the 16-bit ISA bus. Very often the 8-bit data bus was referred to as the IBM PC/XT (extended technology) bus in order to differentiate it from the IBM PC AT (advanced technology). Eventually the IBM PC AT bus became known as the ISA (Industry Standard Architecture) bus since the term "PC AT" was copyrighted by IBM. Throughout this book we use the terms PC/XT and PC interchangeably to refer to the 8-bit portion of the IBM PC AT (ISA) bus. The following is the description of the three main buses of the IBM PC as generated by the 8088 and supporting chips.

Local bus vs. system bus

In the discussion of PC design we often see the terms *local bus* and *system bus*. The system bus not only provides necessary signals to all the chips (RAM, ROM, and peripheral chips) on the motherboard, but also goes to the expansion slot for any plug-in expansion card. In contrast, the local bus is connected directly to the CPU. Any communication with the CPU must go through the local bus. There is a bridge between the local bus and the system bus to make sure they are isolated from each other. Sometimes the system bus is referred to as a *global bus*. We use tri-state buffers to isolate the local bus and system bus. For example, 74LS245 is a widely used chip for the data bus buffer since it is bidirectional. See Figure 9-10. Figure 9-11 shows an example of local and system buses. Figure 9-11 gives an overview of the 8088 and its supporting chips as designed in the original PC. Notice the role of the 74LS245 and 74LS373 in isolating the local and system buses. Everything on the left of the 8288, 74LS373s, and 74LS245 represent the local bus and everything on the right side of those chips are the system bus. The 74LS245 and 74LS373s play the role of bridge to isolate the local and system buses. Now let's look at each of the buses.

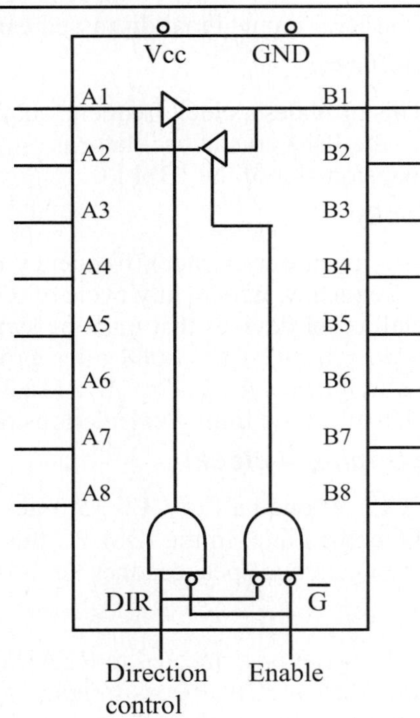

Function Table

Enable $\overline{G}$	Direction Control DIR	Operation
L	L	B data to A bus
L	H	A data to B bus
H	X	Isolation

Figure 9-10. 74LS245 Bidirectional Buffer
(Reprinted by permission of Texas Instruments, Copyright Texas Instruments, 1988)

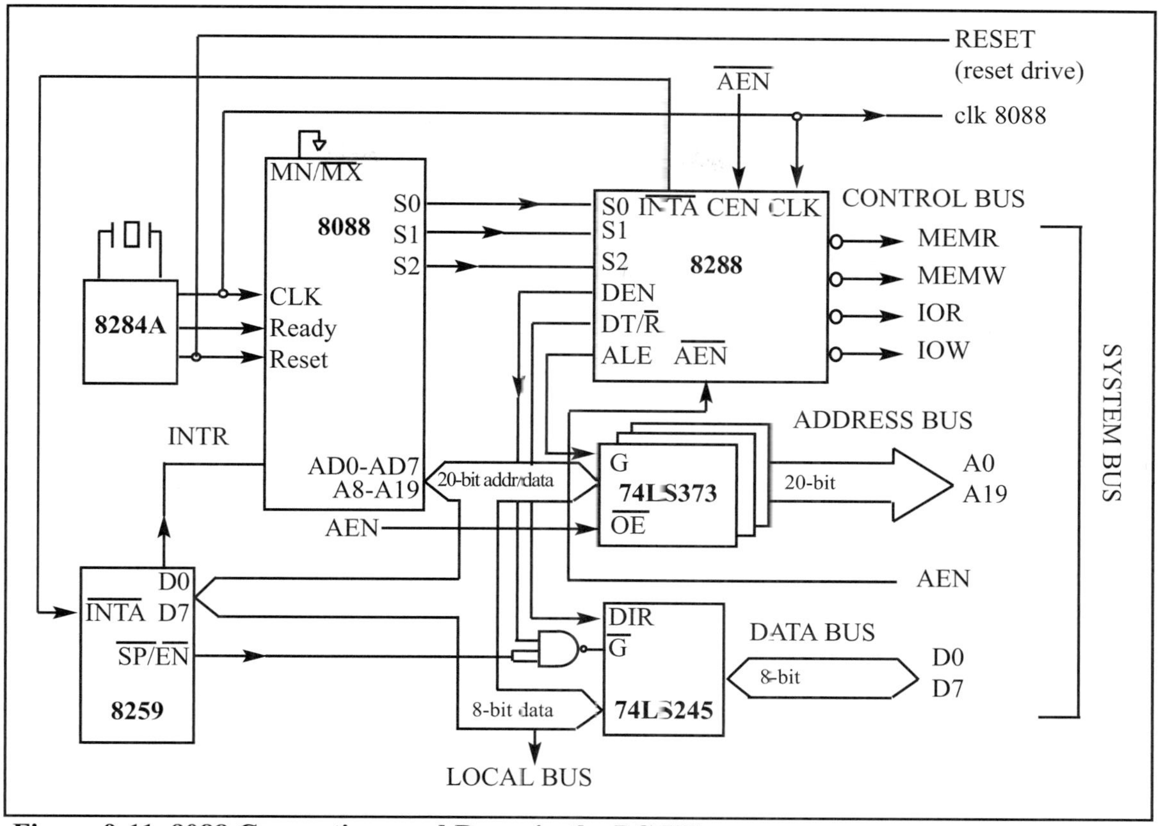

Figure 9-11. 8088 Connections and Buses in the PC/XT

(Reprinted by permission from "IBM Technical Reference" c. 1984 by International Business Machines Corporation)

Address bus

Three 74LS373 chips in Figure 9-11 are used for two functions:

1. To latch the addresses from the 8088 and provide stable addresses to the entire computer. The address bus is a unidirectional bus. The 74LS373 chips are activated by control signals AEN and ALE. When AEN is low, the 8088 provides the address buses to the system. The 8288's ALE (connected to G) enables the 74LS373 to latch the addresses from the CPU, providing a 20-line stable address to memory, peripherals, and expansion slots. Demultiplexing addresses A0–A7 is performed by the 74LS373 connected to pins AD0–AD7 of the CPU. The CPU's A8–A15 is connected to the second 74LS373, and A16–A19 to the third one. Half of the third 74LS373 is unused.
2. To isolate the system address buses from local address buses. The system buses must be allowed to be used by the DMA or any other board through the expansion slot without disturbing the CPU. This is achieved by the 74LS373s through AEN. The AEN signal is described shortly.

Data bus

The bidirectional data bus goes through the 74LS245 transceiver (see Figures 9-10 and 9-11). DT/$\overline{R}$ and DEN are the two signals that activate the 74LS245. DT/$\overline{R}$ goes to DIR of the 74LS245 and makes the transceiver transmit information from the A side to the B side when DT/$\overline{R}$ is high. Conversely, when DT/$\overline{R}$ makes DIR low, the transceiver transfers information from the B side to the A side, thereby receiving information from the system data bus and bringing it to the microprocessor. DEN (an active low signal) enables the 74LS245. This isolates the data buses to make them either a local bus or a system bus. When the 74LS245 is not active, the system data bus is isolated from the local data bus.

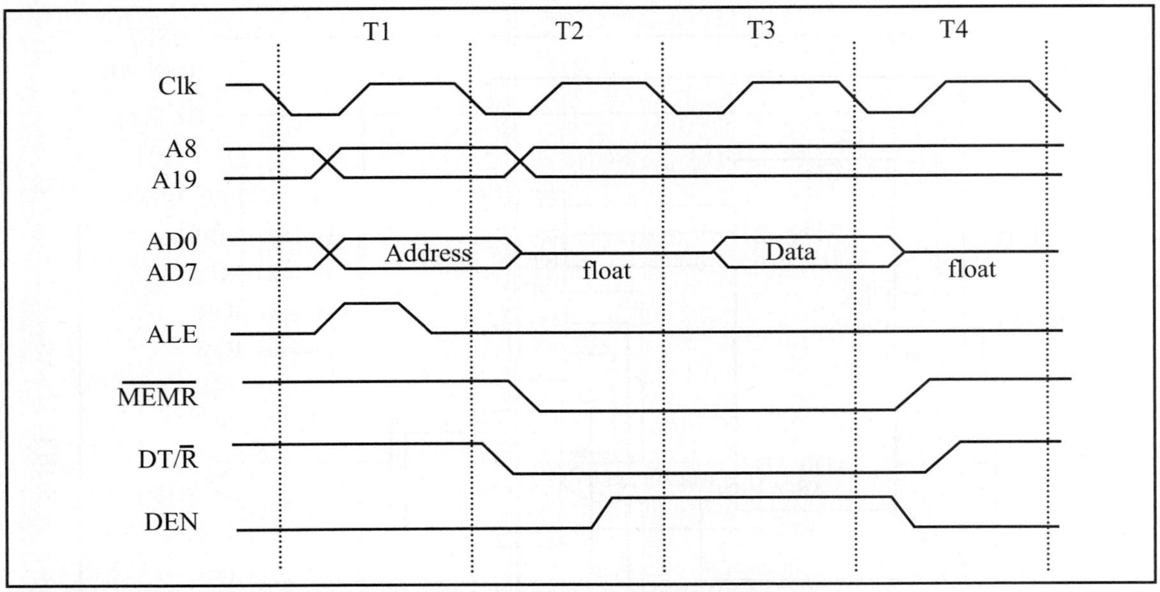

Figure 9-12. ALE, DEN, and DTR Timing for the 8088 System

Control bus

The four most important control signals of the IBM PC are $\overline{\text{IOR}}$ (I/O read), $\overline{\text{IOW}}$ (I/O write), $\overline{\text{MEMR}}$ (memory read), and $\overline{\text{MEMW}}$ (memory write). They are provided by the 8288 chip as shown in Figure 9-11. The timing for the bus activity is shown in Figure 9-12.

One bus, two masters

While the 8088, the main processor, is designed for fetching and executing instructions, it is unacceptably slow for transferring large numbers of bytes of data such as in hard disk data transfers. Instead, the 8237 chip is used for data transfers of large numbers of bytes. The detailed function of this chip is explained in Chapter 15. All that is needed here is to know that the 8237's job is to transfer data and it must have access to all three buses to do that. Since no bus can serve two masters at the same time, there must be a way to allow either the 8088 processor or the 8237 DMA to gain control over the buses. This is called *bus arbitration* and is achieved by the AEN (address enable) generation circuitry.

AEN signal generation

When the system is turned on, the 8088 CPU is in control of all the buses. The CPU maintains control as long as it is fetching and executing instructions. As can be seen from Figure 9-13, AEN is the output signal of the D flip-flop. Since Q is either high or low, depending on the status of this signal, either the CPU or the DMA can access the buses. Table 9-5 shows the role of AEN in bus arbitration.

Control of the bus by DMA

How does AEN become high, handing control of the system buses to DMA? The answer is that when DMA receives a request for service, it will notify the CPU that it needs to use the system buses by putting a LOW on HRQDMA (this is the same as the HOLD signal in minimum mode of the 8088). This in turn will provide a high on the D3 output of the 74LS175, assuming that the current memory cycle is finished and that LOCK is not activated. In the following clock cycle, HLDA (hold acknowledge) is provided to the DMA and AEN becomes high, giving control over the buses to the DMA.

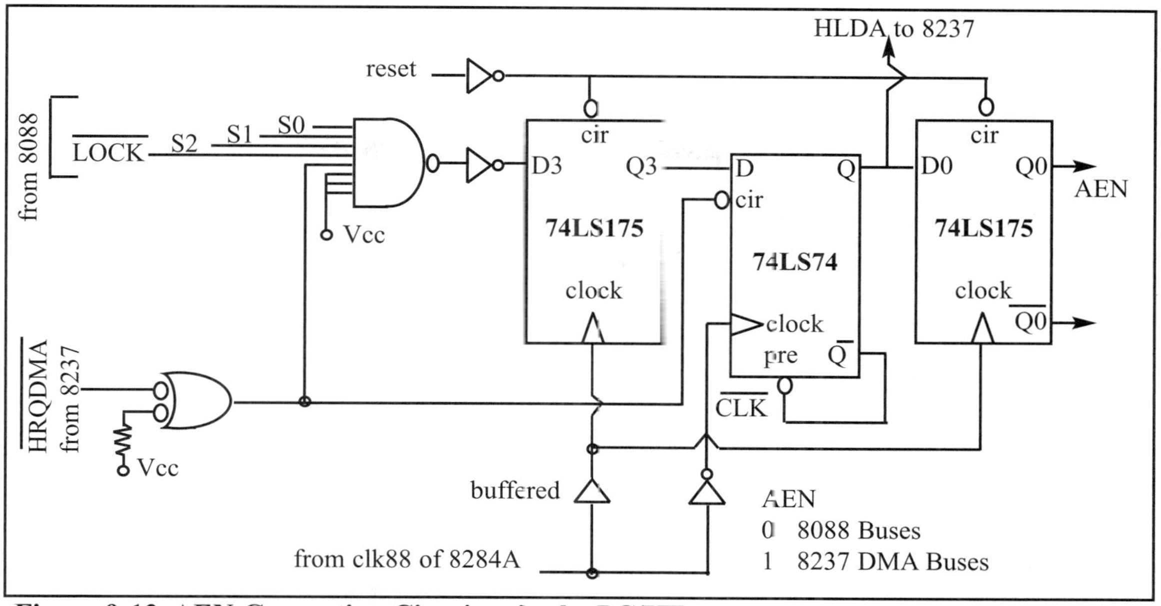

Figure 9-13. AEN Generation Circuitry in the PC/XT

(Reprinted by permission from "IBM Technical Reference" c. 1984 by International Business Machines Corporation)

Table 9-5: AEN Bus Arbitration

AEN	Bus Control
0	Buses controlled by CPU
1	Buses controlled by DMA

Bus boosting

One more point that needs explaining is bus boosting of the control, data, and address buses to provide sufficiently strong signals to drive various IC chips. When a pulse leaves an IC chip it can lose some of its strength, depending on how far away the receiving IC chip is located. In addition, the more pins a signal is connected to, the stronger the signal must be to drive them all. Therefore, the signals must be amplified. Stated another way, every pin connected to a given signal has input capacitance, and the capacitances are in parallel; thus as far as that signal is concerned they are all added together, making one big capacitor load. This requires that the signal be strong enough to drive all the inputs (see Chapter 26 for more details on this topic). It is common to combine the functions of bus isolation and bus boosting into a single chip. For example, 74LS373 chips are used to boost the addresses provided by the 8088 microprocessor in addition to the bus isolation mentioned earlier. The signals provided by the CPU need boosting since the 8088 is a CMOS chip. CMOS has a much lower driving capability than TTL, of which 74LS373s are made. Likewise, the 74LS245 is used for both data bus booster and data bus isolation. Details of IC interfacing and how 74LS245 chips are used for signal amplification (boosting) are shown in Chapter 26.

8-bit section of the ISA bus

As stated earlier, the original IBM PC had an 8-bit data bus. Later with the introduction of the 80286, the 16-bit version of the bus became available. The 80286 bus became known as the ISA bus. The 8-bit bus is a subset of the 16-bit ISA bus and used in many peripheral boards. Figure 9-14 shows the 8-bit portion of the ISA bus expansion slot. From that figure notice that addresses A0–A19 and data signals D0–D7 are on the A side of the expansion slot. On the A side, also notice the AEN pin. On the B side are found control signals IOR, IOW, MEMR, and MEMW. The - sign on these and other control signals implies an active-low signal. In Chapter 11 we use signals A0–A19, D0–D7, AEN, IOR, and IOW to design an I/O interfacing card. The rest of the signals in Figure 9-14 will be covered in subsequent chapters. The signals asscociated with interrupts (IRQs) are covered in Chapter 14; signals associated with DMA (DREQs and DACKs) are covered in Chapter 15.

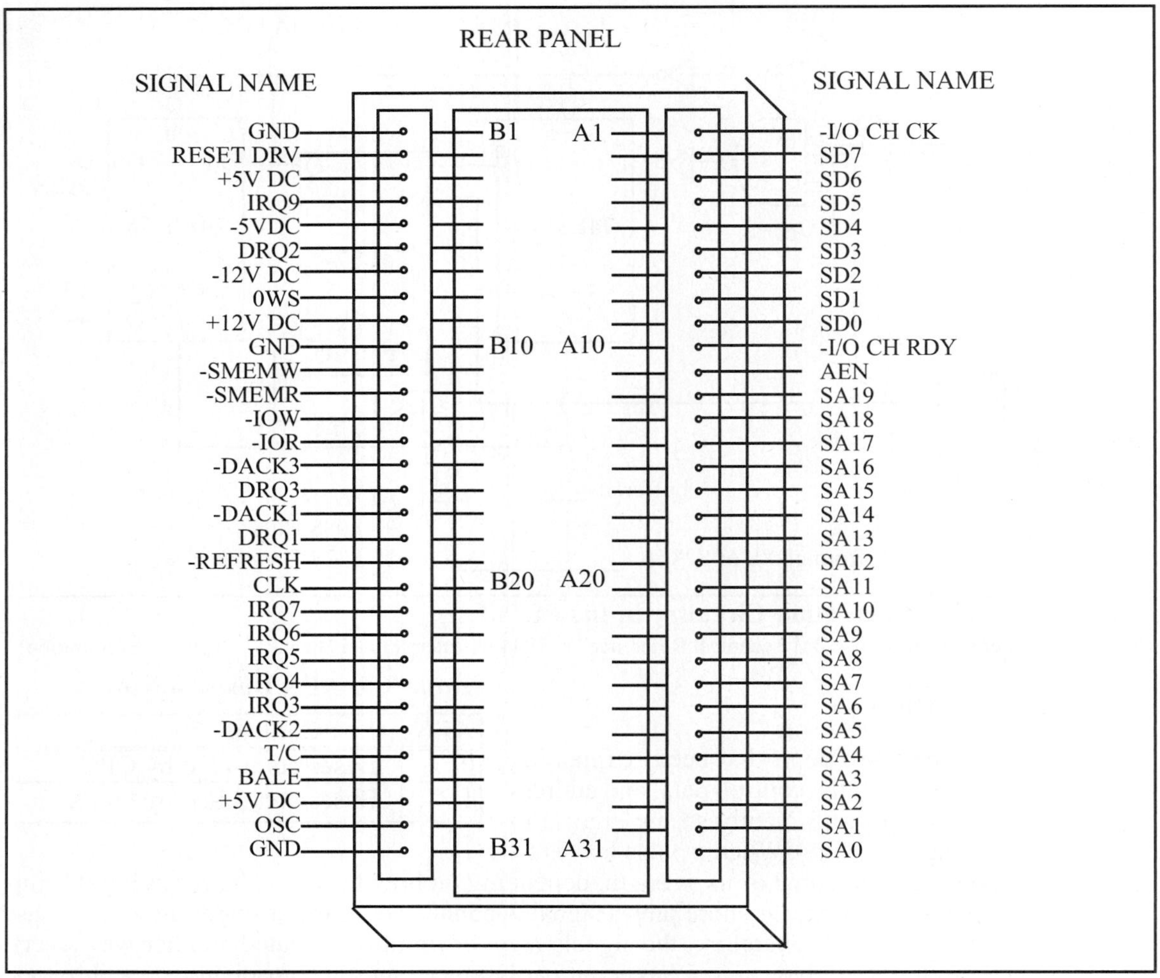

Figure 9-14. ISA Bus Slot Signals Detail (8-bit Section)

(Reprinted by permission from "IBM Technical Reference" c. 1985 by International Business Machines Corporation)

Review Questions

1. The system bus can be accessed either by the CPU or by ____________.
2. The control signal that provides bus arbitration is _______.
3. True or false. After a cold boot, DMA is given control of the buses.
4. The bidirectional data bus goes through the 74LS245 transceiver. Signal _______ determines whether data is flowing from the A to the B side or from the B to the A side.
5. Bus ______________ is required to provide strong signals to various IC chips in the IBM PC.

SECTION 9.4: 80286 MICROPROCESSOR

The 80286 is a 68-pin microprocessor available in either of two packaging formats: LCC (leaded chip carrier) and PGA (pin grid array). This is in contrast to the 8088, which is a 40-pin DIP (dual in-line package). To package the 68-pin IC in DIP packaging would have made it a long IC physically and consequently more fragile. Such packaging would also necessitate a longer path for some signals and as a result make it unsuitable for use in high-frequency systems. Figure 9-15 shows the 80286 in LCC packaging.

The 80286 can work in one of two modes: real mode or protected mode. In real mode, the maximum memory it can access is 1M, 00000H to FFFFFH. To access the entire 16M bytes of memory, 000000H to FFFFFFH, it must work in protected mode. In

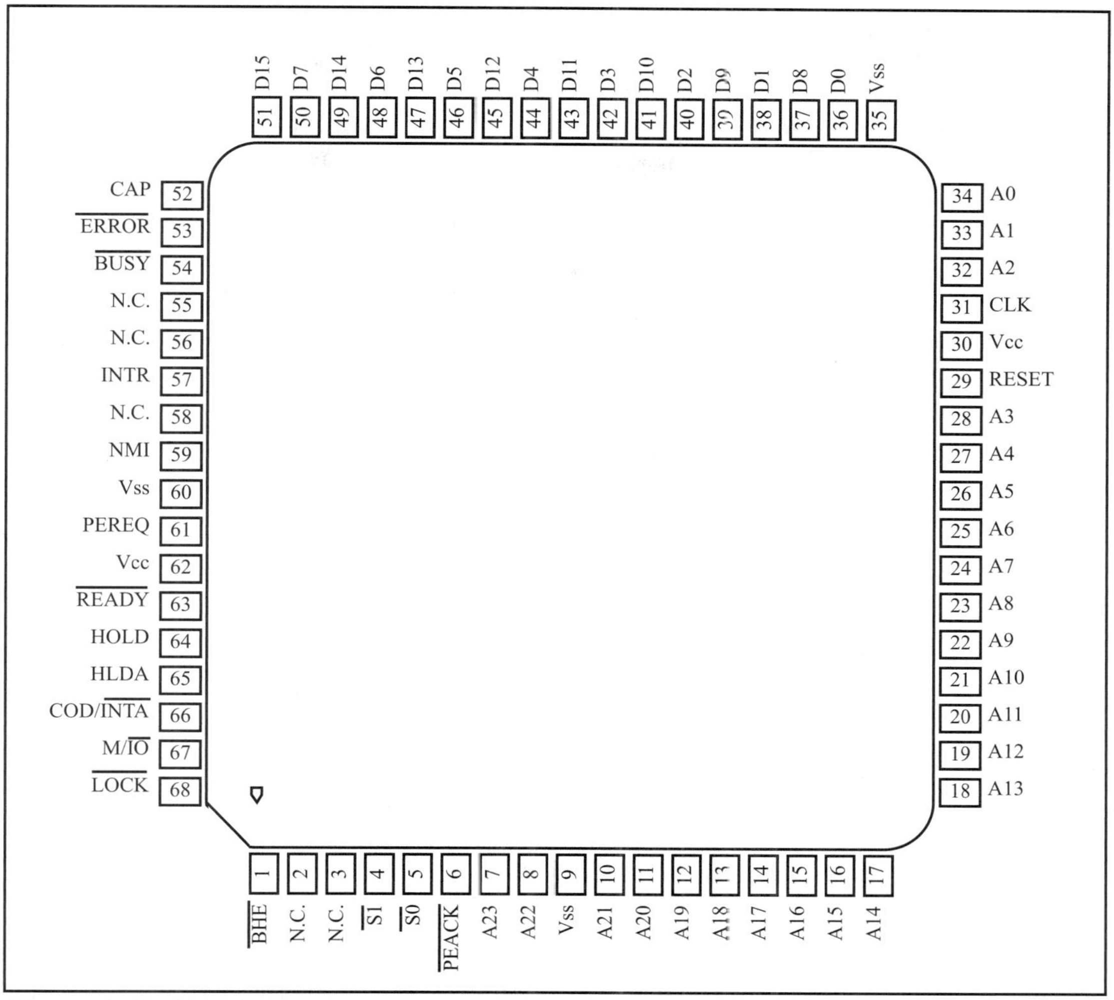

Figure 9-15. 80286 Microprocessor (LCC Packaging)

real mode, the 80286 is a faster version of the 8086 with a few new instructions. When power is applied to the 80286, it starts up in real mode and can be switched to protected mode at any time through a software instruction. However, to use the 286 in protected mode requires an extremely complex memory management system. Since very few systems are using the 286 in protected mode, it is not discussed here (even in protected mode it is still a 16-bit computer, meaning that all registers are 16-bit, as opposed to 32-bit).

Pin descriptions

The following are pin descriptions of the 80286 microprocessor.

Pins A0–A23 (address bus)

These output signals provide a 24-bit address to be used by the decoding circuitry to locate memory or I/O. When providing an address for memory, all 24 pins must be used (A0–A23); therefore, it can access a maximum of 16M bytes of memory (2^{24} = 16M). To access an I/O address, only A0–A15 are used. If the I/O address is a 16-bit address, A0–A15 are used to provide the address, and pins A16–A23 are low. If the I/O address is an 8-bit address, only A0–A7 are used, and A8–A23 are all low.

Pins D0–D15 (data bus)

These pins provide the 16-bit path for data to be transferred in and out of the CPU. It must be noted that unlike the 8088, the data bus is not multiplexed. The use of separate pins for address and data results in higher pin counts, but saves time since it eliminates the need for a demultiplexer. This 2-byte data path to the CPU allows the transfer of data on both bytes or on either byte, depending on the operation. The 80286 coordinates the activity on the D0–D15 data bus with the help of A0 and BHE.

Pin $\overline{BHE}$ (bus high enable)

This is an active-low output signal used to indicate that data is being transferred on D8–D15. Table 9-6 shows how $\overline{BHE}$ and A0 are used to indicate whether the data transfer is on D0–D7, D8–D15, or the entire bus, D0–D15.

Table 9-6: BHE, A0, and Byte Selection in the 80286

BHE	A0	Data Bus Status
0	0	Transferring 16-bit data on D0–D15
0	1	Transferring a byte on the upper half of data bus D8–D15
1	0	Transferring a byte on the lower half of data bus D0–D7
1	1	Reserved (the data bus is idle)

Pin CLK (clock)

CLK is an input providing the working frequency for the 80286. The processor always works on half of this frequency. For example, if CLK = 16 MHz, the system is an 8-MHz system. In other words, for the 80286 computer to be an 8-MHz system, the CLK must be 16 MHz.

Pin M/$\overline{IO}$ (memory I/O select)

M/$\overline{IO}$ is an output signal used by the CPU to distinguish between I/O and memory access. When it is high, memory is being accessed, and when it is low, I/O is being addressed.

Pin COD/$\overline{INTA}$ (code/interrupt acknowledge)

This is an output signal used by the CPU to indicate whether it is performing memory read/write of data or an instruction fetch. It is also used to distinguish between the action of interrupt acknowledge and I/O cycle. This signal, along with the status signals and M/$\overline{IO}$, is used to define the bus cycle.

Pins $\overline{S1}$ and $\overline{S0}$ (status signals)

These status signals for the bus cycle are both output signals used by the CPU along with M/$\overline{IO}$ and COD/$\overline{INTA}$ to define the type of bus cycle.

Pins HOLD and HLDA (hold and hold acknowledge)

HOLD and HLDA allow the CPU to control the buses. HOLD is an input signal to the 80286 and is active high. It is used by devices such as DMA to request permission to use the buses. In response, the CPU activates the output signal HLDA by putting a high on it to inform the requesting device that it has released the buses for the device's use. The DMA has control over the buses as long as HOLD is high, and in response the CPU keeps HLDA high. Whenever the DMA brings HOLD low, the CPU responds by making HLDA low, and regains control over the buses.

RESET pin

This is an input signal and is active high. When there is a low-to-high transition on RESET (and it stays high for at least 16 clocks), the 80286 initializes all registers to their predefined values and the output pins of the 80286 will have the status shown in Table 9-7. Of the above signals, the status of the following pins must be

Table 9-7: Pin State (Bus Idle) During Reset

Pin Name	Signal Level during Reset
D0–D15	High impedance
A0–A23	High
W/$\overline{R}$	Low
M/$\overline{IO}$	High

noted since they are used in the memory design of the IBM PC AT computers: A20 = 1, A21 = 1, A22 = 1, and A23 = 1.

As long as the RESET pin is high, no instruction or bus activity is allowed. The contents of the instruction pointer and segment registers of the 80286 after RESET are shown in Table 9-8.

Table 9-8: IP and Segment Registers After RESET

Register	Contents
CS	F000
IP	FFF0
DS	0000
SS	0000
ES	0000

(Reprinted by permission of Intel Corporation, Copyright Intel Corp. 1983)

It must be noted that when RESET of the 80286 is activated, it forces the 80286 to enter into real mode. In other words, the CPU wakes up in real mode. In real mode the 80286 (indeed, all the x86s from the 80286 to the Pentium 4) processor can address only 1 megabyte since it uses only address lines A0–A19. Since RESET also causes A20–A23 to be high, the first instruction for the 286 must be at physical address FFFFF0H. This is due to the fact that at reset, CS = F000 and IP = FFF0, making the logical address of the first instruction F000:FFF0. This provides the physical address of FFFF0H on A0–A19, and since A20–A23 is high at reset, the physical address of the first instruction must be FFFFF0H. This is 16 bytes from the top of the 16M address range of the 80286. The 80286 expects to have a far jump at location FFFFF0H and when the JMP is executed, the 286 puts 0s on pins A20–A23, making it effectively a 1M range real-mode system. Further implications of these facts are discussed in Chapter 10.

Pin INTR (interrupt request)

INTR is an input signal into the 80286 requesting suspension of the current program execution. It is used for external hardware interrupt expansion along with the 8259 interrupt controller chip. See Chapter 14 for more information.

Pin NMI (nonmaskable interrupt request)

NMI is an active-high input signal. When this pin is activated, the 80286 will automatically perform INT 2, meaning that there is no INTA response since INT 2 is assigned to it. See Chapter 14 for more details of this pin.

$\overline{\text{READY}}$ pin

$\overline{\text{READY}}$ is an active-low input signal used to insert a wait state and consequently prolong the read and write cycle for slow memory and I/O devices.

Figure 9-16 shows the 80286 timing. Notice the 2-clock cycle time for read.

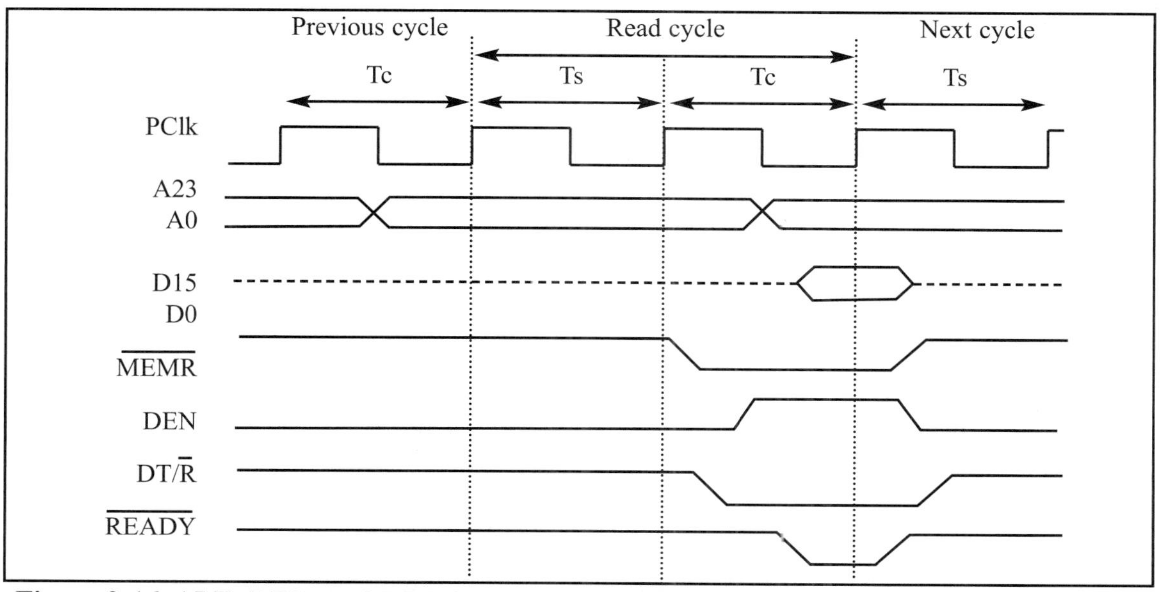

Figure 9-16. ALE, DEN, and DTR Timing for the 80286 CPU

Review Questions

1. When power is applied to the 80286, which mode does it wake up in, real mode or protected mode?
2. The 286 can access __________ of memory in real mode and __________ in protected mode.
3. When RESET is set to high, what are the contents of the CS and IP registers?

SECTION 9.5: 16-BIT ISA BUS

The origin of technical specifications of many of today's x86 PCs is the 80286-based IBM PC/AT. Much of the PC/AT in turn is based on the original 8088-based IBM PC introduced in 1981. A major legacy of those original PCs is the ISA (Industry Standard Architecture) bus slot. Remember that ISA is another name for the PC/AT bus since PC/AT is a trademark copyrighed by IBM Corp. In this section we examine the address, data, and control buses of the ISA expansion bus and some of the issues related to them. Whether the microprocessor used in a PC is Intel's Pentium, 386, 486, or an equivalent AMD processor, if it has an ISA bus slot, the material in this section is relevant and needs to be understood if you want to design expansion cards for ISA slots.

Figure 9-17 shows the 80286 microprocessor, along with supporting chips used in the original PC/AT computers. The address, data, and control buses in this figure are used throughout the motherboard and are also provided to the ISA expansion slot. In today's PC the 80286 is replaced with Intel's Pentium or AMD's Athlon microprocessor, and all the control signals are provided by a chipset. A *chipset* is an IC chip containing all the circuitry needed to support the CPU in a given motherboard. For educational purposes throughout the book, we use simple logic gates from the original PC to discuss some design concepts, even though in the real world the chipset uses CPLDs (Complex Programmable Logic Devices) for design with all the circuitry details buried inside. Next, we examine the major signals of the ISA expansion slot.

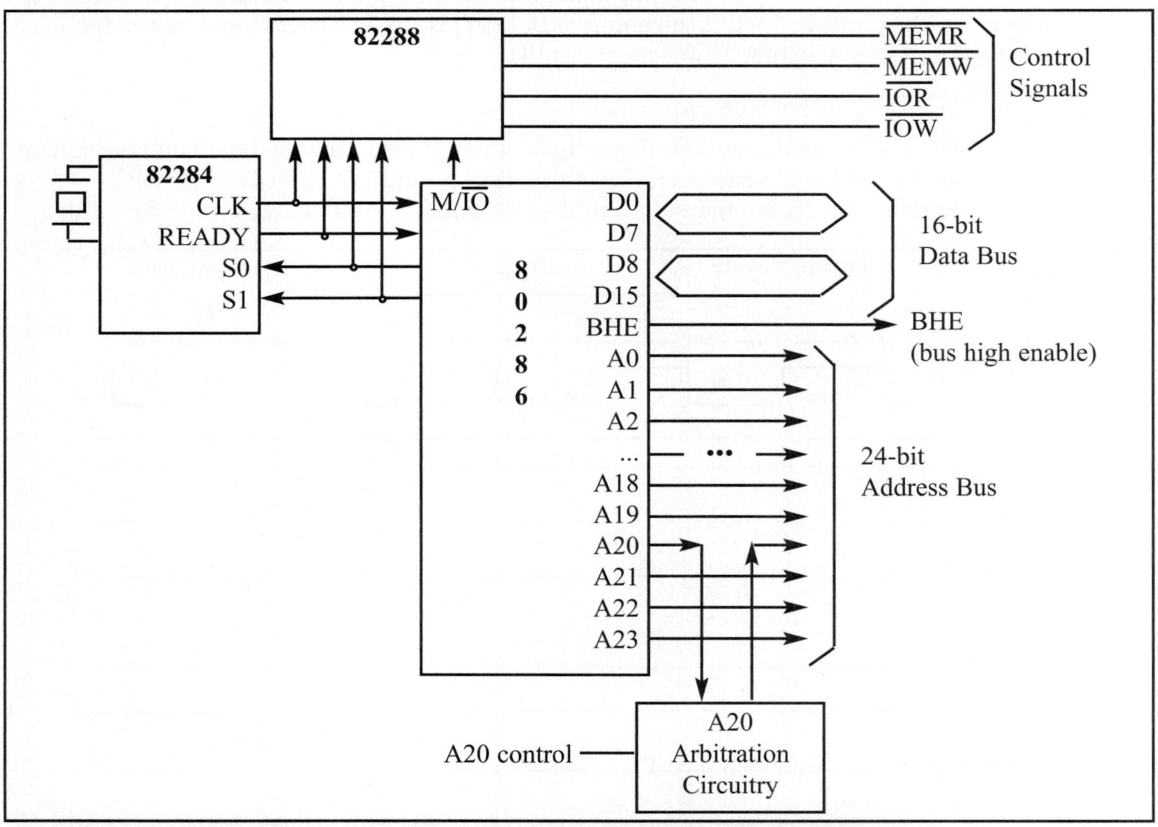

Figure 9-17. 80286 Block Diagram and Supporting Chips in the PC AT

Exploring ISA bus signals

In Section 9.3 we discussed the 8-bit section of the ISA bus. The 8-bit section uses a 62-pin connector to provide access to the system buses. In order to maintain compatibility with the original PC, the 16-bit ISA slot used the 8-bit section as a subset. A 36-pin connector was added to incorporate the new signals as shown in Figure 9-18. In designing a plug-in peripheral card for the ISA slot we need to understand the basic features of the ISA signals. The ISA bus has 24 address pins (A0–A23), 16 data pins (D0–D15), plus many control signals.

Address bus

Addresses A0–A19 are latched using ALE. These addresses are used throughout the motherboard and are also provided to the 62-pin part of the ISA slot as SA0–SA19 (system address). See Figure 9-18. Notice that this is already latched and cannot be latched again by a plug-in card. The A20–A23 part of the address is provided in the 36-pin section. In the 36-pin section of the ISA slot, A17–A23 are also provided as LA17–LA23 (latchable address). We need to use the ALE signal to latch these addresses in the design of plug-in cards. The ALE signal is provided as BALE (buffered ALE) and can be used to latch LA17–LA23.

Data bus

The data bus is composed of pins D0 to D15. The data bus is buffered by a pair of 74ALS245 data bus transceivers that are used throughout the motherboard to access memory and ports. They are also provided at the expansion slot as SD0–SD15 (system data). However, it must be noted that SD0–SD7 are provided at the 62-pin part in order to make it compatible with the original 8088-based PC/XT, while SD8–SD15 show up on the 36-pin part. This allows the 16-bit data bus to access any 16-bit peripheral. To select the upper byte or the lower byte of 16-bit data, we use BHE (bus high enable). BHE is latched and used on the system board and also provided at the expansion slot under SBHE (system bus high enable). We will see how to use this pin below.

Memory and I/O control signals

IOR and IOW are the two control signals used to access ports throughout the system. They show up on the 62-pin section of the ISA expansion slot. This makes them 8088 PC/XT compatible. We will discuss how these signals are used in peripheral interfacing in Chapter 11.

Signals MEMR, MEMW, SMEMR, and SMEMW are used to access memory. There is a reason for duplicate memory read and write signals. To allow access to any memory within the range of 16 megabytes, read/write control signals are provided to the 36-pin section of the ISA expansion slot under the designations of MEMR and MEMW, respectively.

To maintain compatibility with the original 8088-based PC/XT, MEMR and MEMW are designated as SMEMR and SMEMW and are provided on the 62-pin part of ISA on the same strip as the XT bus systems. In other words, MEMR and MEMW can be used to access memory in any location, but to access memory within the 1 megabyte range, we must use SMEMR and SMEMW on the 62-pin part of the ISA bus. In this case they can be used only to address memory locations 0–FFFFFH. Of course, to allow the same signals, MEMR and MEMW, from the support chip to show up in two distinctive places with two different names and functions requires some extra logic circuitry. Such details are buried inside the chipsets in today's PC.

Other control signals

Examining the ISA bus pins in Figure 9-18, we see numerous control signals that we have not seen before. The rest of the control signals in Figure 9-18 are related to the interrupt and DMA chips. IRQ and DMA signals are covered in subsequent chapters.

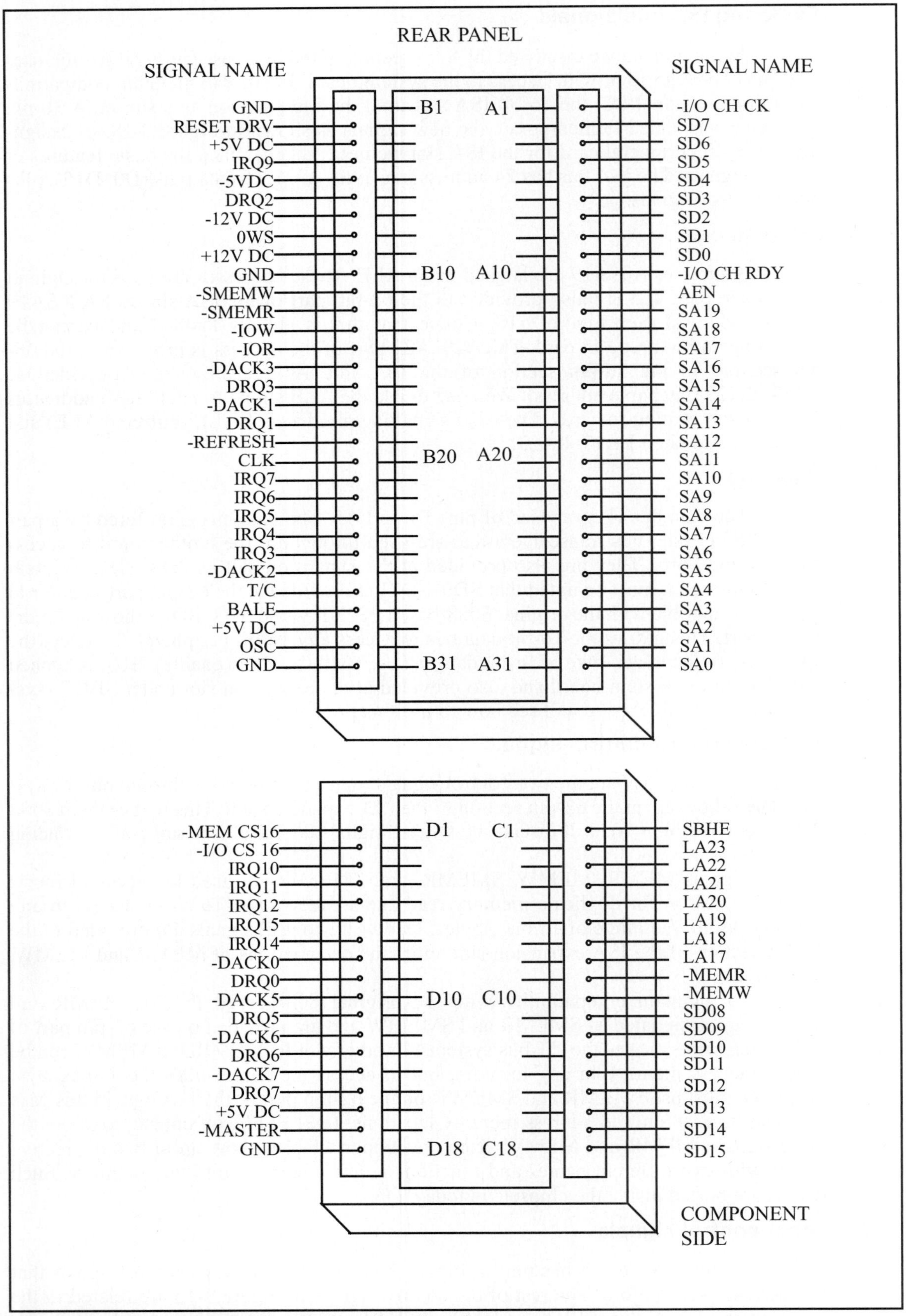

Figure 9-18. ISA (IBM PC AT) Bus Slot Signals

(Reprinted by permission from "IBM Technical Reference" c. 1985 by International Business Machines Corporation)

ODD and EVEN bytes and BHE

In the 36-pin section of the ISA bus there is a pin called SBHE that we explain next. Pin C1 is the same as the BHE pin from the 80286 that we studied in the last section. The BHE pin has to do with the differences between the 8-bit and 16-bit data bus CPUs. Like all general-purpose microprocessors, the memory (and I/O) space of x86 microprocessors is byte addressable. That means that every address location can provide a maximum of one byte of data. If the CPU has an 8-bit data bus, like the 8088, then the addresses are designated as 0 to FFFFFH, as shown in Figure 9-19. Notice in Figure 9-19 that the bus width for the data bus is only 8 bits. In other words, only 8 strips of wire connect the CPU's data bus to devices such as memory and I/O ports. Since the vast majority of memory and I/O devices also have an 8-bit data bus of D0–D7, their interfacing to CPUs with an 8-bit data bus is simple and straightforward. The CPU's D0–D7 data bus is connected directly to the D0–D7 data bus of memory and I/O devices. This is a perfect match. If the CPU has a 16-bit data bus, like 8086/80286/80386SX microprocessors, then the address spaces are designated as odd and even bytes, as shown in Figure 9-20. In such cases, the D0–D7 byte is designated as even and the D8–D15 byte as odd. To distinguish between odd and even bytes, 8086/286/386SX CPUs provide an extra signal called BHE (bus high enable). BHE, in association with the A0 pin, is used to select the odd or even byte according to Table 9-9. In Figure 9-20, notice the odd and even banks. They are called odd and even banks since the memory chips have only an 8-bit data bus of D0–D7 and two IC chips must be used, one for each byte. Although Figure 9-20 shows only 1 megabyte of memory space, the concept of odd and even bytes applies to the entire memory and I/O space of the x86 CPU. This is also the case for the 386 and 486 CPUs with a 32-bit data bus.

Table 9-9: Distinguishing Between Odd and Even Bytes

BHE	A0		
0	0	Even word	D0–D15
0	1	Odd byte	D8–D15
1	0	Even byte	D0–D7
1	1	None	

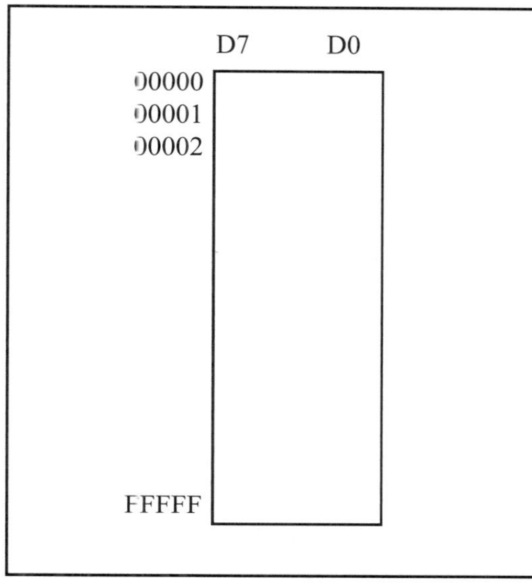

Figure 9-19. Memory Byte Addressing in 8088 (8-bit Data Bus)

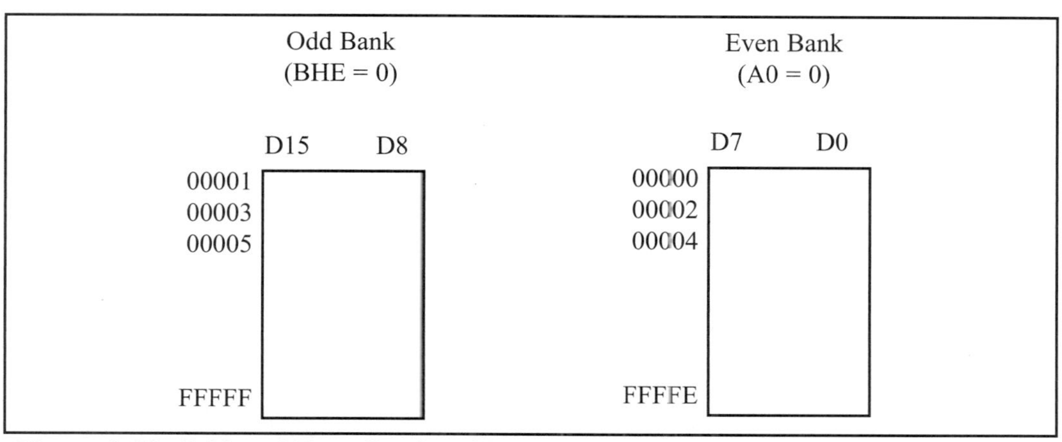

Figure 9-20. Odd and Even Banks of Memory in 16-bit CPUs (80286)

A20 gate and the case of high memory area (HMA)

A20 is an anomaly associated with 286 and higher microprocessors that needs to be discussed. In the 8088, when the segment register added to the offset is more than FFFFFH, it automatically wraps around and starts at 00000H. However, in 80286 and higher processors in real mode, such a wrap-around will not occur. Instead, the result will be 100000H, making A20 = 1. The problem is that A20–A23 is supposed to be activated only when the CPU is in protected mode. To control activation of A20, IBM used a latch controlled by the keyboard in the original PC AT; however, with the introduction of PS computers, control of A20 can also be handled by port 92H. One can use this A20 gate (as it is commonly called) to create a high memory area (HMA). This concept is important for understanding HMA memory in x86 PCs and is discussed in Chapter 25. See Examples 9-1 and 9-2 for clarification on this issue. Notice that the process of enabling and disabling the A20 gate in Figure 9-17 is handled by a piece of software called the *A20 handler*, which is provided with MS-DOS and Windows operating systems.

Example 9-1

(a) If the A20 gate is enabled, show the highest address that 286 (and higher processors) can access while still in real mode.
(b) How far high above 1M is this address?

Solution:

(a) To access the highest physical location in real mode, we must have CS = FFFFH and IP = FFFFH. We shift left the segment register CS and add the offset IP = FFFF:

CS shifted left one hex digit	FFFF0H
adding the offset IP	+ FFFFH
	10FFEFH

Therefore, the addresses FFFF0H–10FFEFH are the range that the CPU can access while it is in real mode. This is a total of 64K bytes.

(b) If the A20 gate is enabled, accessible memory locations above 1M are 100000 to 10FFEF. This is a total of 65,520 bytes, or 16 locations short of 1M + 64K.

Example 9-2

Assume that CS = FF25H. Find the lowest and highest physical addresses for
(a) the 8088 (b) the 80286
Specify the bit on A20.

Solution:

(a) The lowest physical address is FF250H and the highest is 0F24FH (FF250H + FFFFH); since there are only lines A0–A19 in the 8088, the 1 is dropped.
(b) In the 286 the lowest address is the same as in the 8088, but the highest physical address is 10F24F; therefore, A20 = 1.

Review Questions

1. The first IBM PC AT used the ___________ microprocessor.
2. What is the advantage of using the 74LS573 chip address latch over the 74LS373?
3. What is the purpose of the A20 handler circuitry?
4. What address area is called the high memory area in the 80286?
5. Since the control signals MEMR and MEMW are available on the 62-pin part of the expansion slot, why are they duplicated on the 36-pin part?

PROBLEMS

SECTION 9.1: 8088 MICROPROCESSOR

1. State the main differences between the 8088 and 8086 pinouts. Are the two chips interchangeable?
2. ALE is an ____________ (input, output) signal for the 8088.
3. What is the maximum number of bytes of memory addressable by the 8088, and why?
4. RESET is an _____________ (input, output) signal for the 8088.
5. When the 8088 uses the pins for addresses, they are _____________(input, output, both in and out), but when they are used for data, they will be _________ (input, output, both in and out).
6. To use a math coprocessor with the 8088, one must connect the 8088 in ______________ (maximum, minimum) mode.
7. True or false. An address must be latched from pins AD0–AD7 in the 8088.
8. Which of the following signals is provided by the 8088 CPU in minimum mode? (a) INTR (b) ALE (c) WR (d) IO/M
9. What is the advantage of demultiplexing address/data in the 8088 CPU?
10. What is the penalty (disadvantage) in terms of clocks in Problem 9?
11. ALE is activated in which T state?
12. Why are 8086-based systems more expensive compared to 8088-based systems?
13. To use the 8088 with the 8087 math coprocessor, is the minimum/maximum pin connected to low or high?
14. When the input signal RESET in the 8088 is activated, what are the contents of the IP and CS registers?

Use the following for the next three problems.

MEMR*	MEMW*
(a) 0	0
(b) 0	1
(c) 1	0
(d) 1	1

* Active low.

15. Which of the above control signals is activated during the memory read cycle?
16. Which one is activated during memory write?
17. Which of the above absolutely cannot happen at the same time?

SECTION 9.2: 8284 AND 8288 SUPPORTING CHIPS

18. In maximum mode in an 8088-based system, which chip provides the ALE signal, the 8088 or the 8288?
19. Which of the following signals are provided by the 8288 chip?
(a) IOR (b) RESET (c) IOW
(d) NMI (e) MEMR (f) MEMW

Use the following for the next three problems.

MEMR*	MEMW*
(a) 0	0
(b) 0	1
(c) 1	0
(d) 1	1

* Active low.

20. Which of the above control signals is activated during the memory read cycle?
21. Which one is activated during memory write?
22. Which of the above absolutely cannot happen at the same time?

SECTION 9.3: 8-BIT SECTION OF ISA BUS

23. When the computer is RESET, which master takes over, the 8088 or DMA?
24. To latch all the address bits of the 8088, how many 74LS373 chips are needed?
25. In the IBM PC, when AEN = 0 it indicates that the ____________ (8088 CPU, DMA) is in charge of the buses. Which controls the buses when AEN = 1?
26. Which chip is used for the following?
    (a) bidirectional bus buffering
    (b) unidirectional bus buffering
27. In the IBM PC, the 74LS373 is used for which of the following?
    (a) address latch (b) isolating the address bus
    (c) address bus boosting (d) all of the above
28. Draw a block diagram for the 8088 minimum mode connection to the 74LS373 and 74LS245. (Modify Figure 9-5.)
29. To access the buses for interfacing with the CPU, AEN must be _______ (low, high).
30. In the 74LS245, to allow the transfer of data from side A to side B, DIR = _______ and G = __________.
31. Answer Problem 30 if data is transferred from side B to side A.
32. In the 74LS245, what happens if G = 1 and DIR = 0?
33. The 74LS245 chip is used for ________ (address, data) buses.
34. To allow the passage of data through the 74LS373, G = _________ and OE = _________. When is the data actually latched?

SECTION 9.4: 80286 MICROPROCESSOR

35. True or false. The 80286 is available in both LCC and PGA packages.
36. The sales clerk at the local computer store says that the 80286 has 24 bits for address and 16 bits for data; therefore, it has 2^{24} times 2 bytes = 33,554,432 bytes = 32M memory space. Is this person right? Give justification for your answer.
37. When A0 = 0 and BHE = 0, which section of the data bus (the high byte or the low byte or both) is transferring information?
38. When A0 = 0, it makes the address an ___________ (odd, even) address.
39. True or false. If CLK is 20 MHz, the 80286-based system is a 10-MHz system.
40. True or false. The entire 16-megabyte memory space of the 80286 is accessible in real mode.
41. When power is applied to the 80286, it wakes up in _____________ (real, protected) mode.
42. Indicate the contents of CS, IP, DS, SS, and ES when power is applied to the 286.
43. In what physical address does the 286 look for the first opcode?
44. Justify your answer in Problem 43.

SECTION 9.5: 16-BIT ISA BUS

45. The ISA expansion slot of the 80286 has two parts. How many pins does each part have? State also the number of pins for A, B, C. Which side is the component side?
46. In the ISA bus, which part of the expansion slot provides signals A20–A23?
47. In the ISA bus, which part of the expansion slot provides signals D8–D15?
48. Why is D0–D7 provided on the 62-pin part of the expansion slot?
49. The BHE signal is provided on the ______ (62-pin, 36-pin) section of the ISA bus. Why?
50. If CS = FC48H and IP = 7652H, find the status of A20 for each member of the x86 family.
51. Which of the 8088 and 80286 microprocessors have the BHE pin?
52. To access memory anywhere in the 16M range, we must use ______ and _______ for the memory write and memory read control signals. Which part of the ISA bus provides them?

53. True or false. In the 8088, there is no A20 pin.
54. True or false. The 62-pin part of the ISA bus is almost the same as the 62-pin expansion slot of the original PC/XT.

ANSWERS TO REVIEW QUESTIONS

SECTION 9.1: 8088 MICROPROCESSOR

1. In the 8086, pins AD0–AD15 are used for the data bus; the 8088 has an 8-bit external data bus, pins AD0–AD7.
2. The ALE (address latch enable) pin signals whether the information is data or an address.
3. 4
4. Minimum
5. (a) both (b) output (c) output
6. IO/M = 0 and RD = 0
7. IO/M = 0 and WR = 0

SECTION 9.2: 8284 AND 8288 SUPPORTING CHIPS

1. Output, input
2. True
3. False
4. DTR

SECTION 9.3: 8-BIT SECTION OF ISA BUS

1. 8237 DMA
2. AEN
3. False
4. DT/R
5. Buffering

SECTION 9.4: 80286 MICROPROCESSOR

1. Real mode
2. 1 megabyte, 16 megabytes
3. CS = F000H and IP = FFF0

SECTION 9.5: 16-BIT ISA BUS

1. 80286
2. The advantage of the 75LS573 is that all outputs are on one side and all inputs on the other, which reduces noise in high-frequency systems and makes the circuit board easier to design.
3. The A20 handler circuitry allows control of the A20 address bit by software, thereby solving the problem associated with the A20 pin in the 80286.
4. 100000H–FFFFFFH
5. They are provided on the 36-pin part to allow access to extended memory.

CHAPTER 10

MEMORY AND MEMORY INTERFACING

OBJECTIVES

Upon completion of this chapter, you will be able to:

- **>> Define the terms *capacity*, *organization*, and *speed* as used in semiconductor memories**
- **>> Calculate the chip capacity and organization of semiconductor memory chips**
- **>> Compare and contrast the variations of ROM: PROM, EPROM, EEPROM, Flash EPROM, and mask ROM**
- **>> Compare and contrast the variations of RAM: SRAM, DRAM, and NV-DRAM**
- **>> Diagram methods of address decoding for memory chips**
- **>> Diagram the memory map of the IBM PC in terms of RAM, VDR, and ROM allocation**
- **>> Describe the checksum method of ensuring data integrity in ROM**
- **>> Describe the parity bit method of ensuring data integrity in DRAM**
- **>> Describe 16-bit memory design and related issues**

This chapter explores memory and memory interfacing of the x86 PC. We first study the basics of semiconductor memory chips, then in Section 10.2 we present memory address decoding using simple logic gates. The memory map and memory space allocation of the PC are discussed in Section 10.3. Section 10.4 explores the issue of data integrity in RAM and ROM. Section 10.5 discusses the CPU's bus cycle time for memory and shows how to calculate bus bandwidth.

SECTION 10.1: SEMICONDUCTOR MEMORIES

In the design of all computers, semiconductor memories are used as primary storage for code and data. Semiconductor memories are connected directly to the CPU and they are the memory that the CPU first asks for information (code and data). For this reason, semiconductor memories are sometimes referred to as *primary memory.* The main requirement of primary memory is that it must be fast in responding to the CPU; only semiconductor memories can do that. Among the most widely used semiconductor memories are ROM and RAM. Before we discuss different types of RAM and ROM, we discuss terminology common to all semiconductor memories, such as capacity, organization, and speed.

Memory capacity

The number of bits that a semiconductor memory chip can store is called its *chip capacity*. It can be in units of K bits (kilobits), M bits (megabits), and so on. This must be distinguished from the storage capacity of computers. While the memory capacity of a memory IC chip is always given in bits, the memory capacity of a computer is given in bytes. For example, an article in a technical journal may state that the 64M chip has become popular. In that case, although it is not mentioned that 64M means 64 megabits, it is understood since the article is referring to an IC memory chip. However, if an advertisement states that a computer comes with 64M memory, since it is referring to a computer it is understood that 64M means 64 megabytes.

Memory organization

Memory chips are organized into a number of locations within the IC. Each location can hold 1 bit, 4 bits, 8 bits, or even 16 bits, depending on how it is designed internally. The number of bits that each location within the memory chip can hold is always equal to the number of data pins on the chip. How many locations exist inside a memory chip depends on the number of address pins. The number of locations within a memory IC always equals 2^x where x is the number of address pins. Therefore, the total number of bits that a memory chip can store is equal to the number of locations times the number of data bits per location. To summarize:

1. Each memory chip contains 2^x locations, where x is the number of address pins on the chip.
2. Each location contains y bits, where y is the number of data pins on the chip.
3. The entire chip will contain $2^x \times y$ bits, where x is the number of address pins and y is the number of data pins on the chip.
4. The $2^x \times y$ is referred to as the *organization* of the memory chip, where x is the number of address pins and y is the number of data pins on the chip.
5. For 2^x, use Table 10-1 to give the number of locations in K or M units.
6. $2^{10} = 1024 = 1K$. Notice that in common speech, 1K is 1000 (as in discussing salaries or distance), but in computer terminology it is 1024.

Speed

One of the most important characteristics of a memory chip is the speed at which data can be accessed from it. To access the data, the address is presented to the address pins, and after a certain amount of time has elapsed, the data shows up at the data pins.

The shorter this elapsed time, the better, and consequently, the more expensive the memory chip. The speed of the memory chip is commonly referred to as its *access time*. The access time of memory chips varies from a few nanoseconds to hundreds of nanoseconds, depending on the IC technology used in the design and fabrication.

The three important memory characteristics of capacity, organization, and access time will be used extensively in this chapter and throughout the book. Many of these topics will be explored in more detail in the context of applications in this and future chapters. Table 10-1 serves as a reference for the calculation of memory organization. See Examples 10-1 and 10-2 for clarification.

Table 10-1: Powers of 2

x	2^x
10	1K
11	2K
12	4K
13	8K
14	16K
15	32K
16	64K
17	128K
18	256K
19	512K
20	1M
21	2M
22	4M
23	8M
24	16M

ROM (read-only memory)

ROM is a type of memory that does not lose its contents when the power is turned off. For this reason, ROM is also called *nonvolatile memory*. There are different types of read-only memory, such as PROM, EPROM, EEPROM, Flash ROM, and mask ROM. Each is explained next.

Example 10-1

A given memory chip has 12 address pins and 8 data pins. Find:
(a) the organization (b) the capacity

Solution:

(a) This memory chip has 4096 locations ($2^{12} = 4096$), and each location can hold 8 bits of data. This gives an organization of 4096×8, often represented as $4K \times 8$.
(b) The capacity is equal to 32K bits since there is a total of 4K locations and each location can hold 8 bits of data.

Example 10-2

A 512K memory chip has 8 pins for data. Find:
(a) the organization (b) the number of address pins for this memory chip

Solution:

(a) A memory chip with 8 data pins means that each location within the chip can hold 8 bits of data. To find the number of locations within this memory chip, divide the capacity by the number of data pins. 512K/8 = 64K; therefore, the organization for this memory chip is $64K \times 8$.
(b) The chip has 16 address lines since $2^{16} = 64K$.

PROM (programmable ROM) or OTP ROM

PROM refers to the kind of ROM that the user can burn information into. In other words, PROM is a user-programmable memory. For every bit of the PROM, there exists a fuse. PROM is programmed by blowing the fuses. If the information burned into PROM is wrong, that PROM must be discarded since internal fuses are blown permanently. For this reason, PROM is also referred to as *OTP* (one-time programmable). The process of programming ROM is also called *burning* ROM and requires special equipment called a ROM burner or ROM programmer.

EPROM (erasable programmable ROM)

EPROM was invented to allow changes in the contents of PROM after it is burned. In EPROM, one can program the memory chip and erase it thousands of times. This is especially useful during development of the prototype of a microprocessor-based project. The only problem with EPROM is that erasing its contents can take up to 20 minutes. All EPROM chips have a window that is used to shine ultraviolet (UV) radiation to erase the chip's contents. For this reason, EPROM is also referred to as UV-erasable EPROM or simply UV-EPROM. Figure 10-1 shows the pins for a 64K-bit UV-EPROM chip. Notice the A0–A12 address pins and O0–O7 (output) for D0–D7 data pins. The OE (out enable) is for the read signal.

To program a UV-EPROM chip, the following steps must be taken:

1. Its contents must be erased. To erase a chip, remove it from its socket on the system board and place it in EPROM erasure equipment to expose it to UV radiation for 15–20 minutes.
2. Program the chip. To program a UV-EPROM chip, place it in the ROM burner (programmer). To burn code and data into EPROM, the ROM burner uses 12.5 volts or higher, depending on the EPROM type. This voltage is referred to as VPP in the UV-EPROM data sheet.
3. Place the chip back into its socket on the system board.

As can be seen from the above steps, in the same way that there is an EPROM programmer (burner), there is also separate EPROM erasure equipment. The main problem, and indeed the major disadvantage of UV-EPROM, is that it cannot be programmed while in the system board (motherboard). To find a solution to this problem, EEPROM was invented.

EEPROM (electrically erasable programmable ROM)

EEPROM has several advantages over EPROM, such as the fact that its method of erasure is electrical and therefore instant, as opposed to the 20-minute erasure time required for UV-EPROM. In addition, in EEPROM, one can select which byte to be erased, in contrast to UV-EPROM, in which the entire contents of ROM are erased. However, the main advantage of EEPROM is the fact that one can program and erase its contents while it is still in the system board. It does not require physical removal of the memory chip from its socket. In other words, unlike UV-EPROM, EEPROM does not require an external erasure and programming device. To utilize EEPROM fully, the designer must incorporate into the system board the circuitry to program the EEPROM, using 12.5 V for VPP. EEPROM with VPP of 5–7 V is available, but it is more expensive. In general, the cost per bit for EEPROM is much higher than for UV-EPROM.

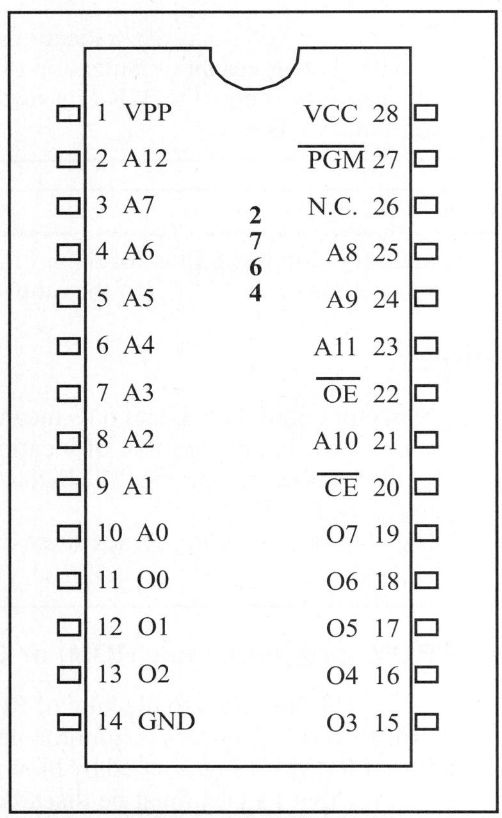

Figure 10-1. UV-EPROM Chip
(Reprinted by permission of Intel Corporation, Copyright Intel Corp., 1987)

Table 10-2 shows examples of some popular ROM chips and their characteristics. Notice the patterns of the IC numbers. For example, 27128-20 refers to UV-EPROM that has a capacity of 128K bits and access time of 200 nanosec-

Table 10-2: Examples of ROM Memory Chips

Type	Part Number	Speed (ns)	Capacity	Organization	Pins	VPP
UV-EPROM	2716	450	16K	2K × 8	24	25
	2716-1	350	16K	2K × 8	24	25
	2716B	450	16K	2K × 8	24	12.5
	2732A-45	450	32K	4K × 8	24	21
	2732A-20	200	32K	4K × 8	24	21
	27C32	450	32K	4K × 8	24	25
	2764A-25	250	64K	8K × 8	28	12.5
	27C64-15	150	64K	8K × 8	28	12.5
	27128-20	200	128K	16K × 8	28	12.5
	27C128-25	250	128K	16K × 8	28	12.5
	27256-20	200	256K	32K × 8	28	12.5
	27C256-20	200	256K	32K × 8	28	12.5
	27512-25	250	512K	64K × 8	28	12.5
	27C512-25	250	512K	64K × 8	28	12.5
	27C010-12	120	1M	128K × 8	32	12.5
	27C201-12	120	2M	256K × 8	32	12.5
	27C401-12	120	4M	512K × 8	32	12.5
EEPROM	28C16A-25	250	16K	2K × 8	24	5
	2864A	250	64K	8K × 8	28	5
	28C256-15	150	256K	32K × 8	28	5
	28C256-25	250	256	32 × 8	28	5
Flash ROM	28F256-20	200	256K	32K × 8	32	12
	28F256-15	150	256K	32K × 8	32	12
	28F010-20	200	1M	128K × 8	32	12
	28F020-15	150	2M	256K × 8	32	12

onds. The capacity of the memory chip is indicated in the part number and the access time is given with a zero dropped. In part numbers, C refers to CMOS technology. While 27xx is for UV-E PROM, 28xx is for EEPROM. See Figure 10-2.

Flash memory

Since the early 1990s, Flash ROM has become a popular user-programmable memory chip, and for good reasons. First, the process of erasure of the entire contents takes only a few seconds, or one might say in a flash, hence its name: Flash memory. In addition, the erasure method is electrical and for this reason it is sometimes referred to as Flash EEPROM. To avoid confusion, it is commonly called Flash ROM. The major difference between EEPROM and Flash memory is the fact that when Flash memory's contents are erased the entire device is erased, in contrast to EEPROM, where one can erase a desired section or byte. Although there are some Flash memories recently made available in which the contents are divided into blocks and the erasure can be done block by block, unlike EEPROM, no byte erasure option is available. Because Flash ROM can be programmed while it is in its socket on the system board, it is widely used to upgrade the BIOS ROM of the PC or the operating system on Cisco routers. Some designers believe that Flash memory will replace the hard disk as a mass storage medium. This would increase the performance of computers tremendously, since Flash memory is semiconductor memory with access time in the range of 100 ns compared with disk access time in the range of tens of milliseconds. For this to happen, Flash memory's program/erase cycles must become infinite, just like hard disks. *Program/erase cycle* refers to the number of times that a chip can be erased and programmed before it becomes unusable. At this time, the program/erase cycle is 500,000 for Flash and EEPROM, 2000 for UV-EPROM, and infinite for RAM and disks. In Table 10-2, notice that the part number for Flash ROM uses the 28Fxx designation, where F indicates the Flash type ROM.

Example 10-3

For ROM chip 27128, find the number of data and address pins, using Table 10-2.

Solution:

The 27128 has a capacity of 128K bits. Table 10-2 also shows that it has 16K × 8 organization, which indicates that there are 8 pins for data, and 14 pins for address (2^{14} = 16K).

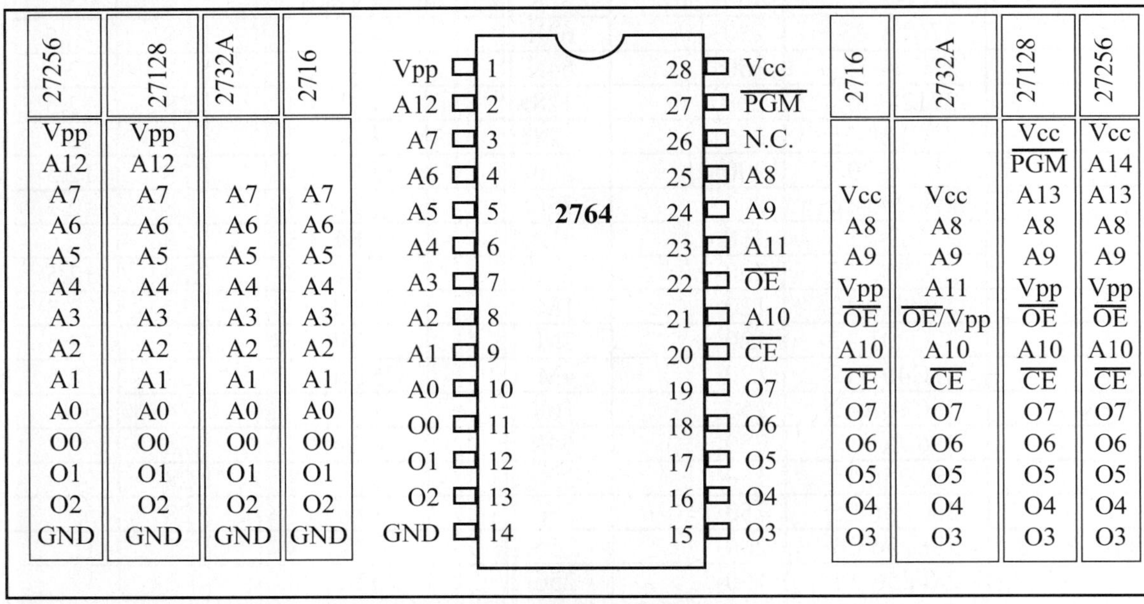

Figure 10-2. Pin Configurations for 27xx ROM Family

Mask ROM

Mask ROM refers to a kind of ROM whose contents are programmed by the IC manufacturer. In other words, it is not a user-programmable ROM. The terminology *mask* is used in IC fabrication. Since the process is costly, mask ROM is used when the needed volume is high and it is absolutely certain that the contents will not change. It is common practice to use UV-EPROM or Flash for the development phase of a project, and only after the code/data have been finalized is mask ROM ordered. The main advantage of mask ROM is its cost, since it is significantly cheaper than other kinds of ROM, but if an error in the data is found, the entire batch must be thrown away.

RAM (random access memory)

RAM memory is called *volatile memory* since cutting off the power to the IC will mean the loss of data. Sometimes RAM is also referred to as RAWM (read and write memory), in contrast to ROM, which cannot be written to. There are three types of RAM: static RAM (SRAM), dynamic RAM (DRAM), and NV-RAM (nonvolatile RAM). Each is explained separately.

SRAM (static RAM)

Storage cells in static RAM memory are made of flip-flops and therefore do not require refreshing in order to keep their data. This is in contrast to DRAM, discussed below. The problem with the use of flip-flops for storage cells is that each cell requires at least 6 transistors to build, and the cell holds only 1 bit of data. In recent years, the cells have been made of 4 transistors, which is still too many. The use of 4-transistor cells plus the use of CMOS technology has given birth to a high-capacity SRAM, but the capacity of SRAM is far below DRAM. Table 10-3 shows some examples of SRAM. SRAMs are widely used for cache memory, which is discussed in Chapter 22. Figure 10-3 shows the pin diagram for the 6116 SRAM chip. The 6116 has an organization of 2K × 8, which gives a capacity of 16K bits, as indicated in the part number. The following is a description of the 6116 SRAM pins.

A0–A10 are for address inputs, where 11 address lines gives $2^{11} = 2K$.
I/O0–I/O7 are for data I/O, where 8-bit data lines give an organization of 2K × 8.
WE (write enable) is for writing data into SRAM (active low).
OE (output enable) is for reading data out of SRAM (active low)
CS (chip select) is used to select the memory chip.
The functional diagram for the 6116 SRAM is given in Figure 10-4.

Figure 10-5 shows the following steps to write data into SRAM.

1. Provide the addresses to pins A0–A10.
2. Activate the CS pin.
3. Make WE = 0 while RD = 1.
4. Provide the data to pins I/O0–I/O7.
5. Make WE = 1 and data will be written into SRAM on the positive edge of the WE signal.

The following are steps to read data from SRAM. See Figure 10-6.

1. Provide the addresses to pins A0–A10. This is the start of the access time (t_{AA}).
2. Activate the CS pin.
3. While WE = 1, a high-to-low pulse on the OE pin will read the data out of the chip.

In the 6116 SRAM, the access time, t_{AA}, is measured as the time elapsed from the moment the address is provided to the address pins to the moment that the data is available at the data pins. The speed for the 6116 chip can vary from 100 ns to 15 ns.

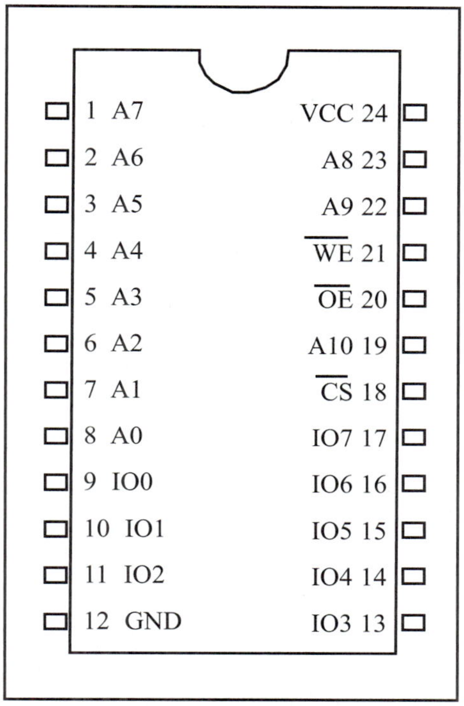

Figure 10-3. 6116 2K × 8 SRAM

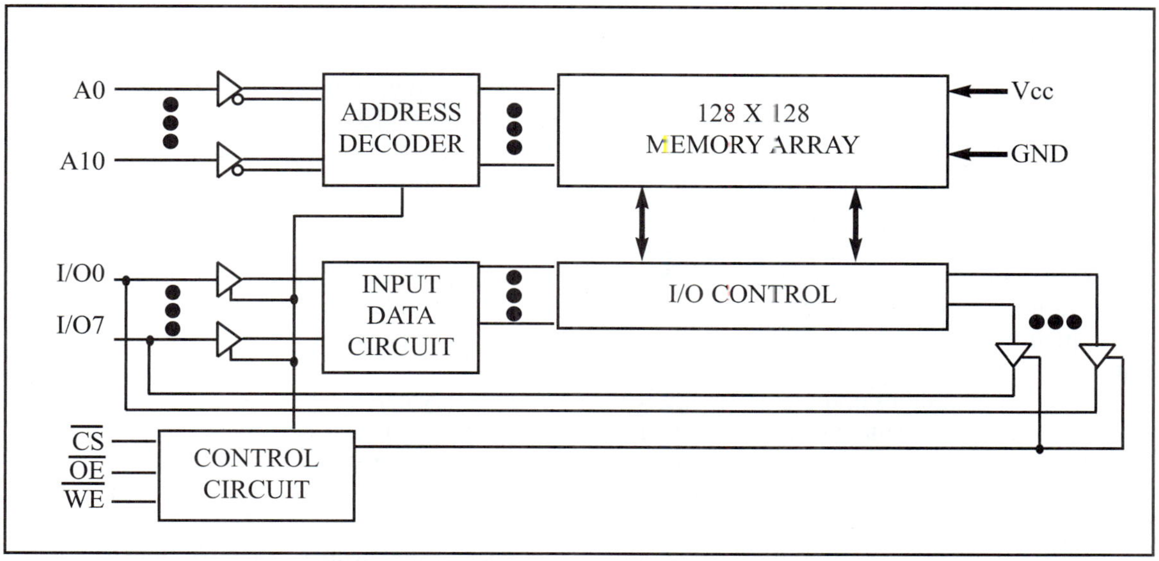

Figure 10-4. Functional Block Diagram for 6116 SRAM

Examine the read cycle time for SRAM in Figure 10-6. The read cycle time (t_{RC}) is defined as the minimum amount of time required to read one byte of data, that is, from the moment we apply the address of the byte to the moment we can begin the next read operation. In SRAM for which t_{AA} = 100 ns, t_{RC} is also 100 ns. This implies that we can read the contents of consecutive address locations with each taking no more than 100 ns. Hence, in SRAM and ROM, $t_{AA} = t_{RC}$. They are not equal in DRAM, as we will find in Chapter 22.

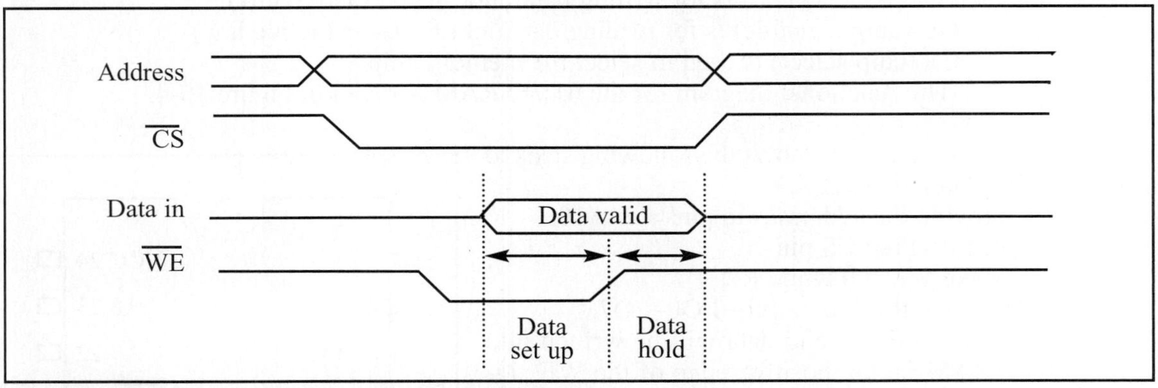

Figure 10-5. Memory Write Timing for SRAM

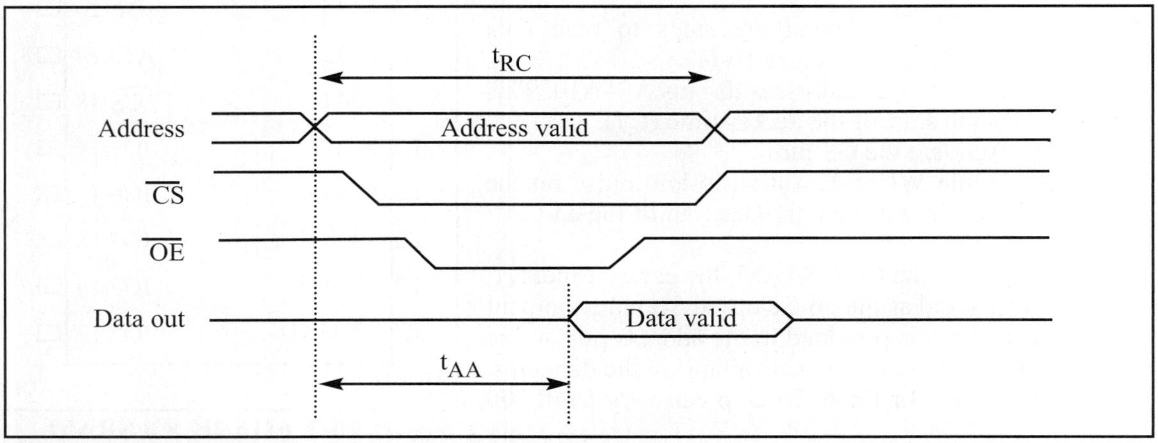

Figure 10-6. Memory Read Timing for SRAM

DRAM (dynamic RAM)

Since the early days of the computer, the need for huge, inexpensive read/write memory was a major preoccupation of computer designers. In 1970, Intel Corporation introduced the first dynamic RAM (random access memory). Its density (capacity) was 1024 bits and it used a capacitor to store each bit. The use of a capacitor as a means to store data cuts down the number of transistors needed to build the cell; however, it requires constant refreshing due to leakage. This is in contrast to SRAM (static RAM), whose individual cells are made of flip-flops. Since each bit in SRAM uses a single flip-flop and each flip-flop requires 6 transistors, SRAM has much larger memory cells and consequently lower density. The use of capacitors as storage cells in DRAM results in much smaller net memory cell size.

The advantages and disadvantages of DRAM memory can be summarized as follows. The major advantages are high density (capacity), cheaper cost per bit, and lower power consumption per bit. The disadvantage is that it must be refreshed periodically, due to the fact that the capacitor cell loses its charge; furthermore, while it is being refreshed, the data cannot be accessed. This is in contrast to SRAM's flip-flops, which retain data as long as the power is on, which do not need to be refreshed, and whose contents can be accessed at any time. Since 1970, the capacity of DRAM has exploded. After the 1K-bit

(1024) chip came the 4K-bit in 1973, and then the 16K chip in 1976. The 1980s saw the introduction of 64K, 256K, and finally 1M and 4M memory chips. The 1990s saw the 16M, 64M, and 256M DRAM chips. In 1980 when the IBM PC was being designed, 16K-bit chips were widely used, but currently motherboards use 64M, 256M, and 1G chips. Keep in mind that when talking about IC memory chips, the capacity is always assumed to be in bits. Therefore, a 1G chip means a 1-gigabit chip and a 256M chip means a 256M-bit chip. However, when talking about the memory of a computer system, it is always assumed to be in bytes. For example, if we say that a PC motherboard has 1G, it means 1G bytes of memory.

Packaging issue in DRAM

In DRAM it is difficult to pack a large number of cells into a single chip with the normal number of pins assigned to addresses. For example, a 64K-bit chip (64K × 1) must have 16 address lines and 1 data line, requiring 16 pins to send in the address if the conventional method is used. This is in addition to VCC power, ground, and read/write control pins. Using the conventional method of data access, the large number of pins defeats the purpose of high density and small packaging, so dearly cherished by IC designers. Therefore, to reduce the number of pins needed for addresses, multiplexing/demultiplexing is used. The method used is to split the address into halves and send in each half of the address through the same pins, thereby requiring fewer address pins. Internally, the DRAM structure is divided into a square of rows and columns. The first half of the address is called the *row* and the second half is called the *column*. For example, in the case of DRAM of 64K × 1 organization, the first half of the address is sent in through the 8 pins A0–A7, and by activating RAS (row address strobe), the internal latches inside DRAM grab the first half of the address. After that, the second half of the address is sent in through the same pins and by activating CAS (column address strobe), the internal latches inside DRAM latch this second half of the address. This results in using 8 pins for addresses plus RAS and CAS, for a total of 10 pins, instead of the 16 pins that would be required without multiplexing. To access a bit of data from DRAM, both row and column addresses must be provided. For this concept to work, there must be a 2-by-1 multiplexer outside the DRAM circuitry while the DRAM chip has its own internal demultiplexer. Due to the complexities associated with DRAM interfacing (RAS, CAS, the need for external multiplexer and refreshing circuitry), many small microprocessor-based projects that do not require much RAM use SRAM instead of DRAM. Figure 10-7 shows the pins for a DRAM chip. Notice the RAS and CAS pins. Also notice the WE (write enable) pin for read and write actions. Table 10-3 provides some examples of DRAM chips.

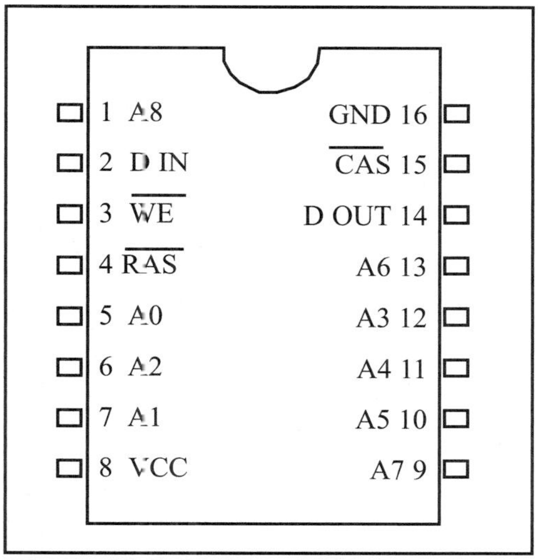

Figure 10-7. 256K × 1 DRAM

DRAM, SRAM, and ROM organizations

Although the organizations for SRAMs and ROMs are always × 8, DRAM can have × 1, × 4, × 8, or even × 16 organizations. In some memory chips (notably SRAM), the data pins are called I/O. In some DRAMs there are separate pins Din and Dout. The DRAMs with x1 organization are widely used for parity bit as we will soon see in this chapter. See Examples 10-4 and 10-5.

Table 10-3: Examples of RAM Chips

Type	Part Number	Speed (ns)	Capacity	Organization	Pins
SRAM	6116-1	100	16K	2K × 8	24
	6116LP-70*	70	16K	2K × 8	24
	6264-10	100	64K	8K × 8	28
	62256LP-10*	100	256K	32K × 8	28
DRAM	4116-20	200	16K	16K × 1	16
	4116-15	150	16K	16K × 1	16
	4116-12	120	16K	16K × 1	16
	4416-12	120	64K	16K × 4	18
	4416-15	150	64K	16K × 4	18
	4164-15	150	64K	64K × 1	16
	41464-8	80	256K	64K × 4	18
	41256-15	150	256K	256K × 1	16
	41256-6	60	256K	256K × 1	16
	414256-10	100	1M	256K × 4	20
	511000P-8	80	1M	1M × 1	18
	514100-7	700	4M	4M × 1	20
NV-SRAM	DS1220	100	16K	2K × 8	24
	DS1225	150	64K	8K × 8	28
	DS1230	70	256K	32K × 8	28

* LP indicates low power.

Example 10-4

Show possible organizations and number of address pins for the: (a) 256K DRAM chip, and (b) 1M DRAM chip.

Solution:

(a) For 256K chips, possible organizations are 256K × 1 or 64K × 4. In the case of 256K × 1, there are 256K locations and each location inside DRAM provides 1 bit. The 256K locations are accessed through the 18-bit address A0–A17 since 2^{18} = 256K. The chip has only A0–A8 physical pins plus RAS and CAS and one pin for data in addition to VCC, ground, and the R/W pin that every DRAM chip must have. For 64K × 4, it requires 16 address bits to access each location (2^{16} = 64K), and each location inside the DRAM has 4 cells. That means that it must have 4 data pins, D0–D3, 8 address pins, A0–A7, plus RAS and CAS.

(b) In the case of a 1M chip, there can be either 1M × 1 or 256K × 4 organizations. For 1M × 1, there are A0–A9, 10 pins, to access 2^{20} = 1M locations with the help of RAS and CAS and one pin for data. The 256K × 4 has 9 (A0–A8) and 4 (D0–D3) pins, respectively, for address and data plus RAS and CAS pins.

Example 10-5

Discuss the number of pins set aside for addresses in each of the following memory chips.
(a) 16K × 4 DRAM (b) 16K × 8 SRAM

Solution:
Since 2^{14} = 16K:
(a) For DRAM we have 7 pins (A0–A6) for the address pins and 2 pins for RAS and CAS.
(b) For SRAM we have 14 pins (A0–A13) for address and no pins for RAS and CAS since they are associated only with DRAM.

NV-RAM (nonvolatile RAM)

While both DRAM and SRAM are volatile, there is a new type of RAM called *NV-RAM*, nonvolatile RAM. Like other RAMs, it allows the CPU to read and write to it; but when the power is turned off, the contents are not lost, just as for ROM. NV-RAM combines the best of RAM and ROM: the read and writability of RAM, plus the nonvolatility of ROM. To retain its contents, every NV-RAM chip internally is made of the following components:

1. It uses extremely power-efficient (extremely low power consumption) SRAM cells built out of CMOS.
2. It uses an internal lithium battery as a backup energy source.
3. It uses an intelligent control circuitry. The main job of this control circuitry is to monitor the VCC pin constantly to detect loss of the external power supply. If the power to the VCC pin falls below out-of-tolerance conditions, the control circuitry switches automatically to its internal power source, the lithium battery. In this way, the internal lithium power source is used to retain the NV-RAM contents only when the external power source is off.

It must be emphasized that all three of the components above are incorporated into a single IC chip, and for this reason nonvolatile RAM is much more expensive than SRAM as far as cost per bit is concerned. Offsetting the cost, however, is the fact that it can retain its contents up to ten years after the power has been turned off and allows one to read and write exactly the same as in SRAM. See Table 10-3 for NV-RAM parts made by Dallas Semiconductor. In the x86 PC, NV-RAM is used to save the system setup. This NV-RAM in PC is commonly referred to as *CMOS* RAM.

Review Questions

1. The speed of semiconductor memory is in the range of ____________.
2. Find the organization and chip capacity for each of the following with the indicated number of address and data pins.
   (a) 11 address, 8 data SRAM (b) 13 address, 8 data ROM
   (c) 8 address, 4 data DRAM (d) 9 address, 1 data DRAM
3. Find the capacity and number of pins set aside for address and data for memory chips with the following organizations.
   (a) 16K × 8 SRAM (b) 32K × 8 EPROM (c) 1M × 1 DRAM
   (d) 256K × 4 DRAM (e) 64K × 8 EEPROM (f) 1M × 4 DRAM
4. Why is Flash memory preferable to UV-EPROM in system development?
5. What kind of memory is used in the CMOS RAM of the x86 PC?

SECTION 10.2: MEMORY ADDRESS DECODING

Current system designs use CPLDs (complex programmable logic devices), in which memory and address decoding circuitry are integrated into one programmable chip. However, it is still important to understand how this task can be performed with common logic gates. In this section we show how to use simple logic gates to accomplish address decoding. The CPU provides the address of the data desired, but it is the job of the decoding circuitry to locate the memory chip where the desired data is stored. To explore the concept of decoding circuitry, we look at the use of NAND and 74LS138 chips as decoders. In this discussion we use SRAM or ROM for the sake of simplicity.

Simple logic gate as address decoder

As seen in the last section, memory chips have one or more CS (chip select) pins that must be activated for the memory's contents to be accessed. Sometimes the chip select is also referred to as chip enable (CE). In connecting a memory chip to the CPU, the data

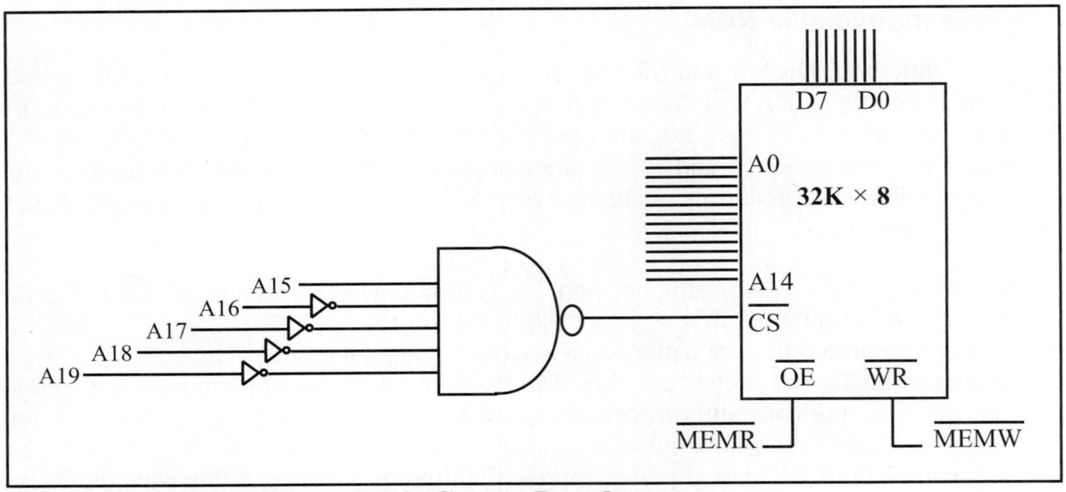

Figure 10-8. Using Simple Logic Gate as Decoder

A19				A0	
0000	1000	0000	0000	0000	= 08000H address of the first location
0000	1111	1111	1111	1111	= 0FFFFH address of the last location

Figure 10-9. Address Range Assigned to Memory Chip in Figure 10-8

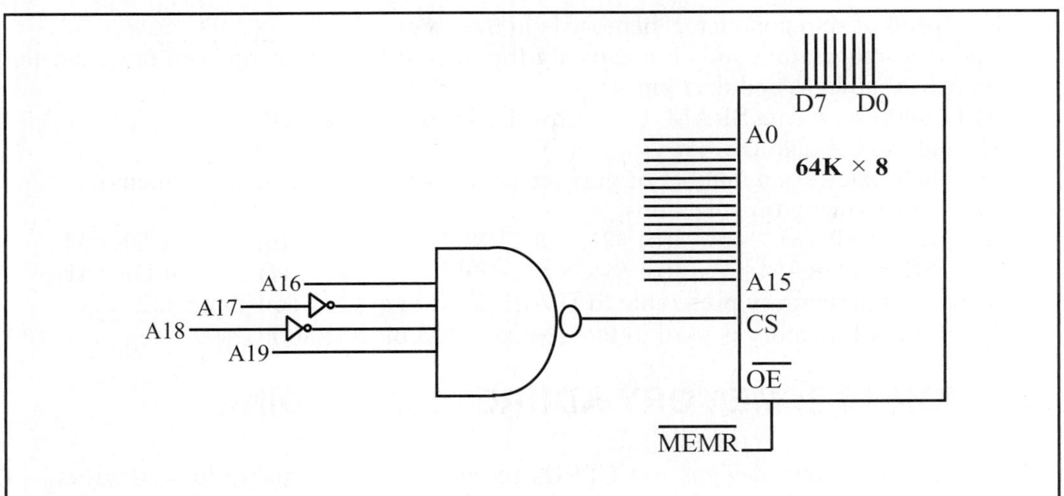

A19				A0	
1001	0000	0000	0000	0000	= 90000H address of the first location
1001	1111	1111	1111	1111	= 9FFFFH address of the last location

Figure 10-10. Decoder and Its Associated Address Range

bus is connected directly to the data pins of the memory. Control signals MEMR and MEMW are connected to the OE and WR pins of the memory chip, respectively (see Figure 10-8). In the case of the address buses, while the lower bits of the address go directly to the memory chip address pins, the upper ones are used to activate the CS pin of the memory chip. It is the CS pin along with RD/WR that allows the flow of data in or out of the memory chip. In other words, no data can be written into or read from the memory chip unless CS is activated. The CS input is active low and can be activated using some simple logic gates, such as NAND and inverters. See Figures 10-8, 10-9, and 10-10. Example 10-6 shows the address range calculation for Figure 10-10.

Example 10-6

Referring to Figure 10-10 we see that the memory chip has 64K bytes of space. Show the calculation that verifies that address range 90000 to 9FFFFH is comprised of 64K bytes.

Solution:

To calculate the total number of bytes for a given memory address range, subtract the two addresses and add 1 to get the total bytes in hex. Then the hex number is converted to decimal and divided by 1024 to get K bytes.

```
 9FFFF          FFFF
-90000       +     1
 0FFFF         10000 hex = 65,536 decimal = 64K
```

In Figure 10-10, notice that the output of the NAND gate is active low and that the CS pin is also active low. That makes them a perfect match. Also notice that A19–A16 must equal 1001 in order for CS to be activated. This results in the assignment of addresses 9000H to 9FFFFH to this memory block. Figures 10-8 and 10-10 show that for every block of memory, we need a NAND gate. The 74LS138 has 8 NAND gates in it; therefore, a single chip can control 8 blocks of memory. This was the method of memory addressing decoding used before the introduction of CPLD, and it is still the best method if you do not have access to CPLD.

Using the 74LS138 as decoder

In the absence of CPLD or FPGA as address decoders, the 74LS138 chip is an excellent choice. The 3 inputs A, B, and C of the 74LS138 generate 8 active-low outputs Y0–Y7, as shown in Figure 10-11. Each Y output is connected to the CS of a memory chip, allowing control of 8 memory blocks by a single 74LS138. This eliminates the need for using NAND and inverter gates. As shown in Figure 10-11, where A, B, and C select which output is activated, there are three additional inputs, G2A, G2B, and G1, that can be used for address or control signal selection. Notice that G2A and G2B are both active low, while G1 is active high. If any one of the inputs G1, G2A,

Block Diagram

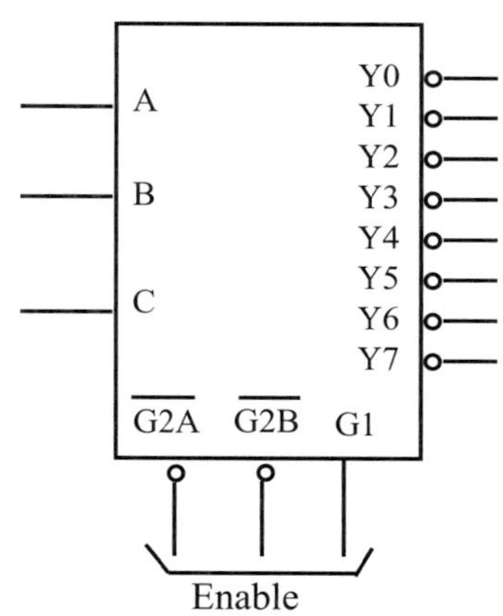

Function Table

Inputs: Enable G1 G2	Inputs: Select C B A	Outputs Y0 Y1 Y2 Y3 Y4 Y5 Y6 Y7
X H	X X X	H H H H H H H H
L X	X X X	H H H H H H H H
H L	L L L	L H H H H H H H
H L	L L H	H L H H H H H H
H L	L H L	H H L H H H H H
H L	L H H	H H H L H H H H
H L	H L L	H H H H L H H H
H L	H L H	H H H H H L H H
H L	H H L	H H H H H H L H
H L	H H H	H H H H H H H L

Figure 10-11. 74LS138 Decoder
(Reprinted by permission of Texas Instruments, Copyright Texas Instruments, 1988)

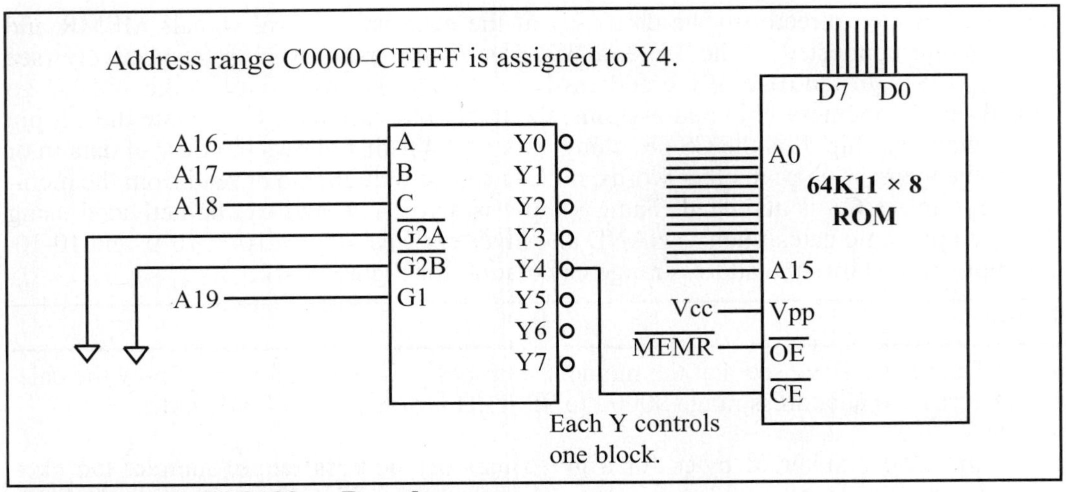

Figure 10-12. 74LS138 as Decoder

or G2B is not connected, they must be activated permanently by either VCC or ground, depending on the activation level.

In Figure 10-12, we have A0–A15 going from the CPU directly to A0–A15 of the memory chip. A16–A18 are used for the A, B, and C inputs of the 74LS138. A19 is controlling the G1 pin of the 74138. For the 74138 to be enabled, we need G2A = 0, G2B = 0, and G1 = 1. G2A and G2B are grounded. When G1 = 1, this 74138 is selected. Depending on the status of pins A, B, and C, one of the Ys is selected. To select Y4, we need CBA = 100 (in binary). That gives us the address range of C0000 to CFFFFH for the memory chip controlled by the Y4 output. For further clarification, see Example 10-7 and Figure 10-13.

Example 10-7

Looking at the design in Figure 10-13, find the address range for (a) Y4, (b) Y2, and (c) Y7, and verify the block size controlled by each Y.

Solution:

(a) The address range for Y4 is calculated as follows.

A19	A18	A17	A16	A15	A14	A13	A12	A11	A10	A9	A8	A7	A6	A5	A4	A3	A2	A1	A0
1	1	1	1	0	0	0	0	0	0	0	0	0	0	0	0	0	0	0	0
1	1	1	1	0	0	1	1	1	1	1	1	1	1	1	1	1	1	1	1

The above shows that the range for Y4 is F0000H to F3FFFH. In Figure 10-13, notice that A19, A18, and A17 must be 1 for the decoder to be activated. Y4 will be selected when A16 A15 A14 = 100 (4 in binary). The remaining A13–A0 will be 0 for the lowest address and 1 for the highest address.

(b) The address range for Y2 is E8000H to EBFFFH.

A19	A18	A17	A16	A15	A14	A13	A12	A11	A10	A9	A8	A7	A6	A5	A4	A3	A2	A1	A0
1	1	1	0	1	0	0	0	0	0	0	0	0	0	0	0	0	0	0	0
1	1	1	0	1	0	1	1	1	1	1	1	1	1	1	1	1	1	1	1

(c) The address range for Y7 is FC000H to FFFFFH. Notice that FFFFF – FC000H = 3FFFH, which is equal to 16,383 in decimal. Adding 1 to it because of the 0 location, we have 16,384. 16,384/1024 = 16K, the block (chip) size.

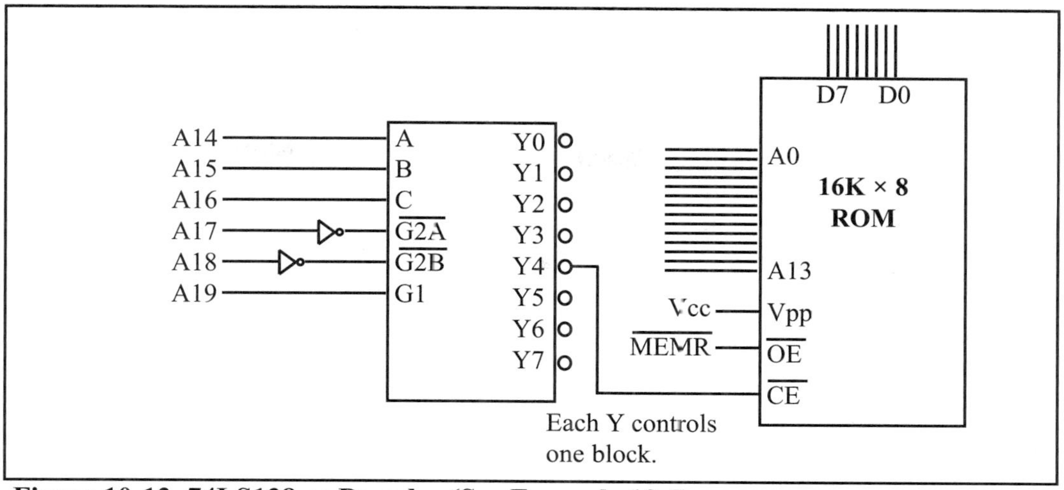

Figure 10-13. 74LS138 as Decoder (See Example 10-7)

Review Questions

1. The MEMR signal from the CPU is connected to the _______ pin of the ROM chip.
2. The MEMW signal from the CPU is connected to the _______ pin of the RAM chip.
3. The CS pin of the memory chip is normally an __________ (active-low, active-high) signal.
4. The 74LS138 has total of _______ outputs.
5. The Y output of the 74138 is ______________ (active low, active high).

SECTION 10.3: IBM PC MEMORY MAP

All x86 CPUs in real mode provide 20 address bits (A0–A19). Therefore, the maximum amount of memory that they can access is one megabyte. How this 1M is allocated in the original PC is the main topic of this section. The 20 lines, A0–A19, of the system address bus can take the lowest value of all 0s to the highest value of all 1s in binary. Converting these values to hexadecimal gives an address range of 00000H to FFFFFH. This is shown in Figure 10-14. Any address that is assigned to any memory block in the 8088-based original PC must fall between these two ranges. This includes all x86 microprocessors in real mode.

The 20-bit address of the 8088 provides a maximum of 1M (1024K bytes) of memory space. Of the 1024K bytes, the designers of the original IBM PC decided to set aside 640K for RAM, 128K for video display RAM (VDR), and 256K for ROM, as shown in Figure 10-15. In today's PC, 640K bytes is not that much, but the standard of the personal computer in 1980 was 64K bytes of memory. At that time, 640K seemed like more than anyone would ever need. Next we discuss the memory map of the PC.

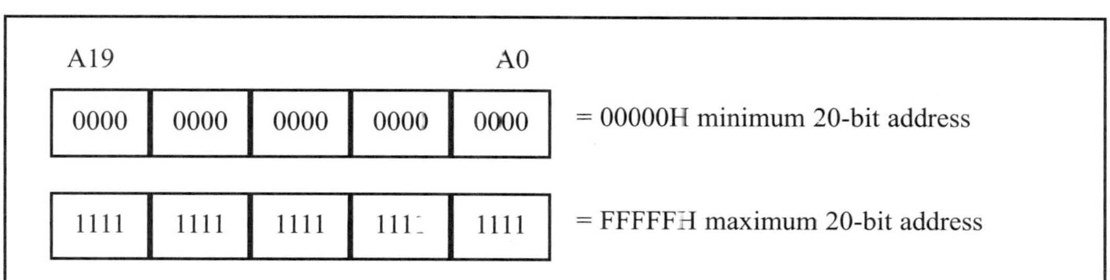

Figure 10-14. 20-bit Address Range in Real Mode for x86 CPUs

Conventional memory: 640K of RAM

In the x86 PC, the addresses from 00000 to 9FFFFH, including location 9FFFFH, are set aside for RAM. In early PCs, only 64K to 256K bytes of RAM were on the motherboard and the rest had to be expanded by adding a memory expansion plug-in card. In those early models, when a RAM memory board was installed, switches had to be set to inform BIOS and DOS of the added memory. In today's x86 PC this is done by the CMOS set-up process and this information is kept by the CMOS NV-RAM for the next cold boot. Of the 640K bytes of memory, some were used by the operating system (the amount depends on the version of DOS) and the rest of the available RAM is used by utilities and application programs. This 640K bytes of memory is commonly referred to as *conventional memory*. Notice that even though the vast majority of PCs use MS Windows for the operating system the above concepts are still valid since DOS is embedded into Windows to run legacy applications.

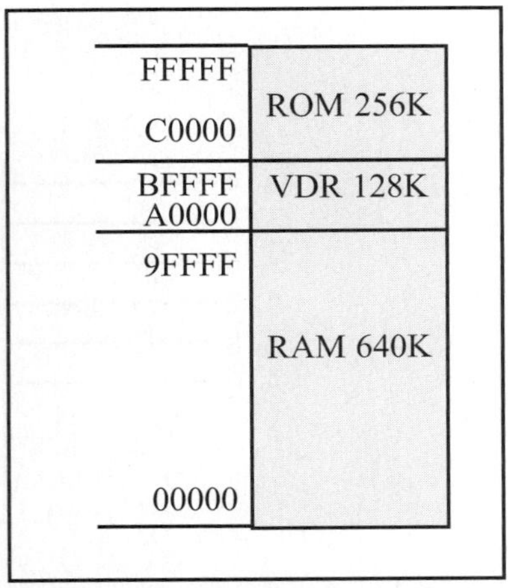

Figure 10-15. Memory Map of the IBM PC

Of the total amount of RAM installed, the first 1K (00000 to 003FF = 1024 bytes) is set aside for the interrupt vector table (see Chapter 15). 00400 to 004FF is set aside for the BIOS temporary data area. Finally, a certain number of kilobytes is occupied by the DOS operating system itself.

Example 10-8

Show the calculation that verifies that addresses 00000 to 9FFFFH comprise 640K bytes.

Solution:

To calculate the total number of bytes for a given memory address range, subtract the two addresses and add 1 to get the total bytes in hex. Then the hex number is converted to decimal and divided by 1024 to get K bytes.

```
  9FFFF          9FFFF
– 00000        + 00001
  9FFFF          A0000 hex = 655,360 decimal = 640K
```

BIOS data area

As mentioned earlier, the BIOS data area is a section of RAM memory used by BIOS to store some extremely important system information. A partial list of that information is given in Figure 10-16. BIOS stores system information in the BIOS data area as it tests each section of the PC. The operating system navigates the system hardware with the help of information stored in the BIOS data area. For example, the BIOS data area tells the operating system how many serial and parallel ports are installed in the PC. We will examine this topic further in the serial and parallel port chapters.

Video display RAM (VDR) map

To display information on the monitor of the PC, the CPU must first store that information in memory called *video display RAM (VDR)*. It is the job of the video controller to display the contents of VDR on the screen. Therefore, the address of the VDR must be within the CPU address range. In the x86 PC, from A0000 to BFFFFH, a total of

Partial Listing of IBM PC RAM Memory Map for Interrupt, BIOS Data		
Memory Location	Bytes	Description
0000:0000 to 0000:03FF	1024	interrupt table
0000:0400 to 0000:0401	2	port address of com1
0000:0402 to 0000:0403	2	port address of com2
0000:0404 to 0000:0405	2	port address of com3
0000:0406 to 0000:0407	2	port address of com4
0000:0408 to 0000:0409	2	port address of lpt1
0000:040A to 0000:040B	2	port address of lpt2
0000:040C to 0000:040D	2	port address of lpt3
0000:040E to 0000:040F	2	port address of lpt4
0000:0410 to 0000:0411	2	list of installed hardware
0000:0412 to 0000:0412	1	initialization flag
0000:0413 to 0000:0414	2	memory size (K bytes)
0000:0415 to 0000:0416	2	memory in I/O channel (if any)
0000:0417 to 0000:0418	2	keyboard status flag
0000:0419 to 0000:0419	1	alternate key entry storage

Figure 10-16. The BIOS Data Area in PC

Table 10-4: Video Display RAM Memory Map

Adapters	Number of Bytes Used	Starting Address
CGA, EGA, VGA	16,384 (16 K)	B8000H
MDA, EGA, VGA	4096 (4K)	B0000H
EGA, VGA	65,536 (64K)	A0000H

128K bytes of the CPU's addressable memory is allocated for video. Of that 128K, only a portion is used for VDR, the amount used depending on the mode in which the video system is being used (text or graphics), and the resolution. For example, the monochrome video mode uses only addresses starting at B0000 up to 4K bytes of RAM, color graphics mode uses addresses starting at B8000, and VGA has a starting address of A0000. See Table 10-4. For more details of each video mode and how many bytes of memory are used in text and graphics modes and their resolution, see Chapter 16.

ROM address and cold boot in the PC

When power is applied to a CPU it must wake up at an address that belongs to ROM. Obviously, the first code executed by the CPU must be stored in nonvolatile memory. The IBM PC is no exception to this design rule. After RESET the 8088 has the values shown in Table 10-5. This means that upon RESET, the 8088 starts to fetch information from CS:IP of FFFF:0000, which gives the physical address FFFF0H. This is the reason that BIOS ROM is located at the upper address range of the memory map. As a result, when the PC is RESET, ROM BIOS is the memory block that is accessed first by the CPU. The ROM BIOS has, among other things, programs that do the testing of the CPU, ROM, and RAM. After those tests, it initializes all peripheral devices, sets up the system, and

Table 10-5: 8088 After RESET

CPU	Contents
CS	FFFFH
DS	0000H
SS	0000H
ES	0000H
IP	0000H
Flags	Clear
Queue	Empty

(Reprinted by permission of Intel Corporation, Copyright Intel 1989)

loads the operating system from hard disk into DRAM and hands over the control of the PC to the operating system. Since the microprocessor starts to fetch and execute instructions from physical location FFFF0H there must be an opcode sitting in that ROM location. In the x86 PC, the CPU finds the opcode for the FAR jump, EA, at location FFFF0H and the target address of the JUMP. You can verify that on your PC regardless of the microprocessor installed on the motherboard. Example 10-9 shows one such case using a simple DEBUG command. Notice in Example 10-9 that the date of ROM BIOS of a PC is stored in locations F000:FFF5 to F000:FFFD of BIOS ROM.

Example 10-9

Using the DEBUG dump command, verify the JMP address for the cold boot and the BIOS date.

Solution:

From the directory containing DEBUG, enter the following:

```
C>DEBUG
-d ffff:0 LF
FFFF:0000 EA 5B E0 00 F0 30 34 2F-30 33 2F 30 37 00 FC .[...04/03/07..
-Q
C>
```

The first 5 bytes showed the jump command "EA" and the destination "F0000:E05B". The next 8 bytes show the BIOS date, 04/03/07.

Example 10-10

Suppose that you buy a software package and encounter a problem installing and running it on your computer. After contacting the technical support department of the manufacturer, you are told that the package is good for the BIOS ROM date of 10/08/07, but you are not told how to find the date. Use DEBUG to find the date for the ROM BIOS of an x86 PC.

Solution:

In the DOS directory (or wherever you keep DEBUG), type the following at the DOS prompt:

```
C>DEBUG
-D F000:FFF5 FFFD
F000:FFF0                30 34 2F-30 33 2F 30 37 00          04/03/07.
-Q
C>
```

The BIOS data is stored at F000:FFF5 through F000:FFFD. In the above case, the BIOS data is 04/03/07, which is earlier than the 10/08/07 date that you hoped to find.

Review Questions

1. What address range is called conventional memory? How many K bytes is that?
2. If the starting physical address of VDR is B0000H, what is the last address if it uses 16K bytes of RAM?
   (a) Show the beginning and ending physical addresses.
   (b) Give the corresponding logical addresses.
3. If the total ROM memory space used by BIOS and other expansion boards is 92K bytes, how many bytes are still unused?
4. What are the contents of CS and IP after the 8088 is reset (cold boot)?
5. What is the implication of Question 4?

SECTION 10.4: DATA INTEGRITY IN RAM AND ROM

When storing data, one major concern is maintaining data integrity. That is, ensuring that the data retrieved is the same as the data stored. The same principle applies when transferring data from one place to another. There are many ways to ensure data integrity depending on the type of storage. The checksum method is used for ROM, and the parity bit method is used for DRAM. For mass storage devices such as hard disks and for transferring data on the Internet, the CRC (cyclic redundancy check) method is employed. In this section we discuss the checksum and parity methods.

Checksum byte

To ensure the integrity of the contents of ROM, every PC must perform a checksum calculation. The process of checksum will detect any corruption of the contents of ROM. One of the causes of ROM corruption is current surge, either when the PC is turned on or during operation. The checksum method uses a checksum byte. This checksum byte is an extra byte that is tagged to the end of a series of bytes of data. To calculate the checksum byte of a series of bytes of data, the following steps can be taken.

1. Add the bytes together and drop the carries.
2. Take the 2's complement of the total sum, and that is the checksum byte, which becomes the last byte of the stored information.

To perform the checksum operation, add all the bytes, including the checksum byte. The result must be zero. If it is not zero, one or more bytes of data have been changed (corrupted). To clarify these important concepts, see Examples 10-11 and 10-12.

Checksum program

When the PC is turned on, one of the first things the BIOS does is to test the system ROM. The code for such a test is stored in the BIOS ROM. Figure 10-17 shows the program using the checksum method. Notice in the code how all the bytes are added together without keeping the track of carries. Then, the total sum is ORed with itself to see if it is zero. The zero flag is expected to be set to high upon return from this subroutine. If it is not, the ROM is corrupted.

```
                ;-----------------------------------
                ;      ROS CHECKSUM SUBROUTINE       :
                ;-----------------------------------
EC4C            ROS_CHECKSUM PROC NEAR    ;NEXT_ROS_MODULE
EC4C B90020                   MOV  CX,8192 ;NUMBER OF BYTES TO ADD
EC4F            ROS_CHECKSUM_CNT:   ;ENTRY PT. FOR OPTIONAL ROS TEST
EC4F 32C0                     XOR  AL,AL
EC51            C26:
EC51 0207                     ADD  AL,DS:[ BX]
EC53 43                       INC  BX     ;POINT TO NEXT BYTE
EC54 E2FB                     LOOP C26    ;ADD ALL BYTES IN ROS MODULE
EC56 0AC0                     OR   AL,AL ; SUM = 0?
EC58 C3                       RET
                ROS_CHECKSUM ENDP
```

Figure 10-17. PC BIOS Checksum Routine

(Reprinted by permission from "IBM Technical Reference" c. 1984 by International Business Machines Corporation)

Example 10-11

Assume that we have 4 bytes of hexadecimal data: 25H, 62H, 3FH, and 52H.
(a) Find the checksum byte.
(b) Perform the checksum operation to ensure data integrity.
(c) If the second byte 62H had been changed to 22H, show how checksum detects the error.

Solution:

(a) The checksum is calculated by first adding the bytes.

```
   25H
+  62H
+  3FH
+  52H
 1 18H
```

The sum is 118H, and dropping the carry, we get 18H. The checksum byte is the 2's complement of 18H, which is E8H.

(b) Adding the series of bytes including the checksum byte must result in zero. This indicates that all the bytes are unchanged and no byte is corrupted.

```
   25H
+  62H
+  3FH
+  52H
+  E8H
 2 00H (dropping the carry)
```

(c) Adding the series of bytes including the checksum byte shows that the result is not zero, which indicates that one or more bytes have been corrupted.

```
   25H
+  22H
+  3FH
+  52H
+  E8H
 1 C0H  dropping the carry, we get C0H.
```

Example 10-12

Assuming that the last byte of the following data is the checksum byte, show whether the data has been corrupted or not: 28H, C4H, BFH, 9EH, 87H, 65H, 83H, 50H, A7H, and 51H.

Solution:
The sum of the bytes plus the checksum byte must be zero; otherwise, the data is corrupted
28H + C4H + BFH + 9EH + 87H + 65H + 83H + 50H + A7H + 51H = 500H
By dropping the accumulated carries (the 5), we get 00. The data is not corrupted. See Figure 10-17 for a program that performs this verification.

Use of parity bit in DRAM error detection

System boards or memory modules are populated with DRAM chips of various organizations, depending on the time they were designed and the availability of a given chip at a reasonable cost. The memory technology is changing so fast that DRAM chips on the boards have a different look every year or two. While early PCs used 64K DRAMs, current PCs commonly use 1G chips. To understand the use of a parity bit in detecting data storage errors, we use some simple examples from the early PCs to clarify some very important design concepts. It must be noted that in today's PCs, these design concepts are still the same, even though the DRAMs have much higher density and FPGAs in the chipset are used in place of TTL logic gates. You may wish to review DRAM organization and capacity, covered earlier in this chapter, before proceeding.

DRAM memory banks

The arrangement of DRAM chips on the system or memory module boards is often referred to as a *memory bank*. For example, the 64K bytes of DRAM can be arranged as one bank of 8 IC chips of 64K × 1 organization, or 4 banks of 16K × 1 organization. The first IBM PC introduced in 1981 used memory chips of 16K × 1 organization. Figure 10-18 shows the memory banks for 640K bytes of RAM using 256K and 1M DRAM chips. Notice the use of an extra bit for every byte of data to store the parity bit. With the extra parity bit, every bank requires an extra chip of × 1 organization for parity check. Figure 10-19 shows DRAM design and parity bit circuitry for a bank of DRAM. First, note the use of the 74LS158 to multiplex the 16 address lines A0–A15, changing them to the 8 address lines of MA0–MA7 (multiplexed address) as required by the 64K × 1 DRAM chip. The resistors are for the serial bus line termination to prevent undershooting and overshooting at the inputs of DRAM. They range from 20 to 50 ohms, depending on the speed of the CPU and the printed circuit board layout.

A few additional observations about Figure 10-19 should be made. The output of multiplexer addresses MA0–MA7 will go to all the banks. Likewise, memory data MD0–MD7 and memory data parity MDP will go to all the banks. The 74LS245 not only buffers the data bus MD0–MD7 but also boosts it to drive all DRAM inputs. Since the banks of the DRAMs are connected in parallel and the capacitance loading is additive, the data line must be capable of driving all the loads. Next we discuss how parity is used to detect RAM defects.

Parity bit generator/checker in the IBM PC

There are two types of errors that can occur in DRAM chips: soft error and hard error. In a *hard error*, some bits or an entire row of memory cells inside the memory chip get stuck to high or low permanently, thereafter always producing 1 or 0 regardless of what you write into the cell(s). In a *soft error*, a single bit is changed from 1 to 0 or from 0 to 1 due to current surge or certain kinds of particle radiation in the air. Parity is used to

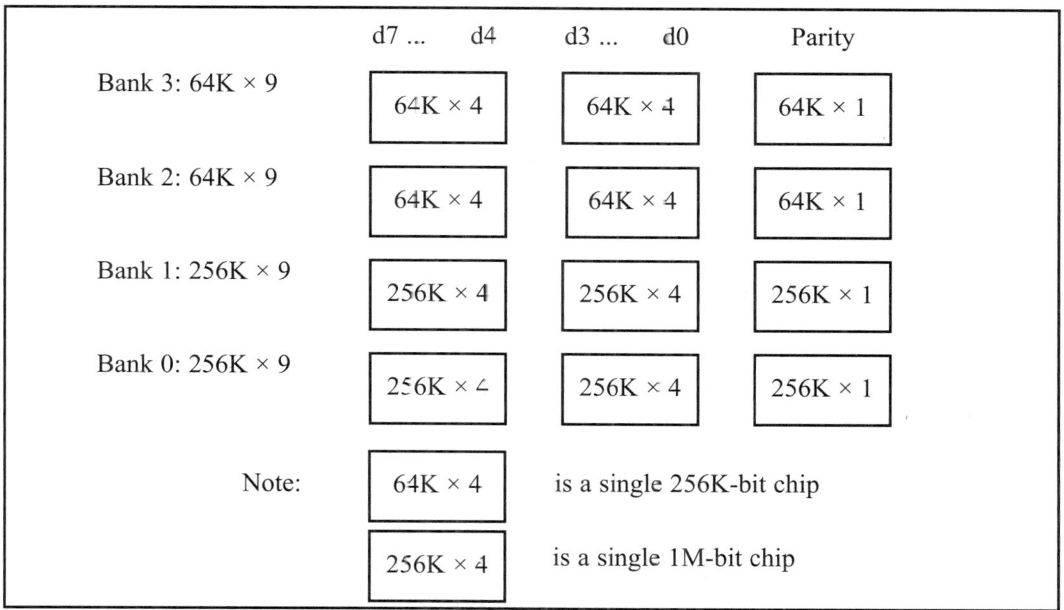

Figure 10-18. A Possible Memory Configuration for 640K DRAM

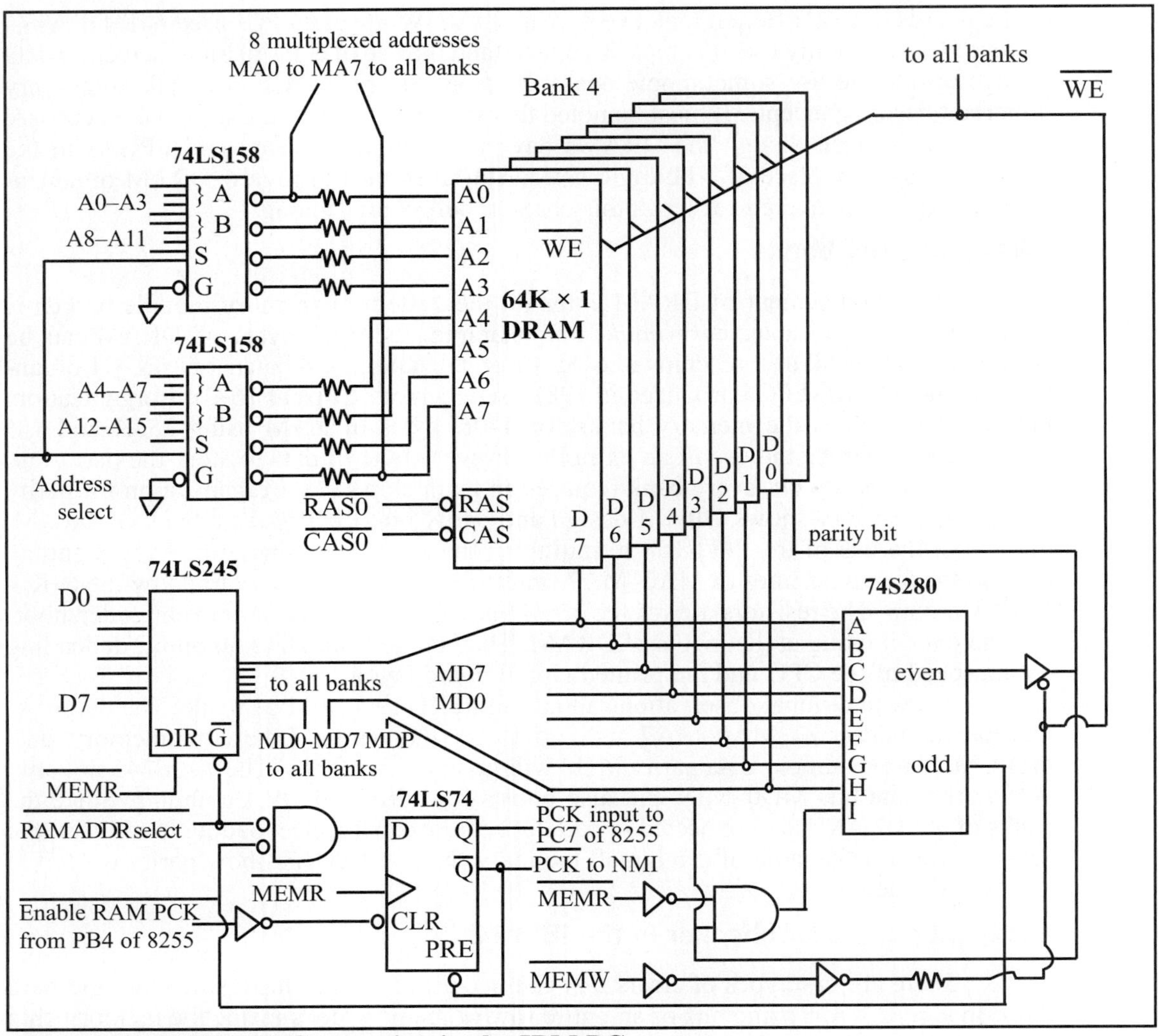

Figure 10-19. DRAM Connection in the IBM PC

(Reprinted by permission from "IBM Technical Reference" c. 1984 by International Business Machines Corporation)

detect such errors. Including a parity bit to ensure *data integrity* in RAM is the most widely used method since it is the simplest and cheapest. This method can only indicate if there is a difference between the data that was written to memory and the data that was read. It cannot correct the error as is the case with some high-performance computers. In those computers and some of the x86-based servers, the EDC (error detection and correction) method is used to detect and correct the error bit. The early IBM PC and compatibles use the 74S280 parity bit generator and checker to implement the concept of the parity bit. The study of that chip should help us to understand the parity bit concept.

74S280 parity bit generator and checker

In order to understand the parity bit circuitry of Figure 10-19 it is necessary first to understand the 74LS280 parity bit generator and checker chip. This chip has 9 inputs and 2 outputs. Depending on whether an even or odd number of ones appears in the input, the even or odd output is activated according to Table 10-6.

As can be seen from Table 10-6, if all 9 inputs have an even number of 1 bits, the even output goes high, as in cases 1 and 4. If the 9 inputs have an odd number of high bits, the odd output goes high, as in cases 2 and 3. The way the IBM PC uses this chip is as follows. Notice that in Figure 10-19, inputs A–H are connected to the data bus, which is 8 bits, or one byte. The I input is used as a parity bit to check the correctness of the byte of data read from memory. When a byte of information is written to a given memory location in DRAM, the even-parity bit is generated and saved on the ninth DRAM chip as a parity bit with use of control signal $\overline{\text{MEMW}}$. This is done by activating the tri-state buffer using $\overline{\text{MEMW}}$. At this point, I of the 74S280 is equal to zero since $\overline{\text{MEMR}}$ is high. When a byte of data is read from the same location, the parity bit is gated into the I input of the 74S280 through $\overline{\text{MEMR}}$. This time the odd output is taken out and fed into a 74LS74. If there is a difference between the data written and the data read, the Q output (called PCK, parity bit check) of the 74LS74 is activated and Q activates NMI, indicating that there is a parity bit error, meaning that the data read is not the same as the data written. Consequently, it will display a parity bit error message. For example, if the byte of data written to a location has an even number of 1s, A to H has an even number of 1s, and I is zero, then the even-parity output of 74280 becomes 1 and is saved on parity bit DRAM. This is case 1 shown in Table 10-6. If the same byte of data is read and there is an even number of 1s (the byte is unchanged), I from the ninth bit DRAM, which is 1, is input to the 74S280, even becomes low, and odd becomes high, which is case 2 in Table 10-6. This high from the odd output will be inverted and fed to the 74LS74, making Q low. This means that $\overline{\text{Q}}$ is high, thereby indicating that the written byte is the same as the byte read and that no errors occurred. If the number of 1s in the byte has changed from even to odd and the 1 from the saved parity DRAM makes the number of inputs even (case 4 above), the odd output becomes low, which is inverted and passed to the 7474 D flip-flop. That makes Q = 1 and $\overline{\text{Q}}$ = 0, which signals the NMI to display a parity bit error message on the screen.

Table 10-6: 74280 Parity Check

	Inputs		Outputs	
Case	**A–H**	**I**	**Even**	**Odd**
1	Even	0	1	0
2	Even	1	0	1
3	Odd	0	0	1
4	Odd	1	1	0

Review Questions

1. Find the checksum byte for the following bytes: 24H, 76H, F5H, 98H, 89H, 7AH, 61H, C2H.
2. Show a simple program in Assembly language to find the checksum byte of the 8 bytes of information (code or data) given in Question 1. Assume that SI equals the offset address of the data.
3. In a given PC we have only 512K of memory on the motherboard. Show possible configurations and the number of chips used to add memory up to the maximum allowed by the limits of conventional memory if we have each of the following. Include the parity bit in your configuration and count.
(a) 64K × 1 (b) 64K × 4 and 64K × 1
4. To detect corruption of information stored in RAM and ROM memories, system designers use the ____________ method for RAM and the _________ method for ROM.
5. Assume that due to slight current surge in the power supply, a byte of RAM has been corrupted while the computer is on. Can the system detect the corruption while the computer is on? Is this also the case for ROM?

SECTION 10.5: 16-BIT MEMORY INTERFACING

In the design of current x86 PCs, a single IC chip called a *chipset* has replaced the 100 or so logic ICs connected together in the original PC. As a result, the details of CPU connection to memory and other peripherals are not visible for educational purposes. The 16-bit bus interfacing to memory chips is one of these details that is now buried within a chipset but still needs to be understood. In this section we explore memory interfacing for 16-bit CPUs. We use the 286 as an example but the concepts can apply to any 16-bit microprocessor. We also discuss the topics of memory cycle time and bus bandwidth.

ODD and EVEN banks

In a 16-bit CPU such as the 80286, memory locations 00000–FFFFF are designated as odd and even bytes as shown in Figure 10-20. Although Figure 10-20 shows only 1 megabyte of memory, the concept of odd and even banks applies to the entire memory space of a given processor with a 16-bit data bus. To distinguish between odd and even bytes, the CPU provides a signal called BHE (bus high enable). BHE in association with A0 is used to select the odd or even byte according to Table 10-7.

Examine Figure 10-20 to see how the odd and even addresses are designated for the 16-bit-wide data buses. Figure 10-21 shows 640 KB of DRAM for 16-bit buses. Figure 10-22 shows the connection for the 16-bit data bus. In Figure 10-22, notice the use of A0 and BHE as bank selectors. Also notice the use of the 74LS245 chip as a data bus buffer.

Table 10-7: Distinguishing Between Odd and Even Bytes

BHE	A0		
0	0	Even word	D0–D15
0	1	Odd byte	D8–D15
1	0	Even byte	D0–D7
1	1	None	

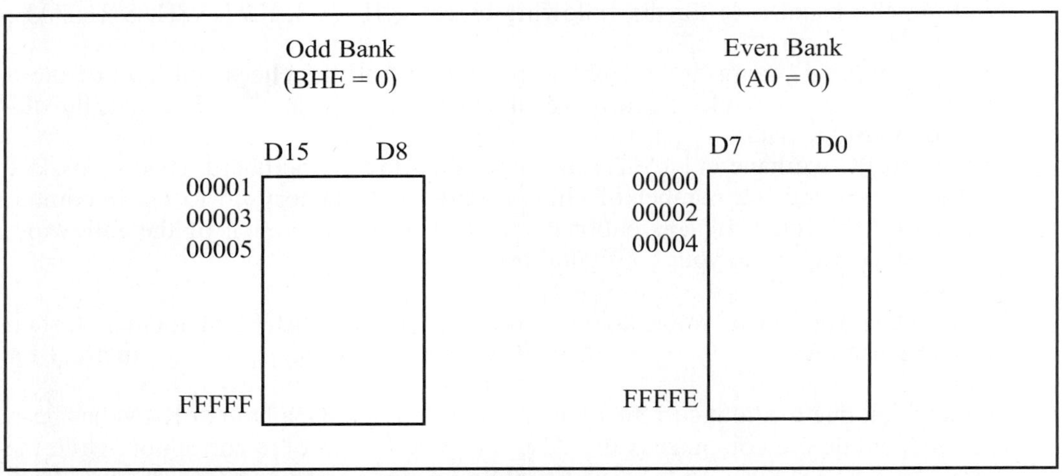

Figure 10-20. Odd and Even Banks of Memory

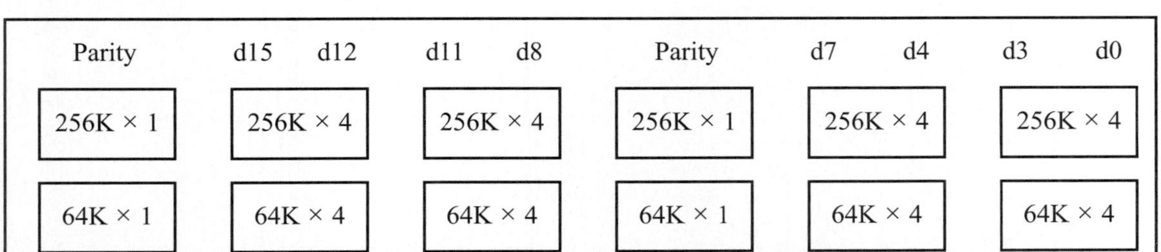

Figure 10-21. 640K Bytes of DRAM with Odd and Even Banks Designation

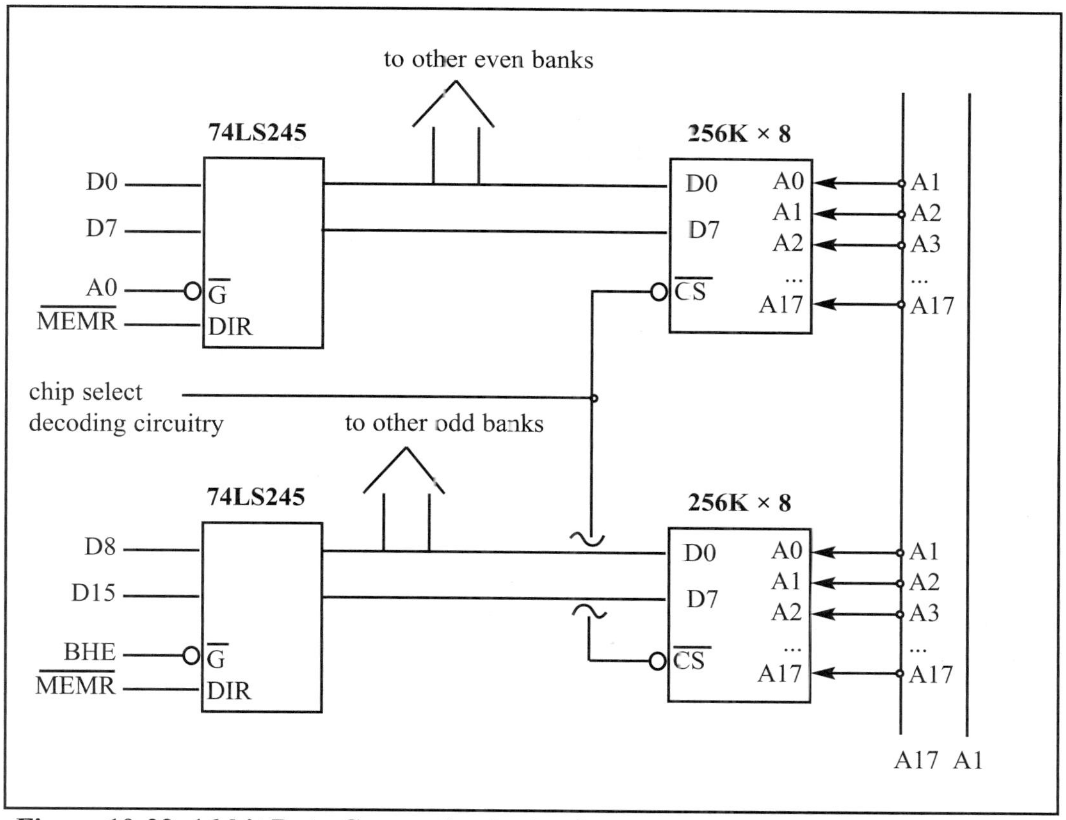

Figure 10-22. 16-bit Data Connection in the Systems with 16-bit Data Bus

Memory cycle time and inserting wait states

To access an external device such as memory or I/O, the CPU provides a fixed amount of time called a *bus cycle time*. During this bus cycle time, the read and write operation of memory or I/O must be completed. Here, we cover the memory bus cycle time. Bus cycle time for I/O devices is given in the next chapter. For the sake of clarity we will concentrate on reading memory, but the concepts apply to write operations as well. The bus cycle time used for accessing memory is often referred to as *MC* (memory cycle) time. The time from when the CPU provides the addresses at its address pins to when the data is expected at its data pins is called *memory read cycle time*. While in older processors such as the 8088 the memory cycle time takes 4 clocks, in the newer CPUs the memory cycle time is 2 clocks. In other words, in all x86 CPUs from the 286 to the Pentium, the memory cycle time is only 2 clocks. If memory is slow and its access time does not match the MC time of the CPU, extra time can be requested from the CPU to extend the read cycle time. This extra time is called a *wait state* (WS). In the 1980s, the clock speed for memory cycle time was the same as the CPU's clock speed. For example, in the 20 MHz 286/386/486 processors, the buses were working at the same speed of 20 MHz. This resulted in 2 × 50 ns = 100 ns for the memory cycle time (1/20 MHz = 50 ns). When the CPU's speed was under 100 MHz, the bus speed was comparable to the CPU speed. In the 1990s the CPU speed exploded to 1 GHz (gigahertz) while the bus speed maxed out at around 133 MHz. The gap between the CPU speed and the bus speed is one of the biggest problems in the design of high-performance computers. To avoid the use of too many wait states in interfacing memory to CPU, cache memory and other high-speed DRAMs were invented. These are discussed in Chapter 22.

It must be noted that memory access time is not the only factor in slowing down the CPU, even though it is the largest one. The other factor is the delay associated with signals going through the data and address path. Delay associated with reading data stored in memory has the following two components:

1. The time taken for address signals to go from CPU pins to memory pins, going through decoders and buffers (e.g., 74LS245). This, plus the time it takes for the data to travel from memory to CPU, is referred to as a *path delay*.
2. The memory access time to get the data out of the memory chip. This is the largest of the two components.

The total sum of these two must equal the memory read cycle time provided by the CPU. Memory access time is the largest and takes about 80% of the read cycle time. See Examples 10-13 and 10-14 for further clarification of these points. These concepts are critical in the design of microprocessor-based products.

Example 10-13

Calculate the memory cycle time of a 20-MHz 8386 system with
(a) 0 WS,
(b) 1 WS, and
(c) 2 WS.
Assume that the bus speed is the same as the processor speed.

Solution:

1/20 MHz = 50 ns is the processor clock period. Since the 386 bus cycle time of zero wait states is 2 clocks, we have:

	80386 20 MHz
Memory cycle time with 0 WS	2 × 50 = 100 ns
Memory cycle time with 1 WS	100 + 50 = 150 ns
Memory cycle time with 2 WS	100 + 50 + 500 = 200 ns

It is preferred that all bus activities be completed with 0 WS. However, if the read and write operations cannot be completed with 0 WS, we request an extension of the bus cycle time. This extension is in the form of an integer number of WS. That is, we can have 1, 2, 3, and so on WS, but not 1.25 WS.

Example 10-14

A 20-MHz 80386-based system is using ROM of 150 ns speed. Calculate the number of wait states needed if the path delay is 25 ns.

Solution:

If ROM access time is 150 ns and the path delay is 25 ns, every time the 80386 accesses ROM it must spend a total of 175 ns to get data into the CPU. A 20-MHz CPU with zero WS provides only 100 ns (2 × 50 ns = 100 ns) for the memory read cycle time. To match the CPU bus speed with this ROM we must insert 2 wait states. This makes the cycle time 200 ns (100 + 50 + 50 = 200 ns). Notice that we cannot ask for 1.5 WS since the number of WS must be an integer. That would be like going to the store and wanting to buy half an apple. You must get one or more complete WS or none at all.

Accessing EVEN and ODD words

As you recall from earlier chapters, Intel defines 16-bit data as a *word*. The address of a word can start at an even or an odd number. For example, in the instruction "MOV AX,[2000]" the address of the word being fetched into AX starts at an even address. In the case of "MOV AX,[2007]" the address starts at an odd address. In systems with a 16-bit data bus, accessing a word from an odd addressed location can be slower. This issue is important and applies to 32-bit and 64-bit systems with 386 and Pentium processors, as we will see in Chapter 22.

As shown in Figure 10-23, in the 8-bit system, accessing a word is treated like accessing two bytes regardless of whether the address is odd or even. Since accessing a byte takes one memory cycle, accessing any word will take 2 memory cycles. In the 16-bit system, accessing a word with an even address takes one memory cycle. That is because one byte is carried on D0–D7 and the other on D8–D15 in the same memory cycle. However, accessing a word with an odd address requires two memory cycles. For example, see how accessing the word in the instruction "MOV AX,[F617]" works as shown in Figure 10-24. Assuming that DS = F000H in this instruction, the contents of physical memory locations FF617H and FF618H are being moved into AX. In the first cycle, the 286 CPU accesses location FF617H and puts it in AL.

In the second cycle, the contents of memory location FF618H are accessed and put into AH. The lesson to be learned from this is to try not to put any words on an odd address if the program is going to be run on a 16-bit system. Indeed this is so important that there is a pseudo-op specifically designed for this purpose. It is the EVEN directive and is used as follows:

```
                EVEN
VALUE1          DW     ?
```

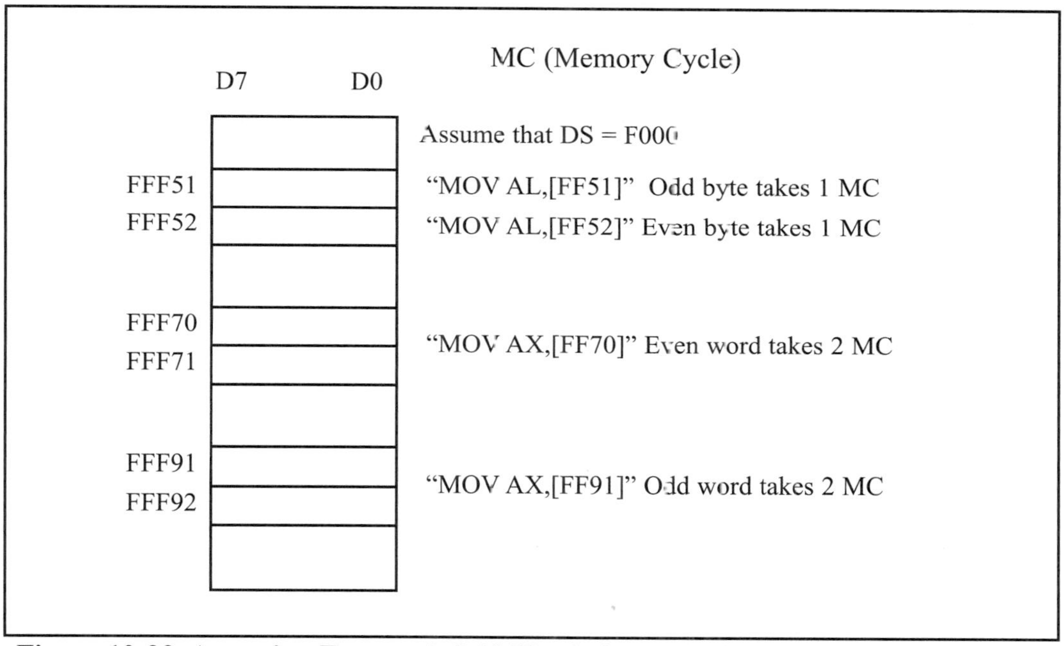

Figure 10-23. Accessing Even and Odd Words in the 8-bit CPU

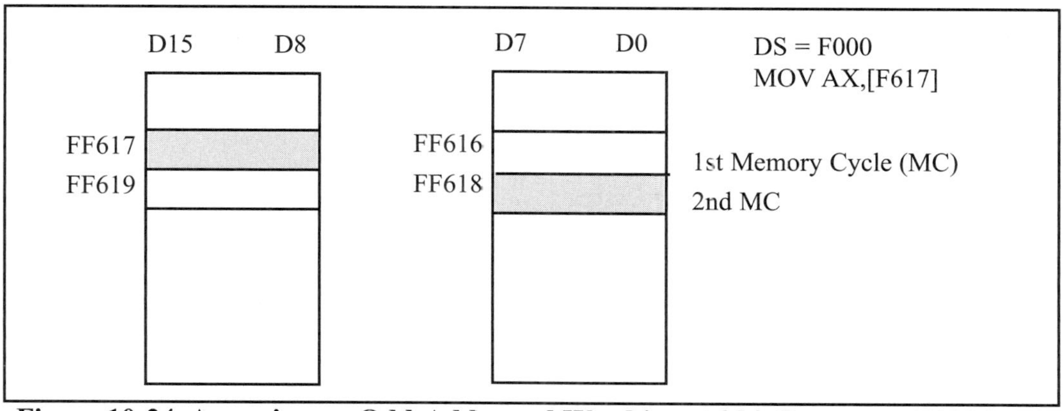

Figure 10-24. Accessing an Odd-Addressed Word in a 16-bit Processor (80286)

This ensures that VALUE1, a word-sized operand, is located in an even address location. Therefore, an instruction such as "MOV AX,VALUE1" or "MOV VALUE1,CX" will take only a single memory cycle.

Bus bandwidth

The main advantage of the 16-bit data bus is a doubling of the rate of transfer of information between the CPU and the outside world. The rate of data transfer is generally called *bus bandwidth*. In other words, bus bandwidth is a measure of how fast buses transfer information between the CPU and memory or peripherals. The wider the data bus, the higher the bus bandwidth. However, the advantage of the wider external data bus comes at the cost of increasing the size of the printed circuit board. Now you might ask why we should care how fast buses transfer information between the CPU and outside, as long as the CPU is working as fast as it can. The problem is that the CPU cannot process information that it does not have. In other words, the speed of the CPU must be matched with the higher bus bandwidth; otherwise, there is no use for a fast CPU. This is like driving a Porsche or Ferrari in first gear; it is a terrible underusage of CPU power. Bus bandwidth is measured in MB (megabytes) per second and is calculated as follows:

bus bandwidth = (1/bus cycle time) × bus width in bytes

In the above formula, bus cycle time can be either memory or I/O cycle time. The I/O cycle time is discussed in Chapter 11. Example 10-15 clarifies the concept of bus bandwidth. As can be seen from Example 10-15, there are two ways to increase the bus bandwidth: Either use a wider data bus or shorten the bus cycle time (or do both). That is exactly what 386, 486, and Pentium processors have done. While the data bus width has increased from 16-bit in the 80286 to 64-bit in the Pentium, the bus cycle time is reaching a maximum of 133 MHz. Again, it must be noted that although the processor's speed can go to 1 GHz or higher, the bus speed is limited to around 133 MHz. The reason for this is that the signals become too noisy for the circuit board if they are above 100 MHz. This is even worse for the ISA expansion slot. The ISA bus speed is limited to around 8 MHz. This is because the ISA slot uses large and bulky connectors and they are too noisy for a speed of more than 8 MHz. In other words, in PCs with a 500 MHz Pentium, the CPU must slow down to 8 MHz when accessing the ISA bus. The PCI bus was introduced to solve this limitation. It can go as high as 133 MHz and its data bus width is 64-bit.

Review Questions

1. True or false. If A0 = 0 and BHE = 1, a byte is being transferred on the D0–D7 data bus from an even-address location.
2. True or false. If we have A0 = 1 and BHE = 0, a byte is being transferred on the D0–D7 data bus from an odd-address location.
3. True or false. If we have A0 = 1 and BHE = 0, a byte is being transferred on the D8–D15 data bus.
4. True or false. If we have A0 = 0 and BHE = 0, a word is being transferred on the D0–D15 data bus.
5. In the instruction "MOV AX,[2000]", the transferring of data into the accumulator takes _______ memory cycles for the 8088 and _____ for the 80286.
6. A 16-MHz 286 has a memory cycle time of _____ ns if it is used with a zero wait state.
7. To interface a 10-MHz 286 processor to a 350-ns access time ROM, how many wait states are needed?

Example 10-15

Calculate memory bus bandwidth for the following microprocessors if the bus speed is 20 MHz.

(a) 286 with 0 WS and 1 WS (16-bit data bus)
(b) 386 with 0 WS and 1 WS (32-bit data bus)

Solution:

The memory cycle time for both the 286 and 386 is 2 clocks, with zero wait states. With the 20 MHz bus speed we have a bus clock of 1/20 MHz = 50 ns.

(a) Bus bandwidth = $(1/(2 \times 50 \text{ ns})) \times 2$ bytes = 20M bytes/second (MB/s)
With 1 wait state, the memory cycle becomes 3 clock cycles
$3 \times 50 = 150$ ns and the memory bus bandwidth is = $(1/150 \text{ ns}) \times 2$ bytes = 13.3 MB/S

(b) Bus bandwidth = $(1/(2 \times 50 \text{ ns})) \times 4$ bytes = 40 MB/s
With 1 wait state, the memory cycle becomes 3 clock cycles
$3 \times 50 = 150$ ns and the memory bus bandwidth is = $(1/150 \text{ ns}) \times 4$ bytes = 26.6 MB/S

From the above it can be seen that the two factors influencing bus bandwidth are:

1. The read/write cycle time of the CPU
2. The width of the data bus

Notice in this example that the bus speed of the 286/386 was given as 20 MHz. That means that the CPU can access memory on the board at this speed. If this 286/386 is used on a PC board with an ISA expansion slot, it must slow down to 8 MHz when communicating with the ISA bus since the maximum bus speed for the ISA bus is 8 MHz. This is done by the chipset circuitry.

PROBLEMS

SECTION 10.1: SEMICONDUCTOR MEMORIES

1. What is the difference between a 4M memory chip and 4M of computer memory as far as capacity is concerned?
2. True or false. The more address pins, the more memory locations are inside the chip.
3. True or false. The more data pins, the more each location inside the chip will hold.
4. True or false. The more data pins, the higher the capacity of the memory chip.
5. True or false. With a fixed number of address pins, the more data pins, the greater the capacity of the memory chip.
6. The speed of a memory chip is referred to as its ________________.
7. True or false. The price of memory chips varies according to capacity and speed.
8. The main advantage of EEPROM over UV-EPROM is ________________.
9. True or false. SRAM has a larger cell size than DRAM.
10. Which of the following, EPROM, DRAM, and SRAM, must be refreshed periodically?
11. Which memory is used for cache?
12. Which of the following, SRAM, UV-EPROM, NV-RAM, DRAM, and cache memory, is volatile memory?
13. RAS and CAS are associated with which memory?
    (a) EPROM (b) SRAM (c) DRAM (d) all of the above
14. Which memory needs an external multiplexer?

(a) EPROM (b) SRAM (c) DRAM (d) all of the above

15. Find the organization and the capacity of memory chips with the following pins.
    (a) EEPROM A0–A14, D0–D7 (b) UV-EPROM A0–A12, D0–D7
    (c) SRAM A0–A11, D0–D7 (d) SRAM A0–A12, D0–D7
    (e) DRAM A0–A10, D0 (f) SRAM A0–A12, D0-D7
    (g) EEPROM A0–A11, D0–D7 (h) UV-EPROM A0–A10, D0–D7
    (i) DRAM A0–A8, D0–D3 (j) DRAM A0–A7, D0–D7
16. Find the capacity, address, and data pins for the following memory organizations.
    (a) 16K × 8 ROM (b) 32K × 8 ROM
    (c) 64K × 8 SRAM (d) 256K × 4 DRAM
    (e) 64K × 8 ROM (f) 64K × 4 DRAM
    (g) 1M × 4 DRAM (h) 4M × 4 DRAM
    (i) 64K × 8 DRAM

SECTION 10.2: MEMORY ADDRESS DECODING

17. Find the address range of the following memory design.

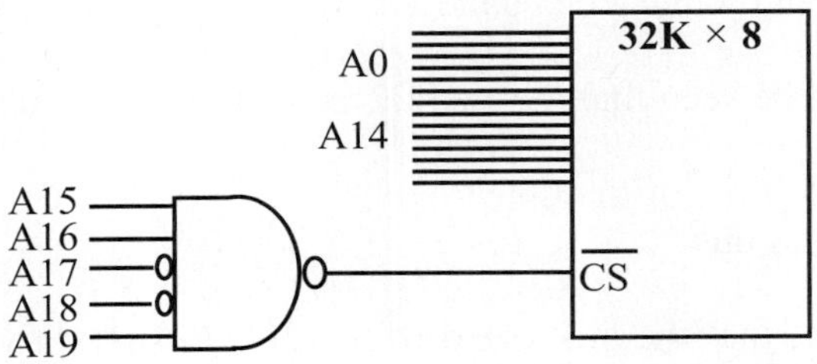

18. Using NAND gates and inverters, design decoding circuitry for the address range 0C0000H–0C0FFFH.
19. Find the address range for Y0, Y3, and Y6 of the 74LS138 for the following design. This is the ROM interfacing with the 8088 CPU in the original PC.

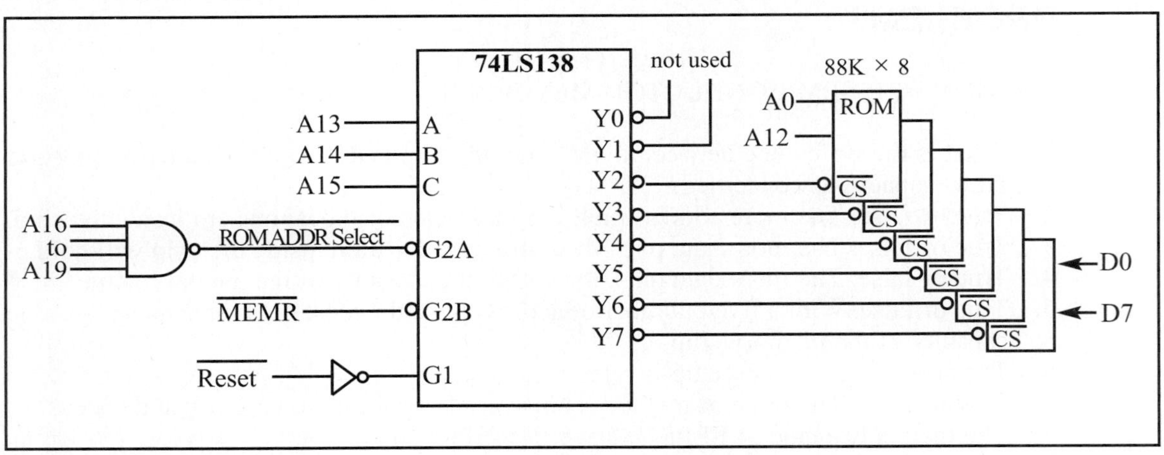

20. Using the 74138, design the memory decoding circuitry in which the memory block controlled by Y0 is in the range 00000H to 03FFFH. Indicate the size of the memory block controlled by each Y.
21. Find the address range for Y3, Y6, and Y7 in Problem 20.
22. Using the 74138 and OR gates, design memory decoding circuitry in which the memory block controlled by Y0 is in the 80000H to 807FFFH space. Indicate the size of the memory block controlled by each Y.

23. Find the address range for Y1, Y4, and Y5 in Problem 22.
24. The CS pin of the memory chip is active _____ (low, high). What about the RD pin?
25. Which one can accommodate more inputs, the 74138 or CPLD?

SECTION 10.3: IBM PC MEMORY MAP

26. Indicate the address range and total kilobytes of memory allocated to the RAM, ROM, and video display RAM of the PC.
27. What address range is called conventional memory, and how many K bytes is it?
28. Can we increase the size of conventional memory? Explain your answer.
29. What are the contents of CS and IP in the 8088 upon RESET?
30. A user wants to add some EPROM to a PC. Can he/she use the address range 00000–9FFFFH? What happens if this range is used?
31. Give the logical and physical location where the BIOS ROM date is stored.
32. Suppose that the memory address range C0000H–C7FFFH is used in a certain plug-in adapter card. How many K bytes is that, and is this memory in the RAM or ROM allocated area?
33. If a video card uses only 4K bytes of VDR and the starting address is B0000H, what is the ending address of this VDR?
34. In a certain video card the starting address is B8000H and it uses only 16K bytes of memory. What is the ending address of this video card?
35. Why is ROM mapped where it is in the PC? Why can't we use addresses starting at 00000?
36. When the CPU is powered up, at what physical address does it expect to see the first opcode? In the PC, what opcode is there normally?

SECTION 10.4: DATA INTEGRITY IN RAM AND ROM

37. Find the checksum byte for the following bytes.
    34H, 54H, 7FH, 11H, E6H, 99H
38. For each of the following sets of data (the last byte is the checksum byte) verify if the data is corrupted.
    (a) 29H, 1CH, 16H, 38H, and 6DH (b) 29H, 1CH, 16H, 30H, and 6DH
39. To maintain data integrity, the checksum method is used for __________ type memory and the parity bit method for __________ memory.
40. True or false. ROM is tested for corruption during a cold boot-up, but data corruption in RAM can be detected any time the system is on.
41. A given PC needs only 320K bytes to reach the maximum allowed conventional memory. Show the memory configuration using 256K × 1 and 64K × 1 memory chips. How many chips are needed? (Include the parity bit.)
42. Repeat Problem 41 if we have 256K × 4 and 64K × 4 chips in addition to 64 × 1 and 256K × 1.
43. Why is it preferable to use higher density memory chips in memory design?
44. True or false. To access DRAM, the RAS address is provided first and then the CAS address.
45. True or false. The 74S280 is both a parity generator and a checker.
46. In the 74S280 we have 10010011 for A–H inputs and I = 0. What is the status of the even and odd output pins?
47. In the 74S280 we have 11101001 for A–H inputs and I = 0. What is the status of the even and odd output pins?
48. In the 74S280 we have 10001001 for A–H inputs and I = 1. What is the status of the even and odd output pins?

SECTION 10.5: 16-BIT MEMORY INTERFACING

49. Odd and even banks are associated with which microprocessor, the 8088 or the 80286?
50. How many memory chips are needed if we use 256K × 8, 256K × 1, 64K × 8, and 64K × 1 memory chips for conventional memory of the 80286 PC/compatible?
51. State the status of A0 and BHE when accessing an odd-addressed byte.
52. State the status of A0 and BHE when accessing an even-addressed word.
53. State the status of A0 and BHE if we only want to access D0–D7 of the data bus.
54. State the status of A0 and BHE if we only want to access D8–D15 of the data bus.
55. What is the use of 74LS245 in memory interfacing?
56. What is the bus bandwidth unit?
57. Give the variables that affect the bus bandwidth.
58. True or false. One way to increase the bus bandwidth is to widen the data bus.
59. True or false. An increase in the number of address bus pins results in a higher bus bandwidth for the system.
60. Calculate the memory bus bandwidth for the following systems.
    (a) 80286 of 10 MHz and 0 WS
    (b) 80286 of 16 MHz and 0 WS

ANSWERS TO REVIEW QUESTIONS

SECTION 10.1: SEMICONDUCTOR MEMORIES

1. Nanoseconds
2. (a) 2K × 8, 16K bits (b) 8K × 8, 64K (c) 64K × 4, 256K (d) 256K × 1, 256K
3. (a) 128K bits, 14 address, and 8 data (b) 256K, 15 address, and 8 data
   (c) 1M, 10 address, and 1 data (d) 1M, 9 address, and 4 data
   (e) 512K, 16, and 8 data (f) 4M, 10 address, and 4 data
4. It takes much less time to erase and does not need to be removed from the system board.
5. NV-RAM

SECTION 10.2: MEMORY ADDRESS DECODING

1. OE
2. WE
3. Active low
4. 8
5. Active low

SECTION 10.3: IBM PC MEMORY MAP

1. 00000–9FFFFH, 640K bytes
2. (a) B0000H–B0FFFH (b) B000:0000–B000:0FFF
3. 256K – 92K = 164K
4. CS = FFFFH, IP = 0000
5. It indicates that the CPU fetches the first opcode at the physical address FFFF0H when the system is turned on. Therefore, no RAM can be mapped into the last segment of the 8088; the memory space must be occupied by a cold boot ROM.

SECTION 10.4: DATA INTEGRITY IN RAM AND ROM

1. Adding the bytes: 24H + 76H + F5H + 98H + 89H + 7AH + 61H + C2H = 44DH. Dropping the carries, we get 4DH, and taking the 2's complement, we have B3H for the checksum byte.

2.

```
        MOV SI,OFFSET DATA      ;LOAD THE OFFSET ADDRESS
        MOV CX,08               ;LOAD THE COUNTER
        SUBAL,AL
BOO:    ADD AL,[ SI]            ;ADD THE BYTE AND IGNORE THE CARRY
        INCSI                   ;POINT TO NEXT BYTE
        LOOP BOO                ;CONTINUE UNTIL COUNT IS ZERO
```

3. Since the maximum limit is 640K bytes, we need add only 128K bytes of RAM (640 – 512 = 128).
   (a) two banks each of nine chips of 64K × 1, total = 18 chips
   (b) two banks each of two 64K × 4 to contain data and one 64K × 1 for parity, total = 6 chips
4. Parity bit generation/checker, checksum
5. While the computer is on, any corruption in the contents of RAM is detected by the parity bit error checking circuitry when that data is accessed (read) again. However, the ROM corruption is not detected since the checksum detection is performed only when the system is booted.

SECTION 10.5: 16-BIT MEMORY INTERFACING

1. True
2. False
3. True
4. True
5. 2 and 1
6. 125 ns
7. 2 WS

CHAPTER 11

8255 I/O PROGRAMMING

OBJECTIVES

Upon completion of this chapter, you will be able to:

- **>> Code Assembly language instructions to read and write data to and from I/O ports**
- **>> Diagram the design of peripheral I/O using the 74LS373 output latch and the 74LS244 input buffer**
- **>> Describe the I/O address map of x86 PCs**
- **>> List the differences in memory-mapped I/O versus peripheral I/O**
- **>> Describe the purpose of the 8255 programmable peripheral interface chip**
- **>> Code Assembly language instructions to perform I/O through the 8255**
- **>> Code I/O programming for Microsoft Visual C/C++**
- **>> Code I/O programming for Linux C/C++**

In addition to memory space, x86 microprocessors also have I/O space. This allows it to access ports. Ports are used either to bring data into the CPU from an external device such as the keyboard or to send data from the CPU to an external device such as a printer. In this chapter we study I/O instructions and I/O design for x86 PCs. In Section 11.1 we discuss I/O instructions and programming. In Section 11.2 we look at ways to design I/O ports for 8088-based systems. In Section 11.3, the I/O map of the x86 IBM PC is given. The 8255 chip and its programming are discussed in Section 11.4. In addition, the details of an 8255 connection to the ISA/PC104 bus will be given along with I/O programming using C/C++.

SECTION 11.1: 8088 INPUT/OUTPUT INSTRUCTIONS

All x86 microprocessors, from the 8088 to the Pentium, can access external devices called *ports*. This is done using I/O instructions. The x86 CPU is one of the few processors that have I/O space in addition to memory space. While memory can contain both opcodes and data, I/O ports contain data only. There are two instructions for this purpose: "OUT" and "IN". These instructions can send data from the accumulator (AL or AX) to ports or bring data from ports into the accumulator. In accessing ports, we can use an 8-bit or 16-bit data port. Since 8-bit data ports in the 8088 are the most widely used, we will concentrate on them and introduce 16-bit data ports only in the last section of this chapter.

8-bit data ports

The 8-bit I/O operation of the 8088 is applicable to all x86 CPUs from the 8088 to the Pentium. The 8-bit port uses the D0–D7 data bus to communicate with I/O devices. In 8-bit port programming, register AL is used as the source of data when using the OUT instruction and as the destination for the IN instruction. This means that to input or output data from any other registers, the data must first be moved to the AL register. Instructions OUT and IN have the following formats:

```
                 Inputting Data          Outputting Data
Format:          IN     dest,source      OUT   dest,source

(1)              IN     AL,port#         OUT   port#,AL


(2)              MOV    DX,port#         MOV   DX,port#
                 IN     AL,DX            OUT   DX,AL
```

In format (1), port# is the address of the port and can be from 00 to FFH. This 8-bit address allows 256 input ports and 256 output ports. In this format, the 8-bit port address is carried on address bus A0–A7. No segment register is involved in computing the address, in contrast to the way data is accessed from memory.

In format (2), the port# is also the address of the port, except that it can be from 0000 to FFFFH, allowing up to 65,536 input and 65,536 output ports. In this case, the 16-bit port address is carried on address bus A0–A15, and no segment register (DS) is involved. This is the way Intel Corporation expanded the number of ports from 256 to 65,536 while maintaining compatibility with the earlier 8085 microprocessors. The use of a register as a pointer for the port address has an advantage in that the port address can be changed very easily, especially in cases of dynamic compilations where the port address can be passed to DX.

How to use I/O instructions

I/O instructions are widely used in programming peripheral devices such as printers, hard disks, and keyboards. The port address can be either 8-bit or 16-bit. For an 8-bit port address, we can use the immediate addressing mode. The following program sends a byte of data to a fixed port address of 43H.

```
        MOV    AL,36H          ;AL=36H
        OUT    43H,AL          ;send value 36H to port address 43H
```

The 8-bit address used in immediate addressing mode limits the number of ports to 256 for input plus 256 for output. To have a larger number of ports we must use the 16-bit port address instruction.

To use the 16-bit port address, register indirect addressing mode must be used. The register used for this purpose is DX. The following program sends values 55H and AAH to I/O port address 300H (a 16-bit port address). In other words, the program below toggles the bits of port address 300H continuously.

```
BACK:  MOV    DX,300H         ;DX = port address 300H
       MOV    AL,55H
       OUT    DX,AL           ;toggle the bits
       MOV    AL,0AAH
       OUT    DX,AL           ;toggle the bits
       JMP    BACK
```

Notice that we can only use register DX for 16-bit I/O addresses; no other register can be used for this purpose. Also notice the use of register AL for 8-bit data. For example, the following code transfers the contents of register BL to port address 378H.

```
        MOV DX,378H            ;DX=378 the port address
        MOV AL,BL              ;load data into accumulator
        OUT DX,AL              ;write contents of AL to port
                               ;whose address is in DX
```

To bring into the CPU a byte of data from an external device (external to the CPU) we use the IN instruction. Example 11-1 shows decision making based on the data that was input.

Just like the OUT instruction, the IN instruction uses the DX register to hold the address and AL to hold the arrived 8-bit data. In other words, DX holds the 16-bit port address while AL receives the 8-bit data brought in from an external port. The following program gets data from port address 300H and sends it to port address 302H.

```
        MOV    DX,300H         ;load port address
        IN     AL,DX           ;bring in data
        MOV    DX,302H
        OUT    DX,AL           ;send it out
```

Example 11-1

In a given 8088-based system, port address 22H is an input port for monitoring the temperature. Write Assembly language instructions to monitor that port continuously for the temperature of 100 degrees. If it reaches 100, then BH should contain 'Y'.

Solution:

```
BACK:          IN     AL,22H   ;get the temperature from port # 22H
               CMP    AL,100         ;is temp = 100?
               JNZ    BACK           ;if not, keep monitoring
               MOV    BH,'Y          ;temp = 100, load 'Y' into BH
```

Review Questions

1. In the x86 system, if we use only the 8-bit address bus A0–A7 for port addresses, what is the maximum number of (a) input, and (b) output ports?
2. The x86 can have a maximum of how many I/O ports?

3. What does the instruction "OUT 24H,AL" do?
4. Write Assembly language instructions to accept data input from port 300H and send it out to port 304H.
5. Write Assembly language instructions to place the status of port 60H in CH.

SECTION 11.2: I/O ADDRESS DECODING AND DESIGN

In this section we show the design of simple I/O ports using TTL logic gates 74LS373 and 74LS244. For the purpose of clarity we use simple logic gates such as AND and inverter gates for decoders. It may be helpful to review the address decoding section for memory interfacing in the preceding chapter before you embark on this section. The concept of address bus decoding for I/O instructions is exactly the same as for memory. The following are the steps:

1. The control signals IOR and IOW are used along with the decoder.
2. For an 8-bit port address, A0–A7 is decoded.
3. If the port address is 16-bit (using DX), A0–A15 is decoded.

Using the 74LS373 in an output port design

In every computer, whenever data is sent out by the CPU via the data bus, the data must be latched by the receiving device. While memories have an internal latch to grab the data, a latching system must be designed for simple I/O ports. The 74LS373 can be used for this purpose. Notice in Figure 11-1 that in order to make the 74LS373 work as a latch, the OC pin must be grounded. For an output latch, it is common to AND the output of the address decoder with the control signal IOW to provide the latching action as shown in Figure 11-2. See Example 11-2.

IN port design using the 74LS244

Likewise, when data is coming in by way of a data bus, it must come in through a three-state buffer. This is referred to as *tristated*, which comes from the term tri-state buffer ("tri-state" is a registered trademark of National Semiconductor Corp.).

As was the case for memory chips, such a tri-state buffer is internal and therefore invisible. For the simple input ports we use the 74LS244 chip. See Figure 11-4 for the internal circuitry of the 74LS244. In Figure 11-4, notice that since 1G and 2G each control only 4 bits of the 74LS244, they both must be activated for the 8-bit input. Examine Figure 11-5 to see the use of the 74LS244 as an entry port to the system data bus. Notice in Figures 11-5 and 11-6 how the address decoder and the IOR control signal together activate the tri-state input.

The 74LS244 not only plays the role of buffer, but also provides the incoming data with sufficient driving capability to travel all the way to the CPU. Indeed, the 74LS244 chip is widely used for buffering and providing high driving capability for unidirectional buses. The 74LS245 is used for bidirectional buses, as seen in Chapter 9.

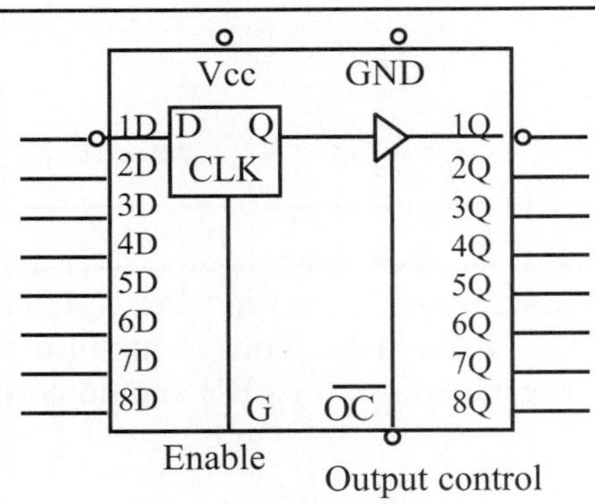

Function Table

Output Control	Enable G	D	Output
L	H	H	H
L	H	L	L
L	L	X	Q0
H	X	X	Z

Figure 11-1. 74LS373 D Latch
(Reprinted by permission of Texas Instruments, Copyright Texas Instruments, 1988)

Memory-mapped I/O

Communicating with I/O devices using IN and OUT instructions is referred to as *peripheral I/O*.

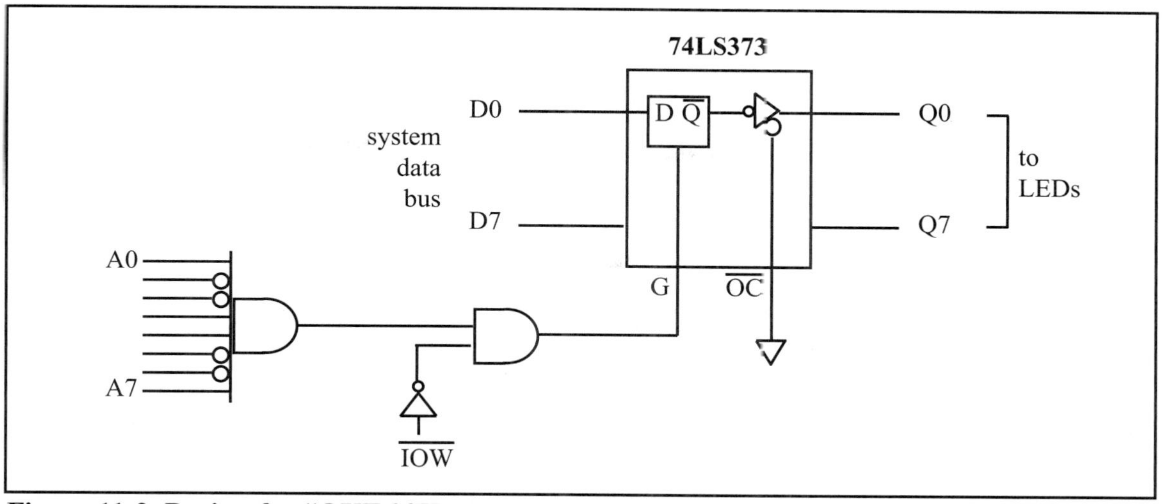

Figure 11-2. Design for "OUT 99H,AL"

Example 11-2

Show the design of an output port with an I/O address of 31FH using the 74LS373.

Solution:

31F9H is decoded, then ANDed with IOW to activate the G pin of the 74LS373 latch. This is shown in Figure 11-3.

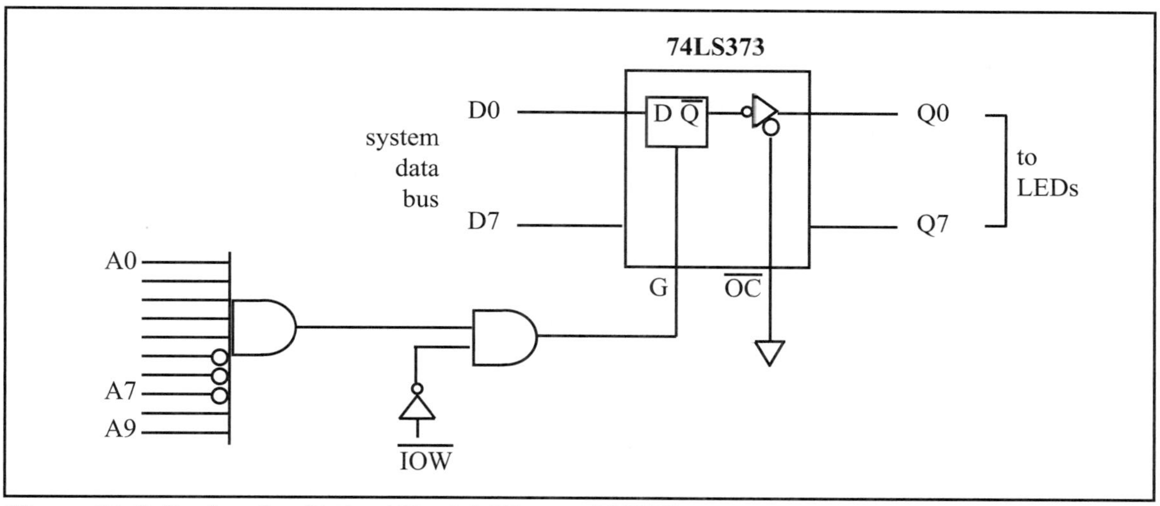

Figure 11-3. Design for Output Port Address of 31FH

Some designers also refer to it as *isolated I/O*. However, there are many microprocessors, such as the new RISC processors, that do not have IN and OUT instructions. In such cases, these microprocessors use what is called *memory-mapped I/O*. In memory-mapped I/O, a memory location is assigned to be an input or output port. The following are the differences between peripheral I/O and memory-mapped I/O in the x86 PC.

1. In memory-mapped I/O, we must use instructions that access memory locations to access the I/O ports instead of IN and OUT instructions. For example, an instruction such as "MOV AL,[2000]" will access an input port of memory address 2000 and "MOV [2010],AL" will access the output port.
2. In memory-mapped I/O, the entire 20-bit address, A0–A19, must be decoded. This is in contrast to peripheral I/O, in which only A0–A15 are decoded. Furthermore, since the 20-bit address involves both the segment and an offset, the DS register must be loaded before memory-mapped I/O is accessed. For example, if physical memory

address 35000H is used for the input port, the following instructions can be used to access the port.

```
MOV AX,3000H ;load the segment value
MOV DS,AX
MOV AL,[ 5000] ;get a byte from loc. 35000H
```

Physical address 35000H is generated by shifting left DS one hex digit and adding it to offset address 5000 (30000 + 5000 = 35000H). Since all 20-bit addresses are decoded, the decoding circuitry for memory-mapped I/O is more expensive.

3. In memory-mapped I/O circuit interfacing, control signals MEMR and MEMW are used. This is in contrast to peripheral I/O, in which IOR and IOW are used.
4. In peripheral I/O we are limited to 65,536 input ports and 65,536 output ports, whereas in memory I/O the number of ports can be as high as 2^{20} (1,048,576). Of course, that many ports are never needed.
5. In memory-mapped I/O, one can perform arithmetic and logic operations on I/O data directly without first moving them into the accumulator. In memory-mapped I/O, data can be transferred into any register, rather than into the accumulator.
6. One major and severe disadvantage of memory-mapped I/O is that it uses memory address space, which could lead to memory space fragmentation.

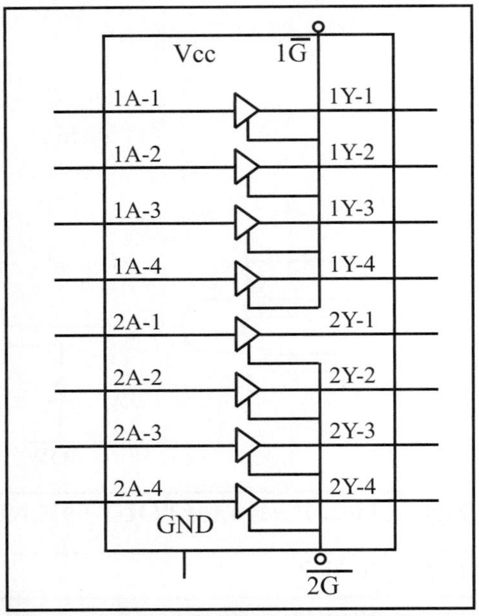

Figure 11-4. 74LS244 Octal Buffer
(Reprinted by permission of Texas Instruments, Copyright Texas Instruments, 1988)

Example 11-3

Show the design of "IN AL,9FH" using the 74LS244 as a tri-state buffer.

Solution:

9FH is decoded, then ANDed with IOR. To activate OC of the 74LS244, it must be inverted since OC is an active-low pin. This is shown in Figure 11-5.

Review Questions

1. Designers use a ____________ (latch, tri-state buffer) for output and a ___________ (latch, tri-state buffer) for input.
2. Why do we use latches in I/O design?
3. Why is the 74LS373 called the transparent latch?
4. To use the 74LS373 as a latch, OC must be set to ______ permanently.
5. True or false. To access the maximum number of ports in the x86, we must decode addresses A0–A15.
6. In memory-mapped I/O, which signal is used to select the (a) output, and (b) input devices?

SECTION 11.3: I/O ADDRESS MAP OF x86 PCs

Designers of the original IBM PC decided to make full use of I/O instructions. This led to assignment of different port addresses to various peripherals such as LPT and COM ports, and other chips and devices. The list of the designated I/O port addresses is referred to as the *I/O map*. Table 11-1 shows the I/O map for the x86 PC. A much more detailed I/O map of the x86 PC see Appendix E. Any system that needs to be compatible with the x86 IBM PC must follow the I/O map of Table 11-1. For example, the map shows that we can use I/O address 300–31F for a prototype card.

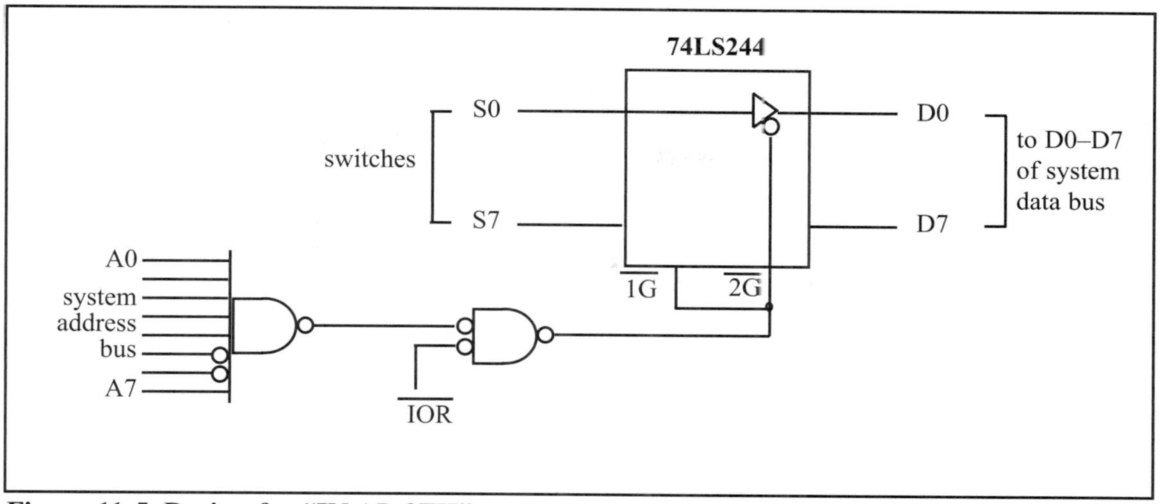

Figure 11-5. Design for "IN AL,9FH"

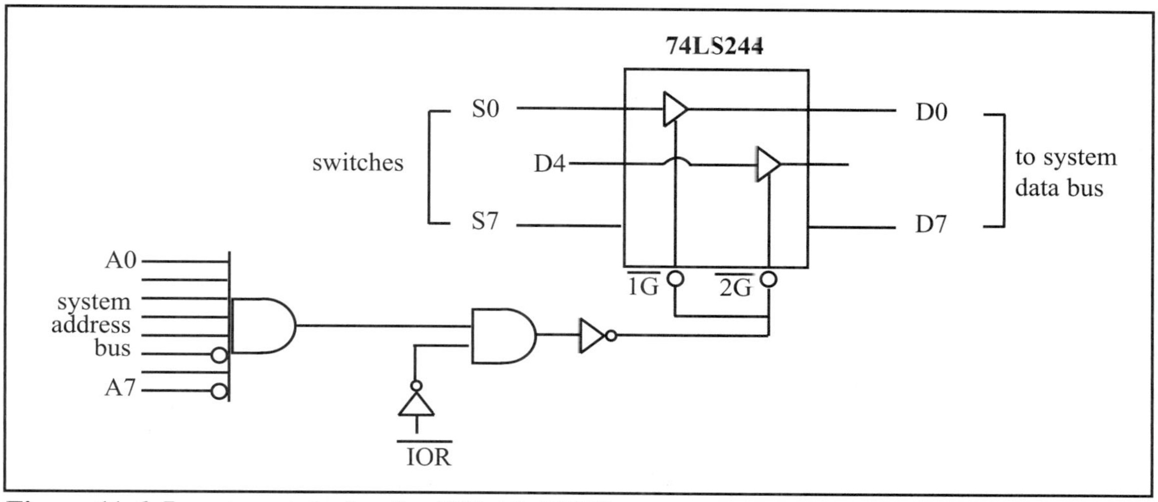

Figure 11-6. Input port design for "IN AL,5FH"

Absolute vs. linear select address decoding

In decoding addresses, either all of them or a selected number of them are decoded. If all the address lines are decoded, it is called *absolute decoding*. If only selected address pins are used for decoding, it is called *linear select decoding*. Linear select is cheaper, since the less input there is, the fewer the gates needed for decoding. The disadvantage is that it creates what are called *aliases*, the same port with multiple addresses. In cases where linear select is used, we must document port addresses in the I/O map thoroughly. In the first IBM PC, linear select decoding was used and that resulted in large numbers of address aliases, as we will see in future chapters. If you see a large gap in the I/O address map of the x86 PC, it is due to the address aliases of the original PC.

Prototype addresses 300–31FH in x86 PC

In the x86 PC, the address range 300H–31FH is set aside for prototype cards to be plugged into the expansion slot. These prototype cards can be data acquisition boards used to monitor analog signals such as temperature, pressure, and so on. Interface cards using the prototype address space use the following signals on the 62-pin section of the ISA expansion slot:

1. IOR and IOW. Both are active low.
2. AEN signal: AEN = 0 when the CPU is using the bus.
3. A0–A9 for address decoding.

Hex Range	Device
000–01F	DMA controller 1, 8237A-5
020–03F	Interrupt controller 1, 8259A, Master
040–05F	Timer, 8254-2
060–06F	8042 (keyboard)
070–07F	Real-time clock, NMI mask
080–09F	DMA page register, 74LS612
0A0–0BF	Interrupt controller 2, 8237A-5
0C0–0DF	DMA controller 2, 8237A-5
0F0	Clear math coprocessor busy
0F1	Reset math coprocessor
0F8–0FF	Math coprocessor
1F0–1F8	Fixed disk
200–207	Game I/O
20C–20D	Reserved
21F	Reserved
278–27F	Parallel printer port 2
2B0–2DF	Alternate enhanced graphics adapter
2E1	GPIB (adapter 0)
2E2 & 2E3	Data acquisition (adapter 0)
2F8–2FF	Serial port 2
300–31F	Prototype card
360–363	PC network (low address)
364–367	Reserved
368–36B	PC network (high address)
36C–36F	Reserved
378–37F	Parallel printer port 1
380–38F	SDLC, bisynchronous 2
390–393	Cluster
3A0–3AF	Bisynchronous 1
3B0–3BF	Monochrome display and printer adapter
3C0–3CF	Enhanced graphics adapter
3D0–3DF	Color/graphics monitor adapter
3F0–3F7	Disk controller
3F8–3FF	Serial port 1
6E2 & 6E3	Data acquisition (adapter 1)
790–793	Cluster (adapter 1)
AE2 & AE3	Data acquisition (adapter 2)
B90–B93	Cluster (adapter 2)
EE2 & EE3	Data acquisition (adapter 3)
1390–1393	Cluster (adapter 3)
22E1	GPIB (adapter 1)
2390–2393	Cluster (adapter 4)
42E1	GPIB (adapter 2)
62E1	GPIB (adapter 3)
82E1	GPIB (adapter 4)
A2E1	GPIB (adapter 5)
C2E1	GPIB (adapter 6)
E2E1	GPIB (adapter 7)

Table 11-1: I/O Map for the x86 PC

Use of simple logic gates as address decoders

Figure 11-7 shows the circuit design for a 74LS373 latch connected to port address 300H of an x86 PC via an ISA expansion slot. Notice the use of signals A0–A9 and AEN. AEN is low when the x86 microprocessor is in control of the buses. After all, it is the job of the CPU to control all the peripheral devices and not the DMA. In Figure 11-7, we are using simple logic gates such as NAND and inverter gates for the I/O address decoder. These can be replaced with the 74LS138 chip because the 74LS138 is a group of NAND gates in a single chip.

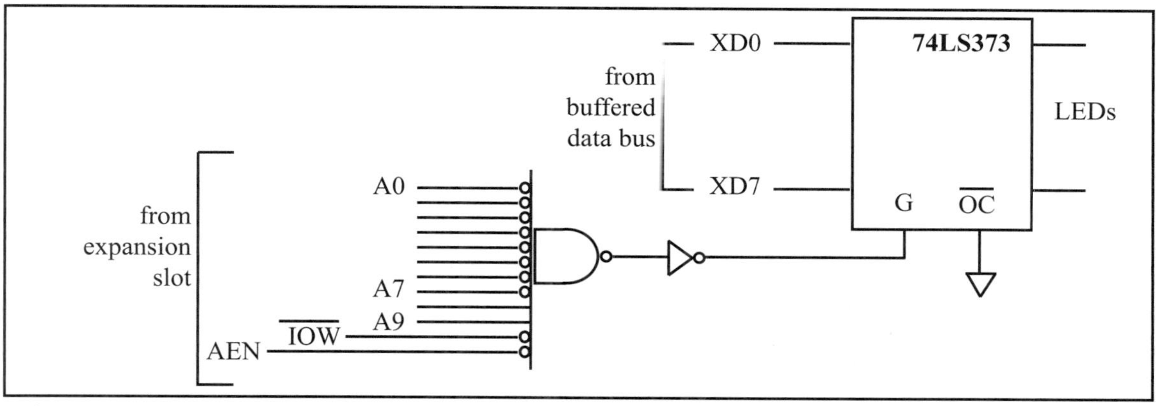

Figure 11-7. Using Simple Logic Gate for I/O Address Decoder (I/O Address 300H)

Use of 74LS138 as decoder

In current system board design, CPLD (complex programmable logic device) chips are used for supporting logics such as decoders. In the absence of CPLD, one could use NANDs, inverters, and 74LS138 chips for decoders as we saw in the preceding chapter for memory address decoding. The same principle applies to I/O address decoding. Figure 11-8 shows the 74LS138. To see an example of the use of a 74LS138 for an I/O address decoder, examine Figure 11-9. Notice how each Y output can control a single device. Figure 11-9 shows the address decoding for an input port located at address 304H. The Y4 output, together with the IOR signal, controls the 74LS244 input buffer. Alternatively, Y0 along with the IOW control signal could be used to control a 74LS373 latch. In other words, each Y output controls a single I/O device. Contrast that with Figure 11-7. The 74LS138 is much more efficient than the use of simple logic gates as decoders.

Block Diagram

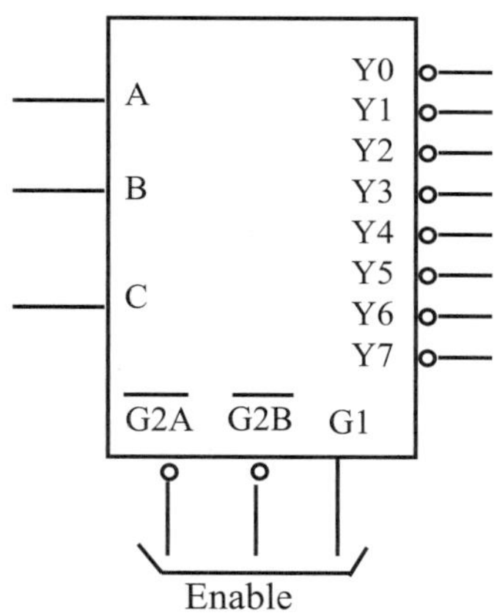

Function Table

Inputs: Enable G1	G2	Select C	B	A	Outputs Y0	Y1	Y2	Y3	Y4	Y5	Y6	Y7
X	H	X	X	X	H	H	H	H	H	H	H	H
L	X	X	X	X	H	H	H	H	H	H	H	H
H	L	L	L	L	L	H	H	H	H	H	H	H
H	L	L	L	H	H	L	H	H	H	H	H	H
H	L	L	H	L	H	H	L	H	H	H	H	H
H	L	L	H	H	H	H	H	L	H	H	H	H
H	L	H	L	L	H	H	H	H	L	H	H	H
H	L	H	L	H	H	H	H	H	H	L	H	H
H	L	H	H	L	H	H	H	H	H	H	L	H
H	L	H	H	H	H	H	H	H	H	H	H	L

Figure 11-8. 74LS138 Decoder

IBM PC I/O address decoder

Figure 11-10 shows a 74LS138 chip used as an I/O address decoder in the original IBM PC. Notice that while A0 to A4 go to individual peripheral input addresses, A5, A6, and A7 are responsible for the output selection of outputs Y0 to Y7. In order to enable the 74LS138, pins A8, A9, and AEN all must be low. While A8 and A9 will directly affect the port address calculations, AEN is low only when the x86 is in control of the system bus.

Since all the peripherals are programmed by the x86, AEN is low during CPU activity. See Table 11-2. We will discuss each Y output of 74LS138 in Figure 11-8 in subsequent chapters.

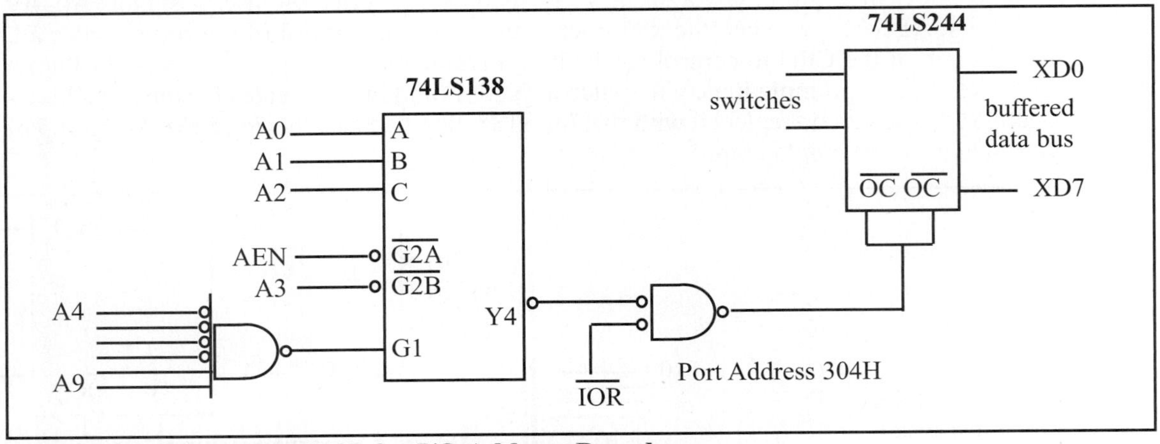

Figure 11-9. Using 74LS138 for I/O Address Decoder

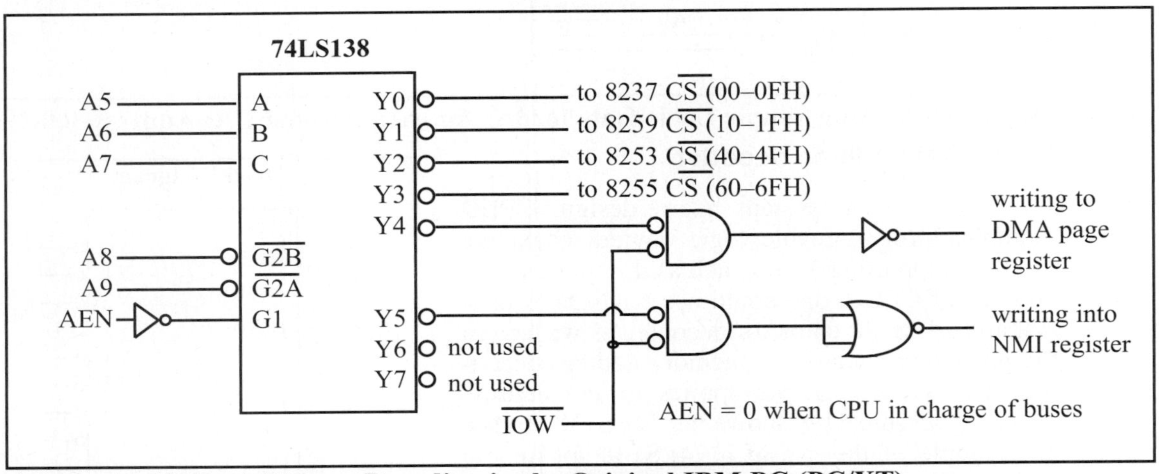

Figure 11-10. Port Address Decoding in the Original IBM PC (PC/XT)

Table 11-2: Port Addresses Decoding Table on the Original PC

G1	G2A	G2B	C B A	
AEN	A9	A8	A7 A6 A5 A4 A3 A2 A1 A0	
0	0	0	0 0 0 0 0 0 0 0	00 Lowest port address
0	0	0	1 1 1 1 1 1 1 1	FF Highest port address

Port 61H and time delay generation

Appendix E provides a detailed I/O map of the PC. In order to maintain compatibility with the IBM PC and run operating systems such as MS-DOS and Windows, the assignment of I/O port addresses must follow the standard set in Appendix E. Port 61H is a widely used port, the details of which are shown in Appendix E. We can use this port to generate a time delay, which will work in any PC with any type of processor from the 286 to the Pentium. I/O port 61H has eight bits (D0–D7). Bit D4 is of particular interest to us. In all 286 and higher PCs, bit D4 of port 61H changes its state every 15.085 microseconds (μs). In other words, it stays low for 15.085 μs and then changes to high and stays high for the same amount of time before it goes low again. This toggling of bit D4 goes on indefinitely as long as the PC is on. Chapter 13 provides more details on this topic.

The following program shows how to use port 61H to generate a delay of 1/2 second. In this program all the bits of port 310H are toggled with a 1/2 second delay in between.

```
;TOGGLING ALL BITS OF PORT 310H EVERY 0.5 SEC
            MOV    DX,310H
HERE:       MOV    AL,55H        ;toggle all bits
            OUT    DX,AL
            MOV    CX,33144      ;delay=33144x15.085 us=0.5 sec
            CALL   TDELAY
            MOV    AL,0AAH
            OUT    DX,AL
            MOV    CX,33144
            CALL   TDELAY
            JMP    HERE

;CX=COUNT OF 15.085 MICROSEC
TDELAY      PROC   NEAR
            PUSH   AX            ;save AX
W1:         IN     AL,61H
            AND    AL,00010000B
            CMP    AL,AH
            JE     W1            ;wait for 15.085 usec
            MOV    AH,AL
            LOOP   W1            ;another 15.085 usec
            POP    AX            ;restore AX
            RET
TDELAY      ENDP
```

In the above program, notice that when port 61H is read, all the bits are masked except D4. The program waits for D4 to change every 15.085 μs before it loops again.

Review Questions

1. What I/O address range is set aside for prototype cards?
2. In the x86 PC, give the status of the AEN signal when I/O ports are being addressed by the CPU.
3. In decoding addresses for I/O instructions, why do we need to include AEN, and what is the activation level?
4. Which bit of port 61H toggles every 15.085 μs?
5. Calculate the time delay if CX = 25,000 for Question 4.

SECTION 11.4: PROGRAMMING AND INTERFACING THE 8255

In this section we study the 8255 chip, a widely used I/O chip. The 8255 is a 40-pin DIP chip (see Figure 11-11). It has three separately accessible ports, A, B, and C, which can be programmed, hence the name PPI (programmable peripheral interface). Each port of the 8255 can be programmed to be input or output. They can also be changed dynamically, in contrast to the 74LS244 and 74LS373, which are hard-wired.

Port A (PA0–PA7)

This 8-bit port A can be programmed all as input or all as output.

Port B (PB0–PB7)

This 8-bit port B can be programmed all as input or all as output.

Port C (PC0–PC7)

This 8-bit port C can be all input or all output. It can also be split into two parts, CU (upper bits PC4–PC7) and CL (lower bits PC0–PC3). Each can be used for input or output. Any of PC0 to PC7 can be programmed individually.

RD and WR

These two active-low control signals are inputs to the 8255. If the 8255 is using peripheral I/O design, IOR and IOW of the system bus are connected to these two pins. If the port uses memory-mapped I/O, MEMR and MEMW activate them.

RESET

This is an active-high signal input into the 8255 used to clear the control register. When RESET is activated, all ports are initialized as input ports. This pin must be connected to the RESET output of the system bus or ground, making it inactive.

A0, A1, and CS

While CS (chip select) selects the entire chip, address pins A0 and A1 select the specific port within the 8255. These three pins are used to access ports A, B, C, or the control register, as shown in Table 11-3.

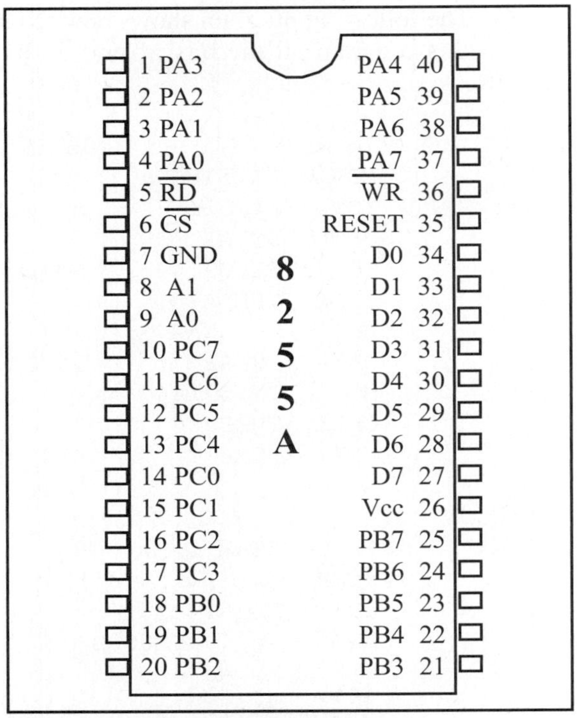

Figure 11-11. 8255 PPI Chip

(Reprinted by permission of Intel Corporation, Copyright Intel, 1983)

Table 11-3: 8255 Port Selection

CS	A1	A0	Selects
0	0	0	Port A
0	0	1	Port B
0	1	0	Port C
0	1	1	Control register
1	x	x	8255 is not selected

(Reprinted by permission of Intel Corporation, Copyright Intel Corp. 1983)

Mode selection of the 8255A

While ports A, B, and C are used for I/O data, it is the control register that must be programmed to select the operation mode of the three ports A, B, and C. The ports of the 8255 can be programmed in various modes, as shown in Figure 11-12. Mode 0, the simple I/O mode, is the most widely used mode and the only one we are concerned with. In this mode, any of the ports A, B, CL, and CU can be programmed as input or output. In this mode, all bits are out or all are in. In other words, there is no control of individual bits. Examples 11-4, 11-5, and 11-6 show programming of 8255 ports in simple I/O mode.

Although the ISA bus has been replaced with the PCI bus, the ISA bus specification lives on in PC104 form. See Chapter 26. For more information on PC104 see:

http://www.pc104.org

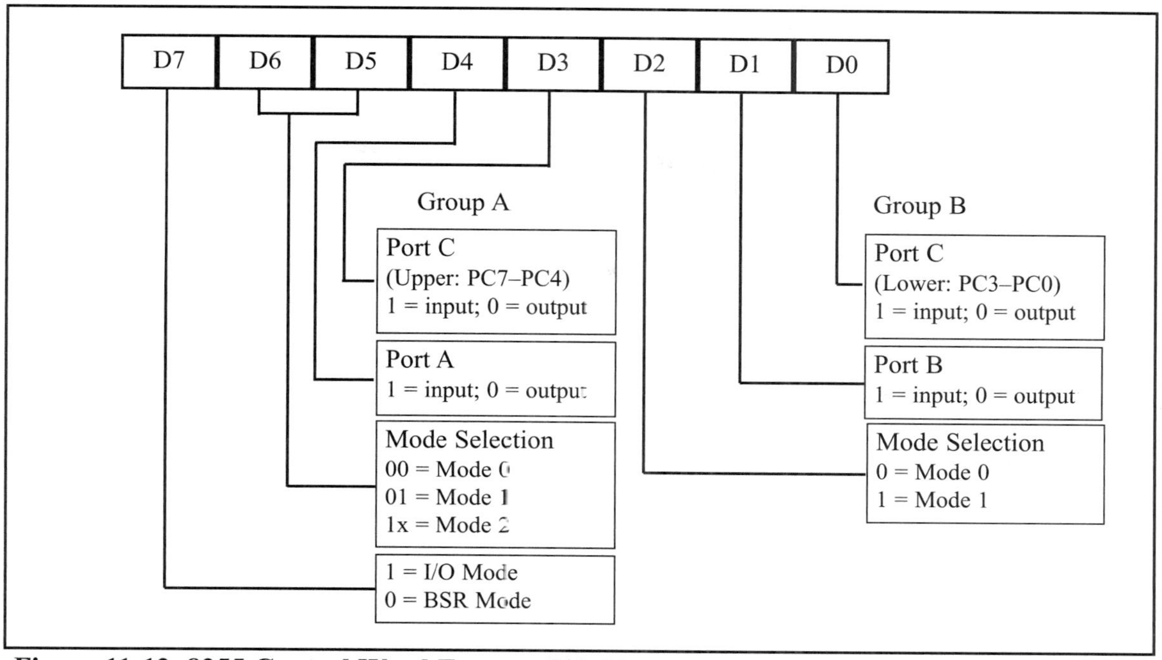

Figure 11-12. 8255 Control Word Format (I/O Mode)
(Reprinted by permission of Intel Corporation, Copyright Intel, 1983)

Example 11-4

(a) Find the control word if PA = out, PB = in, PC0–PC3 = in, and PC4–PC7 = out.
(b) Program the 8255 to get data from port A and send it to port B. In addition, data from PCL is sent out to the PCU.
Use port addresses of 300H–303H for the 8255 chip.

Solution:

(a) From Figure 11-12 we get the control word of 1000 0011 in binary or 83H.

(b) The code is as follows:

```
B8255C EQU   300H   ;Base address of 8255 chip
CNTL    EQU  83H    ;PA=out,PB=in,PCL=in,PCU=out
MOV    DX,B8255C+3;load control reg. address
                    ;(300H + 3 = 303H)
MOV    AL,CNTL      ;load control byte
OUT    DX,AL        ;send it to control register
MOV    DX,B8255C+1 ;load PB address
IN     AL,DX        ;get the data from PB
MOV    DX,B8255C    ;load PA address
OUT    DX,AL        ;send it to PA
MOV    DX,B8255C+2 ;load PC address
IN     AL,DX        ;get the bits from PCL
AND    AL,0FH       ;mask the upper bits
ROL    AL,1
ROL    AL,1         ;shift the bits
ROL    AL,1         ;to upper position
ROL    AL,1
OUT    DX,AL        ;send it to PCU
```

Example 11-5

The 8255 shown in Figure 11-13 is configured as follows: port A as input, B as output, and all the bits of port C as output.
(a) Find the port addresses assigned to A, B, C, and the control register.
(b) Find the control byte (word) for this configuration.
(c) Program the ports to input data from port A and send it to both ports B and C.

Solution:
(a) The port addresses are as follows:

$\overline{CS}$	A1	A0	Address	Port
11 0001 00	0	0	310H	Port A
11 0001 00	0	1	311H	Port B
11 0001 00	1	0	312H	Port C
11 0001 00	1	1	313H	Control register

(b) The control word is 90H, or 1001 0000.
(c) One version of the program is as follows:

```
        MOV     AL,90H          ;control byte PA=in, PB=out, PC=out
        MOV     DX,313H         ;load control reg address
        OUT     DX,AL           ;send it to control register
        MOV     DX,310H         ;load PA address
        IN      AL,DX           ;get the data from PA
        MOV     DX,311H         ;load PB address
        OUT     DX,AL           ;send it to PB
        MOV     DX,312H         ;load PC address
        OUT     DX,AL           ;and to PC
```

Using the EQU directive one can rewrite the above program as follows:

```
CNTLBYTE    EQU     90H     ;PA=in, PB=out, PC=out
PORTA       EQU     310H
PORTB       EQU     311H
PORTC       EQU     312H
CNTLREG     EQU     313H
            ....
            ...
            MOV     AL,CNTLBYTE
            MOV     DX,CNTLREG
            OUT     DX,AL
            MOV     DX,PORTA
            IN      AL,DX
            ;and so on.
```

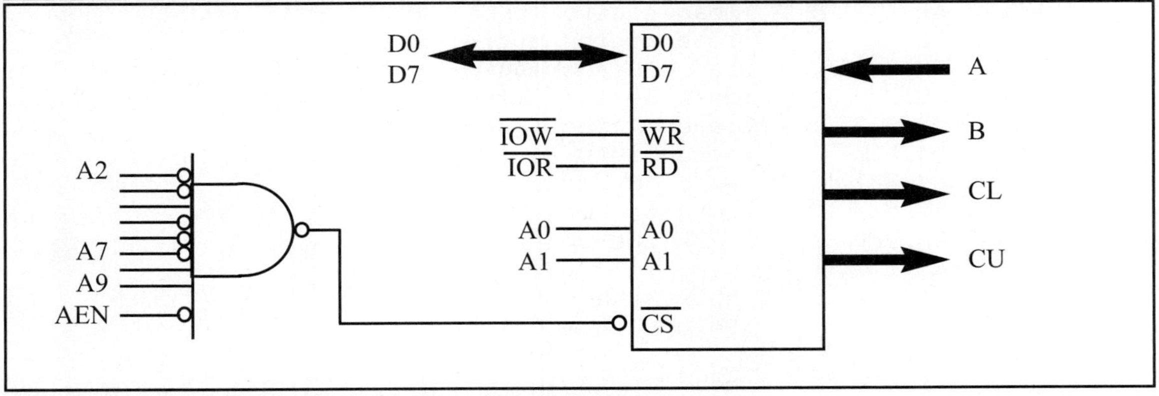

Figure 11-13. 8255 Configuration for Example 11-5

Example 11-6

Show the address decoding where port A of the 8255 has an I/O address of 300H, then write a program to toggle all bits of PA continuously with a 1/4 second delay. Use INT 16H to exit if there is a keypress.

Solution:

The address decoding for the 8255 is shown in Figure 11-14. The control word for all ports as output is 80H. The program below will toggle all bits of PA indefinitely with a delay in between. To prevent locking up the system, we press any key to exit to DOS.

```
            MOV     DX,303H         ;CONTROL REG ADDRESS
            MOV     AL,80H          ;ALL PORTS AS OUTPUT
            OUT     DX,AL
AGAIN:      MOV     DX,300H
            MOV     AL,55H
            OUT     DX,AL
            CALL    QSDELAY         ;1/4 SEC DELAY
            MOV     AL,0AAH         ;TOGGLE BIT
            OUT     DX,AL
            CALL    QSDELAY
            MOV     AH,01
            INT     16H             ;CHECK KEYPRESS
            JZ      AGAIN           ;PRESS ANY KEY TO EXIT
            MOV     AH,4CH
            INT     21H             ;EXIT

QSDELAY     PROC    NEAR
            MOV     CX,16572        ;16,572x15.085 usec=1/4 sec
                    PUSH  AX
W1:         IN      AL,61H
            AND     AL,00010000B
            CMP     AL,AH
            JE      W1
            MOV     AH,AL
            LOOP    W1
            POP     AX
            RET
QSDELAY     ENDP
```

Notice the use of INT 16H option AH = 01 where the keypress is checked. If there is no keypress, it will continue. We must do that to avoid locking up the x86 PC.

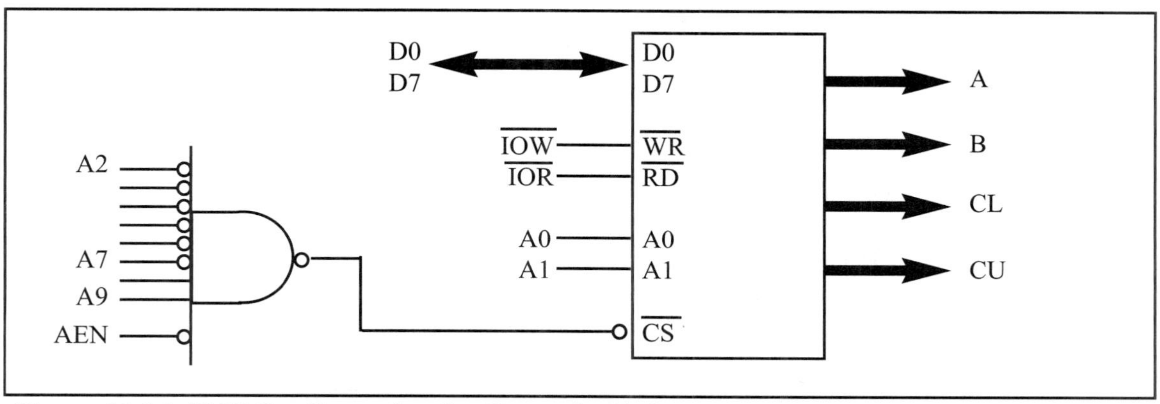

Figure 11-14. 8255 Configuration for Example 11-6

Buffering 300–31FH address range

When accessing the system bus via the expansion slot, we must make sure that the plug-in card does not interfere with the working of system buses on the motherboard. To do that we isolate (buffer) a range of I/O addresses using the 74LS245 chip. In buffering, the data bus is accessed only for a specific address range, and access by any address beyond the range is blocked. Figure 11-15 shows how the I/O address range 300H–31FH is buffered with the use of the 74LS245. Figure 11-16 shows another example of 8255 interfacing using the 74LS138 decoder. As shown in Figure 11-16, Y0 and Y1 are used for the 8255 and 8253, respectively. Table 11-4 shows the 74LS138 address assignment.

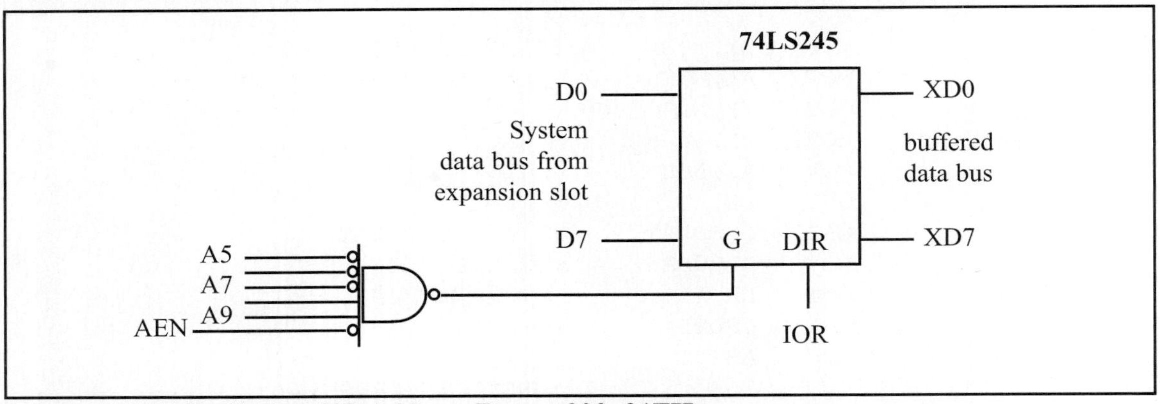

Figure 11-15. Buffering I/O Address Range 300–31FH

Table 11-4: Address Assignment for Figure 11-16

Selector	Address	Assignment
Y0	300–303	Used by 8255
Y1	304–307	Used by 8253
Y2	308–30B	Available
Y3	30C–30F	Available
Y4	310–313	Available

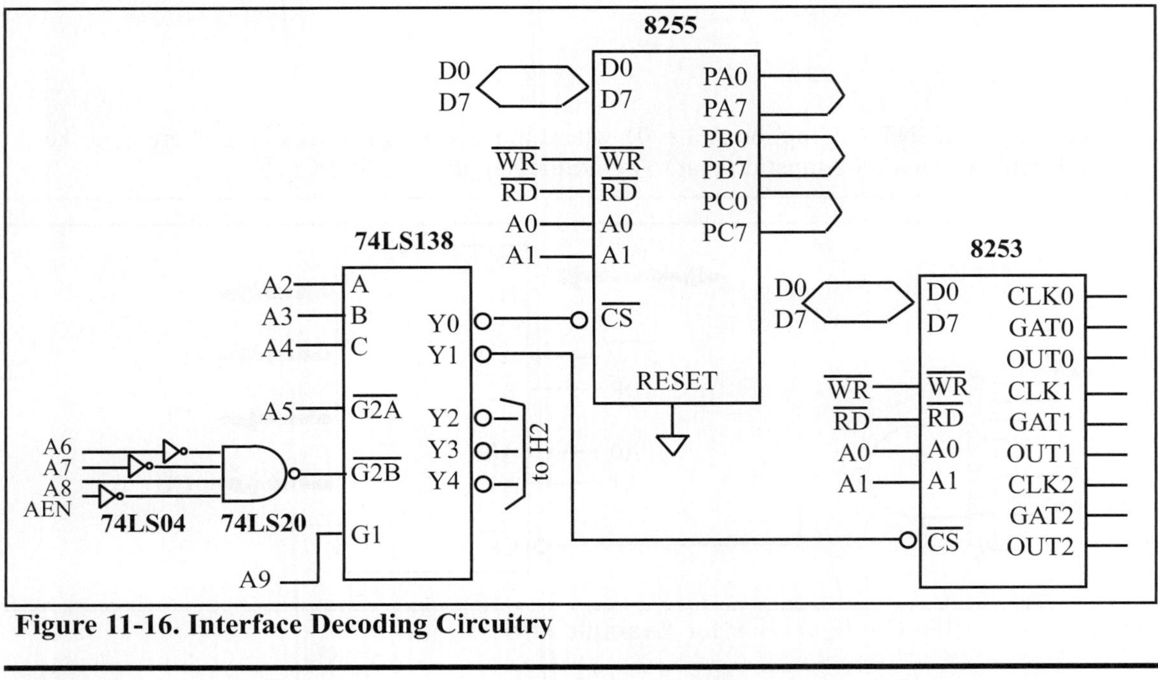

Figure 11-16. Interface Decoding Circuitry

Figure 11-17 shows the circuit for buffering all the buses. Notice the use of the 74LS244 to boost the address and control signals. This ensures the integrity of the signal transmitted to the plug-in board.

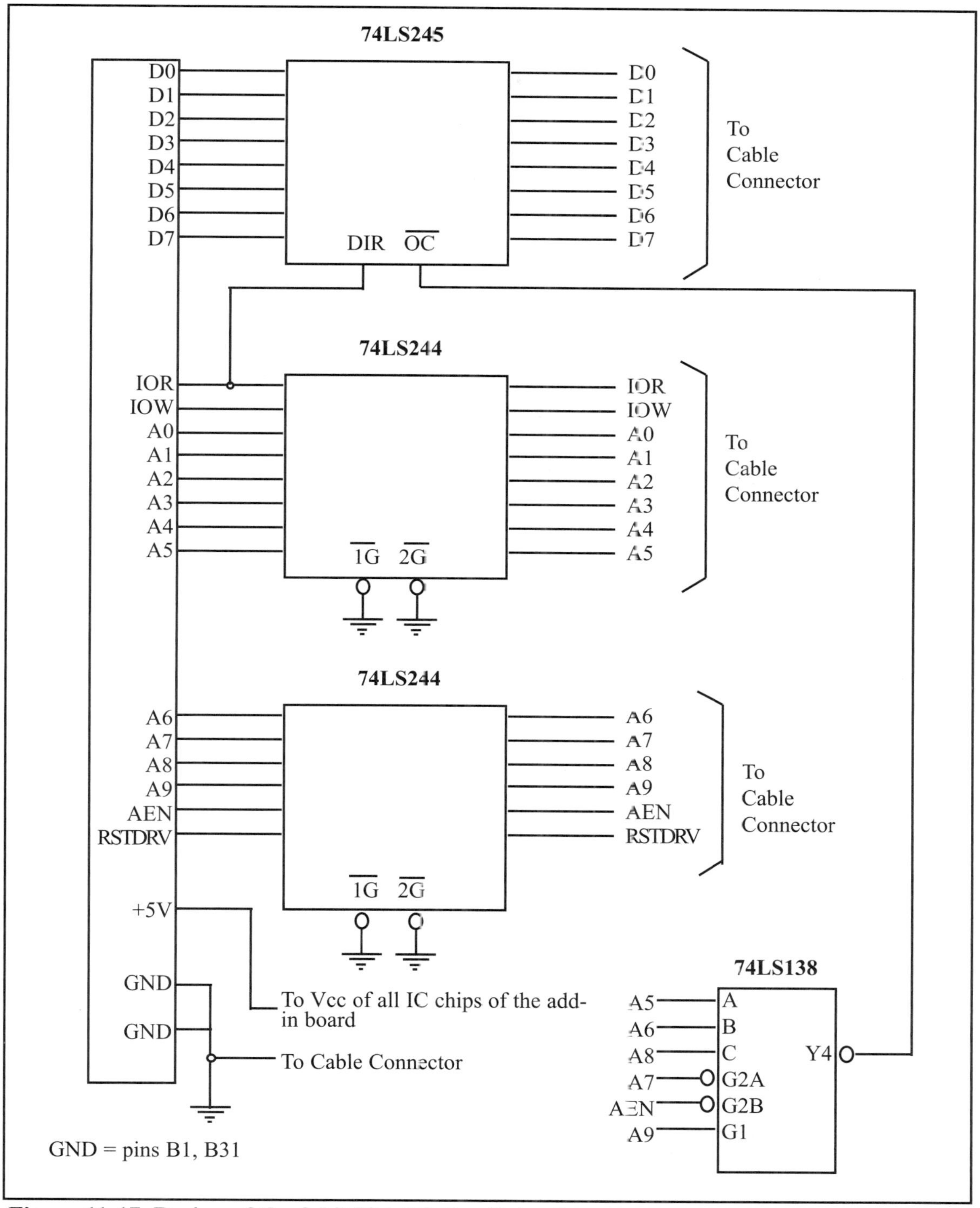

Figure 11-17. Design of the 8-bit ISA PC Bus Extender

Example 11-7 shows a test program to toggle the PA and PB bits. Notice that in order to avoid locking up the system, we use INT 16H to exit upon pressing any key. You can modify Example 11-7 to toggle all bits of PA, PB, and PC. Make sure to put a message on the x86 PC screen to prompt the user to exit by pressing any key.

Example 11-7

Write a program to toggle all bits of PA and PB of the 8255 chip on the PC Trainer. Put a 1/2 second delay in between "on" and "off" states. Use INT 16H to exit if there is a keypress.

Solution:

The program below toggles all bits of PA and PB indefinitely. Pressing any key exits the program.

```
            MOV     DX,303H        ;CONTROL REG ADDRESS
            MOV     AL,80H               ;ALL PORTS AS OUTPUT
            OUT     DX,AL
AGAIN:              MOV    DX,300H       ;PA ADDRESS
            MOV     AL,55H
            OUT     DX,AL
            INC     DX             ;PB ADDRESS
            OUT     DX,AL
            CALL    HSDELAY        ;1/2 SEC DELAY
            MOV     DX,300H        ;PA ADDRESS
            MOV     AL,0AAH
            OUT     DX,AL
            INC     DX             ;PB ADDRESS
            OUT     DX,AL
            CALL    HSDELAY        ;1/2 SEC DELAY
            MOV     AH,01
            INT     16H            ;CHECK KEYPRESS
            JZ      AGAIN                ;PRESS ANY KEY TO EXIT
            MOV     AH,4CH         ;
            INT     21H            ;EXIT

HSDELAY     PROC    NEAR
            MOV     CX,33144       ;33144x15.085 usec=1/2 sec
                    PUSH   AX
W1:         IN      AL,61H
            AND     AL,00010000B
            CMP     AL,AH
            JE      W1
            MOV     AH,AL
            LOOP    W1
            POP     AX
            RET
HSDELAY     ENDP
```

Notice the use of INT 16H option AH = 01 where the keypress is checked. If there is no keypress, it will continue.

Visual C/C++ I/O programming

Microsoft Visual C++ is a programming language widely used on the Windows platform. Since Visual C++ is an object-oriented language, it comes with many classes and objects to make programming easier and more efficient. Unfortunately, there is no object or class for directly accessing I/O ports in the full Windows version of Visual C++. The reason for that is that Microsoft wants to make sure the x86 system programming is under full control of the operating system. This precludes any hacking into the system hardware. This applies to Windows NT, 2000, XP, and higher. In other words, none of the system INT instructions such as INT 21H and I/O operations that we have discussed in previous chapters are applicable in Windows XP and its subsequent versions. To access the I/O and other hardware features of the x86 PC in the XP environment you must use the Windows Platform SDK provided by Microsoft. The situation is different in the Windows 9x (95 and 98) environment. While INT 21H and other system interrupt instructions are blocked in Windows 9x, direct I/O addressing is available. To access I/O directly in Windows 9x, you must program Visual C++ in console mode. The instruction syntax for I/O operations is shown in Table 11-5. Notice the use of the underscore character (_) in both the _outp and _inp instructions. It must also noted that while the x86 Assembly language makes a distinction between the 8-bit and 16-bit I/O addresses by using the DX register, there is no such distinction in C programming, as shown in Table 11-5. In other words, for the instruction "outp(port#,byte)" the port# can take any address value between 0000 and FFFFH. See Examples 11-8 and 11-9.

Table 11-5: I/O Operations in Microsoft Visual C++ (for Windows 98)

x86 Assembly	Visual C++
OUT port#,AL	_outp(port#,byte)
OUT DX,AL	_outp(port#,byte)
IN AL,port#	_inp(port#)
IN AL,DX	_inp(port#)

Example 11-8

Write a Visual C++ program for Windows 98 to toggle all bits of PA and PB of the 8255 chip. Use the kbhit function to exit if there is a keypress.

Solution:

```
//Tested by Dan Bent
#include<conio.h>
#include<stdio.h>
#include<iostream.h>
#include<iomanip.h>
#include<windows.h>
void main()
  {
  cout<<setiosflags(ios::unitbuf);  // clear screen buffer
  cout<<"This program toggles the bits for Port A and Port B.";
  _outp(0x303,0x80);                //MAKE PA,PB of 8255 ALL OUTPUT
  do
    {
    _outp(0x300,0x55);              //SEND 55H TO PORT A
    _outp(0x301,0x55);              //SEND 55H TO PORT B
    _sleep(500);                    //DELAY of 500 msec.
    _outp(0x300,0xAA);              //NOW SEND AAH TO PA, and PB
    _outp(0x301,0xAA);
    _sleep(500);
    }
    while(!kbhit());
  }
```

Example 11-9

Write a Visual C++ program for Windows 98 to get a byte of data from PA and send it to both PB and PC of the 8255 chip in PC Trainer.

Solution:

```
#include<conio.h>
#include<stdio.h>
#include<iostream.h>
#include<iomanip.h>
#include<windows.h>
#include<process.h>
//Tested by Dan Bent
void main()
  {
     unsigned char mybyte;
      cout<<setiosflags(ios::unitbuf);// clear screen buffer
     system("CLS");
     _outp(0x303,0x90);      //PA=in, PB=out, PC=out
     _sleep(5);              //wait 5 milliseconds
     mybyte=_inp(0x300);     //get byte from PA
     _outp(0x301,mybyte);    //send to PB
     _sleep(5);
     _outp(0x302,mybyte);    //send to Port C
     _sleep(5);
     cout<<mybyte;           //send to PC screen also
     cout<<"\n\n";
  }
```

I/O programming in Linux C/C++

Linux is a popular operating system for the x86 PC. You can get a copy of the latest C/C++ compiler from http://gcc.gnu.org. Table 11-6 provides the C/C++ syntax for I/O programming in the Linux OS environment.

Table 11-6: Input/Output Operations in Linux

x86 Assembly	Linux C/C++
OUT port#,AL	outb(byte,port#)
OUT DX,AL	outb(byte,port#)
IN AL,port#	inb(port#)
IN AL,DX	inb(port#)

Compiling and running Linux C/C++ programs with I/O functions

To compile the I/O programs of Examples 11-10 and 11-11, the following points must be noted.

1. To compile with a keypress loop, you must link to library ncurses as follows:
   > gcc -lncurses toggle.c -o toggle
2. To run the program, you must either be root or root must change permissions on executable for hardware port access.
   Example: (as root or superuser)
   > chown root toggle
   > chmod 4750 toggle

   Now toggle can be executed by users other than root. More information on this topic can be found at www.microdigitaled.com.

Example 11-10

Write a C/C++ program for a PC with the Linux OS to toggle all bits of PA and PB of the 8255 chip on the PC Trainer. Put a 500 ms delay between the "on" and "off" states. Pressing any key should exit the program.

Solution:

```
//    This program demonstrates low level I/O
//    using C language on a Linux based system.
//    Tested by Nathan Noel       //
#include <stdio.h>     // for printf()
#include <unistd.h>    // for usleep()
#include <sys/io.h>    // for outb() and inb()
#include <ncurses.h>  // for console i/o functions
int main ()
  {
  int n=0;             // temp char variable
  int delay=5 e5;      // sleep delay variable

  ioperm(0x300,4,0x300);   // get port permission
  outb(0x80,0x303);    // send control word

  //----- begin ncurses setup ----------
  //--- (needed for console i/o) -------

  initscr();           // initialize screen for ncurses
  cbreak();            // do not wait for carriage return
  noecho();            // do not echo input character
  halfdelay(1);        // only wait for 1ms for input
                       // from keyboard
  //-----  end ncurses setup  ----------

  do                   // main toggle loop
    {
    printf("0x55 \n\r");   // display status to screen
    refresh();         // refresh() to update console
    outb(0x55,0x300); // send 0x55 to PortA (01010101B)
    outb(0x55,0x301); // send 0x55 to PortB (01010101B)
    usleep(delay);     // wait for 500ms (5 e5 microseconds)
    printf("0xAA \n\r");   // display status to screen
    refresh();         // refresh() to update console
    outb(0xaa,0x300); // send 0xAA to PortA (10101010B)
    outb(0xaa,0x301); // send 0xAA to PortB (10101010B)
    usleep(delay);     // wait for 500ms
                       // get input from keyboard
    n=getch();         // if no keypress in 1ms, n=0
                       // due to halfdelay()
   }
while(n<=0);           // test for keypress
                       // if keypress, exit program
  endwin();            // close program console for ncurses
  return 0;            // exit program
  }
```

Example 11-11

Write a C/C++ program for a PC with the Linux OS to get a byte of data from port A and send it to both port B and port C of the 8255 in the PC Trainer.

Solution:

```
//    This program gets data from Port A and
//    sends a copy to both Port B and Port C.
//    Tested by: Nathan Noel -- 2/10/2002
//-----------------------------------------------
#include <stdio.h>
#include <unistd.h>
#include <sys/io.h>
#include <ncurses.h>

int main ()
  {
  int n=0;              // temp variable
  int i=0;              // temp variable

  ioperm(0x300,4,0x300);// get permission to use ports
  outb(0x90,0x303);     // send control word for
                // PortA=input, PortB=output, PortC=output

  initscr();            // initialize screen for ncurses
  cbreak();             // do not wait for carriage return
  noecho();             // do not echo input character
  halfdelay(1);         // only wait for 1ms for input

  do                    // main toggle loop
    {
    i=inb(0x300);       // get data from PortA
    usleep(1e5);        // sleep for 100ms

    outb(i,0x301);      // send data to PortB
    outb(i,0x302);      // send data to PortC

    n=getch();          // get input from keyboard
                        // if no keypress in 1ms, n=0
    }while(n<=0);       // test for keypress
                        // if keypress, exit program

  endwin();             // close program window
  return (0);           // exit program
  }
```

Review Questions

1. Find addresses for all 8255 ports if A7–A2 = 111101 is used to activate CS.
2. Find the control word for an 8255 in mode 0 (simple I/O) if all the ports are configured as output ports.
3. Find the control word for an 8255 in mode 0 (simple I/O) if all the ports are configured as input ports.
4. Program an 8255 with the following specifications: All ports are output ports. Write 55H to the ports. After a delay, switch them all to AAH.
5. How are ports configured after the control register is loaded with 89H?

PROBLEMS

SECTION 11.1: 8088 INPUT/OUTPUT INSTRUCTIONS

1. True or false. While memory contains both code and data, ports contain data only.
2. In the instruction "OUT 99H,AL", the port address is:
   (a) 8 bits (b) 16 bits (c) both (a) and (b) (d) none of the above
3. In the instruction "OUT DX,AL", the port address is:
   (a) 8 bits (b) 16 bits (c) either (a) or (b) (d) none of the above
4. True or false. In the instruction "IN AL,78H", register AL is the destination.
5. Explain what the instruction "IN AL,5FH" does.
6. In the instruction "OUT DX,AL", assume that AL = 3BH and DX = 300H. Explain what the instruction does.

SECTION 11.2: I/O ADDRESS DECODING AND DESIGN

7. In the execution of an OUT instruction, which control signal is activated?
8. In the execution of an IN instruction, which control signal is activated?
9. True or false. Segment register DS is used to generate a port's physical address.
10. True or false. In "OUT 65H,AL", only address pins A0–A7 are used by the 8088 to provide the address.
11. True or false. An input port is distinguished from an output port by the port address assigned to it.
12. True or false. An input port is distinguished from an output port by the IOR and IOW control signals.
13. A _______________ (latch, tri-state buffer) is used in the design of input ports.
14. A _______________ (latch, tri-state buffer) is used in the design of output ports.
15. ________ (IOR, IOW) is used in the design of input ports.
16. ________ (IOR, IOW) is used in the design of output ports.
17. Draw a logical design for "OUT 16H,AL" using AND and inverter gates in addition to a 74LS373.
18. Draw a logical design for "IN AL,81H" using AND and inverter gates in addition to a 74LS244.
19. Show one implementation of Problem 17 using NAND and inverter gates. Use as many as you need.
20. Show one implementation of Problem 18 using NAND and inverter gates. Use as many as you need.
21. True or false. Memory-mapped I/O uses control signals MEMR and MEMW.
22. True or false. In memory-mapped I/O, one can perform logical and arithmetic operations on the data without moving it into the accumulator first.
23. Show the logical design of "MOV [0100],AL" for memory-mapped I/O using AND and inverter gates and a 74LS373 latch. Assume that DS = B800H.
24. Why is memory-mapped I/O decoding more expensive?

SECTION 11.3: I/O ADDRESS MAP OF x86 PCs

25. Show the circuit connection to the PC bus for the following instructions. Use simple logic gates 74LS373 and 74LS244.
(a) OUT 309H,AL (b) IN AL,30CH
26. Repeat Problem 25 using a 74LS138 for the decoder.
27. Show the design of an 8255 connection to the PC bus using simple logic gates. Assume port address 304H as the base port address for the 8255.
28. Show the design of an 8255 connection to the PC bus using a 74LS138. Assume base address 31CH.
29. In the IBM PC, how many port addresses are available in the address space commonly referred to as prototype?
30. Which one is more economical, linear address select or absolute address decoding?
31. Explain address aliasing.
32. Which one creates aliases, the linear address select or absolute address decoding?
33. True or false. To design an IBM PC compatible system, one must follow the I/O map of the PC.
34. In accessing ports in the PC, why must the AEN = 0 signal be used in decoding?
35. What port address is used for a fixed delay in the x86 PC?
36. In x86 PC, the ___ bit of port address ___H toggles every ___ microseconds.
37. In Problem 36, to get a 1/4 second delay, we need to load the CX with what value?
38. In Problem 36, calculate the time delay if CX = 38,000.

SECTION 11.4: PROGRAMMING AND INTERFACING THE 8255

39. How many pins of the 8255 are used for ports, and how are they categorized?
40. What is the function of data pins D0–D7 in the 8255?
41. What is the advantage of using the 8255 over the 74LS373 and 74LS244?
42. True or false. All three ports, A, B, and C, can be programmed for simple I/O.
43. True or false. In simple I/O programming of port A of the 8255, we can use PA0–PA3 for output and PA4–PA7 for an input port.
44. Show the decoding circuitry for the 8255 if we want port A to have address 68H. Use NAND and inverter gates.
45. Which of the following port addresses cannot be assigned to port A of the 8255, and why?
(a) 32H (b) 45H (c) 89H (d) BAH
46. If 91H is the control word, indicate which port is input and which is output.
47. Find the control word if PA = input, PB = input, and PC0–PC7 = output.
48. In the 8255, which mode is used if we want to simply send out data?
49. Write a program to monitor PA for a temperature of 100. If it is equal, it should be saved in register BL. Also, send AAH to port B and 55H to port C. Use the port address of your choice.
50. Write a program in Assembly language to get a byte of the data from PA, convert it to ASCII bytes, and store them in registers CL, AH, and AL. For example, an input of FFH will show as 255. (Note: FF in binary becomes 323535 in ASCII.)

ANSWERS TO REVIEW QUESTIONS

SECTION 11.1: 8088 INPUT/OUTPUT INSTRUCTIONS

1. 256 input and 256 output ports
2. 65,536 input and 65,536 output ports

3. It sends the contents of 8-bit register AL to port address 24H.
4.
```
MOV DX,300H        ;LOAD THE PORT ADDR
IN AL,DX           ;GET THE DATA FROM PORT
MOV DX,304H        ;LOAD THE PORT ADDR
OUT DX,AL          ;SEND OUT THE DATA
```
5.
```
IN AL,60H          ;GET DATA FROM PORT ADDRESS 60H
MOV CH,AL          ;GIVE THE COPY TO CH REG
```

SECTION 11.2: I/O ADDRESS DECODING AND DESIGN

1. Latch, tri-state buffer
2. The CPU provides the data on the data bus only for a short amount of time. Therefore, it must be latched before it is lost.
3. Assuming that OC = 0, the input data is transferred from D to Q when G goes from low to high, making it available right away; but it is actually latched when G goes from high to low. This reduces the time delay from D to Q.
4. Low
5. True
6. MEMW* for output and MEMR* for input devices

SECTION 11.3: I/O ADDRESS MAP OF x86 PCs

1. 300–31FH
2. AEN = 0
3. The I/O devices are programmed by the CPU; therefore, with AEN = 0 it will make sure that the I/O device is accessed by the addresses provided by the CPU and not the DMA. AEN is active low when the CPU is using the buses.
4. D4 bit
5. 25,000 × 15.085 ms = 377.125 ms

SECTION 11.4: PROGRAMMING AND INTERFACING THE 8255

1. F4H, F5H, F6H, and F7H for PA, PB, PC, and control register, respectively
2. 80H (see Figure 11-12)
3. 9BH (see Figure 11-12)
4.
```
MOV   AL,80H
OUT   CONTREG,AL
MOV   AL,55H
OUT   PORTA,AL
OUT   PORTB,AL
OUT   PORTC,AL
CALL  DELAY
MOV   AL,0AAH
OUT   PORTA,AL
OUT   PORTB,AL
OUT   PORTC,AL
```
5. All are simple I/O. PA and PB are both out. PC0–PC3 and PC4–PC7 are both in (see Figure 11-12).

CHAPTER 12

INTERFACING TO LCD, MOTOR, ADC, AND SENSOR

OBJECTIVES

Upon completion of this chapter, you will be able to:

- **>> Diagram the interfacing of a PC to an LCD, and code the corresponding programs in Assembly and C/C++**
- **>> Diagram the interfacing of a PC to a stepper motor, and code the corresponding programs in Assembly and C/C++**
- **>> Diagram the interfacing of a PC to a DAC (digital-to-analog converter) device, and code the corresponding programs in Assembly and C/C++**
- **>> Diagram the interfacing of a PC to an ADC (analog-to-digital converter) device, and code the corresponding programs in Assembly and C/C++**
- **>> Show the interfacing of ADC devices to sensors**

In this chapter we show PC interfacing to some real-world devices such as an LCD, stepper motor, ADC and DAC devices, and sensors. Section 12.1 describes interfacing and programming of an LCD. In Section 12.2, stepper motor interfacing is described. DAC (digital-to-analog converter) interfacing to PC is shown in Section 12.3 and ADC (analog-to-digital converter) interfacing to PC is shown in Section 12.4. Sensors, such as temperature sensors, and their interfacing are also described in Section 12.4.

> The Assembly language programs in this chapter can only be used with x86 systems with the PC104 bus and operating systems that support Assembly language programming.
>
> We are in the process of designing a USB-I/O Trainer which works with C++, C#, and Visual Basic. This trainer will have 24 pins for general-purpose I/O and two analog-to-digital converters. See www.MicroDigitalEd.com for more information.

SECTION 12.1: INTERFACING TO AN LCD

This section describes the operation modes of LCDs, then describes how to program and interface an LCD to a PC via an 8255.

LCD operation

In recent years the LCD is replacing LEDs (seven-segment LEDs or other multisegment LEDs). This is due to the following reasons:

1. The declining prices of LCDs.
2. The ability to display numbers, characters, and graphics. This is in contrast to LEDs, which are limited to numbers and a few characters.
3. Incorporation of the refreshing controller into the LCD itself, thereby relieving the CPU of the task of refreshing the LCD. In the case of the LED, it must be refreshed by the CPU (or in some other way) to keep displaying the data.
4. Ease of programming for both characters and graphics.

Table 12-1: Pin Descriptions for LCD

Pin	Symbol	I/O	Description
1	V_{SS}	--	Ground
2	V_{CC}	--	+5V power supply
3	V_{EE}	--	Power supply to control contrast
4	RS	I	RS = 0 to select command register, RS = 1 to select data register
5	R/W	I	R/W = 0 for write, R/W = 1 for read
6	E	I/O	Enable
7	DB0	I/O	The 8-bit data bus
8	DB1	I/O	The 8-bit data bus
9	DB2	I/O	The 8-bit data bus
10	DB3	I/O	The 8-bit data bus
11	DB4	I/O	The 8-bit data bus
12	DB5	I/O	The 8-bit data bus
13	DB6	I/O	The 8-bit data bus
14	DB7	I/O	The 8-bit data bus

LCD pin descriptions

The LCD discussed in this section has 14 pins. The function of each pin is given in Table 12-1. Figure 12-1 shows the pin positions for various LCDs.

VCC, VSS, and VEE: While VCC and VSS provide +5V and ground, respectively, VEE is used for controlling the LCD contrast.

RS, register select: There are two registers inside the LCD and the RS pin is used for their selection as follows. If RS = 0, the instruction command code register is selected, allowing the user to send a command such as clear display, cursor at home, and so on. If RS = 1, the data register is selected, allowing the user to send data to be displayed on the LCD (or data to be retrieved).

R/W, read/write: R/W input allows the user to write information into the LCD or

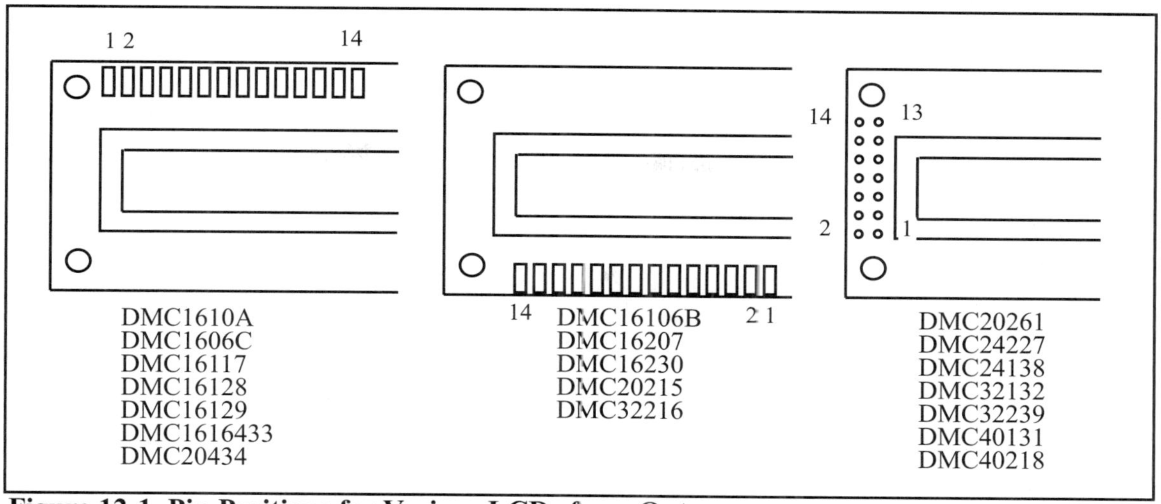

Figure 12-1. Pin Positions for Various LCDs from Optrex

read information from it. R/W = 1 when reading and R/W = 0 when writing.

E, enable: The enable pin is used by the LCD to latch information presented to its data pins. When data is supplied to data pins, a high-to-low pulse must be applied to this pin in order for the LCD to latch in the data present at the data pins. This pulse must be a minimum of 450 ns wide.

D0–D7: The 8-bit data pins are used to send information to the LCD or read the contents of the LCD's internal registers.

To display letters and numbers, we send ASCII codes for the letters A–Z, a–z, and numbers 0–9 to these pins while making RS = 1.

There are also instruction command codes that can be sent to the LCD in order to clear the display, force the cursor to the home position, or blink the cursor. Table 12-2 lists the instruction command codes.

Table 12-2: LCD Command Codes

Code (Hex)	Command to LCD Instruction Register
1	Clear display screen
2	Return home
4	Decrement cursor (shift cursor to left)
6	Increment cursor (shift cursor to right)
5	Shift display right
7	Shift display left
8	Display off, cursor off
A	Display off, cursor on
C	Display on, cursor off
E	Display on, cursor on
F	Display on, cursor blinking
10	Shift cursor position to left
14	Shift cursor position to right
18	Shift the entire display to the left
1C	Shift the entire display to the right
80	Force cursor to beginning of 1st line
C0	Force cursor to beginning of 2nd line
38	2 lines and 5 × 7 matrix

Note: This table is extracted from Table 12-4.

Sending commands to LCDs

To send any of the commands from Table 12-2 to the LCD, make pin RS = 0 and send a high-to-low pulse to the E pin to enable the internal latch of the LCD. The connection of an 8255 to an LCD is shown in Figure 12-2.

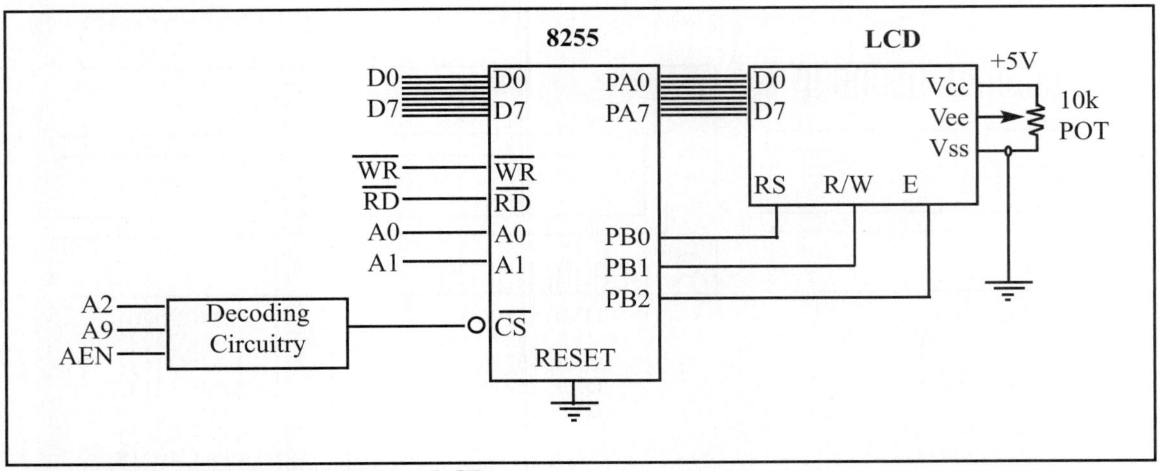

Figure 12-2. 8255 Connection to LCD

Notice the following for the connection in Figure 12-2:

1. The LCD's data pins are connected to Port A of the 8255.
2. The LCD's RS pin is connected to PB0 of Port B of the 8255.
3. The LCD's R/W pin is connected to PB1 of Port B of the 8255.
4. The LCD's E pin is connected to PB2 of Port B of the 8255.
5. Both Ports A and B are configured as output ports.

```
;The following sends all the necessary commands to the LCD
 MOV  AL,38H         ;initialize LCD for 2 lines & 5x7 matrix
 CALL COMNDWRT       ;write the command to LCD
 CALL DELAY          ;wait before issuing the next command
 CALL DELAY          ;this command needs lots of delay
 CALL DELAY
 MOV  AL,0EH         ;send command for LCD on, cursor on
 CALL COMNDWRT       ;write the command to LCD
 CALL DELAY          ;wait before issuing the next command
 MOV  AL,01          ;clear LCD
 CALL COMNDWRT
 CALL DELAY          ;wait
 MOV  AL,06          ;command for shifting cursor right
 CALL COMNDWRT
 CALL DELAY          ;wait

COMNDWRT PROC        ;this procedure writes commands to LCD
  PUSH DX            ;save DX
  MOV DX,PORTA
  OUT DX,AL          ;send the code to Port A
  MOV DX,PORTB       ;Port B address
  MOV AL,00000100B ;RS=0,R/W=0,E=1 for H-TO-L pulse
  OUT DX,AL          ;to Port B
  NOP                ;wait for high-to-low pulse to be
  NOP                ;wide enough
  MOV AL,00000000B;RS=0,R/W=0,E=0 for H-TO-L pulse
  OUT DX,AL          ;
  POP DX             ;restore DX
  RET                ;return to caller
COMNDWRT ENDP
```

In the above program, we must wait before issuing the next command; otherwise, it will jam the LCD. A delay of 20 ms should work fine. We can use the port 61H delay generation shown in Chapter 11. The code is shown below.

Sending data to the LCD

In order to send data to the LCD to be displayed, we must set pin RS = 1, and also send a high-to-low pulse to the E pin to enable the internal latch of the LCD. The following code sends characters to the LCD. Again, it places sufficient time delays between each data issue to ensure that the LCD is ready for new data.

```
   MOV  AL,'Y'      ;display 'Y' letter
   CALL DATWRIT     ;issue it to LCD
   CALL DELAY       ;wait before issuing the next character
   MOV  AL,'E'      ;display 'E' letter
   CALL DATWRIT     ;issue it to LCD
   CALL DELAY       ;wait before issuing the next character
   MOV  AL,'S'      ;display 'S' letter
   CALL DATWRIT     ;issue it to LCD
   CALL DELAY       ;wait

;data write to LCD without checking the busy flag
;AL=char sent to LCD

DATWRIT PROC
   PUSH DX   ;save DX
   MOV  DX,PORTA          ;DX=port A address
   OUT  DX,AL             ;issue the char to LCD
   MOV  AL,00000101B      ;RS=1, R/W=0, E=1 for H-to-L pulse
   MOV  DX,PORTB          ;port B address
   OUT  DX,AL             ;make enable high
   MOV  AL,00000001B  ;RS=1, R/W=0, AND E=0 for H-to-L pulse
   OUT  DX,AL
   POP  DX
   RET
DATWRIT ENDP

;delay generation using the PB4 bit of port 61H

DELAY PROC
      MOV CX,1325         ; 1,325x15.085 usec=20 msec
      PUSH  AX
W1:   IN    AL,61H
      AND   AL,00010000B
      CMP   AL,AH
      JE    W1
      MOV   AH,AL
      LOOP  W1
      POP AX
      RET
DELAY ENDP
```

Checking LCD busy flag

The above programs used a time delay before issuing the next data or command. This allows the LCD a sufficient amount of time to get ready to accept the next data. However, the LCD has a busy flag. We can monitor the busy flag and issue data when it is ready. This will speed up the process. To check the busy flag, we must read the command register (R/W = 1, RS = 0). The busy flag is the D7 bit of that register. Therefore, if R/W = 1, RS = 0. When D7 = 1 (busy flag = 1), the LCD is busy taking care of internal operations and will not accept any new information. When D7 = 0, the LCD is ready to receive new information. It is recommended by the LCD manufacturer's data sheet to monitor the busy flag before sending the data or command codes to the LCD. This ensures that the LCD is ready to receive data. See the code below.

```
        ;writing to LCD with checking the busy flag, AL=char
        MOV   AL,38H       ;initialize LCD for 2 lines & 5x7
        CALL COMNDWRT      ;write the command to LCD
        MOV   AL,0EH       ;send command for LCD on,cursor on
        CALL COMNDWRT      ;write the command to LCD
        MOV   AL,01        ;clear LCD
        CALL COMNDWRT
        MOV   AL,06        ;command for shifting cursor right
        CALL COMNDWRT
        MOV    AL,'Y'      ;display 'Y' letter
        CALL DATWRT        ;issue it to LCD
        MOV    AL,'E'      ;display 'E' letter
        CALL DATWRT        ;issue it to LCD
        MOV    AL,'S'      ;display 'S' letter
        CALL DATWRT        ;issue it to LCD
        ....

;--------------

DATWRT PROC
        CALL   LCDREADY
        PUSH DX   ;save DX
        MOV   DX,PORTA           ;DX=port A address
        OUT   DX,AL              ;issue the char to LCD
        MOV   AL,00000101B ;RS=1, R/W=0, E=1 for H-to-L pulse
        MOV   DX,PORTB           ;port B address
        OUT   DX,AL              ;make enable high
        NOP
        NOP
        MOV   AL,00000001B  ;RS=1, R/W=0, and E=0 for H-to-L
        OUT   DX,AL
        POP   DX
        RET
DATWRT ENDP

;------------

COMNDWRT PROC
        LCDREADY
        PUSH DX            ;save DX
        MOV DX,PORTA
        OUT DX,AL          ;send the code to Port A
        MOV DX,PORTB       ;Port B address
        MOV AL,00000100B   ;RS=0, R/W=0, E=1 for H-to-L pulse
```

```
        OUT DX,AL           ;to Port B
        NOP                 ;wait for high-to-low pulse to be
        NOP                 ;wide enough
        MOV AL,00000000B    ;RS=0, R/W=0, E=0 for H-to-L pulse
        OUT DX,AL
        POP DX              ;restore DX
        RET                 ;return to caller
COMNDWRT ENDP



  ;----------------
LCDREADY PROC
        PUSH AX
        PUSH DX
        MOV AL,90H    ;PA=input to read LCD status,PB=OUT
        MOV DX,CNTPORT      ;DX=control port address
        OUT DX,AL           ;issue to control reg
        MOV AL,00000110B    ;RS=0 busy flag is a
                  ;command R/W=1, E=1 (L-to-H for E)
        MOV DX,PORTB        ;port B address
        OUT DX,AL           ;issue it to port B
        MOV DX,PORTA        ;port A address
AGAIN:IN AL,DX              ;read command reg busy flag is D7
        ROL AL,1            ;send busy flag to carry flag
        JC AGAIN            ;if CF=1 LCD not ready try again
        MOV AL,80H          ;make PA=OUT to send character
        MOV DX,CONTPORT     ;DX=control port address
        OUT DX,AL           ;issue to 8255's control reg
        POP DX
        POP AX
        RET
LCDREADY ENDP
```

LCD cursor position

In the LCD, one can put data at any location. For the 20 × 2 LCD, the address for the first location of line 1 is 80H, and for line 2 it is C0H. The following shows address locations and how they are accessed:

RS	R/W	DB7	DB6	DB5	DB4	DB3	DB2	DB1	DB0
0	0	1	A	A	A	A	A	A	A

where AAAAAAA = 0000000 to 0100111 for line 1 and AAAAAAA = 1000000 to 1100111 for line 2. See Table 12-3. The upper address range can go as high as 0100111 for the 40-character-wide LCD while for the 20-character-wide LCD it goes up to 010011 (19 decimal = 10011 binary). Notice that the upper range 0100111 (binary) = 39 decimal, which corresponds to locations 0 to 39 for the LCDs of 40 × 2 size. From the above discussion we can get the addresses of cursor positions for various sizes of LCDs. See Figure 12-3. Note that all the addresses are in hex.

As an example of setting the cursor at the fourth location of line 1 we have the following:

```
        MOV   AL,83H      ;LINE 1 POSITION 4
        CALL  COMNDWRT
```

and for the sixth location of the second line we have:

```
MOV    AL,0C5H
CALL   COMNDWRT
```

Notice that since the location addresses are in hex, 0 is the first location.

LCD programming in Visual C/C++

In Chapter 11 we showed how to program the x86 PC I/O port using C/C++ for MS Visual C/C++. Example 12-1 shows LCD programming using Visual C/C++.

LCD timing and data sheet

Figures 12-4 and 12-5 show timing diagrams for LCD write and read timing, respectively. Notice that the write operation happens on the H-to-L pulse of the E pin while the read is activated on the L-to-H pulse of the E pin. Table 12-4 provides a more detailed list of LCD instructions.

Table 12-3: LCD Addressing

	DB7	DB6	DB5	DB4	DB3	DB2	DB1	DB0
Line 1 (min)	1	0	0	0	0	0	0	0
Line 1 (max)	1	0	1	0	0	1	1	1
Line 2 (min)	1	1	0	0	0	0	0	0
Line 2 (max)	1	1	1	0	0	1	1	1

16 x 2 LCD	80	81	82	83	84	85	86 through 8F
	C0	C1	C2	C3	C4	C5	C6 through CF
20 x 1 LCD	80	81	82	83	through 93		
20 x 2 LCD	80	81	82	83	through 93		
	C0	C1	C2	C3	through D3		
20 x 4 LCD	80	81	82	83	through 93		
	C0	C1	C2	C3	through D3		
	94	95	96	97	through A7		
	D4	D5	D6	D7	through E7		
40 x 2 LCD	80	81	82	83	through A7		
	C0	C1	C2	C3	through E7		

Note: All data is in hex.

Figure 12-3. Cursor Addresses for Some LCDs

Example 12-1

Write a Visual C/C++ program to display "Hello" on line 1 starting at the sixth position.

Solution:

```
#include<conio.h>
#include<stdio.h>
#include<iostream.h>
#include<iomanip.h>
#include<windows.h>
//tested by Dan Bent
void main()
  {
  unsigned int i;
  char message[ 5] ="Hello";
  cout<<setiosflags(ios::unitbuf);

    _outp(0x303,0x80);          // control word for Port A, B, C
    _outp(0x300,0x38);          //init LCD for 2 lines & 5x7 matrix
    _outp(0x301,0x04);          //RS=0,R/W=0,E=1 for H-to-L pulse
    _outp(0x301,0x00);          //RS=0,R/W=0,E=0 for H-to-L pulse
    _sleep(500);                //delay 500 milliseconds
    _outp(0x300,0x0E);          //send command for LCD on, cursor on
    _outp(0x301,0x04);          //RS=0,R/W=0,E=1 for H-to-L pulse
    _outp(0x301,0x00);          //RS=0,R/W=0,E=0 for H-to-L pulse
    _sleep(250);                //delay 250 milliseconds
    _outp(0x300,0x01);          //clear LCD
    _outp(0x301,0x04);          //RS=0,R/W=0,E=1 for H-to-L pulse
    _outp(0x301,0x00);          //RS=0,R/W=0,E=0 for H-to-L pulse
    _sleep(250);
    _outp(0x300,0x06);          //shift cursor right
    _outp(0x301,0x04);          //RS=0,R/W=0,E=1 for H-to-L pulse
    _outp(0x301,0x00);          //RS=0,R/W=0,E=0 for H-to-L pulse
    _sleep(250);

    _outp(0x300,0x85);          //move cursor to beginning of line
    _outp(0x301,0x04);          //RS=0,R/W=0,E=1 for H-to-L pulse
    _outp(0x301,0x00);          //RS=0,R/W=0,E=0 for H-to-L pulse
    _sleep(250);

//write data to LCD
  for(i=0;i<strlen(message);i++)
    {
     _outp(0x300,(int)message[ i] );
     _outp(0x301,0x05);         //RS=1,R/W=0,E=1 for H-to-L pulse
     _outp(0x301,0x01);         //RS=1,R/W=0,E=0 for H-to-L pulse
     _sleep(250);               //delay 250 milliseconds
    }
  }
```

Review Questions

1. The RS pin is an _______ (input, output) pin for the LCD.
2. The E pin is an ________ (input, output) pin for the LCD.
3. The E pin requires an _______ (H-to-L, L-to-H) pulse to latch in information at the data pins of the LCD.
4. For the LCD to recognize information at the data pins as data, RS must be set to _____ (high, low).
5. Give the command codes for line 1, first character, and line 2, first character.

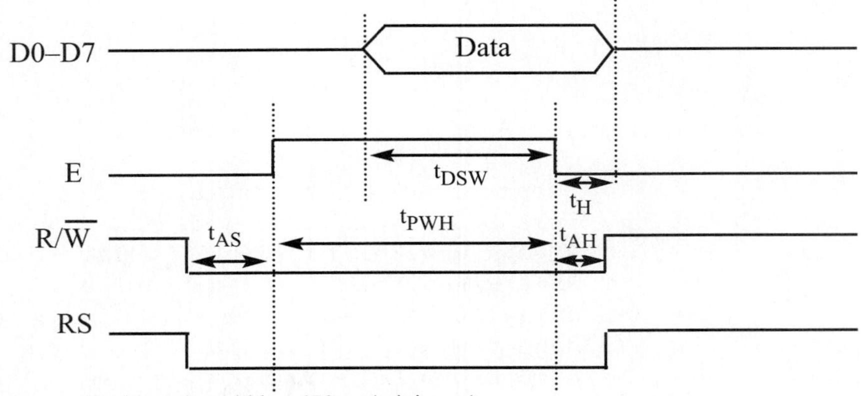

t_{PWH} = Enable pulse width = 450 ns (minimum)
t_{DSW} = Data setup time = 195 ns (minimum)
t_H = Data hold time = 10 ns (minimum)
t_{AS} = Setup time prior to E (going high) for both RS and R/W = 140 ns (minimum)
t_{AH} = Hold time after E has come down for both RS and R/W = 10 ns (minimum)

Figure 12-4. LCD Write Timing

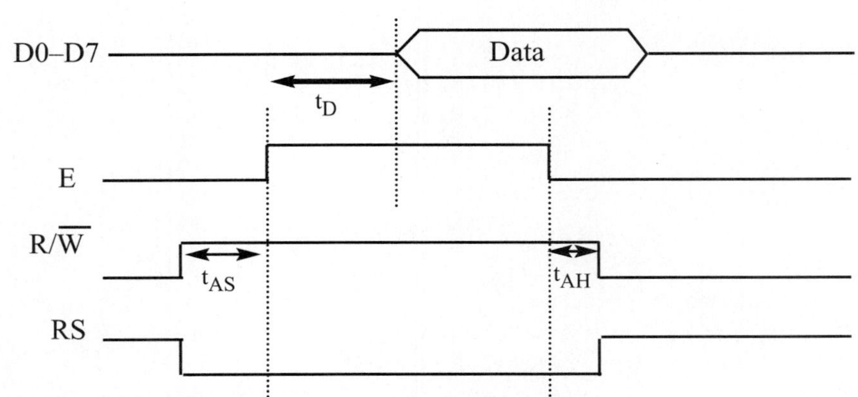

t_D = Data output delay time
t_{AS} = Setup time prior to E (going high) for both RS and R/W = 140 ns (minimum)
t_{AH} = Hold time after E has come down for both RS and R/W = 10 ns (minimum)

Note: Read requires an L-to-H pulse for the E pin.

Figure 12-5. LCD Read Timing

Table 12-4: List of LCD Instructions

Instruction	RS	R/W	DB7	DB6	DB5	DB4	DB3	DB2	DB1	DB0	Description	Execution Time (Max)
Clear Display	0	0	0	0	0	0	0	0	0	1	Clears entire display and sets DD RAM address 0 in address counter	1.64 ms
Return Home	0	0	0	0	0	0	0	0	1	-	Sets DD RAM address 0 as address counter. Also returns display being shifted to original position. DD RAM contents remain unchanged.	1.64 ms
Entry Mode Set	0	0	0	0	0	0	0	1	1/D	S	Sets cursor move direction and specifies shift of display. These operations are performed during data write and read.	40 μs
Display On/ Off Control	0	0	0	0	0	0	1	D	C	B	Sets On/Off of entire display (D), cursor On/Off (C), and blink of cursor position character (B).	40 μs
Cursor or Display Shift	0	0	0	0	0	1	S/C	R/L	-	-	Moves cursor and shifts display without changing DD RAM contents.	40 μs
Function Set	0	0	0	0	1	DL	N	F	-	-	Sets interface data length (DL), number of display lines (L), and character font (F).	40 μs
Set CG RAM Address	0	0	0	1	AGC						Sets CG RAM address. CG RAM data is sent and received after this setting.	40 μs
Set DD RAM Address	0	0	1	ADD							Sets DD RAM address. DD RAM data is sent and received after this setting.	40 μs
Read Busy Flag & Address	0	1	BF	AC							Reads Busy flag (BF) indicating internal operation is being performed and reads address counter contents.	40 μs
Write Data CG or DD RAM	1	0	Write Data								Writes data into DD or CG RAM.	40 μs
Read Data CG or DD RAM	1	1	Read Data								Reads data from DD or CG RAM.	40 μs

Notes:

1. Execution times are maximum times when fcp or fosc is 250 kHz.
2. Execution time changes when frequency changes. Example: When fcp or fosc is 270 kHz: 40 μs × 250 / 270 = 37 μs.
3. Abbreviations:

DD RAM	Display data RAM		
CG RAM	Character generator RAM		
ACC	CG RAM address		
ADD	DD RAM address, corresponds to cursor address		
AC	Address counter used for both DD and CG RAM addresses.		
1/D = 1	Increment	1/D = 0	Decrement
S = 1	Accompanies display shift		
S/C = 1	Display shift;	S/C = 0	Cursor move
R/L = 1	Shift to the right;	R/L = 0	Shift to the left
DL = 1	8 bits, DL = 0: 4 bits		
N = 1	1 line, N = 0: 1 line		
F = 1	5 × 10 dots, F = 0: 5 × 7 dots		
BF = 1	Internal operation;	BF = 0 Can accept instruction	

SECTION 12.2: INTERFACING TO A STEPPER MOTOR

This section begins with an overview of the basic operation of stepper motors. Then we describe how to interface a stepper motor to the PC. Finally, we use Assembly language programs to demonstrate control of the angle and direction of stepper motor rotation.

See www.MicroDigitalEd.com for the USB-I/O version of this section.

Stepper motors

A stepper motor is a widely used device that translates electrical pulses into mechanical movement. In applications such as disk drives, dot matrix printers, and robotics, the stepper motor is used for position control. Every stepper motor has a permanent magnet rotor (also called the *shaft*) surrounded by a stator (see Figure 12-6). The most common stepper motors have four stator windings that are paired with a center-tapped common as shown in Figure 12-7. This type of stepper motor is commonly referred to as a *four-phase* stepper motor. The center tap allows the change of current direction in each of two coils when a winding is grounded, which results in a polarity change of the stator. Notice that while a conventional motor shaft runs freely, the stepper motor shaft moves in a fixed repeatable increment, which allows one to move it to a precise position. This repeatable fixed movement is possible as a result of the basic magnet theory where poles of the same polarity repel and opposite poles attract. The direction of the rotation is dictated by the stator poles. The stator poles are determined by the current sent through the wire coils. As the direction of current is changed, the polarity is also changed causing the reverse motion of the rotor. The stepper motor discussed here has a total of six leads: four leads representing the four stator windings and two commons for the center tapped leads. As the sequence of

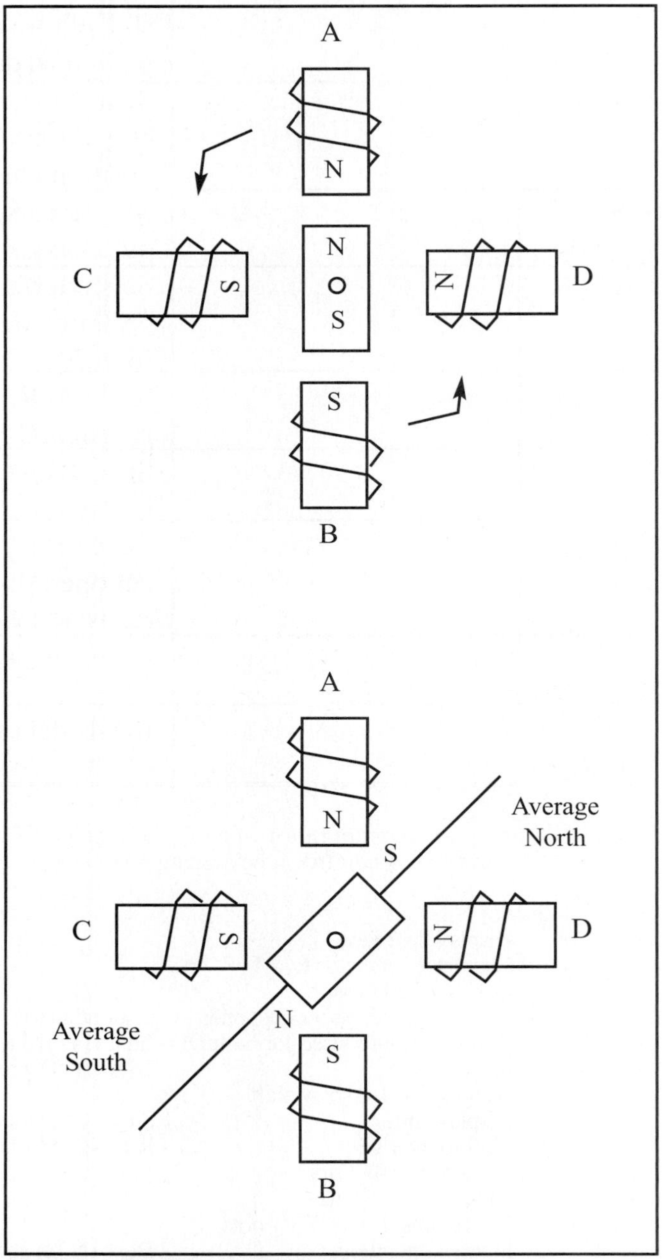

Figure 12-6. Rotor Alignment

power is applied to each stator winding, the rotor will rotate. There are several widely used sequences where each has different degree of precision. Table 12-5 shows the normal 4-step sequence.

It must be noted that although we can start with any of the sequences in Table 12-5, once we start we must continue in the proper order. For example, if we start with step 3 (0110) we must continue in the sequence of steps 4, 1, 2, and so on.

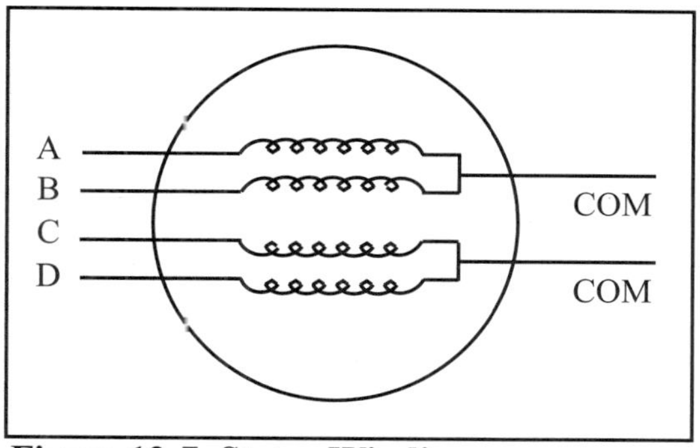

Figure 12-7. Stator Windings Configuration

Table 12-5: Normal 4-Step Sequence

Clockwise	Step #	Winding A	Winding B	Winding C	Winding D	Counter-clockwise
↓	1	1	0	0	1	↑
	2	1	1	0	0	
	3	0	1	1	0	
	4	0	0	1	1	

Table 12-6: Stepper Motor Step Angles

Step Angle	Steps per Revolution
0.72	500
1.8	200
2.0	180
2.5	144
5.0	72
7.5	48
15	24

Step angle

How much movement is associated with a single step? This depends on the internal construction of the motor, in particular the number of teeth on the stator and the rotor. The step angle is the minimum degree of rotation associated with a single step. Various motors have different step angles. Table 12-6 shows some step angles for various motors. In Table 12-6, notice the term *steps per revolution*. This is the total number of steps needed to rotate one complete rotation or 360 degrees (e.g., 180 steps × 2 degrees = 360).

It must be noted that perhaps contrary to one's initial impression, a stepper motor does not need to have more terminal leads for the stator to achieve smaller steps. All the stepper motors discussed in this section have four leads for the stator winding and two com wires for the center tap. Although some manufacturers have set aside only one lead for the common signal instead of two, they always have four leads for the stators.

With this background on stepper motors, next we see how we can interface them with the PC.

Stepper motor connection and programming

Example 12-2 shows the programming of the stepper motor as connected in Figure 12-8. Study this example very carefully since it contains some very important points on motor interfacing.

Example 12-2

Describe the 8255 connection to the stepper motor of Figure 12-8 and code a program to rotate it continuously.

Solution:

The following steps show the 8255 connection to the stepper motor and its programming.

1. Use an ohmmeter to measure the resistance of the leads. This should identify which COM leads are connected to which winding leads.
2. The common wire(s) are connected to the positive side of the motor's power supply. In many motors, +5 V is sufficient.
3. The four leads of the stator winding are controlled by the four bits of port A (PA0–PA3). However, since the 8255 lacks sufficient current to drive the stepper motor windings, we must use a driver such as the ULN2003 to energize the stator. Instead of the ULN2003, we could use transistors as drivers. However, notice that if transistors are used as drivers, we must also use diodes to take care of inductive current generated when the coil is turned off. One reason that the ULN2003 is preferable to the use of transistors as drivers is that the ULN2003 has an internal diode to take care of back EMF.

```
        MOV    AL,80H
        MOV    DX,CNTRLPORT        ;LOAD CONTROL PORT ADDRESS
        OUT    DX,AL               ;PORT AS OUTPUT
        MOV    BL,33H
AGAIN:MOV      AH,01
        INT    16H                 ;CHECK KEY PRESS
        JNZ    EXIT                ;EXIT UPON KEY PRESS
        MOV    AL,BL
        MOV    DX,PORTA
        OUT    DX,AL
        MOV    CX,20000
HERE:   LOOP   HERE                ;DELAY
        ROR    BL,1
        JMP    AGAIN
EXIT:
```

In the above program we are sending the sequence 33H, 66H, CCH, and 99H to the stepper motor continuously. The motor keeps moving unless a key is pressed.

By changing the value of DELAY, we can change the speed of rotation. In your program use a fixed time delay. The fixed time delay generation was shown in Chapter 11.

Steps per second and RPM relation

The relationship between the RPM (revolutions per minute), steps per revolution, and steps per second is intuitive and is as follows.

$$\text{Steps per second} = \frac{\text{RPM} \times \text{Steps per revolution}}{60}$$

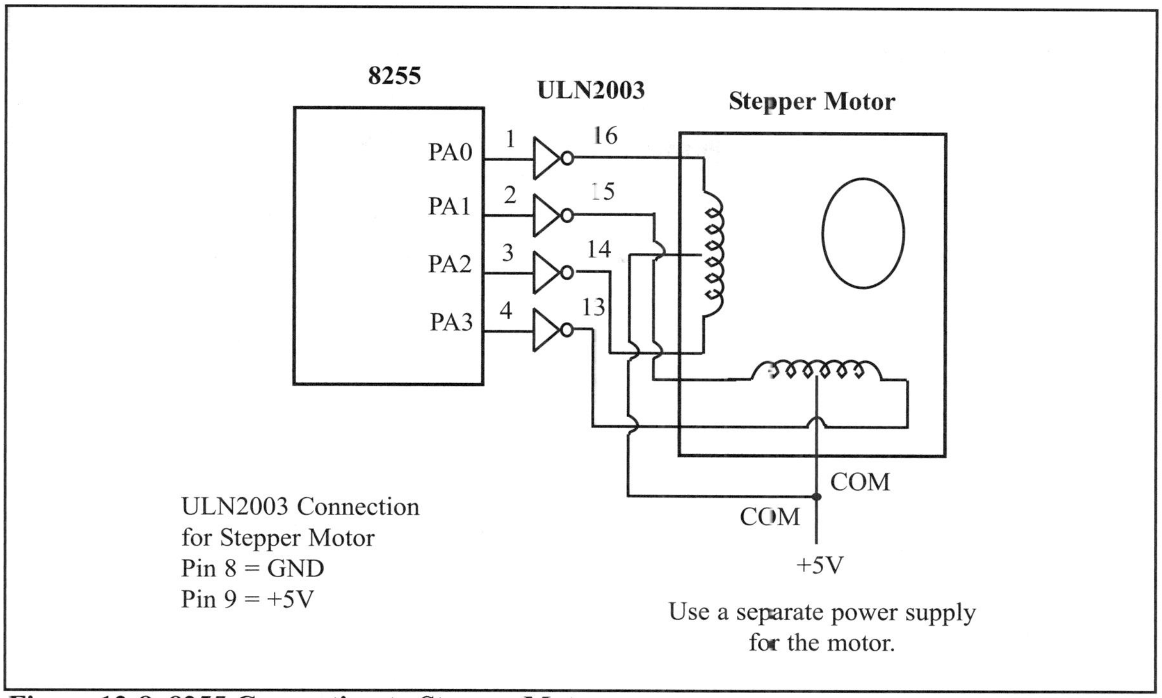

Figure 12-8. 8255 Connection to Stepper Motor

The four-step sequence and number of teeth on rotor

The switching sequence shown above in Table 12-5 is called the 4-step switching sequence since after four steps the same two windings will be "ON". How much movement is associated with these four steps? After completing every four steps, the rotor moves only one tooth pitch. Therefore, in a stepper motor with 200 steps per revolution, its rotor has 50 teeth, since $4 \times 50 = 200$ steps are needed to complete one revolution. This leads to the conclusion that the minimum step angle is always a function of the number of teeth on the rotor. In other words, the smaller the step angle, the more teeth the rotor passes. See Example 12-3.

Example 12-3

Give the number of times the 4-step sequence in Table 12-5 must be applied to a stepper motor to make an 80-degree move if the motor has a 2-degree step angle.

Solution:

A motor with a 2-degree step angle has the following characteristics:

Step angle:	2 degrees	Steps per revolution:	180
Number of rotor teeth:	45	Movement per 4-step sequence:	8 degrees

To move the rotor 80 degrees, we need to send 10 four-step sequences consecutively, since 10×4 steps $\times$ 2 degrees = 80 degrees.

Looking at Example 12-3, one might wonder what happens if we want to move 45 degrees since the steps are 2 degrees each. To allow for finer resolutions, all stepper motors allow what is called an 8-step switching sequence. The 8-step sequence is also called half-stepping since in following the 8-step sequence each step is half of the normal step angle. For example, a motor with a 2-degree step angle can be used as a 1-degree step angle if the sequence of Table 12-7 is applied.

Example 12-4 shows the C version of the program to turn the stepper motor clockwise.

Table 12-7: Half-Step 8-Step Sequence

Clockwise	Step #	Winding A	Winding B	Winding C	Winding D	Counter-clockwise
	1	1	0	0	1	
	2	1	0	0	0	
	3	1	1	0	0	
	4	0	1	0	0	
	5	0	1	1	0	
	6	0	0	1	0	
	7	0	0	1	1	
	8	0	0	0	1	

Example 12-4

Write a Turbo C++ program to turn the stepper motor clockwise continuously. Pressing any key should exit the program.

Solution:

```
//Turning the stepper motor clockwise continuously
#include <conio.h>
#include <stdio.h>
main()
{
  _outp(0x303,0x80);     //CONFIGURE 8255 AS OUT
  printf("\n Turning the Stepper motor clockwise. Press any key to
exit this program\n");
  do
  {
    _outp(0x300,0x99);
    _sleep(500);                  //500 msec
  _outp(0x300,0xcc);
    _sleep(500);                  //500 msec
  _outp(0x300,0x66);
    _sleep(500);                  //500 msec
    _outp(0x300,0x33);
    _sleep(500);                  //500 msec
   }
  while(!kbhit());               //PRESS ANY KEY TO STOP
  return(0);
}
```

Motor speed

The motor speed, measured in steps per second (steps/s), is a function of the switching rate. Notice in Example 12-2 that by changing the length of the time delay loop, we can achieve various rotation speeds.

Holding torque

The following is the definition of the holding torque: "With the motor shaft at standstill or zero RPM condition, the amount of torque, from an external source, required to break away the shaft from its holding position. This is measured with rated voltage and current applied to the motor." The unit is ounce-inch (or kg-cm).

Wave drive 4-step sequence

In addition to the 8-step sequence and the 4-step sequence discussed earlier, there is another sequence called the wave drive 4-step sequence. It is shown in Table 12-8. Notice that the sequence of Table 12-8 is simply the combination of the wave drive 4-step and normal 4-step normal sequences shown in Tables 12-5 and 12-7, respectively. Experimenting with the wave drive 4-step is left to the reader.

Table 12-8: Wave Drive 4-Step Sequence

Clockwise	Step #	Winding A	Winding B	Winding C	Winding D	Counter-clockwise
↓	1	1	0	0	0	↑
	2	0	1	0	0	
	3	0	0	1	0	
	4	0	0	0	1	

Notice that if a given motor requires more current than the ULN2003 can provide, we can use transistors, as shown in Figure 12-9.

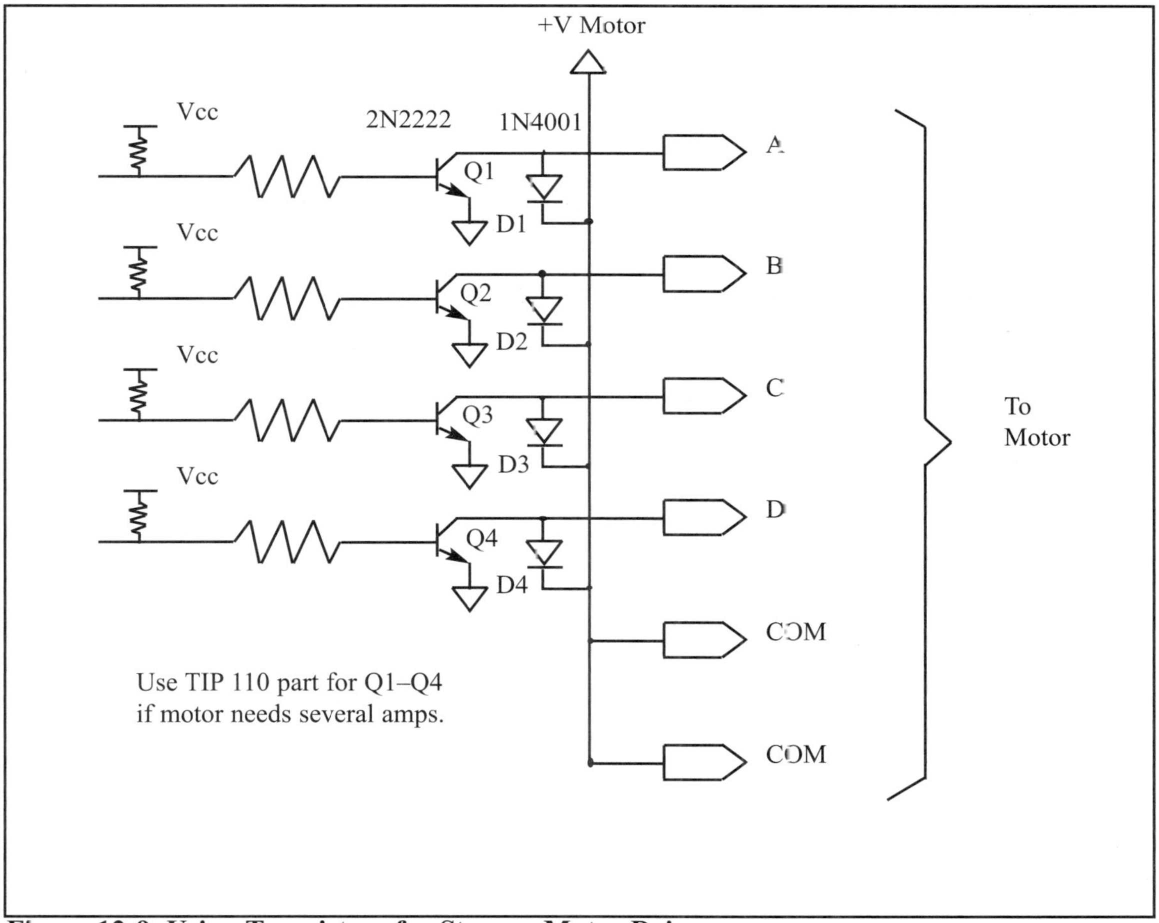

Figure 12-9. Using Transistors for Stepper Motor Driver

Review Questions

1. Give the 4-step sequence of a stepper motor if we start with 0110.
2. A stepper motor with a step-angle of 5 degrees has ____ steps per revolution.
3. Why do we put a driver between the 8255 and the stepper motor?

SECTION 12.3: INTERFACING TO A DAC

This section will show how to interface a DAC (digital-to-analog converter) to a PC via the 8255. Then we demonstrate how to generate a sine wave on the scope using the DAC.

Digital-to-analog (DAC) converter

The digital-to-analog converter (DAC) is a device widely used to convert digital pulses to analog signals. In this section we discuss the basics of interfacing a DAC to a PC.

Recall from your digital electronics book the two methods of making a DAC: binary weighted and R/2R ladder. The vast majority of integrated circuit DACs, including the MC1408 used in this section, use the R/2R method since it can achieve a much higher degree of precision. The first criterion for judging a DAC is its resolution, which is a function of the number of binary inputs. The common ones are 8, 10, and 12 bits. The number of data bit inputs decides the resolution of the DAC since the number of analog output levels is equal to 2^n, where n is the number of data bit inputs. Therefore, the 8-input DAC such as the MC1408 provides 256 discrete voltage (or current) levels of output. Similarly, the 12-bit DAC provides 4096 discrete voltage levels. Although there are 16-bit DACs, they are expensive.

DAC 808

In the DAC808, the digital inputs are converted to current (I_{out}). By connecting a resistor to the I_{out} pin, we convert the result to voltage. The total current provided by the Iout is a function of the binary numbers at the D0–D7 inputs of the 1408 and the reference current (I_{ref}), and is as follows:

$$I_{out} = I_{ref}\left(\frac{D7}{2} + \frac{D6}{4} + \frac{D5}{8} + \frac{D4}{16} + \frac{D3}{32} + \frac{D2}{64} + \frac{D1}{128} + \frac{D0}{256}\right)$$

where D0 is the LSB, D7 is the MSB for the inputs, and I_{ref} is the input current that must be applied to pin 14. The I_{ref} current is generally set to 2.0 mA. Figure 12-10 shows the generation of current reference (setting $I_{ref} = 2$ mA) by using the standard 5-V power supply and 1K, 1.5K ohm standard resistors. Some also use the zener diode (LM336), which overcomes any fluctuation associated with the power supply voltage. Now assuming that $I_{ref} = 2$ mA, if all the inputs to the DAC are high, the maximum output current is 1.99 mA (verify this for yourself).

Converting I_{out} to voltage in 1408 DAC

We connect the output pin I_{out} to a resistor, convert this current to voltage, and monitor the output on the scope. However, in real life this can cause inaccuracy since the input resistance of the load where it is connected will also affect the output voltage. For this reason, the I_{ref} current output is isolated by connecting it to an op amp such as the 741 with $R_f = 5$ kilohms for the feedback resistor. Assuming that R = 5 kilohms, by changing the binary input, the output voltage changes as shown in Example 12-5.

Generating a sine wave

To generate a sine wave, we first need a table whose values represent the magnitude of the sine of angles between 0 and 360 degrees. The values for the sine function vary

Example 12-5

Assuming that R = 5K and I_{ref} = 2 mA, calculate V_{out} for the following binary inputs:

(a) 10011001 binary (99H) (b) 11001000 (C8H)

Solution:

(a) I_{out} = 2 mA (153/255) = 1.195 mA and V_{out} = 1.195 mA × 5K = 5.975 V

(b) I_{out} = 2 mA (200/256) = 1.562 mA and V_{out} = 1.562 mA × 5K = 7.8125 V

from –1.0 to +1.0 for 0 to 360 degree angles. Therefore, the table values are integer numbers representing the voltage magnitude for the sine of theta. This method ensures that only integer numbers are output to the DAC by the x86 processor. Table 12-9 shows the angles, the sine values, the voltage magnitude, and the integer values representing the voltage magnitude for each angle with 30-degree increments. To generate Table 12-9, we assumed the full-scale voltage of 10V for the DAC output. Full-scale output of the DAC is achieved when all the data inputs of the DAC are high. Therefore, to achieve the full-scale 10V output, we use the following equation.

$$V_{out} = 5\text{ V} + (5 \times \sin\theta)$$

To find the value sent to the DAC for various angles, we simply multiply the V_{out} voltage by 25.60 because there are 256 steps and full-scale V_{out} is 10 volts. Therefore, 256 steps / 10 V = 25.6 steps per volt. To further clarify this, look at Example 12-6.

Example 12-6

Verify the values of Table 12-9 for the following angles: (a) 30 (b) 60.

Solution:

(a) V_{out} = 5 V + (5 V × sin θ) = 5 V + 5 × sin 30 = 5 V + 5 × 0.5 = 7.5 V

DAC input values = 7.5 V × 25.6 = 192 (decimal)

(b) V_{out} = 5 V + (5 V × sin θ) = 5 V + 5 × sin 60 = 5 V + 5 × 0.866 = 9.33 V

DAC input values = 9.33 V × 25.6 = 238 (decimal)

The following program sends the values of Table 12-9 to the DAC. See Figure 12-11.

```
        ;in data segment
TABLE   DB 128,192,238,255,238,192,128,64,17,0,17,64,128
        ;in code segment
        ;PA is assumed to be output
A1:     MOV   CX,12          ;count
        MOV   BX,OFFSET TABLE
        MOV   DX,PORTA       ;port A address
NEXT:   MOV   AL,[BX]
        OUT   DX,AL
        INC   BX
        CALL  DELAY          ;let DAC recover
        LOOP  NEXT
        JMP   A1             ;do it again
```

To produce a simple stair-step sine wave, we can use Example 12-7. Example 12-8 uses the Turbo C++ math functions to generate the look-up table values. You can use Visual C++ instead.

Table 12-9: Angle v. Voltage Magnitude for Sine Wave

Angle θ (degrees)	Sin θ	Vout (Voltage Magnitude) 5 V + (5 V × sin θ)	Values Sent to DAC (decimal) (Voltage Mag. × 25.6)
0	0	5	128
30	0.5	7.5	192
60	0.866	9.33	238
90	1.0	10	255
120	0.866	9.33	238
150	0.5	7.5	192
180	0	5	128
210	–0.5	2.5	64
240	–0.866	0.669	17
270	–1.0	0	0
300	–0.866	0.669	17
330	–0.5	2.5	64
360	0	5	128

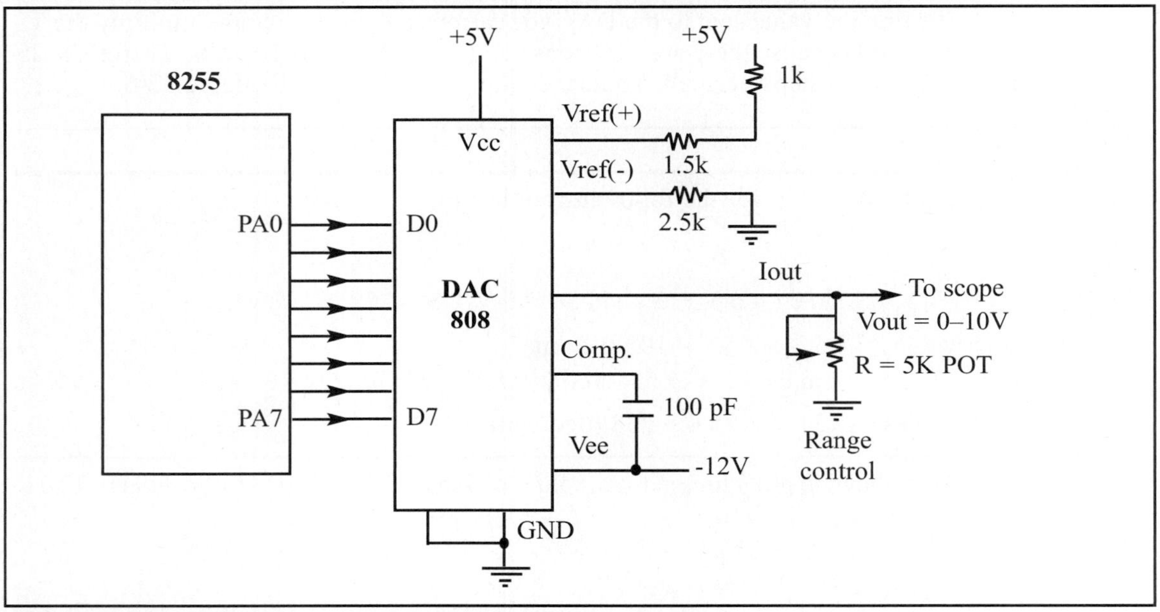

Figure 12-10. 8255 Connection to DAC808

Example 12-7

In order to generate a stair-step ramp, set up the circuit in Figure 12-10 and connect the output to an oscilloscope. Then write a program to send data to the DAC to generate a stair-step ramp.

Solution:

```
          MOV   AL,80H            ;PA=OUT
          MOV   DX,303H
          OUT   DX,AL
          MOV   DX,300H
AGAIN:    MOV   AH,01
          INT   16H               ;CHECK KEY PRESS
          JNZ   STOP              ;EXIT UPON KEY PRESS
          SUB   AL,AL
BACK: OUT DX,AL
          INC   AL
          CMP   AL,0
          JZ    AGAIN
          CALL  DELAY        ;LET DAC RECOVER
          JMP   BACK
STOP:                        ;EXIT
```

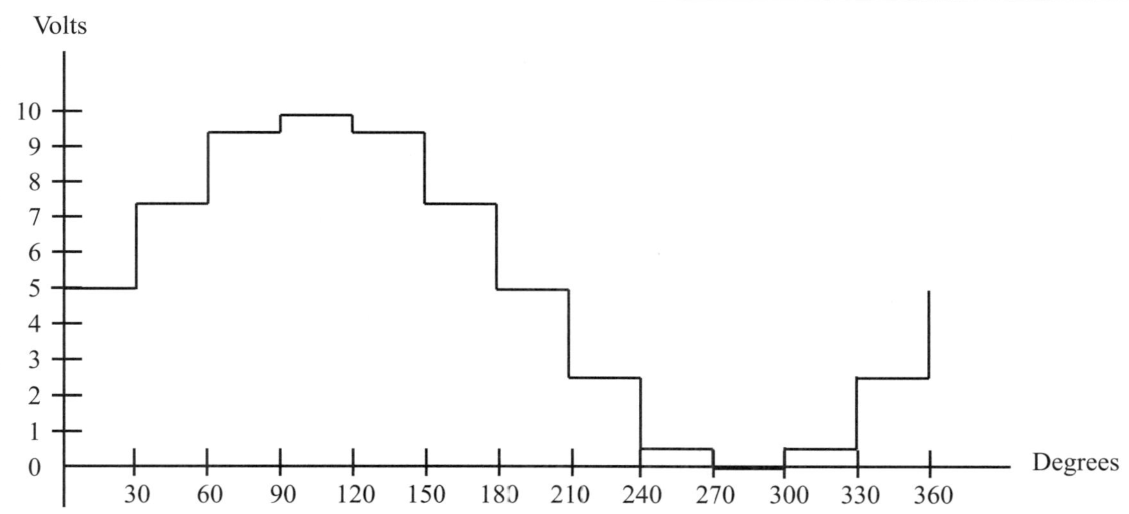

Figure 12-11. Angle v. Voltage Magnitude for Sine Wave

Review Questions

1. In a DAC, input is ______ (digital, analog) and output is _____ (digital, analog). Answer for ADC input and output as well.
2. DAC808 is a(n) ____-bit D-to-A converter.
3. The output of DAC808 is in ______ (current, voltage).

Example 12-8

Write a Turbo C++ program to generate a sine wave on PA. Use the C++ math functions to generate the look-up table values. Pressing any key should exit the program.

Solution:

```
//GENERATING SINE WAVE VIA A DAC CONNECTED TO PORT A
#include <conio.h>
#include <stdio.h>
#include <math.h>
main()
{
 outp(0x303,0x80);              //CONFIGURE 8255 AS OUT
 unsigned char v1;              //v1 IS A BYTE SIZE DATA
 float Vout,magnitude;
 int a;
 printf("\n Press any key to exit this program\n");
 do
 {
      for (a=0;a<360;a++)       //FOR THE FULL 360 DEGREES
     {
     Vout=5.0+(5.0 * sin ((3.14*a)/180));      //VOLTAGE MAGNITUDE
     magnitude=Vout * 25.6;     //VALUE SENT TO DAC
     v1=(char)magnitude;        //MAKE IT A BYTE SIZE
     delay(1);
     outp(0x300,v1);            //OUTPUT IT TO PORT A
     }
 }
 while(!kbhit());               //PRESS ANY KEY TO EXIT
 return(0);
 }
```

SECTION 12.4: INTERFACING TO ADC CHIPS AND SENSORS

This section will explore interfacing ADC (analog-to-digital converter) chips and temperature sensors to a PC. After describing the ADC chips, we show how to interface them to the PC using the PC Interface Trainer. Then we examine the characteristics of the LM3/35 temperature sensor and show how to interface it with proper signal conditioning.

ADC devices

Analog-to-digital converters are among the most widely used devices for data acquisition. Digital computers use binary (discrete) values, but in the physical world everything is analog (continuous). Temperature, pressure (wind or liquid), humidity, and velocity are a few examples of physical quantities that we deal with every day. A physical quantity is converted to electrical (voltage, current) signals using a device called a *transducer*. Transducers are also referred to as *sensors*. There are sensors for temperature, velocity, pressure, light, and many other natural quantities, and they produce an output that is voltage (or current). Therefore, we need an analog-to-digital converter to translate the analog signals to digital numbers so that the PC can read them. Next we describe an ADC chip.

ADC0848 chip

The ADC0848 IC is an analog-to-digital converter in the family of the ADC0800 series from National Semiconductor Corp. Data sheets for this chip can be found at their web site, www.national.com.

The ADC0848 has a resolution of 8 bits. It is an 8-channel ADC, thereby allowing it to monitor up to 8 different analog inputs. See Figures 12-12 and 12-13. The ADC0844 chip in the same family has 4 channels. The following is the discussion of the pins of the ADC0848.

CS: Chip select is an active-low input used to activate the 848 chip. To access the 848, this pin must be low.

RD (read): This is an input signal and is active low. ADC converts the analog input to its binary equivalent and holds it in an internal register. RD is used to get the converted data out of the 848 chip. When CS = 0, if the RD pin is asserted low, the 8-bit digital output shows up at the D0–D7 data pins. The RD pin is also referred to as output enable (OE).

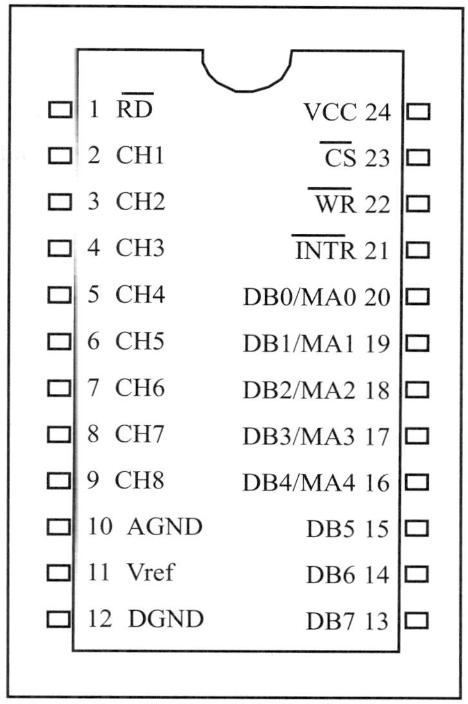

Figure 12-12. ADC0848 Chip

Vref is an input voltage used for the reference voltage. The voltage connected to this pin dictates the step size. For the ADC0848, the step size is $V_{ref}/256$ since it is an 8-bit ADC and 2 to the power of 8 gives us 256 steps. See Table 12-10. For example, if the analog input range needs to be 0 to 4 volts, V_{ref} is connected to 4 volts. That gives 4 V/256 = 15.62 mV for step size. In another case, if we need the step size of 10 mV then V_{ref} = 2.56 V, since 2.56 V/256 = 10 mV.

Table 12-10: ADC0848 V_{ref} vs. Step Size

V_{ref}(V)	Step size (mV)
5	19.53
4	15.62
2.56	10
1.26	5
0.64	2.5

Note: Step size = $V_{ref}/256$.

DB0–DB7 are the digital data output pins. With a D0–D7 output, the 848 must be an 8-bit ADC. The step size, which is the smallest change, is dictated by the number of digital outputs and the V_{ref} voltage. To calculate the output voltage, we use the following formula:

$$D_{out} = \frac{V_{in}}{step\ size}$$

where D_{out} = digital data output (in decimal), V_{in} = analog input voltage, and step size (resolution) is the smallest change, which is $V_{ref}/256$ for an 8-bit ADC. See Example 12-9 for clarification. Notice that D0–D7 are tri-state buffered and that the converted data is accessed only when CS = 0 and a low pulse is applied to the RD pin. Also, notice the dual role of pins D0–D7. They are also used to send in the channel address. This is discussed next.

MA0–MA4 (multiplexed address). The ADC0848 uses multiplexed address/data pins to select the channel. Notice in Figure 12-13 that a portion of the DB0–DB7 pins are also designated as MA0–MA4. The D0–D7 pins are inputs when the channel's address is sent in. However, when the converted data is being read, D0–D7 are outputs. While the use of multiplexed address/data saves some pins, it makes the I/O interfacing more difficult as we will soon see.

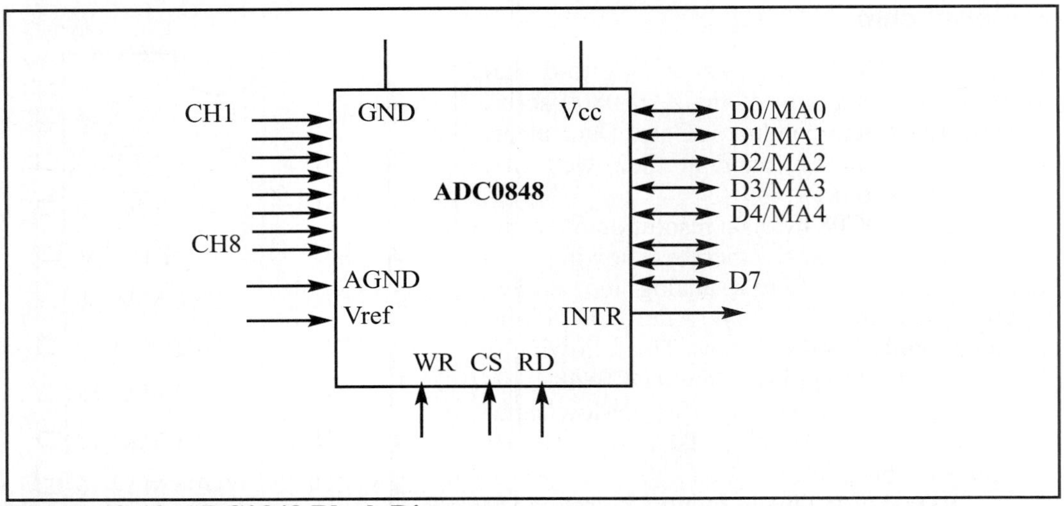

Figure 12-13. ADC0848 Block Diagram

WR (write; a better name might be "start conversion"). This is an input into the ADC0848 chip and plays two important roles: (1) It latches the address of the selected channel present on the D0–D7 pins, and (2) it informs the ADC0848 to start the conversion of analog input at that channel. If CS = 0 when WR makes a low-to-high transition, the ADC0848 latches in the address of the selected channel and starts converting the analog input value to an 8-bit digital number. The amount of time it takes to convert is a maximum of 40 microseconds for ADC0848. The conversion time is set by an internal clock.

CH1–CH8 are 8 channels of the V_{in} analog inputs. In what is called single-ended mode, each of the 8 channels can be used for analog V_{in} where the AGND (analog ground) pin is used as a ground reference for all the channels. These 8 channels of input allow us to read 8 different analog signals, but not all at the same time since there is only a single D0–D7 output. We select the input channel by using the MA0–MA4 multiplexed address pins according to Table 12-11. In Table 12-11, notice that MA4 = low and MA3 = high for single-ended mode. The ADC0848 can also be used in differential mode. In differential mode, two channels, such as CH1 and CH2, are paired together for the $V_{in}(+)$ and Vin(–) differential analog inputs. In that case V_{in} = CH1(+) – CH2(–) is the differential analog input. To use ADC0848 in differential mode, MA4 = don't care, and MA3 is set to low. For more on this, see the ADC0848 data sheet on the www.national.com web site.

VCC is the +5 volt power supply.

AGND, DGND (analog ground and digital ground). Both are input pins providing the ground for both the analog signal and the digital signal. Analog ground is connected to the ground of the analog V_{in} while digital ground is connected to the ground of the VCC pin. The reason that we have two ground pins is to isolate the analog V_{in} signal from transient voltages caused by digital switching of the output D0–D7. Such isolation contributes to the accuracy of the digital data output. Notice that in the single-ended mode the voltage at the channel is the analog input and AGND is the reference for the V_{in}. In our discussion, both the AGND and DGND are connected to the same ground; however, in the real world of data acquisition, the analog and digital grounds are handled separately.

INTR (interrupt; a better name might be "end of conversion"). This is an output pin and is active low. It is a normally high pin and when the conversion is finished, it goes low to signal the CPU that the converted data is ready to be picked up. After INTR goes low, we make CS = 0 and apply a low pulse to the RD pin to get the binary data out of the ADC0848 chip. See Figure 12-14.

Selecting an input channel

The following are the steps for data conversion by the ADC0848 chip.

1. While CS = 0 and RD = 1, provide the address of the selected channel (see Table 12-11) to the DB0–DB7 pins and apply a low-to-high pulse to the WR pin to latch in the address and start the conversion. The channel's addresses are 08H for

Example 12-9

For a given ADC0848, we have V_{ref} = 2.56 V. Calculate the D0–D7 output if the analog input is: (a)1.7 V, and (b) 2.1 V.

Solution:

Since the step size is 2.56/256 = 10 mV, we have the following.

(a)D_{out} = 1.7V/10 mV = 170 in decimal, which gives us 10101011 in binary for D7–D0.

(b)D_{out} = 2.1V/10 mV = 210 in decimal, which gives us 11010010 in binary for D7–D0.

CH1, 09H for CH2, 0AH for CH3, and so on, as shown in Table 12-11. Notice that this process not only selects the channel, but also starts the conversion of the analog input at the selected channel.

2. While WR = 1 and RD = 1, keep monitoring the INTR pin. When INTR goes low, the conversion is finished and we can go to the next step. If INTR is high, keep polling until it goes low, signalling end-of-conversion.
3. After the INTR has become low, we must make CS = 0, WR = 1, and apply a low pulse to the RD pin to get the data out of the 848 IC chip.

Table 12-11: ADC0848 Analog Channel Selection (Single-Ended Mode)

Selected Analog Channel	MA4	MA3	MA2	MA1	MA0
CH1	0	1	0	0	0
CH2	0	1	0	0	1
CH3	0	1	0	1	0
CH4	0	1	0	1	1
CH5	0	1	1	0	0
CH6	0	1	1	0	1
CH7	0	1	1	1	0
CH8	0	1	1	1	1

Note: Channel is selected when CS = 0, RD = 1, and an L-to-H pulse is applied to WR.

ADC0848 connection to 8255

The following is a summary of the connection between the 8255 and the ADC0848 as shown in Figure 12-15.

PA0–PA7 to D0–D7 of ADC:	Channel selection (out), data read (in)
PB0 to INTR	Port B as input
PC0 to WR	Port C as output
PC1 to RD	Port C as output

Notice the following facts about the above connection:

1. Port A is an output when we select a channel, and it is an input when we read the converted data.
2. We must monitor the INTR pin of the ADC for end-of-conversion; therefore, we configure PB as input. Since both WR and RD are inputs into ADC, Port C is configured as an output port.

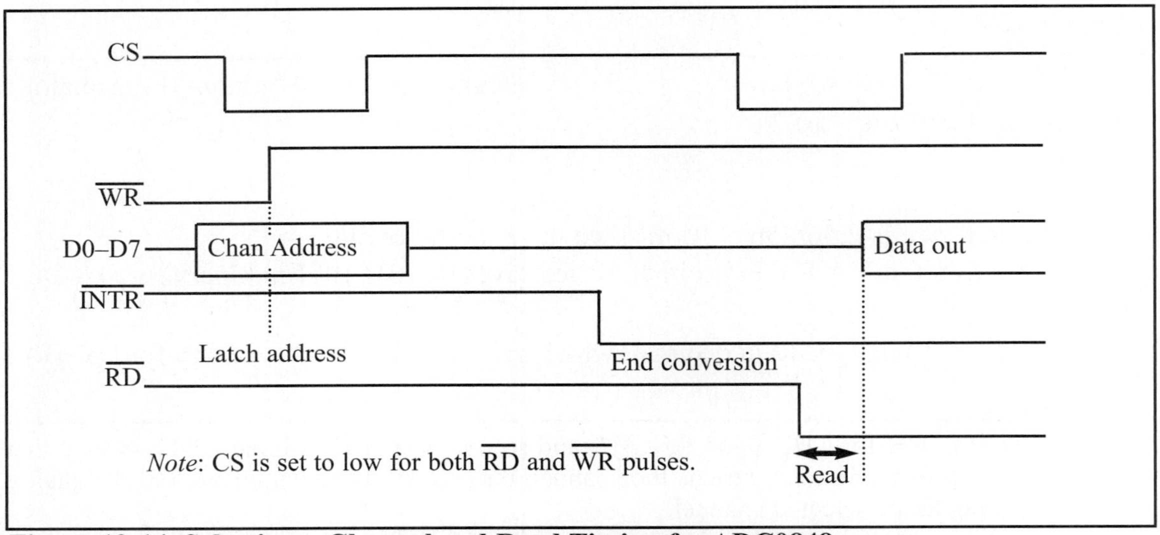

Figure 12-14. Selecting a Channel and Read Timing for ADC0848

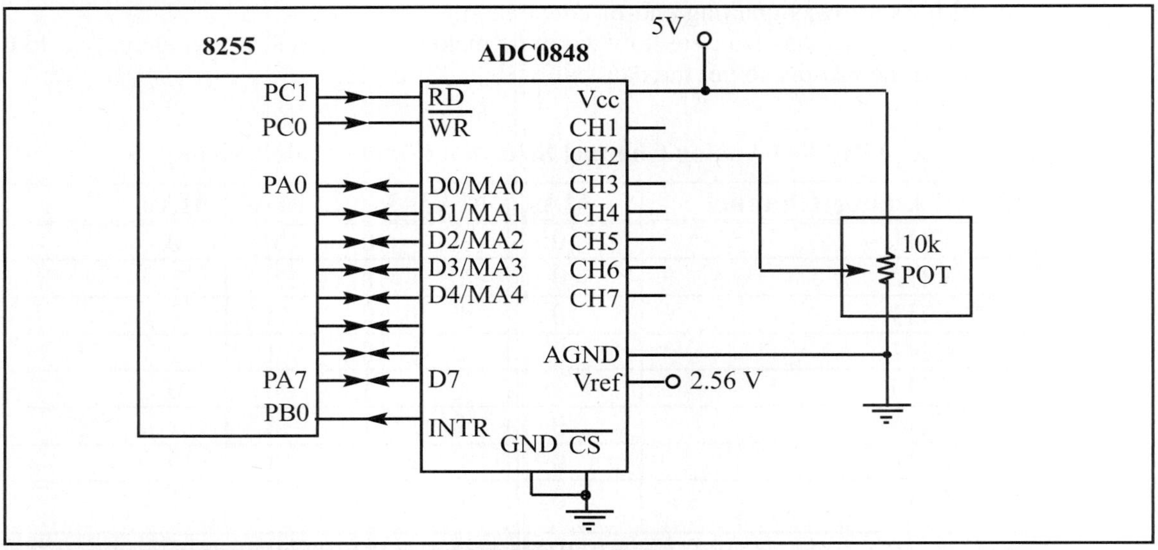

Figure 12-15. 8255 Connection to ADC0848 for CH2

The following program is for Figure 12-15. It selects channel 2. After reading its data, the data is converted from binary (hex) to ASCII. In the program, CL = least significant digit (LSD) and AL = most significant digit (MSD).

```
MOV    AL,82H        ;PA=OUT, PB=IN, PC=OUT
MOV    DX,CNT_PORT
OUT    DX,AL
MOV    AL,09         ;CHANNEL 2 ADDRESS (Table 12-11)
MOV    DX,PORT_A
OUT    DX,AL
MOV    AL,02         ;WR=0,RD=1
MOV    DX,PORT_C
OUT    DX,AL
CALL   DELAY         ;a few microseconds delay
MOV    AL,03         ;WR=1, RD=1
OUT    DX,AL         ;TO LATCH CHANNEL ADDRESS
CALL   DELAY         ;few usec
MOV    AL,92H               ;PA=IN, PB=IN, AND PC=OUT
MOV    DX,CNT_PORT
```

```
        OUT    DX,AL
        MOV    DX,PORT_B     ;GET READY TO MONITOR INTR
B1:     IN     AL,DX         ;MONITOR INTR
        AND    AL,01         ;MASK ALL BITS EXCEPT INTR
        CMP    AL,01         ;IS IT HIGH?
        JNE    B1            ;KEEP MONITORING FOR LOW
        MOV    AL,01         ;WR=1, RD=0 TO READ DATA
        MOV    DX,PORT_C
        OUT    DX,AL
        MOV    DX,PORT_A     ;PORT_A  TO GET  DATA
        IN     AL,DX         ;GET THE CONVERTED DATA
;converting 00-FFH hex value to decimal and then to ASCII.
; AL,AH,CL will have decimal values in ASCII
        MOV    BL,10
        SUB    AH,AH         ;CLEAR AH FOR WORD/BYTE DIV
        DIV    BL            ;AX/BL
        MOV    CL,AH         ;SAVE LSD IN CL REG
        SUB    AH,AH         ;
        DIV    BL            ;AX/BL FOR 2ND DIGIT
;make them all ASCII
        OR     AX,3030H
        OR     CL,30H
```

Notice the conversion of the above data to ASCII. In order to display ADC input on a screen or LCD, it must be converted to ASCII. However, to convert it to ASCII, it must be converted to decimal first. To convert a 00–FF hex value to decimal we keep dividing it by 10 until the remainder is less than 10. Each time we divide it by 10 we keep the quotient as one of our decimal digits. In the case of an 8-bit data, dividing it by 10 twice will do the job. For example, if we have FFH it will become 255 in decimal. To convert from decimal to ASCII format, we OR each digit with 30H. Now all we have to do is to send the digits to the PC screen by using INT 21H or send them to the LCD as was shown in the first section of this chapter. One advantage of using C/C++ programs is that such a conversion is done by the compiler.

Interfacing a temperature sensor to PC

Transducers are used to convert physical quantities such as temperature, light intensity, flow, and speed to electrical signals. Depending on the transducer, the output produced is in the form of voltage, current, resistance, or capacitance. For example, temperature is converted to electrical signals using a transducer called a *thermistor*. The thermistor responds to temperature change by changing its resistance. However, its response is not linear, as shown in Table 12-12.

The complexity associated with writing software for such nonlinear devices has led many manufacturers to market the linear temperature sensor. Simple and widely used linear temperature sensors include the LM34 and LM35 series from National Semiconductor Corp. They are discussed next.

Table 12-12: Thermistor Resistance vs. Temperature

Temperature (C)	Tf (K ohms)
0	29.490
25	10.000
50	3.893
75	1.700
100	0.817

From William Kleitz, *Digital Electronics*

LM34 and LM35 temperature sensors

The sensors of the LM34 series are precision integrated-circuit temperature sensors whose output voltage is linearly proportional to the Fahrenheit temperature. The LM34 requires no external calibration since it is inherently calibrated. It outputs 10 mV for each degree of Fahrenheit temperature. Table 12-13 is the selection guide for the LM34. The LM35 series sensors are also precision integrated-circuit temperature sensors whose output voltage is linearly proportional to the Celsius (centigrade) temperature. The

LM35 requires no external calibration since it is inherently calibrated. It outputs 10 mV for each degree of centigrade temperature. Table 12-14 is the selection guide for the LM35.

Table 12-13: LM34 Temperature Sensor Series Selection Guide

Part Scale	Temperature Range	Accuracy	Output
LM34A	–50 F to +300 F	+2.0 F	10 mV/F
LM34	–50 F to +300 F	+3.0 F	10 mV/F
LM34CA	–40 F to +230 F	+2.0 F	10 mV/F
LM34C	–40 F to +230 F	+3.0 F	10 mV/F
LM34D	–32 F to +212 F	+4.0 F	10 mV/F

Note: Temperature range is in degrees Fahrenheit.

Table 12-14: LM35 Temperature Sensor Series Selection Guide

Part	Temperature Range	Accuracy	Output Scale
LM35A	–55 C to +150 C	+1.0 C	10 mV/F
LM35	–55 C to +150 C	+1.5 C	10 mV/F
LM35CA	–40 C to +110 C	+1.0 C	10 mV/F
LM35C	–40 C to +110 C	+1.5 C	10 mV/F
LM35D	0 C to +100 C	+2.0 C	10 mV/F

Note: Temperature range is in degrees Celsius.

Signal conditioning and interfacing the LM35 to a PC

Signal conditioning is a widely used term in the world of data acquisition. The most common transducers produce an output in the form of voltage, current, charge, capacitance, and resistance. However, we need to convert these signals to voltage in order to send input to an A-to-D converter. This conversion (modification) is commonly called *signal conditioning*. Signal conditioning can be a current-to-voltage conversion or a signal amplification. For example, the thermistor changes resistance with temperature. The change of resistance must be translated into voltages in order to be of any use to an ADC. Look at the case of connecting an LM35 to an ADC0848. Since the ADC0848 has 8-bit resolution with a maximum of 256 steps and the LM35 (or LM34) produces 10 mV for every degree of temperature change, we can condition V_{in} of ADC0848 to produce a V_{out} of 2560 mV (2.56 V) for full-scale output. Therefore, in order to produce the full-scale V_{out} of 2.56 V for the ADC0848, we need to set V_{ref} = 2.56. This makes V_{out} of the ADC0848 correspond directly to the temperature as monitored by the LM35. This is shown in Table 12-15.

Figure 12-17 shows connection of the temperature sensor to CH2 of the ADC0848. Notice that we use the LM336-2.5 zener diode to fix the voltage across the 10k POT at 2.5 volts. The use of the LM336-2.5 should overcome any fluctuations in the power supply. The LM336 has three leads. However, the third lead is unconnected.

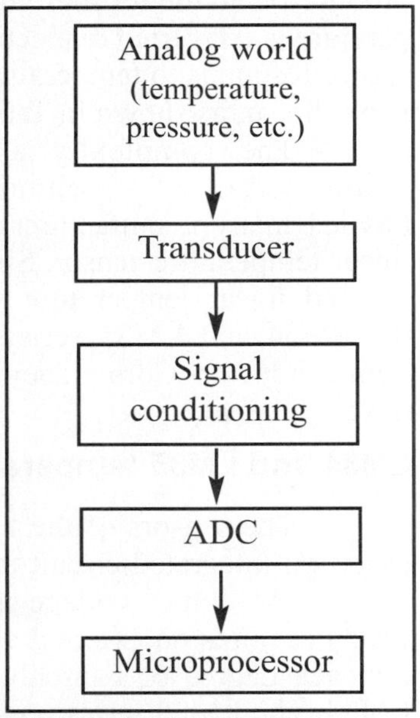

Figure 12-16. Getting Data to the CPU

ADC808/809

Another popular ADC is the ADC808/809 chip. It has eight input channels allowing it to convert 8 different

Table 12-15: Temperature v. Vout of the ADC0848

Temp. (C)	Vin (mV)	Vout (D7–D0)
0	0	0000 0000
1	10	0000 0001
2	20	0000 0010
3	30	0000 0011
10	100	0000 1010
30	300	0001 1110

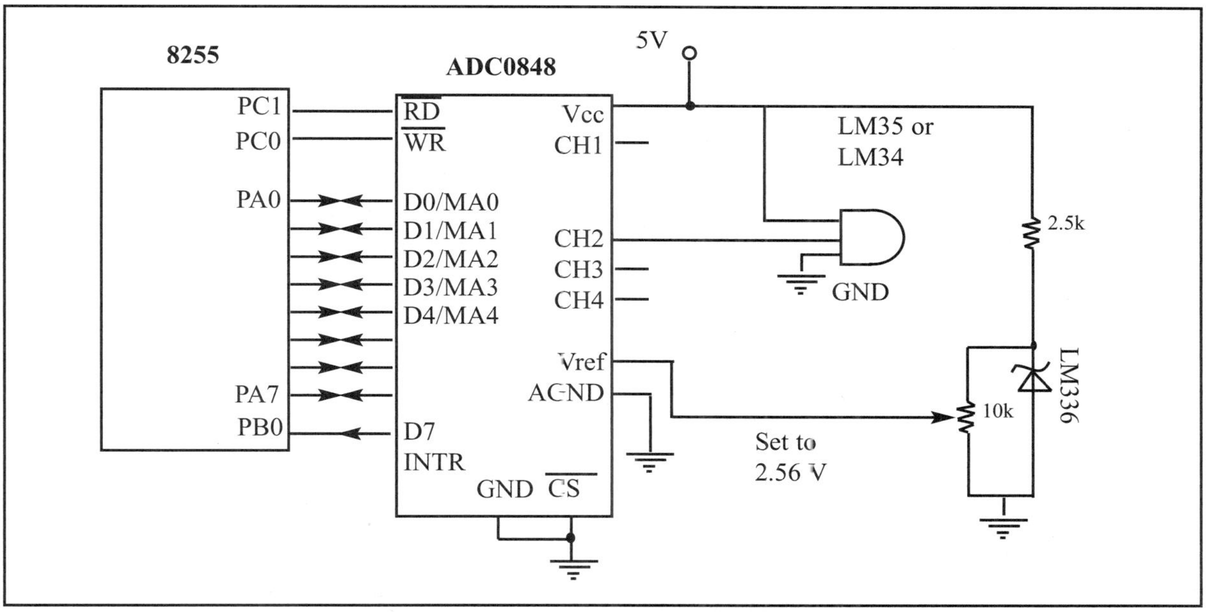

Figure 12-17. 8255 Connection to ADC0848 and Temperature Sensor

analog inputs. See Figure 12-18. It is an 8-bit ADC. The following is the pin description of the ADC808/809 chip.

OE (output enable): This is an input signal and is active high. ADC converts the analog input to its binary equivalent and holds it in an internal register. OE is used to get the converted data out of the ADC808 chip. If a low-to-high pulse is applied to the OE pin, the 8-bit digital output shows up at the D0–D7 data pins. The OE pin is also referred to as RD (read).

SC (start conversion): This is an input pin and is used to inform the ADC808 to start the conversion process. If we apply a low-to-high pulse to this pin, the ADC808 starts converting the analog input value of V_{in} to an 8-bit digital number. The amount of time it takes to convert varies depending on the CLK value. When the data conversion is complete, the EOC (end of conversion) pin is forced low by the ADC808.

CLK is an input pin and is connected to an external clock source. The CLK speed dictates the conversion time. While the ADC0848 uses an internal clock, for the ADC808 the clock source is external. This way one can control the conversion speed.

EOC (end of conversion): This is an output pin and is active low. It is a normally high pin and when the conversion is finished, it goes low to signal the CPU that the converted data is ready to be picked up. After EOC goes low, we send a low-to-high pulse to the OE pin to get the data out of the ADC808 chip.

V_{ref}(+) and V_{ref}(–) are both input voltages used for the reference voltage. The voltage connected to these pins dictates the step size. For the ADC808/809, the step size is $[V_{ref}(+) - V_{ref}(-)]$ /256, since it is an 8-bit ADC and 2 to the power of 8 gives us 256 steps. For example, if the analog input range needs to be 0 to 4 volts, $V_{ref}(+)$ is connected to 4 volts and $V_{ref}(-)$ is grounded. That gives 4 V/256 = 15.62 mV for the step size. In

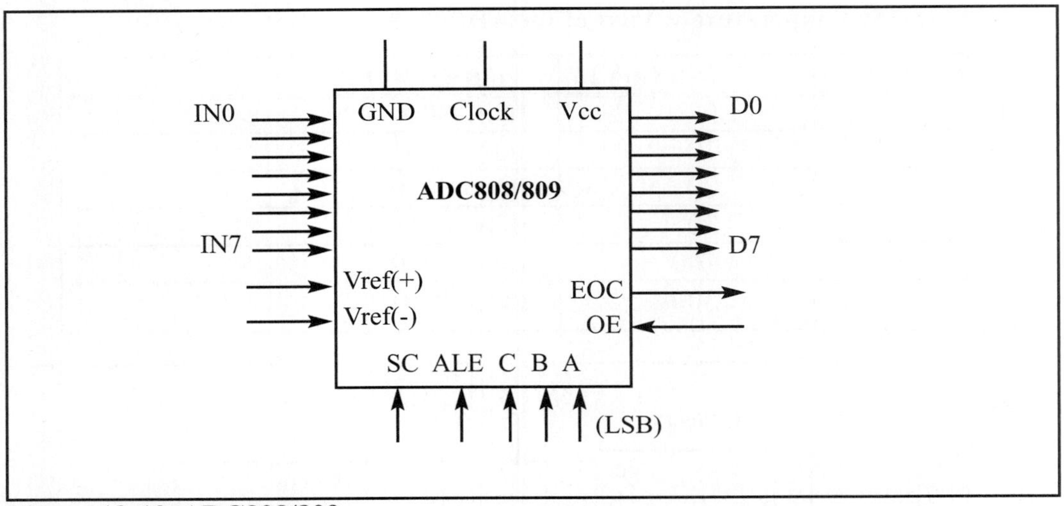

Figure 12-18. ADC808/809

another case, if we need the step size of 10 mV, V_{ref} = 2.56 V since 2.56 V/256 = 10 mV. Notice that if we connect the $V_{ref}(-)$ input to a voltage other than ground, the step size is calculated based on the differential value of the $V_{ref}(+) - V_{ref}(-)$ inputs.

D0–D7 are the digital data output pins. These are tri-state buffered and the converted data is accessed only when OE is forced high. While the analog input voltage is in the range 0 to +5V, the output D0–D7 is given in binary.

IN0–IN7 are the 8 channels of the V_{in} analog inputs. These 8 channels of input allow us to read 8 different analog signals. However, they cannot all be read at the same time since there is only a single D0–D7. We select the input channel by using the A, B, and C address selector pins according to Table 12-16.

A, B, C, and ALE. The input channel is selected by using the A, B, C, and ALE pins. These are input signals into the ADC808/809 and the channel is selected according to Table 12-16. To select a channel, we provide the channel address to the A, B, and C pins according to Table 12-16 and then apply an L-to-H pulse to the ALE pin to latch in the address.

Table 12-16: ADC808/809 Analog Channel Selection

Selected Analog Channel	C	B	A
IN0	0	0	0
IN1	0	0	1
IN2	0	1	0
IN3	0	1	1
IN4	1	0	0
IN5	1	0	1
IN6	1	1	0
IN7	1	1	1

Note: Channel is selected when OE = 0, and an L-to-H pulse is applied to ALE.

How to read ADC808/809 data

Comparing the ADC808/809 with the ADC0848 shows that the ADC808/809 has a clock pin. This means that we must provide an external clock source. Therefore, the conversion speed varies according to the speed of the external clock source. Also, notice that the ADC808/809 has no CS pin.

The following are the steps to select a channel and read its data.

1. Provide the channel address (see Table 12-16) to pins A, B, and C.
2. Apply an L-to-H pulse to the ALE pin to latch in the channel address.
3. Apply an L-to-H pulse to the SC pin to start the conversion of analog input to digital data.
4. After the passage of 8 clocks, the EOC pin will go low to indicate that the data is converted and ready to be picked up. We can either use a small time delay and then read the data out, or monitor the EOC pin and read the data out after it goes low. Notice that if you use a time delay to wait before you read the data, the size of the delay varies depending on the speed of the clock connected to the clock pin of the ADC808/809.
5. Apply an L-to-H pulse to the OE pin and read the data.

Review Questions

1. In the ADC0848, the INTR signal is an _______ (input, output).
2. To begin conversion, send a(n)______ pulse to the ______.
3. Which pin of the ADC0848 indicates end-of-conversion?
4. The LM35 provides ____ mV for each degree of _____ (Fahrenheit, Celsius) temperature.
5. Both the ADC0848 and ADC808 are ________-bit converters.

PROBLEMS

SECTION 12.1: INTERFACING TO AN LCD

1. The LCD discussed in this section has ________ (4, 8) data pins.
2. Describe the function of pins E, R/W, and RS in the LCD.
3. What is the difference between the VCC and VEE pins on the LCD?
4. Clear LCD is a __________ (command code, data item) and its value is ___ hex.
5. What is the hex value of the command code for display on, cursor on?
6. Give the state of RS, E, and R/W when sending a command code to the LCD.
7. Give the state of RS, E, and R/W when sending data character 'Z' to the LCD.
8. Which of the following is needed on the E pin in order for a command code (or data) to be latched in by the LCD?
   (a) H-to-L pulse (b) L-to-H pulse
9. True or false. For the above to work, the value of the command code (data) must be already at the D0–D7 pins.
10. There are two methods of sending streams of characters to the LCD: (1) checking the busy flag, or (2) putting some time delay between each character without checking the busy flag. Explain the difference and advantage and disadvantage of each method. Also explain how we monitor the busy flag.
11. For a 16 × 2 LCD the location of the last character of line 1 is 8FH (its command code). Show how this value came about.
12. For a 16 × 2 LCD the location of the first character of line 2 is C0H (its command code). Show how this value came about.
13. For a 20 × 2 LCD the location of the last character of line 2 is 93H (its command code). Show how this value came about.
14. For a 20 × 2 LCD the location of the third character of line 2 is C2H (its command code). Show how this value came about.
15. For a 40 × 2 LCD the location of the last character of line 1 is A7H (its command code). Show how this value came about.
16. For a 40 × 2 LCD the location of the last character of line 2 is E7H (its command code). Show how this value came about.
17. Show the value (in hex) for the command code for the 10th location, line 1 on a 20 × 2 LCD. Show how you got your value.

18. Show the value (in hex) for the command code for the 20th location, line 2 on a 40 × 2 LCD. Show how you got your value.
19. Rewrite the COMNDWRT procedure (shown in Section 12.1) if port C is used for control signals. Assume that PC4 = RS, PC5 = R/W, PC6 = E.
20. Repeat the above program for a data write procedure. Send the string "Hello" to the LCD without checking the busy flag.

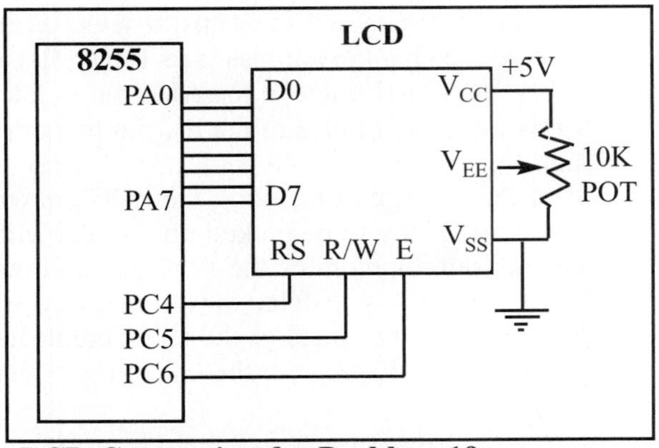

LCD Connection for Problem 19

SECTION 12.2: INTERFACING TO A STEPPER MOTOR

21. If a motor takes 90 steps to make one complete revolution, what is the step angle for this motor?
22. Calculate the number of steps per revolution for a step angle of 7.5 degrees.
23. Finish the normal 4-step sequence clockwise if the first step is 0011 (binary).
24. Finish the normal 4-step sequence clockwise if the first step is 1100 (binary).
25. Finish the normal 4-step sequence counterclockwise if the first step is 1001 (binary).
26. Finish the normal 4-step sequence counterclockwise if the first step is 0110 (binary).
27. What is the purpose of the ULN2003 placed between the 8255 and the stepper motor? Can we use that for 3A motors?
28. Which of the following cannot be a sequence in the normal 4-step sequence for a stepper motor?
    (a) CCH (b) DDH (c) 99H (d) 33H
29. What is the effect of a time delay between issuing each step?
30. In Question 29, how can we make a stepper motor go faster?

SECTION 12.3: INTERFACING TO A DAC

31. True or false. DAC1408 is the same as DAC0808. Are they pin compatible?
32. Find the number of discrete voltages provided by the n-bit DAC for the following.
    (a) $n = 8$ (b) $n = 10$ (c) $n = 12$
33. For DAC1408, if $I_{ref} = 2$ mA show how to get the I_{out} of 1.99 when all inputs are high.
34. Find I_{out} for the following inputs. Assume $I_{ref} = 2$ mA for DAC1408.
    (a) 10011001 (b) 11001100 (c) 11101110
    (d) 00100010 (e) 00001001 (f) 10001000
35. To get a smaller step, we need a DAC with _____ (more, fewer) digital inputs.
36. To get full-scale output, what should be the inputs for DAC?

SECTION 12.4: INTERFACING TO ADC CHIPS AND SENSORS

37. Give the status of CS and WR needed to start conversion for the ADC0848.
38. Give the status of CS and WR needed to get data from the ADC0848.
39. In the ADC0848 what happens to the converted analog data? How do we know that the ADC is ready to provide us the data?
40. In the ADC0848 what happens to the old data if we start conversion again before we pick up the last data?
41. In the ADC0848 INTR is an _______ (input, output) signal. What is its function in the ADC0848?
42. For an ADC0848 chip, find the step size for each of the following V_{ref}.
    (a) $V_{ref} = 1.28$V (b) $V_{ref} = 1$V (c) $V_{ref} = 1.9$V

43. In the ADC0848 what should be the V_{ref} value if we want a step size of 20 mV?
44. In the ADC0848 what should be the V_{ref} value if we want a step size of 5 mV?
45. In the ADC0848 how is the analog channel selected?
46. With a step size of 19.53 mV, what is the analog input voltage if all outputs are 1?
47. With V_{ref} = 1.28V, find the V_{in} for the following outputs.
    (a) D7–D0 = 11111111 (b) D7–D0 = 10011001 (c) D7–D0 = 1101100
48. What does it mean when it is said that a given sensor has a linear output?
49. The LM34 sensor produces ________ mV for each degree of temperature.
50. What is signal conditioning?

ANSWERS TO REVIEW QUESTIONS

SECTION 12.1: INTERFACING TO AN LCD

1. Input
2. Input
3. H-to-L
4. High
5. 80H and C0H

SECTION 12.2: INTERFACING TO A STEPPER MOTOR

1. 0110,0011,1001,1100 for clockwise, and 0110, 1100, 1001, 0011 for counterclockwise
2. 72
3. Because the 8255 does not provide sufficient current to drive the stepper motor

SECTION 12.3: INTERFACING TO A DAC

1. Digital, analog. In ADC the input is analog, the output is digital.
2. 8
3. Current

SECTION 12.4: INTERFACING TO ADC CHIPS AND SENSORS

1. Output
2. L-to-H, WR pin
3. INTR
4. 10, both
5. 8

CHAPTER 13

8253/54 TIMER

OBJECTIVES

Upon completion of this chapter, you will be able to:

- **>> Describe the function of each pin of the 8253/54 PIT (programmable interval timer)**
- **>> Program the three counters of the 8253/54 by use of the chip's control word**
- **>> Diagram how the 8253/54 timer is connected in the IBM PC**
- **>> Write programs to play music notes on the x86 PC speaker**

In the PC there is a single clock used to synchronize activities of all peripheral chips connected to the CPU. That clock, which has the highest frequency in the system, belongs to the x86 CPU. There are functions within the PC that require a clock with a lower frequency. The 8253/54 PIT (programmable interval timer) is used to bring down the frequency to the desired level for various uses such as the beep sound in the PC. This chapter focuses on 8253/54 timer chip and how it is used in the x86 PC. In Section 13.1 we will describe the 8253/54 timer and show the processes of initializing and programming it. The interfacing and the use of the 8253/54 in the x86 PC is discussed in Section 13.2. Section 13.3 will show how the 8253/54 can be used to generate various frequencies, including musical notes on the x86 PC.

SECTION 13.1: 8253/54 TIMER

The 8253 chip was used in the IBM PC/XT, but starting with the IBM PC AT, the 8254 replaced the 8253. The 8254 and 8253 have exactly the same pinout. The 8254 is a superset of the 8253, meaning that all programs written for the 8253 will run on the 8254. The following are pin descriptions of the 8253/54.

A0, A1, CS

Inside the 8253/54 timer, there are three counters. Each works independently and is programmed separately to divide the input frequency by a number from 1 to 65,536. Each counter is assigned an individual port address. The control register common to all three counters has its own port address. This means that a total of 4 ports are needed for a single 8253/54 timer. The ports are addressed by A0, A1, and CS, as shown in Table 13-1. Each of the three counters has three pins associated with it, CLK (clock), GATE, and OUT, as shown in Figure 13-1. See Example 13-1.

Table 13-1: Addressing 8253/54

$\overline{\text{CS}}$	A1	A0	Port
0	0	0	Counter 0
0	0	1	Counter 1
0	1	0	Counter 2
0	1	1	Counter register
1	x	x	8253/54 is not selected

CLK

CLK is the input clock frequency, which can range between 0 and 2 MHz for the 8253. For input frequencies higher than 2 MHz, the 8254 must be used; the 8254 can go as high as 8 MHz, and the 8254-2 can go as high as 10 MHz.

OUT

Although the input frequency is a square wave of 33% duty cycle, the shape of the output frequency coming from the OUT pin after being divided can be programmed. Among the options are square wave, one shot, and other square shape waves of various duty cycles but no sine wave or saw tooth shapes.

GATE

This pin is used to enable or disable the counter. Putting HIGH (5 V) on GATE enables the counter, whereas LOW (0 V) disables it. In some modes a 0 to 1 pulse must be applied to GATE to enable the counter.

D0–D7

The D0–D7 data bus of the 8253/54 is a bidirectional bus connected to D0–D7 of the system data bus. The data bus allows the CPU to access various registers inside the 8253/54 for both read and write operations. RD and WR (both active low) are connected to IOR and IOW control signals of the system bus.

Initialization of the 8253/54

Each of the three counters of the 8253/54 must be programmed separately. In order to program any of the three counters, the control byte must first be written into the control register, which among other things tells the 8253/54 what shape of output pulse is needed. In addition, the number that the input clock should be divided by must be written into that counter of the 8253/54. Since this number can be as high as FFFF (16-bit data) and the data bus for the 8253/54 timer is only 8 bits wide, the divisor must be sent in one byte at a time. The 8253/54 must be initialized before it is used.

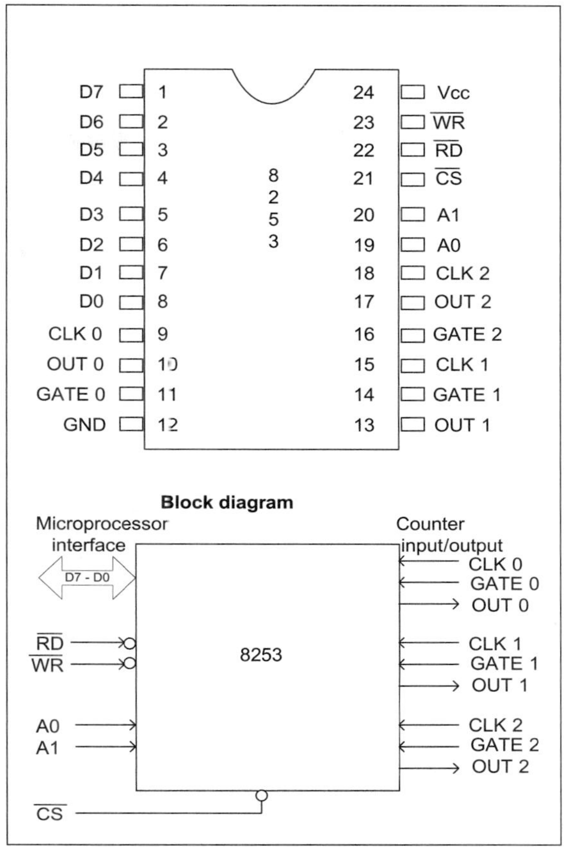

Figure 13-1. 8253 Pin and Function Diagram

Control word

Figure 13-2 shows the one-byte control word of the 8253/54. This byte, which is sent to the control register, has the following bits.

D0 chooses between a binary number divisor of 0000 to FFFFH or a BCD divisor of 0000 to 9999H. The lowest number that the input frequency can be divided by for both options is 0001. The highest number is 216 for binary and 104 for BCD. To get the highest count (65,536 decimal and 10000 BCD), the counter is loaded with zeros.

D1, **D2**, and **D3** are for mode selection. There are six possible modes that determine the shape of the output signal.

Mode 0	Interrupt on terminal count
Mode 1	Programmable one shot
Mode 2	Rate generator
Mode 3	Square wave rate generator
Mode 4	Software triggered strobe
Mode 5	Hardware triggered strobe

D4 and **D5** are for RL0 and RL1. The data bus of the 8253/54 is 8 bits (1 byte), but the number that the input frequency can be divided by (divisor) can be as high as FFFFH. Therefore, RL0 and RL1 are used to indicate the size of the divisor. RL0 and RL1 have three options: (1) read/write the most significant byte (MSB) only, (2) read/write the least significant byte (LSB) only, (3) read/write the LSB first followed immediately by the MSB.

The options for RL0 and RL1 show that programmers can not only write the value of the divisor into the 8253/54 timer but read the contents of the counter at any given time, as well. Since all counters are down counters, and the count register is decremented, the count register's contents can be read at any time, thus using the 8253/53 as an event counter.

D6 and **D7** are used to select which of the three counters, counter 0, counter 1, or counter 2, is to be initialized by the control byte.

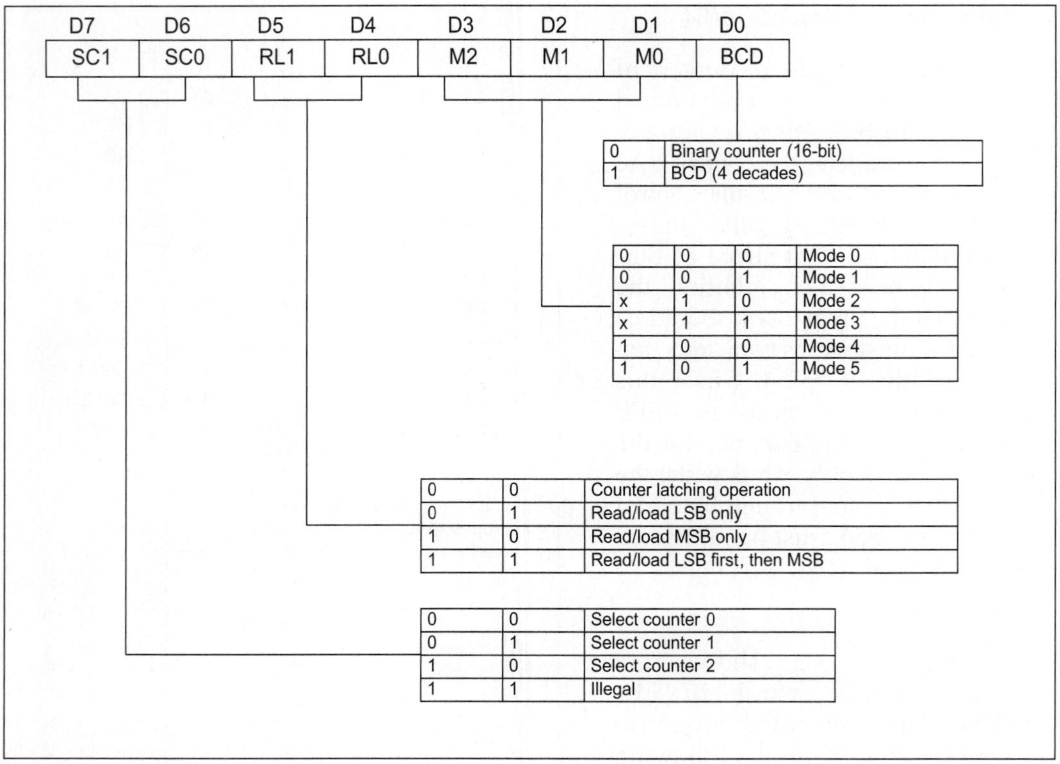

Figure 13-2. 8253/54 Control Word Format

Example 13-1

Pin CS of a given 8253/54 is activated by binary address A7–A2 = 100101.
(a) Find the port addresses assigned to this 8253/54.
(b) Find the configuration for this 8253/54 if the control register is programmed as follows.

```
MOV   AL,00110110
OUT   97H,AL
```

Solution:

(a) From Table 13-1, we have the following:

CS	A1A0	Port	Port address (hex)
1001 01	0 0	Counter 0	94
1001 01	0 1	Counter 1	95
1001 01	1 0	Counter 2	96
1001 01	1 1	Control register	97

(b) Breaking down the control word 00110110 and comparing it with Table 13-1 indicates counter 0 since the SC bits are 00. The RL bits of 11 indicate that the low-byte read/write is followed by the high byte. The mode selection is mode 3 (square wave), and finally binary counting is selected since the D0 bit is 0.

To program a given counter of the 8253/4 to divide the CLK input frequency, one must send the divisor to that specific counter's register. In other words, although all three counters share the same control register, the divisor registers are separate for each counter.

Regarding the options bit D0 of the control byte, it must be noted that in BCD

Example 13-2

Using the port addresses in Example 13-1, show the programming of counter 1 to divide CLK1 by 10,000, producing the mode 3 square wave. Use the BCD option in the control byte.

Solution:

```
MOV   AL,77H        ;counter 1, mode 3, BCD
OUT   97H,AL        ;send it to control register
SUB   AL,AL         ;AL = 0 load the divisor for 10,000
OUT   95H,AL        ;send the low byte
OUT   95H,AL        ;and then the high byte to counter 1
```

Example 13-3

Use the port addresses in Example 13-1 to:
(a) program counter 0 for binary count of mode 3 (square wave) to divide CLK0 by number 4282 (BCD),
(b) program counter 2 for binary count of mode 3 (square wave) to divide CLK2 by number C26A hex,
(c) find the frequency of OUT0 and OUT2 in (a) and (b) if CLK0 = 1.2 MHz, CLK2 = 1.8 MHz.

Solution:

(a) To program counter 0 for mode 3, we have 00110111 for the control word. Therefore,

```
MOV   AL,37H        ;counter 0, mode 3, BCD
OUT   97H,AL        ;send it to control register
MOV   AX,4282H      ;load the divisor (BCD needs H for hex)
OUT   94H,AL        ;send the low byte
MOV   AL,AH         ;to counter 0
OUT   94H,AL        ;and then the high byte to counter 0
```

(b) By the same token:

```
MOV   AL,B6H        ;counter 2, mode 3, binary(hex)
OUT   97H,AL        ;send it to control register
MOV   AX,C26AH      ;load the divisor
OUT   96H,AL        ;send the low byte
MOV   AL,AH         ;to counter 2
OUT   96H,AL        ;send the high byte to counter 2
```

(c) The output frequency for OUT0 is 1.2 MHz divided by 4282, which is 280 Hz. Notice that the program in part (a) used instruction "MOV AX,4282H" since BCD and hex numbers are represented in the same way, up to 9999. For OUT2, CLK2 of 1.8 MHz is divided by 49770 since C26AH = 49770 in decimal. Therefore, OUT2 frequency is a square wave of 36 Hz.

mode, if we program the counter for 9999, the input frequency is divided by that number. However, to divide the frequency by 10,000 we must send in 0 for both high and low bytes. See Examples 13-2 and 13-3.

We can program any of the counters for divisors of up to 65,536 if we use the binary option for D0. To program the counter for the divisor of 65,536, the counter must be loaded with 0 for the low byte and another 0 for the high byte of the divisor. In that case, D0 = 0 for the control byte.

Review Questions

1. True or false. Any code written for the 8253 will work on the 8254.
2. The 8253/54 can be used to ___________ (divide, multiply) a square wave digital frequency.
3. If CS of the 8253/54 is activated by A7–A2 = 0110 00 binary, find the port address for this timer.
4. Find the control byte to program counter 2 for mode 1 (programmable one shot), BCD count, low byte, followed by high byte R/W.
5. True or false. To divide input frequency CLK1 by 5065, we must send the 5065 to the control register.
6. For Question 5, give the port address using the ports in Question 3.
7. To divide the CLK frequency by 52,900, which option for D0 of the control byte must be selected, and why?
8. If D0 = 0 in the control byte, what is the highest number for the divisor?

SECTION 13.2: x86 PC 8253/54 TIMER CONNECTION AND PROGRAMMING

The first IBM PC used a 74LS138 to decode addresses for CS of the 8253 as shown in Figure 13-3. The port addresses are selected as indicated in Table 13-2, assuming zeros for x's. Chapter 11 contains a complete discussion of port selection.

The three clocks of the 8253, CLK0, CLK1, and CLK2, are all connected to a constant frequency of 1.1931817 MHz. This frequency is from PCLK of the 8284 chip after it has been divided by 2 with the use of D flip-flop 72LS175, as shown in Figure 13-4. PCLK of the 8284 (discussed in Chapter 9) is 2.3863633 MHz and must be divided by 2 since the maximum allowed input frequency of CLK of the 8253 is 2 MHz.

Table 13-2: 8253/54 Port Address Calculation in the x86 PC

Binary Address				
$\overline{AEN}$ A9 A8	A7 A6 A5 A4	A3 A2 A1 A0	Hex Address	Function
1 0 0	0 1 0 x	x x 0 0	40	Counter 0
1 0 0	0 1 0 x	x x 0 1	41	Counter 1
1 0 0	0 1 0 x	x x 1 0	42	Counter 2
1 0 0	0 1 0 x	x x 1 1	43	Counter register

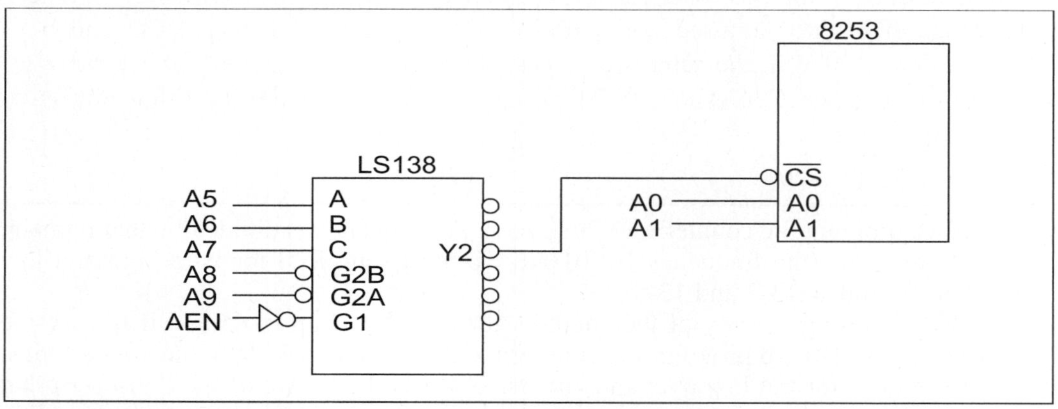

Figure 13-3. 8253 Port Selection in the x86 PC

GATE0 and GATE1, which enable counter 0 and counter 1, respectively, are connected to HIGH (5 V), thereby making those two counters enabled permanently. GATE2 of counter 2 can be enabled or disabled through PB0 of port B of the 8255. Now that the input frequency to each timer is known, programming and applications of each counter in the PC can be explained.

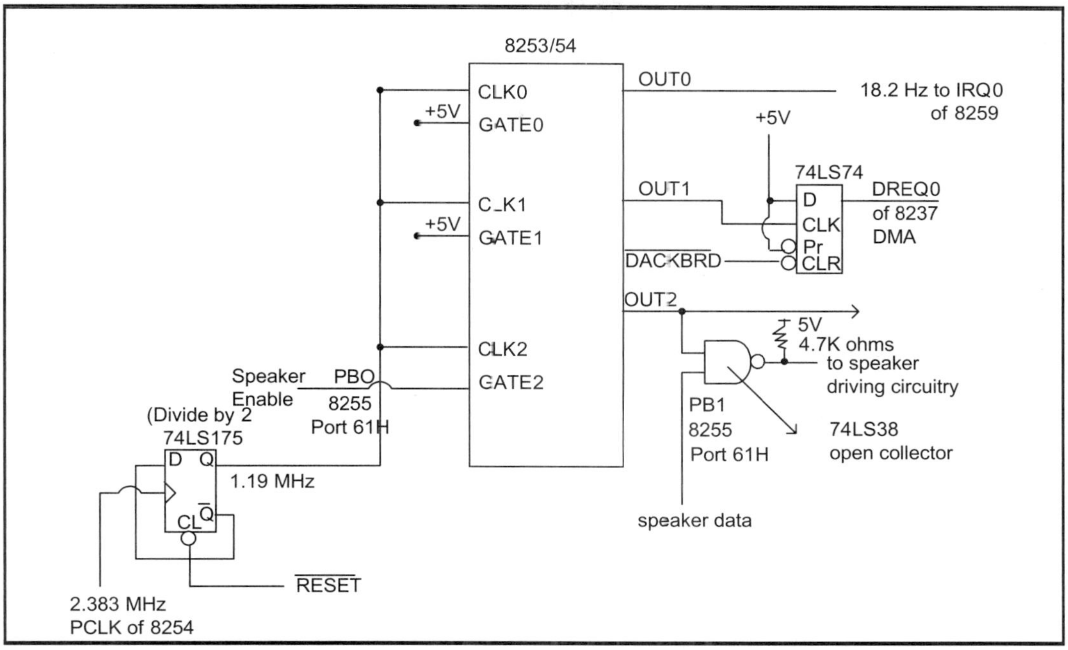

Figure 13-4. 8253/54 Chip Connection in the x86 PC

Using counter 0

CLK0 of counter 0 is 1.193 MHz, and GATE0 is connected to high permanently. OUT0 of counter 0 is connected to IRQ0 (the highest priority interrupt) of the 8259 interrupt controller. The next question is: How often is IRQ0 activated, or in other words, what is the output frequency? IRQ0 is activated 18.2 times per second, or put another way, the OUT0 frequency is 18.2 Hz. If the frequency of CLK0 is 1.193 MHz and the output frequency should be 18.2 Hz, the counter must be programmed to divide 1.193 MHz by 65,536. The wave shape is a square wave (mode 3 of the 8253) in order to trigger IR0 on the positive edge of each pulse of the square wave so that a high pulse will not be mistaken for a multiple interrupt. Using the above information and Figure 13-2, the control word can be calculated in the following way:

D0 = 0 for the binary (or hex) value of the counter divisor. The timer is decremented after every input pulse until it reaches zero and then the original value is loaded again. Therefore, to divide the input frequency by 65,536, the timer is programmed with 0s for both high and low bytes.

D3 D2 D1 = 011, mode 3, for the square wave output of 18.2 Hz frequency.
D4 D5 = 11 for reading/writing the LSB first, followed by the MSB.
D7 D6 = 00 for counter 0.

Summarizing the above gives the following control word:

```
D7  D6  D5  D4        D3  D2  D1  D0
 0   0   1   1         0   1   1   0  = 36H
```

The programming of counter 0 is as follows:

```
MOV   AL,36H        ;control word
OUT   43H,AL        ;to control register of 8253
MOV   AL,00         ;00 LSB and MSB of the divisor
OUT   40H,AL        ;LSB to timer 0
OUT   40H,AL        ;MSB  to timer 0
```

The IBM PC BIOS shows the same process as follows:

```
                 TIMER EQU 40H
     ..          ...
E277 B036        MOV   AL,36H       ;SET TIM 0,LSB,MSB,MODE 3
E279 E643        OUT   TIMER+3,AL   ;WRITE TIMER MODE REG
E27B B000        MOV   AL,0
E27D E640        OUT   TIMER,AL     ;WRITE LSB TO TIMER 0 REG
     ..          ...
E284 E640        OUT   TIMER,AL     ;WRITE MSB TO TIMER 0 REG
```

At the rate of 18.2 Hz (or every 54.94 ms), BIOS will make this interrupt available by going to the vector table of INT 1CH. The user can define CS:IP of a service routine at the vector location belonging to INT 1CH and use it for any purpose, as will be seen in Chapter 14. If the user is not using this interrupt, control will automatically be returned to BIOS.

Using counter 1

In counter 1, CLK1 is connected to 1.193 MHz and GATE is high permanently. OUT1 generates a periodic pulse required to refresh DRAM memory of the computer. This refreshing must be done at least every 15 us for each cell. As will be discussed in Chapter 15, in the IBM PC the task of refreshing DRAM is performed by the 8237 DMA. It is up to the 8253's counter 1 to inform DMA periodically, lest the allowed time pass. To achieve this, OUT1 will provide DMA a pulse of approximately 15 us duration or 66,278 Hz. This means that counter 1 must divide the input frequency 1.19318 MHz by 18 (1.19318 MHz divided by 18 = 66,278 Hz). Using Figure 13-2, the control byte can be figured out as follows:

D0 = 0 for binary option

D3 D2 D1 = 010 for mode 2 shape output. In this mode, OUT1 stays high for a total of 18 pulses and goes low for one pulse. This action is repeated continuously.

D5 D4 = 01 for the LSB only, since the byte is less than FF. CLK1 is divided by 18; therefore, 18 is the LSB and there is no need for the MSB.

D7 D6 = 01 for counter 1

```
D7 ... D0
0101 0100 = 54H for the control word
```

The programming of the 8253 counter 1 in the IBM BIOS is listed as follows, with slight modifications for the sake of clarity:

```
MOV   AL,54H          ; the control word
OUT   43H,AL          ; to control register
MOV   AL,18           ;18 decimal, the divisor
OUT   41H,AL          ; to counter 1
```

Using counter 2

The output of counter 2 is connected to two different devices: the speaker and PC5 of the 8255. In early models of the IBM PC/XT, it was also connected to the cassette circuitry. That option has been eliminated in all the IBM PC and PS/2 computers built in recent years. Since counter 2 in the IBM PC is used to play music, it is important to understand counter 2 programming thoroughly.

Use of timer 2 by the speaker

In the IBM PC, CLK2 is connected to a frequency of 1.19318 MHz, and GATE2 is programmed by PB0 of port 61H (port B). The IBM PC uses counter 2 to generate the beep sound. Although BIOS uses timer 2 for the beep sound, it can be changed to play any musical note, as will be shown in the next section. The beep sound has a frequency of 896 Hz of mode 3 (square wave). Dividing the input frequency of 1.19318 MHz by 896 Hz gives 1331 (0533 hex) for the value to be loaded to counter 2. This gives the following control word:

D7 ... D0

1011 0110 = B6H for binary option, mode 3 (square wave), LSB first, then MSB, counter 2. The program would be as follows:

```
MOV    AL,0B6H               ;control word
OUT    43H,AL
MOV    AL,33H                ;low byte
OUT    42H,AL
MOV    AL,05                 ;high byte
OUT    42H,AL
```

or it can be written as follows:

```
TIMER EQU    40H
.........    ........
MOV    AL,10110110B          ;SET TIM 2,LSB,MSB,BINARY
OUT    TIMER+3,AL            ;WRITE THE TIMER MODE REG
MOV    AX,533H               ;DIVISOR FOR 1000 HZ (896 HZ)
OUT    TIMER+2,AL            ;WRITE TIMER 2 CNT    LSB
MOV    AL,AH
OUT    TIMER+2,AL            ;WRITE TIMER 2 CNT    MSB
```

Turning on the speaker via PB0 and PB1 of port 61H

The process of turning on the speaker is the same for all the x86 PCs regardless of the microprocessor used. As can be seen from Figure 13-4, GATE2 must be high to provide the CLK to timer 2. This function is performed by PB0 of port 61H. Again from Figure 13-4, OUT2 of timer 2 is ANDed with PB1 of port 61H, then is input to the driving circuitry of the speaker. Therefore, to allow OUT2 to go to the speaker, PB1 of port 61H must be set to high as well. The following is the code to turn the speaker on, which is exactly the same as the IBM BIOS's code to sound the BEEP.

```
IN     AL,61H                ;GET THE CURRENT SETTING OF PORT B
MOV    AH,AL                 ;SAVE IT
OR     AL,00000011B          ;MAKE PB0=1 AND PB1=1
OUT    61H,AL                ;TURN THE SPEAKER ON.
{HOW LONG THE BEEP SHOULD SOUND GOES HERE}
MOV    AL,AH                 ;GET THE ORIGINAL SETTING OF PORT B
OUT    61H,AL                ;TURN OFF THE SPEAKER
```

The amount of time that a musical note is played is referred to as its time delay and is produced with the help of the CPU in the x86 PC.

Time delay for x86 PCs

Time delays are often needed for various applications. Using the instructions of the x86 CPU to generate the delay is unreliable since the CPU speed varies among the x86 PCs. For example, the following delay subroutine is dependent on CPU speed, and therefore unacceptable:

```
        SUB    CX,CX

G7:     LOOP   G7
        DEC    BL
        JNZ    G7
```

This is the reason that the x86 PC provides a scheme to create a time delay using hardware that is CPU speed independent. To create a CPU-independent delay, x86 makes PB4 of port 61H toggle every 15.085 microseconds. That means that by monitoring PB4 of port 61H, a fixed time delay can be obtained, as shown next. Upon entering this subroutine called WAITF, register CX must hold the number of 15.085-microsecond time delays needed.

```
;(CX) = COUNT OF 15.085 MICROSECONDS
WAITF PROC  NEAR
      PUSH  AX
WAITF1:
      IN    AL,61H
      AND   AL,10H          ;CHECK PB4
      CMP   AL,AH       ;DID IT JUST CHANGE?
      JE    WAITF1      ;WAIT FOR CHANGE
      MOV   AH,AL       ;SAVE THE NEW PB4 STATUS
      LOOP  WAITF1      ;CONTINUE UNTIL CX BECOMES 0
      POP   AX
      RET
WAITF ENDP
```

Now a time delay of any duration can be created regardless of the x86 CPU frequency. For example, to create a half-second delay, set CX = 33,144 (33,144 × 15.085 μs = 1/2 second), and then call the above routine:

```
      MOV   CX,33144    ;1/2-second delay
      CALL  WAITF
```

See Example 13-4.

Example 13-4

Using the BIOS WAITF routine, show how to create a 1.5-second time delay.

Solution:

Since the 1.5-second delay requires the counter to be set to 99,436 (1.5 s/15.085 μs = 99,436) and the maximum value of CX is 65,536, the following method is used to generate the 1.5-second delay.

```
            MOV   BL,03
BACK:       MOV   CX,33144    ;1/2-second delay
            CALL  WAITF
            DEC   BL
            JNZ   BACK
```

Review Questions

1. What port addresses are assigned to the 8253/54 timer on the x86 PC motherboard?
2. Of the three counters of the 8253/54 timer on the PC motherboard, which one is used for the speaker, and what port address belongs to it?
3. True or false. In the PC, counters 0 and 1 are used for internal system use.
4. True or false. While the user can program counter 2, users cannot program counters 0 and 1 since they are for system use only.

5. True or false. In the PC, while GATE0 and GATE1 are high permanently, GATE2 can be controlled by the user.
6. In the PC, how is GATE2 controlled by the user?
7. Find the time delay generated by the following code using the method of monitoring PB4 of port 61H in x86 PCs and compatibles.

```
      MOV   DL,200
BACK: MOV   CX,16572      ;delay=16572 x 15.085 microsec
WAIT: IN    AL,61H
      AND   AL,10H        ;check PB4
      CMP   AL,AH         ;did it just change?
      JE    WAIT          ;wait for change
      MOV   AH,AL         ;save the new PB4 status
      LOOP  WAIT          ;continue until CX becomes 0
      DEC   DL
      JNZ   BACK          ;try until DL is 0
```

SECTION 13.3: GENERATING MUSIC ON THE x86 PC

As mentioned earlier, counter 2 is connected to the speaker and it can be programmed to output any frequency that is desired. First, look at the list of piano notes and their frequencies given in Figure 13-5. Since the input frequency to counter 2 is fixed at 1.1931817 MHz for all x86 PCs, programs for playing music found in this section can run on any of them without modification. To play music, the input frequency of 1.1931817 MHz is divided by the desired output frequency to get the value that must be loaded into counter 2. See Examples 13-5 and 13-6.

Example 13-5

Show the values to be loaded into counter 2 in order to have the output frequency for the notes (a) D3, (b) A3, and (c) A4.

Solution:

From Figure 13-5, notice that the frequency for note D3 is 147. The value that must be loaded into counter 2 is 1.1931 MHz divided by 147, which is 8116. Going through this procedure for each note gives the following:

		Value Loaded into Counter 2	
Note	Frequency	Decimal	Hex
D3	147 Hz	8116	1FB4
A3	220 Hz	5423	152F
A4	440 Hz	2711	0A97

Now that the values to be loaded into counter 2 are known, the program for getting the speaker to sound the notes for a certain duration is shown in Example 13-6.

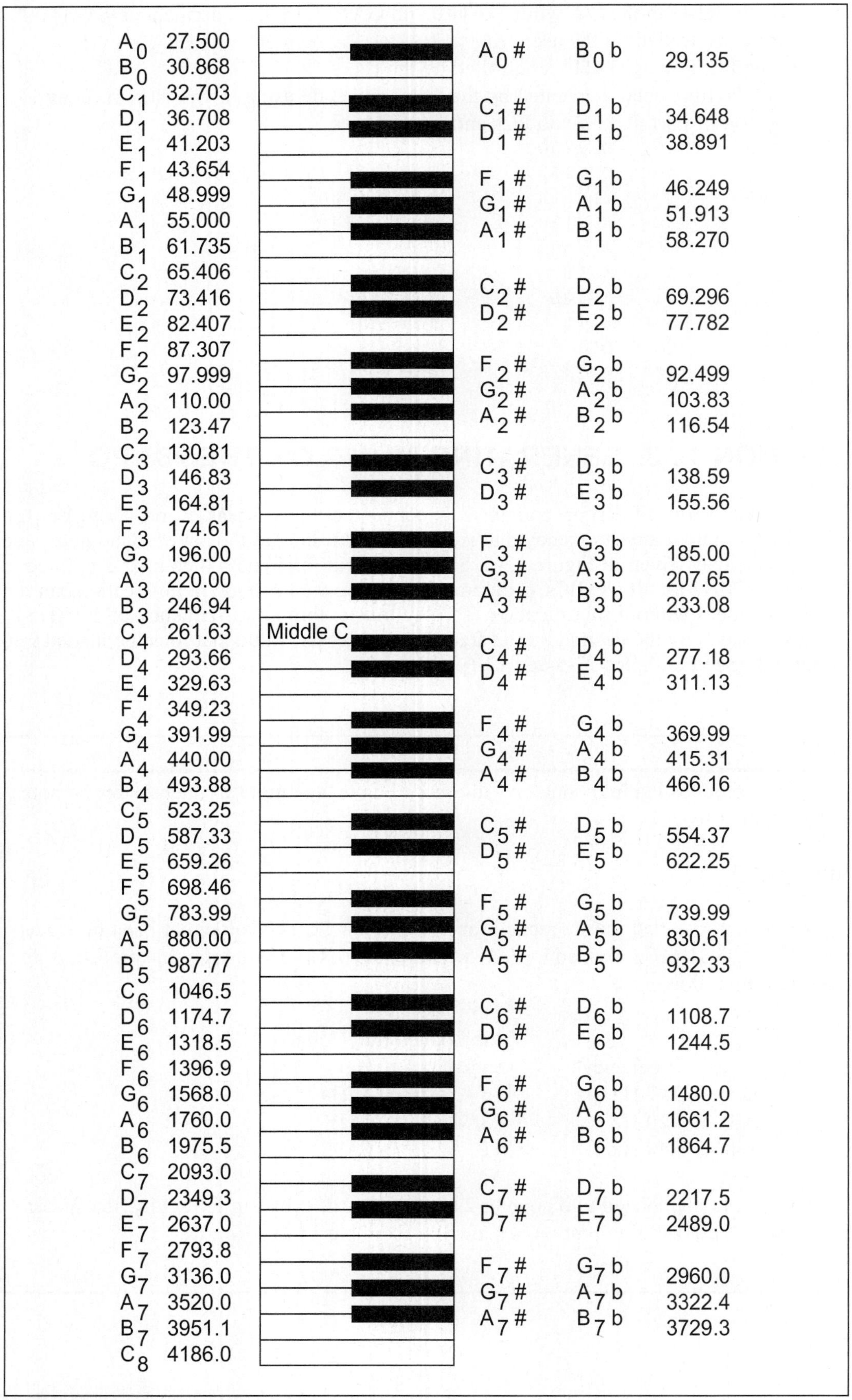

Figure 13-5. Piano Note Frequencies

Example 13-6

Program counter 2 to play the following notes: D3, A3, A4, for durations of 250, 500, and 500 ms, respectively. Place a 5-ms silence between each note.

Solution: This program uses the values calculated in Example 13-5.

```
      MOV   AL,0B6H      ;control byte:counter 2, LSM, MSB, binary
      OUT   43H,AL       ;send the control byte to control reg
;load the counter 2 value for D3 and play it for 250 ms
      MOV   AX,1FB4H     ;for D3 note
      OUT   42H,AL       ;the low byte
      MOV   AL,AH
      OUT   42H,AL       ;the high byte
;turn the speaker on
      IN    AL,61H       ;get the current setting of port B
      MOV   AH,AL        ;save it
      OR    AL,00000011B;make PB0 = 1 and PB1 = 1
      OUT   61H,AL       ;turn the speaker on
      CALL  DELAY        ;play this note for 250 ms
      MOV   AL,AH        ;get the original setting of port B
      OUT   61H,AL       ;turn off the speaker
      CALL  DELAY_OFF    ;speaker off for this duration
;load the counter 2 value for A3 and play it for 500 ms
      MOV   AX,152FH     ;for A3 note
      OUT   42H,AL       ;the low byte
      MOV   AL,AH
      OUT   42H,AL       ;the high byte
;turn the speaker on
      IN    AL,61H       ;get the current setting of port B
      MOV   AH,AL        ;save it
      OR    AL,00000011B;make PB0 = 1 and PB1 = 1
      OUT   61H,AL       ;turn the speaker on
      CALL  DELAY        ;play for 250 ms
      CALL  DELAY        ;play for another 250 ms
      MOV   AL,AH        ;get the original setting of port B
      OUT   61H,AL       ;turn off the speaker
      CALL  DELAY_OFF    ;speaker off for this duration
;load the counter 2 value for A4 and play it for 500 ms
      MOV   AX,0A97H     ;for A4 note
      OUT   42H,AL       ;the low byte
      MOV   AL,AH
      OUT   42H,AL       ;the high byte
;turn the speaker on
      IN    AL,61H       ;get the current setting of port B
      MOV   AH,AL        ;save it
      OR    AL,00000011B;make PB0 = 1 and PB1 = 1
      OUT   61H,AL       ;turn the speaker on
      CALL  DELAY        ;play for 250 ms
      MOV   AL,AH        ;get the original setting of port B
      OUT   61H,AL       ;turn off the speaker
      CALL  DELAY_OFF    ;speaker off for this duration
```

For a delay of 250 ms in x86 PCs, the following routine can be used.

```
;this 250 ms delay for x86 PCs
;
      DELAY        PROC   NEAR
                   MOV    CX,16578      ;16578 x 15.08 microsec = 250
ms
                   PUSH   AX
      WAIT:
                   IN     AL,61H
                   AND    AL,10H        ;check PB4
                   CMP    AL,AH         ;did it just change?
                   JE     WAIT          ;wait for change
                   MOV    AH,AL         ;save the new PB4 status
                   LOOP   WAIT          ;decrement CX and continue
                                        ;until CX becomes 0
                   POP    AX
                   RET
      DELAY        ENDP
```

A delay of 5 ms between notes can be achieved in the same way.

```
      DELAY_OFF    PROC   NEAR
                   MOV    CX,331        ;331 x 15.08 microsec = 5 ms
                   PUSH   AX
      WAIT:        IN     AL,61H
                   AND    AL,10H        ;check PB4
                   CMP    AL,AH         ;did it just change?
                   JE     WAIT          ;wait for change
                   MOV    AH,AL         ;save the new PB4 status
                   LOOP   WAIT          ;continue until CX becomes 0
                   POP    AX
                   RET
      DELAY_OFF    ENDP
```

Playing "Happy Birthday" on the PC

This background should be sufficient to develop a program to play any song. The tune for the song "Happy Birthday" is given below.

Lyrics	Notes	Freq. (Hz)	Duration
hap	C4	262	1/2
py	C4	262	1/2
birth	D4	294	1
day	C4	262	1
to	F4	349	1
you	E4	330	2
hap	C4	262	1/2
py	C4	262	1/2
birth	D4	294	1
day	C4	262	1
to	G4	392	1
you	F4	349	2
hap	C4	262	1/2
py	C4	262	1/2
birth	C5	523	1
day	A4	440	1
dear	F4	349	1
so	E4	330	1

so	D4	294	3
hap	B4b	466	1/2
py	B4b	466	1/2
birth	A4	440	1
day	F4	349	1
to	G4	392	1
you	F4	349	2

The following program plays the first seven notes of the "Happy Birthday" song on Windows NT/2000/XP. Notice we are using arrays for notes and duration. This program was tested using MASM.

```
;Tested by Danny Causey and Hanani Bonda
;This program plays the first 7 notes of "Happy Birthday" song
      .MODEL SMALL
      .STACK 64
      .DATA
NOTES        DW     11CAH,11CAH,0FDAH,11CAH,0D5BH,0E1FH,11CAH
Duration     DB     2,2,4,4,4,8,2
      .CODE
START PROC  FAR
      MOV   AX,@DATA
      MOV   DS,AX
      MOV   AL,0B6H     ;control byte: counter 2, LSB, MSB, binary
      OUT   43H,AL       ;send the control byte to control reg
      MOV   BX,7         ;set up counter
      LEA   SI,NOTES
      LEA   DI,Duration
AGAIN:MOV   AX,[ SI]     ;set up pointer
      OUT   42H,AL
      MOV   AL,AH
      OUT   42H,AL
      MOV   DL,[ DI]
      CALL  SpkON
      INC   SI           ;increments NOTES pointer
      INC   SI
      INC   DI
      DEC   BX           ;decrements the counter
      JNZ   AGAIN  ;if CX is not zero run the loop until CX=0
      MOV   AH,4CH
      INT   21H
START ENDP

SpkON PROC
      IN     AL,61H      ;get the current setting of port B
      MOV    AH,AL       ;save it
      OR     AL,00000011B;make PB0 = 1 and PB1 = 1
      OUT    61H,AL      ;turn the speaker on
BACK: CALL   DELAY       ;play this note for 250 ms
      DEC    DL
      CMP    DL,00
      JNE    BACK
      MOV    AL,AH       ;get the original setting of port B
      OUT    61H,AL      ;turn off the speaker
      CALL   DELAY_OFF   ;speaker off for this duration
      RET
SpkON ENDP
```

```
DELAY_OFF PROC    NEAR
      MOV    CX,331       ;5 ms
      PUSH   AX
WAIT0:IN     AL,61H
      AN     DAL,10H      ;check PB4
      CMP    AL,AH        ;did it just change?
      JE     WAIT0        ;wait for change
      MOV    AH,AL        ;save the new PB4 status
      LOOP   WAIT0        ;continue until CX becomes 0
      POP    AX
      RET
DELAY_OFF    ENDP

DELAY   PROC  NEAR
      MOV    CX,33156     ;250 ms
      PUSH   AX
WAIT1:IN     AL,61H
      AND    AL,10H       ;check PB4
      CMP    AL,AH        ;did it just change?
      JE     WAIT1        ;wait for change
      MOV    AH,AL        ;save the new PB4 status
      LOOP   WAIT1        ;continue until CX becomes 0
      POP    AX
      RET
DELAY ENDP
      END    START
```

In all examples concerning counter 2, values loaded into that counter were calculated by dividing 1.1931817, the input to CLK2, by the desired OUT2 frequency. One can use the x86 to do the calculation as well, by loading 1.1931817 MHz into registers DX:AX and then dividing it by the desired output frequency using the DIV instruction.

Generating music using C#

The x86 compilers restrict access to I/O ports using the I/O instructions in Assembly language so widely used in the previous progams. This prevents us from generating music directly on the timer. However, Microsoft's .NET architecture provides us with an interface to achieve the same result as follows:

```
//C# of the first seven notes of the "Happy Birthday" song
using System;
using System.Threading;
namespace Music
{
  class Program
  {
    static void Main(string[] args)
    {
      Console.Beep(262, 500);        //hap for 500 ms
      Thread.Sleep(5);               //speaker off 5 ms
      Console.Beep(262, 500);        //py for 500 ms
      Thread.Sleep(5);               //speaker off 5 ms
      Console.Beep(294, 1000);       //birth for 1 s
      Thread.Sleep(5);               //speaker off 5 ms
      Console.Beep(262, 1000);       //day for 1 s
      Thread.Sleep(5);               //speaker off 5 ms
      Console.Beep(349, 1000);       //to for 1 s
      Thread.Sleep(5);               //speaker off 5 ms
```

```
            Console.Beep(330, 2000);          //you for 2 s
            Thread.Sleep(5);                  //speaker off 5 ms
            Console.Beep(262, 500);           //hap for 500 ms
        }
    }
}
```

Console.Beep(frequency, duration) will play the frequency for the duration in milliseconds. Note that the frequencies can range from 37 to 32767. Anything else will create an exception. Thread.Sleep(duration) will pause execution for the duration in milliseconds. Use these functions to generate the rest of the song for practice.

Review Questions

1. Find the frequency and the value that must be loaded into the register for counter 2 to play the following notes.
   (a) C4 (b) D3 (c) E4 (d) F4
2. Write pseudocode to program counter 2 to play a note.
3. Of the steps in Question 2, which must the x86 be involved in, and why?

PROBLEMS

SECTION 13.1: 8253/54 TIMER
Note: Problems 1–10 are not necessarily x86 PC compatible.

1. True or false. Each of the 8253/54 counters must be programmed independently.
2. CLK of the 8253/54 is an _______ (input, output) _________ (square, sine) wave.
3. Design the decoder for the 8253/54, where A7–A2 = 0010 11 is used to activate CS. Use NAND and inverters only. Give the port address for each port of this design.
4. Which of the following addresses cannot be assigned to counter 0 of the 8253/54, and why? 23H, 54H, 97H, 51H, FCH, 59H
5. Give the highest number by which a single counter of the 8253/54 can divide the input frequency, and what value is loaded into the counter. Give your answer for both binary and BCD options.
6. True or false. If the divisor is larger than 255, we must send the low byte first, then the high byte to the counter.
7. Find the control word to program counter 1 for mode 3, binary count, low byte first, followed by high byte.
8. Write a program for Problem 7 if CLK1 = 1.6 MHz and OUT1 = 1200 Hz. Use the port addresses in Problem 3.
9. Repeat Problem 8 for OUT1 = 250 Hz.
10. In Problem 8, what would be the OUT1 frequency if it is programmed for the maximum divisor? What if the maximum divisor BCD option were used?

SECTION 13.2: x86 PC 8253/54 TIMER CONNECTION AND PROGRAMMING

11. State the CLK frequency of all three counters of the 8253/54 in the IBM PC.
12. State the source of CLK in Problem 11.
13. What port addresses are assigned to the 8253/54 in the PC? Can they be changed?
14. State the function of each counter in the 8253/54 of the PC.
15. True or false. A PC user can program counter 2 only and should not program counters 0 and 1.

16. State the status of the GATE input for each of the counters of 8253/54 in the PC.
17. Why is a time delay based on the microprocessor's instruction clock count not widely used?
18. Write a program to generate a 10-second delay using a fixed hardware delay.

SECTION 13.3: GENERATING MUSIC ON THE x86 PC

19. To generate the following notes, state the value programmed into the divisor of counter 2 in the PC. A3, G5, B6
20. Write a program to play the song "Mary Had a Little Lamb," shown at the right.

Lyrics	Note	Freq (Hz)	Length
Mar	E4	330	1
y	D4	294	1
had	C4	262	1
a	D4	294	1
lit	E4	330	1
tle	E4	330	1
lamb	E4	330	2
lit	D4	294	1
tle	D4	294	1
lamb	D4	294	2
lit	E4	330	1
tle	G4	392	1
lamb	G4	392	2
Mar	E4	330	1
y	D4	294	1
had	C4	262	1
a	D4	294	1
lit	E4	330	1
tle	E4	330	1
lamb	E4	330	1
whose	E4	330	1
fleece	D4	292	1
was	D4	292	1
white	E4	330	1
as	D4	294	1
snow.	C4	262	4

"Mary Had a Little Lamb"

ANSWERS TO REVIEW QUESTIONS

SECTION 13.1: 8253/54 TIMER

1. True
2. Divide
3. 60H is the base address and 63H is the address for the control register.
4. B3H
5. False; it must be sent to the counter 1 register.
6. The port address of 61H
7. D0 = 0 since the maximum BCD number is 10,000 but the binary (hex) option goes as high as 65,536.
8. 65,536

SECTION 13.2: x86 PC 8253/54 TIMER CONNECTION AND PROGRAMMING

1. 40H, 41H, 42H, and 43H
2. Counter 2 at port 42H
3. True
4. True
5. True
6. Using PB0 of port address 61H
7. Monitoring of PB4 of port address 61 provides us $16{,}572 \times 15.085\ \mu s = 0.25$ s 200×0.25 s = 50 s.

SECTION 13.3: GENERATING MUSIC ON THE x86 PC

1. Since CLK2 = 1.193187 MHz, we must divide this input frequency by the desired OUT2 frequency of each note to get the value to be loaded into counter 2. Therefore, we have: (a) 262, 4554 (b) 147, 8116 (c) 330, 3616 (d) 349, 3419
2. The sequence is as follows:
   (a) Load the control byte for the 8253/54.
   (b) Load the divisor into port 42H.
   (c) Get the status of port 61 and save it.
   (d) Turn the speaker on by setting high both PB0 and PB1.
   (e) Let the 8253/54 play the note.
   (f) Use the x86 to generate a time delay for the duration of the note.
   (g) Turn off the speaker by restoring the original status of port 61.
3. All of them except step (e) since playing of the notes is performed by the 8253/54, independent of the x86.

CHAPTER 14

INTERRUPTS IN x86 PC

OBJECTIVES

Upon completion of this chapter, you will be able to:

- **>> Explain how the x86 PC executes interrupts by using the interrupt vector table and interrupt routines**
- **>> List the differences between interrupts and CALL instructions**
- **>> Describe the differences between hardware and software interrupts**
- **>> Examine the ISR for any interrupt, given its interrupt number**
- **>> Describe the function of each pin of the 8259 programmable interrupt controller (PIC) chip**
- **>> Explain the purpose of each of the four control words of the 8259 and demonstrate how they are programmed**
- **>> Examine the interrupts in x86 PCs**

This chapter examines the interrupts in x86 PCs. We also discuss sources of hardware interrupts in the x86 PC. In Section 14.1 we discusses the concept of interrupts in the 8088/86 CPU, then in Section 14.2 we look at the interrupt assignment of the original IBM PC. Section 14.3 examines the 8259 interrupt controller chip in detail. Use of the 8259 chip in the x86 PC is discussed in Section 14.4. In Section 14.5, hardware interrupts and interrupt assignments in the x86 PC are discussed.

SECTION 14.1: 8088/86 INTERRUPTS

An interrupt is an external event that informs the CPU that a device needs its service. In the 8088/86 there are a total of 256 interrupts: INT 00, INT 01, ... , INT FF (sometimes called TYPEs). When an interrupt is executed, the microprocessor automatically saves the flag register (FR), the instruction pointer (IP), and the code segment register (CS) on the stack, and goes to a fixed memory location. In x86 PCs, the memory location to which an interrupt goes is always four times the value of the interrupt number. For example, INT 03 will go to address 0000CH ($4 \times 3 = 12 =$ 0CH). Table 14-1 is a partial list of the interrupt vector table.

Table 14-1: Interrupt Vector

INT Number	Physical Address	Logical Address
INT 00	00000	0000-0000
INT 01	00004	0000-0004
INT 02	00008	0000-0008
INT 03	0000C	0000-000C
INT 04	00010	0000-0010
INT 05	00014	0000-0014
...	...	...
INT FF	003FC	0000-03FC

Interrupt service routine (ISR)

For every interrupt there must be a program associated with it. When an interrupt is invoked it is asked to run a program to perform a certain service. This program is commonly referred to as an *interrupt service routine* (ISR). The interrupt service routine is also called the *interrupt handler*. When an interrupt is invoked, the CPU runs the interrupt service routine. Now the question is, where is the address of the interrupt service routine? As can be seen from Table 14-1, for every interrupt there are allocated four bytes of memory in the interrupt vector table. Two bytes are for the IP and the other two are for the CS of the ISR. These four memory locations provide the addresses of the interrupt service routine for which the interrupt was invoked. Thus the lowest 1024 bytes ($256 \times 4 = 1024$) of memory space are set aside for the interrupt vector table and must not be used for any other function. Figure 14-1 provides a list of interrupts and their designated functions as defined by Intel Corporation.

Example 14-1

Find the physical and logical addresses in the interrupt vector table associated with:
(a) INT 12H (b) INT 8

Solution:

(a) The physical addresses for INT 12H are 00048H–0004BH since ($4 \times$ 12H = 48H). That means that the physical memory locations 48H, 49H, 4AH, and 4BH are set aside for the CS and IP of the ISR belonging to INT 12H. The logical address is 0000:0048H–0000:004BH.
(b) For INT 8, we have $8 \times 4 = 32 =$ 20H; therefore, memory addresses 00020H, 00021H, 00022H, and 00023H in the interrupt vector table hold the CS:IP of the INT 8 ISR. The logical address is 0000:0020H–0000:0023H.

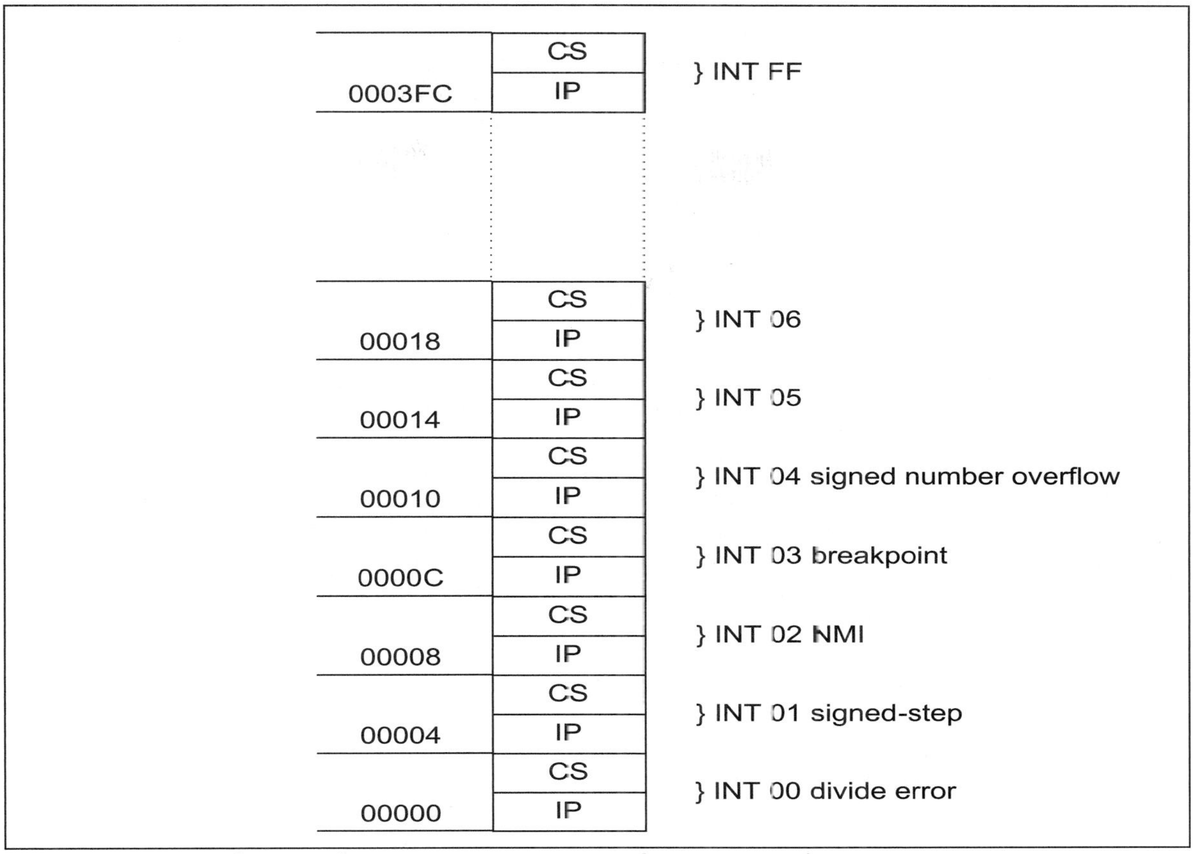

Figure 14-1. Intel's List of Designated Interrupts for the 8088/86

Difference between INT and CALL instructions

If the INT instruction saves the CS:IP of the following instruction and jumps indirectly to the subroutine associated with the interrupt, what is the difference between that and a CALL FAR instruction, which also saves the CS:IP and jumps to the desired subroutine (procedure)? The differences can be summarized as follows:

1. A "CALL FAR" instruction can jump to any location within the 1-megabyte address range of the 8088/86 CPU, but "INT nn" goes to a fixed memory location in the interrupt vector table to get the address of the interrupt service routine.
2. A "CALL FAR" instruction is used by the programmer in the sequence of instructions in the program but an externally activated hardware interrupt can come in at any time, requesting the attention of the CPU.
3. A "CALL FAR" instruction cannot be masked (disabled), but "INT nn" belonging to externally activated hardware interrupts can be masked. This is discussed in a later section.
4. A "CALL FAR" instruction automatically saves only CS:IP of the next instruction on the stack, while "INT nn" saves FR (flag register) in addition to CS:IP of the next instruction.
5. At the end of the subroutine that has been called by the "CALL FAR" instruction, the RETF (return FAR) is the last instruction, whereas the last instruction in the interrupt service routine (ISR) for "INT nn" is the instruction IRET (interrupt return). The difference is that RETF pops CS and IP off the stack but the IRET pops off the FR (flag register) in addition to CS and IP.

Categories of interrupts

"INT nn" is a 2-byte instruction where the first byte is for the opcode and the second byte is the interrupt number. This means that we can have a maximum of 256 (INT 00 INT FFH) interrupts. Of these 256 interrupts, some are used for software interrupts and some are for hardware interrupts.

Hardware interrupts

As we saw in Chapters 9 and 10, there are three pins in the x86 that are associated with hardware interrupts. They are INTR (interrupt request), NMI (nonmaskable interrupt), and INTA (interrupt acknowledge). The use of INTA will be discussed in Section 14.3. INTR is an input signal into the CPU, which can be masked (ignored) and unmasked through the use of instructions CLI and STI. However, NMI, which is also an input signal into the CPU, cannot be masked and unmasked using instructions CLI and STI, and for this reason it is called a *nonmaskable* interrupt. INTR and NMI are activated externally by putting 5 V on the pins of NMI and INTR of the x86 microprocessor. When either of these interrupts is activated, the x86 finishes the instruction that it is executing, pushes FR and the CS:IP of the next instruction onto the stack, then jumps to a fixed location in the interrupt vector table and fetches the CS:IP for the interrupt service routine (ISR) associated with that interrupt. At the end of the ISR, the IRET instruction causes the CPU to get (pop) back its original FR and CS:IP from the stack, thereby forcing the CPU to continue at the instruction where it left off when the interrupt came in.

Intel has embedded "INT 02" into the x86 microprocessor to be used only for NMI. Whenever the NMI pin is activated, the CPU will go to memory location 00008 to get the address (CS:IP) of the interrupt service routine (ISR) associated with NMI. Memory locations 00008, 00009, 0000A, and 0000B contain the 4 bytes of CS:IP of the ISR belonging to NMI. In contrast, this is not the case for the other hardware pin, INTR. There is no specific location in the vector table assigned to INTR. The reason is that INTR is used to expand the number of hardware interrupts and should be allowed to use any "INT nn" that has not been previously assigned. The 8259 programmable interrupt controller (PIC) chip can be connected to INTR to expand the number of hardware interrupts to 64. In the case of the IBM PC, one Intel 8259 PIC chip is used to add a total of 8 hardware interrupts to the microprocessor. IBM PC AT, PS/2 80286, 80386, 80486, and Intel Pentium computers use two 8259 chips to allow up to 16 hardware interrupts. The design of hardware interrupts and the use of the 8259 in the IBM PC are covered in Sections 14.3 and 14.4, while ISA bus interrupts are covered in Section 14.5.

Software interrupts

If an ISR is called upon as a result of the execution of an x86 instruction such as "INT nn", it is referred to as a software interrupt since it was invoked from software, not from external hardware. Examples of such interrupts are DOS "INT 21H" function calls and video interrupts "INT 10H", which were covered in Chapter 4. These interrupts can be invoked in the sequence of code just like a CALL or any other x86 instruction. Many of the interrupts in this category are used by the MS DOS operating system and IBM BIOS to perform essential tasks that every computer must provide to the system and the user. Within this group of interrupts there are also some predefined functions associated with some of the interrupts. They are "INT 00" (divide error), "INT 01" (single step), "INT 03" (breakpoint), and "INT 04" (signed number overflow). Each is described below. These interrupts are shown in Figure 14-1. Aside from "INT 00" to "INT 04", which have predefined functions, the rest of the interrupts from "INT 05" to "INT FF" can be used to implement either software or hardware interrupts.

Interrupts and the flag register

Among bits D0 to D15 of the flag register, there are two bits that are associated with the interrupt: D9, or IF (interrupt enable flag), and D8, or TF (trap or single step flag). In addition, OF (overflow flag) can be used by the interrupt. See Figure 14-2.

The 16 bits of the flag register:

15	14	13	12	11	10	9	8	7	6	5	4	3	2	1	0
R	R	R	R	OF	OF	IF	TF	SF	ZF	U	AF	U	PF	U	CF

R = reserved
U = undefined
OF = overflow flag
DF = direction flag
IF = interrupt flag
TF = trap flag
SF = sign flag
ZF = zero flag
AF = auxiliary carry flag
PF = parity flag
CF = carry flag

Figure 14-2. Flag Register

The interrupt flag is used to mask (ignore) any hardware interrupt that may come in from the INTR pin. If IF = 0, all hardware interrupt requests through INTR are ignored. This has no effect on interrupts coming from the NMI pin or "INT nn" instructions. The instruction CLI (clear interrupt flag) will make IF = 0. To allow interrupt requests through the INTR pin, this flag must be set to one (IF = 1). The STI (set interrupt flag) instruction can be used to set IF to 1. Section 14.3 will show how to use STI and CLI to mask or allow interrupts through the INTR pin. The trap flag (TF) is explained below in the discussion of "INT 01", the single step interrupt.

Processing interrupts

When the 8088/86 processes any interrupt (software or hardware), it goes through the following steps:

1. The flag register (FR) is pushed onto the stack and SP is decremented by 2, since FR is a 2-byte register.
2. IF (interrupt enable flag) and TF (trap flag) are both cleared (IF = 0 and TF = 0). This masks (causes the system to ignore) interrupt requests from the INTR pin and disables single stepping while the CPU is executing the interrupt service routine. Depending on the nature of the interrupt procedure, a programmer can unmask the INTR pin by the STI instruction.
3. The current CS is pushed onto the stack and SP is decremented by 2.
4. The current IP is pushed onto the stack and SP is decremented by 2.
5. The INT number (type) is multiplied by 4 to get the physical address of the location within the vector table to fetch the CS and IP of the interrupt service routine.
6. From the new CS:IP, the CPU starts to fetch and execute instructions belonging to the ISR program.
7. The last instruction of the interrupt service routine must be IRET, to get IP, CS, and FR back from the stack and make the CPU run the code where it left off.

Functions associated with INT 00 to INT 04

As mentioned earlier, interrupts INT 00 to INT 04 have predefined tasks (functions) and cannot be used in any other way. The function of each is described next.

INT 00 (divide error)

This interrupt belongs to the category of interrupts referred to as *conditional* or *exception* interrupts. Internally, they are invoked by the microprocessor whenever there are conditions (exceptions) that the CPU is unable to handle. One such situation is an attempt to divide a number by zero. Since the result of dividing a number by zero is undefined, and the CPU has no way of handling such a result, it automatically invokes the divide error exception interrupt. In the 8088/86 microprocessor, out of 256 interrupts, Intel has set aside only INT 0 for the exception interrupt. There are many more exception handling interrupts in x86 CPUs, which are discussed in Section 14.5. INT 00 is invoked by the microprocessor whenever there is an attempt to divide a number by zero. In the x86 PC, the service subroutine for this interrupt is responsible for displaying the message "DIVIDE ERROR" on the screen if a program such as the following is executed:

```
MOV   AL,92        ;AL=92
SUB   CL,CL        ;CL=0
DIV   CL           ;92/0=undefined result
```

INT 0 is also invoked if the quotient is too large to fit into the assigned register when executing a DIV instruction. Look at the following case:

```
MOV   AX,0FFFFH    ;AX=FFFFH
MOV   BL,2         ;BL=2
DIV   BL           ;65535/2 = 32767 larger than 255
                   ;maximum capacity of AL
```

Put INT 3 at the end of the above two programs in DEBUG and see the reaction of the PC. For further discussion of divide error interrupts due to an oversized quotient, see Chapter 3.

INT 01 (single step)

In executing a sequence of instructions, there is a need to examine the contents of the CPU's registers and system memory. This is often done by executing the program one instruction at a time and then inspecting registers and memory. This is commonly referred to as *single-stepping*, or performing a trace. Intel has designated INT 01 specifically for implementation of single-stepping. To single-step, the trap flag (TF), D8 of the flag register, must be set to 1. Then after execution of each instruction, the 8088/86 automatically jumps to physical location 00004 to fetch the 4 bytes for CS:IP of the interrupt service routine, whose job is, among other things, to dump the registers onto the screen. Now the question is, how is the trap flag set or reset? Although Intel has not provided any specific instruction for this purpose (unlike IF, which uses STI and CLI instructions to set or reset), one can write a simple program to do that. The following shows how to make TF = 0:

```
PUSHF
POP   AX
AND   AX,1111111011111111B
PUSH  AX
POPF
```

Recall that TF is D8 of the flag register. The analysis of the above two programs is left to the reader. To make TF = 1, one simply uses the OR instruction in place of the AND instruction above.

INT 02 (nonmaskable interrupt)

All Intel x86 microprocessors have a pin designated NMI. It is an active-high input. Intel has set aside INT 2 for the NMI interrupt. Whenever the NMI pin of the x86 is activated by a high (5 V) signal, the CPU jumps to physical memory location 00008 to fetch the CS:IP of the interrupt service routine associated with NMI. Section 14.4 contains a detailed discussion of its purpose and application.

INT 03 (breakpoint)

To allow implementation of breakpoints in software engineering, Intel has set aside INT 03 solely for that purpose. Whereas in single-step mode, one can inspect the CPU and system memory after the execution of each instruction, a breakpoint is used to examine the CPU and memory after the execution of a group of instructions. One interesting point about INT 3 is the fact that it is a 1-byte instruction. This is in contrast to all other interrupt instructions of the form "INT nn", which are 2-byte instructions.

INT 04 (signed number overflow)

This interrupt is invoked by a signed number overflow condition. There is an instruction associated with this, INTO (interrupt on overflow). For a detailed discussion of signed number overflow, see Chapter 6. If the instruction INTO is placed after a signed number arithmetic or logic operation such as IMUL or ADD, the CPU will activate INT 04 if OF = 1. In cases where OF = 0, the INTO instruction is not executed but is bypassed and acts as a NOP (no operation) instruction. To understand that, look at the following example.

```
MOV    AL,DATA1
MOV    BL,DATA2
ADD    AL,BL;add BL to AL
INTO
```

Suppose in the above program that DATA1 = +64 = 0100 0000 and DATA2 = +64 = 0100 0000. The INTO instruction will be executed and the 8088/86 will jump to physical location 00010H, the memory location associated with INT 04. The carry from D6 to D7 causes the overflow flag to become 1.

```
     + 64   0100 0000
+    + 64   0100 0000
     +128   1000 0000   OF=1 and the result is not +128
```

The above incorrect result causes OF to be set to 1. INTO causes the CPU to perform "INT 4" and jump to physical location 00010H of the vector table to get the CS:IP of the service routine. Suppose that the data in the above program was DATA1 = +64 and DATA2 = +17. In that case, OF would become 0; the INTO is not executed and acts simply as a NOP (no operation) instruction.

Review Questions

1. True or false. When any interrupt (software or hardware) is activated, the CPU jumps to a fixed and unique address.
2. There are _____ bytes of memory in the interrupt vector table for each "INT nn".
3. How many K bytes of memory are used by the interrupt vector table, and what are the beginning and ending addresses of the table?
4. The program associated with an interrupt is referred to as ___________.
5. What is the function of the interrupt vector table?
6. What physical memory locations in the interrupt vector table hold the CS:IP of INT 10H?
7. The 8088/86 has assigned INT 2 to NMI. Can that be changed?
8. Which interrupt is assigned to divide error exception handling?

SECTION 14.2: x86 PC AND INTERRUPT ASSIGNMENT

Of the 256 possible interrupts in the x86, some are used by the PC peripheral hardware (BIOS), some are used by the Microsoft operating system, and the rest are available for programmers of software applications. Table 14-2 lists many of the PC interrupts. Example 14-2 shows the physical address as it relates to the logical address.

Example 14-2

For a given ISR, the logical address is F000:FF53. Verify that the physical address is FFF53H.

Solution:

Since the logical address is F000:FF53, this means that CS = F000H and IP = FF53H. Shifting left the segment register one hex digit and adding it to the offset gives the physical address FFF53H.

Examining the interrupt vector table of your PC

Example 14-3 shows how to use DEBUG's dump command to examine the interrupt vector table of a x86 PC, regardless of which x86 CPU it contains.

From the CS:IP address of the ISR, it is possible to determine which source provides the service: DOS or BIOS. This is shown in Example 14-4.

Analyzing an x86 PC BIOS interrupt service routine

To understand the structure of an ISR, we examine the interrupt service routine of INT 12H from IBM BIOS. The interrupt 12H service is available on any PC with an x86 microprocessor.

INT 12H: checking the size of RAM on the IBM PC

IBM PC BIOS uses INT 12H to provide the amount of installed conventional (0 to 640K bytes) RAM memory on the system. By system is meant both the motherboard and expansion boards. One of the functions of the BIOS POST (power on self test) is to

Example 14-3

(a) Use the DEBUG dump command to dump the contents of memory locations 00000–0002FH.
(b) Find the CS:IP of divide error, NMI, and INT 8.

Solution:
(a) It is very possible that the data you get on your PC will be different from the following dump, depending on the OS version and the BIOS chip date of your PC.

```
C>debug
-D 0000:0000-002F
0000:0000 E8 56 2B 02 56 07 70 00-C3 E2 00 F0 56 07 70 00 .V+.V.p.....V.p.
0000:0010 56 07 70 00 54 FF 00 F0-47 FF 00 F0 47 FF 00 F0 V.p.T...G...G...
0000:0020 A5 FE 00 F0 87 E9 00 F0-DD E6 00 F0 DD E6 00 F0 ................
```

(b) For the divide error interrupt (INT 0), CS:IP is located at addresses 0, 1, 2, 3. Remember that because of the little endian convention, the low address has the low value; therefore, IP = 56E8H and CS = 022BH. By the same token, NMI's INT 2 is located at the vector table addresses of 8, 9, A, and B. Therefore, we have IP = E2C3H and CS = F000H. The CS:IP of INT 8 ISR is located in addresses starting at 00020 since 8 × 4 = 32 = 20H, IP = FEA5, and CS = F000.

Table 14-2: IBM PC Interrupt System

Interrupt	Logical Addr.	Physical Addr.	Purpose
0	00E3:3072	03EA2	Divide error
1	0600:08ED	068ED	Single-step (trace command in DEBUG)
2	F000:E2C3	FE2C3	Nonmaskable interrupt
3	0600:08E6	068E6	Breakpoint
4	0700:0147	07147	Signed number arithmetic overflow
5	F000:FF54	FFF54	Print screen (BIOS)
6, 7			Reserved
8	F000:FEA5	FFEA5	IRQ0 of 8259 (BIOS timer interrupt)
9	F000:E987	FE987	IRQ1 of 8259 (BIOS keyboard interrupt)
A			IRQ2 of 8259 (reserved)
B			IRQ3 of 8259 (reserved for serial com 2)
C			IRQ4 of 8259 (reserved for serial com 1)
D			IRQ5 of 8259 (reserved for hard disk XT)
E	F000:EF57	FEF57	IRQ6 of 8259 (floppy disk)
F	0070:0147	00847	IRQ7 of 8259 (parallel printer LPT 1)
10	F000:F065	FF065	Video I/O (BIOS)
11	F000:F84D	FF84D	Equipment configuration check (BIOS)
12	F000:F841	FF841	Memory size check (BIOS)
13	F000:EC59	FEC59	Disk I/O (BIOS)
14	F000:E739	FE739	RS-232 I/O (BIOS)
15	F000:F859	FF859	Cassette I/O (BIOS)
16	F000:E82E	FE82E	Keyboard I/O (BIOS)
17	F000:EFD2	FEFD2	Parallel printer I/O (BIOS)
18	F600:0000	F6000	Load ROM BASIC
19	F000:E6F2	FE6F2	Load boot-strap (BIOS)
1A	F000:FE6E	FFE6E	Time-of-day (BIOS)
1B	0070:0140	00840	Ctrl-Brk control (BIOS)
1C	F000:FF53	FFF53	Timer control
1D	F000:F0A4	FF0A4	Video parameters table
1E	0000:0522	00522	Floppy disk parameters table
1F	00E3:0B07	01937	Graphics character table (DOS 3.0 and up)
20	PSP:0000	-----	DOS program terminate
21	Relocatable	-----	DOS function calls
22	PSP:000A	-----	DOS terminate address
23	PSP:000E	-----	DOS Ctrl-Brk exit address
24	PSP:0012	-----	DOS critical error-handling vector
25	Relocatable	-----	DOS absolute disk read
26	Relocatable	-----	DOS absolute disk write
27	Relocatable	-----	DOS terminate but stay resident (TSR)
28–2E		-----	Reserved for DOS
2F	Relocatable	-----	Multiplex interrupt
30–3F		-----	Reserved for DOS
40		FEC59	Disk I/O (XT)
41		FE401	Fixed (hard) disk parameters (XT)
42–5F		-----	Reserved for DOS
60–66		-----	User defined
67		-----	Expanded memory manager
68–7F		-----	Not used
80–85		-----	Reserved for BASIC
86–F0		-----	BASIC interpreter
F1–FF		-----	Not used

Example 14-4

Examine the answers for Example 14-3(b) to determine whether DOS or BIOS provides the ISR for the divide error and NMI.

Solution:

In the case of INT 0 (divide error), the logical address for CS:IP is 022B:56E8. This results in a physical address of 07998H for the divide error interrupt service routine. This area of memory belongs to DOS, as discussed in Chapter 10. For NMI interrupt 2, we have a logical address of CS:IP = F000:E2C3, which corresponds to physical address FE2C3H. This is the BIOS ROM area.

test and count the total K bytes of conventional RAM memory installed on the system and write it in memory locations 00413H and 00414H, which have been set aside for this purpose in the BIOS data area. In Chapter 10, we showed the data area used by BIOS. The job of INT 12H is to copy that value from memory locations 00413H and 00414H into AX and return. In other words, after executing INT 12H, AX will contain the total K bytes of conventional RAM memory on the system. This value is in hex and must be converted to decimal to get values of 1 to 640K bytes. The interrupt service routine for INT 12H looks as follows in the IBM PC Technical Reference.

```
MEM_SIZE    PROC  FAR
            STI                 ;interrupt back on
            PUSH  DS            ;save segment
            SUB   AX,AX         ;set DS = 0
            MOV   DS,AX         ;for BIOS data area
            MOV   AX,[0413]     ;conv mem size in 413,414
            POP   DS            ;recover segment
            IRET                ;return to caller
MEM_SIZE    ENDP
```

See Examples 14-5 and 14-6.

Example 14-5

Execute INT 12H followed by INT 3 (breakpoint) in DEBUG. Verify the memory size.

Solution:

```
C>DEBUG
-a
1131:0100    INT 12
1131:0102    INT 3
1131:0103
-G
AX=0280 BX=0000 CX=0000 DX=0000 SP=FFEE BP=0000 SI=0000 DI=0000
DS=1132 ES=1132 SS=1132 CS=1132 IP=0102 NV UP EI PL NZ NA PO NC
1131:0102 CC            INT   3
-Q
```

AX = 0280, which is in hex format. Converting it to decimal gives a size of 640K bytes of memory installed on this computer.

Example 14-6

Use the DEBUG D (dump) command to dump memory locations 0040:0000 to 0040:001FH and inspect the contents of locations 0040:13 and 0040:14. Does this match the result of Example 14-5?

Solution:

```
C>DEBUG
-D 0040:0000 1F
0040:0000 F8 03 F8 02 00 00 00 00-78 03 78 02 00 00 00 00 ........x.x.....
0040:0010 6F 94 00 80 02 40 02 00-00 00 22 00 22 00 66 21 o...@....".".f!
-Q
```

Location 13 contains 80 and 14 contains 02. 0280 matches the result of Example 14-5.

Review Questions

1. Find the logical address, CS, and IP values for the ISR if it is located at the ROM physical address of FF065H.
2. In Question 1, if the ISR belongs to INT 10H, find the exact contents of memory locations in the interrupt vector table.
3. Assume that after using the dump command of DEBUG to dump a section of the interrupt vector table, we have the following:
   0000:0030 .... -57 EF 00 FF 00 00 00 00
   To what interrupt number does the above dump belong?
4. For Question 3, show the logical address and exact values for CS and IP of the ISR.
5. In a given embedded x86 PC, the documentation states that the motherboard comes with 512K bytes of RAM. Running the INT 12H shows the value we expect in the AX register.
6. Which one is the last instruction in the ISR of INT 12H: RET or IRET?

SECTION 14.3: 8259 PROGRAMMABLE INTERRUPT CONTROLLER

The x86 has only pins INTR and INTA for interrupts, but one can use these two pins to expand the number of interrupts. Intel Corporation has provided an IC chip called the 8259 programmable interrupt controller (PIC) to make the job of expanding the number of hardware interrupts much easier. See Figures 14-3 and 14-4. This section covers the 8259 IC chip pins and programming options. Note that this section is about the 8259 chip. The ports and programs covered in this section are not related to the PC.

CAS0–CAS2

CAS0, CAS1, and CAS2 can be used to set up several 8259 chips to expand the number of hardware interrupts of the 8088/86 to 64 by cascading the 8259 chips in a master/slave configuration. This section will focus on slave mode. Section 14.5 discusses both the master and slave configurations as used in PC/AT type computers. To use the 8259 in slave mode, the chip must be programmed and CAS0 to CAS2 are ignored.

$\overline{SP}/\overline{EN}$

$\overline{SP}/\overline{EN}$ (slave programming/enable) in buffered mode is an output signal from the 8259 to activate the transceiver ($\overline{EN}$). In nonbuffered mode it is an input signal into the 8259, $\overline{SP}$ = 1 for the master and $\overline{SP}$ = 0 for the slave.

INT

INT is an output that is connected to INTR of the x86.

$\overline{\text{INTA}}$

$\overline{\text{INTA}}$ is input to the 8259 from INTA of the x86.

IR0–IR7

Inputs IR0 to IR7 (interrupt request) are used as hardware interrupts. When a HIGH is put on any interrupt from IR0 to IR7, the 8088/86 will jump to a vector location. For each IR there exists a physical memory location in the interrupt vector table. The x86 has 256 hardware or software interrupts (INT 00–INT FF).

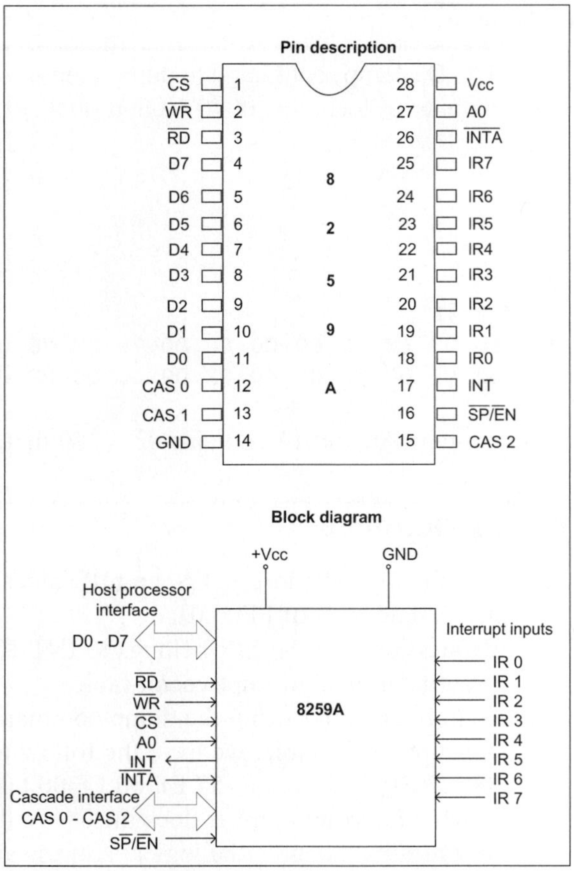

Figure 14-3. 8259A Programmable Interrupt Controller

8259 control words and ports

The four control words associated with the 8259 are ICW1 (initialization command word), ICW2, ICW3, and ICW4. ICW3 is used in master mode only and is discussed in Section 14.5. As can be seen from the pins of the 8259, there is only one address line A0 to communicate with the chip. Table 14-3 and Example 14-7 show the values that A0 and CS must take to initialize the 8259.

Table 14-3: 8259 Initialization

$\overline{\text{CS}}$	A0	Initialization
0	0	ICW1
0	0	ICW2,ICW3,ICW4
1	x	8259 is not addressed

ICW1 (initialization command word 1)

In looking at Table 14-3, the question might arise: How can the 8259 make a distinction between ICW2, ICW3, and ICW4 when they are sent to the same address? This is one of the functions of ICW1. D0, the LSB of ICW1, will tell the 8259 if it should look for ICW4 or not. In a similar manner, if D1 is high it knows that the system is configured in slave mode and it should not expect any ICW3 in the initialization sequence. The initialization sequence must always start with ICW1, followed by ICW2,

Example 14-7

Find the address for ICW1–ICW4 if chip select is activated by A7–A1 = 0010011.

```
A7 A6 A5 A4    A3 A2 A1 A0
 0  0  1  0     0  1  1  0
 0  0  1  0     0  1  1  1
```

Solution:

The above shows 26H to be the port for ICW1 and 27H the port for ICW2, ICW3, and ICW4.

and finally the last one, if needed. There is no jumping ahead. D2 is always set low (= 0) for the x86. D3 chooses between level triggering or edge triggering of the input signals IR0–IR7. In edge triggering, a low-to-high input is recognized as an interrupt request. In level triggering, a high on the IR is recognized as an interrupt request. D4 must always be high. D5, D6, and D7 are all low for the x86 microprocessors (they are used only for the 8080/85). See Figure 14-5.

ICW2 (initialization command word 2)

It is the function of ICW2 to assign interrupt numbers to IR0–IR7. While the lower three bits, D0, D1, and D2, vary from 000 to 111, they, along with D3–D7 (T3 through T7), form the 8-bit INT type number assigned to the corresponding IR0 through IR7. That means that D3–D7 can only be programmed according to the assignment of the INT type, with the lower bits being provided by the 8259, depending on which interrupt of IR0 to IR7 is activated. See Figure 14-5.

ICW3 (initialization command word 3)

ICW3 is used only when two or more 8259s are cascaded. In this mode, a single 8259 can be connected to eight slave 8259s, thereby providing up to 64 hardware interrupts. In cascade mode, there are separate ICW3 words for the master and the slave. For the master, it indicates which IR has a slave connected to it, and a separate ICW3 informs the slave which IR of the master it is connected to. See Figure 14-6.

ICW4 (initialization command word 4)

D0 indicates the processor mode (PM), the choice of microprocessor. D0 equals 1 for the 8088/86 and 0 for the 8080/8085. When D1, which is AEOI (automatic end of interrupt), is high it eliminates the need for an EOI instruction to be present before the IRET (interrupt return) instruction in the interrupt service routine. When D1 is zero, the EOI must be issued using the OCW (operation command word) to the 8259. In other words, if D1 = 0, the last three instructions of the interrupt service routine for IR0–IR7 must be issuing the EOI followed by IRET. The significance of this is discussed shortly when the OCW is discussed. D2 and D3 are for systems where data buses are buffered with the use of bidirectional transceivers. The 8259 can work in either buffered or non-buffered mode. D4 is for SFNM (special fully nested mode). This mode must be used when the 8259 is in master mode, and then D4 = 1; otherwise, it is 0. D5–D7 must be zero, as required by the 8259. See Figure 14-6 and Example 14-8.

Masking and prioritization of IR0–IR7 interrupts

One might ask what happens if more than one of interrupts IR0–IR7 is activated at the same time? Can we mask any of the interrupts? What about responding to another interrupt while an interrupt is being serviced? To answer all these questions, the function of the OCW (operation command word) must be understood. This is discussed next.

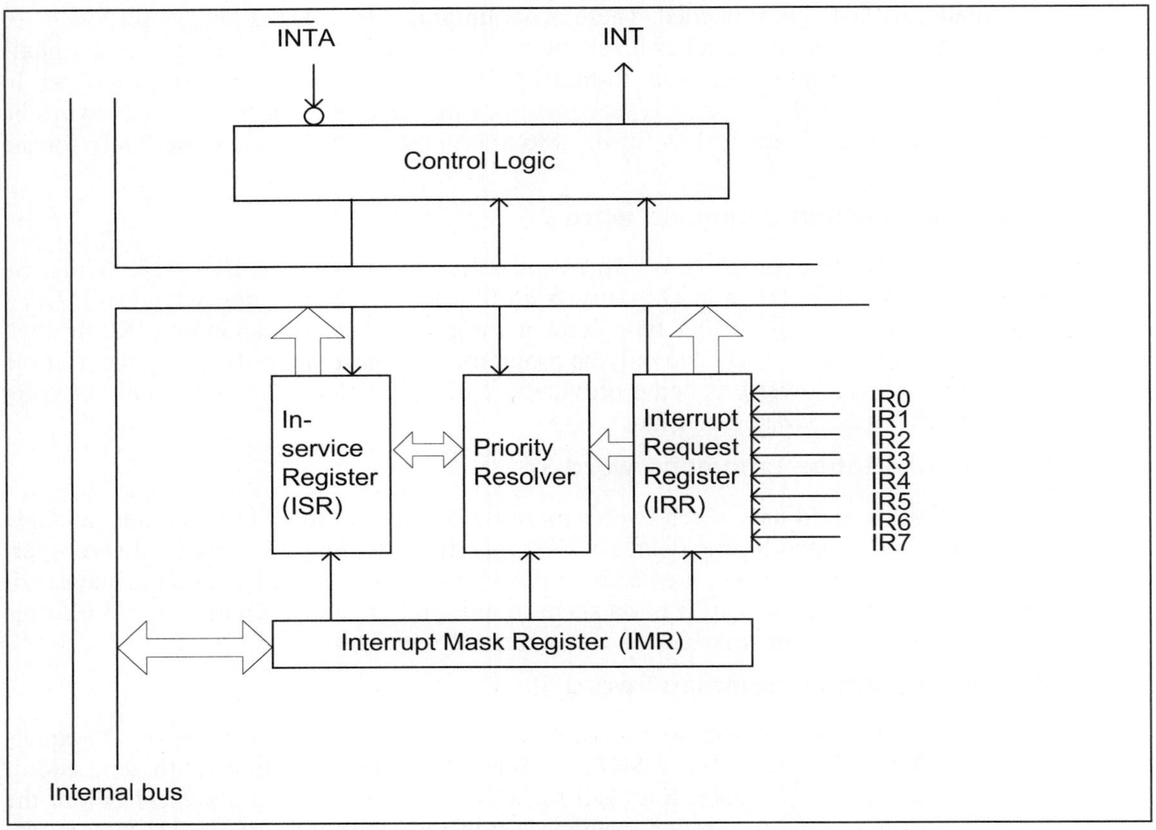

Figure 14-4. Partial Block Diagram of the 8259A

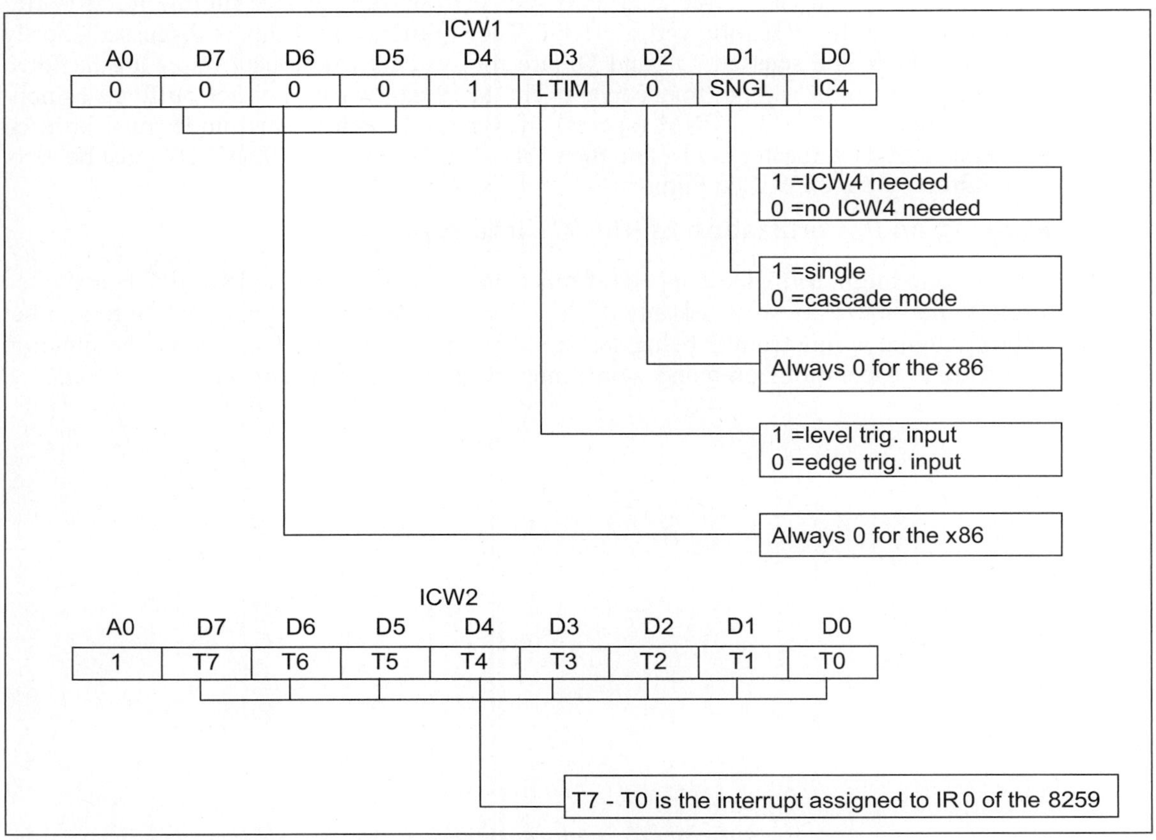

Figure 14-5. ICW Formats (ICW1 and ICW2) for the 8259

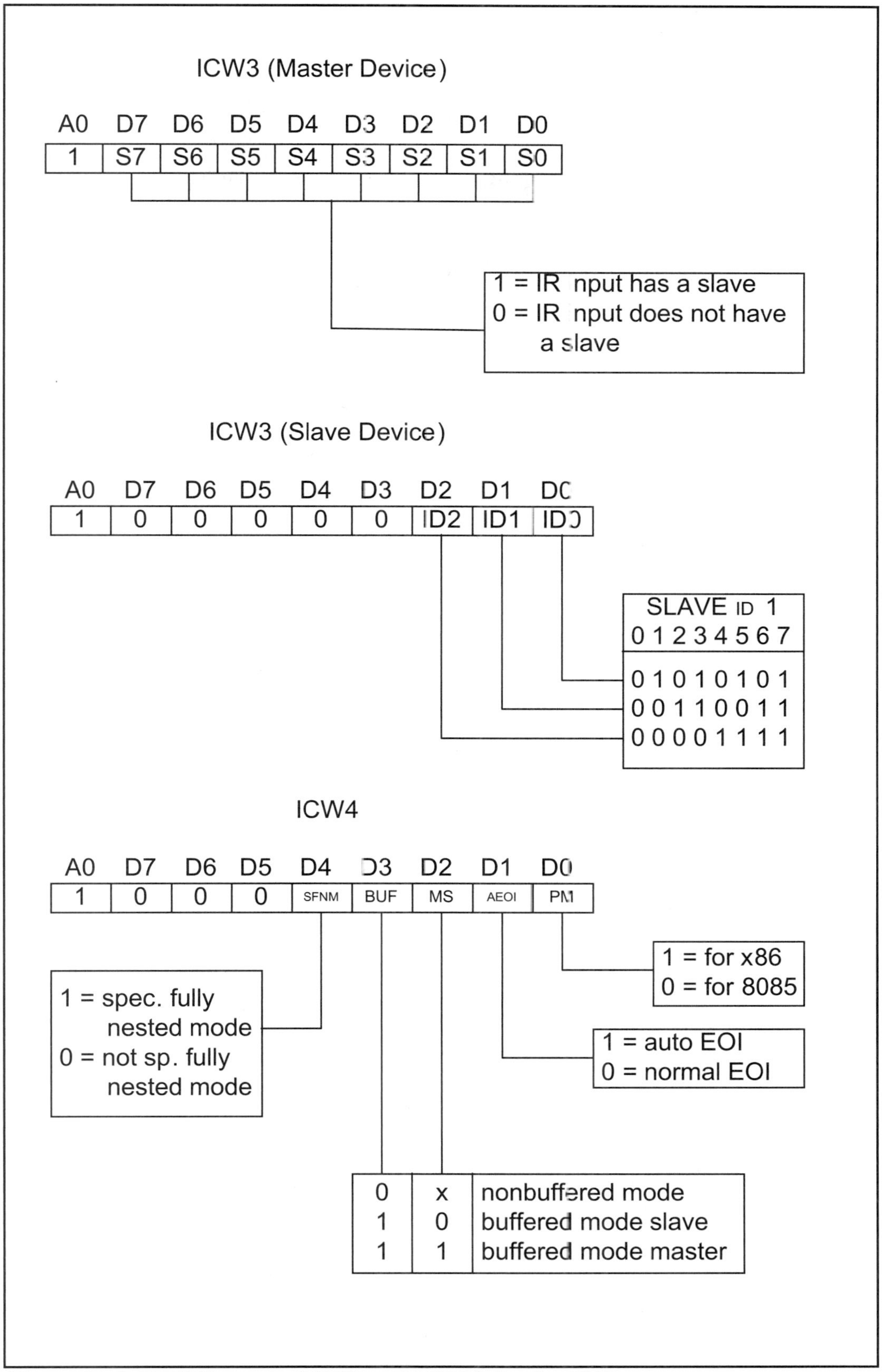

Figure 14-6. ICW Formats (ICW3 and ICW4) for the 8259

Example 14-8

(a) Find the ICWs of the 8259 if it is used with an 8088/86 CPU, single, level triggering IRs, and IR0 is assigned "INT 50H". The 8259 is in slave buffered mode with normal EOI.
(b) Show the program to initialize the 8259 using the port addresses in Example 14-7.
(c) Find the addresses associated with IR0, IR1, and IR2 in the interrupt vector table.
Note: This example is not PC-compatible and is given only for an exercise.

Solution:

(a) From Figure 14-5, we get the following for each of the ICWs:

ICW1	
D0 = 1	ICW4 needed
D1 = 1	single
D2 = 0	this is always zero for x86 CPUs
D3 = 1	level triggering
D4 = 1	required by the ICW1 itself
D5 = D6 = D7 = 0	this is always zero for x86 CPUs

This gives ICW1 = 00011011 = 1BH. To get ICW2, look at Table 14-4. Always equate ICW2 to the INT # assigned to IR0: ICW2 = 01010000 = 50H. Notice that "INT nn", assigned to IR0, can decide only bits D7–D3 (T7–T3 in Figure 14-5) of ICW2. This means that the "INT nn" assigned to IR0 must have the lower three bits = 000; therefore, it can take either values of X0H or X8H, where X is a hex number. For example "INT 45H" cannot be assigned to IR0.

No ICW3 is needed since it is single and not cascaded.

ICW4	
D0 = 1	8088/86
D1 = 0	normal (we must issue EOI before IRET instruction)
D2 = 0; D3 = 1	slave buffered mode
D4 = 0	not nested
D5 = D6 = D7 = 0	required by the ICW4

We get ICW4 = 00001001 = 09H.

(b) The program is as follows:

```
MOV   AL,1BH        ;ICW1
OUT   26H,AL        ;TO PORT 26H
MOV   AL,50H        ;ICW2
OUT   27H,AL        ;TO PORT 27H
MOV   AL,09         ;ICW4
OUT   27H,AL        ;TO PORT 27H
```

(c) If "INT 50H" is assigned to IR0, then IR1 and IR2 have "INT 51H" and "INT 52", respectively, and so on. The vector memory locations associated with the IRs are as follows:

		Vector Location	
IRQ (Pin of 8259)	INT	Logical Address	Physical Address
IR0	50H	0000:0140H-0143	00140H-00143
IR1	51H	0000:0144H-0147	00144H-00147
IR2	52H	0000:0148H-014B	00148H-0014B

and so on.

Table 14-4: INT Number for Hardware Interrupts in Example 14-8

Binary Data for ICW2		8259 Interrupt	
D7 D6 D5 D4	D3 D2 D1 D0	Input	INT Type
0 1 0 1	0 0 0 0	IR0	INT 50
0 1 0 1	0 0 0 1	IR1	INT 51
0 1 0 1	0 0 1 0	IR2	INT 52
0 1 0 1	0 0 1 1	IR3	INT 53
0 1 0 1	0 1 0 0	IR4	INT 54
0 1 0 1	0 1 0 1	IR5	INT 55
0 1 0 1	0 1 1 0	IR6	INT 56
0 1 0 1	0 1 1 1	IR7	INT 57

OCW (operation command word)

After ICW1, ICW2, and ICW4 have been issued in sequence to the 8259 chip in order to initialize it, the 8088/86 is ready to receive hardware interrupts through the 8259's IR0–IR7 pins. After the process of initialization, the OCW (operation command word) can be sent to mask any of IR0–IR7, or change the priority assigned to each IR. There are three operation command words: OCW1, OCW2, and OCW3. See Table 14-5. With the help of OCWs, a programmer can dynamically change the priority associated with each of IR0–IR7, or mask any of them. Example 14-9 shows how the OCWs are sent to the 8259.

Table 14-5: Addresses for 8259 OCWs

$\overline{CS}$	A0	Operation Command Word
0	0	OCW2,OCW3
0	1	OCW1
1	x	8259 is not addressed

Figure 14-7 shows the OCWs for the 8259. Below is a discussion of each. Before discussing the OCWs, the existence of three registers inside the 8259 must be noted. They are the ISR (in-service register), IRR (interrupt request register), and IMR (interrupt mask register). See Figure 14-4.

Example 14-9

Find the port addresses for the OCWs of the 8259 in Example 14-7.

Solution:

From Table 14-5 and Example 14-7:

A7	A6	A5	A4	A3	A2	A1	A0		
0	0	1	1	0	1	1	0	= 26H	Port for OCW2 and OCW3
0	0	1	1	0	1	1	1	= 27H	Port for OCW1

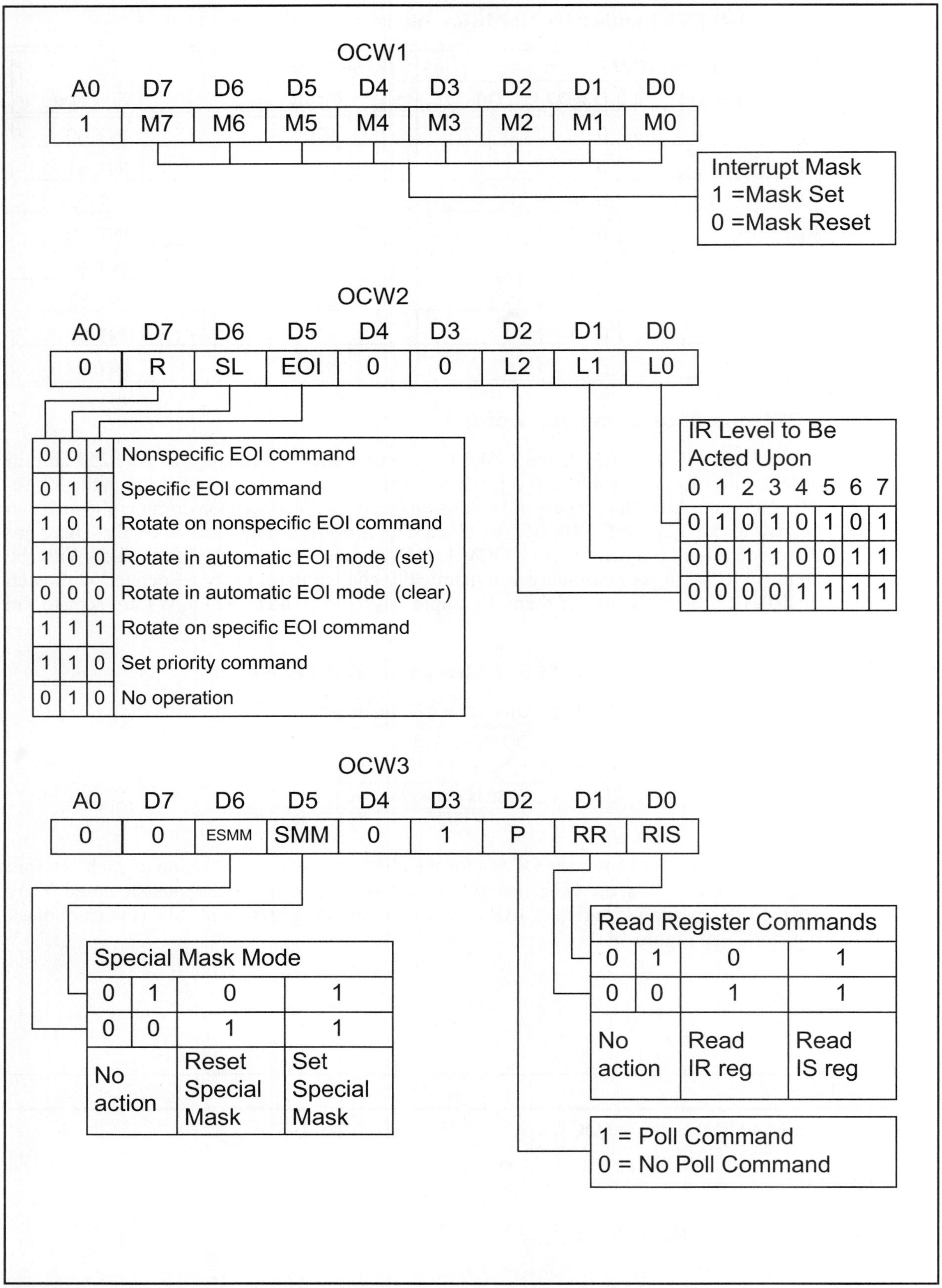

Figure 14-7. OCW Format for the 8259

OCW1 (operation command word 1)

OCW1 is used to mask any of IR0–IR7. Logic 1 is for masking (disabling) and 0 is for unmasking (enabling). For example, 11111000 is the OCW1 to enable (unmask) IR0, IR1, IR2, and disable (mask) the rest (IR3 through IR7). When this byte is written to the 8259 (by making A0 = 1 and CS = low), it goes into the internal register called IMR (interrupt mask register).

There are occasions when one needs to know which IRs are disabled and which ones are enabled. In that case, simply read OCW1, which is the contents of IMR. For example, to read OCW1 using the port addresses in Example 14-10, code "IN AL,27H". By examining the contents of AL, one can find out which IRs are enabled and which ones are disabled.

Example 14-10

Write the code to unmask (enable) IR0–IR7. Use the ports in Example 14-9.

Solution:

To enable IR0–IR7, use 0 for M0–M7 in OCW1 of Figure 14-7.
OCW1 = 0000 0000 = 00H

```
        MOV   AL,00        ;OCW1 to unmask IR0 IR7
        OUT   27,AL        ;issue OCW1 to IMR
```

OCW2 (operation command word 2)

This command word is used to assign a specific priority to the IRs. Three methods for assigning priority to IR0–IR7 are discussed below.

Fully nested mode

This assigns the highest priority to IR0 and the lowest to IR7. In this case, if IR3 and IR5 are activated at the same time, first IR3 is served and then IR5. What happens if IR3 is being served when both IR2 and IR4 request service? In that case, IR3 is put on hold, then IR2 is served. After IR2 is served, IR3 is completed and finally IR4 is served. This is the default mode when the 8259 is initialized. The 8259 can be programmed to change the default mode to assign the highest priority to any IR. For example, the following shows OCW2 if IR6 has been assigned the highest priority, then IR7 has the next priority, and so on.

IR0	IR1	IR2	IR3	IR4	IR5	IR6	IR7	interrupt pin on 8259
2	3	4	5	6	7	0	1	priority 0 = highest and 7 = lowest

Automatic rotation mode

In this scheme, when an IR has been served it will take the lowest priority and will not be served until every other request has had a chance. This prevents interrupt starvation, where one device monopolizes the interrupt service.

Specific rotation mode

In this scheme the 8259 can be programmed to make the rotation follow a specific sequence rather than IR0 to IR7, which is the case for the automatic rotation mode. In this mode, the IR served will be stamped as the lowest priority, meaning that it will not be served until every other request has had a chance. The only difference between this mode and automatic rotation is the sequence of rotation.

Having concluded this brief description of priority schemes, the following will discuss the OCW2 bits. D2–D0 are used to assign a new priority scheme to an IR other than the default. For example, to assign the highest priority to IR6, use D2–D0 = 110. D4–D3 must always be 0 for OCW2. D5, EOI (end of interrupt), is used to issue an end-of-interrupt command to the 8259. This is a very important function and is widely used in IBM BIOS. The following paragraph discusses this at length. D6 (SL, select) and D7 (R, rotation) bits are used to program the 8259 for the various priority schemes discussed earlier. A frequently used bit combination for OCW2 in the x86 PCs is 0010000 = 20H.

Importance of the EOI (end-of-interrupt) command

Why is it necessary to issue an EOI command to the 8259? To understand the answer to this question, consider the following case. Assume that an 8259 has been initialized and is in the default fully nested mode (where IR0 has the highest priority and IR7 the lowest). Now assume that IR3 is activated and the CPU acknowledges the interrupt by sending back a signal through INTA. Then the CPU goes to the vector table and gets CS:IP of the interrupt service routine and starts to execute the routine. When the CPU acknowledges IR3, the 8259 marks (sets to 1) the bit associated with IR3 in its ISR (in-service register) to indicate that this is being serviced now. Issuing EOI to the 8259 indicates that servicing of IR3 is now complete and the bit associated with IR3 in register ISR can be reset to zero, thereby allowing IR3 to come in again. Of course, the EOI must be issued at the end of the service routine; otherwise, IR3 might keep interrupting itself again and again. If it is not issued and the CPU goes back to the main program after it finishes servicing IR3, it will not be able to be serviced again since the bit in the ISR register indicates that IR3 is being serviced. The important point of this is that the last three instructions of any interrupt service routine for IR0–IR7 must be issuing the EOI, followed by IRET (see Example 14-11). It should be noted that while IR3 is being serviced, IR2, IR1, and IR0 all are allowed to come in and interrupt it since they have higher priority, but no lower priority interrupts of IR4–IR7 are responded to. For example, if IR3 is being serviced and IR1 is activated, only IR0 can interrupt, and IR2 and IR4–IR7 will not be responded to. If the programmer has failed to issue the EOI at the end of IR1 and IR3, he has simply put these two IRs out of circulation in addition to IR2, IR4, IR5, IR6, and IR7. Only IR0 will be responded to by the 8259, since the ISR has marked IR3 and consequently all the lower priority interrupts IR4–IR7. Then IR1 puts the lower priority IR2 out of circulation. See Example 14-11.

Example 14-11

Show the last three instructions of the interrupt service routine for IR1 of the 8259 in Example 14-8.

Solution:

Before the interrupt service routine returns control to the main program, it issues OCW2 to the 8259.

```
INT_SERV     PROC  FAR          ;routine for IR1
             ...   ...
             MOV   AL,20H       ;the EOI byte for OCW2
             OUT   26H,AL       ;to port designated for OCW2
             IRET               ;return from ISR to the main program
```

OCW3 (operation command word 3)

OCW3 is used, among other functions, to read the 8259's internal registers IRR (interrupt request register) and ISR (in-service register). D0 and D1 allow the program to read these registers in order to see which of IR0–IR7 is pending for service and which one is being served. (As mentioned earlier, OCW1 is used to peek into the IMR.) The rest of the bits are for changing the masking mode and other advanced functions of the 8259. Interested readers should refer to Intel manuals.

Review Questions

1. A single 8259 can add up to _______ hardware interrupts to an x86 CPU.
2. INTR is an ________ (input, output) signal for the 8259 but it is an __________ (input, output) signal for the x86.
3. True or false. CAS0, CAS1, and CAS2 are used for master/slave mode only.
4. Indicate the logic level (high or low) on input pins A0 and CS needed to send ICW1 to the 8259.
5. True or false. ICWs can be sent to the 8259 in random order.
6. The 8259 can receive ICWs in the sequence ICW1, ICW2, ICW3, ICW4 or ICW1, ICW2, ICW4. How does it know which option is being programmed?
7. True or false. When the ISR (interrupt service routine) of IR5 is being executed, the 8259 prevents requests from the same interrupt by marking bit IR5 in its in-service register (ISR).
8. True or false. Fully nested mode is the default mode.
9. In fully nested mode, which IR has the highest priority?
10. Assume that an 8259 is configured in fully nested mode and the CPU is executing the interrupt service routine for IR5. During the execution of IR5, which interrupts can come in and which ones are blocked?

SECTION 14.4: USE OF THE 8259 CHIP IN x86 PCs

The original 8088-based PC used only one 8259 chip to extend the number of hardware interrupts to eight, but in subsequent PCs, two 8259 chips were used to extend the hardware interrupts up to 15. Since the original PC is a subset of today's x86 PC, we discuss that first and examine the rest of interrupts in a later section.

Interfacing the 8259 to the 8088 in IBM PC

To interface the 8259 to the 8088, there must be two port addresses assigned to the 8259. One is for ICW1 and the second one is for ICW2 and ICW4. Figure 14-8 shows the address decoding for the 8259 in the IBM PC. Since the chip select is activated by Y1 and all the x's for don't care must be zero, the addresses can be calculated in the manner indicated as shown in Table 14-6.

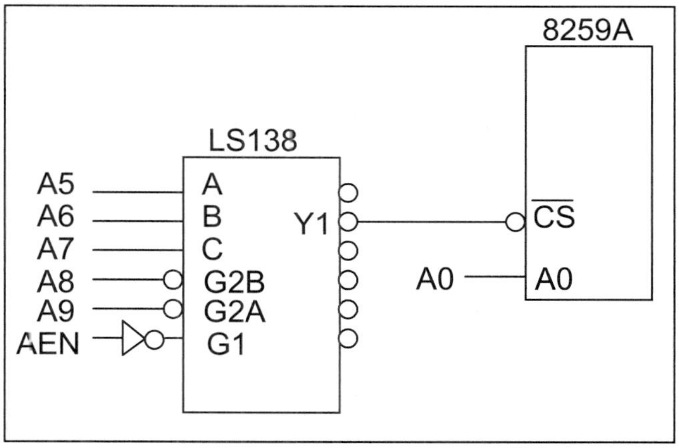

Figure 14-8. Chip Select Decoder of the 8259A

Table 14-6: Port Addresses of ICWs and OCWs

Binary Address											Hex	Function
AEN	A9	A8	A7	A6	A5	A4	A3	A2	A1	A0	Address	
1	0	0	0	0	1	X	X	X	X	0	20	Port address ICW1
1	0	0	0	0	1	X	X	X	X	1	21	Port addresses ICW2,ICW3,ICW4
1	0	0	0	0	1	X	X	X	X	0	20	Port addresses OCW2,OCW3
1	0	0	0	0	1	X	X	X	X	1	21	Port address OCW1

Initialization words of the 8259 in the IBM PC

Next the IBM PC initialization words for the 8259 will be explained. From earlier discussions and Figure 14-5, the following configuration for the control words ICW1, ICW2, and ICW4 can be calculated:

ICW1	IBM PC Configuration
D0 = 1	ICW4 needed
D1 = 1	Single
D2 = 0	For x86 this must be zero
D3 = 0	Edge triggering
D4 = 1	Required by ICW1 itself
D5 = D6 = D7 = 0	0s for 8088/86 based systems

ICW1 = 00010011 = 13H

As was explained in Section 14.2, out of the 256 interrupts of the 8088/86, IBM PC designers assigned INT 08–INT 0F for expansion of hardware interrupts. These interrupts are used by IR0–IR7 of the 8259 and commonly referred to as IRQ0–IRQ7. INT 08 is for IRQ0, INT 09 is for IRQ1, and so on. It is the function of ICW2 to inform the 8259 which interrupt numbers are assigned to IRQ0–IRQ7. This is done by equating ICW2 of the 8259 to the interrupt number assigned to IRQ0. In other words, ICW2 is the interrupt number for IR0, which in the case of the IBM PC and compatibles is INT 08. The 8259 is only programmed for the value of IRQ0, so the 8259 generates the INT numbers for IR1 through IR7. These are listed in Table 14-7. Summarizing the above discussion gives ICW2 = 00001000 = 08H.

Table 14-7: IBM PC Hardware Interrupts

Binary Data ICW2		8259 Interrupt	
D7 D6 D5 D4	D3 D2 D1 D0	Input	INT Type
0 0 0 0	1 0 0 0	IR0	INT 08
0 0 0 0	1 0 0 1	IR1	INT 09
0 0 0 0	1 0 1 0	IR2	INT 0A
0 0 0 0	1 0 1 1	IR3	INT 0B
0 0 0 0	1 1 0 0	IR4	INT 0C
0 0 0 0	1 1 0 1	IR5	INT 0D
0 0 0 0	1 1 1 0	IR6	INT 0E
0 0 0 0	1 1 1 1	IR7	INT 0F

ICW3 is used only when multiple 8259 chips are connected in master/slave mode, which is discussed in Section 14.5. Next the ICW4 configuration will be examined.

ICW4	IBM PC Configuration
D0 = 1	8088/86
D1 = 0	Normal (issue EOI before IRET)
D2 = 0, D3 = 1	Slave buffered mode
D4 = 0	Not nested
D5 = D6 = D7 = 0	Required by ICW4

ICW4 = 00001001 = 09H

This gives the following code for 8259 initialization.

```
MOV    AL,13H              ;the ICW1
OUT    20H,AL
MOV     AL,8               ;the ICW2
OUT    21H,AL
MOV    AL,9                ;the ICW4
OUT    21H,AL
```

The IBM PC BIOS version of the above program is as follows:

```
LOC OBJ        LINE       SOURCE
0020           19  INTA00 EQU    20H        ;8259 PORT
0021           20  INTA01 EQU    21H        ;8259 PORT
....           ...        ...... ...                    .............
               553;              INITIALIZE THE 8259 INTERRUPT CONTROLLER
CHIP
               554;--------------------------------------------------------
------------
E1B4           555 C21:
E1B4    B013   556        MOV    AL,13H     ; ICW1 - EDGE, SNGL, ICW4
E1B6    E620   557        OUT    INTA00,AL
E1B8    B008   558        MOV    AL,8       ;SETUP ICW2    INT TYPE 8 (8-F)
E1BA    E621   559        OUTI   NTA01,AL
E1BC    B009   560        MOV    AL,09      ;SETUP ICW3 - BUFFERED,8086 MODE
E1BE    E621   561        OUT    INTA01,AL
....           ...        ...... ...                    .............
```

(Reprinted by permission from "IBM BIOS Technical Reference" c. 1984 by International Business Machines Corporation)

Now that the 8259 is initialized, it is ready to accept an interrupt on any of the inputs IRQ0–IRQ7, thereby expanding the number of hardware interrupts for the 8088/86. What if the 8259 IC is defective? The 8259 is tested by a program in BIOS. There is a collection of programs in BIOS that is responsible for testing and initialization of the CPU and peripheral chips. This is commonly referred to as the POST (power on self test).

Since it is possible that one of the bits of the IMR (interrupt mask register) has become stuck to zero or one during the fabrication of the chip and escaped detection, the following program from IBM BIOS tests the IMR of the 8259 chip by writing 0s and 1s to it and reading them back. Reminder: To access the IMR, use OCW1, which has the port address 21H in the IBM PC (see Table 14-6). In the following program, if the test fails, the system will beep.

```
               620; TEST THE IMR REGISTER
               621
E217    BA2100 622        MOV  DX,0021    ;POINT INTR. CHIP ADDR 21
E21A    B000   623        MOV  AL,0       ;SET IMR TO ZERO
E21C    EE     624        OUT  DX,AL
E21D    EC     625        IN   AL,DX            ;READ IMR
E21E    0AC0   626        OR   AL,AL            ;IMR = 0?
E220    7515   627        JNZ  D6         ;GO TO ERR ROUTINE IF NOT 0
E222    BOFF   628        MOV  AL,0FF     ;DISABLE DEVICE INTERRUPTS
E224    EE     629        OUT  DX,AL            ;WRITE TO IMR
E225    EC     630        IN   AL,DX            ;READ IMR
E226    0401   631        ADD  AL,1       ;ALL IMR BIT ON?
E228    750D   632        JNZ  D6         ;NO GO TO ERR ROUTINE
...     ...    ...        .... ..         ........................
```

(Reprinted by permission from "IBM BIOS Technical Reference" c. 1984 by International Business Machines Corporation)

Sequences of hardware interrupts with the 8259

When a high is put on any of IR0–IR7, how does the microprocessor become involved? As mentioned earlier, INTA of the 8088/86 is connected to INTA of the 8288 bus controller, and INTR of the 8259 is connected to INTR of the 8088/86. The following is the sequence of events after an IR of the 8259 is activated.

1. After an IR is activated, the 8259 will respond by putting a high on INTR, thereby signaling the CPU for an interrupt request.
2. The 8088/86 puts the appropriate signals on S0, S1, and S2 (S0 = 0, S1 = 0, and S2 = 0), indicating to the 8288 that an interrupt has been requested.
3. The 8288 issues the first INTA to the 8259.
4. The 8259 receives the first INTA and does internal housekeeping, which includes resolution of priority (if more than one IR has been activated) and resolution of cascading.
5. The 8288 issues the second INTA to the 8259.
6. The second INTA pulse makes the 8259 put a single interrupt vector byte on the data bus in which the 8088/86 will latch. The value of the single byte depends on ICW2 and which IR has been activated, as discussed earlier.
7. The 8088/86 uses this byte to calculate the vector location, which is four times the value of the INT type.
8. The 8088/86 pushes the flag register onto the stack, clears IF (Interrupt Flag) and TF (Trap Flag), thereby disabling further external interrupt requests and disabling single-step mode, and finally pushes the present CS:IP registers onto the stack.
9. The 8088/86 reads CS:IP of the interrupt service routine from the vector table and begins execution of the interrupt routine.

Next we see which devices in the IBM PC use the eight hardware interrupts, IRQ0 to IRQ7, of the 8259.

Sources of hardware interrupts in the IBM PC

With the use of the 8259, the IBM PC has eight interrupts, IR0 to IR7, plus NMI of the 8088/86. First the assignment of IR0 to IR7 will be discussed, then NMI and its use in the IBM PC.

Of the eight interrupts for the 8259, IBM has used two, IR0 and IR1, for internal use by the system. The other six, IR2 through IR7, are available through the expansion slots. Of those used internally, IR0 is for channel 0 of the 8253 timer to update the time of day (TOD) clock, and IR1 is dedicated to the keyboard. IR1 is activated whenever the serial-in, parallel-out shift register of the keyboard has a byte of data. IR2 to IR7 are generally used with the following assignments. The following two are used on the motherboard:

INT	08	IRQ0	Channel 0 of 8253 timer to update TOD
INT	09	IRQ1	Keyboard input data

The following are available through the expansion slot bus and used widely in industry, as indicated. Figure 14-9 summarizes the hardware interrupt assignment in the IBM PC.

INT	0AH	IRQ2	Reserved
INT	0BH	IRQ3	Serial COM2
INT	0CH	IRQ4	Serial COM1
INT	0DH	IRQ5	Alternative printer
INT	0EH	IRQ6	Floppy disk
INT	0FH	IRQ7	Parallel printer LPT1

Sources of NMI in the IBM PC

The last hardware interrupt to be discussed for the 8088/86 computer is the NMI (nonmaskable interrupt). This interrupt is actually one of the pins of the CPU (similar to the TRAP pin in the 8080/85), and unlike INTR there is no need for the INTA pin to acknowl-

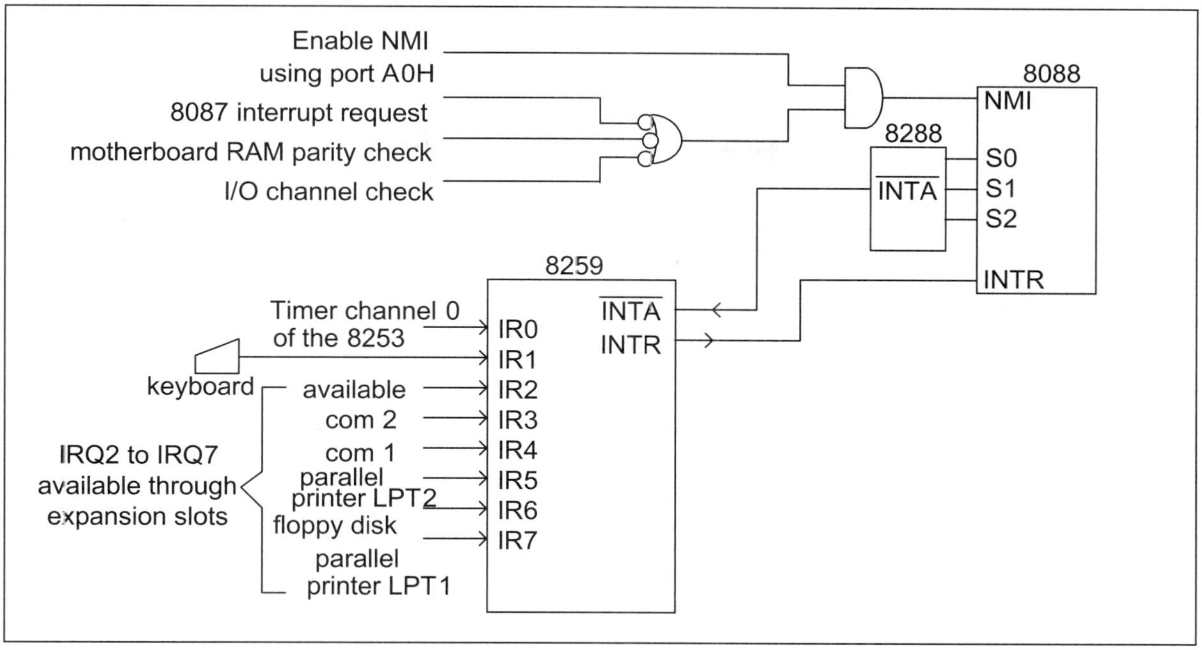

Figure 14-9. PC Sources of Hardware Interrupts

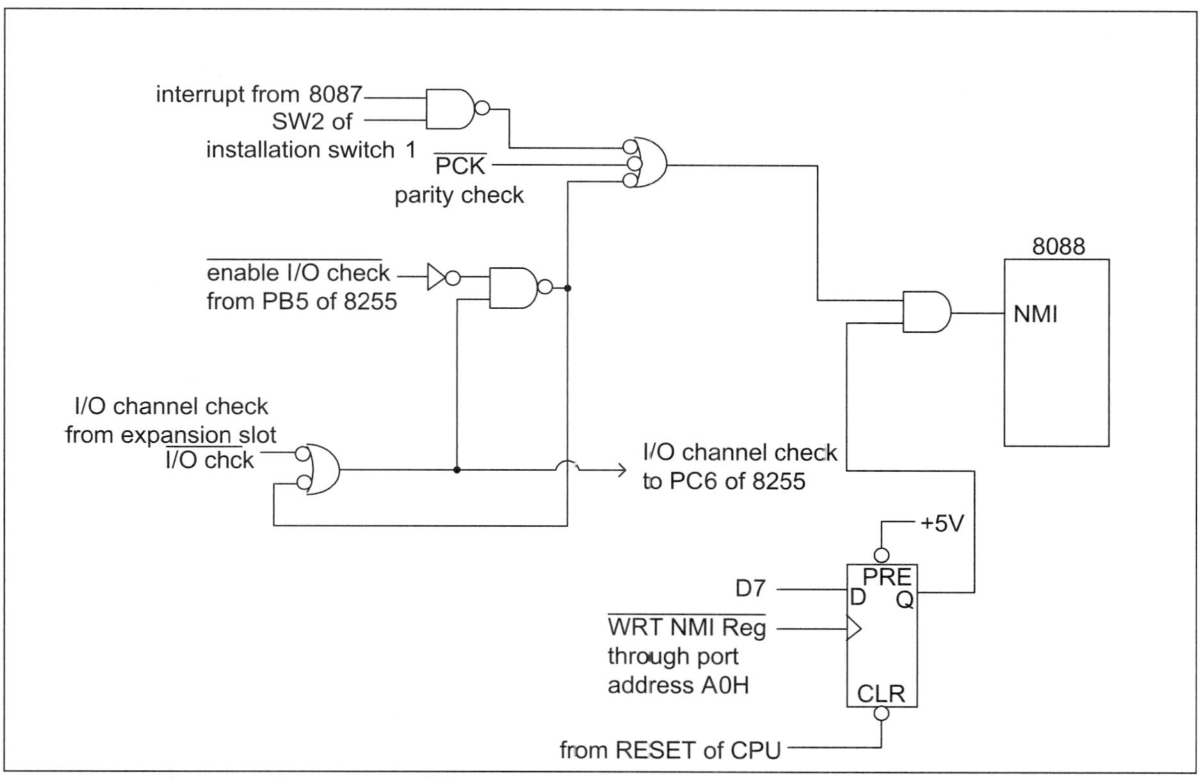

Figure 14-10. Sources of NMI in the PC

edge it. Furthermore, it cannot be masked (disabled) by software as is the case for INTR, which can be masked at any time through use of the instruction CLI (clear interrupt flag). It is for this reason that the IBM PC has used the NMI for parity bit checking of DRAM to make sure that all read/write memory is working properly. In the absence of RAM memory, the operating system would not be loaded and the computer could not function.

If the NMI is so important to the system, which devices can activate it, and can they be masked at all?

First, as can be seen from Figure 14-10, there are three sources of activation of the NMI:

1. NPIRQ (numerical processor interrupt request)
2. Read/write PCK (parity check)
3. IOCHK (input/output channel check)

Since three different sources can activate NMI, how does the system know which one is requesting interrupt service at any given time? The IBM PC system recognizes which of these interrupt requests has been activated by checking input port C of the 8255. It looks at PC6 of the 8255 to see if it is IOCHK and at PC7 to see if it is PCK. The NMI service routine software must check PC6 and PC7 and determine which one has requested service. If neither of these two is requesting service, the request must have come from the 8087 coprocessor on the motherboard (in IBM terminology, planer). IBM BIOS checks the source of each and, as it finds them, displays an appropriate message on the video screen.

The following code shows how BIOS detects the source of the NMI interrupt.

```
E2C3                     ORG    0E2C3H
E2C3          NMI_INT    PROC   NEAR
E2C3 50                  PUSH   AX             ;SAVE ORIG CONTENTS OF AX
E2C4 E462                IN     AL,PORT_C
E2C6 A8C0                TEST   AL,0C0H        ;PARITY CHECK?
E2C8 7415                JZ     D14            ;NO, EXIT FROM ROUTINE

.........                ...           ;ADDR OF ERROR MSG
E2CE A840                TEST   AL,40H         ;I/O PARITY CHECK
E2D0 7504                JNZ    D13            ;DISPLAY ERROR MEG
...  ....                ...    ...    ;MUST BE PLANER
E2D6          D13:
...  ....                ...    ...    } sends the message to
...  ....                       ...    ...    } video and halts the
system.
E2DF          D14:
E2DF 58                  POP    AX             ;RESTORE ORIGINAL AX
E2E0 CF                  IRET
              NMI_INT    ENDP
```

(Reprinted by permission from "IBM BIOS Technical Reference" c. 1984 by International Business Machines Corporation)

Is there any way that NMI can be masked? The answer is yes. As can be seen from Figure 14-10, NMI is masked by a RESET signal from the CPU with CLR of the D flip-flop when the computer is first turned on. It can also be unmasked or masked through port A0H by setting D7 of the data bus to 1 (unmask) or 0 (mask). Again from the IBM PC BIOS we see the following:

```
                  ;       ENABLE NMI INTERRUPTS

E5BC B080                 MOV    AL,80H        ;ENABLE NMI INTERRUPTS
E5BE E6A0                 OUT    0A0H,AL
.... ....    ....         ...    ........      .....................
```

(Reprinted by permission from "IBM BIOS Technical Reference" c. 1984 by International Business Machines Corporation)

Review Questions

1. True or false. The original IBM PC used only one 8259.
2. What ports are assigned to ICWs in the PC?
3. In the PC, the IRQs are ______________ (edge, level) triggered.
4. Of the 256 possible interrupts of the 8088, which ones are assigned to IRQ0–IRQ7 of the 8259?
5. True or false. IRQ0 and IRQ1 can be used by the system but not by the user.

6. Which IRQs of the 8259 are available on the expansion slot?
7. True or false. The x86 can mask and unmask the NMI by using the STI and CLI instructions.
8. True or false. If there is a problem with the memory of the PC, NMI is activated.

SECTION 14.5: MORE ON INTERRUPTS IN x86 PCs

When the first PC was introduced, only the six hardware interrupts, IRQ2–IRQ7, were available through the 8-bit section of expansion slot. The other two, IRQ0 and IRQ1, were used by the motherboard. With the introduction of the 80286-based PC AT, another eight interrupts, IRQ8–IRQ15, were added. IBM implemented the additional hardware interrupts with the use of a second 8259 programmable interrupt controller. To make their computers IBM compatible, all subsequent x86 PCs have remained faithful to the original IBM PC. In this section we study the hardware interrupt assignment for x86 PCs.

x86 PC hardware interrupts

In the design of the second generation of IBM PC, IBM designers had to make sure that it was compatible with the 8088-based original PC. This led to the use of IRQ0 and IRQ1 for the system timer and keyboard, respectively, as was the case in the original PC. IBM made the first 8259 a master, and added the second 8259 in slave mode. To do that, it connected the INT pin of the slave 8259 to IRQ2 of the master 8259. The master and slave 8259s communicate with each other through pins IRQ2, INT, CAS0, CAS1, and CAS2. See Figure 14-11. Table 14-8 shows the interrupt assignment for x86 PCs.

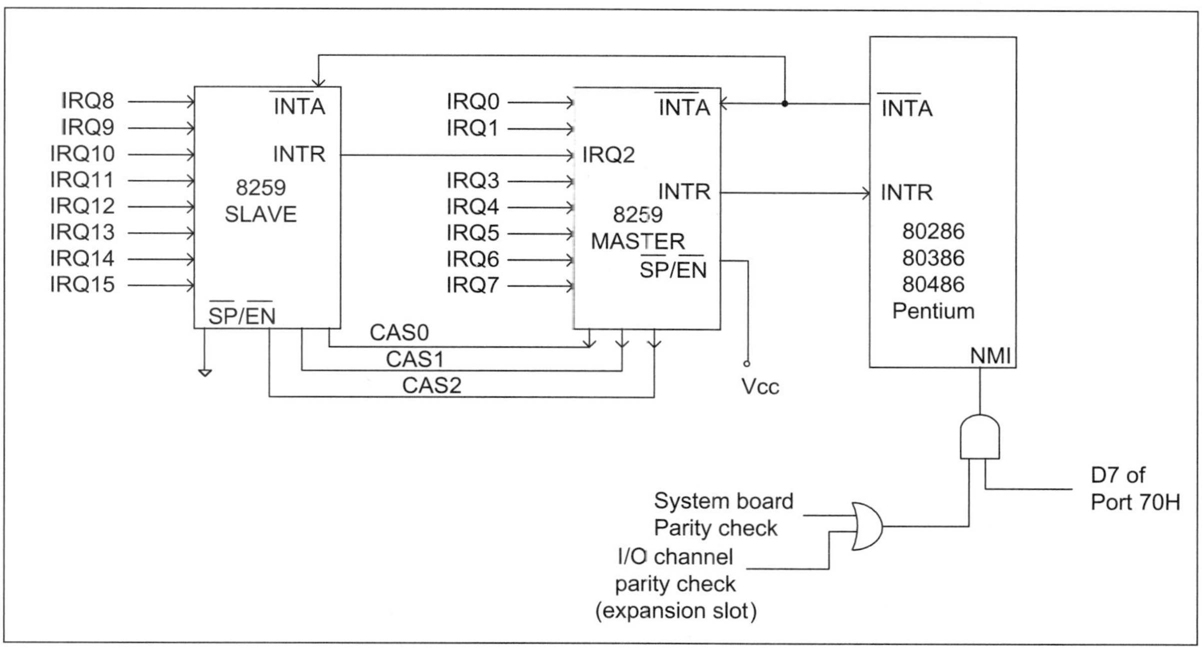

Figure 14-11. 8259 Chips in Master/Slave Relation for 286 and x86 PCs

The hardware interrupts on the ISA bus are shown in Figure 14-12. Notice on the ISA expansion slot that IRQ10, IRQ11, IRQ12, IRQ14, and IRQ15 are on the 32-pin section and IRQ9, IRQ3, IRQ4, IRQ5, IRQ6, and IRQ7 are on the 62-pin section.

The x86 microprocessor generated interrupts (exceptions)

As mentioned in Section 14.1, when the CPU encounters an unusual situation such as dividing a number by zero, it generates an exception. The 8088/86 had only one exception, divide error or INT 0. In the 8088/86, Intel Corporation left the first 32 interrupts (INT 00 to INT 1FH) reserved for future microprocessors. However, designers of

Table 14-8: Hardware Interrupt Assignment for ISA PCs

IRQ	INT Number	Use
IRQ0	INT 08	8254 TIMER OUT0
IRQ1	INT 09H	Keyboard
IRQ2	INT 0AH	Interrupt from 8259 #2
IRQ3	INT 0BH	Serial port COM2 (and COM4)
IRQ4	INT 0CH	Serial port COM1 (and COM3)
IRQ5	INT 0DH	Parallel port 2: LPT2
IRQ6	INT 0EH	Floppy disk controller
IRQ7	INT 0FH	Parallel port 1: LPT1
IRQ8	INT 70H	CMOS real-time clock
IRQ9	INT 71H	Software redirected to INT 0AH
IRQ10	INT 72H	Available
IRQ11	INT 73H	Available
IRQ12	INT 74H	PS/2 mouse
IRQ13	INT 75H	Math coprocessor
IRQ14	INT 76H	Hard disk
IRQ15	INT 77H	Available

the first IBM PC ignored this and assigned many of these interrupts to hardware and software interrupts on the system. By not adhering to Intel's specifications, IBM has created a massive headache for software designers of protected mode 386 and later systems. This is due to the fact that Intel continued to assign the processor exception cases generated by the x86 CPU to INT 5 and higher with each new member of the x86 family. These are shown in Table 14-9. Many of the interrupts in Table 14-9 are used by the x86 in protected mode. Since the Microsoft Windows NT/2000/XP and Vista operating systems use the x86 in protected mode, they have mapped all these interrupts to new interrupts to avoid interrupt conflict with the IRQs of the PC.

Interrupt priority

The next topic in this section is the concept of priority for INT 00 to INT FFH. What happens if two interrupts want the attention of the CPU at the same time? Which has priority? As far as the x86 is concerned, the INTR pin is considered a single interrupt. Therefore, the resolution of priority among the IRQs is up to the 8259. Assume that the INT instruction (such as INT 21H) and INTR both want to be processed. The INT instruction has a higher priority than either INTR or NMI. If both NMI and INTR are activated at the same time, NMI is responded to first since NMI has a higher priority than INTR. Table 14-10 shows the interrupt processing order for the 80286 microprocessor from Intel's manual (1 is the highest priority).

For the IRQs coming through INTR, the 8259 resolves priority depending on the way the 8259 is programmed. In the x86 IBM PC, PS, and compatibles, IRQ0 has the highest priority and IRQ7 is assigned the lowest priority. It must be noted that since IRQ8 to IRQ15 of the slave 8259 are connected to IRQ2 of the master 8259, they have higher priority than IRQ3 to IRQ7 of the master 8259. Figure 14-13 shows the IRQ0–IRQ15 priority.

More about edge- and level-triggered interrupts

As discussed previously, in the 8259 there are two ways to activate the interrupt input IRQ, depending upon how it is programmed. One is level-triggered mode and the other is edge-triggered mode.

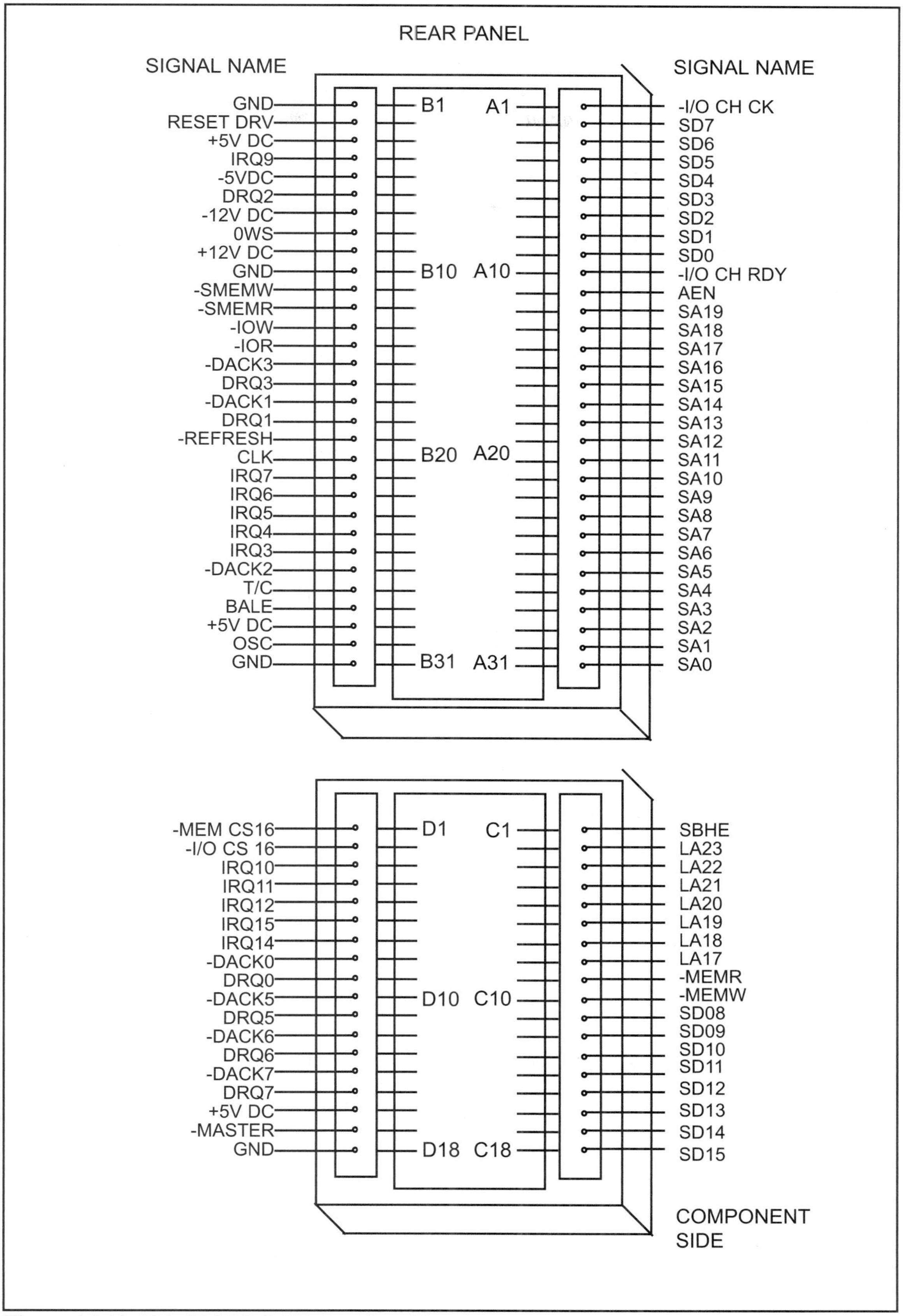

Figure 14-12. ISA (IBM PC AT) Bus Slot Signals

Table 14-9: x86 Microprocessor Interrupt Assignment

Interrupt	8086/88	286	386	486
00	Divide error	Divide error	Divide error	Divide error
01	Single step	Single step	Single step debugging exceptions	Single step debugging exceptions
02	Nonmaskable interrupt	Nonmaskable interrupt	Nonmaskable interrupt	Nonmaskable interrupt
03	Breakpoint	Breakpoint	Breakpoint	Breakpoint
04	INTO detected overflow	INTO detected overflow	INTO detected overflow	INTO detected overflow
05		Bound range exceeded	Bound range exceeded	Bound range exceeded
06		Invalid instruction	Invalid instruction	Invalid instruction
07		Coprocessor not available	Coprocessor not available	Coprocessor not available
08		Double exception detected	Double exception detected	Double exception detected
09		Coprocessor	Coprocessor protection error	Coprocessor protection error
0A		Invalid task state segment	Invalid task state segment	Invalid task state segment
0B		Segment not present	Segment not present	Segment not present
0C		Stack fault	Stack fault	Stack fault
0D		Protection fault	Protection fault	Protection fault
0E			Page fault	Page fault
10		Coprocessor error	Coprocessor error	Coprocessor error
11				Alignment check

Note: Pentium has assigned a new exception called machine check with INT 12H.

Level-triggered mode

In level triggered mode, the 8259 will recognize a high on the IRQ input as an interrupt request. The request on the IRQ line must remain high until the first INTA is acknowledged from the 8259. It is only then that the high must be removed from the IRQ input immediately. If the IRQ input remains high after the end of interrupt (EOI) command has been issued, the 8259 will generate another interrupt on the same IRQ input. Therefore, to avoid multiple interrupt generation, the IRQ input must be brought low before the EOI is issued.

Edge-triggered mode

In edge-triggered mode, the 8259 will recognize an interrupt request only when a low-to-high pulse is applied to an IRQ input. This means that after the low-to-high transition on the IRQ input, the 8259 will acknowledge the interrupt request by activating INTA and the high level input will not generate further interrupts even after the EOI is issued. Therefore, the designer does not need to worry about quickly removing the high to avoid generating multiple interrupts, as is the case for level-triggered mode. This is due to the fact that in edge-triggered mode, before another interrupt can be requested, the IRQ input must be brought back to low. Notice in both edge- and level-triggered modes that the IRQ must stay high until after the falling edge of the first INTA pulse in order to acknowledge the interrupt request. It is interesting to note the role of the IRR (interrupt request register) of the 8259. In level-triggered mode, the IRR latch is always ready to recognize a high on the IRQ as a request for interrupt. But in edge-triggered mode, the IRR latch is disabled after the request is acknowledged and will not latch another interrupt until that IRQ input goes back to low. The disadvantage of edge-triggered mode is the problem of false interrupt caused by a good sized spike as a result of noise on the IRQ line, especially in high-speed systems.

Table 14-10: 80286 Interrupt Priority

Order	Interrupt
1	INT instruction or exception
2	Single step
3	NMI
4	Processor extension segment overrun
5	INTR

IRQ0	HIGHEST PRIORITY
IRQ1	
IRQ8	
IRQ9	
IRQ10	
IRQ11	
IRQ12	
IRQ13	
IRQ14	
IRQ15	
IRQ3	
IRQ4	
IRQ5	
IRQ6	
IRQ7	LOWEST PRIORITY

Figure 14-13. IRQ Priority in the x86 PC

Review Questions

1. True or false. IRQ13 is used for math processor error detection in x86 PCs.
2. Which portion of the ISA bus has the IRQ10?
3. Which has the higher priority, IRQ10 or IRQ7?
4. Which has the higher priority, IRQ1 or IRQ13?
5. With x86 processors, an invalid instruction causes an exception. Which interrupt is assigned to it by Intel?

PROBLEMS

SECTION 14.1: 8088/86 INTERRUPTS

1. Assume that the 8088/86 is executing an instruction with 17 clock counts. Meanwhile, the INTR pin is activated. Does the CPU finish the current instruction before it responds to INTR? How does the CPU resume from where it left off?
2. Give the logical and physical addresses in the interrupt vector table associated with each of the following interrupts.
(a) INT 5 (b) INT 21H
3. What does ISR stand for, and what is it? Give another name for ISR.
4. Where is the address of each ISR kept?
5. Compare the number of bytes of stack memory used by each of the following.
(a) CALL FAR (b) interrupt activation
6. Vector table addresses 003F8H–003FB belong to which interrupt?
7. Give the logical and physical addresses used by the interrupt vector table.

8. How many bytes are used by the interrupt vector table, and why?
9. Why should we not use the first 1K of address space in 8088/86-based systems?
10. Indicate the interrupt(s) set aside for exception handling in the 8088/86.
11. Give the interrupt number (type) assigned to each of the following. (a) divide error (b) single step (c) NMI
12. True or false. When an interrupt through INTR is executed, IF = 0 and TF = 0.
13. True or false. CLI blocks both INTR and NMI.
14. Show how to set TF to high.
15. Show how to set IF to each of the following. (a) low (b) high
16. True or false. Instruction INTO is executed only if the overflow flag is high (OF = 1).
17. The last instruction in the ISR is _________ , whereas the last instruction in a FAR subroutine is __________.
18. What is the difference between RETF and IRET in terms of stack activity?
19. Show the stack frame where CS, IP, and FR are stored for both an interrupt and a CALL FAR routine. Assume that SP = FFE0H.
20. In which of the following sequences are the stack contents popped off by IRET?
    (a) IP, FR, CS (b) FR, IP, CS (c) FR, CS, IP (d) none of the above

SECTION 14.2: x86 PC AND INTERRUPT ASSIGNMENT

21. Answer the following questions, assuming that vector table locations 0000:001C to 0000:001F have the contents indicated below.
    0000:0010 .... 47 FF 00 F0
    (a) Which interrupt does this belong to?
    (b) What is the logical address and physical address of the ISR?
22. In Problem 21, does BIOS or DOS provide the service?
23. INT 12H provides the size of which of the following memories?
    (a) high memory area (b) extended memory
    (c) conventional memory (d) expansion memory installed in the expansion slot
24. If the start of an ISR is located in BIOS ROM at FFE6EH, what are the values of CS and IP in the vector table?
25. In Problem 24, if the ISR belongs to INT 1CH, show the exact contents of the vector table.
26. In what BIOS data area location is the size of conventional memory stored?

SECTION 14.3: 8259 PROGRAMMABLE INTERRUPT CONTROLLER

Note: These problems do not necessarily apply to IBM PC compatibles.

27. True or false. In the 8259, to program ICW1, we must have A0 = 1 and CS = 0.
28. For the 8259, indicate which of the following is input and which is output.
    (a) IR0 IR7 (b) INT (c) INTA
    (d) A0 (e) CS (f) RD
29. Find the addresses for each ICW of the 8259 if CS is activated by A7–A1 = 1001 010.
30. Find the ICW1 and ICW2 if the 8259 is used with an 8088/86, single, edge triggering, no ICW4, and IR0 is assigned INT 88H.
31. Show the programming of ICW0 and ICW2 in Problem 30. Use the port addresses of Problem 29.
32. Which of the following interrupts cannot be assigned to IR0 of the 8259, and why?
    (a) 99H (b) 98H (c) CCH
    (d) 22H (e) 10H (f) F8H
33. Find the INT type number assigned to IR0 and IR7 if IR3 is assigned INT 1BH.
34. Find the INT number assigned to IR0, IR4, and IR6 if IR2 is assigned INT 32H.
35. Which of the OCWs is used to mask a given IR of the 8259?
36. EOI is issued by which of the OCWs?

37. What is the default mode for the prioritization of IR0 to IR7?
38. Find the port addresses assigned to each of the OCWs in Problem 29.
39. Show the program to enable IR2 and IR4 and mask the rest of the IRs. Use the port addresses in Problem 38.
40. OCW2, OCW3, and ICW1 go to the same port address in the 8259 when A0 = 0 and CS = 0. How does the 8259 distinguish between them? How does it distinguish between OCW2 and OCW3 since both go to the same port address?

SECTION 14.4: USE OF THE 8259 CHIP IN x86 PCs

41. Why is signal AEN used in accessing the 8259 in the PC?
42. The PC uses the 8259 in ____________ (single, cascade) mode.
43. Indicate the IRQs level of triggering in the IBM PC (edge triggered, level triggered).
44. What interrupt numbers are assigned to the 8259 in the PC?
45. What port addresses are assigned to the 8259 in the PC?
46. True or false. IRQ0 and IRQ1 are used by the system board and are not available.
47. Which of the IRQs of the 8259 are available on the expansion slot?
48. Indicate on which side, A or B, of the expansion slot IRQs are located.
49. What are the binary and hex values for the EOI, and to which port is it issued in the IBM PC?
50. Which IRQ has the highest priority, and why?
51. True or false. The 8288 chip issues two INTAs to the 8259 when INTR of the 8088 is activated.
52. True or false. In the PC, there is more than one source of NMI activation.
53. In the PC, can NMI be blocked? If yes, how?
54. True or false. In the PC, the parity bit error from both the memory of the system board and the memory board of the expansion slot can activate NMI.

SECTION 14.5: MORE ON INTERRUPTS IN x86 PCs

55. True or false. In the x86 PC, INTR of the x86 comes from the primary (master) 8259 chip.
56. True or false. INTA from the x86 goes to both 8259 chips in the x86 PC.
57. In the x86, what port addresses are assigned to the 8259s? See Appendix E.
58. In the x86 PC, what interrupt numbers are assigned to the second 8259 chip?
59. What IRQs are available on the ISA expansion slot?
60. Why is there no IRQ2 on the ISA expansion slot? Is there any replacement for it?
61. Of IRQ10 and IRQ4, which has the higher priority, and why?
62. True or false. With every generation of the x86, more exception interrupts are added but they are downward compatible.
63. True or false. The INT instruction and the exception interrupt have a lower priority than NMI.
64. True or false. The NMI has a higher priority than INTR.
65. If NMI, IRQ10, and IRQ6 are all activated at the same time, explain the sequence when the system responds and executes them.
66. If IRQ3, IRQ7, and IRQ15 are all activated at the same time, in what order are they serviced?

ANSWERS TO REVIEW QUESTIONS

SECTION 14.1: 8088/86 INTERRUPTS

1. True
2. 4
3. 1K byte beginning at 00000 and ending at 003FFH
4. Interrupt service routine (ISR) or interrupt handler
5. To hold the CS:IP of each ISR

6. 00040H, 41H, 42H, and 43H
7. No; it is internally embedded into the CPU.
8. INT 0

SECTION 14.2: x86 PC AND INTERRUPT ASSIGNMENT

1. F000:F065, CS = F000 and IP = F065
2. INT 10H is assigned memory locations 00040H, 41H, 42H, and 43H in the interrupt vector table. That means that we have 00040 = (65), 00041 = (F0), 00042 = (00), and 00043 = (F0).
3. The dash (-) tells us this is the 8th boundary; therefore, the 0000:0038 address in the interrupt vector table belongs to INT 14 (0E hex).
4. The logical address is F000:EF57; therefore, CS = F000 and IP = EF57.
5. AX = 200
6. IRET

SECTION 14.3: 8259 PROGRAMMABLE INTERRUPT CONTROLLER

1. 8
2. Output, input
3. True
4. CS = 0 and A0 = 0
5. False
6. The bit D1 in ICW1 indicates if it is for single or cascade. If it is cascade, it expects to receive all ICWs, from 1 to 4, but if it is single, it does not expect ICW3.
7. True
8. True
9. IR0
10. Higher priority interrupts IR0, IR1, IR2, IR3, IR4 can come in, but IR6 and IR7 are blocked since they have lower priority.

SECTION 14.4: USE OF THE 8259 CHIP IN x86 PCs

1. True
2. 20H and 21H
3. Edge triggered
4. INT 08 to INT 0FH
5. True
6. IRQ2 through IRQ7
7. False
8. True

SECTION 14.5: MORE ON INTERRUPTS IN x86 PCs

1. True
2. 62-pin section
3. IRQ10
4. IRQ1
5. INT 06

CHAPTER 15

DIRECT MEMORY ACCESS AND DMA CHANNELS IN x86 PC

OBJECTIVES

Upon completion of this chapter, you will be able to:

- **>> Describe the concept of DMA, direct memory accessing**
- **>> List the pins of the 8237A DMA chip and describe their functions**
- **>> Explain how bus arbitration is achieved between DMA and the CPU**
- **>> Explain how the channels of the 8237 are used in the PC**
- **>> List the DMA signals in the ISA bus**

For a computer to work efficiently, there must be a way to transfer a large amount of data in a short amount of time. In the IBM PC, this is accomplished with the help of what is called direct memory access (DMA), and that is the subject of this chapter.

SECTION 15.1: CONCEPT OF DMA

In computers there is often a need to transfer a large number of bytes of data between memory and peripherals such as disk drives. In such cases, using the microprocessor to transfer the data is too slow since the data first must be fetched into the CPU and then sent to its destination. In addition, the process of fetching and decoding the instructions themselves adds to the overhead. For this reason, Intel created the 8237 DMAC (direct memory access controller) chip, whose function is to bypass the CPU and provide a direct connection between peripherals and memory, thus transferring the data as fast as possible. While the 8237 can transfer a byte of data between an I/O peripheral and memory in only 4 clocks, the 8088 would take 39 clocks:

```
                                Number of Clocks
BACK: MOV  AL,[SI]                    10
      OUT PORT,AL                     10
      INC  SI                         2
      LOOP  BACK                      17
;total clocks                         39
```

One problem with using DMA is that there is only one set of buses (one set of each bus: data bus, address bus, control bus) in a given computer and no bus can serve two masters at the same time. The buses can be used either by the main x86 CPU or the 8237 DMA. Since the x86 has primary control over the buses, it must give permission to DMA to use them. How is this done? The answer is that any time the DMA needs to use the buses to transfer data, it sends a signal called HOLD to the CPU and the CPU will respond by sending back the signal HLDA (hold acknowledge) to indicate to the DMA that it can go ahead and use the buses. While the DMA is using the buses to transfer data, the CPU is sitting idle, and conversely, when the CPU is using the bus, the DMA is sitting idle. After DMA finishes its job it will make HOLD go low and then the CPU will regain control over the buses. See Figure 15-1.

For example, if the DMA is to transfer a block of data from memory to an I/O device such as a disk, it must know the address of the beginning of the block (address of the first byte of data) and the number of bytes (count) it needs to transfer. Then it will go through the following steps.

1. The peripheral device (such as the disk controller) will request the service of DMA by pulling DREQ (DMA request) high.
2. The DMA will put a high on its HRQ (hold request), signaling the CPU through its HOLD pin that it needs to use the buses.
3. The CPU will finish the present bus cycle and respond to the DMA request by putting high on its HLDA (hold acknowledge), thus telling the 8237 DMA that it can go ahead and use the buses to perform its task. HOLD must remain active high as long as DMA is performing its task.
4. DMA will activate DACK (DMA acknowledge), which tells the peripheral device that it will start to transfer the data.
5. DMA starts to transfer the data from memory to the peripheral by putting the address of the first byte of the block on the address bus and activating MEMR, thereby reading the byte from memory into the data bus; it then activates IOW to write the data to the peripheral. Then DMA decrements the counter and increments the address pointer and repeats this process until the count reaches zero and the task

is finished.

6. After the DMA has finished its job it will deactivate HRQ, signaling the CPU that it can regain control over its buses.

This above discussion indicates that DMA can only transfer information; unlike the CPU, it cannot decode and execute instructions. Notice also that when the CPU receives a HOLD request from DMA, it finishes the present bus cycle (but not necessarily the present instruction) before it hands over control of the buses to the DMA. This is in contrast to a hardware interrupt, in which the CPU finishes the present instruction before it responds with INTA. One could look at the DMA as a kind of CPU without the instruction decoder/executer logic circuitry. For the DMA to be able to transfer data it is equipped with the address bus, data bus, and control bus signals IOR, IOW, MEMR, and MEMW.

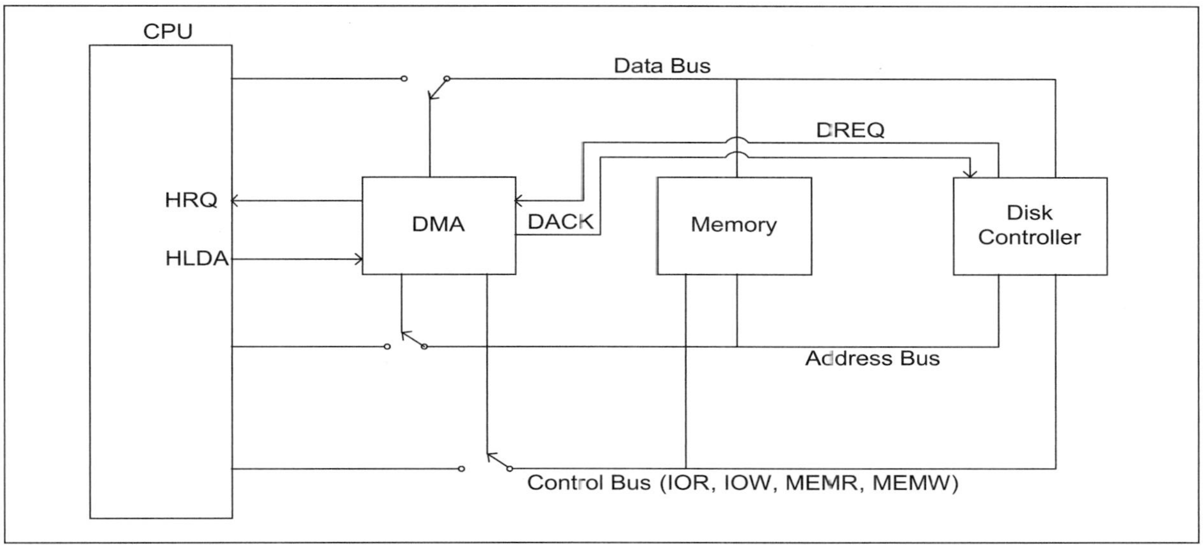

Figure 15-1. DMA Usage of System Bus

Review Questions

1. True or false. When the DMA is working, the CPU is sitting idle.
2. True or false. When the CPU is working, the DMA is sitting idle.
3. True or false. No bus can serve two masters at the same time.
4. True or false. The main CPU (x86) has control over all the system buses.
5. To get control over the system bus the ______ (INTR, HOLD) pin of the x86 is activated.
6. The x86 CPU informs the peripheral that it relinquishes control over the system bus through its ____________ pin.
7. The HOLD is an ___________ (input, output) for the x86 CPU.
8. The HLDA is an ___________ (input, output) for the x86 CPU.

SECTION 15.2: 8237 DMA CHIP PROGRAMMING

The Intel 8237 DMA controller is a 40-pin chip. It has four channels for transferring data, and each must be used for one device. For example, one is used for the floppy disk, one for the hard disk, and so on. Of course, only one device can use the DMA to transfer data at a given time. With every channel there are two associated signals, DREQ (DMA request) and DACK (DMA acknowledge). DREQ is an input to DMA coming from the peripheral device (such as the hard disk controller) and DACK is an output signal from the 8237 going to the peripheral device. From the 8237 DMA, there is only one HOLD and one HLDA that are connected to HOLD and HLDA of the x86. This means that four channels from four different devices can request use of the system buses, but DMA decides who gets control based on the way its priority register has been programmed. Every channel of the 8237 DMA must be initialized separately for the address of the data block and the count (the size of the block) before it can be used. This initialization involves writing into each channel:

1. The address of the first byte of the block of data that must be transferred (called the *base address*).
2. The number of bytes to be transferred (called the *word count*).

After initialization, each channel can be enabled and controlled with the use of a control word. There are many modes of operation, and these various modes and options must be programmed into the 8237's internal registers. To access these registers, the 8237 provides four address pins, A0–A3, along with the CS (chip select) pin. Since each channel needs separate addresses for the base address and the word count, a total of eight ports is set aside for those alone. Table 15-1 shows the internal addresses of the 8237 registers for each channel. Example 15-1 shows how these addresses are generated.

Example 15-1

Find the port addresses for the base address and word count of each channel of the 8237 for Figure 15-2 (CS is activated by A7–A4 = 1001 binary).

Solution:

From Table 15-1, one can get the addresses found in Table 15-3.

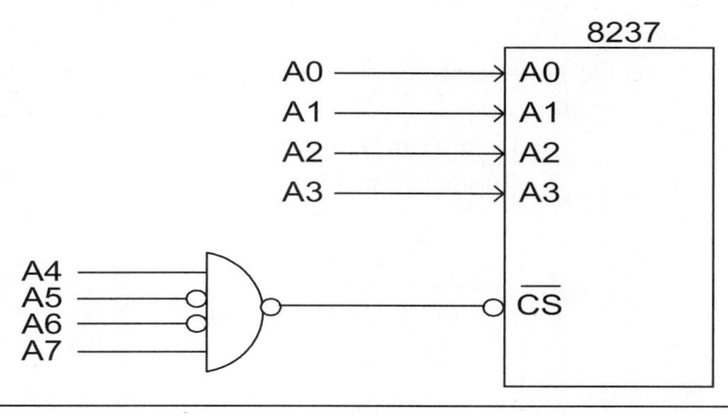

Figure 15-2. Diagram for Example 15-1

Table 15-1: 8237 Internal Addresses for Writing Transfer Addresses and Counts

CH	Register	R/W	$\overline{CS}$	$\overline{IOR}$	$\overline{IOW}$	A3	A2	A1	A0
0	Base and current address	Write	0	1	0	0	0	0	0
	Current address	Read	0	0	1	0	0	0	0
	Base and current word count	Write	0	1	0	0	0	0	1
	Current word count	Read	0	0	1	0	0	0	1
1	Base and current address	Write	0	1	0	0	0	1	0
	Current address	Read	0	0	1	0	0	1	0
	Base and current word count	Write	0	1	0	0	0	1	1
	Current word count	Read	0	0	1	0	0	1	1
2	Base and current address	Write	0	1	0	0	1	0	0
	Current address	Read	0	0	1	0	1	0	0
	Base and current word count	Write	0	1	0	0	1	0	1
	Current word count	Read	0	0	1	0	1	0	1
3	Base and current address	Write	0	1	0	0	1	1	0
	Current address	Read	0	0	1	0	1	1	0
	Base and current word count	Write	0	1	0	0	1	1	1
	Current word count	Read	0	0	1	0	1	1	1

Table 15-2: 8237 Internal Addresses for Commands/Status

A3	A2	A1	A0	$\overline{IOR}$	$\overline{IOW}$	Operation
1	0	0	0	0	1	Read status register
1	0	0	0	1	0	Write command register
1	0	0	1	0	1	Illegal
1	0	0	1	1	0	Write request register
1	0	1	0	0	1	Illegal
1	0	1	0	1	0	Write single mask register bit
1	0	1	1	0	1	Illegal
1	0	1	1	1	0	Write mode register
1	1	0	0	0	1	Illegal
1	1	0	0	1	0	Clear byte pointer flip-flop
1	1	0	1	0	1	Read temporary register
1	1	0	1	1	0	Master clear
1	1	1	0	0	1	Illegal
1	1	1	0	1	0	Clear mask register
1	1	1	1	0	1	Illegal
1	1	1	1	1	0	Write all mask register bits

Table 15-3: 8237 Address Selection for Example 15-1

Binary address		Hex		Read/
A7 A6 A5 A4	A3 A2 A1 A0	Address	Function	Write
1 0 0 1	0 0 0 0	90	CHAN0 memory address register	R/W
1 0 0 1	0 0 0 1	91	CHAN0 count register	R/W
1 0 0 1	0 0 1 0	92	CHAN1 memory address register	R/W
1 0 0 1	0 0 1 1	93	CHAN1 count register	R/W
1 0 0 1	0 1 0 0	94	CHAN2 memory address register	R/W
1 0 0 1	0 1 0 1	95	CHAN2 count register	R/W
1 0 0 1	0 1 1 0	96	CHAN3 memory address register	R/W
1 0 0 1	0 1 1 1	97	CHAN3 count register	R/W

The two sets of information needed in order to program a channel of the 8237 DMA to transfer data are (1) the address of the first byte of data to be transferred, and (2) how many bytes of data are to be transferred.

For set 1, the channel's memory address register must be programmed. Since the memory address register of the 8237 is 16 bits and the data bus of the 8237 is 8 bits, one byte at a time, consecutively, is sent in to the same port address. For set 2, the channel's count register is programmed. The count can go as high as FFFFH. Since the count register is 16 bits and the data bus of the DMA is only 8 bits, it takes two consecutive writes to program that register. This is shown in Example 15-2.

Example 15-2

Assume that channel 2 of the DMA in Example 15-1 is to transfer a 2K (2048) byte block of data from memory locations starting at 53400H. Use the port addresses of Example 15-1 for the DMA to program the memory address register and count register of channel 2.

Solution:

The port addresses for the channel 2 memory address register and count register in Example 15-1 are 94H and 95H, respectively. The initialization will look as follows:

```
MOV    AX,3400H     ;load lower 4 digits of start address
OUT    94H,AL       ;send out the low byte of the address
MOV    AL,AH        ;
OUT    94H,AL       ;send out the high byte of the address
MOV    AX,2048      ;load block size into AX
OUT    95H,AL       ;send out the low byte of the count
MOV    AL,AH        ;
OUT    95H,AL       ;send out the high byte of the count
```

The contents of the memory address and count registers can be read in the same manner (low byte first, then high byte) to monitor these registers at any time. From looking at the above program one might ask, since the system address bus is 20 bits and the memory address is 53400H, why does this program use 16-bit addresses? This is a limitation of the 8237 DMA. In the 8237, not only is the register holding the address of the block 16 bits, but in addition there are only 16 address pins that carry the addresses. The x86 PC solves this problem by using an external *n*-bit register to hold the upper bits of the addresses.

8237's internal control registers

Although the 8237 has four channels and each channel must be programmed separately for the base address and count, there is only one set of control/command registers used by all channels. These registers are shown in Table 15-1. To understand how to access those registers, look at Example 15-3.

Of these eight registers, only the most essential ones will be explained in detail here. The reader can refer to Intel manuals for information concerning others. The functions of 8237 pins are described in Section 15.4 in the context of some real-life designs, such as the x86 PC.

Example 15-3

Use the circuit in Example 15-1 to find the address of the 8237 DMA control registers.

Solution:

Using Table 15-2 and substituting for A7–A3 gives the information in Table 15-4.

Table 15-4: Address Selection for Example 15-3

Binary address				
A7 A6 A5 A4	A3 A2 A1 A0	Hex	Register Name	Read/ Write
1 0 0 1	1 0 0 0	98	Status/command register	R/W
1 0 0 1	1 0 0 1	99	Request register	W
1 0 0 1	1 0 1 0	9A	Single mask register bit	W
1 0 0 1	1 0 1 1	9B	Mode register	W
1 0 0 1	1 1 0 0	9C	Clear byte pointer	W
1 0 0 1	1 1 0 1	9D	Master clear/temporary register	R/W
1 0 0 1	1 1 1 0	9E	Clear mask register	W
1 0 0 1	1 1 1 1	9F	Mask register bits	W

Command register

This is an 8-bit register used for controlling the operation of the 8237 (see Figure 15-3). It must be programmed (written into) by the CPU. It is cleared by the RESET signal from the CPU or the master clear instruction of the DMA. The function of each bit is described below.

The 8237 is capable of transferring data (1) from a peripheral device to memory (reading from disk), (2) from memory to a peripheral device (writing the file into disk), or (3) from memory to memory. One example of the use of the memory-to-memory option is what is called *shadow RAM*. In computers such as 386- and 486-based systems, the access time of ROM is too long. However, the system can copy the ROM into a portion of RAM and allow the CPU to access it from RAM, which has a much shorter access time than ROM.

D0 gives the option to use only channels 0 and 1 for transferring a block of data from memory to memory. Why the need for two channels? Channel 0 must be used for the source and channel 1 for the destination. Channel 0 reads the byte into a temporary register inside the 8237, and then channel 1 will write it to the destination. This is in contrast to I/O-to-memory or memory-to-I/O transfers, in which the data is read into the data bus and transferred to the destination, all without being saved anywhere temporarily.

D1 is used only when the memory-to-memory option is enabled and can be used to disable the memory incrementation/decrementation of channel 0 in order to write a fixed value into a block of memory.

D2 is used to enable or disable DMA.

D3 gives the option to choose between the normal memory cycle of 4 clock pulses and compressed timing of 2 clock pulses per memory cycle. There are 4 clock pulses per byte after the initial delay, assuming that the high byte address is already latched. If every byte of transfer requires both high byte and low byte addresses, an extra clock pulse for the address latch is required, which makes the bus cycle 5 clock pulses. The same is the case for the compressed option, making it 3 clock ticks per bus cycle.

D4 gives the option of using the four channels on fixed priority or rotating priority. If fixed priority is chosen, DREQ0 has the highest priority and DREQ3 has the lowest priority. If more than one DREQ is activated at the same time, it will always respond to the one with the highest priority. In rotating mode, DREQ0 again has the highest priority and DREQ3 the lowest, but the system rotates through DREQ0, DREQ1, DREQ2, and DREQ3 in that order, servicing one request from each if present. In other words, when DREQ0 is served it will not be given a chance until the rest of the DREQs are given a chance. This prevents monopolization by the DREQ with the highest priority.

D5 allows time for the write signal to be extended for slow devices.

D6 gives the option of programming the activation level of DREQ. It can be an active-high or active-low signal.

D7 gives the option of programming the activation level of DACK. It can be an active-high or active-low signal.

The command byte is issued to this register through port address X8H, where X is the combination provided to activate CS, as shown in Examples 15-4 and 15-5.

Example 15-4

Program the command register of the 8237 in Example 15-3 for the following options: no memory-to-memory transfer, normal timing, fixed priority, late write, and DREQ and DACK both active-high.

Solution:

From Figure 15-3, the command byte would be 1000 0000 = 80H and the program is

```
MOV   AL,80H ;load the command byte into AL
OUT   98H,AL       ;issue the command byte to port 98H
```

Example 15-5

Assume that the CPU is doing some very critical processing and that the 8237 DMA should be disabled. Use the ports in Example 15-3 to show the program.

Solution:

To disable the 8237, send 0000 0100 = 04H to the command register as follows:

```
MOV   AL,04H
OUT   98H,AL
```

Status register

This is an 8-bit register that can only be read by the CPU through the same port address as the command register. This register is often referred to as RO (read only) in PC documentation. As mentioned above, the port is X8 hex, where X is for CS. It contains various information about the operating state of the four channels. The lower four bits, D0 D3, are used to indicate if channels 0–3 have reached their TC (terminal count). TC is set high when the count register has been decremented to zero. This gives the option to monitor the count register by software. This monitoring also can be done by hardware through the EOP pin of the 8237, as we will see in the next section. The upper four bits, D4–D7, of the status register keep count of pending DMA requests. This information can be used by the CPU to see which channel has a pending DMA request. See Figure 15-4.

Mode register

This register can only be written to by the CPU through port address XBH, where X is the address combination for CS activation. Of the 8 bits of the mode register, the lower two, D0 and D1, are used for channel selection. The other 6 bits are used to select

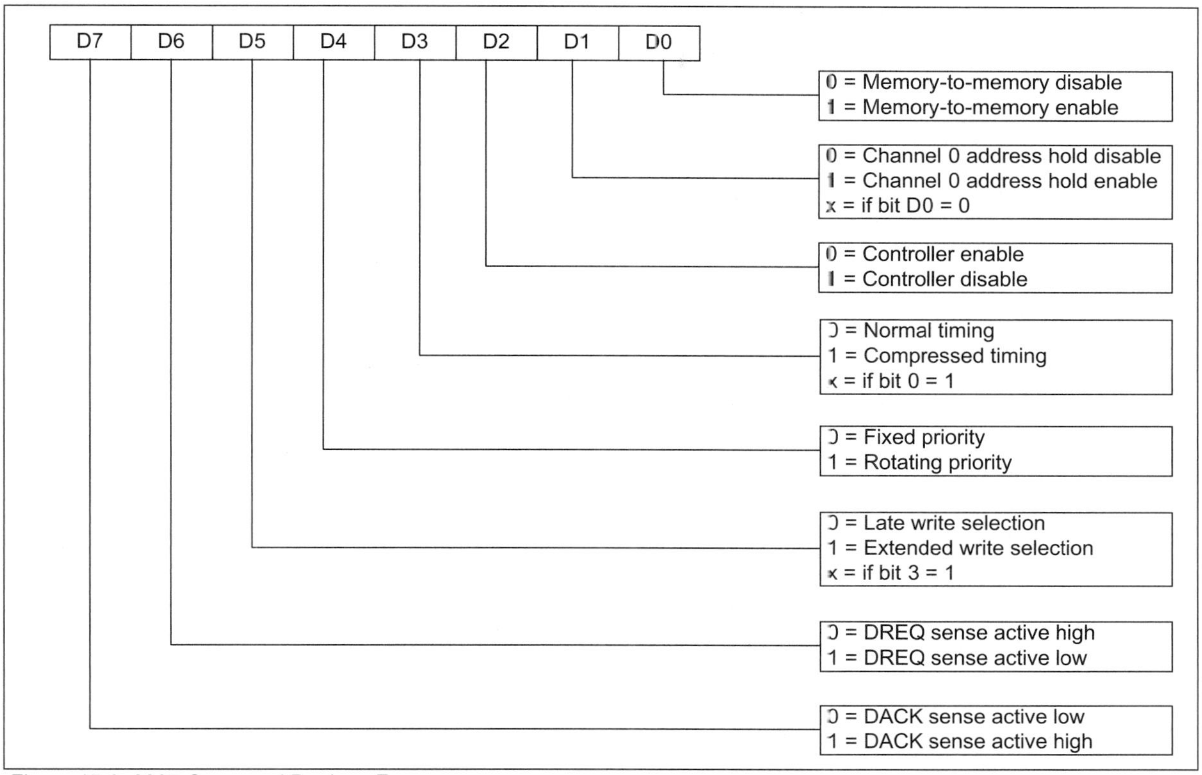

Figure 15-3. 8237 Command Register Format

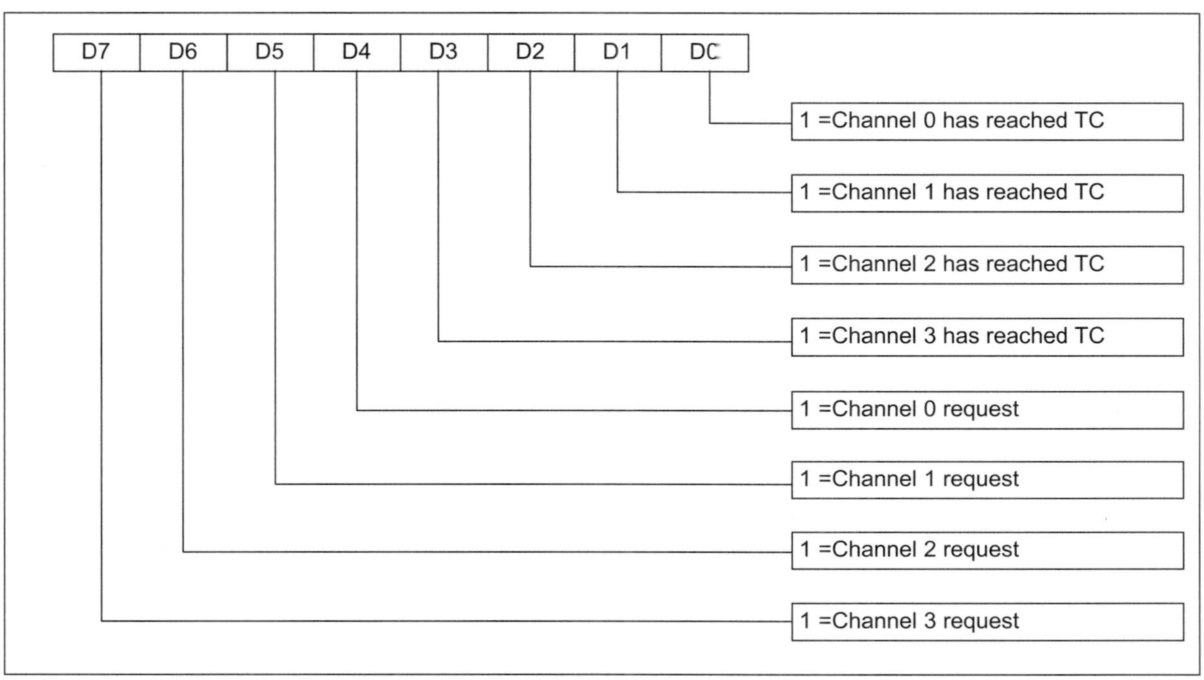

Figure 15-4. 8237 Status Register Format

various operation modes to be used for the channel selected by bits D0 and D1. D2 and D3 specify data transfer mode. In the write transfer option, the DMA transfers from an I/O device (such as a disk) to memory by activating IOR and MEMW. Reading from memory to an I/O is a read transfer and is achieved by activating MEMR and IOW. The verify transfer is called *pseudo* and is like a read or write except that it does not generate any control signals, such as IOR, MEMR, and so on. D4 is used for autoinitialization. If enabled, the memory address register and the count register are reloaded with their original values at the end of a DMA data transfer (when the count register becomes zero). In this way those registers are programmed only once and the original values are saved internally. D5 gives the option to increment or decrement the address. D6 and D7 determine the way the 8237 is used. The options are:

1. *Demand mode*, where the transfer of data continues until DREQ is deactivated or the terminal count has been reached. This ensures that the DMA can finish the job without interruption even though it means monopolization of the system buses by the DMA for the duration of the transfer of the entire block of data.
2. *Block mode*, which is the same as demand mode except that DREQ can be deactivated after the DMA cycle starts and the process of data transfer will go on until the TC (terminal count) state has been reached. In other words, there is no need to keep the DREQ high for the duration of the data transfer.
3. *Single mode*, where if DREQ is held active, the DMA transfers one byte of data, then allows the x86 to gain control of the system bus by deactivating its HRQ for one bus cycle. This process goes on alternating access to the system bus between the CPU and DMA until the TC has been reached, and then autoinitialization will happen if that choice has been made in the control word. This is the option used in all x86 PCs and compatibles since the DMA and CPU alternately share the system buses, allowing both to do their job without either monopolizing the buses.
4. *Cascade mode*, in which several DMAs can be cascaded to expand the number of

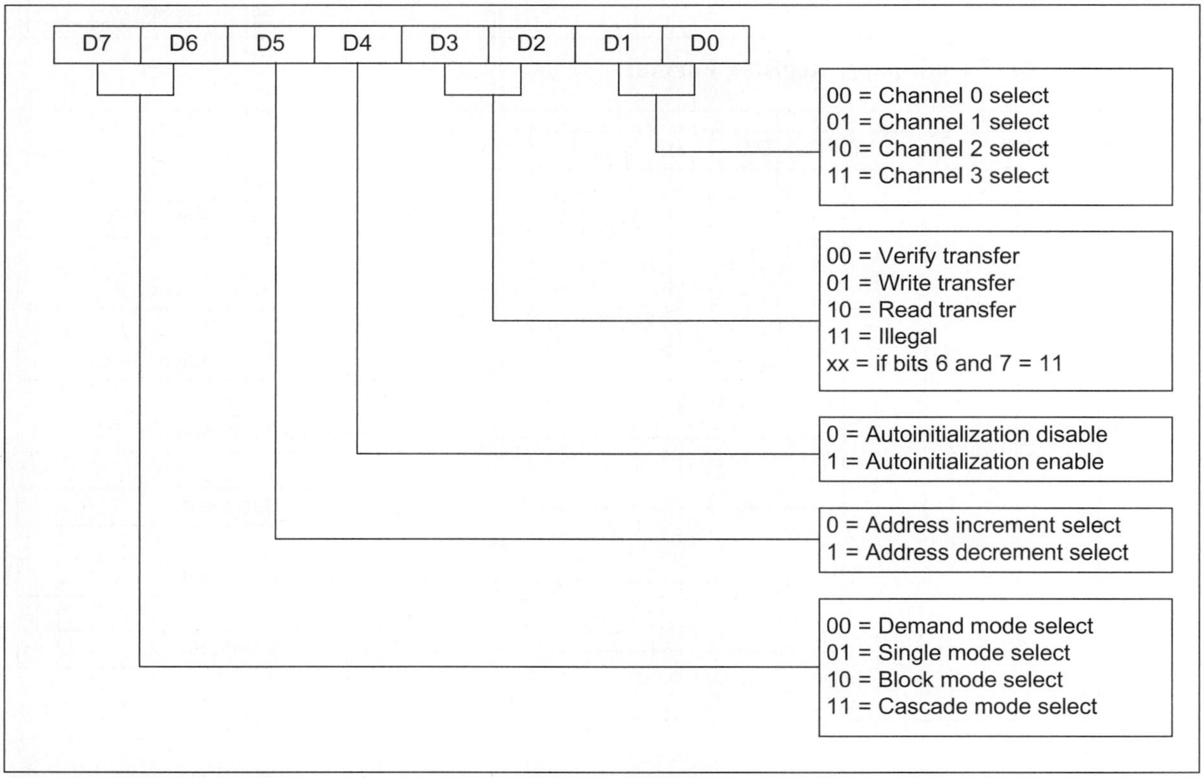

Figure 15-5. Mode Register Format

DREQs to more than 4. This option is used in x86 PC, as we will see in Section 15.5. The original PC used only one 8237. Examples 15-6 and 15-7 show the programming of the mode register.

Single mask register

This register can only be written to by the CPU through port address XA hex, where X is for CS. Of the 8 bits of this register, only three are used. D0 and D1 select the channel. D2 clears or sets the mask bit for that channel. It is through this register that the DREQ input of a specific channel can be masked (disabled) or unmasked (enabled). For example, if the value 00000101 is written to this register, it will mask (block) DREQ1 and the DMA will not respond to DREQ of channel 1 when DREQ1 is activated. While the command register can be used to disable the whole DMA chip, this register allows the programmer to disable or enable a specific channel. The only problem is that only one channel can be masked or unmasked at a time. To mask or unmask more than one channel, the all mask register is used. Figure 15-6 shows the single mask register format.

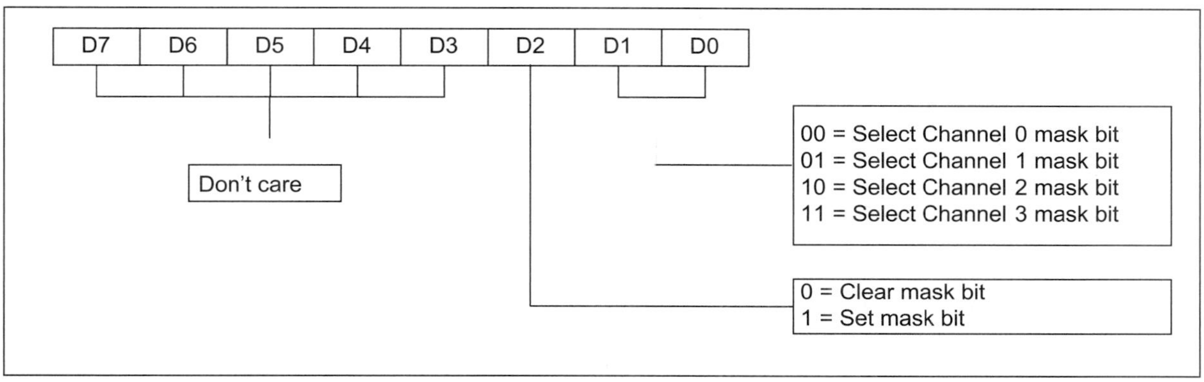

Figure 15-6. 8237 Single Mask Register Format

All mask register

In function, this register is similar to the single mask register except that all 4 channels can be masked or unmasked with one write operation. For example, if 00000010 is written to this register, it will mask the OUT of channel 1 and unmask (enable) the other channels. See Figure 15-7. Again this register can only be written to by the CPU through the port address XFH, where X is for CS activation.

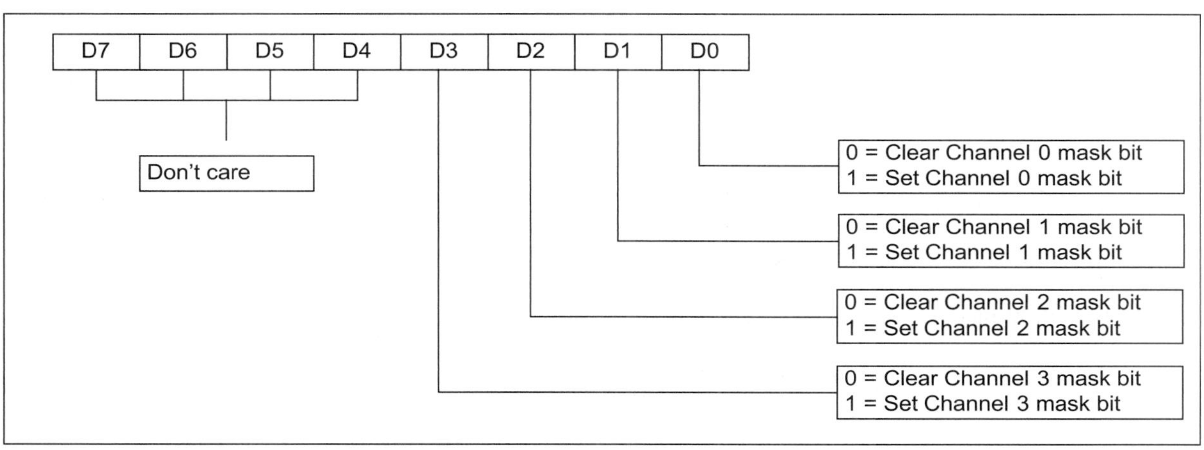

Figure 15-7. 8237 All Mask Register Format

Example 15-6

Program the 8237's mode register of Example 15-3 to select channel 2 to transfer from memory to I/O using autoinitialization, address increment, and single-byte transfer.

Solution:

From Figure 15-5, with these options, the mode register must have 01011010 = 5AH. The port address for the command register is 9BH, which results in

```
MOV   AL,5AH
OUT   9BH,AL
```

Example 15-7

Program the 8237 of Example 15-3 to enable channel 2.

Solution:

From Figure 15-6, the value for the single mask register to enable (unmask) channel 2 is 0000 0010 = 02H and is sent to port 9AH as follows:

```
MOV   AL,02
OUT   9AH,AL
```

Master clear/temporary register

This register must only be written to by the CPU through port address XDH, where X is for CS activation. The byte sent to this register does not matter since it simply clears the status, command, request, and mask registers and forces the DMA to the idle cycle. This is the same as activating the hardware RESET of the 8237. If an attempt is made to read from this register, the DMA will provide the last byte of data that was transferred during the memory-to-memory transfer. Note that when the DMA is doing an I/O-to-memory or memory-to-I/O transfer, it transfers the data directly between these two sections of the computer without bringing it into the DMA, but in memory-to-memory transfers it must bring each byte into the DMA before it sends it to the destination, since it has to switch the contents of the address bus for the source and destination. This is similar to string instruction MOVSB, except that it is performed by the DMA instead of the CPU.

Clear mask register

This register can be written to by the CPU only through port address XEH, where X is for CS. The bit patterns written to it do not matter. Its function is to clear the mask bits of all four channels, thereby enabling them to accept the DMA request through the DREQs.

Review Questions

1. Which address bits are used to select a register inside the 8237?
2. For an 8237, why are addresses X0H to XFH used to access its internal registers?
3. True or false. To use a channel to transfer data, both the memory address register and count register for that channel must be programmed.
4. State the functions of the memory address register and the count register.
5. Show instructions to program channel 0 memory address and count registers to transfer 4K starting from offset 1440H. Use port addresses from Example 15-1.
6. In the fixed priority scheme, which channel has the highest priority?
7. True or false. Programming some control registers is optional (depending on how the 8237 is used), but the command register must always be programmed.
8. True or false. The command register is accessed by CS = 0 and A3–A0 = 1000.
9. True or false. The mode register is accessed by CS = 0 and A3–A0 = 1011.
10. True or false. The level of activation, high or low, for DREQ and DACK of each channel can be programmed.

SECTION 15.3: 8237 DMA INTERFACING IN THE IBM PC

As shown in Figure 15-8, the 8237 DMA has eight addresses, A0–A7. Four of these, A0–A3, form a bidirectional address bus, sending addresses into the 8237 to select one of the 16 possible registers, assuming that chip select is activated. In the IBM PC, chip select is activated by Y0 of the 74LS138 as shown in Figure 15-9. The address selection of the registers inside the 8237 is summarized as shown in Table 15-5, assuming zero for each x. The conditions for A6, A7, A8, A9, and AEN were discussed in Chapter 12 and will not be repeated here. From Table 15-5 it can be seen that port addresses 0 to 7 are assigned to the four channels, and 08–0F are assigned to the control registers commonly used by all the channels.

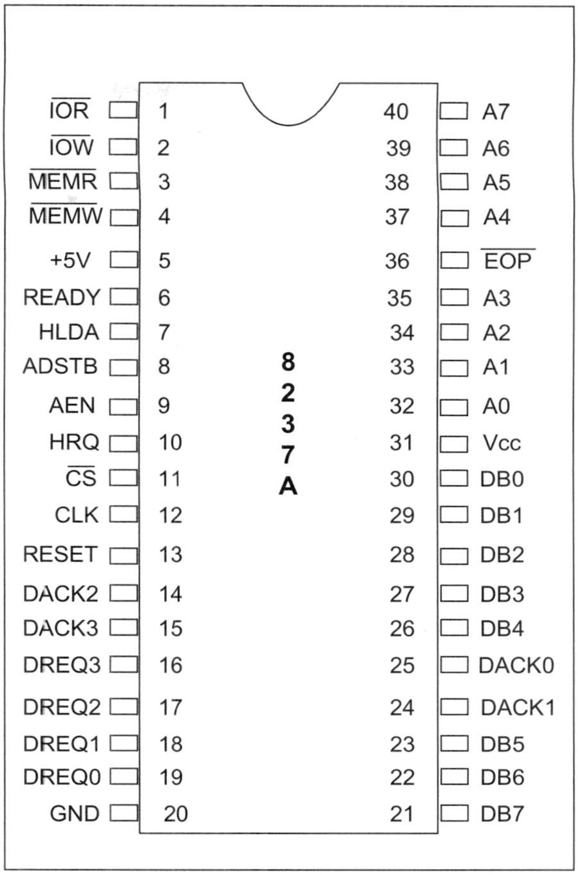

Figure 15-8. 8237A DMA Pin Layout

8237 and 8088 connections in the IBM PC

Since the DMA must be capable of transferring data between I/O and memory without any interference from the CPU, it must have all the required control, data, and address buses. Looking at Figure 15-10, one can see that the 8237 has its own data bus, D0–D7. This is a bidirectional bus connected to the system bus D0–D7. It also has all four control buses, IOR, IOW, MEMR, and MEMW. However, its address bus, A0–A7, is only 8 bits. If the 8237 can transfer up to 64K bytes of data between I/O and memory, it must have 16 address lines, A0–A15. Where are the other eight address pins, A8–A15? The answer is that the high byte of the 16-bit address changes only once, while D0–D7 are

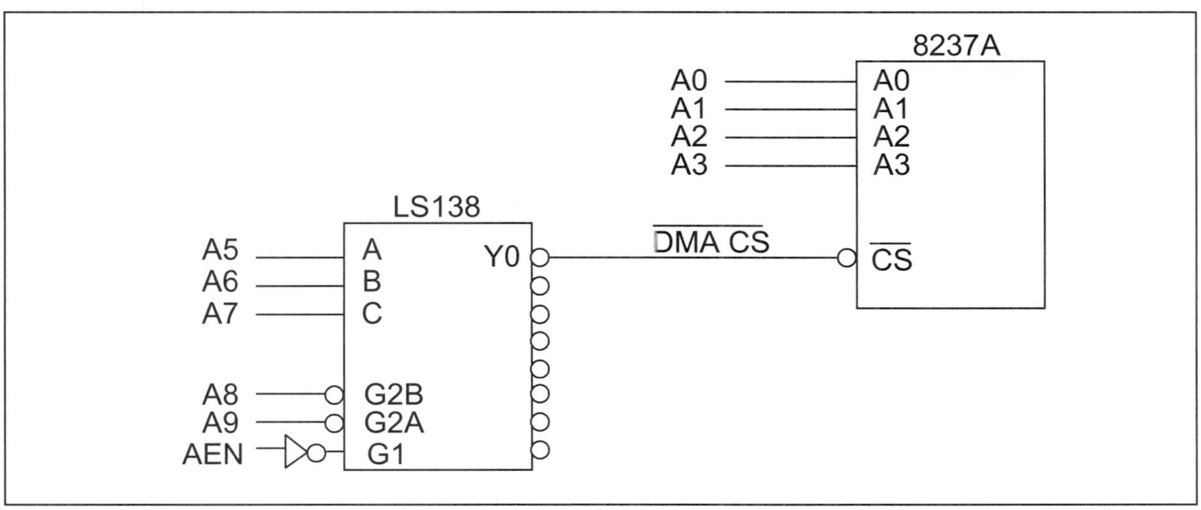

Figure 15-9. Chip Selection of the 8237A in the PC

used by the 8237 to send out the upper part of the address whenever A0–A7 rolls over from FF to 00. There must be a device to latch and hold the A8–A15 part of the address from the D0–D7 data bus. This is the function of the 74LS373. The function of the ADSTB (address strobe) is to activate the latch whenever the 8237 provides the upper 8-bit address through the data bus. Similar to ALE, the ADSTB goes high only when D0–D7 are used to provide the upper address, meaning that as long as ADSTB stays low, D0–D7 are a normal data bus. Figure 15-11 diagrams the 8237 circuit connection.

It should be noted that A0–A3 and all control buses are bidirectional, so that when programming the 8237 they can be used to communicate with the internal registers. As long as the CPU is idle and the DMA is in control of transferring data, A0–A7 and all the control buses are unidirectional.

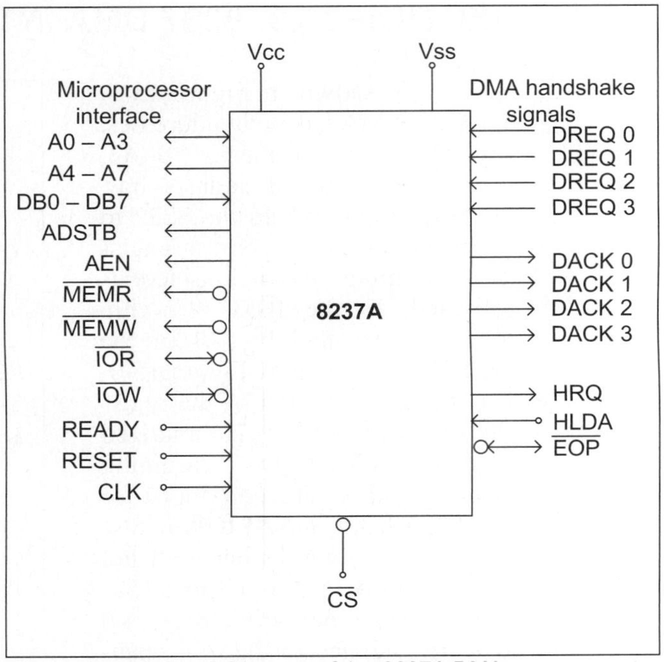

Figure 15-10. Block Diagram of the 8237A DMA

One last point about A0–A15 from the 8237 is that the system bus can be used by the CPU only when the 8237 is not functioning. This is ensured through the AEN signal. This signal was discussed in Chapter 9 and is summarized here.

AEN	
0	x86 is in control of the system bus
1	8237 DMA is in control of the system bus

The rest of the pins in Figure 15-10 are described below.

Table 15-5: PC8237 Internal Register Port Addresses

Binary address		Hex		Read/
A7 A6 A5 A4	A3 A2 A1 A0	Address	Function	Write
0 0 0 x	0 0 0 0	00	CHAN0 memory address register	R/W
0 0 0 x	0 0 0 1	01	CHAN0 count register	R/W
0 0 0 x	0 0 1 0	02	CHAN1 memory address register	R/W
0 0 0 x	0 0 1 1	03	CHAN1 count register	R/W
0 0 0 x	0 1 0 0	04	CHAN2 memory address register	R/W
0 0 0 x	0 1 0 1	05	CHAN2 count register	R/W
0 0 0 x	0 1 1 0	06	CHAN3 memory address register	R/W
0 0 0 x	0 1 1 1	07	CHAN3 count register	R/W
0 0 0 x	1 0 0 0	08	Status/command register	R/W
0 0 0 x	1 0 0 1	09	Request register	W
0 0 0 x	1 0 1 0	0A	Single mask register bit	W
0 0 0 x	1 0 1 1	0B	Mode register	W
0 0 0 x	1 1 0 0	0C	Clear byte pointer	W
0 0 0 x	1 1 0 1	0D	Master clear/temporary register	R/W
0 0 0 x	1 1 1 0	0E	Clear mask register	W
0 0 0 x	1 1 1 1	0F	Mask register bits	W

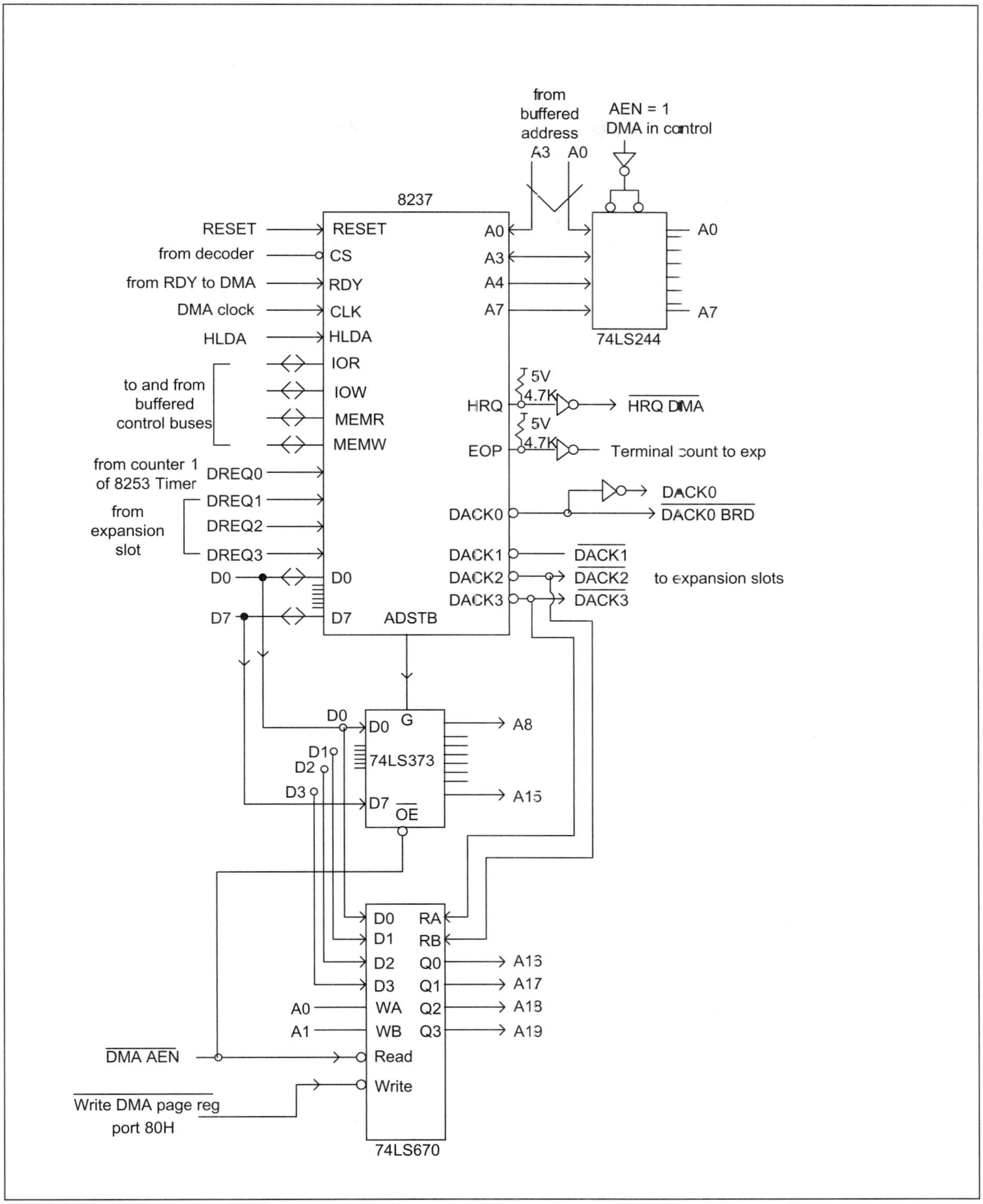

Figure 15-11. 8237 DMA Circuit Connection in the PC

RESET is the input coming from the RESET of 8284.

CS is from the 74LS138 decoder, as shown earlier.

READY input is from the RDYDMA of the wait state generation circuitry. The purpose of this is to extend the memory cycle of the DMA.

HOLD and HLDA are connected to the pins with the same name on the x86 CPU.

EOP (end of process) is inverted and becomes TC (terminal count). This signal is activated whenever the count register of any of the four channels is decremented to zero. This signal could be used with the DACK of a specific channel to prevent multiple DMA requests from that channel at the same time or could be used to inform the requesting device that the DMA has finished the job and it should deactivate its DREQ. In other words, EOP is a hardware pin indicating that the counter has reached zero. Using software one can monitor the count register of each channel by reading the status register, as was shown in Section 15.2.

DREQ0 and DACK0 are the signals for channel 0 and are used for refreshing DRAM as explained in Section 15.4. While the DREQ is active high, the DACK0 is programmed to be active low by BIOS, as shown in Section 15.4.

DREQ1–DREQ3 and DACK1–DACK3 are the signals for channel 1 to channel 3, and are available through the expansion slot. The assignment of these channels is discussed next.

Channel assignment of the 8237 in the IBM PC

In the original IBM PC, each of the four channels of the 8237 is assigned in the following fashion.

1. Channel 0 for refreshing DRAM. In the later x86 PC this practice was abandoned.
2. Channel 1 is unused, but in many implementations it is used for networks.
3. Channel 2 usually is used for the floppy disk controller.
4. Channel 3 normally is used for the hard disk controller.

Inspecting IBM BIOS shows that 8237 channels 1, 2, and 3 have been initialized by programming the mode register. The mode register, which must be sent to port address 0BH, is as follows for channel 1 (from Figure 15-5):

D1,D0 = 01 for channel 1
D3,D2 = 00 for verify transfer
D4 = 0 autoinitialization disable
D5 = 0 for address increment
D7,D6 = 01 for single byte mode

D7 D0
0100 0001 = 41H mode register for channel 1

For channels 2 and 3, the value for the mode register is the same except that D0 and D1 are changed to 10 and 11, respectively. Therefore, channels 2 and 3 have mode register values of 42 and 43. The program could look like the following code.

```
MOV   DX,000BH      ;load the mode register address
MOV   AL,41H        ;chan1 mode reg value
OUT   DX,AL
MOV   AL,42H        ;chan2 mode reg value
OUT   DX,AL
MOV   AL,43H        ;chan3 mode reg value
OUT         DX,AL
```

The way IBM BIOS does the initialization is slightly more compact:

```
E136 B20B                 MOV    DL,0BH         ;DX=000B
....                      ...    ...........
E142 B103                 MOV    CL,3
E144 B041                 MOV    AL,41H         ;SET MODE FOR CHANNEL 1
E146            C18A:
```

```
E146 EE          OUT    DX,AL
E147 FEC0        INC    AL             ;POINT TO NEXT CHANNEL
E149 E2FB        LOOP   C18A
```


These channels are programmed by the device that uses them when the device is installed. For example, the hard disk controller ROM programs channel 3 according to its specifications.

Review Questions

1. What port addresses are assigned to the 8237 in the original PC?
2. What port address is assigned to the command register of the 8237 in the original PC?
3. If the 8237 has only A0–A7, how is the 16-bit address A0–A15 provided?
4. The 8237 in the original IBM PC is programmed to have channel ___ as the highest priority and channel ___ as the lowest priority.

SECTION 15.4: DMA IN x86 PCs

The original IBM PC had only three DMA channels available through the expansion slot. All these channels were designed for 8-bit data transfer. To expand the capability of the PC, designers of the x86 PC added the second 8237 and made it a 16-bit data transfer DMA. This is shown in Figure 15-12.

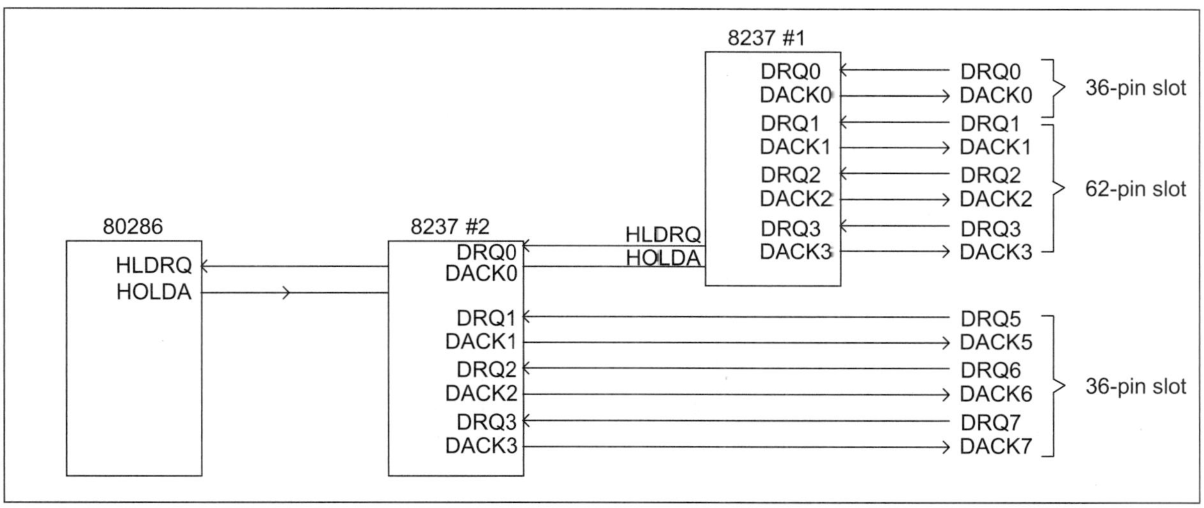

Figure 15-12. 80286 (and Higher) PCDMA

8237 DMA #1

To maintain compatibility with the original PC, DREQ1, DREQ2, and DREQ3 of DMA #1 are available through the expansion slot and are for 8-bit data transfer between 8-bit I/O and the 16 MB memory range of the x86. The ports assigned to DMA#1 are exactly the same as in the PC. The x86 PCs abandoned the idea of refreshing DRAM using DMA channel 0 and instead replaced it with DRAM refresher circuitry. This made channel 0 available through the ISA expansion slot. The signal associated with channel 0 is DREQ0 and DACK0 and is accessed though the 36 edge of the ISA. This is shown in Figure 15-13.

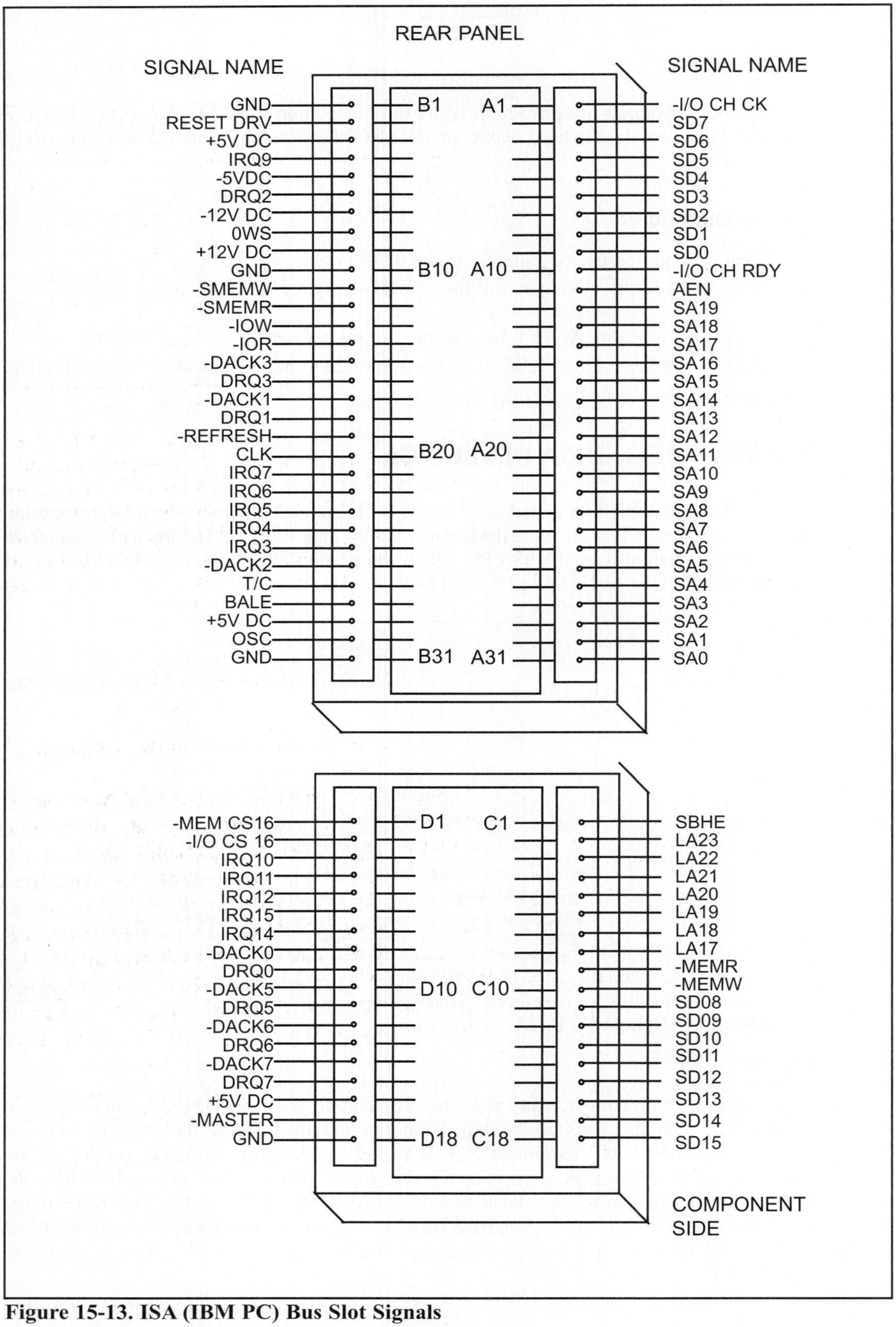

Figure 15-13. ISA (IBM PC) Bus Slot Signals

The following points must be noted regarding channels 0, 1, 2, and 3 of DMA #1 in x86 PC computers.

1. Channels 0, 1, 2, and 3 can be used only for data transfer between 8-bit I/O and system memory. The system memory address can be on an odd-byte or even-byte boundary.
2. Since the count register is a 16-bit register, each of channels 0, 1, 2, and 3 can transfer up to a 64K-byte block of data.
3. Each channel, 0, 1, 2, or 3, can transfer data in 64K-byte blocks throughout the 16M system memory address space.

8237 DMA #2

The second 8237 DMA is connected as master (level 1) and its channel 0 is used for cascading of DMA#1 as shown in Figure 15-12. The other three channels of this DMA are available through the expansion slot (36 edge) under DREQ5 and DACK5, DREQ6 and DACK6, and DREQ7 and DACK7 designations. These three channels must be used for 16-bit data transfer.

Points to be noted regarding 16-bit DMA channels

Channels 5, 6, and 7 of DMA #2 are used exclusively for 16-bit data transfer between the 16 MB memory address space and I/O peripherals. The following points must be noted regarding their use.

1. Channels 5, 6, and 7 must be used for 16-bit data transfers between 16-bit system memory and 16-bit I/O adapters. Notice that the I/O must support 16-bit data.
2. The number of 16-bit (2-byte) words to be transferred is programmed into the count register of channels 5, 6, and 7. Since the count register is a 16-bit register, each channel can transfer up to 65,536 words or 128K bytes between I/O and memory.
3. The memory address for a DMA memory transfer must be on an even-byte address boundary.
4. Channels 5, 6, and 7 transfer data in blocks that have a maximum size of 128K bytes throughout the 16M system memory.
5. Since channels 5, 6, and 7 cannot transfer data on an odd-byte boundary, A0 and BHE are both forced to 0.
6. DMA #2 can be accessed (programmed) by another master from the expansion slot using the MASTER input signal on the 36-pin part of the ISA bus.

DMA channel priority

The BIOS of the x86 PC programs both DMAs to have channel 0 as the highest priority. This means that of the seven DMA channels available through the expansion slot of the x86 ISA bus PC, channel 0 has the highest priority and channel 7 the lowest priority. This is due to the fact that the master DMA (8237 #2) has channel 0 as the highest priority and since the slave 8237 #1 is connected to it, channels 0 through 3 have higher priority than channels 5, 6, and 7. Therefore, we have the following:

	x86 ISA DMA Channel Priority
channel 0	Highest priority
channel 1	
channel 2	
channel 3	
channel 5	
channel 6	
channel 7	Lowest priority

Figure 15-14 shows an example of DMA and I/O write cycle timing.

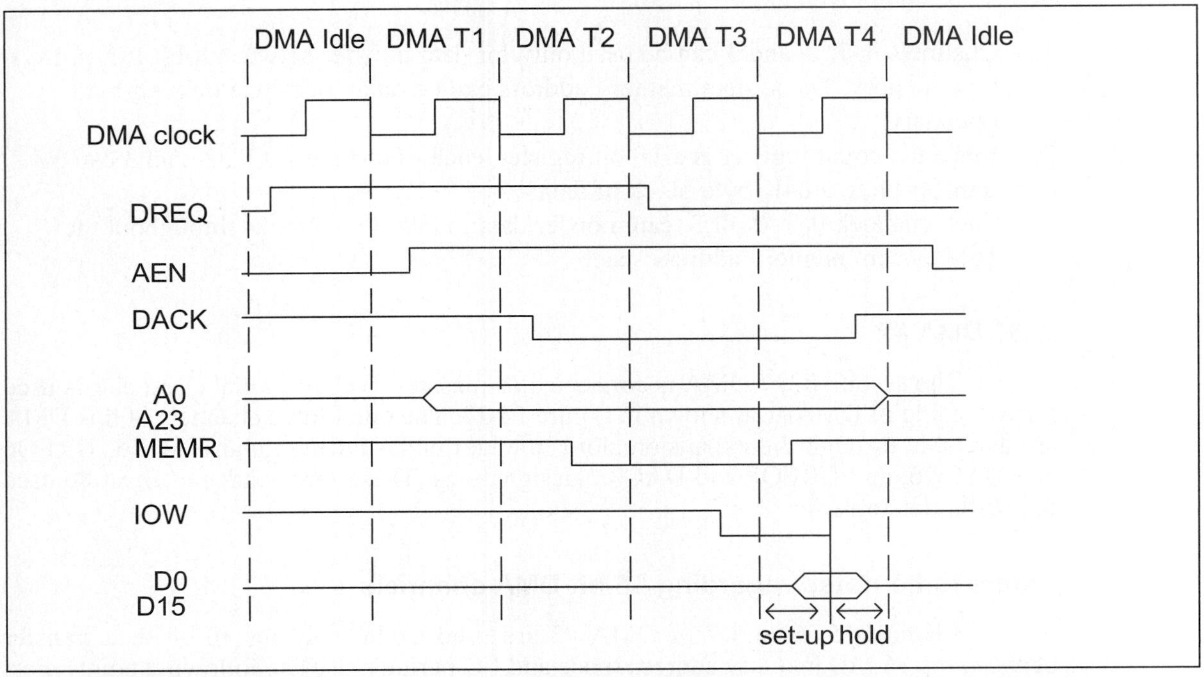

Figure 15-14. DMA Memory Read and I/O Write Bus Cycle for Many x86 and Later PCs

Review Questions

1. True or false. In the ISA bus, all channels of the DMA #1 are available on the expansion slot.
2. How many channels of the DMA are available on the ISA bus?
3. Indicate on what part of the ISA (62- or 36-pin) bus the channels of the DMA are accessible.
4. True or false. Channels 5, 6, and 7 can be used for 8- or 16-bit data transfers.
5. Why is bus control in the IBM PC, PS, and compatibles alternated between DMA and the CPU?

PROBLEMS

SECTION 15.1: CONCEPT OF DMA

1. Compare the rate of data transfer between the 8088 CPU and DMA. For DMA, assume that it takes 4 clocks to transfer a byte. How many times faster is DMA?
2. Calculate the time needed to transfer 512 bytes by the 8088 and by DMA in Problem 1. Assume 200 ns for each clock period.
3. Explain the difference between the CPU's response to signals INTR and HOLDR.
4. For the CPU, HOLDR is an _______ (input, output) signal and HOLDA is an _______ (input, output) signal.
5. In response to activation of HOLDR, the CPU finishes the current __________ before handing the buses to DMA.
   (a) instruction (b) bus cycle (c) subroutine
6. The DMA cannot take over the buses until signal ______ is activated by the CPU.
7. Why it is much less expensive to design a DMA chip than a CPU chip?
8. At what point does the CPU regain control over the buses?

SECTION 15.2: 8237 DMA CHIP PROGRAMMING

9. There are total of ______ port addresses assigned to an 8237.
10. Which of the following port addresses cannot be assigned to the 8237 DMA, and why? (This question is not PC compatible.)
    (a) 88H (b) 80H (c) 92H (d) F0H
11. For a DMA channel to transfer data it must have two sets of information. State these. How many port addresses are assigned to each set?
12. Explain why a total of eight port addresses are set aside for the 8237 channels.
13. If CS is activated by A7–A4 = 0101, give the port addresses assigned to the four channels of the 8237.
14. In Problem 13, what are the port addresses of the 8237 internal control registers?
15. Which register inside the 8237 is used to program the activation level (low or high) of the DREQ and DACK pins? In Problem 14, what port address is that?
16. In fixed priority, which channel has the highest priority? Which has the lowest? How is this different from rotating priority?
17. Assume that the 8237 is programmed for fixed priority. If DREQ2 and DREQ4 are activated at the same time, who gets serviced first?
18. The ____________ (x86 CPU, 8237 DMA) resolves channel priority.
19. State the function of the status register TC bits in the 8237. Can we write into it?
20. Program the mode register in Problem 14 for I/O-to-memory transfer, autoinitialization, address decrement, and block mode for channel 2
21. Program the single mask register in Problem 14 to enable channel 3.
22. Show the programming of the memory address and count registers of channel 3 to transfer 8K bytes of data from I/O to memory starting at address 1500H.

SECTION 15.3: 8237 DMA INTERFACING IN THE IBM PC

23. To access a memory block, explain how many address bits are provided by the 8237, and how they are provided.
24. The 8237 can access 64K-byte blocks of memory. Explain how the 8237 in the IBM PC can access the entire 1M address range.
25. True or false. For every DREQ there is a DACK in the 8237.
26. True or false. For every DREQ there is a HOLD in the 8237.
27. State if each of the following pins is input, output, or both for the 8237.
    (a) HOLD (b) HOLDA (c) DREQs (d) DACKs (e) A3–A0
    (f) ADDSTB (g) IOR (h) IOW (i) MEMR (j) MEMW

SECTION 15.4: DMA IN x86 PCs

28. True or false. In the ISA bus, the use of channel 0 for DRAM refreshing was abandoned.
29. In the ISA bus, how many channels are available through the expansion slot? Indicate on what part (62-pin or 36-pin) they are available.
30. Why do you think there are two separate performance benchmarks for memory-intensive and disk-intensive applications?
31. Although 386 and higher PCs have 32-bit address buses, the DMA uses only 24-bit addresses in ISA type PCs. Explain why and state the implication.

ANSWERS TO REVIEW QUESTIONS

SECTION 15.1: CONCEPT OF DMA

1. True
2. True
3. True
4. True
5. HOLD
6. HLDA
7. Input
8. Output

SECTION 15.2: 8237 DMA CHIP PROGRAMMING

1. A0, A1, A2, A3, and CS
2. Because A0–A3 gives rise to only 16 possibilities, 0–F hex
3. True
4. The memory address of the first byte of the block of data to be transferred is loaded in the memory address register and the number of bytes to be transferred is loaded into the count byte register.
5.
```
MOV AX,1440H     ;LOAD LOWER 4 DIGITS OF START ADDRESS
OUT 90H,AL       ;SEND OUT THE LOW BYTE OF THE ADDRESS
MOV AL,AH
OUT 90H,AL       ;SEND OUT THE HIGH BYTE OF THE ADDRESS
MOV AX,4048      ;LOAD BLOCK SIZE INTO AX  \
OUT 91H,AL       ;SEND OUT THE LOW BYTE OF THE COUNT
MOV AL,AH
OUT 91H,AL       ;SEND OUT THE HIGH BYTE OF THE COUNT
```
6. Channel 0
7. True
8. True
9. True
10. True

SECTION 15.3: 8237 DMA INTERFACING IN THE IBM PC

1. 00–0F hex
2. 0B hex
3. It is provided through the D0–D7 data pins of the 8237 to the 74LS373 latch only when A0–A7 rolls over from FF to 00.
4. 0, 3

SECTION 15.4: DMA IN x86 PCs

1. True
2. Only 7
3. Channels 1–3 on the 62-pin section and channels 0, 5, 6, and 7 on the 36-pin section
4. False; they can be used only for 16-bit data transfers.
5. This is because in the mode register initialization, the DMA is programmed for the single mode. This means that for every DMA cycle there is a CPU cycle in between, making the DMA bus bandwidth much lower than if the DMA had control over the buses for the entire duration of the data transfer. The real reason is that the buses must be released by the DMA in order to allow refreshing of DRAM before it loses the data.

CHAPTER 16

VIDEO AND VIDEO ADAPTERS

OBJECTIVES

Upon completion of this chapter, you will be able to:

- >> Determine the quality of a monitor by technical features such as resolution, dot rate, horizontal and vertical frequency, and dot pitch
- >> Describe how images are produced on the screen by the method called raster scanning
- >> Explain the function of the video adapter board and its two components: video display RAM and the video controller
- >> Describe the differences in text and graphics modes
- >> State the purpose of the attribute byte and how it affects storage space in video display RAM
- >> Write Assembly language programs to manipulate text data on the screen using INT 10H
- >> Describe the relation between the number of colors available for a monitor and the amount of video memory needed
- >> Write an Assembly language program to program pixels on the screen

Although the quality of video monitors has improved dramatically since the introduction of the first IBM PC in 1981, the principles behind them have remained the same. This chapter will look at the video system of the x86 PCs.

SECTION 16.1: PRINCIPLES OF MONITORS AND VIDEO MODES

Video monitors use a method called *raster scanning* to display images on the monitor screen. This method uses a beam of electrons to illumine phosphorus dots, called *pixels*, on the screen. This electron gun rasters from the top left corner of the screen to the bottom right, one line at a time. As the gun turns on and off, it moves from left to right toward the end of the line, at which time it is turned off to move back to the beginning of the next line. This moving back while the gun is off is called *horizontal retrace*. When it reaches the bottom right of the screen, the gun is turned off and moves to the top left of the screen. This turning off and moving back to the top is called *vertical retrace*. Figure 16-1 shows two methods of scanning. One is noninterlaced (normal) scanning and the other is interlaced scanning. The concept for both is the same, but in interlaced scanning (which is the same method used in television sets), each frame is scanned twice. First the odd lines are scanned and then the gun comes back to scan the even lines. This method can create flicker, but allows better vertical resolution at a cheaper cost. Noninterlaced monitors provide much better flicker-free images than do interlaced monitors and for this reason are widely used as monitors of most PCs.

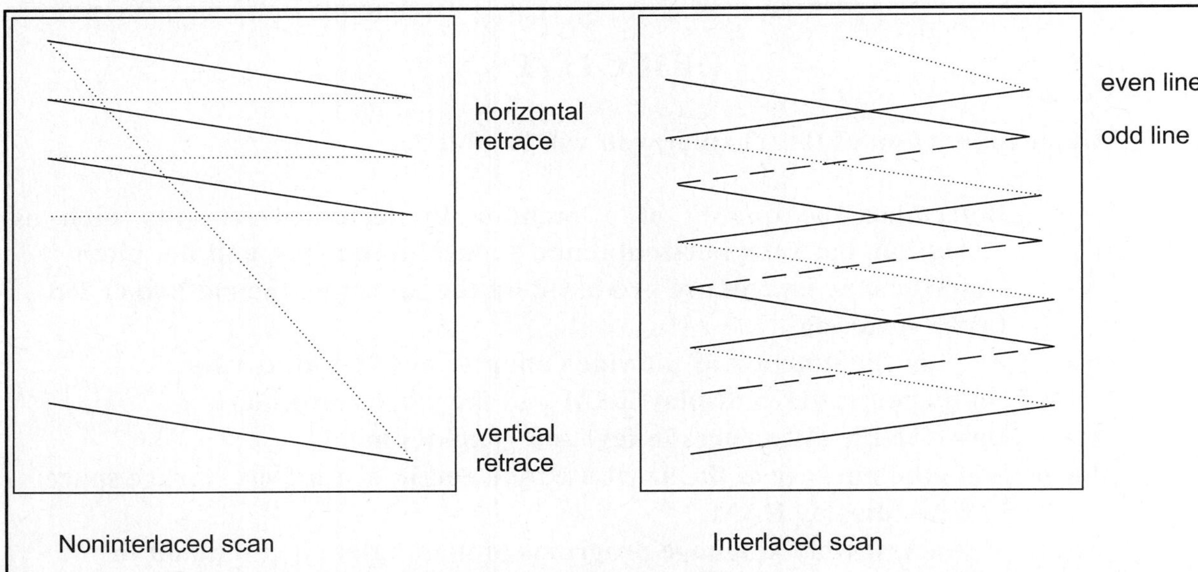

Figure 16-1. CRT Scanning Methods

How to judge a monitor

The resolution of the screen depends upon the following factors:

1. The number of pixels (dots) per scanned line
2. The speed at which the gun can turn on and off the phosphors on the surface of the tube
3. The speed at which it can scan and retrace a horizontal line
4. The number of scan lines per screen (frame)
5. The speed at which it finishes one frame and performs the vertical retrace.

While in a television set, horizontal scanning is done at the rate of 15,750 times per second (15.75 kHz) and vertical scanning at 60 times per second (60 Hz), on the IBM

PC monochrome monitor using the monochrome display adapter (MDA) the frequencies are 18.432 kHz and 50 Hz, respectively. Knowing these two frequencies enables one to calculate the maximum number of scan lines per screen by dividing the horizontal frequency by the vertical one as follows:

number of scanned lines per screen = horizontal freq. (HF) / vertical freq. (VF)
(not all visible)

Example 16-1

In a IBM PC monochrome monitor with HF = 18.432 kHz and VF = 50 Hz, calculate the number of scanned lines per screen.

Solution:

The number of scanned lines = HF/VF; therefore, 18,432 divided by 50 = 368 lines per screen.

Not all 368 horizontal lines in Example 16-1 are visible on the screen since some lines are for overscan and some are used for vertical retrace time. Overscan refers to the lines above or below the visible portion of the screen; these lines ensure clear edges at the top and bottom of the screen. In the IBM PC MDA, only 350 lines are visible on screen and of the remaining 18 (368–350), some (about 3 or 4) are used for overscanning. The time that would have been taken for scanning the rest (approximately 14) is used for the vertical retrace. Now that the number of scan lines is known, the next question is, how many pixels (dots) are there per line? This is calculated by dividing the video frequency (sometimes called *dot frequency*) by the horizontal frequency:

number of pixels per scan line = dot frequency / horizontal frequency
(not all visible)

In the IBM monochrome display adapter (MDA), dot frequency is 16.257 MHz, which, when divided by HF =18.43 kHz, gives 882 pixels per line. Again, all 882 pixels are not visible. With the IBM PC monochrome adapter, only 720 pixels are visible for each scan line. The time set aside for the remaining 162 is used for the time taken by horizontal retrace and overscanning on the left and right sides of the screen. Again, this overscanning allows sharp edges on the right and left sides of the screen. From the above discussion it can be seen that the three most critical factors in a monitor are:

1. The video frequency (also referred to as dot rate, pixel rate, or video bandwidth)
2. The horizontal frequency
3. The vertical frequency

From these three parameters, the number of pixels per line and the number of lines per screen can be calculated, keeping in mind that not all pixels and lines are visible on the screen, due to overscanning and retrace times. The number of visible pixels is given by the manufacturer of the adapters or monitors. Looking at Table 16-1 for the IBM monochrome adapter, the number of pixels is 720 × 350, which means that there are 720 pixels per line and 350 lines per screen, giving a total of 252,000 pixels. The total number of pixels (dots) per screen is a major factor in assessing a monitor's resolution, which is one of its most critical characteristics. The total number of pixels per screen is determined by the size of the pixel and how far apart pixels are spaced. For this reason one must look at what is called the *dot pitch* in monitor specifications.

Table 16-1. Adapter Characteristics

Adapter	Year Intro.	Dot Rate MHz	Horizontal Rate kHz	Vertical Rate Hz	Monitor Type	
CGA	1981	14.318	15.75	60	COMP. dig. RGB	
MDA	1981	16.257	18.43	50	Digital mono	
EGA	1984	14.318	15.75	60	Digital RGB	
		16.257	18.43	50	Digital mono	
		16.257	21.85	60	Digital RGB	
VGA	1987	25.175	31.5	70	Analog	
		28.175	31.5	70	Analog	
		25.175	31.5	70	Analog	
		25.175	31.5	60	Analog	

Dot pitch

Dot pitch is the distance between adjacent pixels (dots) and is given in millimeters. For example, a dot pitch of 0.31 means that the distance between pixels is 0.31 mm. Consequently, the smaller the size of the pixel itself and the smaller the space between them, the higher the total number of pixels and the better the resolution. Dot pitch varies from 0.6 inch in some low-resolution monitors to 0.2 inch in higher-resolution monitors. In some video monitor specifications, it is given in terms of the number of dots per square inch, which is the same way it is given for laser printers, for example, 300 DPI (dots per inch).

Dot pitch and monitor size

Monitors, like televisions, are advertised according to their diagonal size. For example, a 14-inch monitor means that its diagonal measurement is 14 inches. There is a relation between the number of horizontal and vertical pixels, the dot pitch, and the diagonal size of the image on the screen. The diagonal size of the image must always be less than the monitor's diagonal size. The following simple equation can be used to relate approximately these three factors to the diagonal measurement. It is derived from the Pythagorean theorem:

$$(\text{image diagonal size})^2 = (\text{number of horizontal pixels} \times \text{dot pitch})^2 + (\text{number of vertical pixels} \times \text{dot pitch})^2$$

Since the dot pitch is in millimeters, the size given by the equation above would be in mm, so it must be multiplied by 0.039 to get the size of the monitor in inches. See Example 16-2.

As can be seen from the above discussion, one can use a lower vertical frequency to get a higher number of vertical pixels. However, this can result in flickering, as hap-

Example 16-2

A manufacturer has advertised a 14-inch monitor of 1024 × 768 resolution with a dot pitch of 0.28. Calculate the diagonal size of the image on the screen. It must be less than 14 inches.

Solution:

The calculation is as follows:

$(\text{diagonal size})^2 = (1024 \times 0.28 \text{ mm})^2 + (768 \times 0.28 \text{ mm})^2$

diagonal size (inches) = 358 mm × 0.039 inch per mm = 13.99 inches pens in 1024 × 768 interlaced monitors. The interlaced method is a cheap way of increasing the vertical pixels by halving the vertical frequency and making the frame be scanned in two successive sweeps of odd and even fields.

Phosphors

Another factor that determines the quality of the monitor is the type of phosphors used, since the brightness of the pixels depends on two factors:

1. The intensity of the electron beam, which decides how bright or dark each pixel should be. This can be controlled by software, as will be seen in Section 16.4.
2. The phosphor material used. After the pixel has been illumined, some phosphor materials retain their brightness for longer periods of time than others. This characteristic, called *persistence*, is fixed in the monitor and cannot be changed. In the early IBM PC monochrome monitors, a high-persistence type of phosphor was used to create a better look on the monitor in case there was a need for frequency compensation. In situations where the frequency cannot be adjusted, such as the IBM monochrome monitors, a phosphor with longer persistence is used to compensate.

Color monitors

In color monitors the principles are the same, except that every phosphor dot is made of three colors: red, green, and blue, hence the name *RGB* (red/green/blue) monitors. Color monitors require three different wires to carry three electronic beams, one for each color, unless the monitor is of the composite brand. In composite monitors there is one single wire that carries all three colors. The process of combining, then separating, the three colors diminishes the quality of the image compared to true RGB monitors. In older RGB monitors, for every color wire there was a separate gun, but in newer models a single gun carries three wires for the three beams. Another difference between color and monochrome monitors is the presence of the shadow mask on color monitors. The shadow mask is a metal plate with many holes that is placed just before the phosphor-coated screen in order to coordinate the shooting of the electron beam of each gun through a single hole. This ensures that the red gun illumines the red dot only, the blue gun illumines the blue dot only, and the green gun illumines the green dot only.

Since each pixel has three dots of color, red, green, and blue, how is the dot pitch measured? The answer is that the dot pitch on color monitors is the distance between two dots of the same color, or as some manufacturers advertise, the distance between two consecutive holes of the shadow mask. While in monochrome monitors, it is changing the intensity of the electron gun that generates the shades of gray-black-white, in color monitors it is the combination of the three primary colors that generates all other colors. In other words, by changing the intensity of the red, green, and blue triad, one can create all the colors (see the next section for some examples).

Analog and digital monitors

Another monitor characteristic to be explained is digital versus analog. Digital monitors, such as the MDA- and CGA-based monitors, use a number of bits to specify variations of color and intensity. To increase these variations one must employ large numbers of bits. Analog monitors, because they can accommodate many more variations, have much better quality pictures. To understand the difference between digital and analog systems, imagine that we have defined a temperature of 20 degrees as cold and 100 degrees as hot. These would be represented in digital as 0 and 1 for cold and hot, but an analog system could accommodate temperature variations of 20 to 100 in increments of one, allowing many more variations. Another example is the state of a light bulb. In digital it is represented as 0 and 1 for on and off. In analog, one can accommodate many more variations, similar to the concept of using a dimmer switch.

Video display RAM and video controller

Communication between the system board (motherboard) and the video display monitor is through the video adapter board. Among the components every video board must have is a video controller and video display RAM. The information displayed on the monitor (either text or graphics) is stored in memory called *video display RAM* (VDR), also called video buffer. In order for the information to be displayed, it must be written first into video RAM by the CPU; then it is the job of the video adapter's controller (processor) to read the information from video RAM and convert it to the appropriate signals to be displayed on the screen. In other words, there is a separate controller, often called a *CRT controller* or *video processor*, apart from the main x86 CPU. The video controller's sole job is to take care of the video section of the computer. Since the CRT controller is built specifically for that purpose, it can perform the tasks associated with video much more efficiently than can a CPU such as the x86. That also means that video RAM must be accessible to both the main processor and the video processor. In the IBM PC and compatibles, 128K bytes of the 1 megabyte of addressable memory, from address A0000H to BFFFFH, is set aside for video display RAM. Of this 128K bytes of memory, only some is used; the amount depends on the resolution of the video adapter and the selected mode: text mode or graphics. For example, when displaying text the IBM monochrome adapter uses only 4K bytes of memory, starting at memory address B0000H. Of this 4K bytes, 2K bytes are for the full screen of characters (80 characters per line and 25 lines per screen = 2000 bytes) and another 2000 bytes are for the attributes. The attribute byte provides various information, such as color, intensity, and blinking, to the video circuitry. As will be seen later, every time a byte of character data is accessed, its attribute is automatically fetched as well. The memory requirement of each video board and the number of colors that it can handle will be given in the next section.

If the same video display RAM is accessed by both the microprocessor and the video controller, how can two masters access the same RAM at the same time? There are several solutions to this dilemma.

1. The CPU can access the video RAM only during the time when the video controller is doing the retrace.
2. To use a more expensive, specially designed kind of RAM called *VRAM* (video RAM). This kind of RAM allows the transfer of data by the video controller at a much higher rate than is allowed by normal DRAM.
3. Another approach used in some high-performance graphics system is to use dual-port RAMs. This kind of RAM has two sets of data pins, allowing both the CPU and video controller to access the video RAM with much less conflict, since it eliminates the time wasted by a multiplexer.

It must be noted that if the CPU tries to access the VDR while the video controller is accessing it, the CPU is blocked since the screen must be refreshed by the video controller before it is lost. In other words, the video controller has a higher priority than the main CPU in accessing the VDR. If by software manipulation one blocks the video controller's access to the VDR, it will result in snow on the screen.

Video systems have improved dramatically in recent years, due to the fact that the speed of the CPU has reached 5 GHz and can therefore transfer data from (or to) the VDR at a much faster rate during the retrace time.

Character box

Video boards can be programmed in two modes: text and graphics. While in graphics mode the individual pixels are accessed and manipulated, in text mode characters, which are a group of pixels, are accessed. In text mode, horizontal and vertical pixels are grouped into what are called *character boxes*. Each character box can display a single character. The size of the character box matrix varies from adapter to adapter. For example, IBM's MDA (monochrome display adapter) has a 9 pixel by 14 pixel character

Example 16-3

If the MDA character box is 9 × 14 (9 pixels wide and 14 pixels high) and the resolution of MDA is 720 × 350, verify the fact that MDA in text mode can display 80 × 25 characters per screen.

Solution:

720 horizontal scan lines divided by 9, the width of character box, gives 80 columns of characters. Dividing 350 vertical pixels by 14, the height of the character box, results in 25 rows of characters.

Example 16-4

In a given adapter, the character box is 8 × 14 and the adapter in text mode displays 80 × 25 characters. Calculate the pixel resolution.

Solution:

The total number of horizontal pixels is 640 (8 × 80) and the vertical number is 350 (14 × 25). Therefore, it has 640 × 350 resolution.

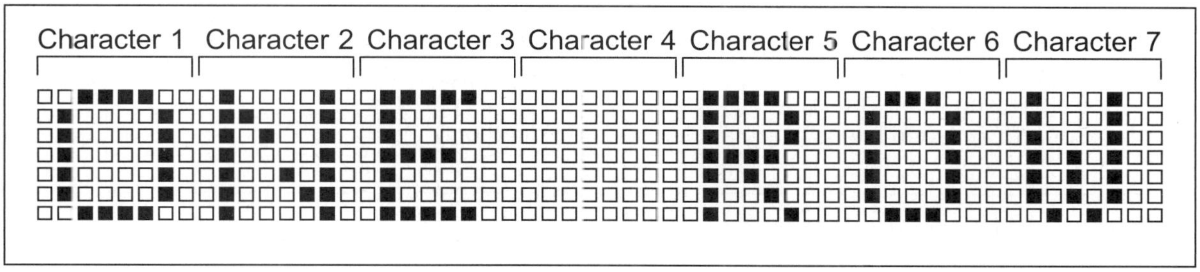

Figure 16-2. Character Boxes

box. Since every character is 9 pixels wide and 14 pixels high, one can calculate the number of character columns per screen by dividing the number of pixels on a horizontal line by 9, and can calculate the number of rows per screen by dividing the number of pixels on a vertical line by 14. Conversely, one can calculate the horizontal and vertical pixels by using the size of the character box, the number of rows, and the number of columns per screen:

pixels per scan line = number of character columns × pixel width of char. box
raster lines = number of rows per screen × pixel height of char. box

See Examples 16-3 and 16-4. To get better-looking characters, the character box size must be increased, which translates to more pixels horizontally and vertically. From the above discussion, one can conclude that for a fixed-size monitor to display a fixed number of rows and columns of characters, the number of horizontal and vertical pixels is the most important factor. Since the number of horizontal and vertical pixels is directly proportional to the horizontal and vertical dot frequencies, in judging CRT monitors one must look for higher HF (horizontal frequency), VF (vertical frequency), and DF (dot frequency). Figure 16-2 illustrates the character box.

x86 PC video modes

When the original IBM PC was introduced in 1981, it had two video monitor options: MDA (monochrome display adapter) and CGA (color graphics adapter). While CGA allowed both graphics and text mode options, MDA allowed only text mode. Consequently, CGA had color for both graphics and text, but the text was not very crisp. On the other hand, MDA had excellent text, but did not support graphics. Not until 1985 was EGA (enhanced graphics adapter) introduced in order to provide both graphics and text on the same monitor. In 1987, IBM introduced the new video standards called VGA (video graphics adapter) and MCGA (multicolor graphics array). In MDA and CGA, IBM used the Motorola 6845 CRT controller to design the adapter board, and in EGA used a set of proprietary LSI chips, but in all three cases the adapter board had to be plugged into one of the expansion slots. Later on, IBM put the video circuitry on the motherboard. Currently, some manufacturers make high-performance VGA-compatible adapter plug-in boards for graphic-intensive video games.

VGA (video graphics array)

VGA is a single-chip video controller designed by IBM that performs many tasks previously done by several chips in EGA. However, many use the term VGA to refer to the entire adapter. It has excellent resolution of up to 720 × 400 for text modes and 640 × 480 for graphics modes. In today's PC, VGA is already on the motherboard, but one can also purchase an adapter board for high-performance video games to be plugged into one of the expansion slots of the x86 PC. Table 16-1 shows the history of early video modes.

D7	D6	D5	D4	D3	D2	D1	D0
B	R	G	B	I	R	G	B
	Background			Foreground			

B = blinking, I = intensity
Both blinking and intensity are applied to foreground only .

Figure 16-3. CGA Text Mode Attribute Byte

Table 16-2 shows the various video modes. Notice that VGA emulates all the modes of CGA, MDA, and EGA, plus some new modes that are not available with the earlier adapters. Figure 16-3 shows the various color modes available in text mode. Figure 16-4 shows the text mode attributes in black and white. See Examples 16-5 and 16-6.

Video memory and attributes in VGA

Up to 1 megabyte of DRAM can be installed on VGA boards. This extra memory is used to store pixels and their attributes. Since VGA can display up to 256 colors out of 262,144 possible colors at once in graphics mode, it requires 1 megabyte of DRAM to store them. How the 1M of memory is mapped into the 128K-byte address space A00000–BFFFFH is discussed in Section 16.3, which covers text mode programming.

When VGA is programmed to emulate CGA text, the address for the video is B8000H, but for MDA emulation the address is B0000H. This is in order to be compatible with previous adapters.When VGA is in text mode it uses mode 3, as we will see in the next section.

Table 16-2: The 16 Possible Colors

I	R	G	B	Color
0	0	0	0	Black
0	0	0	1	Blue
0	0	1	0	Green
0	0	1	1	Cyan
0	1	0	0	Red
0	1	0	1	Magenta
0	1	1	0	Brown
0	1	1	1	White
1	0	0	0	Gray
1	0	0	1	Light blue
1	0	1	0	Light green
1	0	1	1	Light cyan
1	1	0	0	Light red
1	1	0	1	Light magenta
1	1	1	0	Yellow
1	1	1	1	High-intensity white

Table 16-3: Video Modes and Their Definitions

AL	Pixels	Chars	Box	T/G	Colors	Mode	Buf	Start
00H	320 × 200	40 × 25	8 × 8	Text	16*	CGA	8	B8000h
	320 × 350	40 × 25	8 × 14	Text	16*	EGA	8	B8000h
	360 × 400	40 × 25	9 × 16	Text	16*	VGA	8	B8000h
	320 × 400	40 × 25	8 × 16	Text	16*	MCG	8	B8000h
01H	320 × 200	40 × 25	8 × 8	Text	16	CGA	8	B8000h
	320 × 350	40 × 25	8 × 14	Text	16	EGA	8	B8000h
	360 × 400	40 × 25	9 × 16	Text	16	VGA	8	B8000h
	320 × 400	40 × 25	8 × 16	Text	16	MCGA	8	B8000h
02H	640 × 200	80 × 25	8 × 8	Text	16*	CGA	8	B8000h
	640 × 350	80 × 25	8 × 14	Text	16*	EGA	8	B8000h
	720 × 400	80 × 25	9 × 16	Text	16*	VGA	8	B8000h
	640 × 400	80 × 25	8 × 16	Text	16*	MCGA	8	B8000h
03H	640 × 200	80 × 25	8 × 8	Text	16	CGA	8	B8000h
	640 × 350	80 × 25	8 × 14	Text	16	EGA	8	B8000h
	720 × 400	80 × 25	9 × 16	Text	16	VGA	8	B8000h
	640 × 400	80 × 25	8 × 16	Text	16	MCGA	8	B8000h
04H	320 × 200	40 × 25	8 × 8	Graphics	4	CGA,EGA,VGA,MCGA	1	B8000h
05H	320 × 200	40 × 25	8 × 8	Graphics	4*	CGA,EGA,VGA,MCGA	1	B8000h
06H	640 × 200	80 × 25	8 × 8	Graphics	2	CGA,EGA,VGA,MCGA	1	B8000h
07H	720 × 350	80 × 25	9 × 14	Text	Mono	MDA	8	B0000h
	720 × 350	80 × 25	9 × 14	Text	Mono	EGA	4	B0000h
	720 × 400	80 × 25	9 × 16	Text	Mono	VGA	8	B0000h
08H – 0CH not used								
0DH	320 × 200	40 × 25	8 × 8	Graphics	16	EGA	2/4	A0000h
	320 × 200	40 × 25	8 × 8	Graphics	16	VGA	8	A0000h
0EH	640 × 200	80 × 25	8 × 8	Graphics	16	EGA	1/2	A0000h
	640 × 200	80 × 25	8 × 8	Graphics	16	VGA	4	A0000h
0FH	640 × 350	80 × 25	9 × 14	Graphics	Mono	EGA	1	A0000h
	640 × 350	80 × 25	8 × 14	Graphics	Mono	VGA	2	A0000h
10H	640 × 350	80 × 25	8 × 14	Graphics	4	EGA	1/2	A0000h
	640 × 350	80 × 25	8 × 14	Graphics	16	VGA	2	A0000h
11H	640 × 480	80 × 30	8 × 16	Graphics	2	VGA,MCGA	1	A0000h
12H	640 × 480	80 × 30	8 × 16	Graphics	16	VGA	1	A0000h
13H	320 × 200	40 × 25	8 × 8	Graphics	256	VGA,MCGA	1	A0000h

* Color burst off.

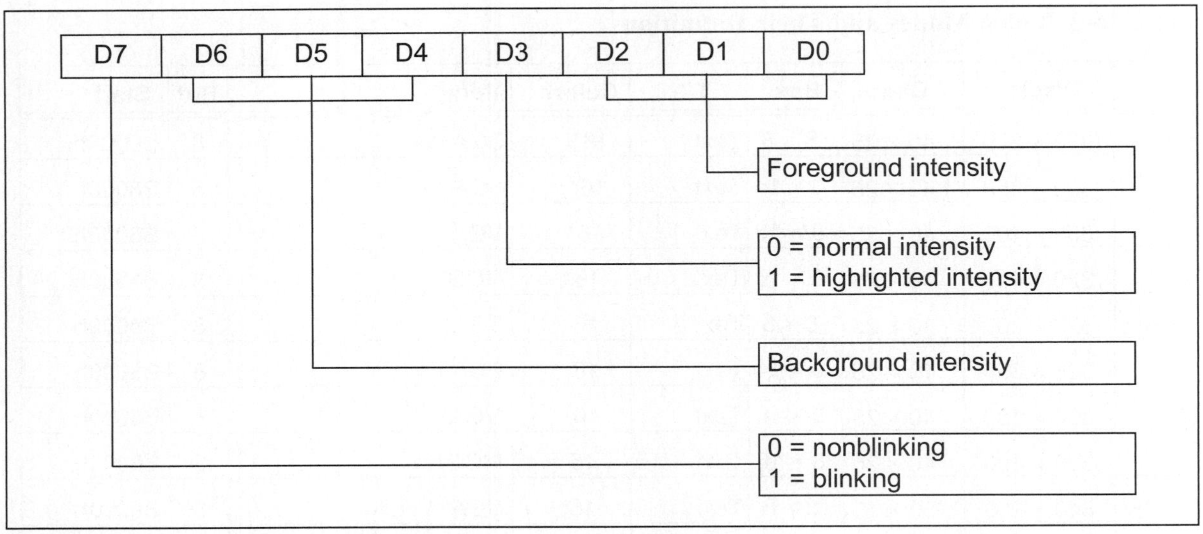

Figure 16-4. Attribute Byte in MDA

Example 16-5

Using Figure 16-3, find the attribute byte (in binary and hex) for the following color options:

(a) blue on black (b) green on blue (c) high-intensity white on blue
(d) red on blue

Solution:

	Binary	Hex	Color Effect
(a)	0000 0001	01	Blue on black
(b)	0001 0010	12	Green on blue
(c)	0001 1111	1F	High-intensity white on blue
(d)	0001 0100	14	Red on blue

Example 16-6

Find the attributes associated with the following attribute bytes in MDA.
(a) 07H (b) 0FH (c) 70H
Solution:
(a) 07H = 00000111 gives background black, foreground normal intensity, nonblinking.
(b) 0FH = 00001111 gives the same as (a) except with foreground highlighted.
(c) 70H = 01110000 gives black on white, a reverse video screen mode in which the foreground is black and the background is white, nonblinking.

Review Questions

1. The way images are displayed on the monitor screen is referred to as __________.
2. Increasing the dot frequency (DF) but keeping HF and VF constant will increase the number of ________________ (horizontal, vertical) scan lines.
3. True or false. The smaller the dot pitch, the better the monitor.
4. Of the three frequencies DF, HF, and VF, state which has the:

(a) highest frequency (b) lowest frequency

5. True or false. For a fixed-size monitor design, one must increase the dot pitch to get more pixels.
6. True or false. For any information to be displayed, it must be stored in the VDR.
7. True or false. The VDR memory address range (memory space) must be accessible to both the main CPU and the video processor.
8. True or false. To display crisper characters, one must design more pixels into the character box.

SECTION 16.2: TEXT MODE PROGRAMMING AND VIDEO RAM

As we showed in Chapter 4, we can use BIOS INT 10H to program video modes of the x86 using Assembly language instructions. In this section we revisit text and graphic mode programming. At the end of this section, character generator ROM is discussed.

Finding the current video mode

To find the current video mode, set AH = 0F and use INT 10H as follows:

```
MOV    AH,0FH ;AH=0F
INT    10H
```

Reminder: DEBUG assumes that numbers are in hex. If you are assembling the above program in DEBUG, make sure to remove the H, and also put INT 3 as the last instruction.

Changing the video mode

To change the video mode, use INT 10H with AH = 00 and AL = video mode. A list of video modes is given in Table 16-3. Regardless of what mode is selected, MDA, CGA, EGA, and MCGA all are supported by the VGA monitor. Examples 16-7 and 16-8 shows the use of INT 10H functions.

Setting the cursor position (AH = 02)

The cursor position is set by making AH = 2, DH = row, and DL = column. The following sets the cursor to row 12, column 28.

```
MOV    AH,02 ;set the cursor
MOV    BH,0        ;page 0
MOV    DH,12       ;row 12
MOV    DL,28       ;col 28
INT    10H   ;invoke interrupt
```

Getting the current cursor position (AH = 03)

To get the current cursor position, AH = 03 of INT 10H must be used:

```
MOV    AH,03 ;get cursor position
MOV    BH,0  ;page 0
INT    10H
```

After running the above code, registers DH and DL have the row and column

Example 16-7

Write the following program using INT 10H to:

(a) change the video mode to 03
(b) display letter "A" in 80 locations with the attributes of red on blue
(c) use the DEBUG utility to run and verify the program
(d) use DEBUG to dump the video RAM contents

Solution:

(a)
```
        MOV   AH,00        ;set mode option
        MOV   AL,03        ;CGA text mode of 80x25
        INT   10H          ;monitor is monochrome
```

(b) Using INT 10H function AH = 09, one can display a character a certain number of times with specific attributes.

```
        MOV   AH,09        ;display option
        MOV   BH,00        ;page 0
        MOV   AL,41H       ;ASCII for letter "A"
        MOV   CX,50H       ;repeat it 80 (50h) times
        MOV   BL,14H             ;red on blue
        INT   10H
```

(c) Reminder: DEBUG assumes that all the numbers are in HEX.

```
C:\>debug
-A
1131:0100   MOV   AH,00
1131:0102   MOV   AL,03
1131:0104   INT   10
1131:0106   MOV   AH,09
1131:0108   MOV   BH,00
1131:010A   MOV   AL,41
1131:010C   MOV   CX,50
1131:010F   MOV   BL,14
1131:0111   INT   10
1131:0113   INT   3
1131:0114
-
```

Now see the result by typing in command -G. Make sure that IP = 100 before you run it.

(d) When EGA and VGA monitors emulate CGA in text mode, the video memory address starts at B8000H. Dumping memory immediately after running the above program gives the following. Notice the character and the attribute byte stored in even and odd addresses.

```
-D B800:0 4F
B800:0000 41 14 41 14 41 14 41 14-41 14 41 14 41 14 41 14 A.A.A.A.A.A.A.A.
B800:0010 41 14 41 14 41 14 41 14-41 14 41 14 41 14 41 14 A.A.A.A.A.A.A.A.
B800:0020 41 14 41 14 41 14 41 14-41 14 41 14 41 14 41 14 A.A.A.A.A.A.A.A.
B800:0030 41 14 41 14 41 14 41 14-41 14 41 14 41 14 41 14 A.A.A.A.A.A.A.A.
B800:0040 41 14 41 14 41 14 41 14-41 14 41 14 41 14 41 14 A.A.A.A.A.A.A.A.
```

Example 16-8

Modify Example 16-7 using mode 7 to emulate MDA. Use attribute 87H (white on black blinking).

Solution:

The following shows the code as it would be entered in DEBUG.

```
1131:0100    MOV    AH,00
1131:0102    MOV    AL,07
1131:0104    INT    10
1131:0106    MOV    AH,09
1131:0108    MOV    BH,00
1131:010A    MOV    AL,41
1131:010C    MOV    CX,50
1131:010F    MOV    BL,87
1131:0111    INT    10
1131:0113    INT    3
1131:0114
-
```

After running the above program, we dump the MDA video buffer starting at memory address B000:0 and see the data and attribute bytes.

```
-D B000:0 F
B000:0000 41 87 41 87 41 87 41 87-41 87 41 87 41 87 41 87 A.A.A.A.A.A.A.A.
```

positions of the cursor (in hex), respectively.

Scrolling the window up to clear the screen (AH = 06)

The options 06 and 07 are called scroll functions. They are used to scroll a part or all of the screen up or down. One of the most widely used applications of option AH = 06 is to clear the screen, as shown next.

```
MOV    AH,06          ;scroll up option
MOV    AL,0           ;the entire screen
MOV    BH,07          ;normal attribute
MOV    CL,0           ;col 0 (top left col)
MOV    CH,0           ;row 0 (top left row)
MOV    DL,79          ;col 79 (bottom right col)
MOV    DH,24          ;row 24 (bottom right row)
INT    10H
```

A more efficient version of the above code is

```
MOV    AX,0600H
MOV    BH,07
MOV    CX,0
MOV    DX,184FH       ;18H=24 AND 4FH=79
INT    10H
```

Character generator ROM

To display characters on the screen, every video board must have access to the pixel patterns of the characters. In CGA, the patterns are burned into BIOS ROM starting at F000:FA60H. To decipher the patterns, first remember that the CGA character box is 8 × 8. Therefore, for every ASCII character there must be 64 (8 × 8 = 64) bits for each pattern. This means that every 8 bytes of the ROM provides the pattern for one character. Now let's see the patterns by using DEBUG.

```
A>DEBUG
-D F000:FA6E L4F
F000:FA60   FF FF FF FF FF FF FF FF-FF FF FF FF FF FF 00 00
F000:FA70   00 00 00 00 00 00 7E 81-A5 81 BD 99 81 7E 7E FF
F000:FA80   DB FF C3 E7 FF 7E 6C FE-FE FE 7C 38 10 00 10 38
F000:FA90   7C FE 7C 38 10 00 38 7C-38 FE FE 7C 38 7C 10 38
F000:FAA0   38 7C FE 7C 38 7C 00 00-18 3C 3C 18 00 00 FF FF
```

Starting with FA6E, the first 8 bytes are for blank (null), resulting in all 00s. The second 8 bytes are for the patterns for happy face, and so on. Inspecting the contents of ROM BIOS reveals the following character definition table.

```
Address     Patterns in Hex              ASCII         Hex    Dec
F000:FA6E   00,00,00,00,00,00,00,00      NULL          00     00
F000:FA76   7E,81,A5,81,BD,99,81,7E      HAPPY FACE    01     01
F000:FA7E   7E,FF,DB,FF,C3,E7,FF,7E      ...           ..     ..
.........   ........................     ...           ..     ..
F000:FBEE   7C,C6,CE,DE,F6,E6,7C,00      0             30     48
F000:FBF6   30,70,30,30,30,30,FC,00      1             31     49
.........   ........................     ...           ..     ..
F000:FC36   78,CC,CC,7C,0C,18,70,00      9             39     57
........    ........................     ...           ..     ..
F000:FC76   30,78,CC,CC,FC,CC,CC,00      A             41     65
F000:FC7E   FC,66,66,7C,66,66,FC,00      B             42     66
.........   ........................     ...           ..     ..
F000:FD3E   FE,C6,8C,18,32,66,FE,00      Z             5A     90
.........   ........................     ...           ..     ..
F000:FD76   00,00,78,0C,76,CC,76,00      a             61     97
F000:FD7E   E0,60,60,7C,66,66,DC,00      b             62     98
.........   ........................     ...           ..     ..
F000:FE36   00,00,CC,CC,CC,7C,0C,F8      y             79     121
F000:FE3E   00,00,FC,98,30,64,FC,00      z             7A     122
.........   ........................     ...           ..     ..
F000:FE66   00,10,38,6C,C6,C6,FE,00      DELTA         7F     127
```

For example, for the happy face character we have 7E, 81, A5, 81, BD, 99, 81, and 7E for the hex patterns of the bits that form the character of the screen. Example 16-9 demonstrates this.

How characters are displayed in text mode

To see how characters are displayed on the screen, analyze the MDA block diagram shown in Figure 16-6. For CGA, the process of generating the signals is exactly the same as in MDA except that the video circuitry generates R (red), B (blue), G (green), as well as vertical, horizontal, intensity, and dot pattern signals.

Example 16-9

Draw the patterns for the happy face (02H) and letter "A" (41H) on an 8 × 8 box.

Solution:

See Figure 16-5.

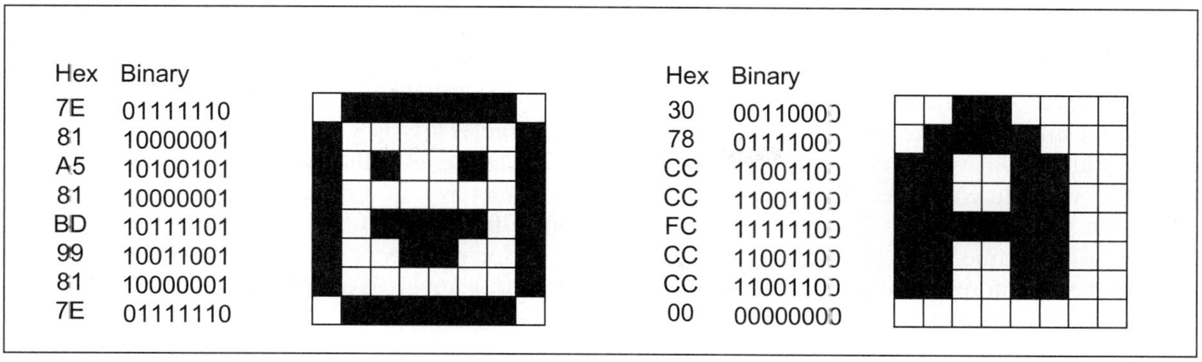

Figure 16-5. Diagram for Example 16-9

The function of the multiplexer in Figure 16-6 is to allow access to the video RAM by both the CPU and the 6845 CRT controller. The character generator ROM has patterns for all the ASCII characters. We just examined the content of this ROM holding the CGA patterns. First, the 6845 CRT controller is initialized by the CPU. Then to display characters, the x86 CPU writes the characters and their attributes into the video RAM. The job of the CRT controller is to fetch the characters and send them to the character generator ROM to get the patterns of every character on each row for the scan lines. As shown in Figure 16-6, the RA0–RA4 (row address) output pins from the 6845 fetch the specific row of the character and send it to a parallel-in-serial-out register (often referred to as a serializer). The job of this register is to provide the patterns to the video circuitry, one at a time, to be output serially to the monitor with the attribute.

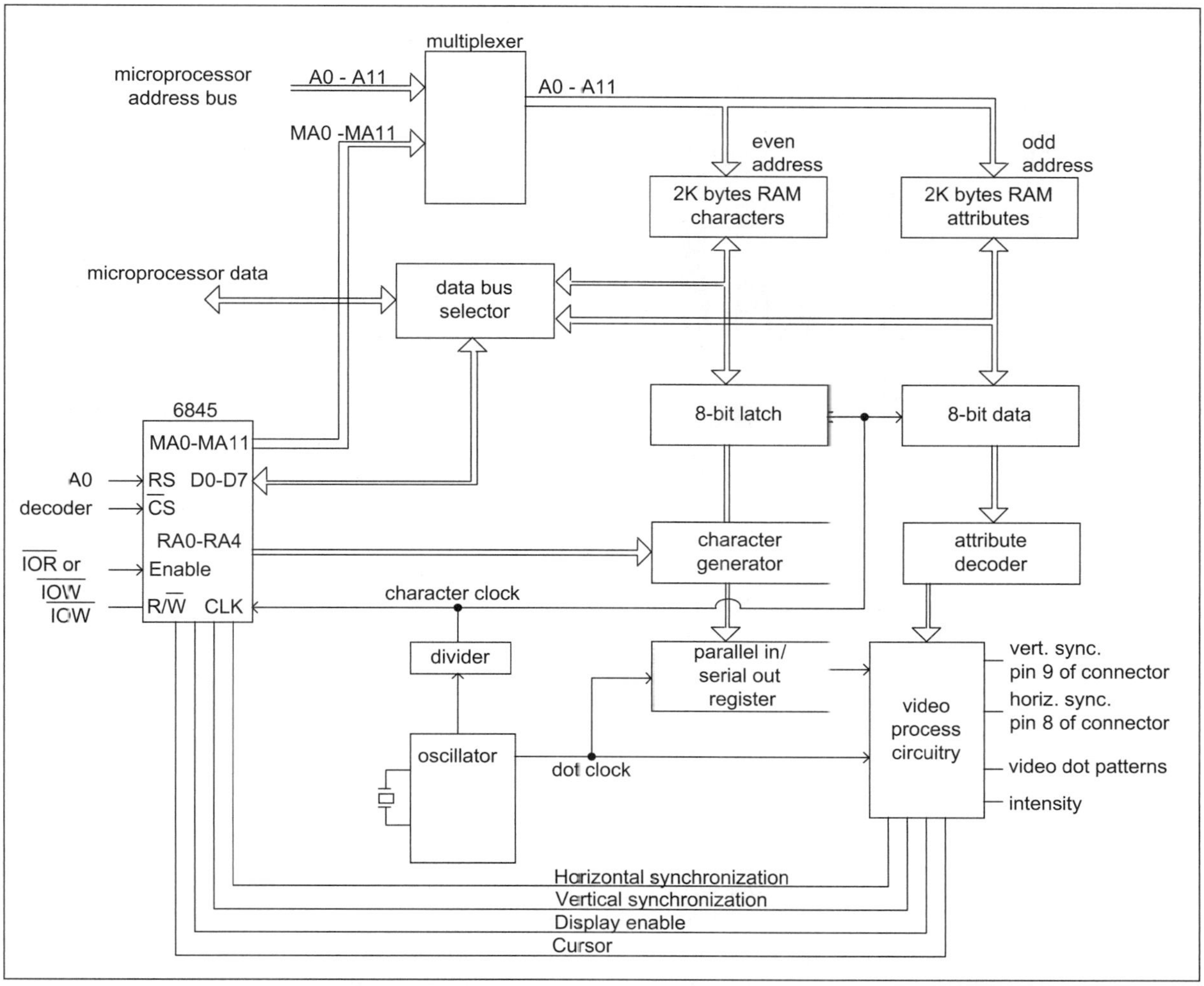

Figure 16-6. Original IBM PC Monochrome Block Diagram

Character definition table in VGA

In the discussion of CGA, the character definition table was examined by inspecting the contents of character generator BIOS ROM. The character box in CGA is 8 × 8 and as a result, the text is not very sharp. In VGA, the character box is 8 × 16 and the patterns for all the characters are stored in ROM memory. The address for that memory varies from computer to computer. To get the address of the character definition table, use INT 10H with AH = 11H, AL = 30H, and BH = 06. On return from INT 10H, ES:BP has the address. The number of bytes used to form patterns for each character is given in CX, and DL has the row number minus one. This is shown in Example 16-11. Example 16-10 diagrams two characters for VGA.

Example 16-10

Draw the patterns for VGA characters of happy face (02H) and letter "A" (41H) on a 9 × 16 box. Contrast this with CGA in Example 16-9.

Solution:

The patterns for these characters are as follows. See Example 16-13 for how to get the patterns. Figure 16-7 shows the diagram.

00,00,7E,81,A5,81,81,BD,99,81,81,7E,00,00,00,00 happy face
00,00,10,38,6C,C6,C6,FE,C6,C6,C6,C6,00,00,00,00 A

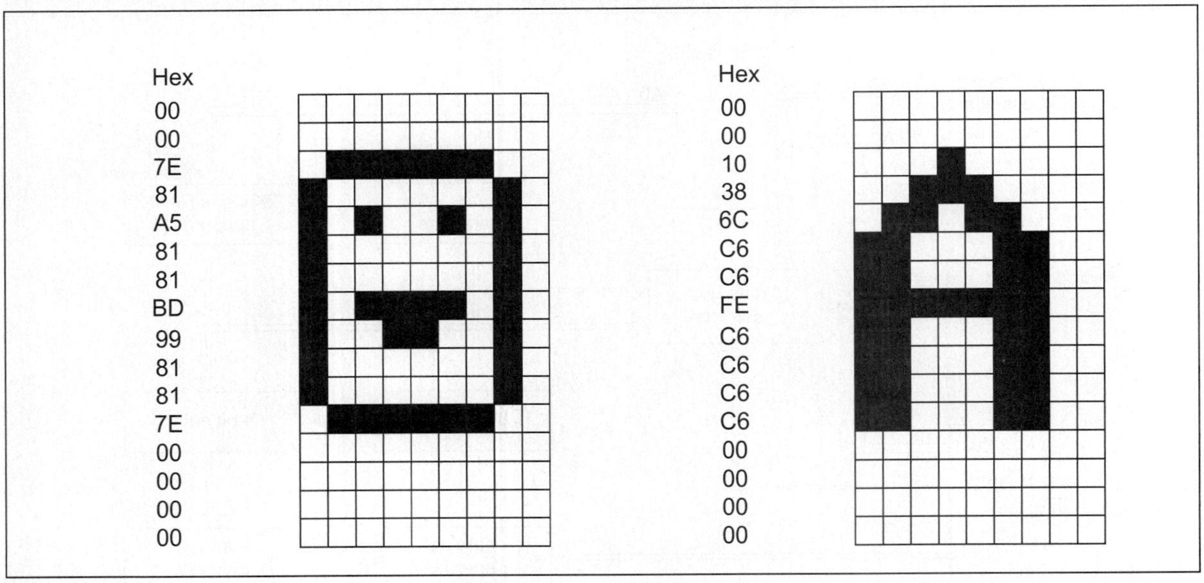

Figure 16-7. Diagram for Example 16-10

Review Questions

1. What colors are defined by the following attribute bytes in CGA text mode?
   (a) 0 (b) 14H
2. The characters displayed are the ___________(foreground, background).
3. When VGA emulates CGA, it uses what starting address for the video buffer?
4. When VGA emulates MDA, it uses what starting address for the video buffer?
5. True or false. The patterns for CGA characters are provided in BIOS ROM.
6. True or false. The patterns for VGA characters are provided in BIOS ROM.

Example 16-11

Use DEBUG to find the address of the character definition table of a given VGA board.

Solution:

```
C:\>DEBUG
-A
17D9:0100    MOV AH,11
17D9:0102    MOV AL,30
17D9:0104    MOV BH,06
17D9:0106    INT 10
17D9:0108    INT 3
17D9:109
-G
AX=XXXX BX=XXXX CX=0010 DX=0018 SP=XXXX BP=61E7 SI=XXXX DI=XXXX
DS=17D9 ES=E000 SS=XXXX .........................................
17D9:0108 CC             INT 3
-
```

CX = 10H gives 16 bytes used for each character. DL = 18H = 24, which indicates that there are a total of 25 rows of characters per screen. Dumping the contents of memory location at ES:BP, which is E000:61E7, shows the following pattern:

```
-D E000:61E7
E000:61E7                               00-00 00 00 00 00 00 00 00   .......
E000:61F0   00 00 00 00 00 00 00 00-00 7E 81 A5 81 81 BD 99   ......
E000:6200   81 81 7E 00 00 00 00 00-00 7E FF DB FF FE C3 E7   .....
E000:6210   FF FF FE 00 00 00 00 00-00 00 00 6C FE FE FE FE   .....
..........................................................................
```

and so on.
Going through these memory locations reveals:

BYTE PATTERNS	ASCII	HEX
00,00,00,00,00,00,00,00,00,00,00,00,00,00,00,00	null	00
00,00,7E,81,A5,81,81,BD,99,81,81,7E,00,00,00,00	happy face	01
....	....	..
00,00,38,6C,C6,C6,D6,D6,D6,C6,C6,6C,38,00,00,00	0	30
00,00,18,38,78,18,18,18,18,18,18,7E,00,00,00,00	1	31
00,00,7C,C6,06,0C,18,30,60,C0,C6,FE,00,00,00,00	2	32
....	...	..
....	...	..
00,00,10,38,6C,C6,C6,FE,C6,C6,C6,C6,00,00,00,00	A	41
00,00,FC,66,66,66,7C,66,66,66,66,FC,00,00,00,00	B	42
....	..	..
....	..	..

and so on.

SECTION 16.3: GRAPHICS AND GRAPHICS PROGRAMMING

In all the video programming examples given so far, characters have been used as units to be addressed and a character was treated as a group of pixels. In this section, programming individual pixels will be discussed. In graphics mode, pixel accessing is also referred to as *bit-mapped graphics*. First, the relationship between pixel resolution, the number of colors supported, and the amount of video memory in a given video board is clarified.

Graphics: pixel resolution, color, and video memory

There are two facts associated with every pixel on the screen:

1. The location of the pixel
2. Its attributes: color and intensity

These two facts must be stored in the video RAM. The higher the number of pixels and colors options, the larger the amount of memory that is needed to store them. In other words, the memory requirement goes up as the resolution and the number of colors supported go up. The number of colors displayed at one time is always 2^n where n is the number of bits set aside for the color. For example, when 4 bits are assigned for the color of the pixel, this allows 16 combinations of colors to be displayed at one time because $2^4 = 16$. See Example 16-12. The relation between the video memory, resolution, and color for each video adapter is discussed separately.

Example 16-12

In certain video graphics, a maximum of 256 colors can be displayed at one time. How many bits are set aside for the color of the pixels?

Solution:

To display 256 colors at once, we must have 8 bits set for color since $2^8 = 256$.

The case of CGA

The CGA board can have a maximum of 16K bytes of video memory since the 6845 chip used in the video design of the original PC had only 14 address pins (2^{14} = 16K). This 16K bytes of memory can hold up to 4 pages of data, where each page represents one full screen of 80 × 25 characters. In graphics mode, the number of colors supported varies depending on the resolution, as shown next.

320 × 200 (Medium resolution)

In this mode there are a total of 64,000 pixels (320 columns × 200 rows = 64,000). Dividing the total video RAM memory of 128K bits (16K × 8 bits = 128K bits) by the 64,000 pixels gives 2 bits for the color of each pixel. These 2 bits give rise to 4 colors since $2^2 = 4$. Therefore, the 320 × 200 resolution CGA can support only up to 4 different colors at a time. See Figure 16-8. These 4 colors can be selected from a palette of 16 possible colors. To select this mode, use set mode option AH = 0 of INT 10H with AL = 04 for mode. After setting the video mode to AL = 04, we must use option 0BH of INT 10H to select the color of the pixel displayed on the screen.

640 × 200 (High resolution)

In this mode there are a total of 128,000 pixels (200 × 640 = 128,000). Dividing the 16K bytes of memory by this gives 1 bit (128,000/128,000 = 1) for color. The bit can be on (white) or off (black). Therefore, the 640 × 200 high-resolution CGA can support

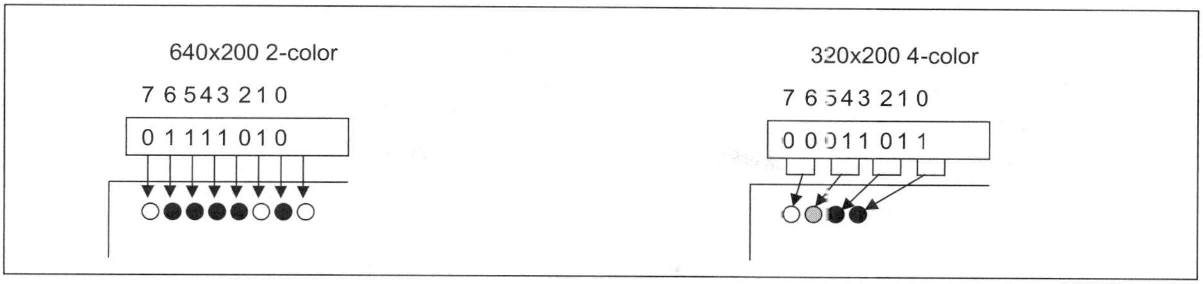

Figure 16-8. CGA Pixel Mapping

only two colors: black and white. To select this mode, use set mode option AH = 0 of INT 10H with AL = 06 for mode.

Note that for a fixed amount of video RAM, as the resolution increases, the number of supported colors decreases. The 16-color 160 × 100 low resolution is used with color TV sets.

The case of EGA

In the EGA board, the memory buffer was increased to a maximum of 256K bytes. This allowed both the number of colors and the number of pixels supported in graphics mode to increase. Although EGA can have up to 64 colors, only 16 of them can be displayed on the screen at a time. This is in contrast to CGA, which displayed only 4 colors of a 16-color palette. EGA graphics memory starts at A0000H and goes to a maximum of AFFFFH, using only 64K bytes of the PC's memory space. How is this 256K bytes of memory accessed through a 64K-byte address window? To solve this problem, IBM designers used four parallel planes, each 64K bytes, to access the entire 256K bytes of video RAM.In this scheme, each plane holds one bit of the 4-bit color. The assignment of 4 bits for color allows a maximum of 16 colors to be displayed at any given time. In the EGA card, IBM introduced what are called palette registers. There are a total of 16 palette registers in the EGA, each holding 8 bits. EGA uses only 6 bits out of the 8 bits of the palette register, giving rise to a maximum of 64 hues.

Video memory size and color relation for EGA

In EGA, to support 640 × 350 pixels with 16 colors requires a minimum of 640 × 350 × 4 = 896,000 bits of memory, but because of the concept of the plane and the 64K-byte address space of A0000H–AFFFFH, the memory must be 256K bytes, although some portions of video memory are unused.

In EGA, one can use 64K bytes for the video RAM, making only 16K bytes available for each color plane, but this reduces the number of colors supported. EGA is downwardly compatible with CGA in graphics mode, the same as in text mode. To program the palette registers of the EGA, use option AL = 0 of INT 10H. INT 10H has many options for pixel programming of EGA and VGA.

The case of VGA

In VGA, the number of pixels was increased to 640 × 480 with support for 256 colors displayed at one time. The color palette was increased to 2^{18} = 262,144 hues. The number of palette registers was also increased to 256. Each palette register holds 18 bits, 6 bits for each of the red, green, and blue colors.VGA was the first analog monitor introduced by IBM. All previous monitors were digital. In analog VGA, the analog colors of red, green, and blue replace the digital red, green, and blue of the digital display, allowing substantial increases of the number of colors supported. This gives rise to the use of what is called a *video DAC* (digital-to-analog converter). Each color of red, green, and blue has a 6-bit D/A converter, allowing 64 combinations for each color, making a total of 18 bits used for the palette, which gives rise to a total of 262,144 (2^{18}) hues. If the video DAC size is expanded from 6 to 8, the number of combinations for the three signals will be 256 × 256 × 256 = 2^{24} = 16,777,216 hues for the color palette, which is referred to as 16.7 million colors in many advertisements.

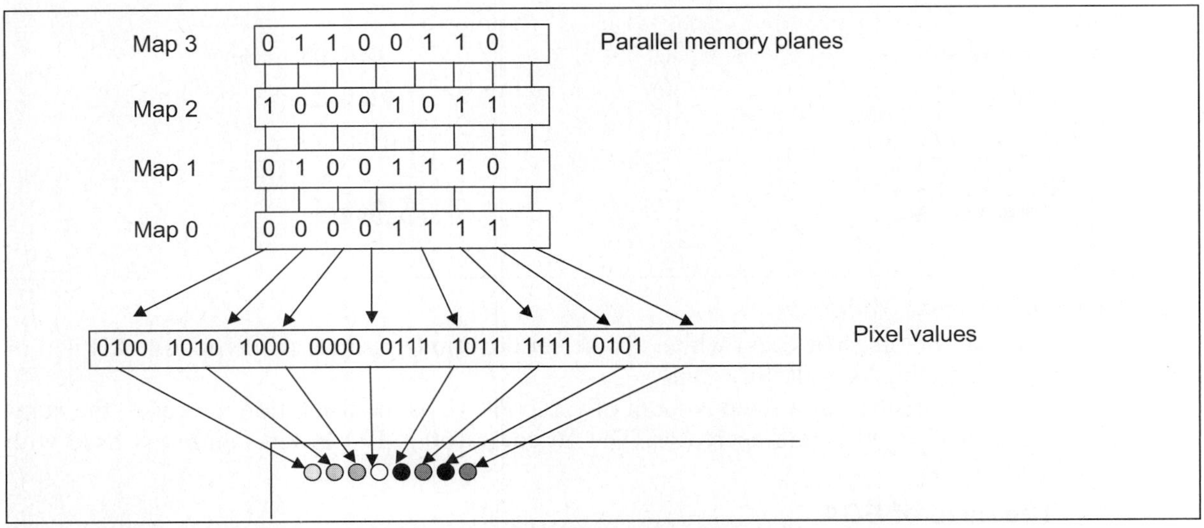

Figure 16-9. 16-Color EGA and VGA Mode Pixel Mapping

Video memory size and color relation for VGA

In VGA, 640 x 480 resolution with support for 256 colors displayed at one time will require a minimum of 640 × 480 × 8 = 2,457,600 bits of memory, but due to the architectural design of VGA, there must be 256K bytes of memory available on the video board. Using the concept of planes means that each plane has 64K bytes. See Figure 16-9. VGA is downward compatible with both CGA and EGA in graphics mode. To access one of the 256 palette registers of VGA, set AH = 10H and use option AL = 10H of INT 10H. As mentioned earlier, for the AH = 10H mode there are many options available for pixel programming of both EGA and VGA. These options are selected through register AL.

The case of SVGA

In SVGA, all the resolutions of 800 × 600, 1024 × 768, and 1024 × 1024 are supported. The memory requirement for these boards can reach millions of bytes, depending on the number of colors supported. For example, SVGA of 800 × 600 pixels with 256 colors displayed at the same time requires a minimum of 800 × 600 × 8 = 3,840,000 bits of memory, or a total of 480,000 bytes. Due to the use of bit planes, a total of 512K bytes of DRAM is needed. See Table 16-4. Another example is the total memory required by 800 × 600 resolution with 16 million colors. In this case we need 800 × 600 × 24 = 11,520,000 bits or 1,440,000 bytes, or 1406K bytes. Due to the use of bit planes, it uses 1.5M bytes of DRAM (see Table 16-4).

INT 10H and pixel programming

To address a single pixel on the screen, use INT 10H with AH = 0CH. To do that, X and Y coordinates of the pixel must be known. The values for X (column) and Y (row) vary depending on the resolution of the monitor. CX holds the column point (the X coordinate) and DX the row point (Y coordinate). If the display mode supports more than one page, then BH = page number; otherwise, it is ignored. To turn the pixel on or off, AL = 1 or AL = 0 for black and white. The value of AL can be modified for various colors.

Drawing horizontal or vertical lines in graphics mode

To draw a horizontal line, choose values for the row and column points at the beginning of the line and then continue to increment the column until it reaches the end of the line as shown in Example 16-13.

Table 16-4: Video Memory Requirements by Resolution

Resolution	16 Colors (4 bits)	256 Colors (8 bits)	65,536 Colors (16 bits)	16,777,216 Colors (24 bits)
640 x 480	256K	512K	1M	1M
800 x 600	256K	512K	1M	1.5M
1024 x 768	512K	1M	1.5M	2.5M
1280 x 1024	1M	1.5M	2.5M	4M
1600 x 1200	1M	2M	4M	6M

Example 16-13

Using INT 10H, write a program to:

(a) clear the screen

(b) set the mode to CGA of 640 × 200 resolution

(c) draw a horizontal line starting at column 50 and row 50 and ending at column 200, row 50

Solution:

```
            ;clear the screen
                  MOV    AX,0600H
                  MOV    BH,07
                  MOV    CX,0
                  MOV    DX,184F
                  INT    10H
            ;set the mode to 06 (CGA high resolution)
                  MOV    AH,00
                  MOV    AL,06
                  INT    10H
            ;draw the horizontal line from (50,50) to (200,50)
                  MOV    CX,50                 ;col pixel=50
                  MOV    DX,50                 ;row pixel=50
      BACK:       MOV    AH,0C                 ;0CH option to write a pixel
                  MOV    AL,01                 ;turn on the pixel
                  INT    10H
                  INC    CX                    ;increment horizontal position
                  CMP    CX,200                      ;check for the last posi-
tion
                  JNZ    BACK                  ;if not, continue
```

Review Questions

1. The 320 × 200 resolution CGA can support ______ colors.
2. True or false. In 640 × 200 resolution, the pixel color can be black or white.
3. As the number of pixels goes up, there is _______ (more, less) video memory for storage of color bits.
4. If a total of 24 bits is set aside for color, how many colors are available?
5. Calculate the total video memory needed for 1024 × 768 resolution with 16 colors displayed at the same time.

PROBLEMS

SECTION 16.1: PRINCIPLES OF MONITORS AND VIDEO MODES

1. Calculate the number of scan lines and dots per line for each of the following.
   (a) 640 × 200 EGA, DF = 14.318 MHz, HF = 15.75 kHz, and VF = 60 Hz
   (b) 640 × 350 EGA, DF = 16.257 MHz, HF = 21.85 kHz, and VF = 60 Hz
   (c) 640 × 480 VGA, DF = 25.175 MHz, HF = 31.5 kHz, and VF = 60 Hz
2. In Problem 1, find the number of scan lines and pixels used for overscan and retrace.
3. The following table is from PC Magazine, July 1993, showing the recommended dot pitch for various resolutions and monitor size. Calculate the diagonal size used by the image on screen for each case.

Monitor size	640 × 480	800 × 600	1,024 × 768	1,280 × 1,024
14	0.35	0.28	0.22	0.18
15"	0.38	0.30	0.24	0.19
17"	0.43	0.34	0.27	0.22
20"	0.50	0.40	0.31	0.25

4. A person wants to use a 14-inch monitor of 0.5 mm dot pitch for 640 × 480 VGA color resolution applications. Show by mathematics why he (she) cannot do that.
5. Dot pitch refers to the size of the ___________ (pixel, distance between pixels).
6. True or false. In color monitors each pixel has its own unique color.
7. For 640 × 200 resolution used for the 80 × 25 characters per screen, find the size of the character box.
8. A 320 × 200 resolution used for the 8 × 8 character box allows only ________ characters per screen.

SECTION 16.2: TEXT MODE PROGRAMMING AND VIDEO RAM

9. True or false. VGA supports all previous video standards.
10. Give the starting memory address used by VGA emulating MDA.
11. Rewrite Example 16-7, to display letter "B" in 80 locations with the attributes of blue on red.
12. Rewrite Example 16-8 for white on black blinking.
13. Draw the character boxes for the letter B and digits 2 and 9, both in CGA and VGA.

SECTION 16.3: GRAPHICS AND GRAPHICS PROGRAMMING

14. True or false. The more color a given video board supports, the more DRAM it needs.
15. True or false. The more pixels a given video board supports, the more DRAM it needs.
16. The 16-bit color depth can support how many color hues?
17. Why can 620 × 200 CGA support only black and white?
18. In a given VGA board with 256K bytes of memory, how can it fit into space A0000–AFFFFH?
19. To support 16,777,216 colors, the number of bits set aside for color depth must be _____.
20. Verify the memory requirements of the video board resolution and color depth of Table 16-9 for the following cases.
    (a) 640 × 480 of 16,256, 65,536 and 16 million colors
    (b) 1024 × 768 of 16,256, 65,536 and 16 million colors
    (c) 1600 × 1200 of 256 and 65,536 colors
21. Write a program to change the video mode to 620 × 200 CGA high-resolution graphics and then draw two vertical lines splitting the screen into three equal sections.

ANSWERS TO REVIEW QUESTIONS

SECTION 16.1: PRINCIPLES OF MONITORS AND VIDEO MODES

1. Raster scan
2. Horizontal
3. True
4. (a) DF (b) VF
5. False
6. True
7. True
8. True

SECTION 16.2: TEXT MODE PROGRAMMING AND VIDEO RAM

1. (a) 00H = 0000 0000B is black on black. (b) 14H = 0001 0100 is red on blue.
2. Foreground
3. B8000H
4. B0000H
5. True
6. False

SECTION 16.3: GRAPHICS AND GRAPHICS PROGRAMMING

1. 4
2. True
3. Less
4. 16.7 million
5. $1024 \times 768 \times 4 = 3{,}145{,}728$ bits = 384K bytes, but it uses 512 KB due to bit planes.

CHAPTER 17

SERIAL PORT PROGRAMMING WITH ASSEMBLY AND C#

OBJECTIVES

Upon completion of this chapter, you will be able to:

- **>> List the advantages of serial communication over parallel communication**
- **>> Explain the difference between synchronous and asynchronous communication**
- **>> Define the terms *simplex*, *half duplex*, and *full duplex* and diagram their implementation in serial communication**
- **>> Describe how start and stop bits frame data for serial communication**
- **>> Contrast and compare the measures *baud rate* and *bps* (bits per second)**
- **>> Describe the RS232 standard**
- **>> Contrast and compare DTE (data terminal) versus DCE (data communication) equipment**
- **>> Describe the purpose of handshaking signals such as DTR, RTS, and CTS**
- **>> Code Assembly language instructions to perform serial communication using BIOS INT 14H**
- **>> Describe the function of UART and USART chips**

Computers transfer data in two ways: parallel and serial. In parallel data transfers, often eight or more lines (wire conductors) are used to transfer data to a device that is only a few feet away. Examples of parallel transfers are printers and hard disks using cables with many wire strips. Although in such cases a lot of data can be transferred in a short amount of time by using many wires in parallel, the distance cannot be great. To transfer data to a device located many meters away, the serial method is used. In serial communication, the data is sent one bit at a time, in contrast to parallel communication, in which the data is sent a byte or more at a time. Serial communication and the study of associated chips are the topics of this chapter.

SECTION 17.1: BASICS OF SERIAL COMMUNICATION

When a microprocessor communicates with the outside world it provides the data in byte-sized chunks. In some cases, such as printers, the information is simply grabbed from the 8-bit data bus and presented to the 8-bit data bus of the printer. This can work if the cable is not too long since long cables diminish and even distort signals. In addition, an 8-bit data path is expensive. For these reasons, serial communication is used for transferring data between two systems located at distances of hundreds of feet to millions of miles apart. Figure 17-1 diagrams serial versus parallel data transfers.

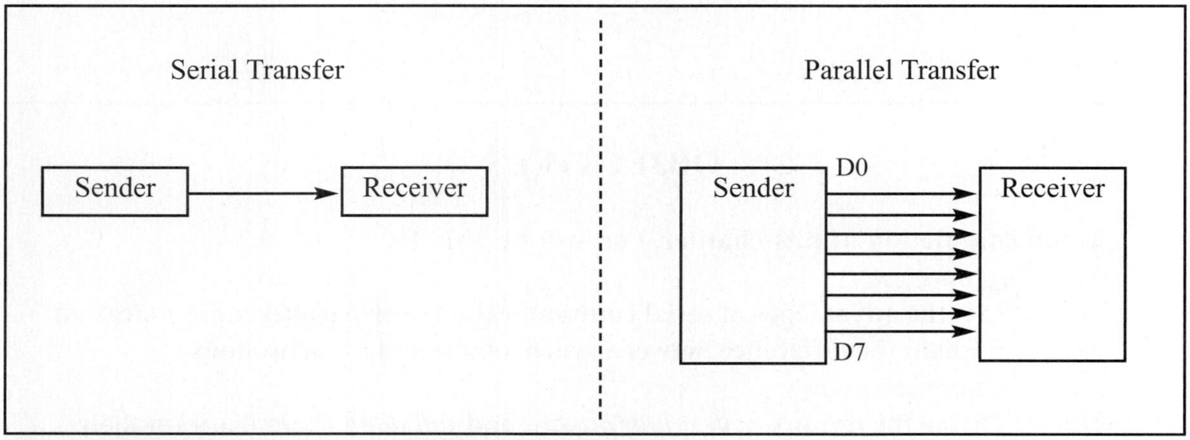

Figure 17-1. Serial versus Parallel Data Transfer

The fact that in serial communication a single data line is used instead of the 8-bit data line of parallel communication not only makes it much cheaper but also makes it possible for two computers located in two different cities to communicate over the telephone.

For serial data communication to work, the byte of data must be grabbed from the 8-bit data bus of the microprocessor and converted to serial bits using a parallel-in-serial-out shift register; then it can be transmitted over a single data line. This also means that at the receiving end there must be a serial-in-parallel-out shift register to receive the serial data, pack it into a byte, and present it to the system at the receiving end. Of course, if data is to be transferred on the telephone line, it must be converted from 0s and 1s to audio tones, which are sinosoidal-shaped signals. This conversion is performed by a peripheral device called a *modem*, which stands for "modulator/demodulator."

When the distance is short, the digital signal can be transferred as it is on a simple wire and requires no modulation. This is how IBM PC keyboards transfer data between the keyboard and the motherboard. However, for long-distance data transfers using communication lines such as a telephone, serial data communication requires a modem to modulate (convert from 0s and 1s to audio tones) and demodulate (convert from audio tones to 0s and 1s).

Serial data communication uses two methods, asynchronous and synchronous.

The synchronous method transfers a block of data (characters) at a time while the asynchronous transfers a single byte at a time.

It is possible to write software to use either of these methods, but the programs can be tedious and long. For this reason, special IC chips are made by many manufacturers for serial data communications. These chips are commonly referred to as UART (universal asynchronous receiver-transmitter) and USART (universal synchronous-asynchronous receiver-transmitter). The COM port in the x86 PC uses the UART.

Half- and full-duplex transmission

In data transmission a duplex transmission is one in which the data can be transmitted and received. This is in contrast to simplex transmissions such as printers, in which the computer only sends data. Duplex transmissions can be half or full duplex, depending on whether or not the data transfer can be simultaneous. If data is transmitted one way at a time, it is referred to as half duplex. If the data can go both ways at the same time, it is full duplex. Of course, full duplex requires two wire conductors for the data lines (in addition to ground), one for transmission and one for reception, in order to transfer and receive data simultaneously. See Figure 17-2.

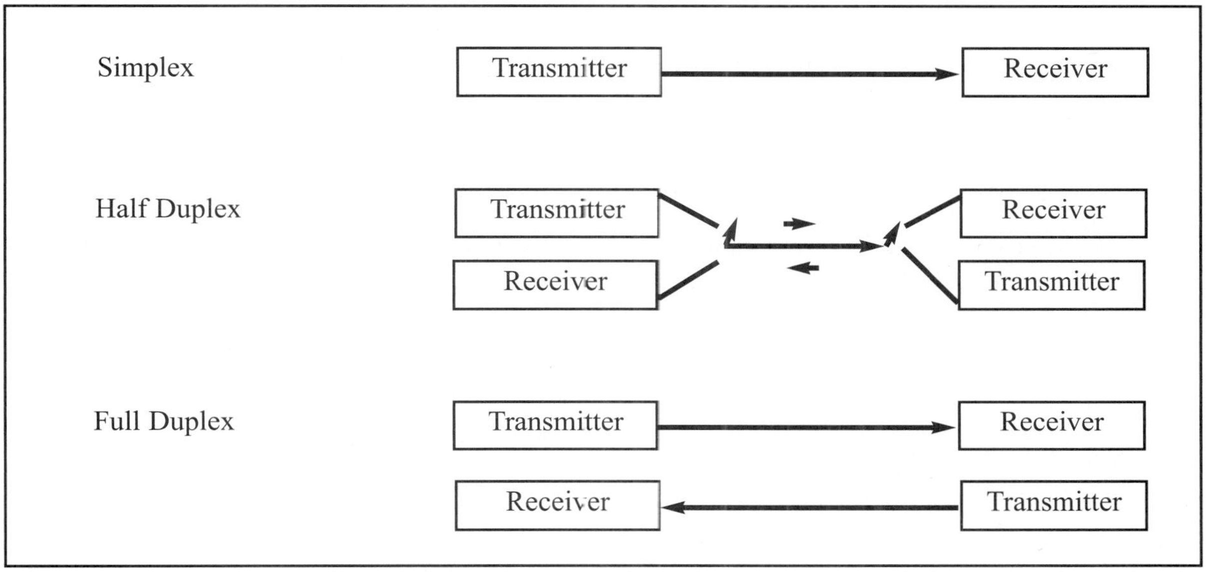

Figure 17-2. Simplex, Half-, and Full-Duplex Transfers

Asynchronous serial communication and data framing

The data coming in at the receiving end of the data line in a serial data transfer is all 0s and 1s; it is difficult to make sense of the data unless the sender and receiver agree on a set of rules, a protocol, on how the data is packed, how many bits constitute a character, and when the data begins and ends.

Start and stop bits

Asynchronous serial data communication is widely used for character-oriented transmissions, and block-oriented data transfers use the synchronous method. In the asynchronous method, each character is put between start and stop bits. This is called *framing*. In data framing for asynchronous communications, the data, such as ASCII characters, are packed between a start bit and a stop bit. The start bit is always one bit but the stop bit can be one or two bits. The start bit is always a 0 (low) and the stop bit(s) is 1 (high). For example, look at Figure 17-3 where the ASCII character "A", binary 0100 0001, is framed between the start bit and 2 stop bits. Notice that the LSB is sent out first.

In Figure 17-3, when there is no transfer the signal is 1 (high), which is referred to as *mark*. The 0 (low) is referred to as *space*. Notice that the transmission begins with a start bit followed by D0, the LSB, then the rest of the bits until the MSB (D7), and finally, the 2 stop bits indicating the end of the character "A".

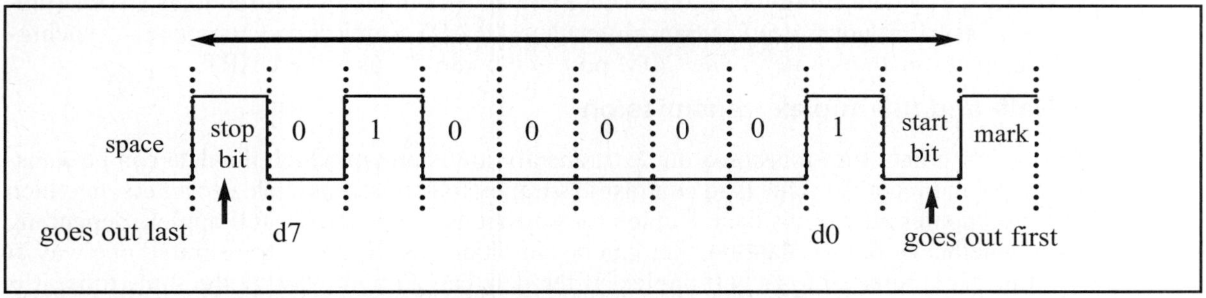

Figure 17-3. Framing ASCII "A" (41H)

In asynchronous serial communications, peripheral chips and modems can be programmed for data that is 5, 6, 7, or 8 bits wide. This in addition to the number of stop bits, 1 or 2. While in older systems ASCII characters were 7-bit, due to extended ASCII characters, 8 bits are required for each character. Small non-ASCII keyboards use 5- and 6-bit characters. In some older systems, due to the slowness of the receiving mechanical device, 2 stop bits were used to give the device sufficient time to organize itself before transmission of the next byte. However, in modern PCs the use of 1 stop bit is common. Assuming that we are transferring a text file of ASCII characters using 2 stop bits, we have a total of 11 bits for each character since 8 bits are for the ASCII code, and 1 and 2 bits are for start and stop bits, respectively. Therefore, for each 8-bit character there are an extra 3 bits, or more than 30% overhead.

In some systems in order to maintain data integrity, the parity bit of the character byte is included in the data frame. This means that for each character (7- or 8-bit, depending on the system) we have a single parity bit in addition to start and stop bits. The parity bit is odd or even. In the case of an odd-parity bit the number of data bits, including the parity bit, has an odd number of 1s. Similarly, in an even-parity bit the total number of bits, including the parity bit, is even. For example, the ASCII character "A", binary 0100 0001, has 0 for the even-parity bit. UART chips allow programming of the parity bit for odd-, even-, and no-parity options, as we will see in the next section. If a system requires the parity, the parity bit is transmitted after the MSB, and is followed by the stop bit.

Data transfer rate

The rate of data transfer in serial data communication is stated in bps (bits per second). Another widely used terminology for bps is baud rate. However, the baud and bps rates are not necessarily equal. This is due to the fact that baud rate is the modem terminology and is defined as number of signal changes per second. In modems, there are occasions when a single change of signal transfers several bits of data. As far as the conductor wire is concerned, the baud rate and bps are the same, and for this reason in this book we use the terms bps and baud interchangeably.

The data transfer rate of a given computer system depends on communication ports incorporated into that system. For example, the early IBM PC could transfer data at the rate of 100 to 9600 bps. However, today's x86 PC can transfer data at rates as high as 115,200 bps. It must be noted that in asynchronous serial data communication, the baud rate is generally limited to less than 100,000 bps when using a telephone line and modem. See Examples 17-1, 17-2, and 17-3.

RS232 and other serial I/O standards

To allow compatibility among data communication equipment made by various manufacturers, an interfacing standard called RS232 was set by the Electronics Industries Association (EIA) in 1960. In 1963 it was modified and called RS232A. RS232B and RS232C were issued in 1965 and 1969, respectively. In this book we refer to it simply as

Example 17-1

Calculate the total number of bits used in transferring 50 pages, each with 80 × 25 characters. Assume 8 bits per character and 1 stop bit.

Solution:

For each character a total of 10 bits is used, 8 bits for the character, 1 stop bit, and 1 start bit. Therefore, the total number of bits is 80 × 25 × 10 = 20,000 bits per page. For 50 pages, 1,000,000 bits will be transferred.

Example 17-2

Calculate the time it takes to transfer the entire 50 pages of data in Example 17-1 using a baud rate of:
(a) 9600 (b) 57,600

Solution:

(a) 1,000,000 / 9600 = 104 seconds
(b) 1,000,000 / 57,600 = 17 seconds

Example 17-3

Calculate the time it takes to download a movie of 2 gigabytes using a telephone line. Assume 8 bits, 1 stop bit, no parity, and 57,600 baud rate.

Solution:

2 × 1,000,000,000 × 10 / 57,600 = 347,222 seconds = 4 days

RS232. Today, RS232 is the most widely used serial I/O interfacing standard. However, since the standard was set long before the advent of the TTL logic family, the input and output voltage levels are not TTL compatible. In the RS232 a 1 is represented by –3 to –25 V, while the 0 bit is +3 to +25 V, making –3 to +3 undefined. For this reason, to connect any RS232 to a TTL-level chip (microprocessor or UART) we must use voltage converters such as MAX232 or MAX233 to convert the TTL logic levels to the RS232 voltage level, and vice versa. MAX232 and MAX233 IC chips are commonly referred to as line drivers. This is shown in Figures 17-4 and 17-5. The MAX232 has two sets of line drivers for transferring and receiving data, as shown in Figure 17-4. The line drivers used for TxD are called T1 and T2, while the line drivers for RxD are designated as R1 and R2. In many applications only one of each is used. Notice in MAX232 that the T1 line driver has a designation of T1in and T1out on pin numbers 11 and 14, respectively. The T1in pin is the TTL side and is connected to TxD of the USART, while T1out is the RS232 side that is connected to the RxD pin of the RS232 DB connector. The R1 line driver has a designation of R1in and R1out on pin numbers 13 and 12, respectively. The R1in (pin 13) is the RS232 side that is connected to the TxD pin of the RS232 DB connector, and R1out (pin 12) is the TTL side that is connected to the RxD pin of the USART. See Figure 17-4. Notice the null modem connection where RxD for one is TxD for the other.

MAX232 requires four capacitors ranging from 1 to 22 mF. The most widely used value for these capacitors is 22 mF. To save board space, some designers use the MAX233 chip from Maxim. The MAX233 performs the same job as the MAX232 but eliminates the need for capacitors. However, the MAX233 chip is much more expensive than the MAX232. See Figure 17-5 for MAX233 with no capacitor used.

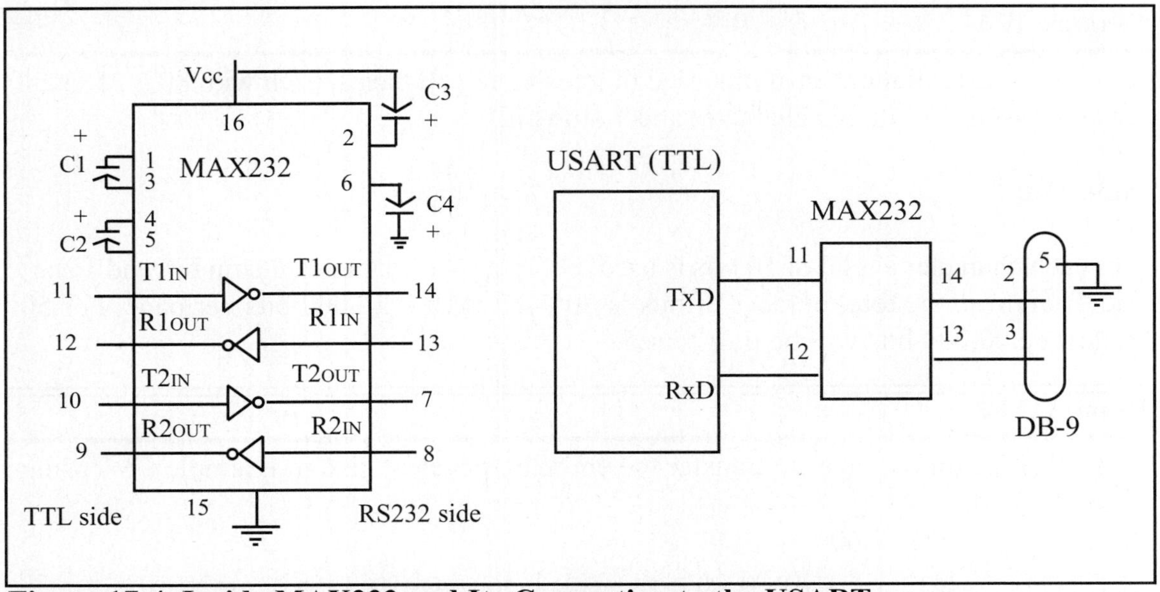

Figure 17-4. Inside MAX232 and Its Connection to the USART

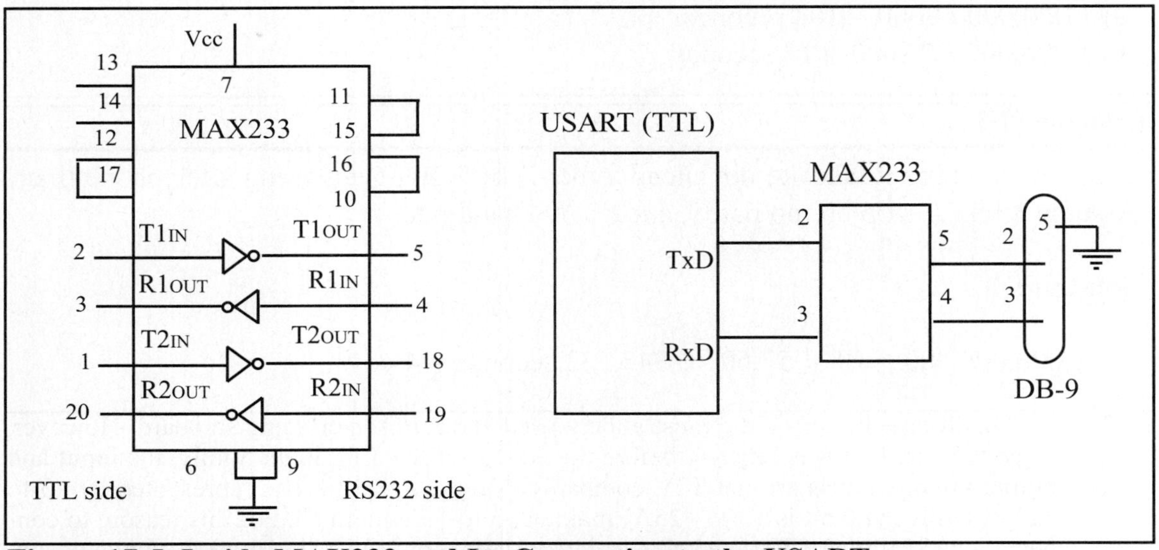

Figure 17-5. Inside MAX233 and Its Connection to the USART

RS232 pins

Table 17-1 provides the pins and their labels for the RS232 cable, commonly referred to as the DB-9 connector. The x86 PC 9-pin serial port is shown in Figure 17-6.

Data communication classification

Current terminology classifies data communication equipment as DTE (data terminal equipment) or DCE (data communication equipment). DTE refers to terminals and computers that send and receive data, while DCE refers to communication equipment, such as modems, that is responsible for transferring the data. Notice that all the RS232 pin function definitions of Table 17-1 are from the DTE point of view.

The simplest connection between two PCs (DTE and DTE) requires a minimum of three pins, TxD, RxD, and ground, as shown in Figure 17-7. Notice that the connection between two DTE devices, such as two PCs, requires pins 2 and 3 to be interchanged as shown in Figure 17-7. In looking at Figure 17-7, keep in mind that the RS232 signal definitions are from the point of view of DTE.

Examining the RS232 handshaking signals

To ensure fast and reliable data transmission between two devices, the data transfer must be coordinated. Due to the fact that in serial data communication the receiving device may have no room for the data there must be a way to inform the sender to stop sending data.

Some of the pins of the RS-232 are used for handshaking signals. They are described below.

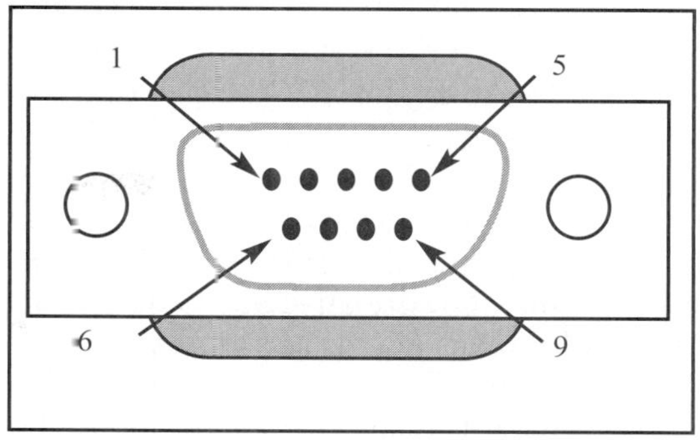

Figure 17-6. DB-9 9-Pin Connector

Table 17-1: IBM PC DB-9 Signals

Pin	Description
1	Data carrier detect ($\overline{DCD}$)
2	Received data (RxD)
3	Transmitted data (TxD)
4	Data terminal ready (DTR)
5	Signal ground (GND)
6	Data set ready ($\overline{DSR}$)
7	Request to send ($\overline{RTS}$)
8	Clear to send ($\overline{CTS}$)
9	Ring indicator (RI)

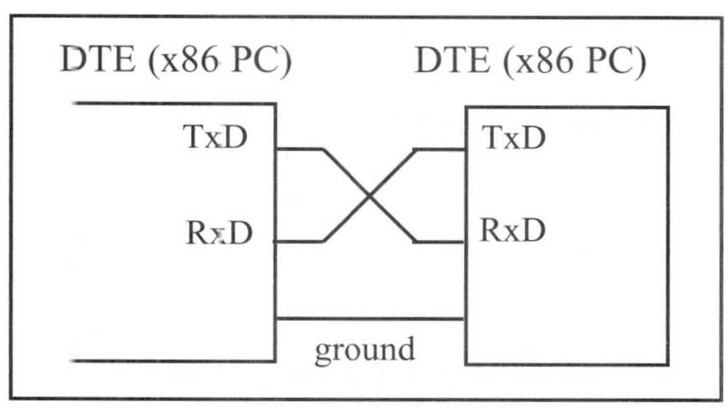

Figure 17-7. Null Modem Connection For Data Lines

1. DTR (data terminal ready). When the terminal (or a PC COM port) is turned on, after going through a self-test, it sends out signal DTR to indicate that it is ready for communication. If there is something wrong with the COM port, this signal will not be activated. This is an active-low signal and can be used to inform the modem that the computer is alive and kicking. This is an output pin from DTE (PC COM port) and an input to the modem.
2. DSR (data set ready). When a DCE (modem) is turned on and has gone through the self-test, it asserts DSR to indicate that it is ready to communicate. Therefore, it is an output from the modem (DCE) and an input to the PC (DTE). This is an active-low signal. If for any reason the modem cannot make a connection to the telephone, this signal remains inactive, indicating to the PC (or terminal) that it cannot accept or send data.
3. RTS (request to send). When the DTE device (such as a PC) has a byte to transmit, it asserts RTS to signal the modem that it has a byte of data to transmit. RTS is an active-low output from the DTE and an input to the modem.
4. CTS (clear to send). In response to RTS, when the modem has room for storing the data it is to receive, it sends out signal CTS to the DTE (PC) to indicate that it can receive the data now. This input signal to the DTE is used by the DTE to start transmission.
5. CD (carrier detect, or DCD, data carrier detect). The modem asserts signal CDC to inform the DTE (PC) that a valid carrier has been detected and that contact between it and the other modem is established. Therefore, CDC is an output from the modem and an input to the PC (DTE).
6. RI (ring indicator). An output from the modem (DCE) and an input to a PC (DTE)

indicates that the telephone is ringing. It goes on and off in synchronization with the ringing sound. Of the six handshake signals, this is the least often used, due to the fact that modems take care of answering the phone. However, if in a given system the PC is in charge of answering the phone, this signal can be used.

From the above description, PC and modem communication can be summarized as follows: While signals DTR and DSR are used by the PC and modem, respectively, to indicate that they are alive and well, it is RTS and CTS that actually control the flow of data. When the PC wants to send data it asserts RTS, and in response, if the modem is ready (has room) to accept the data, it sends back CTS. If, for lack of room, the modem does not activate CTS, the PC will deassert DTR and try again. RTS and CTS are also referred to as *hardware control flow signals*. See Figure 17-8.

This concludes the description of the most important pins of the RS232 handshake signals plus TxD, RxD, and ground. Ground is also referred to as SG (signal ground). In the next section we will see serial communication programming in the x86 PC.

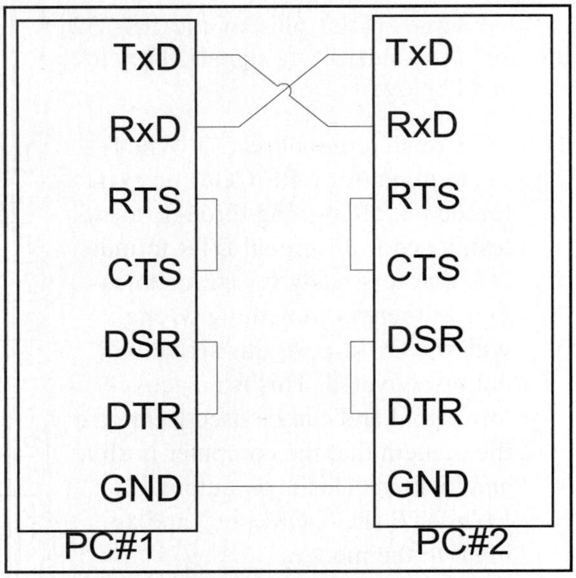

Figure 17-8. Null Modem Connection with Control Signals

Review Questions

1. The transfer of data using parallel lines is __________ (faster, slower) but ______________ (more expensive, less expensive).
2. In communications between two PCs in New York and Dallas, we use ___________ (serial, parallel) data communication.
3. In serial data communication, which method fits block-oriented data?
4. True or false. Sending data to a printer is duplex.
5. True or false. In duplex we must have two data lines.
6. The start and stop bits are used in the __________ (synchronous, asynchronous) method.
7. Assuming that we are transmitting letter "D", binary 100 0100, with odd-parity bit and 2 stop bits, show the sequence of bits transferred.
8. In Question 7, find the overhead due to framing.
9. Calculate the time it takes to transfer 400 characters as in Question 7 if we use 1200 bps.What percentage of time is wasted due to overhead?
10. True or false. RS232 is not TTL-compatible.

SECTION 17.2: PROGRAMMING x86 PC COM PORTS USING ASSEMBLY AND C#

To relieve users and programmers from the tedious details of the USART chip, both Windows and BIOS provide means of accessing the x86 PC serial COM ports. In this section we show how to use Assembly language and Visual C++ to program the COM ports. First some introductory comments on the number of ports in the PC will be given.

IBM PC COM ports

In the x86 PC, as many as four COM ports can be installed. They are numbered 1, 2, 3, and 4 (BIOS numbers them as 0, 1, 2, and 3). When the PC is turned on, it is the job of the POST (power-on self-test) to test the USART chip for each of the four COM ports. If they are installed, their I/O port addresses are written to memory locations 0040:0000–0040:0007. Since the I/O address assigned to each UART is a 16-bit address, it takes 2 bytes for each installed UART. The BIOS data area memory locations 0040:0000 and 0040:00001 will have the I/O port address for COM 0, and the 0040:0002 and 0040:0003 locations have the I/O port address for COM 1, and so on. We can also use the System Tools option in the Accessories menu in Windows to see the COM ports installed in the x86 PC. See Example 17-4.

Example 17-4

A nationally known computer columnist is asked by a reader how he/she can find the number of COM ports installed in a PC and which one is installed. What do you think the answer should be?

Solution:

Dumping memory locations 0040:0000–0040:0007 in DEBUG on the computer, as shown below, showed that there is only one COM port with starting address 03F8H.

```
C:\>DEBUG
-d 0040:0000 L08
0040:0000 F8 03 00 00 00 00 00 00
```

Using HyperTerminal on x86 PC

HyperTerminal is a widely used utility that comes with Windows 2000/XP. It allows communication with the x86 PC via the serial COM port. Some of the baud rates suppoerted by HyperTerminal shown in Table 17-2. The only problem with HyperTerminal is that it is not dynamic, which means we cannot incorporate it into our program. To do that, we can use INT 14H in Assembly language or the COM port statement in C#.

Table 17-2: Some HyperTerminal Baud Rates

Baud Rates
300
600
1,200
2,400
4,800
9,600
19,200
38,400
57,600
115,200

Programming COM ports using BIOS INT 14H with Assembly

The serial communication ports of the x86 PC can be accessed using the BIOS-based INT 14H. Various options of INT 14H are chosen with the AH registers as shown in Figure 17-9. Using BIOS INT 14H we can send and receive characters with another PC via a COM port. The process is as follows.

1. To send a character we use INT 14H, AH = 1, AL = character.

2. To receive a character we use INT 14H, AH = 3 to get the COM port's status in register AH. Notice that this is the status of the COM port and not the status of the MODEM, which is given in AL. Then check D0 of the status port, which is called *received data ready*. If it is high, a character has been received via the COM port and is sitting inside the USART. To read the received character we use INT 14H, AH = 2 where AL holds the character upon return.

AH   INT 14H Function

00   Initialize COM Port

Additional Call Registers	**Result Registers**
AL = parameter (see below)	AH = port status (see below)
DX = port number (0 if COM1, 1 if COM2, etc.)	AL = modem status (see below)

Note 1: The parameter byte in AL is defined as follows:

7	6	5	4	3	2	1	0	Indicates
x	x	x						Baud rate (000 = 110, 001 = 150, 010 = 300, 011 = 600, 100 = 1200, 101 = 2400, 110 = 4800, 111 = 9600)
			x	x				Parity (01 = odd, 11 = even, x0 = none)
					x			Stop bits (0 = 1, 1 = 2)
						x	x	Word length (10 = 7 bits, 11 = 8 bits)

Note 2: The port status returned in AH is defined as follows:

7	6	5	4	3	2	1	0	Indicates
1								Timed-out
	1							Transmit shift register empty
		1						Transmit holding register empty
			1					Break detected
				1				Framing error detected
					1			Parity error detected
						1		Overrun error detected
							1	Received data ready

Note 3: The modem status returned in AL is defined as follows:

7	6	5	4	3	2	1	0	Indicates
1								Received line signal detect
	1							Ring indicator
		1						DSR (data set ready)
			1					CTS (clear to send)
				1				Change in receive line signal detect
					1			Trailing edge ring indicator
						1		Change in DSR status
							1	Change in CTS status

Figure 17-9. BIOS INT 14H Functions *(continued on the following page)*

01 *Write Character to COM Port*

Additional Call Registers	**Result Registers**
AL = character	AH bit 7 = 0 if successful, 1 if not
DX = port number (0 if COM1, 1 if COM2, etc.)	AH bits 0–6 = status if successful
	AL = character

Note: The status byte in AH, bits 0–6, after the call is as follows:

6	5	4	3	2	1	0	Indicates
1							Transmit shift register empty
	1						Transmit holding register empty
		1					Break detected
			1				Framing error detected
				1			Parity error detected
					1		Overrun error detected
						1	Receive data ready

02 *Read Character from COM Port*

Additional Call Registers	**Result Registers**
DX = port number (0 if COM1, 1 if COM2, etc.)	AH bit 7 = 0 if successful, 1 if not
	AH bits 0–6 = status if successful
	AL = character read

Note: The status byte in AH, bits 1–4, after the call is as follows:

4	3	2	1	Indicates
1				Break detected
	1			Framing error detected
		1		Parity error detected
			1	Overrun error detected

03 *Read COM Port Status*

Additional Call Registers	**Result Registers**
DX = port number (0 if COM1, 1 if COM2, etc.)	AH = port status
	AL = modem status

Note: The port status and modem status returned in AH and AL are the same format as in INT 14H function 00H, described above.

Program 17-1 shows the steps needed to read from and write to the COM port:

(1) Check for a key press. If a key has been pressed, get it and write it to the COM port to be transferred. Also check for ESC to exit.

(2) If there is no key pressed, go check the status of the COM port. If a character has been received, read it and display it on the screen.

(3) Go to step (1).

Figure 17-9. BIOS INT 14H Functions *(continued from the previous page)*

```
TITLE SERIAL DATA COMMUNICATION BETWEEN TWO PCS
            .MODEL SMALL
            .STACK
            .DATA
MESSAGE     DB  'Serial communication via COM1, 4800 ,'
            DB  'No P,1 Stop,8-BIT DATA.',0AH,0DH'
            DB ' ANY KEY PRESS IS SENT TO OTHER PC.',0AH,0DH
            DB ' PRESS ESC TO EXIT','$'
            .CODE
MAIN  PROC
            MOV AX,@DATA
            MOV DS,AX
            MOV AH,09
            MOV DX,OFFSET MESSAGE
            INT 21H
            ;initializing COM 1
            MOV AH,0            ;initialize COM port
            MOV DX,0            ;COM 1
            MOV AL,0C3H         ;4800 ,NO P,1 STOP,8-BIT DATA
            INT 14H
    ;checking key press and sending key to COM2 to be transfered
AGAIN:      MOV AH,01  ;check for key press using INT 16H ,AH=01
            INT 16H             ;if ZF=1, there is no key press
            JZ  NEXT            ;If no key go check COM port
            MOV AH,0            ;yes, there is a key press, get it
            INT 16H   ;notice we must use INT 16H twice,2nd time
    ;with AH=0 to get the char itself. AL=ASCII char pressed
            CMP AL,1BH          ;is it esc key?
            JE EXIT             ;yes EXIT
            MOV AH,1            ;no. send the char to COM 1 port
            MOV DX,0
            INT 14H
    ;check COM1 port for a char. if so get it and display it
NEXT:       MOV AH,03           ;get COM 2 status
            MOV DX,0
            INT 14H
            AND AH,01  ;AH has COM port status, mask all but D0
            CMP AH,01          ;check D0 to see if there is a char
            JNE AGAIN           ;no data, go to monitor keyboard
            MOV AH,02           ;yes, COM1 has data: get it
            MOV DX,0
            INT 14H             ;get it
            MOV DL,AL           ;and display it using INT 21H
            MOV AH,02           ;DL has char to be displayed
            INT 21H
            JMP AGAIN           ;keep monitoring keyboard
EXIT:       MOV AH,4CH          ;exit to DOS
            INT 21H
MAIN        ENDP
            END
```

Program 17-1: Serial Communication Between Two PCs

Programming x86 COM port using C# 2005

To program the COM port we can use System.IO.Ports namespace in C# 2005. This is shown in Programs 17-2 and 17-3. Program 17-2 writes data to the COM port and Program 17-3 reads data from the COM port.

```
//Program 17-2 : Send data through the COM port.
//Must be compiled in Visual C# 2005 Express,
// which is availabe for free from Microsoft website.
using System;
using System.IO.Ports;

namespace SerialComm {
  class SerialOut {
    static void Main ()
    {
  // The following line will set the COM port parameters in C#.
      SerialPort com1 = new SerialPort ( "COM1", 9600,
                               Parity.None, 8, StopBits.One );
      com1.Open (); // Open the COM port 1.
      do
      {
        // Send the data through the COM port.
        com1.WriteLine ( "Hello World!!!" );
      }
      while (!Console.KeyAvailable); // Press any key to close.
      com1.Close (); // Close the COM port 1.
    }
  }
}
```

Program 17-2: Sending Data Through the COM Port

```
//Program 17-3 : Get data from COM port and display it on
// the screen.
//Must be compiled in Visual C# 2005 Express,
// which is availabe for free from Microsoft website.
using System;
using System.IO.Ports;

namespace SerialIn {
  class SerialIn {
    static void Main ()
    {
  // The following line will set the COM port parameters in C#.
      SerialPort com1 = new SerialPort ( "COM1", 9600,
                               Parity.None, 8, StopBits.One );
      com1.Open (); // Open the COM port 1.
      do
      {
        // Read from COM port and display on screen.
        Console.WriteLine ( com1.ReadLine () );
      }
      while (!Console.KeyAvailable); // Press any key to close.
      com1.Close (); // Close the COM port 1.
    }
  }
}
```

Program 17-3: Getting Data from the COM Port and Displaying on Screen

Review Questions

1. The maximum number of COM ports allowed in the PC is ___________.
2. Give the minimum and maximum baud rates supported by HyperTerminal.
3. Give the minimum and maximum baud rates supported by INT 14H using AH = 0 and AH = 14.

PROBLEMS

SECTION 17.1: BASICS OF SERIAL COMMUNICATION

1. Which is more expensive, parallel or serial data transfer?
2. True or false. 0- and 5-V digital pulses can be transferred on the telephone without being converted (modulated).
3. Show the framing of the letter ASCII "Z" (0101 1010), even parity, 1 stop bit.
4. If there is no data transfer and the line is high, it is called ___________ (mark, space).
5. What is space?
6. Calculate the overhead percentage if the data size is 6 bits, 2 stop bits, even parity.
7. True or false. The RS232 voltage specification is TTL compatible.
8. What is the function of the MAX232 chip?
9. True or false. The COM port on the back of an x86 PC uses an RS232 male connector.
10. How many pins of the RS232 are used by the null modem connection?
11. True or false. The longer the cable, the higher the data transfer baud rate.
12. The function definition of the RS232 pins is stated from the point of view of __________ (DTE, DCE).
13. If two PCs are connected through the RS232 without the modem, they are both configured as a ________ (DTE, DCE) -to- _______ (DTE, DCE) connection.
14. State the most important signals of the RS232 used for the null modem connection.
15. Calculate the total number of bits transferred if 200 pages of ASCII data are sent using asynchronous serial data transfer. Assume a data size of 8 bits, 1 stop bit, no parity.

SECTION 17.2: PROGRAMMING x86 PC COM PORTS USING ASSEMBLY AND C#

16. In the IBM PC, what is the maximum number of COM ports that can be installed? Use System Tools to find which COM port is installed on your PC.
17. What are the highest baud rates supported by the x86 PC COM ports using BIOS programming?
18. Show the COM2 port setting of 9600 baud rate, 8 data bits, 1 stop bit, no parity bit, using BIOS INT 14H.
19. Modify Program 17-1 to display the key pressed not only on the receiving PC's monitor but also on the monitor of the PC that sent it.

ANSWERS TO REVIEW QUESTIONS

SECTION 17.1: BASICS OF SERIAL COMMUNICATION

1. Faster, more expensive
2. Serial
3. Synchronous
4. False; it is simplex.
5. True
6. Asynchronous

7. With 100 0100 binary we have 1 as the odd-parity bit. The bits as transmitted in the sequence are:
   (a) 0 (start bit)
   (b) 0
   (c) 0
   (d) 1
   (e) 0
   (f) 0
   (g) 0
   (h) 1
   (i) 1 (parity)
   (j) 1 (first stop bit)
   (k) 1 (second stop bit)
8. 4 bits
9. 400 × 11 = 4400 bits total bits transmitted. 4400/1200 = 3.667 seconds, 4/7 = 58%.
10. True

SECTION 17.2: PROGRAMMING x86 PC COM PORTS USING ASSEMBLY AND C#

1. 4
2. 110 to 9600 (can be higher if we bypass BIOS)
3. 110 to 19,200 (can be higher if we bypass BIOS)

CHAPTER 18

KEYBOARD AND PRINTER INTERFACING

OBJECTIVES

Upon completion of this chapter, you will be able to:

>> Diagram how a keyboard matrix is connected to the I/O ports of a PC
>> Describe the processes of key press detection and key identification performed by a microprocessor with pseudocode or flowchart
>> Describe the respective functions of the keyboard microcontroller, INT 09, and the motherboard in keyboard input
>> Code Assembly language instructions using INT 16H to get and check the keyboard input buffer and status bytes
>> Diagram how data from the keyboard is stored in the keyboard buffering
>> State the differences between hard contact and capacitance keyboards
>> List the 36-pin assignments of the Centronics printer interface
>> Describe the BIOS programming for the four parallel printer ports LPT1–LPT4
>> Diagram the I/O port assignment for printer data, status, and control
>> Code Assembly language instructions using INT 17H to check printer status, initialize printers, and send data to the printer
>> Describe the printer time-out problem and how it can be alleviated
>> Discuss the evolution of the PC's parallel port
>> Contrast and compare parallel port types, including SPP, PS/2, EPP, and ECP

Along with video monitors, keyboards and printers are the most widely used input/output devices of the PC, and a basic understanding of them is essential. In this chapter, we discuss fundamentals of the keyboard and printer.

SECTION 18.1: INTERFACING THE KEYBOARD TO THE CPU

At the lowest level, keyboards are organized in a matrix of rows and columns. The CPU accesses both rows and columns through ports; therefore, with two 8-bit ports, an 8 × 8 matrix of keys can be connected to a microprocessor. When a key is pressed, a row and a column make a contact; otherwise, there is no connection between rows and columns. In IBM PC keyboards, a single microcontroller (consisting of a microprocessor, RAM and EPROM, and several ports all on a single chip) takes care of hardware and software interfacing of the keyboard. In such systems, it is the function of programs stored in the EPROM of the microcontroller to scan the keys continuously, identify which one has been activated, and present it to the main CPU on the motherboard. More details of the IBM PC keyboard design are presented in Section 18.2. In this section we look at the mechanism by which the microprocessor scans and identifies the key. For clarity we use 8088/86 Assembly language instructions in examples.

Scanning and identifying the key

Figure 18-1 shows a 4 × 4 matrix connected to two ports. The rows are connected to an output port and the columns are connected to an input port. If no key has been pressed, reading the input port will yield 1s for all columns since they are all connected to high (VCC). If all the rows are grounded and a key is pressed, one of the columns will

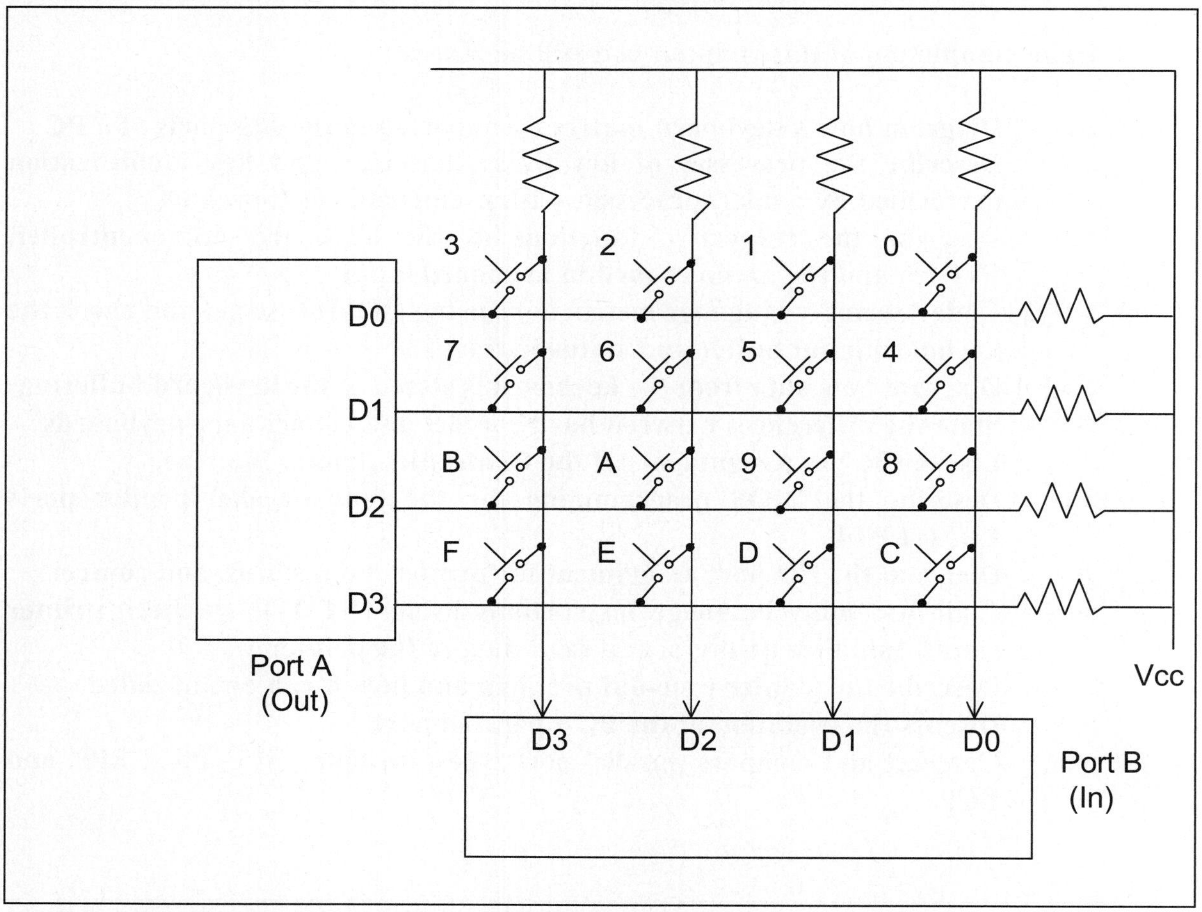

Figure 18-1. Matrix Keyboard Connection to Ports

have 0 since the key pressed provides the path to ground. It is the function of the microprocessor to scan the keyboard continuously to detect and identify the key pressed. How it is done is explained next.

Grounding rows and reading the columns

To detect the key pressed, the microprocessor grounds all rows by providing 0 to the output latch, then it reads the columns. If the data read from the columns is D3–D0 = 1111, no key has been pressed and the process continues until a key press is detected. However, if one of the column bits has a zero, this means that a key press has occurred. For example, if D3–D0 = 1101, this means that a key in the D1 column has been pressed. After a key press is detected, the microprocessor will go through the process of identifying the key. Starting with the top row, the microprocessor grounds it by providing a low to row D0 only; then it reads the columns. If the data read is all 1s, no key in that row is activated and the process is moved to the next row. It grounds the next row, reads the columns, and checks for any zero. This process continues until the row is identified. After identification of the row in which the key has been pressed, the next task is to find out which column the pressed key belongs to. This should be easy since the CPU knows at any time which row and column are being accessed. Look at Example 18-1.

Example 18-1

From Figure 18-1, identify the row and column of the pressed key for each of the following.
(a) D3–D0 = 1110 for the row, D3–D0 = 1011 for the column
(b) D3–D0 = 1101 for the row, D3–D0 = 0111 for the column

Solution:

From Figure 18-1 the row and column can be used to identify the key.
(a) The row belongs to D0 and the column belongs to D2; therefore, the key number 2 was pressed.
(b) The row belongs to D1 and the column belongs to D3; therefore, the key number 7 was pressed.

Program 18-1 is the Assembly language program for the detection and identification of the key activation. In this program, it is assumed that PORT_A and PORT_B are initialized as output and input, respectively. Program 18-1 goes through the following four major stages:

1. To make sure that the preceding key has been released, 0s are output to all rows at once, and the columns are read and checked repeatedly until all the columns are high. When all columns are found to be high, the program waits for a short amount of time before it goes to the next stage of waiting for a key to be pressed.
2. To see if any key is pressed, the columns are scanned over and over in an infinite loop until one of them has a 0 on it. Remember that the output latches connected to rows still have their initial zeros (provided in stage 1), making them grounded. After the key press detection, it waits 20 ms for the bounce and then scans the columns again. This serves two functions: (a) it ensures that the first key press detection was not an erroneous one due to a spike noise, and (b) the 20 ms delay prevents the same key press from being interpreted as a multiple key press. If after the 20 ms delay the key is still pressed, it goes to the next stage to detect which row it belongs to; otherwise, it goes back into the loop to detect a real key press.
3. To detect which row the key press belongs to, it grounds one row at a time, reading the columns each time. If it finds that all columns are high, this means that the key

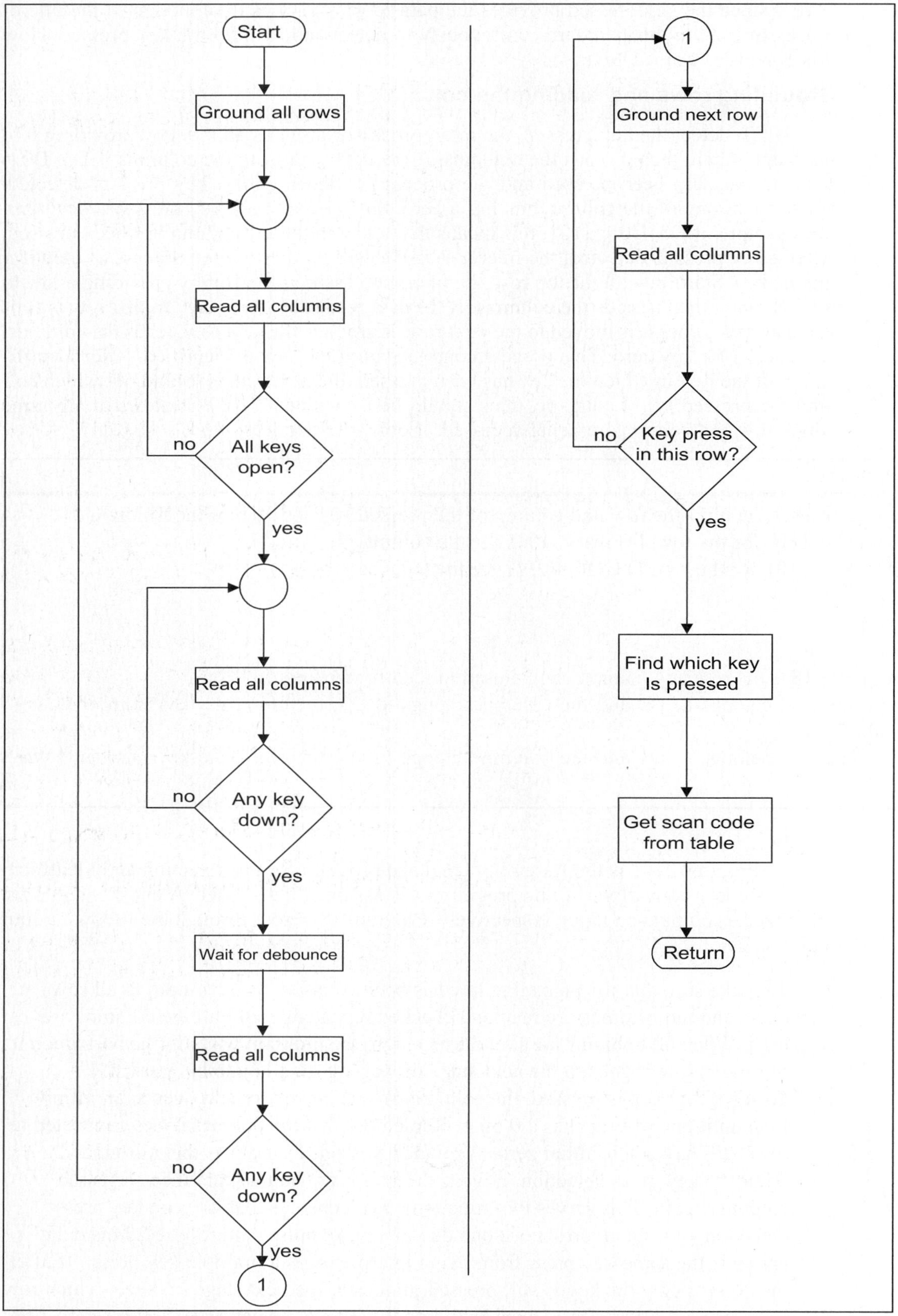

Figure 18-2. Flowchart for Program 18-1

press cannot belong to that row; therefore, it grounds the next row and continues until it finds the row the key press belongs to. Upon finding the row that the key press belongs to, it sets up the starting address for the lookup table holding the scan codes for that row and goes to the next stage to identify the key.

4. To identify the key press, it rotates the column bits, one bit at a time, into the carry flag and checks to see if it is low. Upon finding the zero, it pulls out the scan code for that key from the lookup table; otherwise, it increments the pointer to point to the next element of the lookup table.

While the key press detection is standard for all keyboards, the process for determining which key is pressed varies. The lookup table method shown in Program 18-1 can be modified to work with any matrix up to 8 × 8. Figure 18-2 provides the flowchart for Program 18-1 for scanning and identifying the pressed key.

```
;the following look-up scan codes are in the data segment
KCOD_0 DB 0,1,2,3          ;key codes for row zero
KCOD_1 DB 4,5,6,7          ;key codes for row one
KCOD_2 DB 8,9,0AH,0BH           ;key codes for row two
KCOD_3 DB 0CH,0DH,0EH,0FH       ;key codes for row three

;the following is from the code segment
       PUSH BX                  ;save BX
       SUB AL,AL   ;AL=0 to ground all rows rows at once
       OUT PORT_A,AL;to ensure all keys are open (no contact)
K1:    IN   AL,PORT_B           ;read the columns
       AND AL,00001111B         ;mask the unused bits (D7-D4)
       CMP AL,00001111B         ;are all keys released
       JNE K1              ;keep checking for all keys released
       CALL DELAY               ;wait for 20 ms
K2:    IN   AL,PORT_B           ;read columns
       AND AL,00001111B         ;mask D7-D4
       CMP AL,00001111B         ;see if any key pressed?
       JE   K2                  ;if none keep checking
       CALL DELAY               ;wait 20 ms for debounce
  ;after the debounce see if still pressed
       IN   AL,PORT_B           ;read columns
       AND AL,00001111B         ;mask D7-D4
       CMP AL,00001111B         ;see if any key closed?
       JE   K2                  ;if none keep polling
  ;now ground one row at a time and read columns to find the
key
       MOV AL,11111110B         ;ground row 0 (D0=0)
       OUT PORT_A,AL
       IN   AL,PORT_B           ;read all columns
       AND AL,00001111B         ;mask unused bits (D7-D4)
       CMP AL,00001111B         ;see which column
       JE   RO_1                ;if none go to grounding row 1
       MOV BX,OFFSET KCOD_0    ;BX=start of table for column 0 key
       JMP FIND_IT              ;identify the key
RO_1:  MOV AL,11111101B         ;ground row 1 (D1=0)
       OUT PORT_A,AL
       IN AL,PORT_B             ;read all columns
       AND AL,00001111B         ;mask unused bits (D7-D4)
       CMP AL,00001111B         ;see which column
       JE RO_2                  ;if none go to grounding row 2
```

Program 18-1 *(continued on the following page)*

```
        MOV BX,OFFSET KCOD_1 ;BX=Start of table for column 1 keys
        JMP FIND_IT             ;identify the key
RO_2:   MOV AL,11111011B        ;ground row 2 (D2=0)
        OUT PORT_A,AL
        IN AL,PORT_B            ;read all columns
        AND  AL,00001111B       ;mask unused bits (D7-D4)
        CMP  AL,00001111B       ;see which column
        JE         RO_3         ;if none go to grounding row 3
        MOV  BX,OFFSET KCOD_2 ;BX=start of table for column 2
keys
        JMP  FIND_IT            ;identify the key
RO_3:   MOV  AL,11110111B       ;ground row 3 (D3=0)
        OUT  PORT_A,AL
        IN         AL,PORT_B    ;read all columns
        AND  AL,00001111B       ;mask unused bits (D7-D4)
        CMP  AL,00001111B       ;see which column
        JE    K2    ;if none then false input repeat the process
        MOV  BX,OFFSET KCOD_3 ;BX=start of table for column 3
keys
;A key press has been detected and the row identified.
;Now find which key.
FIND_IT:RCR AL,1        ;rotate the column input to search for 0
        JNC MATCH               ;if zero, go get the code
        INC BX                  ;if not point at the next code
        JMP FIND_IT             ;and keep searching
;GET THE CODE FOR THE KEY PRESSED AND RETURN
MATCH:  MOV AL,[ BX]            ;get the code pointed by BX
        POP BX          ;return with AL=code for pressed key
        RET
;FOR THE DELAY GENERATION SEE CHAPTER 13
```

Program 18-1 *(continued from the previous page)*

There are IC chips such as National Semiconductor's MM74C923 that incorporate keyboard scanning and decoding all in one chip. Such chips use combinations of counters and logic gates (no microprocessor) to implement the underlying concepts presented in Program 18-1.

Review Questions

1. True or false. To see if any key is pressed, all rows are grounded.
2. If D3–D0 = 0111 is the data read from the columns, which column does the key pressed belong to?
3. True or false. Key press detection and key identification require two different processes.
4. In Figure 18-1, if the row has D3–D0 = 1110 and the columns are D3–D0 = 1110, which key is pressed?
5. True or false. To identify the key pressed, one row at a time is grounded.

SECTION 18.2: PC KEYBOARD INTERFACING AND PROGRAMMING

In the IBM PC and compatibles, a microcontroller is used for both detection and identification of keys. This microcontroller has a microprocessor in addition to a few hundred bytes of RAM, a few kilobytes of EPROM, and a few I/O ports, all on one chip. The microcontroller used widely in the IBM PC and compatibles is Intel's 8042 (or some variation). The 8042 is programmed to detect and identify the key press. A scan code is assigned to each key and the microcontroller provides the scan code for the pressed key to the motherboard. To allow the keyboard to be detachable from the system board, the keyboard is connected to the system board through a cable. Such an arrangement necessitates the use of serial data communication to transfer the scan code to the main CPU (serial data transfer was covered in Chapter 17). IBM PC AT keyboards use the following data frame when sending the scan code serially to the motherboard. For each scan code, a total of 11 bits are transferred from the keyboard to the motherboard.

one start bit (always 0)
8 bits for scan code
odd parity bit
one stop bit (always 1)

In the original PC motherboard, a serial-in-parallel-out shift register, 74LS322, was used to receive the serial data coming in through the keyboard cable. The 74LS322 strips away the framing portion, makes an 8-bit scan code, and presents it to port A of the 8255 with I/O port address of 60H. In the subsequent x86 PCs, the 74LS322 and supporting logic were replaced by another 8042. This allows the option of programming the keyboard itself. Therefore, two 8042 microcontrollers, one on the keyboard and one on the motherboard, are responsible for keyboard bidirectional communication in the x86 PC.

Make and break

In the IBM PC, the key press and release are represented by two different scan codes. The key press is referred to as a *make*, and the release of the same key is called a *break*. When a key is pressed (a make), the keyboard sends one scan code, and when it is released (a break), it sends another scan code. The scan code for the break is always 127 decimal (80H) larger than the make scan code. For example, if a given key produces a scan code of 06 on make, the scan code for the break is 86H (06 + 80H = 86H).

IBM PC scan codes

The original IBM PC keyboard had 83 keys, arranged in three major groupings:

1. The standard typewriter keys
2. Ten function keys, F1 to F10
3. 15-key keypad

These 83 keys are shown in Table 18-1. With the introduction of the next generation of PC, IBM added one more key, "Sys Rq", to make a total of 84 keys. The locking shift keys were made more noticeable by providing LED indicators for them. Later, IBM introduced what is called the *advanced* keyboard, known more commonly as the *enhanced* keyboard. The number of keys was increased to 101 for the U.S. market. Tables 18-1, 18-2, and 18-3 provide the scan codes for both the original PC and enhanced keyboards.

In Table 18-1, notice that the same scan code is used for a given lowercase letter and its capital. The same is true for all the keys with dual labels. If the scan code is the same for both of them, how does the system distinguish between them? This is taken care of by the keyboard shift status byte. The data area location 0040:0017H holds the shift status byte. The meaning of each bit is given in Figure 18-3.

The BIOS data area location 0040:0018H holds the second keyboard status byte. The meaning of each bit is given in Figure 18-4. Notice that some of the bits are used for the 101-key enhanced keyboards.

When a key is pressed, the interrupt service routine of INT 9 receives the scan code and stores it in a memory location called a *keyboard buffer*, located in the BIOS data area. However, to relieve programmers from the details of keyboard interaction with the motherboard, IBM has provided INT 16H. We first look at the services provided by the BIOS INT 16H and then we study the details of how the keyboard interacts with the motherboard through hardware INT 09.

Table 18-1: PC Scan Codes for 83 PC Keys

Hex	Key	Hex	Key	Hex	Key	Hex	Key
01	Esc	15	Y and y	29	~ and `	3D	F3
02	! and 1	16	U and u	2A	Left Shift	3E	F4
03	@ and 2	17	I and i	2B	| and \	3F	F5
04	# and 3	18	O and o	2C	Z and z	40	F6
05	$ and 4	19	P and p	2D	X and x	41	F7
06	% and 5	1A	{ and [	2E	C and c	42	F8
07	^ and 6	1B	} and]	2F	V and v	43	F9
08	& and 7	1C	enter	30	B and b	44	F10
09	* and 8	1D	ctrl	31	N and n	45	Num Lock
0A	(and 9	1E	A and a	32	M and m	46	Scroll Lock
0B	) and 0	1F	S and s	33	< and ,	47	7 and Home
0C	and -	20	D and d	34	> and .	48	8 and Up Arrow
0D	+ and =	21	F and f	35	? and /	49	9 and Pg Up
0E	backspace	22	G and g	36	RightShift	4A	- (gray)
0F	tab	23	H and h	37	PrtSc and *	4B	4 and Left Arrow
10	Q and q	24	J and j	38	Alt	4C	5 (keypad)
11	W and w	25	K and k	39	Spacebar	4D	6 and Right Arrow
12	E and e	26	L and l	3A	Caps Lock	4E	+ (gray)
13	R and r	27	: and ;	3B	F1	4F	1 and End
14	T and t	28	" and '	3C	F2	50	2 and Down Arrow
						51	3 and Pg Dn
						52	0 and Ins
						53	. and Del

(Reprinted by permission from "IBM BIOS Technical Reference" c . 1987 by International Business Machines Corporation)

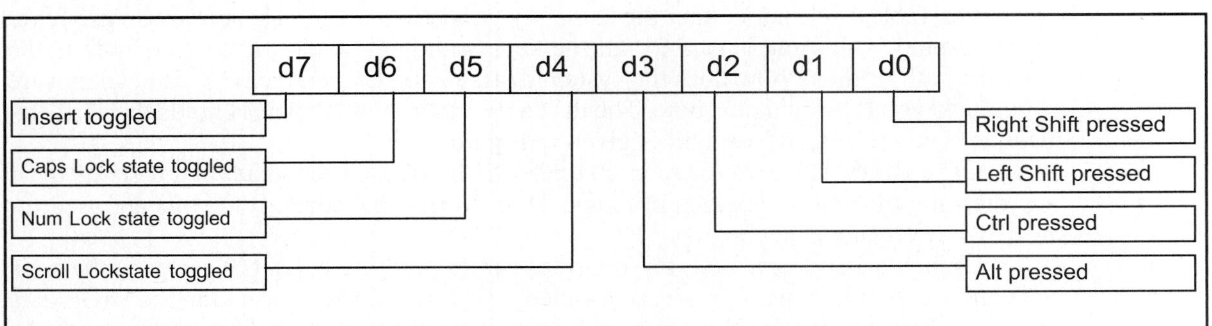

Figure 18-3. First Keyboard Status Byte

Table 18-2: Combination Key Scan Codes

Hex	Keys	Hex	Keys	Hex	Keys	Hex	Keys
54	Shift F1	60	Ctrl F3	6C	Alt F5	78	Alt 1
55	Shift F2	61	Ctrl F4	6D	Alt F6	79	Alt 2
56	Shift F3	62	Ctrl F5	6E	Alt F7	7A	Alt 3
57	Shift F4	63	Ctrl F6	6F	Alt F8	7B	Alt 4
58	Shift F5	64	Ctrl F7	70	Alt F9	7C	Alt 5
59	Shift F6	65	Ctrl F8	71	Alt F10	7D	Alt 6
5A	Shift F7	66	Ctrl F9	72	Ctrl PrtSc	7E	Alt 7
5B	Shift F8	67	Ctrl F10	73	Ctrl Left Arrow	7F	Alt 8
5C	Shift F9	68	Alt F1	74	Ctrl Right Arrow	80	Alt 9
5D	Shift F10	69	Alt F2	75	Ctrl End	81	Alt 10
5E	Ctrl F1	6A	Alt F3	76	Ctrl Pg Dn		
5F	Ctrl F2	6B	Alt F4	77	Ctrl Home		

(Reprinted by permission from "IBM BIOS Technical Reference" c . 1987 by International Business Machines Corporation)

Table 18-3: Extended Keyboard Scan Codes

	Keys	Hex	Keys	Hex	Keys	Hex	Keys
	F11	8E	Ctrl -	97	Alt Home	A0	Alt Down Arrow
	F12	8F	Ctrl 5	98	Alt Up Arrow	A1	Alt Pg Dn
	Shift F11	90	Ctrl +	99	Alt Pg Up	A2	Alt Insert
	Shift F12	91	Ctrl Down Arrow	9A		A3	Alt Delete
	Ctrl F11	92	Ctrl Insert	9B	Alt Left Arrow	A4	Alt /
	Ctrl F12	93	Ctrl Delete	9C		A5	Alt Tab
	Alt F11	94	Ctrl Tab	9D	Alt Right Arrow	A6	Alt Enter
	Alt F12	95	Ctrl /	9E			
	Ctrl UpArrow	96	Ctrl *	9F	Alt End		

(Reprinted by permission from "IBM BIOS Technical Reference" c . 1987 by International Business Machines Corporation)

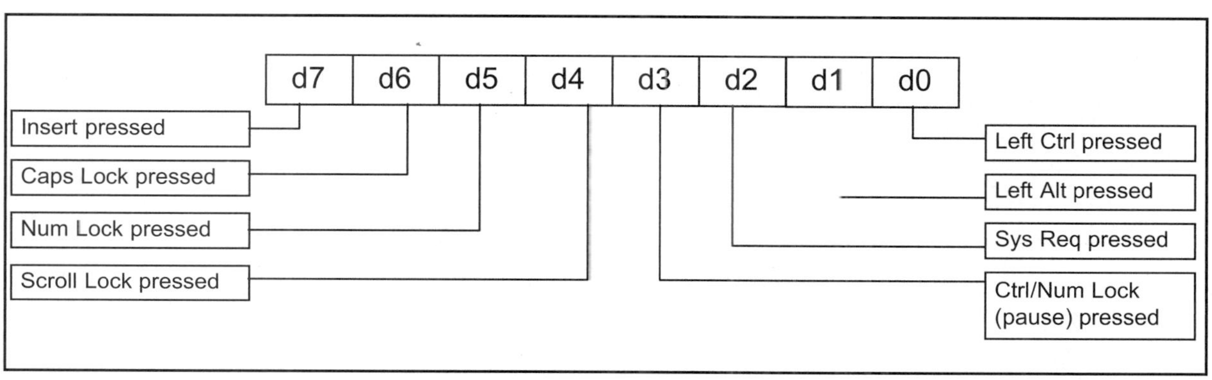

Figure 18-4. Second Keyboard Status Byte

BIOS INT 16H keyboard programming

INT 16H, AH = 0 (read a character)

This option checks the keyboard buffer for a character. If a character is available, it returns its scan code in AH and its ASCII code in AL. If no character is available in the buffer, it waits for a key press and returns it. For characters such as F1–F10 for which there is no ASCII code, it simply provides the scan code in AH and AL = 0. Therefore, if AL = 0, a special function key was pressed. This option simply provides the code for the character and does not display it. See Example 18-2.

Example 18-2

Run the following program in DEBUG. Interpret the result after typing each of the following. Run the program for each separately. (a) Z (b) F1 (c) ALT

```
MOV    AH,0
INT    16H
INT    3
```

Solution:

(a) AX = 2C7A AH = 2C, the scan code, and AL = 7A, ASCII for 'Z'

(b) AX = 3B00 AH = 3B, the scan code for F1, and AL = 00, because F1 is not an ASCII key

(c) Nothing happens because there is no scan code for the Alt key. The status of keys such as Alt is found in the keyboard status byte stored in BIOS at 40:17 and 40:18.

INT 16H, AH = 01 (find if a character is available)

This option, which is similar to the AH = 0 option, checks the keyboard buffer for a character. If a character is available, it returns its scan code in AH and its ASCII code in AL and sets ZF = 0. If no character is available in the buffer, it does not wait for a key press and simply makes ZF = 1 to indicate that.

INT 16H, AH = 02 (return the current keyboard status byte)

This option provides the keyboard status byte in the AL register. The keyboard status byte (also referred to as the *keyboard flag byte*) is located in the BIOS data area memory location 0040:0017H. For the meaning of each bit of the shift status byte, see Figure 18-3 and Example 18-3.

Example 18-3

Run the following program in DEBUG while the right shift key is held down and the Caps Lock key light is on. Verify it also by dumping the 0040:0017 location using DEBUG.

```
MOV    AH,02
INT    16H
INT    3
```

Solution:

Running the program while the Right Shift and Caps Lock keys are activated gives
AH = 41H = 0100 0001 in binary, which can be checked against Figure 18-3. In DEBUG,
-d 0:417 418
will provide the keyboard status byte 41-00.

Due to additional keys on the IBM extended keyboard, BIOS added the following additional services to INT 16H.

INT 16H, AH = 10H (read a character)

This is the same as AH = 0 except that it also accepts the additional keys on the IBM extended (enhanced) keyboard.

INT 16H, AH = 11H (find if a character is available)

This is the same as AH = 1 except that it also accepts the additional keys on the IBM extended (enhanced) keyboard.

INT 16H, AH = 12H (return the current status byte)

This is the same as AH = 2 except that it also provides the shift status byte of the IBM extended (enhanced) keyboard to AH. See Figure 18-5 and Example 18-4.

Example 18-4

Run the following program in DEBUG. Interpret the result after typing each of the following. Run the program for each separately. (a) F11 (b) ALT F11 (c) ALT TAB

```
MOV   AH,10H
INT   16H
INT   3
```

Solution:

After running the program above in DEBUG for each case, we have the following:

(a) AX = 8500, where 85H is the scan code for F11
(b) AX = 8B00, where 8BH is the scan code for Alt-F11
(c) AX = A500, where A5H is the scan code for Alt-Tab

All of the cases above have AL = 00 since there is no ASCII code for these keys.

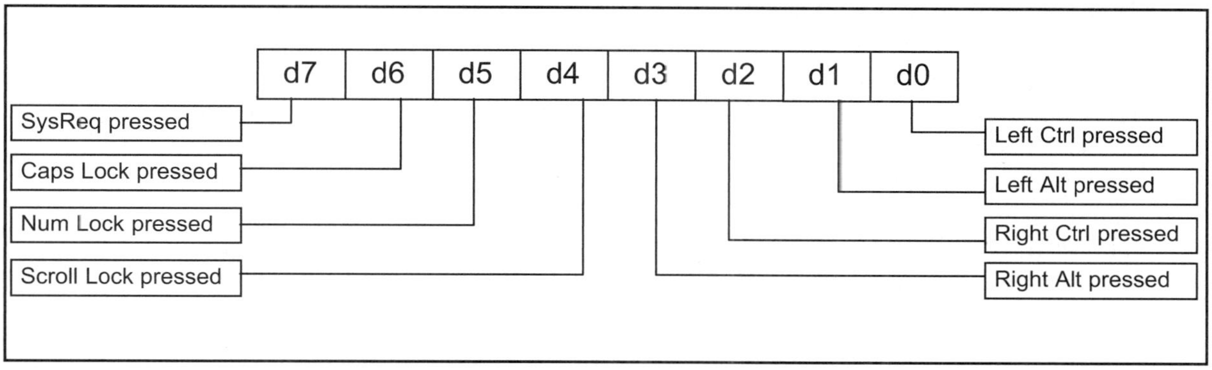

Figure 18-5. Enhanced Keyboard Shift Status Byte

Example 18-5

Write and test a program in DEBUG that increments counter CX whenever Shift-F7 is activated; otherwise, it should exit.

Solution:

The unassembled program in DEBUG follows.

```
-U 100 113
16B7:0100 B402          MOV   AH,02
16B7:0102 CD16          INT   16
16B7:0104 F6C403        TEST  AH,03
16B7:0107 740A          JZ    0113
16B7:0109 B400          MOV   AH,00
16B7:010B CD16          INT   16
16B7:010D 80FC5A        CMP   AH,5A
16B7:0110 7501          JNZ   0113
16B7:0112 41                  INC   CX
16B7:0113 CC                  INT   3
-
```

Hardware INT 09 role in the IBM PC keyboard

To understand fully the principles underlying the IBM PC keyboard, it is necessary to know how INT 09 works. The IBM PC keyboard communicates with the motherboard through hardware interrupt IRQ1 of the 8259. As mentioned in Chapter 14, IRQ1 (INT 09) of the 8259 is used by the keyboard. The way the INT 09 interrupt service works is as follows.

1. The keyboard microcontroller scans the keyboard matrix continuously. When a key is pressed (a make), it is identified and its scan code is sent serially to the motherboard through the keyboard cable (see Figure 18-6). The circuitry on the motherboard receives the serial bits, gets rid of the frame bits, and makes one byte (scan code) with the help of its serial-in-parallel-out shift register, then presents this 8-bit scan code to port A of the 8255 at I/O address of 60H, and finally activates IRQ1.
2. Since IRQ1 is set to INT 09, its interrupt service routine (ISR) residing in BIOS ROM is invoked.
3. The ISR of INT 09 reads the scan code from port 60H.
4. The ISR of INT 09 tests the scan code to see if it belongs to one of the shift keys (Right Shift and Left Shift), Alt, Ctrl keys, and so on. If it does, the appropriate bit of the keyboard status bytes in BIOS memory locations 0040:0017H and 0018H are set. However, it will not write the scan code to the keyboard buffer. If the scan code belongs to any key other than a special key (Shift, Alt, Ctrl, and so on), INT 09 checks to see if there is an ASCII code for the key. If there is one, it will write both the ASCII and scan codes into the keyboard buffer. If there is no ASCII code for the key, it puts 00 in place of ASCII code and the scan code in the keyboard buffer.
5. Before returning from INT 09, the ISR will issue EOI to unmask IRQ1, followed by the IRET instruction. This allows IRQ1 activation to be responded to again.
6. When the key is released (a break), the keyboard generates the second scan code by adding 80H to it and sends it to the motherboard.
7. The ISR of INT 09 checks the scan code to see if there is 80H difference between the last code and this one. This is easy since all it has to do is to test D7 (80H = 10000000 binary). If D7 is high, this is interpreted as meaning that the key has been released and the system ignores the second scan code. However, if the key is held down more than 0.5 seconds, it is interpreted as a new key and INT 09 will write it into the keyboard buffer next to the preceding one. Holding down the key for more than 0.5 seconds is commonly referred to as *typematic* in IBM literature, which means repeating the same key.

From the above steps the following points must be emphasized:

1. The keyboard sends two separate scan codes for make and break to the motherboard.
2. It is the function of the ISR of INT 09 to read the scan code sent by the keyboard and convert it to an ASCII code (if any), then save both the scan code and ASCII code in the keyboard buffer of the motherboard.
3. If any of the special keys, such as Shift, Alt, or Ctrl, is pressed, INT 09 sets the appropriate bits to 1 in the BIOS data area of 0040:0017H and 0018H, but it will not deliver the scan code to the keyboard buffer.
4. If any undefined combination of keys is pressed, INT 09 is activated but it will ignore it since there is no associated scan code. If such key combinations are used by a given program, it is the job of the programmer to intercept them by hooking into INT 09.

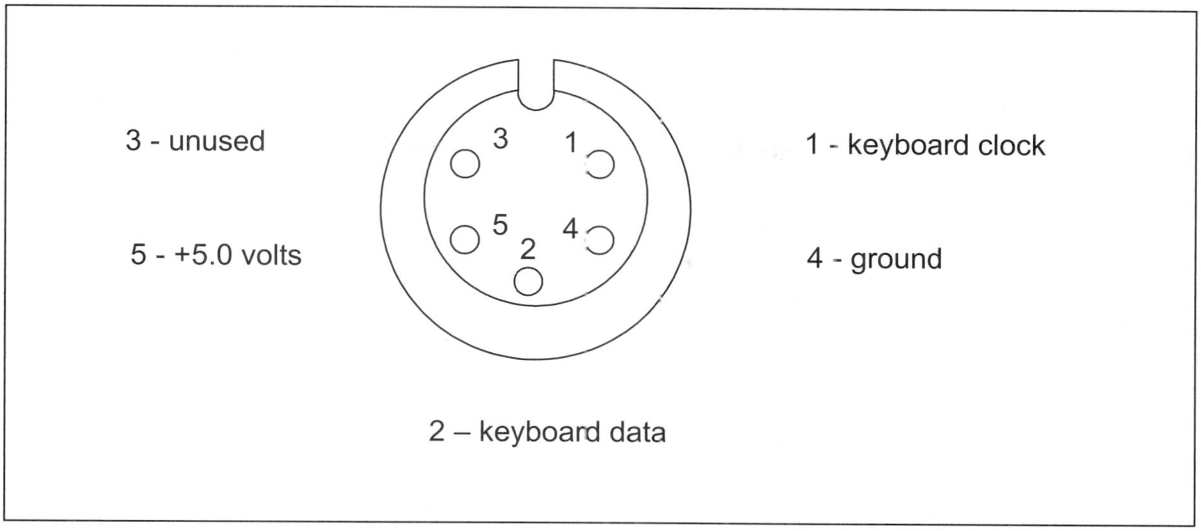

Figure 18-6. Keyboard Cable Jack for the PC

Keyboard overrun

On the keyboard side, the 8042 circuitry must serialize the scan code and send it through the cable to the motherboard. On the motherboard side, there is circuitry responsible for getting the serial data and making a single byte of scan code out of the streams of bits, and holding it for the CPU to read. What happens if the CPU falls behind and cannot keep up with the number of keystrokes? Such a situation is called *keyboard overrun*. The motherboard beeps the speaker when an overrun occurs. The beeping process works as follows. The circuitry on the keyboard has a buffer of its own to store a maximum of 20 key strokes. When this buffer becomes full, it stops receiving keystrokes and sends a special byte called an *overrun* byte (which is FFH in the PC) to the motherboard. After getting the scan code, INT 09 first checks to see if the scan code received is the overrun byte, FFH. If it is, it will sound the speaker; otherwise, it tests for the shift keys and so on, as explained earlier. In other words, the BIOS ROM on the motherboard is responsible for beeping the speaker in the event of keyboard overrun. The following program shows this process. It provides the beginning and ending codes for the INT 09 interrupt service routine, taken from the IBM PC BIOS with some modification for the sake of clarity.

```
;KEYBOARD INT 09 INTERRUPT ROUTINE for PC
KB_INT PROC FAR
      STI                   ;ALLOW FURTHER INTERRUPT
      PUSH   AX             ;SAVE ALL THESE
      PUSH   BX             ;REGISTERS
      PUSH   CX
      PUSH   DX
      PUSH   SI
      PUSH   DI
      PUSH   DS
      PUSH   ES
      ...
      ...
      IN     AL,60H         ;READ IN THE CHARACTER FROM PORT
      ...
      ...
      CMP    AL,0FFH        ;OVERRUN CHARACTER?
      JNZ    K16            ;NO. TEST FOR SHIFT KEY
      JMP    K62            ;SOUND THE BEEPER FOR BUFFER FULL
      ...
      ...
```

```
        CLI                     ;TURN OFF INTERRUPTS
        MOV     AL,20H          ;ISSUE EOI (END OF INTERRUPT) TO 8259
        OUT     20H,AL          ;AT PORT ADDRESS 20H
        POP     ES
        POP     DS
        POP     DI
        POP     SI
        POP     DX
        POP     CX              ;RESTORE ALL THE
        POP     BX              ;REGISTERS
        POP     AX
        IRET                    ;RETURN FROM INTERRUPT
KB_INT ENDP
```

Keyboard buffer in BIOS data area

As mentioned above, the INT 09 interrupt routine gets the scan code from the keyboard and stores it in some memory locations in the BIOS data area. These memory locations are referred to as the *BIOS keyboard buffer*. This keyboard buffer in the BIOS data area should not be confused with the buffer inside the keyboard itself, whose overrun causes the speaker to beep.

If there is an ASCII code, INT 09 also stores the ASCII code for the key in the keyboard buffer; otherwise, it puts 0 there instead. Where this keyboard buffer is located and how it is used by INT 09 are discussed next.

BIOS keyboard buffer

A total of 32 bytes (16 words) of memory in the BIOS data area is set aside for the keyboard buffer. It starts at memory address 40:001EH and goes to 40:003DH, which corresponds to physical addresses 0041EH and 0043DH. Each two consecutive locations are used for a single character, one for the scan code and the other one for the ASCII code (if any) of the character. How does INT 9 know in which word of this 16-word buffer it should put the next character, and how does INT 16H know which of the characters in the keyboard buffer to extract? To answer these questions, we must explain the role of keyboard buffer pointers. There are two keyboard buffer pointers: the head pointer and the tail pointer. See Table 18-4.

Table 18-4: BIOS Data Area Used by Keyboard Buffer

Address of Head Pointer	Address of Tail Pointer	Keyboard Buffer
41A and 41B	41C and 41D	41E to 43D

Tail pointer

Memory locations 0040:001CH and 0040:001DH (physical addresses 0041CH and 0041DH) hold the address for the tail. This means that at any given time, memory locations 0041CH and 0041D hold the address where INT 09 should store the next character. It is the job of INT 09 to put the character in the keyboard buffer and advance the tail by incrementing the word contents of memory location 0041C, where the tail pointer is held.

Head pointer

INT 16H gets the address of where to extract the next character from memory locations 41AH and 41BH, the head pointer. As INT 16H reads each character from the keyboard buffer, it advances the head pointer held by memory locations 41AH and 41BH.

The above discussion can be summarized as follows. As INT 09 inserts the character into the keyboard buffer it advances the tail, and as INT 16H reads the character from the keyboard buffer it advances the head. When they come to the end of the keyboard buffer they both wrap around, creating a ring of 16 words where the head is continuously chasing the tail. This is shown in Figure 18-7.

Notice in Figure 18-7 that if the keyboard buffer is empty, the head address is equal to the tail address. As INT 09 inserts characters into the buffer, the tail is moved. If the buffer is not read by INT 16H, it becomes full, which causes the tail to be right behind the head. Look at Example 18-6.

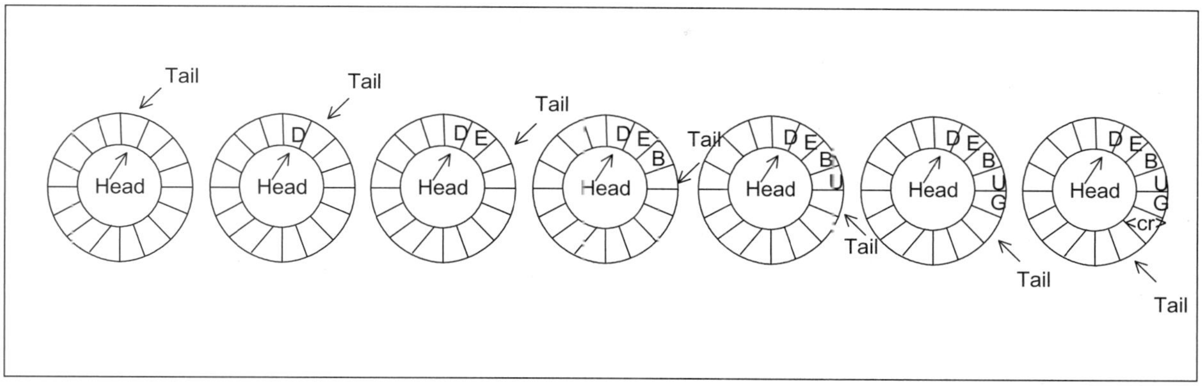

Figure 18-7. Keyboard Buffer Head and Tail

Example 18-6

Using DEBUG, dump the location where the head and tail pointers are held. Compare them. Is the buffer full or empty?

Solution:

```
C>DEBUG
-D 0:410 41F
0000:0410 63 44 F0 80 02 00 01 40-00 00 3C 00 3C 00 20 39 cD.....@..<.<. 9
-
```

In this case, the head and tail pointers point to the same location; therefore, the buffer is empty.

PC keyboard technology

The kind of keyboard shown in Figure 18-1 is referred to as a hard contact keyboard. When a key is pressed, a physical contact between the row and column causes the column to be pulled to ground. Although these kinds of keyboards are cheaper to make, they have the disadvantage of deteriorating at the contact points of rows and columns, eventually becoming too stiff to use, due to metal oxidation of the contact points. The alternative to hard contact keyboards are capacitive keyboards. In such keyboards there is no physical contact between the rows and columns; instead, there is a capacitor for each point of the matrix. In capacitive keyboards when a key is pressed, the change in capacitance is detected by a sense amplifier and produces the logic level, indicating that a key has been pressed. Capacitive keyboards last much longer than do hard contact keyboards. All IBM PC and PS machines use capacitive keyboards except the now abandoned PC Jr. models, which used hard contact keyboards.

Review Questions

1. Show the bits transferred from the keyboard to the motherboard when "J" is pressed.
2. How does the PC recognize the difference between the key press and key release?
3. Find the make and break scan codes for the letter "X" in both hex and binary.

4. True or false. The CPU is notified through INT 09 only for key press, not for key release.
5. Does the Alt key have a scan code? If yes, what is it?
6. True or false. The CPU stores the scan code for the right SHIFT in the buffer.
7. Find the contents of the keyboard status byte if CapsLock and Alt are pressed.
8. True or false. INT 09 is responsible for finding the ASCII code for a given key if there is one.
9. True or false. The keyboard buffer holds both the scan code and the ASCII code for a given key.
10. True or false. The beep sound indicates that the BIOS keyboard buffer is full.
11. What does it mean when the head and tail pointers have the same values?
12. As INT 09 puts the scan code into keyboard buffer it advances the _________ (tail, head) pointer.

SECTION 18.3: PRINTER AND PRINTER INTERFACING IN THE IBM PC

In this section we describe the standard printer interface, called the *Centronics printer interface*. Then we study the IBM PC printer interfacing and provide some examples of printer programming using BIOS INT 17H.

Centronics printer interface pins

The Centronics type parallel printer interface is the printer interface standard in the x86 PC. It is also referred to as Epson FX 100 standard. It is a 36-pin interface connector where the pins are labeled as 1 to 36. Many of the 36 pins are used for ground, allowing many signals to have their own ground return lines, which reduces electrical noise. The 36 pins can be grouped as follows.

1. The data lines, which carry the data sent by the PC to the printer.
2. Printer status signals, which indicate the status of the printer at any given time.
3. Printer control signals, which are used to tell the printer what to do.
4. Ground signals, which provide an individual ground return line for each data line and for certain control and status lines.

Table 18-5: DB-25 Printer Pins

Pin	Description
1	$\overline{\text{Strobe}}$
2	Data bit 0
3	Data bit 1
4	Data bit 2
5	Data bit 3
6	Data bit 4
7	Data bit 5
8	Data bit 6
9	Data bit 7
10	$\overline{\text{Acknowledge}}$
11	Busy
12	Out of paper
13	Select
14	$\overline{\text{Auto feed}}$
15	$\overline{\text{Error}}$
16	Initialize printer
17	$\overline{\text{Select input}}$
18	Ground
19	Ground
20	Ground
21	Ground
22	Ground
23	Ground
24	Ground
25	Ground

Data lines and grounds

Input pins DATA 1 to DATA 8 provide a parallel pathway for 8-bit data sent by the PC to the printer. Table 18-5 describes the DB-25 printer pins. Notice in Table 18-6 that pins 20 to 28 are used for individual ground return lines, one for each data pin. Figure 18-8 shows the connector.

Printer status signals

These are all output pins from the printer to the PC used by the printer to indicate its own status. They are as follows.

PE (pin 12) is used by the printer to indicate that it is out of paper.

BUSY (pin 11) is high if the printer is not ready to accept a new character. This pin is high when the printer is off line or when it is printing and cannot accept any data. The PC monitors this pin continuously and as long as this pin is high, it will not transfer data to the printer.

Table 18-6: Centronics Printer Specification

Serial #	Return #	Signal	Direction	Description
1	19	$\overline{\text{STROBE}}$	IN	STROBE pulse to read data in. Pulse width must be more than 0.5 μs at receiving terminal. The signal level is normally "high"; read -in of data is performed at the "low" level of this signal
2	20	DATA 1	IN	These signals represent information of the 1st to 8th bits of parallel data, respectively. Each signal is at "high" level when data is logical "1", and "low" when logical "0".
3	21	DATA 2	IN	
4	22	DATA 3	IN	
5	23	DATA 4	IN	
6	24	DATA 5	IN	
7	25	DATA 6	IN	
8	26	DATA 7	IN	
9	27	DATA 8	IN	
10	28	$\overline{\text{ACKNLG}}$	OUT	Approximately 0.5 μs pulse; "low" indicates data has been received and printer is ready for data
11	29	BUSY	OUT	A "high" signal indicates that the printer cannot receive data. The signal becomes "high" in the following cases : (1) during data entry, (2) during printing operation, (3) in "off-line" status, (4) during printer error status.
12	30	PE	OUT	A "high" signal indicates that printer is out of paper .
13	--	SLCT	OUT	Indicates that the printer is in the state selected .
14	--	AUTOFEEDXT	IN	With this signal being at "low" level , the paper is fed automatically one line after printing . (The signal level can be fixed to "low" with DIP SW pin 2-3 provided on the control circuit board.)
15	--	NC		Not used.
16	--	0V		Logic GND level.
17	--	CHASISGND	--	Printer chassis GND. In the printer, chassis GND and the logic GND are isolated from each other.
18	--	NC	--	Not used.
19 - 30	--	GND	--	"Twisted-pair return" signal'; GND level.
31	--	$\overline{\text{INIT}}$	IN	When this signal becomes "low" the printer controller is reset to its initial state and the print buffer is cleared . Normally at "high" level; its pulse width must be more than 50 μs at receiving terminal.
32	--	$\overline{\text{ERROR}}$	OUT	The level of this signal becomes "low" when printer is in "paper end" , "off-line" and "error" state.
33	--	GND	--	Same as with pin numbers 19 to 30.
34	--	NC	--	Not used.
35	--			Pulled up to +5 v dc through 4.7 K ohms resistance.
36	--	SLCT IN	IN	Data entry to the printer is possible only when the level of this signal is "low" . (Internal fixing can be carried out with DIP SW 1-8. The condition at the time of shipment is set "low" for this signal.)

(Reprinted by permission from "IBM Technical Reference Options and Adapters" c . 1981 by International Business Machines) Corporation)

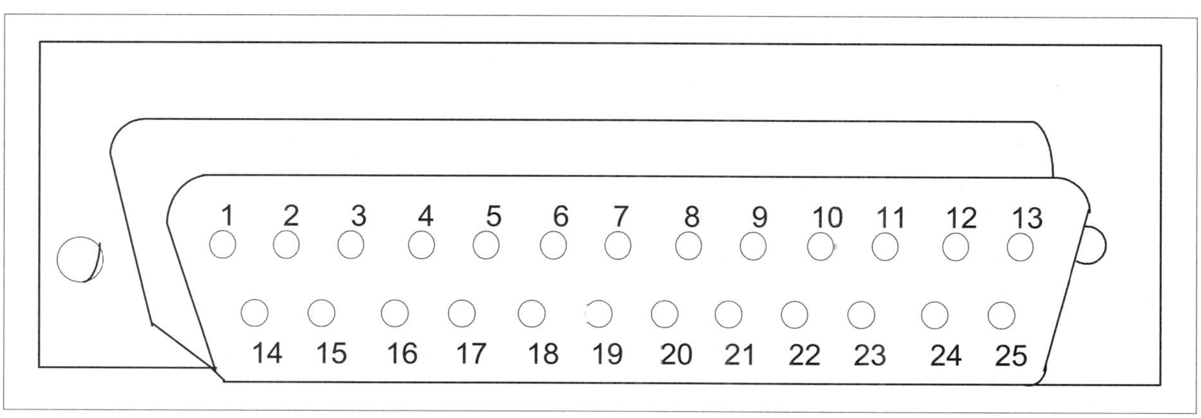

Figure 18-8. DB-25 (Male) Printer Connector

ERROR (pin 32) is normally a high output and is activated (goes low) when there are conditions such as out of paper, off line state, or jammed printhead in which the printer cannot print.

SLCT (pin 13) is active high and goes from the printer to the PC when the printer is turned on and online, indicating that the printer is being selected.

ACKNLG (pin 10) is used by the printer to acknowledge receipt of data and that it can accept a new character.

Printer control signals

STROBE (pin 1) and ACKNLG are the most widely used signals among control and status pins. When the PC presents a character to the data pins of the printer, it activates the STROBE pin of the printer, telling it that there is a byte sitting at the data pins. When the printer picks up the data and is ready for another byte, it sends back the ACKNLG signal. While the STROBE is used by the CPU to tell the printer that there is a byte of data, it is the printer that must acknowledge the data receipt and its readiness for accepting another byte through the ACKNLG line. The ACKNLG signal can be used by the CPU to go and get another byte of data to be presented to the printer.

INIT (pin 31) is an input into the printer and is normally high. When it is activated (active low) it resets the printer. Upon receiving this signal, the printer goes through a sequence of internal initialization, including clearing its own internal buffer.

There are two other control signals in the printer: AUTO FEED XT and SLCT IN. See Table 18-6 for their descriptions. The following are the steps in computer and printer communication.

1. The computer checks to see if a BUSY signal from the printer indicates that the printer is ready (not busy).
2. The computer puts 8-bit data on the data line connected to the printer data pins.
3. The computer activates the STROBE pin by making it low. Prior to asserting the printer input STROBE pin, the data must be at the printer's data pins at least for 0.5 µs. This is data setup time.
4. The STROBE must stay low for at least 0.5 µs before the computer brings it back to high. The data must stay at the printer's data pins at least 0.5 µs after the STROBE pin is deasserted (brought back to high).
5. The activation of STROBE causes the printer to assert its BUSY output pin high, indicating to the computer to wait until it finishes taking care of the last byte.
6. When the printer is ready to accept another byte, it sends the ACKNLG signal back to the computer by making it low. The printer keeps the ACKNLG signal low only for 5 µs. At the rising edge of ACKNLG, the printer makes the BUSY (not BUSY = ready) pin low to indicate that it is ready to accept the next byte.

The CPU can use either the ACKNLDG or BUSY signals from the printer to initiate the process of sending another byte to printer. Some systems use BUSY and some use ACKNLG.

IBM PC printer interfacing

In the IBM PC, the POST (power-on self-test) portion of BIOS is programmed to check for printers connected to parallel ports. As they are identified, the base I/O port address of each is written into the BIOS data area 0040:0008 to 0040:000FH just like the COM port discussed in Chapter 17. A total of 8 bytes of memory in the BIOS data area can store the base I/O address of four printers, each taking 2 bytes.

Memory locations 00408 to 0040FH can be checked to see which LPT (line printer) port is available. Memory locations 0040:0008H and 0040:0009H (physical locations 00408H and 00409H) hold the base I/O address of LPT1, and so on, as shown in Table 18-7. If no printer port is available, 0s are found.

Table 18-7: BIOS I/O Base Addresses for LPT

I/O Base Address	LPT
0040:0008 – 0040:0009	LPT1
0040:000A – 0040:000B	LPT2
0040:000C – 0040:000D	LPT3
0040:000E – 0040:000F	LPT4

It must be emphasized that the base I/O port addresses assigned to LPTs can vary from system to system. This is due to the fact that the POST (power-on self-test) portion of BIOS will check for the existence of a printer port first at I/O address 03BCH, then at 0378H, and finally, at 0278H. Whichever is found first will be written into BIOS data area 408H, where the base I/O address for LPT1 is expected; the second one found is written to the 40AH address for LPT2; and so on.

Table 18-8: IBM PC Printer Ports and Their Functions

Line Printer	Data Port (R/W)	Status Port (Read Only)	Control Port (R/W)
LPT 1	03BCH	03BDH	03BEH
LPT 2	0378H	0379H	037AH
LPT 3	0278H	0279H	027AH

Printer interfacing circuitry uses only three I/O ports starting at the base address: one I/O port for the LPT's data lines, one for the LPT's status lines, and one for the LPT's control lines. For example, if the base I/O port address for LPT1 is 378H, the I/O port address 378H is used for the data, 379H for the status, and 37AH for the control signals. See Table 18-8 for the assignments. Also see Example 18-7.

Figure 18-9 shows the printer's data, status, and control ports.

Example 18-7

Using DEBUG, determine which printer port(s) are available.

Solution:

```
C:\>DEBUG
-D 40:08 L8
0040:0008   78 03 00 00 00 00 00 00        x.......
```

This shows that the base I/O address of LPT1 is 0378H. No other printers are connected to the parallel ports.

What is printer time-out?

Occasionally, the printer time-out message will appear on the screen. This means that the printer port is installed but the printer is not ready to print. This could be due to the fact that the printer is turned off, the printer is not on line, or some other condition in which the printer is connected to the PC but not ready to print. Upon detecting that the printer port is installed, BIOS tries repeatedly for a period of 20 seconds to see if it is ready to accept data. If the printer is not ready, the PC gives up (time-out) and displays a message to indicate that. Can the PC be forced not to give up so soon and try a little bit longer? The answer is yes. The amount of time that BIOS tries to get a response from the printer is stored in BIOS data area 0040:0078 to 0040:007B. Location 0040:0078 holds the time-out time for LPT1, 0040:0079 the time for LPT2, and so on. At boot time, these locations are initialized to 20 seconds.

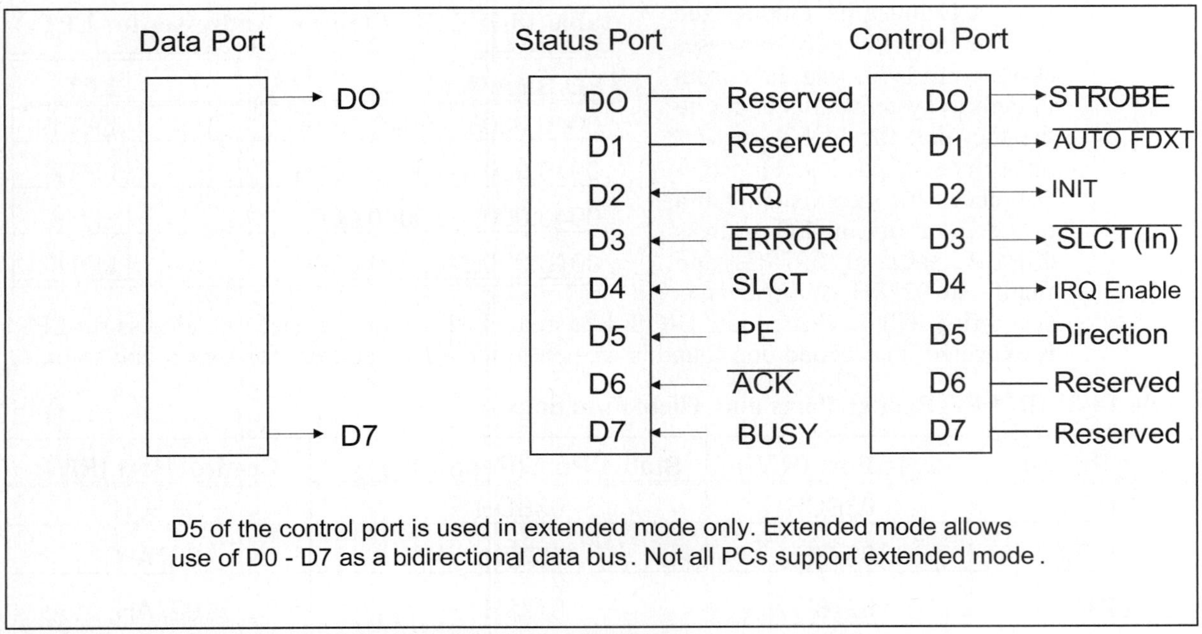

Figure 18-9. Printer's Data, Status, and Control Ports

ASCII control characters

Certain characters in ASCII are used to control the printer. Table 18-9 shows the most commonly used printer control characters in ASCII.

Table 18-9: ASCII Printer Control Characters

ASCII Symbol	Hex Code	Function
BS	08	Backspace
HT	09	Horizontal tab
LF	0A	Line feed (advances one line)
VT	0B	Vertical tab
FF	0C	Form feed (advances to next Page)
CR	0D	Carriage return (return to left margin)

Programming the IBM PC printer with BIOS INT 17H

BIOS INT 17H provides three services: printing a character, initializing the printer port, and getting the printer status port. These options are selected according to the value set in the AH register. This is described as follows.

INT 17H, AH = 0 (print a character)

If this option is selected, INT 17H expects to have the LPT number in register DX (0 for LPT1, 1 for LPT2, and 2 for LPT3) and the ASCII character to be printed in the AL register. Upon return, INT 17H provides the status of the selected printer port as follows.

Bit No. Function
71 = Not BUSY(ready), 0 = BUSY
61 = Acknowledge
51 = Out of paper
41 = Printer selected
31 = I/O error
2,1 unused
01 = Printer time-out

See Example 18-9.

INT 17H, AH = 02 (get the printer port status)

This option allows a programmer to check the status of the printer. Before calling the function, AH is set to 2 and DX holds the printer number (0 = LPT1, 1 = LPT2, and 2 = LPT3). After calling, AH = status, the situation is the same as shown under option 0. See Example 18-10.

Example 18-10

Run the following program in DEBUG to check the LPT1 printer states. Run it once with the printer off line, then run it again with the printer on line. Interpret the AH register upon return.

```
        MOV    AH,2
        MOV    DX,0
        INT    17H
```

Solution:

```
C:\>DEBUG
-A
16B7:0100 MOV AH,2
16B7:0102 MOV DX,0
16B7:0105 INT 17
16B7:0107 INT 3
16B7:0108
-G
AX=0800  BX=0000  CX=0000  DX=0000  SP=CFDE  BP=0000  SI=0000  DI=0000
DS=16B7  ES=16B7  SS=16B7  CS=16B7  IP=0107   NV UP DI PL NZ NA PO NC
16B7:0107 CC             INT     3
-G=100
AX=9000  BX=0000  CX=0000  DX=0000  SP=CFDE  BP=0000  SI=0000  DI=0000
DS=16B7  ES=16B7  SS=16B7  CS=16B7  IP=0107   NV UP DI PL NZ NA PO NC
16B7:0107 CC             INT     3
-Q
```

The first execution of the program occurred when the printer was off line. It returned AH = 08. This indicates an I/O error. The program was run again with the printer on line. This time it returned AH = 90, which indicates "not busy."

Inner working of BIOS INT 17H for printing a character

Below is a listing of a portion of BIOS INT 17H with some modification for the sake of clarity. It shows how a character is printed by monitoring the BUSY signal from the printer and issuing a STROBE. Figure 18-10 diagrams printer timing.

```
;This skeleton of BIOS INT 17H shows how a character is issued
; to printer is taken from
;the IBM PC Technical Reference. The instructions not shown here
; include:
;(a) Loading DX with base  I/O address of printer port from BIOS
; data area 0040:0008 000F
;Reminder: The base I/O address is the address of the printer
; data bus.
;(b) Loading of time out value into BL reg from BIOS data area
; 0040:0078H 007B
;Therefore we have the following upon going into this portion of
; BIOS,
;BL=time out value
;DX=has the base I/O address, which is LPT's data port
;AL=character to be printed

        OUT    DX,AL         ;OUTPUT CHARACTER TO BE PRINTED
        INC    DX            ;POINT TO STATUS PORT
```

```
B3:     SUB    CX,CX          ;TIMER VALUE FOR BUSY
B3_1:   IN     AL,DX          ;GET STATUS
        MOV    AH,AL          ;SAVE IT IN AH
        TEST   AL,80H ;IS BUSY LINE HIGH? (SEE FIG. 18-9, D7 = BUSY)
        JNZ    B4             ;IF READY THEN OUTPUT THE STROBE
        LOOP   B3_1           ;TRY AGAIN
        DEC    BL             ;DROP LOOP COUNT
        JNZ    B3             ;GO UNTIL TIME OUT ENDS
        OR     AH,01          ;SET ERROR FLAG
        AND    AH,0F9H        ;TURN OFF OTHER BITS
.....
        IRET                  ;RETURN WITH ERROR FLAG BIT SET
....
B4:     MOV    AL,0DH         ;SET STROBE HIGH
        INCDX                 ;DX=I/O PRINTER CNTL REG
        OUTDX,AL              ;STROBE IS BIT 0 OF CNTR REG
        MOVAL,0CH             ;SET STROBE LOW
        OUTDX,AL        ;AND SEND IT TO PRINTER CONTROL PORT
```

Notice the steps taken to print a character in the above listing.

1. Send the character to the D7–D0 latch connected to the data pins of the printer.
2. Test to see if BUSY is low (NOT BUSY). If ready (NOT BUSY), issue the STROBE to ask the printer to grab the data by making STROBE = high and then STROBE = low.
3. If the printer is BUSY, try again until the time-out is finished. The time-out forces the CPU to check the printer for a period of time before it gives up.
4. If for whatever reason the printer does not respond after trying repeatedly, go back and set the time-out bit to indicate that.

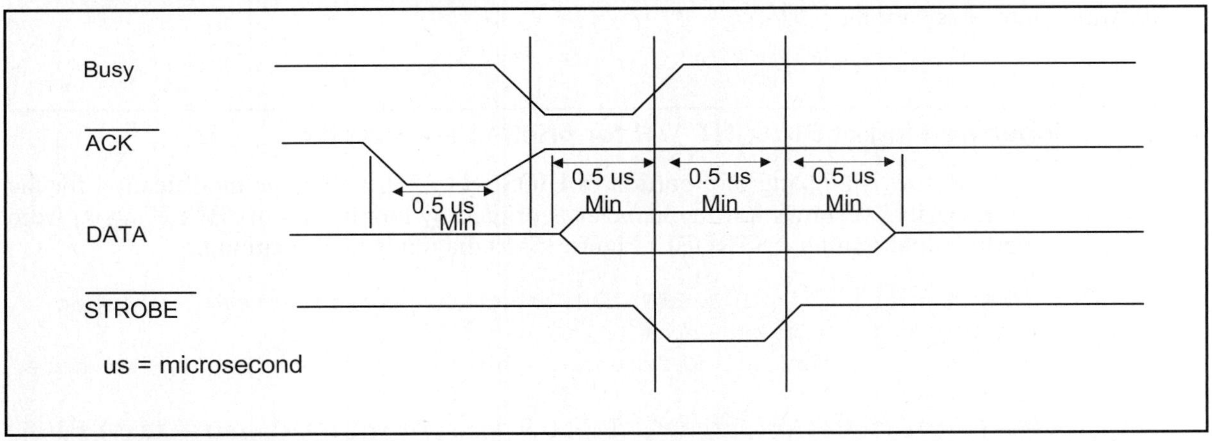

Figure 18-10. Printer Timing

Bidirectional LPT port

Since the introduction of the first IBM PC in 1981, the PC's parallel port has gone through various changes. Next we give an overview of SPP, PS/2, EPP, and ECP parallel port types and provide some parallel port interfacing tips. First, we discuss the characteristics of each parallel port type.

SPP

SPP stands for standard parallel port. This refers to the parallel port of the first IBM PC introduced in 1981. The data bus in SPP is unidirectional and is designed to send data from the PC to the printer. At that time, designers never thought that someone might want to use the LPT's data bus for input. In SPP, the internal logic circuitry is set for data output only and any attempt to use the data bus for input can damage the LPT port. For this reason you should never try to modify the LPT port unless you know what you are doing. Some designers use the status and control port of the SPP to send data in. In such cases, the pull-up resistors are used to prevent damage to the LPT's parallel port. For further information, refer to web page http://www.lvr.com.

PS/2

The first change in the data bus portion of the LPT port occurred in 1987 with the introduction of PS/2 computers by IBM. By then, designers had seen the potential use of parallel ports for fast data acquisition. Therefore, internal circuitry of the data section of the LPT port in the x86 PC was changed to make it bidirectional. However, upon bootup, BIOS configured the LPT port as SPP, meaning that it was to be used only for data output. At the same time, the C5 bit of the control port (base +2) was modified to allow the user to change the data port direction. At bootup, C5 is low (C5 = 0) meaning that the data port is for output. By making C5 = 1, we can make the data port an input port. Recall from the last section that control port C0–C4 was already used by the SPP. Therefore, in the LPT port, C5 of the control port is used for data port direction, while C6 and C7 are reserved.

Example 18-11

Assume that the I/O base address = 278H for LPT2 in a PS/2-compatible PC. Show how to change the control bit C5 to make the data port an input port.

Solution:

The I/O base address = 278H is for the data port. This means that we have 279H for the status port and 27AH for the control port.

```
MOV    DX,27AH          ;DX=control port address
IN     AL,DX            ;get the current information
OR     AL,00100000      ;make C5=1 without changing anything else
OUT    DX,AL            ;now data port is an input port
```

Examine the I/O addresses 278H, 279H, 27AH in Appendix E. It says that the data port is RW (read/write), the status port is RO (read only), and the control port is RW.

How to detect a bidirectional data bus

The following are steps in detecting if your LPT data port is bidirectional.

1. Put the data port in bidirectional mode by writing 1 to C5 of the control port (C5 = 1).
2. Write a known value (such as 55H, AAH, or 99H) to the data port.
3. Read back the value from the data port.
4. If the read value matches the value written to the data port, the data port is not bidirectional.

By making the data port bidirectional in the PS/2, IBM set a new standard, allowing many devices such as tape backup, scanners, and data acquisition instruments to use the LPT port instead of a PC expansion slot. However, there is one problem with LPT ports: They are too slow. This slowness led Intel and Xircom, along with other companies, to set a new LPT standard called EPP.

EPP

EPP stands for *enhanced parallel port*. It is the same, but much faster. Recall from Section 18.3 that handshaking signals such as the strobe signal are generated by software. In EPP, a higher speed was achieved by delegating the handshaking signals to the hardware circuitry on the LPT port itself. The EPP standard also added new registers to the I/O port address space beyond base address +2. In EPP, the I/O space goes from base to base+7. For example, if the base address is 278H, 279H and 27AH are the same as SPP. However, I/O addresses 27BH through 27FH are also used or reserved.

ECP

ECP stands for *extended capability port*. The need for an even faster LPT port led to ECP. The ECP has all the features of EPP plus DMA (direct memory address) capability, allowing it to transfer data via the DMA channel. It also has data compression capability. The DMA and data compression capabilities make ECP an ideal port for high-speed peripherals such as laser printers and scanners. This is the reason that Hewlett Packard joined with Microsoft in developing the ECP standard. While the ECP type LPT port is supposed to support SPP, PS/2, and EPP, not all of them can emulate EPP.

To see if a given ECP supports EPP, examine the PC technical documentation or check the CMOS setup on your PC.

In order to unify these various types of LPT ports, a committee of the IEEE has put together specification IEEE 1284. Refer to the IEEE web page: http://www.ieee.org.

Review Questions

1. The Centronics printer standard uses __________ (serial, parallel) data transfer.
2. Give one reason why there are 8 bits for the data lines in the Centronics standard.
3. The status signals of the printer are _____ (in, out) for the printer and ________ (in, out) for the computer.
4. The control signals of the printer are ____ (in, out) for the printer and ________ (in, out) for the computer.
5. STROBE is an ____ (in, out) signal for the printer and ____ (in, out) for the computer.
6. ACKNLG is an ____ (in, out) signal for the printer and ____ (in, out) for the computer.
7. D1–D8 are _____ (in, out) signals for the printer and ____ (in, out) for the computer.
8. BUSY is an ____ (in, out) signal for the printer and ____ (in, out) for the computer.
9. Out of paper is an ____ (in, out) signal for the printer and ____ (in, out) for the computer.
10. How does the computer know if the printer got the last byte sent and is ready for the next one?
11. State the role and level of activation for the STROBE signal.
12. If the base I/O address of a given LPT is 3BCH, give the I/O address for each of the following lines of the printer.
    (a) control (b) status (c) data
13. Assuming that the I/O base address of LPT1 is 378H, show a simple Assembly language program that monitors the BUSY line of the printer.
14. What is a time-out in IBM PC terminology?
15. Give the ASCII codes for the carriage return and line feed in hex.

PROBLEMS

SECTION 18.1: INTERFACING THE KEYBOARD TO THE CPU

1. In reading the columns of a keyboard matrix, if no key is pressed we should get all ______ (1s, 0s).
2. In Figure 18-1, to detect the key press, which of the following is grounded? (a) all rows (b) one row at time (c) both (a) and (b)
3. In Figure 18-1, to identify the key pressed, which of the following is grounded? (a) all rows (b) one row at time (c) both (a) and (b)
4. For Figure 18-1, indicate the column and row for each of the following. (a) D3–D0 = 0111 (b) D3–D0 = 1110
5. Indicate the steps to detect the key press.
6. Indicate the steps to identify the key pressed
7. Modify Program 18-1 and Figure 18-1 for a 4 × 5 keyboard (4 rows, 5 columns).
8. Modify Program 18-1 and Figure 18-1 for a 5 × 6 keyboard.
9. Indicate an advantage and a disadvantage of using an IC chip for keyboard scanning and decoding instead of using a microprocessor.
10. What is the best compromise for the answer to Problem 9?

SECTION 18.2: PC KEYBOARD INTERFACING AND PROGRAMMING

11. In the IBM PC for each key press (make), _____ bits are transferred to the main CPU. What are these bits?
12. Find the break code for the following make codes.
    (a) 34H (b) 1AH (c) 5FH
13. Identify make and break among the following codes.
    (a) 9BH (b) 89H (c) 17H (d) C2H (e) 79H
14. Since keys "5" and "%" have the same scan code, how are they distinguished?
15. Find the scan code for the following.
    (a) ALT F2 (b) SHIFT F4 (c) &
    (d) V (e) Pg Up (f) F6
16. Which option of INT 16H is used to get the status bytes of enhanced keyboards?
17. Write a program to display a prompt such as "I will play for you the 'Happy Birthday' music if you guess the key I am thinking of. The key is one of the ALT Fs." Using INT 16H to monitor the scan codes continuously, if Alt F9 is activated the PC should play "Happy Birthday" and exit to DOS (for "Happy Birthday" music, see Chapter 13). If any other key is pressed it should display a message such as "Try again" and continue. The Esc key should exit to DOS.
18. INT 09 is assigned to which IRQ of the 8259?
19. True or false. INT 09 is activated for both the make and break scan codes.
20. True or false. If CapsLock is pressed, INT 09 saves it in the keyboard buffer.
21. What value does the BIOS keyboard subroutine save in the keyboard buffer if there is no ASCII code for a given key?
22. When there is keyboard overrun, which of the following generates the sound beep: the circuitry inside the keyboard or the motherboard?
23. The keyboard shift status byte indicates the status of which keys?
24. Which of the following keys are non-ASCII keys?
    (a) HOME (b) ! (c) Arrow (d) *
25. Give the content (in hex and binary) of the keyboard shift status byte if only the NumLock and CapsLock are on.
26. If the content of the first keyboard shift byte is 10000001, what does it mean?
27. Give the physical address of memory locations in the BIOS data area set aside for each of the following keyboard components.
    (a) shift status byte (the first one) (b) buffer
    (c) buffer's tail address (d) buffer's head address

28. When the addresses of the head and tail are the same, what does it mean?
29. The key buffer is ______ (empty, full) if the address of the tail is one number higher than the address of the head.
30. What keyboard technology is used in IBM enhanced keyboards?

SECTION 18.3: PRINTER AND PRINTER INTERFACING IN THE IBM PC

31. State the four categories of Centronics printer pins.
32. Of the following pins, which belongs to the printer's status signal and which belongs to the printer's control signal categories? Indicate which is input and which is output from the point of view of the PC.
    (a) BUSY
    (b) STROBE
    (c) ACKNLG
    (d) SLCT
    (e) INIT
    (f) PE
33. Which pin is used by the printer to indicate that it is out of paper?
34. True or false. Each data line has its own ground return line.
35. What is the function of the BUSY signal?
36. In response to STROBE, the printer makes ACKNLG ________ (low, high).
37. When does the BUSY signal go low?
38. True or false. The base I/O address for the printer port can be 378H or 3BCH.

ANSWERS TO REVIEW QUESTIONS

SECTION 18.1: INTERFACING THE KEYBOARD TO THE CPU

1. True
2. Column 3
3. True
4. 0
5. True

SECTION 18.2: PC KEYBOARD INTERFACING AND PROGRAMMING

1. The scan code is 24H. This has the odd parity bit of 1; therefore, the following bits are transferred from the keyboard to the motherboard: 1 1 00100100 0.
2. The scan code for break is always 80H larger than the scan code for make.
3. 2DH = 00101101 and ADH = 10101101
4. False, for both
5. Yes, it is 38H.
6. False
7. 48H
8. True
9. True
10. False; the buffer inside the keyboard is full.
11. The keyboard buffer on the motherboard is empty.
12. Tail

SECTION 18.3: PRINTER AND PRINTER INTERFACING IN THE IBM PC

1. Parallel
2. Since the characters are 8-bit ASCII code
3. Out, in
4. In, out

5. In, out
6. Out, in
7. In, out
8. Out, in
9. Out, in
10. Through the ACKNLG or BUSY signals; either one can be used.
11. It must be high normally. When the computer has a byte of data for printer it makes it go low to inform the printer.
12. (a) 3BEH for control (b) 3BDH for status (c) 3BCH for data
13.

```
          MOV    DX,379H            ;LPT1 STATUS PORT ADDRESS
A1:              IN     AL,DX       ;GET THE LPT1 STATUS
          TES    TAL,80H            ;IS D7=1 (BUSY SIGNAL)
          JNZ...                    ;NOT BUSY
          JMP    A1                 ;TRY AGAIN
```

14. When the PC tests the printer status port and cannot get any response from the printer, it tries again for a certain time period; then if it does not get any response, it sets the time-out bit.
15. 0DH and 0AH

CHAPTER 19

HARD DISKS

OBJECTIVES

Upon completion of this chapter, you will be able to:

- **>> Contrast and compare the terms primary storage and secondary storage**
- **>> Discuss hard disk organization in terms of the boot record, FAT, and the directory**
- **>> Analyze the capacity of hard disks in terms of sectors, tracks, clusters, cylinders, and platters**
- **>> Define hard disk terminology: partitioning, interleaving, low-level and high-level formatting, parking the head, and MTBF**
- **>> Define the components of hard disk access time: seek time, settling time, and latency time**

This chapter will examine the characteristics of hard disk storage. The chapter includes a discussion of data encoding techniques, interfacing standards, and definitions of hard disk terminology. Hard disk organization, characteristics, and terminology are covered in this chapter.

SECTION 19.1: HARD DISK ORGANIZATION AND PERFORMANCE

In the early days of the personal computer, cassette tape was used to store information. Due to its long access time, it was abandoned as a secondary storage medium. The term *secondary storage* refers to memory other than RAM. RAM is called *primary storage* since the CPU asks for the information that it needs from RAM first. This section will look at the characteristics of hard disks and their organization with emphasis on performance factors such as access time and interfacing standards. The hard disk, referred to sometimes as *fixed* disk, or *Winchester* disk in IBM literature, is judged according to three major criteria: capacity, access time (speed of accessing data), and interfacing standard. Before delving into each category, an explanation should be given for the use of different names such as fixed disk, Winchester disk, and hard disk to refer to the same device. The term hard disk comes from the fact that it uses hard solid metal platters to store information instead of plastic as is the case in floppy disks. It is also called fixed disk because it is mounted (fixed) at a place on the computer and is not portable like the floppy disk (although some manufacturers make removable hard disks). Why is it also called the Winchester disk? When IBM made the first hard disk for mainframes it was capable of storing 30 megabytes on each side and therefore was called a 30/30 disk. The 30/30 began to be called the Winchester 30/30, after the rifle, and soon it came to be known simply as the Winchester disk.

Capacity of the disk

In order to store data on the disk, both sides are coated with magnetic materials. The principles behind the process of reading and writing (storing) digital data on disks is the same as that used in any magnetic-based medium. Each side of the disk is organized into tracks and sectors as shown in Figure 19-1. Tracks are organized as concentric circles and their number per disk varies from disk to disk, depending on the size and technology. Each track is divided into a number of sectors, and again the number of sectors per track varies, depending on the density of the disk. Each sector stores 512 bytes of information.

Hard disk capacity and organization

One of the most important factors in judging a hard disk is its capacity, the number of bytes it can store. The capacity of hard disks ranges from a few hundred megabytes to many thousands of gigabytes (a gigabyte is 1024 megabytes). Regardless of the capacity of the hard disk, they all use hard metal platters to store data. In general, the higher the number of platters, the higher the capacity of the disk. Just as in the floppy disk, both sides of each platter in the disk are coated with magnetic material. Likewise, it uses a storage scheme that divides the area into sectors and tracks. There is one read/write head for each side of every platter, and these heads all move together. For example, a hard disk with 4 plates might have 8 read/write heads, one for each side, and they all move from the outer tracks into the inner tracks by the same arm. Hard disks give rise to more complex organization and hence a new term, the *cylinder*, which consists of all the tracks of the same radius on each platter. Since all the read/write heads move together from track to track it is logical to talk about cylinders in addition to tracks in the hard disk. Why do all the heads move together? The answer is that it is too difficult and expensive to design a hard disk controller that controls the movement of so many different heads. In addition, it would prolong the access time since it must stop one head and then activate a different head continuously until it reaches the end of the file. Using the concept of the cylinder, all the tracks of the same radius are accessed at the same time, and if the end of the file is not reached, all the heads move together to the next track. The number of read/write heads

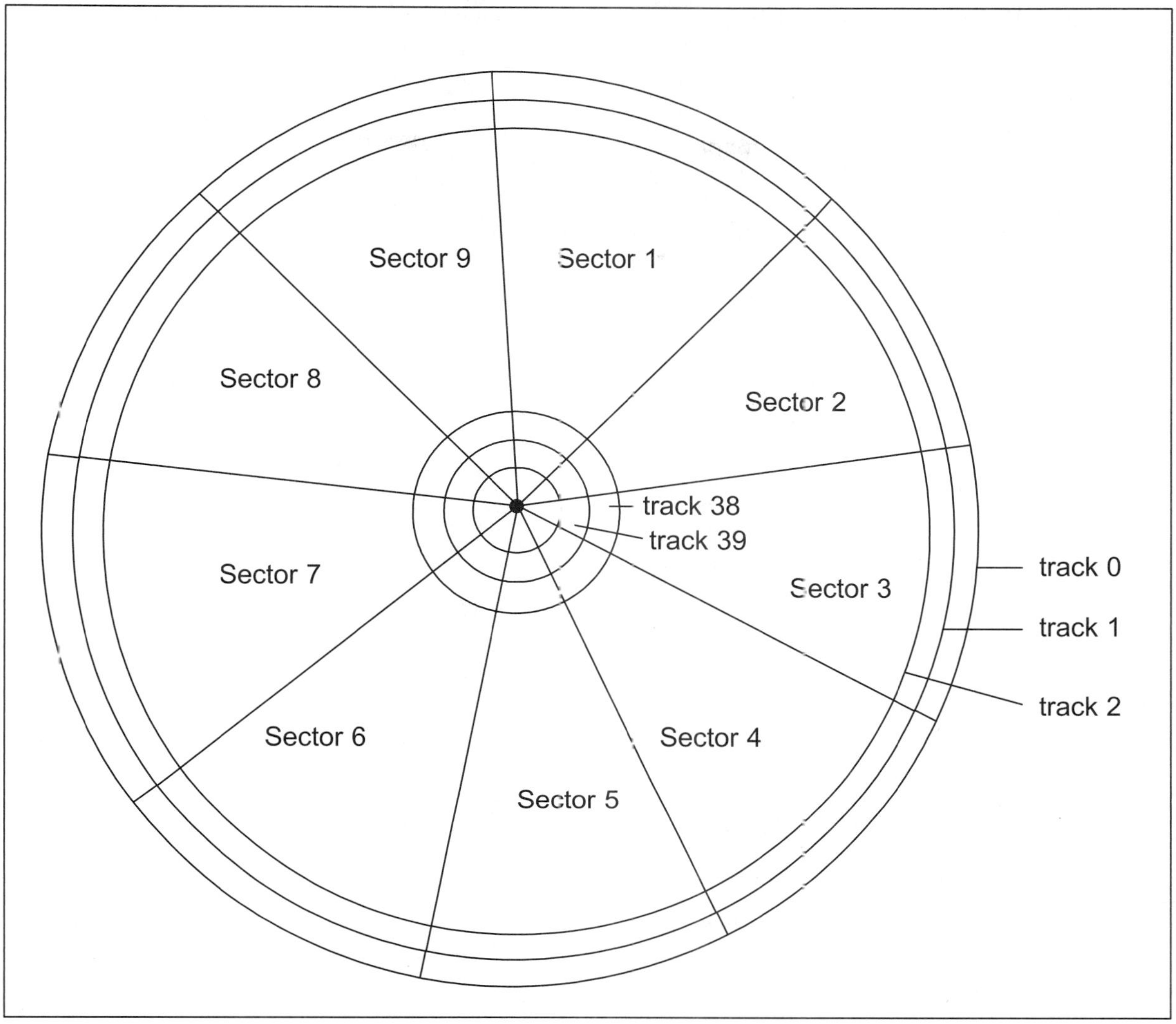

Figure 19-1. Sectors and Tracks of a Disk

varies from one hard disk to another. Knowing the number of cylinders makes it possible to calculate the total capacity of the hard disk. The total capacity of a disk is calculated as follows:

number of tracks = number of cylinders × tracks per cylinder
HD capacity = number of tracks × sectors per track × sector density

See Example 19-1.

The sectors of a disk are grouped into clusters. Cluster size varies among disk sizes, but it is always a power of 2 (2, 4, 8, 16, 32, ...) sectors per cluster. The file allocation table, or FAT, keeps track of what clusters are used to store which files.

Formatting disks

Once a floppy disk is formatted, the computer can read from or write to that disk. Formatting organizes the sectors and tracks in a way that makes it possible for the disk controller to access the information on the disk. When a disk is formatted, a number of sectors are set aside for various functions and the remaining sectors are used to store the user's files. The formatting process sets aside a specific number of sectors for the boot record, directory, and FAT (file allocation table), each of which is explained in detail below. It also copies some system files onto the disk if it was formatted with the "/s"

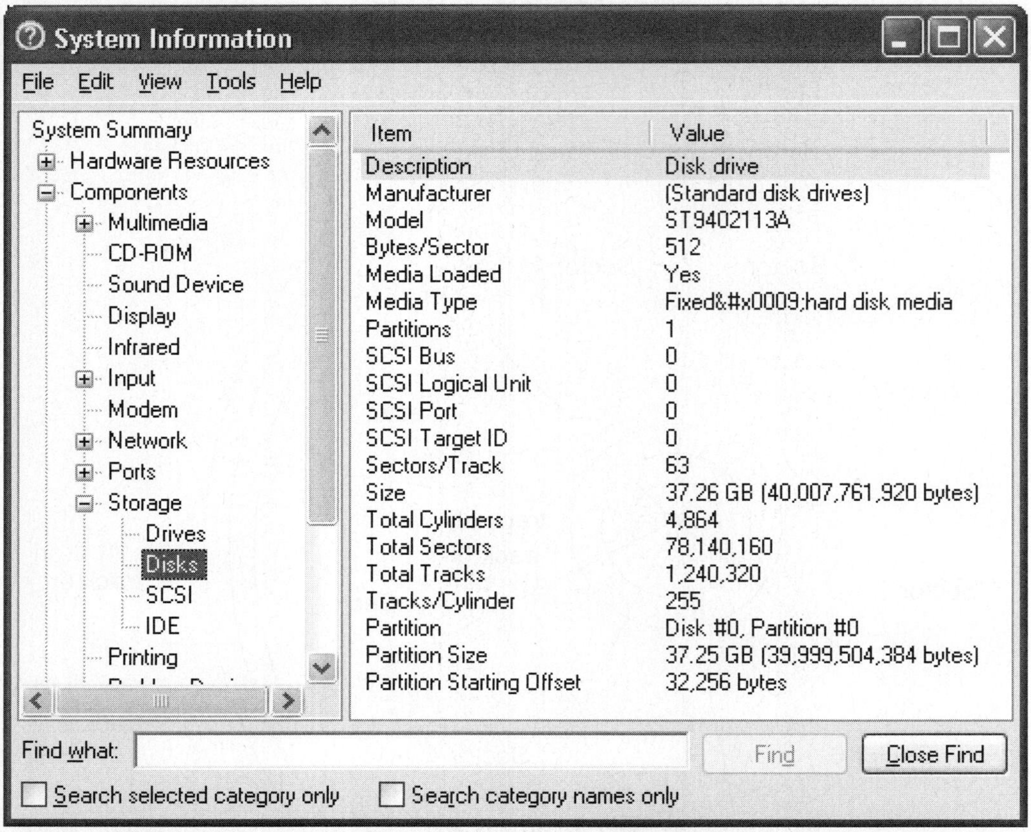

Figure 19-2. Screenshot of a Hard Disk in x86 Taken from System Information

Example 19-1

Verify the capacity of the hard disk using the data in Figure 19-2.

Solution:

As shown in Figure 19-2, the hard disk has 4864 cylinders, 255 tracks per cylinder, 63 sectors per track, and 512 bytes per sector:
Total tracks = 4864 cylinders × 255 tracks per cylinder = 1,240,320 tracks
Total sectors = 1,240,320 tracks × 63 sectors per track = 78,140,160 sectors
Total bytes = 78,140,160 sectors × 512 bytes per sector = 40,007,761,920 bytes

Note that in Figure 19-2 the capacity of the drive is expressed as 37.26 GB, where GB = 2^{30}.

option, which makes it a bootable disk. The difference between bootable and nonbootable disks will be explained later.

Disk organization

Regardless of the type of disk that is used, the first sector of the disk (side 0, track 0, sector 0) is always assigned to hold the boot record; then some sectors are used for storage of the FAT (file allocation table) copies 1 and 2. The number of sectors set aside for FAT depends upon the disk density. After the FAT, the directory is stored in consecutive sectors. Again, the number of sectors used by the directory depends on disk density. In assigning sectors for the FAT and the directory, the operating system uses all the sectors of track 0 of side 0, then goes to side 1 and uses all the sectors of track 0 of side 1, then comes back to side 0 and uses track 1, then goes to side 1 of track 1, and so on. See Figure 19-2. Before moving to the next topic it should be noted that the number of sectors assigned for the boot record, FAT, and directory are fixed for a given kind of disk and operating system version and it is only after assigning these sectors to those essential functions that the OS uses the remaining sectors to store files. The number of sectors set aside for each of the above can be calculated from the information in the boot record. The boot record, FAT, and directory are explained in detail next.

Looking into the boot record

When a disk is formatted, the first sector is used for the boot record. It is from the boot record that the computer will know the disk type, sector density, total number of sectors in the disk, and other essential information needed by BIOS and the operating system.

To examine the boot sector use the WinHex utility available from many web sites. Google WinHex for available sites.

Bootable and nonbootable disks

If the disk is formatted as a system disk (bootable), the first two files are IO.SYS and MSDOS.SYS, which are followed by COMMAND.COM. The first two are hidden files; therefore, they will not be listed when DIR is used. However, if the disk is formatted as a nonbootable disk, it will not have those three files on it after it is formatted. The job of IO.SYS is to provide low-level (hardware) communication (interface) between BIOS and DOS. The high-level (software) interface is provided by the MSDOS.SYS file. This is the section of DOS that contains INT 21H, among other things. Among the functions of COMMAND.COM is to provide the DOS prompt ">", read, interpret, and execute commands typed in by the user. The SYS command can be used to copy these files to a nonbootable disk to make it bootable.

FAT (file allocation table)

If the boot record tells BIOS and the operating system the kind of disk, and the directory provides the lists of all the files contained on the disk, how does the operating system locate a given file? Does it check every one of the hundreds of sectors to see if the file is there? This would obviously take an inordinate amount of time. It is the function of the FAT to provide a road map for the operating system to find where each file is located. In fact, the FAT is so critical to the operating system's ability to locate files that two copies of the FAT are kept on the disk, one for use and another one for backup in case something happens to the first one. If both are damaged, the operating system cannot find any file on that disk. The FAT is always located in the sectors following the boot record sector. The number of sectors used by the FAT varies depending on the size and density of the disk.

Partitioning

Partitioning the disk is the process of dividing the hard disk into many smaller disks. For example, a given hard disk of 80 gigabytes capacity can be partitioned into three smaller logical disks. They are called *logical* disks since it is the same physical disk, but as far as the operating system is concerned, it will be labeled disks C, D, and E.

After the hard disk has been partitioned, high-level formatting should be performed next. The C drive must be formatted with the system option (FORMAT C: /S) so that the system can boot from drive C.

Clusters

In the x86 IBM PC, the sector size is always 512 bytes but the size of the cluster varies among disks of various sizes. The cluster size is always a power of 2: 2, 4, 8, and so on. The fact that a 1-byte file takes a minimum of one cluster is important and must be emphasized. This means that a number of small files on a disk with a large number of sectors per cluster will result in wasted space on the hard disk. Let's look at an example. In a hard disk with a cluster size of 16 sectors (16 × 512 = 8192 bytes), storing a file of 26,000 bytes requires 4 clusters. The result is a waste of 6768 bytes since 4 × 8192 = 32,768 bytes, and 32,768 – 26,000 bytes = 6768. One can use the WinHex utility to find the cluster size.

Speed of the hard disk

One of the most important and widely cited hard disk performance factors is its speed, or how fast the requested data is available to the user. The hard disk access time is in the range of 10–80 ms and is still dropping. This access time is much longer than that of primary DRAM memory, which is in the range of 50–250 ns and lower, which is the reason that disk caching is used, as will be discussed later in this chapter. The access time of the hard disk given by manufacturers is broken down into several smaller times indicating the speed of different sections of the hard disk. The components of access time are seek time, settling time, and latency time. Seek time is the amount of time that the read/write head takes to find the desired cylinder or track. The outer tracks (the outer cylinder) obviously take less time to find since the head is parked on the outermost track and moves into inner tracks. Manufacturers always give the average seek time in the data sheet. To reduce seek time, computers use hard disks with several heads parked at different cylinders (tracks), which reduces seek time drastically but also increases the cost.

Settling time, the second factor in access time, is the time it takes the head to stop vibrating before it can begin reading the data. Some manufacturers include settling time when they give the average seek time.

Rotational latency is the time it takes for the head to locate on the specific sector. In other words, after the head is settled, the platter is rotating at a certain RPM (revolutions per minute) rate. Rotational latency time depends on the distance between the head and the desired sector, but in no case is it more than the time for one revolution. This means that rotational latency is directly proportional to the RPM of the hard disk. The RPM for various disks varies between 2,400 and 15,000 (and as high as 18,000 in some recent ones). Again, the average rotational latency must be taken into consideration. For example, if a given disk has 3600 RPM, which is 60 rotations per second, this is 16.6 ms (1/60 = 16.6 ms) for each full revolution. Since the desired sector could be directly beneath the head or at the end of the track, the average latency due to rotation is 8.3 ms [(16.6 + 0) / 2 = 8.3].

Interfacing standards in the hard disk

To ensure that hard disks made by different manufacturers are compatible, common standards for interfacing the hard disk and personal computers have been devised. These standards are ESDI and SCSI, which are explained below.

ESDI (enhanced small device interface)

The ESDI standard was developed by a group of disk drive manufacturers in 1983. There are some differences between ESDI and ST412:

1. ESDI can achieve a data transfer rate of up to 20 Mbits per second in contrast to the 7.5 Mbits/second of the ST412.
2. With the same RPM as ST412, it can have more sectors per track. The number of sectors for ESDI can vary between the 20s and the 50s.
3. While in ST412 the defect information must be provided manually during low-level formatting, for ESDI the defect map is already stored on the drive.

4. In the ST412 standard, the number of cylinders, heads, and sectors is stored either in the CMOS RAM of the system or in the ROM of the hard disk controller, in contrast to ESDI, where the configuration information is already provided and there is no need to store it externally.

IDE (integrated device electronics)

IDE is the standard for current PCs. In IDE, the controller is part of the hard disk. In other words, there is no longer a need to buy a hard disk and a separate controller as is often the case for ST412. One of the reasons that the IDE drives have a better data transfer rate is the integration of many of the controller's functions into the drive itself with the use of VLSI chips. For example, in the ST412 standard the hard disk read/write heads would read the data and transfer it to the controller through the cable, and then the data is separated from the clock pulses by what is called *data separator circuitry*. By eliminating cable degradation, IDE and SCSI (discussed next) reach a much higher external data transfer rate.

SCSI (small computer system interface)

SCSI (pronounced "scuzzy") is one of the most widely used interface standards not only for high-performance IBMs and compatibles but also for non-80x86 computers by other manufacturers, such as Apple and Sun Micro. The main reason is that unlike IDE, SCSI is the standard for all kinds of peripheral devices, not just hard disks. One can daisy chain up to seven devices, such as CD-ROMs, optical disk, tape drives, floppy disk drives, networks, and other I/O devices, using the SCSI standard. See Figure 19-3.

All the characteristics discussed for ESDI and IDE apply equally to SCSI. In addition, SCSI can have an internal data transfer rate of up to 80 Mbits/second. It must be noted that SCSI hard drives always have the controllers embedded into them and there is no need for a separate controller. The only thing needed is an adapter to convert the SCSI signals to signals compatible with the bus expansion slot of the host computers.

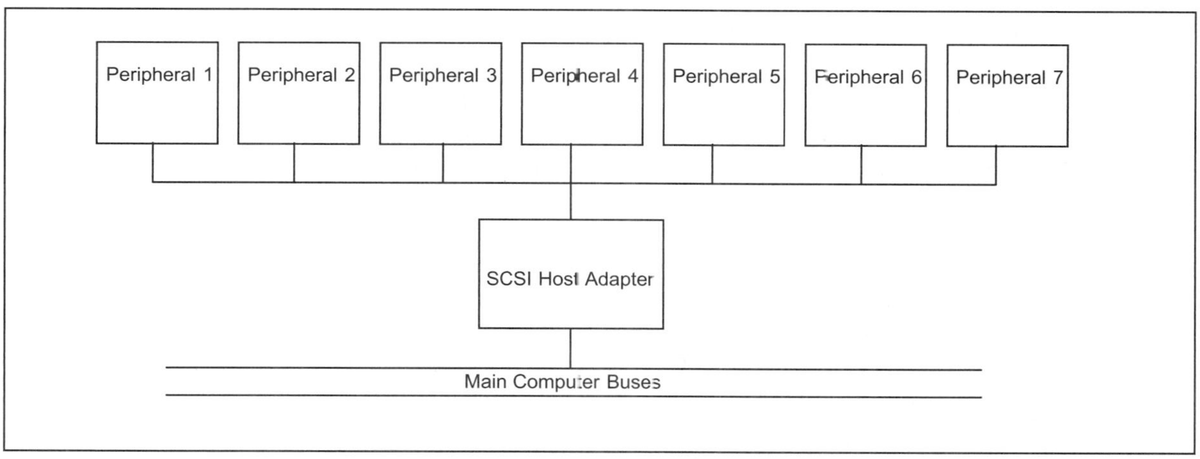

Figure 19-3. Peripheral Devices in SCSI "Daisy Chain"

Interleaving

As the read/write head moves along the track, it must read each sector and pass it to the controller. The controller in turn will deliver this data to the host computer through the buses. If the head and the controller cannot keep up with the stream of data passing under the head, there are two choices: either the rotation should be slower or interleaving should be used. Using a slower rotation, say, 600 RPM instead of 3600, will give an unacceptably long access time. That brings us to interleaving While common sense tells us that the sectors should be numbered on each track sequentially, the head and controller cannot process the data in sector 1 in time to be ready for sector 2 by the time it is under

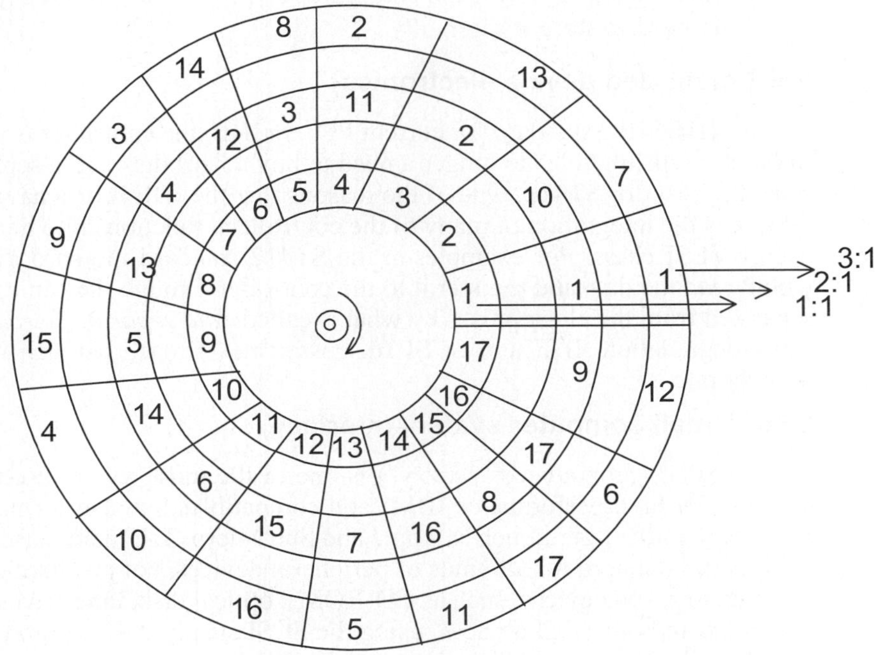

Figure 19-4. Hard Disk Interleaving

the head, so it would have to wait for the next rotation to read sector 2. Likewise, while sector 2 is being processed, sector 3 has already passed under the head, so to read that sector it must wait for the next revolution. This means that reading all 17 sectors of each track will take 17 revolutions. This is 1:1 interleaving and is as bad as slowing down the RPM. In 2:1 interleaving the sectors are numbered and accessed alternately. If the controller is not fast enough, 3:1 can be used. In 3:1 interleaving, every third sector is numbered and accessed. It will take 2 complete revolutions to access all the sectors in 2:1, and 3 revolutions for 3:1 interleaving. This is much better than all the other choices discussed above. In 3:1 interleaving, the computer accesses sector 1, and by the time it finishes processing it, sector 2 is under the head. The two sectors between sectors 1 and 2 give the controller time to get ready for accessing the next sector. Note that today's high-performance computers using IDE and SCSI controllers and 1:1 interleaving can read the entire track with one revolution, due to their fast controllers and wide data buses. Figure 19-4 shows the concept of interleaving.

Disk caching

Due to the long access time of the hard disk, disk caching is used to speed up the disk access time. There are two types of disk caching. In one type, the disk manufacturer puts some (several megabytes) fast memory on the disk. This is called *hardware disk cache* (see Table 19-1). In the other type, a section of memory on the PC motherboard is set aside for disk caching. Obviously, the larger the size of this memory, the more files can be stored there and accessed by the CPU, assuming that there is extra memory to spare. Using a section of motherboard DRAM for disk caching is done by the operating system. This kind of disk caching is called *software disk cache*.

Table 19-1: Seagate LD25 Series Disk Drive Datasheet Data

Specifications	**40 GB** **GB = 1 billion**	**20 GB** **GB = 1 billion**
Model Number	ST9402115AS	ST920217AS
	ST9402115A	ST920217A
Interface Options	SATA/150	SATA/150
	Ultra ATA/100	Ultra ATA/100
Performance		
Spindle Speed (RPM)	5400	5400
Average Latency (mse c)	5.6	5.6
Seek Time		
Average Read/Write (msec)	<16	<16
Transfer Rate		
Maximum Internal (Mbytes/sec)	57.6	57.6
Maximum Sustained (Mbytes/sec)	150/100	150/100
Cache, Multisegmented (Mbytes)	2	2
Configuration/Organization		
Bytes per Sector	512	512
Reliability/Data integrity		
Nonrecoverable Read Errors per Bits Read	1 sector per 1014	1 sector per 1014
Power Management		
Startup current (5V, typical Amps)	1	1
Seek (typical, watts)	2.9	2.9
Idle Average (watts)	2.5	2.5
Standby Average (watts)	<1.5	<1.5
Environmental		
Temperature, Operating (°C)	5 to 55	5 to 55
Temperature, Nonoperating (°C)	–40 to 70	–40 to 70
Shock, Operating: 2 msec (Gs)	100	100
Shock, Nonoperating: 1 msec (Gs)	500	500
Acoustics Idle (bels —sound power)	2.4	2.4
Rotational Vibration (Rad/sec, Gs)	.5	.5
Physical		
Height (mm)	9.5	5.0 +/–0.1
Width (mm)	69.85	69.85
Depth (mm)	100.51	100.51
Weight (g)	100	100

Disk reliability

MTBF (mean time between failures) is a measure of reliability and durability of the disk when the power is on. This factor is given in hours. For example, the ST225 has a MTBF of 100,000 hours. Dividing it by 24 hours gives an MTBF value of 4166.6 days or 11.4 years (4116.6/365). Of course, manufacturers will not power on the disk for that long and then test it since they would be out of business by then. They use extremely reliable statistical analysis to figure out the MTBF.

Review Questions

1. True or false. All sectors have the same capacity (total number of bytes that can be stored).
2. The very first sector always belongs to the ___________ (FAT, boot record).
3. The sectors belonging to the __________ (FAT, directory) are located next to the boot sector.
4. Why are there two copies for each FAT?
5. True or false. The number of sectors set aside for the FAT varies among the various-sized disks.
6. The boot record provides the total ____________________ (byte capacity, number of sectors) per disk.
7. How does the operating system know how many sectors each track is divided into?
8. Sectors set aside for directories are always _______ (before, after) the FAT sector.
9. In hard disks, which sector is set aside for the boot record?
10. True or false. The SCSI interfacing standard is also used for devices other than hard disks, such as CD ROM.
11. How many rotations does it take to read all the sectors of a given track in a 3:1 interleaved hard disk?
12. True or false. Each file begins on a new cluster even if the previous cluster has some empty space.
13. How many sectors are in a cluster?
14. What does "MTBF" stand for, and what does it measure?

PROBLEMS

SECTION 19.1: HARD DISK ORGANIZATION AND PERFORMANCE

1. Why is the hard disk called secondary storage?
2. The first sector of every disk is set aside for _______________.
3. After the boot sector, sectors are assigned to the ___________(FAT, directory).
4. For a disk to be bootable, which file must it contain?
5. Why doesn't the DIR command list MSDOS.SYS and IO.SYS on the screen?
6. True or false. The terms hard disk and fixed disk refer to the same thing.
7. What is a cylinder, and how is used in the hard disk?
8. The total number of tracks in a given hard disk is equal to ____________.
9. Calculate the total number of sectors and the capacity of the hard disk on your computer.
10. True or false. The number of sectors per cluster is always a power of 2.
11. Discuss seek time, settling time, latency time, and how they relate to disk access time.
12. Which of the following has the shortest access time? Which has the longest access time?
    (a) 2400 RPM (b) 3600 RPM (c) 4800 RPM
13. True or false. The SCSI interfacing standard is used only for hard disks.
14. State the number of peripheral devices that can be daisy chained if SCSI is used.
15. To read all the sectors of a given track, how many times must the track rotate under the head for each of the following interleaving factors?
    (a) 1:1 (b) 1:3 (c) 1:5
16. What does MTBF stand for, and what is its use?

ANSWERS TO REVIEW QUESTIONS

SECTION 19.1: HARD DISK ORGANIZATION AND FERFORMANCE

1. True
2. Boot record
3. FAT
4. To use the second for backup in case something happens to the first one, since the FAT is the road map for finding where data is located on the disk.
5. True
6. Number of sectors
7. This information is provided in the boot sector.
8. After
9. Sector 0
10. True
11. 3 rotations
12. True
13. 1, 2, 4, 8, 16, 32, or 64; it is always a power of 2.
14. Mean time between failures; it is a measure of disk longevity and reliability.

CHAPTER 20

THE IEEE FLOATING POINT AND x87 MATH PROCESSORS

OBJECTIVES

Upon completion of this chapter, you will be able to:

- **>> Diagram the bit assignment of floating-point IEEE standards for single- and double-precision data**
- **>> Convert data from real numbers to IEEE floating-point format**
- **>> Diagram the bit assignment for x87 data types: word integer, short integer, long integer, packed decimal, short real, long real, and temporary real**
- **>> Code Assembly language data directives for x87 data types**
- **>> List the registers of the x87**
- **>> Write Assembly language programs using x87 instructions**

This chapter will examine the x87 math coprocessor. In Section 20.1, we study the IEEE standard for floating-point numbers and the Intel x87 math coprocessor's data format. In Section 20.2, the x87 instructions are discussed along with some sample programs using MASM. In Section 20.3, we provide an overview of x87 instructions.

SECTION 20.1: MATH COPROCESSOR AND IEEE FLOATING-POINT STANDARDS

Using a general-purpose microprocessor such as the 8088/86 to perform mathematical functions such as log, sine, and others is very time consuming, not only for the CPU but also for programmers writing such programs. In the absence of a math coprocessor, programmers must write subroutines using 8088/86 instructions for mathematical functions. Although some of these subroutines are already written and can be purchased at a small cost, no matter how good the subroutine, its CPU run time (8088/86, 286, 386) will still be quite long. Table 20-1 provides a comparison of the number of clocks used by the 8087 and 8086 to perform some mathematical functions. One can appreciate the advantage of having a coprocessor by comparing the run time of some programs, such as SPICE (a package for circuit analysis which uses floating-point operations extensively), on a computer with a coprocessor on one without a coprocessor. In some cases the difference is hours.

Table 20-1: Comparison of 8087 and 8086 Clock Times

	Approximate Execution	Time (μs) (5-MHz clock)
Instruction	**8087**	**8086 Emulation**
Multiply (single precision)	19	1,600
Multiply (double precision)	27	2,100
Add	17	1,600
Divide (single precision)	39	3,200
Compare	9	1,300
Load (single precision)	9	1,700
Store (single precision)	18	1,200
Square root	36	19,600
Tangent	90	13,000
Exponentiation	100	17,100

IEEE floating-point standard

Up to the late 1970s, real numbers (numbers with decimal points) were represented differently in binary form by different computer manufacturers. This made many programs incompatible for different machines. In 1980, an IEEE committee standardized the floating-point data representation of real numbers. This standard, much of which was contributed by Intel based on the 8087 math coprocessor, recognized the need for different degrees of precision by different applications; therefore, it established single precision and double precision. Since almost all software and hardware companies, including IBM, Intel, and Microsoft, now abide by these standards, each one is explained thoroughly. RISC processors also use IEEE floating-point standards.

IEEE single-precision floating-point numbers

IEEE single-precision floating-point numbers use only 32 bits of data to represent any real number in the range 2^{128} to 2^{-126}, for both positive and negative numbers. This translates approximately to a range of 1.2×10^{-38} to $3.4 \times 10^{+38}$ in decimal numbers, again for both positive and negative values. In Intel coprocessor terminology, these single-precision 32-bit floating-point numbers are referred to as *short real*. Assignment of the 32 bits in the single-precision format is

Bit	Assignment
31	Sign bit: 0 for positive (+) and 1 for negative (–)
23–30	Biased exponent
22–0	The fraction, also called *significand*

To make the hardware design of the math processors much easier and less transistor consuming, the exponent part is added to a constant of 7FH (127 decimal). This is referred to as a *biased exponent*. Conversion from real to floating point involves the following steps.

1. The real number is converted to its binary form.
2. The binary number is represented in scientific form: 1.xxxx E yyyy
3. Bit 31 is either 0 for positive or 1 for negative.
4. The exponent portion, yyyy, is added to 7F to get the biased exponent, which is placed in bits 23 to 30.
5. The significand, xxxx, is placed in bits 22 to 0.

Examples 20-1, 20-2, and 20-3 demonstrate this process. In Section 20.2 we will verify all the above examples using an assembler.

Example 20-1

Convert 9.75_{10} to single-precision (short real) floating point.

Solution:

decimal 9.75 = binary 1001.11 = scientific binary 1.00111 E 3

Sign bit 31 is 0 for positive.

Exponent bits 30 to 23 are 1000 0010 (3 + 7F = 82H) after biasing.

Significand bits 22 to 0 are 00111000000000000000 ...00.

Putting it all together gives the following binary form, under which is written the hex form:

0100	0001	0001	1100	0000	0000	0000	0000
4	1	1	C	0	0	0	0

This can be verified by using an assembler, such as MASM, as will be seen later in this chapter.

IEEE double-precision floating-point numbers

Double-precision FP (called *long real* by Intel) can represent numbers in the range 2.3×10^{-308} to 1.7×10^{308}, both positive and negative. A total of 53 bits (bits 0 to 52) are for the significand, 11 bits (bits 53 to 62) are for the exponent, and finally, bit 63 is for the sign. The conversion process is the same as for single precision in that the real number must first be represented as 1.xxxxxxx E YYYY, then YYYY is added to 3FF to get the biased exponent.

Other data formats of the 8087

In addition to short real (single-precision) and long real (double-precision) representations for real numbers, the 8087 also supports 16-, 32-, and 64-bit integers. They are referred to as *word integers*, *short integers*, and *long integers*, respectively, and are shown in Figure 20-1. These forms are sometimes referred to as *signed integer numbers*. No decimal points are allowed in integers, in contrast to real numbers, in which decimal points are allowed. There are also two 80-bit data formats in the 8087 coprocessor, packed decimal and temporary real. The packed decimal format has 18 packed BCD numbers, which require a total of 72 bits ($18 \times 4 = 72$). Bits 71 to 0 are used for the numbers, bits 73 to 78 are always 0, and bit 79 is for the sign. The temporary real format is used internally by the 8087 and is shown in Figure 20-1. In the temporary real format, the conversion goes through the same process as shown above, except that the biased exponent is calculated by adding the constant 3FFFH.

Example 20-2

Convert 0.078125_{10} to short IEEE floating-point standard real FP (single precision).

Solution:

decimal 0.078125 = binary 0.000101 = scientific binary 1.01 E –4
Sign bit 31 is 0 for positive.
Exponent bits 30–23 are 0111 1011 (–4 + 7F = 7B) after biasing.
Significand bits 22–0 are 01000000....000.
This number will be represented in binary and hex as

0011	1101	1010	0000	0000	0000	0000	0000
3	D	A	0	0	0	0	0

Example 20-3

Convert -96.27_{10} to single-precision FP format.

Solution:

decimal 96.27 = binary 1100000.01000101000111101 =
scientific binary 1.100000010001010001111101E 6
Sign bit 31 is 1 for negative.
Exponent bits 30–23 are 1000 0101 (6 + 7F = 85H) after biasing,
Fraction bits 22–0 are 10000001000101000111101,
The final form in binary and hex is

1100	0010	1100	0000	1000	1010	0011	1101
C	2	C	0	8	A	3	D

It must be noted that conversion of the decimal portion 0.27 to binary can be continued beyond the point shown above, but because the fraction part of the single precision is limited to 23 bits, this was all that was shown. For that reason, double-precision FP numbers are used in some applications to achieve a higher degree of accuracy.

Example 20-4

Convert 152.1875_{10} to double-precision FP.

Solution:

decimal 152.1875 = binary 10011000.0011 =
scientific binary 1.00110000011 E 7
Bit 63 is 0 for positive.
Exponent bits 62–53 are 10000000110 (7 + 3FF = 406) after biasing.
Fraction bits 52–0 are 00110000011000.....000.

0100	0000	0110	0011	0000	0110	0000	0000	0000	...	...	0000
4	0	6	3	0	6	0	0	0	...	...	0

This example will be verified by an assembler in the next section.

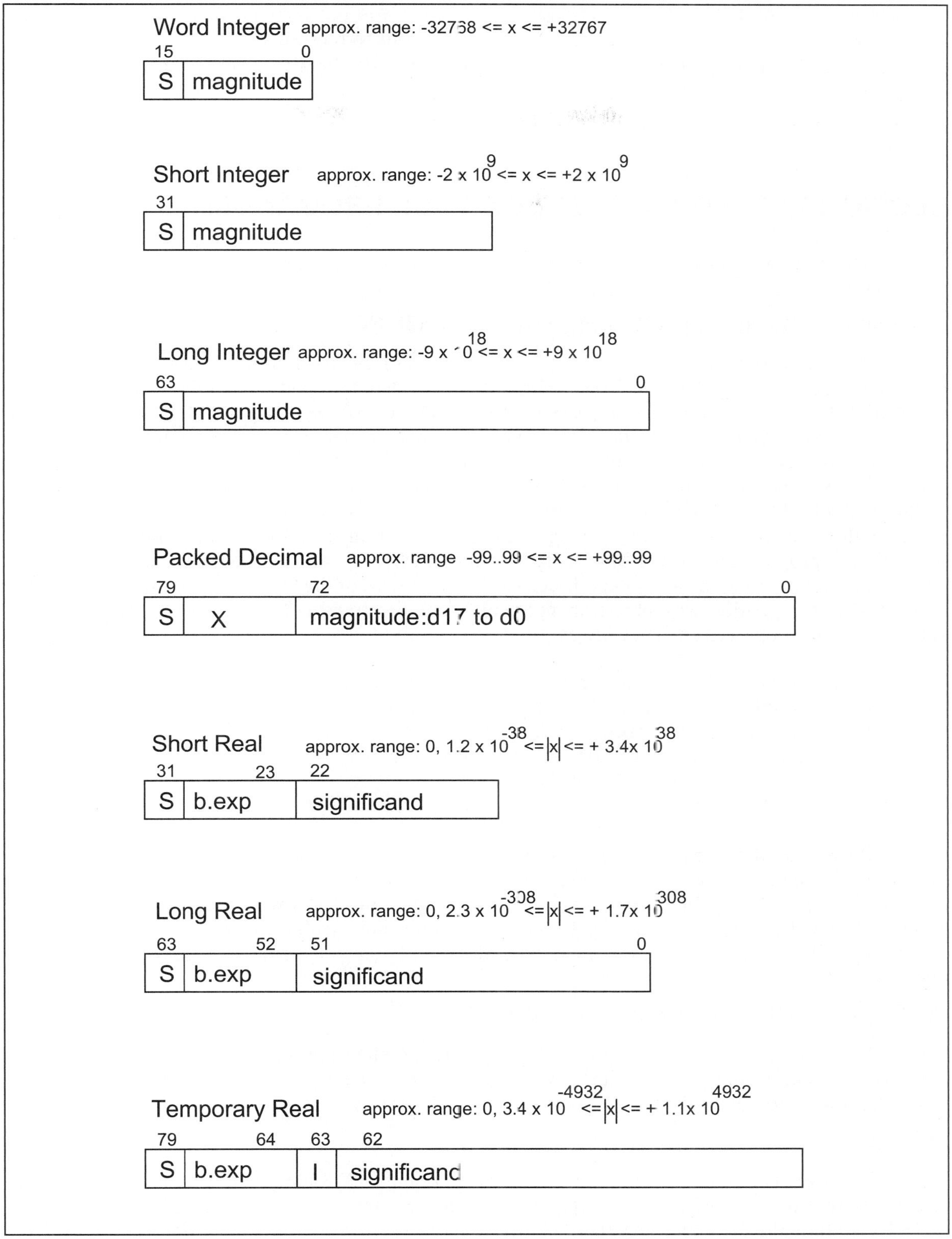

Figure 20-1. x87 Data Format

Review Questions

1. True or false. In the absence of a math processor, the general-purpose processor must perform all math calculations.
2. True or false. The x87 follows the IEEE floating-point standard.

3. Single-precision IEEE FP standard uses _________ bits to represent data.
4. Double-precision IEEE FP standard uses ________ bits to represent data.
5. To get the biased exponent portion of IEEE single-precision floating-point data we add ________.
6. To get the biased exponent portion of IEEE double-precision floating-point data we add ________.

SECTION 20.2: x87 INSTRUCTIONS AND PROGRAMMING

This section examines the x87 registers and instructions and provides some sample programs.

Assembling and running x87 programs on the x86 PCs

The math coprocessors found in x86 (486, Pentium I-IV, and Itanium) processors have their origin in the 8087 coprocessor chip. As far as data types and instructions are concerned, there have been few changes since the introduction of the original 8087, aside from the fact that x86 processors run 8087 instructions much faster. Originally, the 8087, 80287, and 80387 math coprocessors were separate chips. In IBM PC and Compatible computer motherboards there was a socket for these coprocessors right next to the 8088, 80286, and 80386 microprocessors. Starting with the 80486, the coprocessor was integrated with the main processor into a single chip. This section shows how to assemble several 8087 programs using the Microsoft Assembler, MASM, run them on a PC, and analyze the result. First, the assembler directives for data types of the 8087 are explained. In MASM and compatible assemblers, there are different directives to define the different data types of the coprocessor. They are as follows:

DD	(Define double word) for short real (single precision)
DQ	(Define quad word) for long real (double precision)
DD	(Define double word) for short integer
DQ	(Define quad word) for long integer
DT	(Define ten bytes) for packed decimal
DT	(Define ten bytes) for temporary real

Recall that the word size in the x86 family is 16 bits. Therefore, when using DD to define a double word, the result is 32 bits. This is different from some other processors, notably RISC processors, in which a word is defined as 32 bits. It is worth repeating a point made in Chapter 0: Although a byte is defined as 8 bits universally, a word is defined differently by different companies. For example, the Cray computer defines a word as 64 bits.

Verifying the solution for Examples 20-1 to 20-4

Program 20-1 is a portion of the .LST file produced when a program is assembled. It verifies the conversion from decimal to the internal machine representation given in Examples 20-1 through 20-4.

x87 registers

There are only eight general-purpose registers in the x87. Rather than having different-size registers for different-size operands, all the registers of the 8087 are 80 bits wide. Every time the 8087 loads an operand, it automatically converts it to this 80-bit format. This gives uniformity to the registers and makes programming, as well as 8087 hardware design, much easier. Although these eight registers have been numbered from 0 to 7, they are accessed like a stack, meaning that a last-in-first-out policy is used. At any given time, the top of the stack is referred to as ST(0), or simply ST, and all other registers, regardless of their number, are referred to according to their positions compared to the top of the stack, ST. The programming examples below will demonstrate the use of registers

```
                               .8087
                               PAGE 60,132
0000                           .MODEL    SMALL
0000           20  [           .STACK   32
                          ????
                                     ]
0040
                                     ;---------------------------
0000                           .DATA
0000                           ORG 00H
0000   00 00 1C 41 EX1                    DD   9.75          ;example 1
0010                           ORG 10H
0010   00 00 A0  3D EX2                   DD   0.078125      ;example 2
0020                           ORG 20H
0020   3D 8A C0  C2 EX3                   DD   -96.27        ;example 3
0030                           ORG 30H
0030   00 00 00 00 00 06 EX4              DQ   152.1875      ;example 4
        63 40
                               ;-----------------------------
```

Example 20-1 data:

hex: 41 1C 00 00
binary: 0100 0001 0001 1100 0000 0000 0000 0000
sign: 0 for positive
biased exp: 1000 0010 normalize: 82 – 7F = 3
significand: 0011100..00
scientific binary: 1.00111000..00 E3 = 1001.11000...00
decimal: 9.75

Example 20-2 data:

hex: 3D A0 00 00
binary: 0011 1101 1010 0000 0000 0000 0000 0000
sign: 0 for positive
biased exp: 0111 1011 normalize: 7B – 7F = –4
significand: 01000.00
scientific binary: 1.01 E–4 = .000101
decimal: 0.078125

Example 20-3 data:

hex: C2 C0 8A 3D
binary: 1100 0010 1100 0000 1000 1010 0011 1101
sign: 1 for negative
biased exp: 1000 0101 normalize: 85 – 7F = 6
significand: 10000001000101000111101
scientific binary: 1.10000001000101000111101 E6 = 1100000.001000101000111101
decimal: –96.2700078

Example 20-4 data:

hex: 40 63 06 00 00 00 00 00
binary: 0100 0000 0110 0011 0000 0110 000..00
sign: 0 for positive
biased exp: 10000000110 normalize: 406 – 3FF = 7
significand: 00110000011
scientific binary: 1.00110000011 E7 = 10011000.0011
decimal: 152.1875

Program 20-1

in the 8087. Example 20-5 will show a complete Assembly language program using the 8087 coprocessor. First, a few points should be noted:

1. All x87 mnemonics start with the letter "f" to distinguish them from x86 instructions.
2. The x87 must be initialized to make sure that the top of the stack will be register number 7.
3. Whenever a register is not identified specifically, ST [which is ST(0)] is assumed automatically.
4. ST(0) is the top of the stack, ST(1) is one register below that, and ST(2) is two registers below ST(0), and so on. In other words, for register ST(*m*), the number in parentheses, *m*, has nothing to do with the register number. There is a way to find out which register number, 0–7, is ST(0), the top of the stack.
5. In the following programming examples, all values of X, Y, and Z have been defined in the data segment and allocated memory locations. The same is true for variables, such as SUM, for storing the result.

Example 20-5

Write an 8087 program that loads three values for X, Y, and Z, adds them, and stores the result.

Solution:

```
finit          ;initialize the 8087 to start at the top of stack
fld    X       ;load X into ST(0). now ST(0)=X
fld    Y       ;load Y into ST(0). now ST(0)=Y and ST(1)=X
fld    Z       ;load Z into ST(0). now ST(0)=Z,ST(1)=Y,ST(2)=X
fadd   ST(1)          ;add Y to Z and save the result in ST(0)
fadd   ST(2)          ;add X to (Y+Z) and save it in ST(0)
fst    sum     ;store ST(0) in memory location called sum.
```

Now the same program can be written as follows:

```
finit
fld    X              ;load x, now ST(0) = x
fld    Y              ;load y, now ST(0)= y, ST(1) = x
fld    Z              ;load z, now ST(0)=z, ST(1)=y, ST(2)=x
fadd                  ;adds y to z
fadd   ST(2)          ;adds x to (y + z)
fst    sum
```

Program 20-2 shows the actual MASM code and execution. Figure 20-2 shows the registers.

Often, an application requires the use of both real and integer numbers. Real numbers can be rounded into integers by using the x87 instruction FRNDINT, as shown in Program 20-3, which includes a procedure to round real numbers. The sample data used is the real number 5.5. In addition, the data is analyzed to see how the number was rounded. FRNDINT rounds real numbers to integers by rounding up, rounding down, truncating, or rounding to the nearest integer. How real numbers are rounded is determined by the RC (round control) bits in the control word (see Figure 20-3).

```
;Program for Example 20-5 to load 3 numbers and compute their sum
.8087
PAGE  60,132
.MODEL      SMALL
.STACK      32
;----------------------------
            .DATA
            ORG   00H
X           DD         9.75
            ORG   10H
Y           DD         13.09375
            ORG   20H
Z           DD         29.0390625
            ORG   30H
SUM         DD    ?
;----------------------------
            .CODE
START       PROC        FAR
            MOV   AX,@DATA
            MOV   DS,AX
            CALL  CSUM
            MOV   AH,4CH
            INT         21H
START       ENDP
;-----------------------------
CSUM        PROC        NEAR
            FINIT                   ;initialize 8087 stack
            FLD   X                 ;load X into ST(0)
            FLD   Y                 ;load Y into ST(0)
            FLD   Z                 ;load Z into ST(0)
            FADD  ST(0),ST(1)       ;ST(0) = Y + Z
            FADD  ST(0),ST(2)       ;ST(0) = X + (Y + Z)
            FST   SUM               ;store ST(0) in sum
            RET
CSUM        ENDP
            END   START

C:\>DEBUG EX5.EXE
-U CS:0 C
16EF:0000 B8B617        MOV     AX,17B6
16EF:0003 8ED8          MOV     DS,AX
16EF:0005 E80400        CALL    00CC
16EF:0008 B44C          MOV     AH,4C
16EF:000A CD21          INT     21
16EF:000C 9B            WAIT
-G
Program terminated normally
-D 17B6:0 3F
17B6:0000 00 00 1C 41 00 00 00 00-00 00 00 00 00 00 00 00 ...A............
17B6:0010 00 80 51 41 00 00 00 00-00 00 00 00 00 00 00 00 ..QA............
17B6:0020 00 50 E8 41 00 00 00 00-00 00 00 00 00 00 00 00 .PhA............
17B6:0030 00 88 4F 42 00 00 00 00-00 00 00 00 00 00 00 00 ..@.............
-Q
C:\>
```

Program 20-2

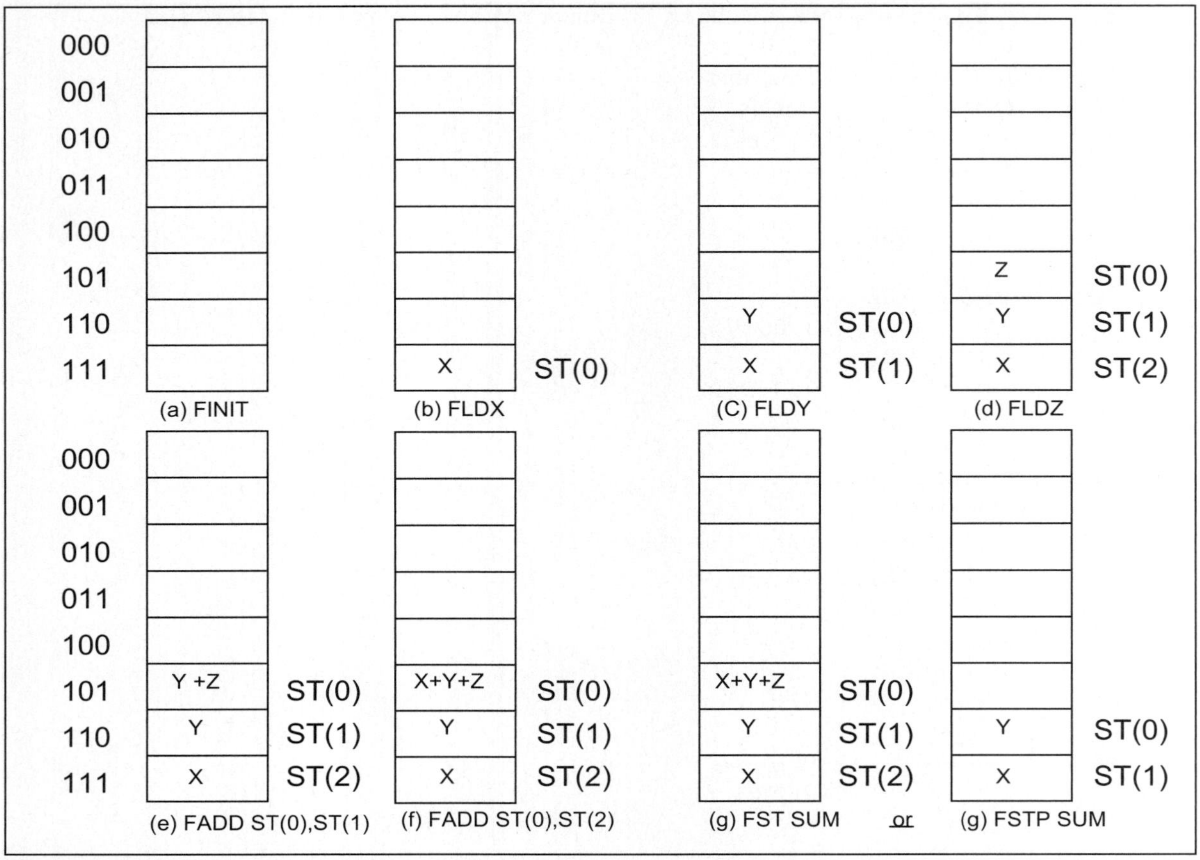

Figure 20-2. Stack Diagram for Example 20-5

Example 20-5 Data Analysis

X:	hex:	41 1C 00 00	binary:	0100 0001 0001 1100 0000 0000 0000 0000
	sign:	0 for positive	biased exp:	1000 0010 normalize: 82 – 7F = 3
	significand:	0011 1000 00..00		
	sci. binary:	1.00111000..00 E3 = 1001.11000.00		
	decimal:	9.75		
Y:	hex:	41 51 80 00	binary:	0100 0001 0101 0001 1000 0000 0000 0000
	sign:	0 for positive	biased exp:	1000 0010 normalize: 82 – 7F = 3
	significand:	1010001100..00		
	sci. binary:	1.1010001100..00 E3 = 1101.00011		
	decimal:	13.09375		
Z:	hex:	41 E8 50 00	binary:	0100 0001 1110 1000 0101 0000 0000 0000
	sign:	0 for positive	biased exp:	1000 0011 normalize: 83 – 7F = 4
	significand:	1101000010100..00		
	sci. binary:	1.1101000010100.00 E4 = 11101.0000101		
	decimal:	20.0390625		
SUM:	hex:	42 4F 88 00	binary:	0100 0010 0100 1111 1000 1000 0000 0000
	sign:	0 for positive	biased exp:	1000 0100 normalize: 84 – 7F = 5
	significand:	10011111000100..00		
	sci. binary:	1.100111110001 E5 = 110011.1110001		
	decimal:	51.8828125		

Program 20-4 calculates the area of a circle. The 8087 has instructions that load the top of the stack, ST(0), with a constant. For example, FLDPI loads PI into ST. To calculate the square of a number, the register is multiplied by itself with FMUL. FMUL can have two operands, such as "FMUL ST(2),ST(4)", where ST(4) is multiplied by ST(2) and the result is placed in ST(2). If no operands are given, the operation is assumed to be "FMUL ST(0),ST(1)", so that the first two stack registers are multiplied together and the result is stored in ST(0).

```
.8087
PAGE 60,132
            .MODEL SMALL
; PROGRAM TO ROUND A REAL NUMBER TO AN INTEGER
            .STACK      32
;-----------------------------
            .DATA
            ORG 00H
REALNUM     DD              5.5
            ORG 10H
INTNUM      DD              ?
;-----------------------------
            .CODE
START       PROC  FAR
            MOV    AX,@DATA
            MOV    DS,AX
            CALL   RND_NUM
            MOV    AH,4CH
            INT          21H
START       ENDP
;------------------------------
;PROCEDURE TO ROUND A REAL NUMBER TO AN INTEGER
RND_NUM    PROC     NEAR
            FINIT                             ;initialize 8087
            FLD          REALNUM              ;load real
            FRNDINT                     ;round to integer
            FIST         INTNUM               ;store integer
            RET
RND_NUM    ENDP
;------------------------------
                  END    START


C>debug ex6.exe
-g
Program terminated normally
...
...
-d 1065:0 1f
1065:0000 00 00 B0 40 00 00 00 00-00 00 00 00 00 00 00 00..0@...........
1065:0010 06 00 00 00 00 00 00 00-00 00 00 00 00 00 00 00...............
-q
```

The verification of the above dump is left to the reader as an exercise. It must be noted that the control word can be accessed in order to see which rounding method was used.

Program 20-3

```
.8087
PAGE 60,132
            .MODEL        SMALL
;PROGRAM TO CALCULATE AREA OF A CIRCLE (radius 91.67)
            .STACK        32
;----------------------------
            .DATA
            ORG    00H
R           DD     91.67
            ORG    10H
AREA        DD     ?
;----------------------------
            .CODE
START       PROC   FAR
            MOV    AX,@DATA
            MOV    DS,AX
            CALL   CIRC_AREA
            MOV    AH,4CH
            INT    21H
START       ENDP
;-----------------------------
;PROCEDURE TO CALCULATE THE AREA OF A CIRCLE
CIRC_AREA   PROC         NEAR
            FINIT                            ;initialize 8087
            FLD          R                   ;load radius
            FMUL         ST(0),ST(0)         ;square R
            FLDPI                            ;load PI
            FMUL         ST(0),ST(1)         ;multiply PI by R squared
            FSTP         AREA                ;store AREA
            RET
CIRC_AREA   ENDP
;------------------------------
            END    START
```

The data dumped in DEBUG looked as follows:

```
-d 1065:0 1f
1065:0000  0A 57 B7 42 00 00 00 00-00 00 00 00 00 00 00 00   .W7B............
1065:0010  0C 40 CE 46 00 00 00 00-00 00 00 00 00 00 00 00   .@NF............
```

r = 42 B7 57 0A Hex

binary =	0100 0010 1011 0111 0101 0111 0000 1010
sign:	0 for positive
biased exp:	1000 0101 normalize: 85 – 7F = 6
significand:	01101110101011100001010
sci. binary:	1.01101110101011100001010 E6 = 1011011.10101011100001010
decimal:	91.6702481689454

area = 46 CE 40 0C

binary =	0100 0110 1100 1110 0100 0000 0000 1100
sign:	0 for positive
biased exp:	1000 1101 normalize: 8D – 7F = 14
significand:	10011100100000000001100
sci. binary:	1.10011100100000000001100 E 14 = 110011100100000.000001100
decimal:	26400.0232375

Program 20-4

Example 20-6

Write, run, and analyze an 8087 program to calculate sin, cosine, tan, and cotan of a 30-degree angle.

Solution:

First the 30-degree angle must be converted to radians: (PI/180) × 30 = 0.523598776 radian.
The program is Program 20-5. The data dump in DEBUG, after Program 20-5 is run, is as follows:

```
-d  1065:0 8F
1065:0000   92 0A 06 3F 00 00 00 00-00 00 00 00 00 00 00 00   ...?............
1065:0010   1B 06 BA 3F 00 00 00 00-00 00 00 00 00 00 00 00   ..:?............
1065:0020   45 CD 56 3F 00 00 00 00-00 00 00 00 00 00 00 00   EMV?............
1065:0030   45 CD D6 3F 00 00 00 00-00 00 00 00 00 00 00 00   EMV?............
1065:0040   00 00 00 3F 00 00 00 00-00 00 00 00 00 00 00 00   ...?............
1065:0050   D8 B3 5D 3F 00 00 00 00-00 00 00 00 00 00 00 00   X3]?............
1065:0060   3A CD 13 3F 00 00 00 00-00 00 00 00 00 00 00 00   :M.?............
1065:0070   D8 B3 DD 3F 00 00 00 00-00 00 00 00 00 00 00 00   X3]?............
1065:0080   1E 2B C0 50 B8 65 10 8E-D8 E8 10 00 E8 23 00 E8   .+@P8e..Xh..h#.h
```

Data analysis:

sin = 3F 00 00 00

binary:	0011 1111 0000 0000 0000 0000 0000 0000
sign:	0 for positive
biased exp:	0111 1110 normalize: 7E – 7F = –1
significand:	000000..00
sci. binary:	1.00..00 E–1 = .1
decimal:	.5

cos = 3F 5D B3 D8

binary:	0011 1111 0101 1101 1011 0011 1101 1000
sign:	0 for positive
biased exp:	0111 1110 normalize: 7E – 7F = –1
significand:	10111011011001111011000
sci. binary:	1.10111011011001111011000 E–1 = .110111011011001111011000
decimal:	.8660254476

tan = 3F13CD3A

binary:	0011 1111 0001 0011 1100 1101 0011 1010
sign:	0 for positive
biased exp:	0111 1110 normalize: 7E – 7F = –1
significand:	00100111100110100111010 E–1
sci. binary:	1.00100111100110100111010 E–1 = .100100111100110100111010
decimal:	.577350259

cot = 3F DD B3 D8

binary:	0011 1111 1101 1101 1011 0011 1101 1000
sign:	0 for positive
biased exp:	0111 1111 normalize: 7F – 7F = 0
significand:	10111011011001111011000 E0
sci. binary:	1.10111011011001111011000
decimal:	1.732050896

```
.8087
PAGE 60,132
          .MODEL SMALL
;program to calculate SIN, COS, TAN, and COT of a 30-degree angle
          .STACK 32
;--------------------------
          .DATA
          ORG    00H
ANGLE     DD     0.523598776 ;angle in radians for 30 degrees
          ORG    10H
X         DD     0
          ORG    20H
Y         DD     0
          ORG    30H
R         DD     0
          ORG    40H
SIN       DD     0
          ORG    50H
COS       DD     0
          ORG    60H
TAN       DD     0
          ORG    70H
COT       DD     0
;--------------------------
          .CODE
START     PROC   FAR
          MOV    AX,@DATA
          MOV    DS,AX
          CALL   CALC_X_Y
          CALL   CALC_R
          CALL   CALC_SIN
          CALL   CALC_COS
          CALL   CALC_TAN
          CALL   CALC_COT
          MOV    AH,4CH
          INT    21H
START     ENDP
;---------------------------
;procedure to calculate X and Y given an angle
CALC_X_Y     PROC        NEAR
             FINIT                           ;initialize 8087
             FLD         ANGLE               ;load ANGLE onto stack
             FPTAN                           ;calculate X and Y
             FSTP        X                   ;store X and POP
             FSTP        Y                   ;store Y and POP
             RET
CALC_X_Y     ENDP
;----------------------------
;procedure to calculate hypotenuse given X and Y
CALC_R       PROC   NEAR
             FINIT                           ;initialize 8087
             FLD         X                   ;load X onto stack
             FMUL        ST(0),ST(0)         ;square X
             FLD         Y                   ;load Y onto stack
             FMUL   ST(0),ST(0)        ;square Y
             FADD        ST(0),ST(1)         ;calculate X**2 + Y**2
             FSQRT                     ;take square root
             FST         R                   ;store R
             RET
CALC_R       ENDP
;----------------------------
```

Program 20-5 *(continued on the following page)*

```
;procedure to calculate SIN, given R and X
CALC_SIN     PROC   NEAR
             FINIT                           ;initialize 8087
             FLD    R                  ;load R onto stack
             FLD           Y                 ;load Y onto stack
             FDIV          ST(0),ST(1)       ;SIN = Y/R
             FST           SIN               ;store SIN
             RET
CALC_SIN     ENDP
;------------------------------
;procedure to calculate COS, given R and X
CALC_COS     PROC   NEAR
             FINIT                           ;initialize 8087
             FLD           R                 ;load R onto stack
             FLD           X                 ;load X onto stack
             FDIV          ST(0),ST(1)       ;COS = X/R
             FST           COS               ;store COS
             RET
CALC_COS   ENDP
;------------------------------
;procedure to calculate TAN, given X and Y
CALC_TAN     PROC   NEAR
             FINIT                           ;initialize 8087
             FLD           X                 ;load X onto stack
             FLD           Y                 ;load Y onto stack
             FDIV          ST(0),ST(1)       ;TAN = Y/X
             FST           TAN               ;store TAN
             RET
CALC_TAN     ENDP
;------------------------------
;procedure to calculate COT, given X and Y
CALC_COT     PROC   NEAR
             FINIT                           ;initialize 8087
             FLD           Y                 ;load Y onto stack
             FLD           X                 ;load X onto stack
             FDIV          ST(0),ST(1)       ;COT = X/Y
             FST           COT               ;store COT
             RET
CALC_COT     ENDP
;------------------------------
                   END    START
```

Program 20-5 *(continued from the previous page)*

Trig functions

Example 20-6 uses trig functions. The instruction FPTAN (partial tangent) calculates Y/X = TAN Z, where Z is the angle in radians and must be 0 < Z < PI/4. Z is stored in ST(0) prior to execution of FPTAN. After the execution, ST(0) = X and ST(1) = Y. Then X and Y are used to calculate the hypotenuse R. After that, it is easy to calculate the sine, cosine, tangent, and cotangent. This process is shown in Program 20-5.

Note in Example 20-6 that in order to calculate sine and cosine we had to use the tangent; however, starting with the 80387 coprocessor there are specific instructions such as FSIN (sine) and FCOS (cosine) for these purposes. To invoke these instructions one must use the .387 directive for the assembler.

Integer numbers

Although performance of real numbers in the x87 is very impressive, integer operations should not be overlooked. One way to appreciate this performance is to compare the addition of two multibyte numbers, each 64 bits, on the 8088/86/286 and on the 8087. Since in the 8086/88/286, AX is only 16 bits wide, it will take a loop of 4 iterations to add a 64-bit number plus the overhead of moving the four 16-bit words in for each number and moving the result in and out of the CPU. The same addition can be performed by

```
;This program adds two positive integer numbers and stores the result.
.8087
PAGE 60,132
       .MODEL SMALL
       .STACK 32
;--------------
       .DATA
       ORG    00H
INT1   DD     -50000
       ORG    10H
INT2   DD     25000000
       ORG    20H
SUM    DD     ?
;--------------
       .CODE
START PROC    FAR
       MOV    AX,@DATA
       MOV    DS,AX
       CALL   ADD_INT
       MOV    AH,4CH
       INT    21H
START ENDP
;--------------
;procedure to add two integers
ADD_INT       PROC   NEAR
       FINIT         ;initialize 8087
       FILD   INT1   ;load integer 1 onto stack
       FIADD INT2    ;add the second integer
       FIST   SUM    ;store result in SUM
       RET
ADD_INT       ENDP
;--------------
       END    START

-d 1065:0 1f
1065:0000  0A 57 B7 42 00 00 00 00-00 00 00 00 00 00 00 00   .W7B............
1065:0010  0C 40 CE 46 00 00 00 00-00 00 00 00 00 00 00 00   .@NF............
```

r = 42 B7 57 0A Hex

binary =	0100 0010 1011 0111 0101 0111 0000 1010
sign:	0 for positive
biased exp:	1000 0101 normalize: 85 – 7F = 6
significand:	01101110101011100001010
sci. binary:	1.01101110101011100001010 E6 = 1011011.10101011100001010
decimal:	91.6702481689454

area = 46 CE 40 0C

binary =	0100 0110 1100 1110 0100 0000 0000 1100
sign:	0 for positive
biased exp:	1000 1101 normalize: 8D – 7F = 14
significand:	10011100100000000001100
sci. binary:	1.10011100100000000001100 E 14 = 110011100100000.000001100
decimal:	26400.0232375

Program 20-6

the x87 with only 4 instructions. In the x87, integer number instructions are distinguished from real number instructions by the letter "I". For example, the instruction FILD loads an integer number into ST(0) while the FLD would do the same thing for real numbers. One important point about differences between real and integer negative numbers is that integer negative numbers are stored in 2's complement. In real negative numbers, the only difference between a number and its negative is the sign bit. These numbers are not stored in 2's complement. Program 20-6 will show the assembler representation of a negative integer.

Review Questions

1. In Assembly language programming, which data directive is used for single-precison data?
2. In Assembly language programming, which data directive is used for double-precison data?
3. State the number of general-purpose registers in the x87.
4. x87 registers are accessed according to _________ (LIFO, FIFO).
5. True or false. In the 8086, "AX" always refers to the same physical register, but in the x87 "ST(2)" could be assigned to different physical registers at different times.
6. The top of stack is referred to as _______.
7. ST(1) is the register _________ (above, below) ST(0).
8. What does "FADD ST(4)" do? What are the operands, and where is the result kept?
9. What is the purpose of the instruction "FINIT"?
10. Instructions using integer data have the letter_____ as part of their mnemonics.
11. What is the difference between the ADD and FADD instructions?
12. True or false. The 8087 has an instruction named "FSIN" to calculate sine.

SECTION 20.3: x87 INSTRUCTIONS

This section provides an overview of the x87 instructions.

Real transfers

```
FLD src      ;pushes source operand onto ST(0)
             ;source may be ST(i) or memory
             ;Tip:  FLD ST(0) duplicates stack top

FST dest     ;copies ST(0) to destination
             ;dest may be ST(i), short or long real variable

FSTP dest    ;copies ST(0) to dest then pops ST(0)
             ;dest may be ST(i), short, long or temporary
             ;real memory
             ;Tip: "FSTP ST(0)" is equivalent to popping
             ;the stack with no data transfer

FXCH dest    ;swaps contents of ST(0) and destination
             ;FXCH with no operands swaps ST(0) and ST(1)
             ;Tip: frequently used to move a register to
             ;the top before
             ;using an instruction which assumes ST(0)
```

Integer transfers

```
FILD src     ;converts source to temporary real and pushes
             ;onto ST(0)

FIST dest    ;rounds ST(0) to integer and copies to
             ;destination
             ;dest may be a word or short integer

FISTP dest   ;functions the same as FIST but then pops
             ;ST(0)
             ;dest may be any binary integer data type
```

Packed decimal transfers

```
FBLD src   ;converts source contents to temporary real
           ;then pushes onto ST(0)

FBSTP dest ;converts ST(0) to BCD and stores at
           ;destination, then pops stack
```

Addition

```
FADD dest,src;adds src to dest, storing result in dest
           ;If no operands are given, ST(0) becomes
           ;ST(0) + ST(1)
           ;If one operand is given, destination is
           ;ST(0), source is operand
           ;source may be ST(i) or real data variable
           ;dest may be ST(i)

FADDP dest,src;adds src to dest then pops ST(0)
           ;dest may be ST(i), src is ST(0)

FIADD src  ;adds src to ST(0)
```

Subtraction

```
FSUB dest,src;subtracts src from dest, stores result in
dest
           ;If no operands are given, ST(1) becomes
           ;ST(1)-ST(0)
           ;If one operand is given, it will be src with
           ;ST(0) as the destination

FSUBP dest,src;subtract src from dest and store in dest
           ;src is ST(0) dest is ST(i)

FISUB src  ;subtract source from ST(0) and store in ST(0)
```

Reversed subtraction

```
FSUBR dest,src;functions the same as FSUB but subtracts
           ;dest from src
           ;and stores the result in dest (R is for
           ;reverse)

FSUBRP dst,src;functions the same as FSUBP but subtracts
           ;dest from src
           ;instead of src from destination

FISUBR src ;operates the same as FISUB but subtracts
           ;ST(0) from source
           ;and stores in ST(0)
```

Multiplication

```
FMUL dest,src;multiplies dest by src and stores result in
           ;dest
           ;If no operands are given, ST(0) becomes ST(0)
           ;x ST(1)
           ;If one operand is given, destination is
           ;ST(0), source is operand
```

```
              ;source may be ST(i) or real data variable
              ;dest may be ST(i)

FMULP dest,src;multiplies dest by src, stores result in
              ;dest and pops src
              ;src is ST(0), dest is ST(i)

FIMUL src     ;multiplies ST(0) by src and stores result in
              ;ST(0)
```

Division

```
FDIV dest,src;divides dest by src and stores result in
              ;dest
              ;If no operands are given, ST(0) becomes
              ;ST(0) / ST(1)
              ;If one operand is given, ST(0) is divided by
              ;the src operand
              ;source may be ST(i) or real data variable
              ;dest is ST(i)

FDIVP dest,src;divides dest by src and stores result in
              ;dest, then pops src
              ;src is ST(0),dest is ST(i)

FIDIV src     ;divides ST(0) by src and stores result in
              ;ST(0)
```

Reversed division

```
FDIVR dest,src;functions identical to FDIV except src is
              ;divided by dest

FDIVRP dst,src;functions the same as FDIVP except src is
              ;divided by dest

FIDIVR src    ;functions the same as FIDIV except src is
              ;divided by dest
```

Other arithmetic instructions

```
FSQRT         ;replaces ST(0) with its square root

FSCALE        ;replaces ST(0) with ST(0) x 2n, where n is
              ;the integer in ST(1)
              ;this provides fast method of multiplying by
              ;integral powers of 2

FPREM         ;ST(1) is repeatedly subtracted from ST(0)
              ;until ST(0) <ST(1)
              ;same as ST(0) mod ST(1)

FRNDINT       ;rounds ST(0) to an integer
              ;rounds according to RC (round control) bits
              ;in the control word
FXTRACT       ;extracts the exponent from ST(0) and places
              ;in ST(1)
              ;extracts the significand from ST(0) and places
              ;in ST(0)
```

```
FABS        ;replaces ST(0) with its absolute value

FCHS        ;reverses sign bit in ST(0)
```

Compare instructions

The following instructions compare ST(0) with the source operand and set condition code bits C3, C2, and C0 of the status word as follows:

C3	C2	C0	
0	0	0	ST(0) > source
0	0	1	ST(0) < source
1	0	0	ST(0) = source
1	1	1	numbers cannot be compared

```
FCOM        ;compare ST(0) with real operand
FICOM       ;compare ST(0) with integer operand
```

The source operand may be ST(i) or a real number. If no source operand is given, ST(0) is assumed.

The condition codes are set as follows:

C3	C2	C1	C0	Meaning
0	0	0	0	+unnormal
0	0	0	1	+NAN (not a number)
0	0	1	0	–unnormal
0	0	1	1	–NAN
0	1	0	0	+normal
0	1	0	1	+infinite
0	1	1	0	–normal
0	1	1	1	–infinite
1	0	0	0	+0
1	0	0	1	empty
1	0	1	0	–0
1	0	1	1	empty
1	1	0	0	+denormal
1	1	0	1	empty
1	1	1	0	–denormal
1	1	1	1	empty

Transcendental instructions

```
FPTAN       ;computes tangent of theta = y/x
            ;theta is in ST(0) and must be between 0 and
            ;pi/4
            ;after the ratio is computed, y replaces theta
            ;in ST(0)
            ;and x is pushed onto the stack, becoming the
            ;new top of stack

FPATAN      ;computes theta = arctan (y/x)
            ;x is in ST(0) and y in ST(1)
            ;ST(0) is popped, theta is written over y in
            ;ST(1), the new stack top

F2XM1       ;computes y = 2^x-1
```

```
              ;x is taken from ST(0) and must be in the
              ;range: -1 to +1
              ;y replaces x in ST(0)

FYL2X         ;computes z = y times log2 x
              ;x is taken from ST(0) and y from ST(1)
              ;x must be greater than 0
              ;z replaces y, which becomes the new stack top
              ;as x is popped off

FYL2XP1       ;computes z = y times log2 (x+1)
              ;x is from ST(0) and is in the range:
              ;0 <|x| <(1 - x1/2 / 2)
              ;y is from ST(1)
              ;z replaces y, which becomes the new stack top
              ;as x is popped off
```

The following instructions are available only in 387 and later coprocessors.

```
FSIN          ;computes sin of ST(0) but provides x = ST(0)
              ;and y = ST(1)
              ;to get the sin of ST(0), perform y/x using
              ;FDIV or to get the cos of ST(0), perform x/y
              ;using FDIVR

FCOS          ;same as FSIN, takes cos of ST(0) and places
              ;x = ST(0) and y = ST(1)
              ;to get cos of ST(0), perform y/x using FDIV
              ;or to get sin of ST(0), perform x/y using
              ;FDIVR

FSINCOS       ;computes both sin and cos of ST(0)
              ;places sin in ST(0) and cos in ST(1)
```

Constant instructions

```
FLDZ          ;pushes +0.0 onto the stack
FLD1          ;pushes +1.0 onto the stack
FLDPI         ;pushes pi onto the stack
FLD2T         ;pushes log2 10 onto the stack
FLDL2E        ;pushes log2 e onto the stack
FLDLG2        ;pushes log10 2 onto the stack
FLDLN2        ;pushes loge 2 onto the stack
```

Many mathematical equations can be implemented using constant and transcendental functions. For example, $x^y = 2^{y \log x}$. If $z = y \log_2 x$, FYL2X can be used to calculate z. Then F2XM1 can be used to calculate $2^z - 1$. Then 1 can be added to this to get 2^z, which is equal to x^y.

The instruction sequence would be

```
FLD   Y
FLD   X
FYL2X
F2XM1
FLD1
```

```
FADD
FST   SUM
```

Other frequently used functions can likewise be calculated, for example e^y and 10^y, substituting e or 10 for x in the above equations. In addition, a little creativity will allow a programmer to use the constant and transcendental functions frequently. For example, if the calculation $\log_2 x$ is needed, the FYL2X ($y \log_2 x$) function can be used by making $y = 1$. Likewise, if 2^x is needed, the F2XM1 function can be used, after which 1 can be added to the result.

Processor control instructions

Many of the following instructions have two mnemonics; the second one has an extra N. This N instructs the CPU not to prefix the instruction with a wait state. The no-wait forms should be used when CPU interrupts are disabled and the 8087 might generate an interrupt, which would create an endless wait. Wait forms are used when the CPU interrupts are enabled.

```
FINIT or     ;resets the processor
FNINIT

FDISI or     ;sets the interrupt enable mask in the control
             ;word
FNDISI       ;thereby disabling interrupts in the 8087

FENI or      ;clears the interrupt enable mask in the
             ;control word
FNENI        ;thereby enabling interrupts in the 8087

FLDCW src    ;replaces the control word with the contents
             ;of src

FSTCW dest or;writes the control word to dest
FNSTCW

FSTSW dest or;writes status word to dest
FNSTSW

FCLEX or     ;clears exception flags, busy flag
FNCLEX       ;and interrupt request flag in the status word

FSAVE dest or;writes to dest the 94-byte save area, which
FNSAVE       ;includes the environment and the stack

FRSTOR src   ;restores the 94-byte save area from src

FSTENV dest or;stores environment (control, status
FNSTENV      ;and tag words and exception pointers) to dest

FLDENV src   ;restores environment previously saved with
             ;FSTENV instruction

FINCSTP      ;increments status word's stack pointer

FDECSTP      ;decrements status word's stack pointer

FFREE dest   ;marks dest as an empty register
```

```
FNOP        ;stores ST(0) to ST(0), therefore performs no
            ;operation

FWAIT       ;same as CPU's wait instruction
            ;used to synchronize CPU and 8087
```

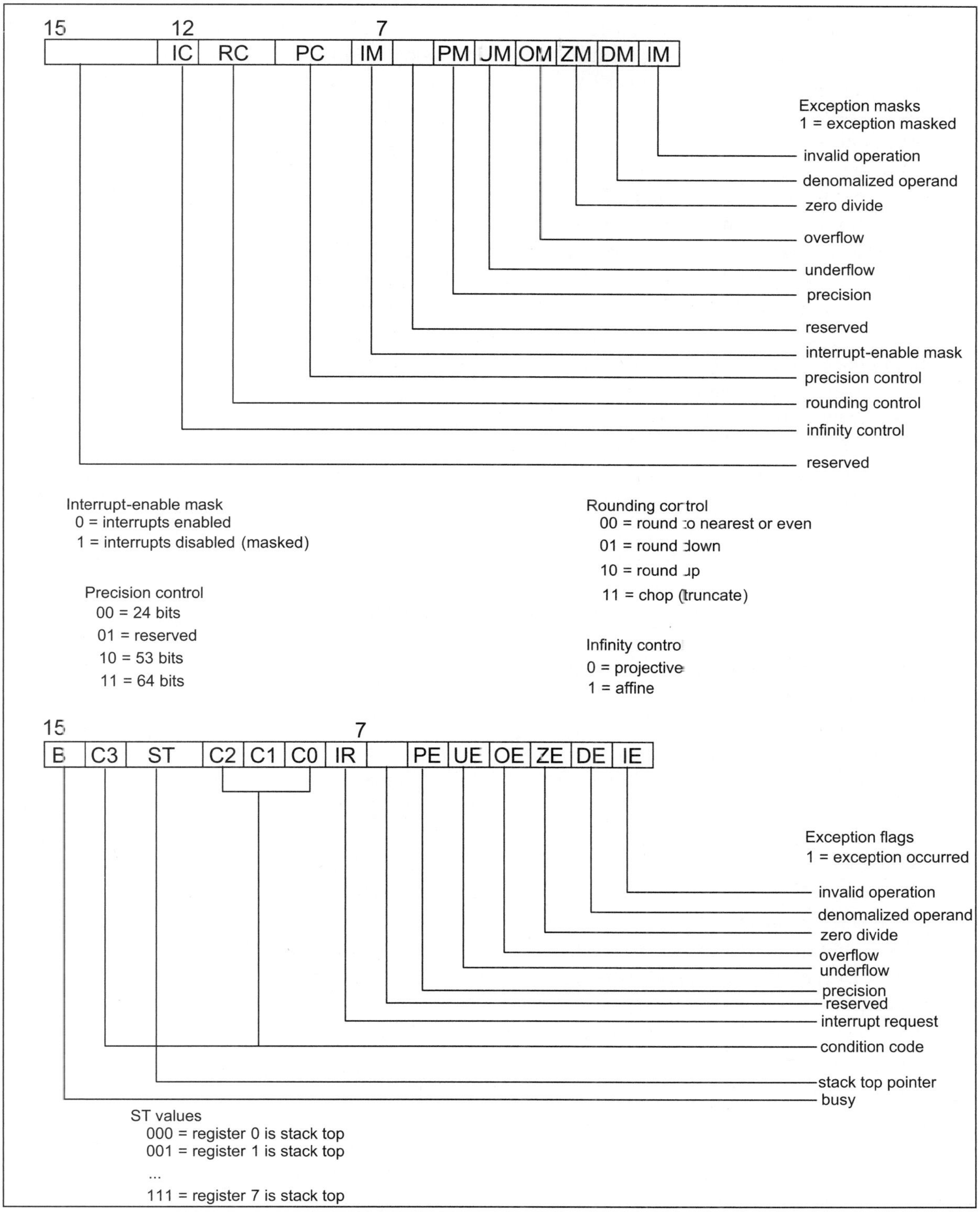

Figure 20-3. 8087 Control and Status Word

PROBLEMS

SECTION 20.1: MATH COPROCESSOR AND IEEE FLOATING-POINT STANDARDS

1. What is the disadvantage of using a general-purpose processor to perform math operations?
2. The IEEE single-precision standard uses _____ bytes to represent a real number.
3. The IEEE double-precision standard uses ______ bytes to represent a real number.
4. Show the bit assignment of the IEEE single-precision standard.
5. Convert (by hand calculation) each of the following real numbers to IEEE single-precision standard.
   (a) 15.575 (b) 89.125 (c) –1022.543 (d) –0.00075
6. Use the last four digits of your ID number and put the decimal point in the middle. Convert it to single-precision IEEE standard (e.g., 9823 is 98.23).
7. What data types are called short real and long real in Intel's literature?
8. Show the bit assignment of the IEEE double-precision standard.
9. In single-precision FP (floating point), the biased exponent is calculated by adding ________ to the _________ portion of a scientific binary number.
10. In double-precision FP, the biased exponent is calculated by adding ________ to the _________ portion of a scientific binary number.
11. Convert the following to double-precision FP.
    (a) 12.9823 (b) 98.76123
12. How many bits are set aside for the magnitude portion of Intel's long integer?
13. Which bits of packed decimal are used for the sign?
14. Packed decimal uses only _____ bits of an 80-bit operand.
15. In Intel's temporary real, the data type is _____ bytes wide.

SECTION 20.2: x87 INSTRUCTIONS AND PROGRAMMING

16. Indicate the data directive used for the following data types.
    (a) single-precision FP
    (b) double-precision FP
    (c) packed decimal
17. Using the assembler of your choice, verify your calculation of Problems 5 and 11.
18. Write and run an x87 program to calculate $z = (x^2 + y^3)^{1/2}$, where $x = 3.12$ and $y = 5.43$.
19. Write and run an x87 program to calculate $y = 2x^2 + 5x + 12.34$, where $x = 1.25$.
20. Write and run an x87 program to calculate the area of a circle if $r = 25.5$.
21. Write and run an x87 program to calculate $3(\pi r^3)/4$ if $r = 25.5$.
22. Write and run an x87 program to calculate the sine of a 45-degree angle.
23. Which of the following processors have an on-chip coprocessor?
    (a) 80286
    (b) 80386
    (c) 80486
    (d) Pentium
24. For the x87 to calculate a trig function, the angle must be in ____________ (degrees, radians).
25. Write the 80387 program to calculate the sin and cos of a 30-degree angle. Run and analyze your program.

ANSWERS TO REVIEW QUESTIONS

SECTION 20.1: MATH COPROCESSOR AND IEEE FLOATING-POINT STANDARDS

1. True
2. True
3. 32
4. 64
5. 7FH
6. 3FFH

SECTION 20.2: x87 INSTRUCTIONS AND PROGRAMMING

1. DD (define double word)
2. DQ (define quad word)
3. 8
4. LIFO
5. True
6. ST(0)
7. Below
8. It means ST(4) + ST(0) and the result is placed in ST(0).
9. To initialize the registers to the top of the stack
10. I
11. The assembler generates the opcode for the ADD instruction to be used by the x86 while it produces the opcode for the FADD instruction to be used by the x87.
12. False. 387 and later coprocessors have the FSIN instruction.

CHAPTER 21

386 MICROPROCESSOR: REAL vs. PROTECTED MODE

OBJECTIVES

Upon completion of this chapter, you will be able to:

- >> List the additional features implemented on the 80186, 80286, and 80386
- >> State the purpose of designing two modes, real and protected, into the 80386
- >> Code Assembly language instructions using the new scaled index addressing mode of the 386
- >> Code Assembly language instructions using the new instructions of the 386
- >> State the purpose of each pin of the 80386 microprocessor
- >> Describe the data misalignment problem with 386 programs and discuss how to resolve the problem
- >> Describe how protection of user and system programs is accomplished in the 386 in protected mode
- >> Contrast and compare the two methods of virtual memory implementation: paging and segmentation
- >> Describe the methods of converting from logical to physical addresses in 386 protected mode

This chapter emphasizes unique features of the 80386 microprocessor, from both hardware and software perspectives. In Section 21.1 we look at the 386 in real mode. The hardware of the 386 is examined in Section 21.2. Section 21.3 provides an introduction to protected mode of the 386.

SECTION 21.1: 80386 IN REAL MODE

In this section we look first at Intel's 80186 microprocessor and then unique features of the 286 and 386 from the perspective of real mode programming.

Evolution of x86 from 80186 to 80386

Intel has a very successful product called the 80186 (and 80188). This chip is alive and doing very well in the embedded controller market, where it is used to replace multiple devices with a single component. Prior to the introduction of the 80186/88, Intel did a survey and found that many are using the 8088/86 along with other peripheral chips, such as the 8237 DMA controller, the 8254 timer, and the 8259 interrupt controller. This led to putting a portion of these chips along with the 8088/86 microprocessor on a single chip and calling it the 80186/88. Internally, the 80186 and 80188 are identical, but externally the 80186 has a 16-bit data bus and the 80188 has an 8-bit external data bus. In this regard they are similar to the 8086 and 8088. The address bus is still 20-bit, making a 1-megabyte memory system. The data bus is multiplexed with the address bus. The 80186/88 is a 68-pin chip that includes the following on-chip functions: (1) clock generator, (2) two 20-bit DMA channels, (3) three 16-bit programmable counters, (4) interrupt controller, (5) programmable wait-state generator, and (6) programmable chip select decoder unit. Although very few 80186/88 microprocessors are used in PCs, millions of them are found in embedded systems. The 80186/88 microprocessor supports all 8088/86 instructions in addition to some new ones. The new instructions of the 80186/88 are as follows:

```
BOUND dest,source
ENTER disp,level
LEAVE
IMUL  result,source,immediate data
INS   dest,port
OUTS  port,dest
SAR   dest,immediate count
SHR
SAL
RCR
ROR
RCL
ROL
PUSH  immediate data
PUSHA
POPA
```

Some of the above instructions, such as ENTER and LEAVE, are intended for implementation in high-level languages but many others can be used in everyday Assembly language programs. For example, look at the shift and rotate instructions. In the 8088/86 to shift or rotate an operand more than once required putting the count in CL; immediate operands could not be used. Starting with the 80186/88, immediate counts are allowed. Look at Example 21-1.

Other useful new instructions of the 80186/88 are PUSHA (push all) and POPA (pop all). Very often in writing a procedure (subroutine) all the registers need to be saved on the stack. In the 8088/86 one must code PUSH and POP for each 16-bit register separately; however, in the 80186/88 the use of PUSHA and POPA can save a lot of coding.

Example 21-1

Show Assembly language code to shift operand right 26H by 5 bits in each of the following systems.
(a) 8088/86 (b) 80186/88

Solution:

(a) 8088/86	(b) 80186/88
MOV AL,26H	MOV AL,26H
MOV CL,5	SHR AL,5
SHR AL,CL	

Example 21-2

Show a sequence of x86/88 instructions equivalent to 80186/88 PUSHA and POPA.

Solution:

8088/86	80186/88
PUSH AX	PUSHA
PUSH CX	
PUSH DX	
PUSH BX	
PUSH SP	
PUSH BP	
PUSH SI	
PUSH DI	
...	...
POP DI	
POP SI	
POP BP	
POP SP	
POP BX	
POP DX	
POP CX	
POP AX	POPA
RET	RET

Note that POPA restores all registers except SP, which is ignored. In other words, it does not disturb the present stack frame.

See Example 21-2.

One can test these new instructions on 286 PCs and later machines. However, using these instructions means that the program will not run on 8088/86 machines. There are two dominant trends in software for x86-based systems.

1. Software that runs on any x86 machine, including 8088/86-based systems.
2. Software that is 32-bit 386 based and must be run on 386 and higher machines.

Also note that the DEBUG utility does not support the new 80186/88 instructions since it is intended to run on any x86 PC, including the 8088/86.

80286 microprocessor

The demand for a more powerful CPU led Intel Corporation to use more than 100,000 transistors to design a new microprocessor called the 80286. This processor is downwardly compatible with the 80186/88 processor. The 80286 microprocessor was a major improvement over the core 8086 in the following ways.

1. There are separate pins for the address and data buses and thus no need for demultiplexing the buses as was the case in the 8088/86 microprocessor. This increased number of pins required abandoning DIP (dual in-line packaging). Instead, PGA (pin grid array) packaging was chosen.
2. The memory cycle time was reduced to 2 clocks from 4 clocks in the 8088/86. This made memory interfacing quite a challenge, especially for frequencies of 20 MHz and beyond. Memory design of high-performance computers is discussed in Chapter 22.
3. Introduction of virtual memory in the 80286 was the most drastic change over the 8088/86. The 80286 works in two different modes: real mode and protected mode. In real mode, the 80286 is simply a faster 8086, capable of handling only 1M byte of memory. It executes all the instructions of the 8086 with fewer clock cycles, as shown in Appendix E. In order to use the entire 16 megabytes of memory space, the 80286 must work in protected mode. When the 80286 microprocessor is turned on, it automatically starts from real mode and can be switched to protected mode. When in protected mode, all the address buses A0–A23 can be used, thereby giving a total of 16M bytes of addressable physical memory (RAM and ROM). It is in this mode that most of the changes over the 8086 have been introduced into the 80286. Due to the declining price of the 80386, very few systems use the 286 in protected mode; therefore, we bypass any discussion of the 80286 in protected mode. However, since protected mode of the 80286 is a subset of the 80386, many of the concepts of 386 protected mode apply to the 286.

Major changes in the 80386

The 80386 microprocessor started a new trend in the x86 family. Although it is downwardly compatible with the 8088/86 and 80286, there are some major changes in its architecture. The following are some of the major changes that have been introduced in the 80386.

1. The data bus was increased from 16 bits to 32 bits, both internally and externally.
2. All the registers were extended to 32 bits, thereby making the 80386 a 32-bit microprocessor.
3. The address bus was increased to 32 bits, thus providing 4 gigabytes (232) of physical memory addressing capability.
4. The paging virtual memory mechanism was introduced, making the 80386 capable of using both segmentation and paging. More about paging and segmentation is provided in Section 21.3.
5. A new addressing mode called scaled index was added.
6. Many new bit-manipulation instructions were added. These instructions work in both real mode and protected mode.
7. The 386 can be switched from protected to real mode by software. This is a major improvement over the 286, which had to be reset to switch back to real mode.

To reduce the cost of board design, Intel made the 80386SX microprocessor available with a 16-bit external data bus, but internally it remained a 32-bit processor, 100% compatible with the 80386. In terms of memory bandwidth, it is slower than the 80386

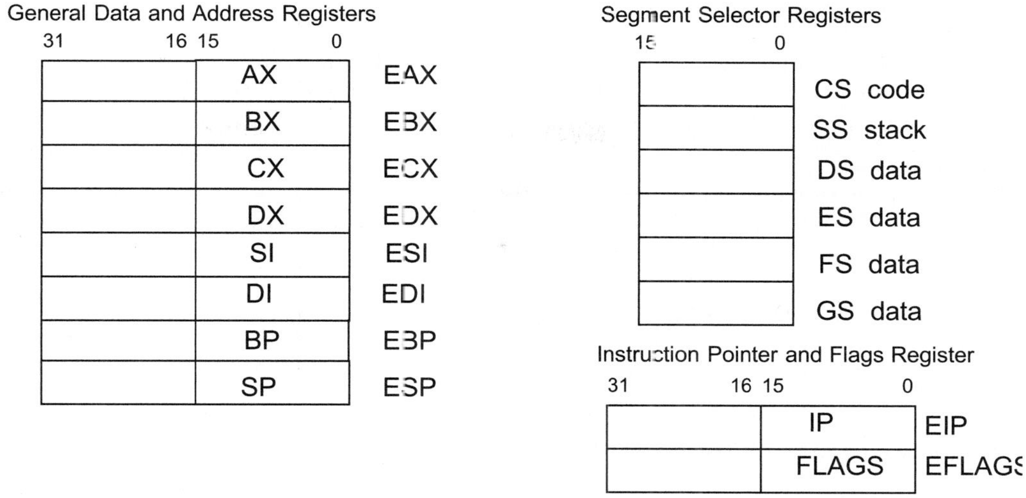

Figure 21-1. Selected Intel 386 Registers

since it takes two memory cycles (each 2 clock cycles) to address a 32-bit word instead of only 1 memory cycle, as is the case in the 80386. Intel also made the 80386SX with only 24-bit address buses, the same as the 80286. In other words, the 80386SX is the same as the 80286 externally, but internally it is a 32-bit processor fully compatible with all 386 computers.

In real mode, the 80386 can access a maximum of 1 megabyte using address pins A19–A0. However, in protected mode the 80386 can access 4 gigabytes of memory through using the 32-bit address bus.

80386 real mode programming

In the design of the 80386, Intel made such massive design changes that it is radically different from the 80286, yet it is still capable of running all the code written for the 286 and 8088/86. Next we describe some of these new features that are available in both real and protected modes.

32-bit registers

In the 80386, register sizes were extended from 16 bits to 32 bits and register names were changed to reflect this. For example, EAX is the extended AX, EBX is the extended BX, and so on. See Figure 21-1. In order to access 32-bit registers, the letter E must be included in the coding. The four general-purpose registers AX, BX, CX, and DX are still accessible in their 8086 formats in addition to the extended format. For example, register EAX is accessible as AL, AH, AX, and EAX. Notice that the upper 16-bit part is not accessible as a separate register. One way to access it is to shift EAX right. See Example 21-3.

Example 21-3

Load EAX with 7698E35FH and move it among the 8-, 16-, and 32-bit registers of the 386.

Solution:

```
MOV   EAX,7698E35FH     ;EAX=7698E35F (AX=E35F,AH=E3,AL=5F)
MOV   EDX,EAX           ;EDX=EAX=7698E35F
MOV   CH,AL             ;CH=AL=5F
MOV   DI,AX             ;DI=AX=E35F
MOV   ESI,EDX           ;ESI=EDX=7698E35F
ROR   EAX,16            ;rotate EAX right 16 times (EAX=E35F7698H)
MOV   BX,AX             ;BX=AX=7698H
MOV   CL,AL             ;CL=AL=98H
```

Which end goes first?

In storing data, the 386 followed the tradition of the 8086/286 in placing the least significant byte (little end of the data) in the low address. As discussed previously, this is referred to as little endian. See Example 21-4.

Example 21-4

Show how data is placed after execution of the following code.

```
        MOV    EAX,7698E39FH       ;EAX=7698E39F
        MOV    [4524],AX
        MOV    [8000],EAX
```

Solution:

For "MOV [4524],AX" we have

```
DS:4524 =(9F)
DS:4525 =(E3)
```

and for "MOV [8000],EAX" we have

```
DS:8000 = (9F)
DS:8001 = (E3)
DS:8002 = (98)
DS:8003 = (76)
```

In Example 21-4, notice how the least significant byte (the little end of the data) 9FH goes to the low address 8000, and the most significant byte of the data 76H goes to the high address 8003. This means that the little end of the data goes in first, hence the name little endian. In the Motorola 68000 family, data is stored the opposite way: The big end (most significant byte) goes into the low address first, and for this reason it is called big endian. Some recent RISC processors, such as Power PC (developed jointly by IBM and Motorola), allow selection of mode, big endian or little endian. The software overhead of converting from one camp to the other led Intel to introduce a new instruction called SWAP in the 80486, specifically to take care of this problem (see Chapter 23).

General registers as pointers

Another major change introduced in the 80386 is the use of general registers such as EAX, ECX, and EDX as pointers. As you might recall from Chapter 1, the 8088/86/286 can use only BX, SI, and DI as pointers into the data segments. But starting with the 386, all 32-bit general-purpose registers can be used for pointers into data segments. Look at the following cases for valid and invalid instructions.

```
MOV   BX,WORD PTR [EAX]    ;move into BX word pointed to by EAX
MOV   BX,WORD PTR [AX]     ;invalid AX can't be used as pointer
MOV   EAX,DWORD PTR [ECX]  ;move into EAX DWORD pointed to by ECX
MOV   AL,BYTE PTR [EDX]    ;move into AL BYTE pointed to by EDX
MOV   EBX,WORD PTR [CX]    ;invalid CX can't be used as pointer
MOV   EAX,DWORD PTR [EDI]  ;move into EAX DWORD pointed to by EDI
```

The 386 also allows the use of displacement for 32-bit register pointers. Therefore, instructions such as "MOV AL,[ECX+100]" are perfectly valid. Of course, the 386 supports all the addressing modes of the 8086/286 discussed in Chapter 1. Table 21-1 shows some of the addressing modes supported by the 386.

Scaled index addressing mode

One of the most powerful addressing modes introduced in the 386 is *scaled index addressing* mode. It allows access of multidimensional arrays with ease. In scaled index

Table 21-1: Addressing Modes for the 80386

Addressing Mode	Operand	Default Segment
Register	Register	None
Immediate	Data	None
Direct	[OFFSET]	DS
Register indirect	[BX] [SI] [DI] [EAX] [EBX] [ECX] [EDX] [ESI] [EDI]	DS DS DS DS DS DS DS DS DS
Based relative	[BX]+disp [BP]+disp [EAX]+disp [EBX]+disp [ECX]+disp [EDX]+disp [EBP]+disp	DS SS DS DS DS DS SS
Indexed relative	[DI]+disp [SI]+disp [EDI]+disp [ESI]+disp	DS DS DS DS
Based indexed relative	[R1][R2]+disp R1 and R2 are any of the above	If BP is used, segment is SS; otherwise, DS is the segment

Note: In based indexed relative addressing, disp is optional.

addressing mode, any of the 32-bit registers, except ESP, can be used as a pointer that is multiplied by a scaled factor of 1, 2, 4, or 8. The scaling (multiplication) factors 1, 2, 4, and 8 correspond to byte, word, doubleword, and quadword operands, respectively. Look at Example 21-5 to see how the effective address is calculated in cases where the scaled index addressing mode is used. Only the 32-bit register pointers can be used for this mode. They are shown in Table 21-2.

Example 21-6 shows how the scaled index addressing mode is used. It must be noted that we cannot use a 16-bit register as a scaled index. In other words, the instruction "MOV AL,[ESI+BX*4]" is invalid.

It must be noted that for an Assembly language program to be run under DOS, the effective address should not exceed FFFFH. In other words, if EBX is used as a pointer, you must make sure that the upper 16 bits of the EBX register are all zero, since DOS works in real mode.

Table 21-2: 386 Scaled Index Addressing Mode

Scaled Index	Default Segment
[EAX]	DS
[EBX]	DS
[ECX]	DS
[EDX]	DS
[ESI]	DS
[EDI]	DS
[EBP]	SS
[ESP]	SS

Example 21-5

Find the effective address in each of the following cases. Assume that ESI = 200H, ECX = 100H, EBX = 50H, and EDI = 100H.

(a) MOV AX,[2000+ESI*4]
(b) MOV AX,[5000+ECX*2]
(c) MOV ECX,[2400+EBX*4]
(d) MOV DX,[100+EDI*8]

Solution:

(a) EA (effective address) is 2000H + 200H × 4 = 2000 + 800H = 2800H. Therefore, the logical address of the operand moved into AX is DS:2800H.

(b) By the same token we have EA = 5000H + 100H × 2 = 5000H + 200 = 5200H.

(c) EA = 2400H + 4 × 50H = 2400H + 140H = 2540H.

(d) 100H + 8 × 100H = 100H + 800H = 900H.

Example 21-6

Using the scaled index addressing mode, write an Assembly language program to add 5 operands of 32-bit size and save the result.

Solution:

```
            .MODEL SMALL
            .386
            .STACK 300H
            .DATA
MYDATA      DD     234556H,0F983F5H,6754AE2H,0C5231239H,0AF34ACB4H
RESULT      DQ     ?
            .CODE
            MOV    AX,@DATA
            MOV    DS,AX
            SUB    EBX,EBX                ;EBX=0
            MOV    EDX,EBX                ;clear EDX
            MOV    EAX,EBX                ;clear EAX
            MOV    CX,5                   ;set the counter to 5
BACK:       ADD    EAX,[ MYDATA+EBX* 4]   ;add the 32-bit operand
            ADC    EDX,0                  ;save the carry
            INC    EBX                    ;point to next 32-bit data
            DEC    CX                     ;decrement the counter
            JNZ    BACK                   ;repeat until counter is zero
            MOV    DWORD PTR RESULT,EAX   ;save the lower 32 bits
            MOV    DWORD PTR RESULT+4,EDX ;save the upper 32 bits
            ;place code here to return to OS
```

In this program, we first define the 32-bit data using the DD directive, and the RESULT is defined as 64-bit using the DQ directive. Notice that EBX is initially zero; therefore, the instruction "ADD EAX,[MYDATA +EBX*4]" adds the first 32-bit operand to EAX since the effective address is MYDATA. "INC EBX" makes EBX = 1; therefore, in the next iteration the effective address is [MYDATA+1 * 4], and likewise in the next iteration the effective address is [MYDATA+2 * 4], which is MYDATA+8, and so on. For example, if the offset address for MYDATA is 2000H, the effective address is 2000H for the first iteration, 2004H for the second iteration, 2008H for the third iteration, 200CH for the fourth iteration, and so on.

Some new 386 instructions

There are many new instructions in the 386 that work in both real and protected modes. A detailed look at each new instruction and how it is used is beyond the scope of this volume. Here are some of the new instructions with examples.

MOVSX and MOVZX instructions

As we discussed in Chapter 6, the 8086 has sign-extend instructions such as CBW (D7 of AL is copied into all the AH bits) and CWD (D15 of the AX is copied into all bits of the DX). In the 386, there is a new instruction CDQ (convert doubleword to quadword) in which the sign bit of EAX, D31, is copied to all the bits of EDX. Notice that in all sign-extend instructions, the accumulator sign is extended. To overcome this limitation, Intel introduced the MOVSX and MOVZX instructions. In the MOVSX, the sign bit of any register (or even a memory location) can be extended (copied) into any register. Similarly, MOVZX zero-extends the contents of a register or memory location. The MOVSX instruction is used to sign-extend the operand in signed number arithmetic to prevent overflow problems. The MOVZX instruction is used in unsigned arithmetic. Look at Example 21-7.

Example 21-7

Find the contents of destination registers after execution of the following code.

```
(a)  MOV    BL,-5          (b) MOV    DL,+9
     MOVSX  CX,BL              MOVSX  EBX,DL
(c)  MOV    AL,95H         (d) MOV    BH,83H
     MOVZX  ECX,AL             MOVZX  AX,BH
```

Solution:

MOVSX copies the source register into the lower bits of the destination register and copies the sign bit into all upper bits of the destination register. Therefore, we have the following.

```
(a)  MOV      BL,-5          ;BL=1111 1011B =FBH (2's complement)
     MOVSX CX,BL ;CL=FBH,CH=FF since BL is copied into
                             ;CL and the sign bit (D7) is copied into
                             ;all CH bits. BL is unchanged
```

(b) DL = 0000 1001B = 09H. Then BL = 09 and D8–D31 of EBX are all zero, the sign bit of DL. Therefore, EBX = 00000009.

```
(c)  MOV      AL,95H     ;AL =1001 0101B =95H
     MOVZX ECX,AL        ;AL =CL =95H and D8-D31 of  ECX are all zeros
                         ;therefore, ECX =00000095H
```

(d) BH = 1000 0011B = 83H. Then AL = BH = 83H and D8–D15 of AX are all zeros. Therefore, AX = 0083H.

Bit scan instructions

The 386 has new instructions allowing a program to scan an operand from LSB to MSB or from MSB to LSB, to find the first high bit (= 1). If the scanning is done from the least significant bit (D0) toward higher bits, the BSF (bit scan forward) instruction is used. If the scanning is done from the most significant bit (D31) toward the lower bits, the BSR (bit scan reverse) instruction is used. In these instructions whenever the first high is found, the scanning is stopped and the position of the bit is written into the destination register. The bit position is numbered from D0 (LSB) to D31 (MSB), regardless of the direction of scanning. See Example 21-8.

Example 21-8

Find the register contents after the execution of the following code.

```
(a)     MOV    BX,4578H
        BSF    DX,BX ;scan BX and put the position of the first high into DX
(b)     MOV    ECX,3A9H
        BSR    EAX,ECX ;scan ECX from D31 down and put position of
                       ;first high into EAX
```

Solution:

(a) DX = 03 since scanning 4578H = 0100 0101 0111 1000B from right to left yields 1 in D3.
(b) EAX = 9 since scanning 000003A9H = 0000 0000 0000 0000 0000 0011 1010 1001 from D31 toward D0 yields the first high in D9; therefore, EAX = 9.

Review Questions

1. The 80188/86 is a(n) _____ (8-bit, 16-bit) processor.
2. What is the size of the external data bus on the 80186?
3. In which x86 was the concept of virtual memory introduced?
4. In which x86 was the protected mode concept introduced?
5. The 80286 works in which of the following?
   (a) real mode (b) protected mode
   (c) both (a) and (b) (d) 8086 virtual mode
6. True or false. The 32-bit registers of the 386 can be accessed only in protected mode.
7. Find the contents of BL, BH, BX, and EBX after execution of instruction "MOV EBX,99FF77AAH".
8. The 80386 uses the _____________ (little endian, big endian) convention.
9. List all the 32-bit registers that can be used as pointers into the data segment.
10. In the instruction "MOV EBX,[EAX+ESI*8]", find the effective address if EAX = 2000 and ESI = 100 (both in hex).
11. Scaled index addressing mode can be used with which of the following registers?
    (a) SI (b) EDI (c) EAX
    (d) DX (e) ECX (f) CX
12. Find the contents of EDX after execution of the following code.
    ```
    MOV     DL,-9
    MOVSX   EDX,DL
    ```
13. Find the contents of ECX after execution of the following code.
    ```
    MOV     DL,-5
    MOVZX   ECX,DL
    ```
14. Find the contents of DX and AX after execution of the following code.
    ```
    MOV    BX,1998H
    BSF    DX,BX
    BSR    AX,BX
    ```

SECTION 21.2: 80386: A HARDWARE VIEW

We present a hardware view of the 386 in this section. To avoid confusion, Intel calls the 80386 with a 32-bit external data bus the 80386 and 80386SX refers to the 386 with a 16-bit external data bus. Figures 21-2 and 21-3 provide a block diagram and pin layout of the 80386, respectively. Signal functions are provided in Table 21-3.

Overview of pin functions of the 80386

D31–D0 (data bus)

These provide the 32-bit data path to the system board. They are grouped into 8-bit data chunks, D0–D7, D8–D15, D16–D23, and D24–D31. Each 8-bit data bus is accessed by a separate byte enable pin (BE).

A31–A2 and BE0, BE1, BE2, BE3

These provide the 32-bit address path to the system board. Notice the absence of A0, A1, or BHE seen in earlier generations of the x86. Since the 80386 supports data types of byte (8 bits), word (16 bits), and double word (32 bits), the external buses must be able to access any of the 4 banks of memory connected to the 32-bit data bus. BE0–BE3 are used to access each bank independently. BE, which stands for byte enable, is active low and used for bank selection. According to Table 21-4, to select D7–D0, BE0 is used, BE1 is for D15–D8, and so on. See Figure 21-4.

Table 21-3: Data Bus Selection and BE

Data Bus	Byte Enable
D7 – D0	BE0
D15 – D8	BE1
D23 – D16	BE2
D31 – D24	BE3

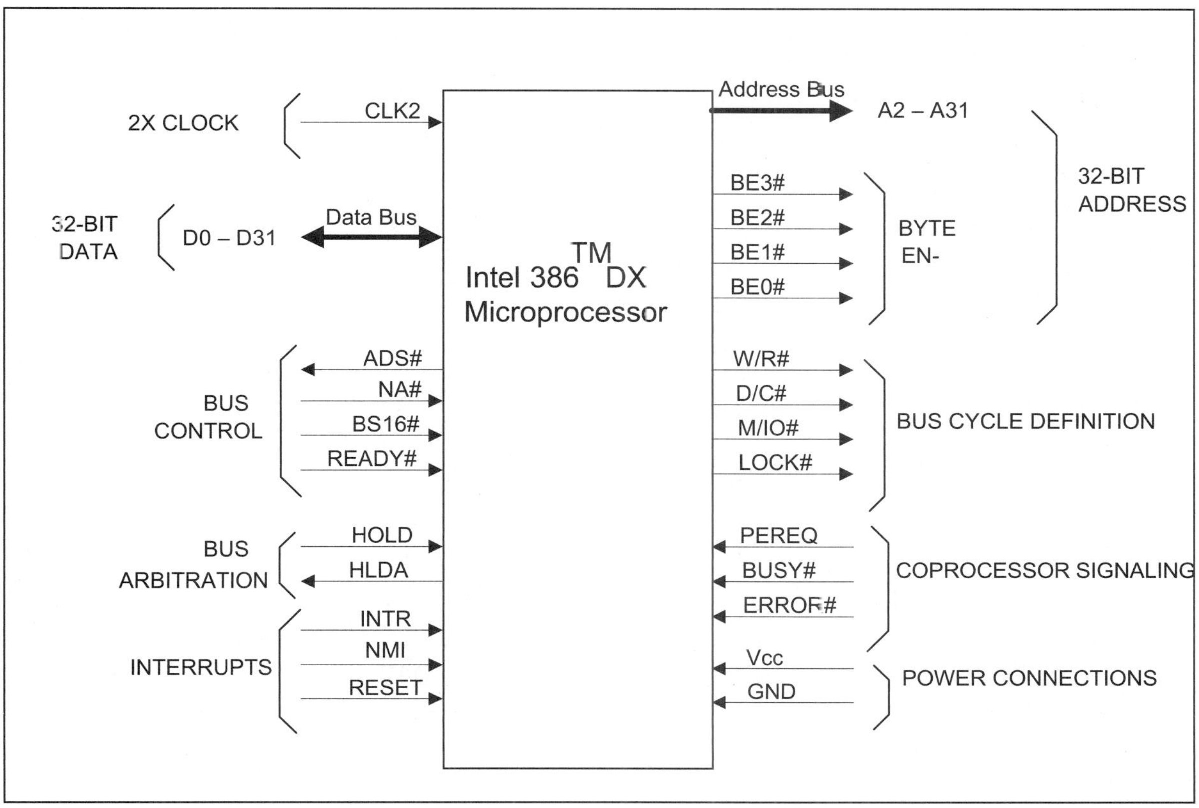

Figure 21-2. 80386 Block Diagram (# indicates active low)

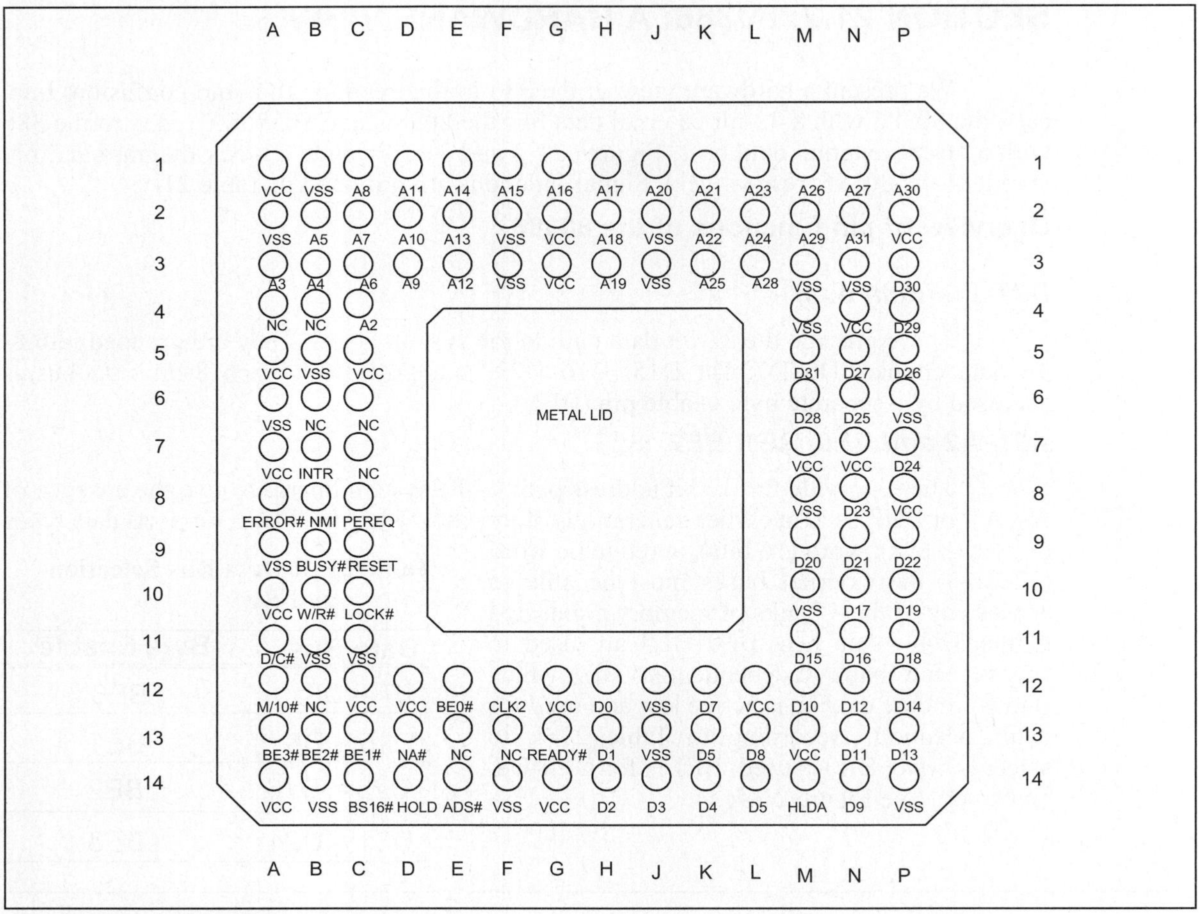

Figure 21-3. 80386 PGA Pins

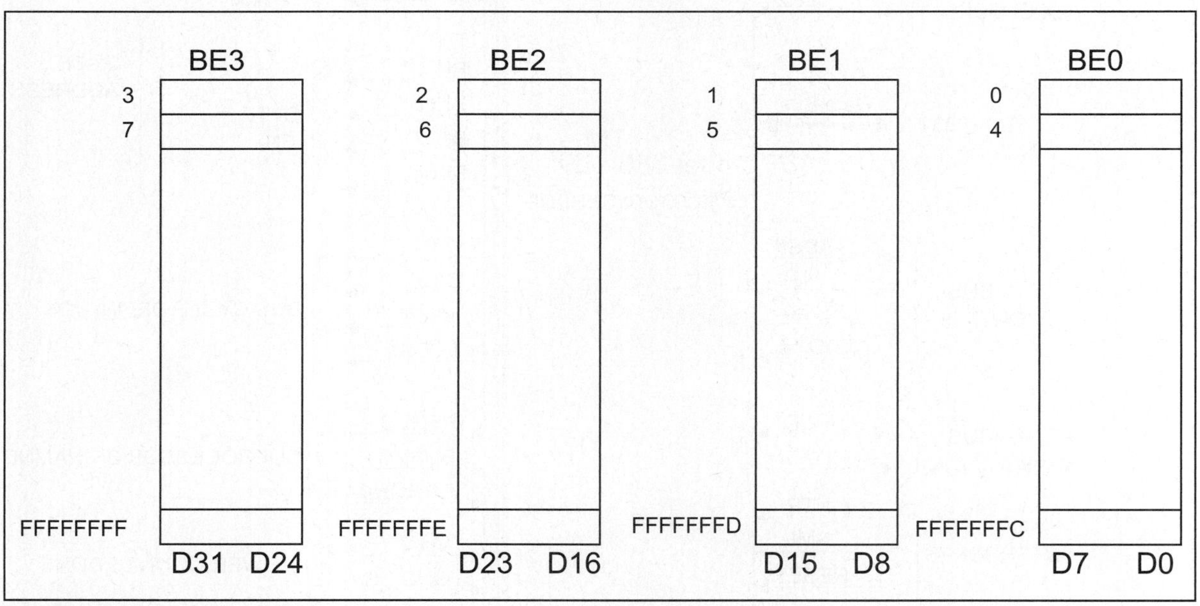

Figure 21-4. 80386 Banks

ADS, BS16, NA, and READY

ADS (address status), BS16 (bus size), NA (next address request), and READY are bus control signals. These signals allow the implementation of an efficient bus control

Table 21-4: Intel 386 DX PGA Pinout

Signal/Pin		Signal/Pin		Signal/Pin		Signal/Pin		Signal/Pin		Signal/Pin	
A2	C4	A24	L2	D6	L14	D28	M6	Vcc	C12	Vss	F2
A3	A3	A25	K3	D7	K12	D29	P4	Vcc	D12	Vss	F3
A4	B3	A26	M1	D8	L13	D30	P3	Vcc	G2	Vss	F14
A5	B2	A27	N1	D9	N14	D31	M5	Vcc	G3	Vss	J2
A6	C3	A28	L3	D10	M12	D/C#	A11	Vcc	G12	Vss	J3
A7	C2	A29	M2	D11	N13	ERROR#	A8	Vcc	G14	Vss	J12
A8	C1	A30	P1	D12	N12	HLDA	M14	Vcc	L12	Vss	J13
A9	D3	A31	N2	D13	P13	HOLD	D14	Vcc	M3	Vss	M4
A10	D2	ADS#	E14	D14	P12	INTR	B7	Vcc	M7	Vss	M8
A11	D1	BE0#	E12	D15	M11	LOCK#	C10	Vcc	M13	Vss	M10
A12	E3	BE1#	C13	D16	N11	M/IO#	A12	Vcc	N4	Vss	N3
A13	E2	BE2#	B13	D17	N10	NA#	D13	Vcc	N7	Vss	P6
A14	E2	BE3#	A13	D18	P11	NMI	B8	Vcc	P2	Vss	P14
A15	F1	BS16#	C14	D19	P10	PEREQ	C8	Vcc	P8	W/R#	B10
A16	G1	BUSY#	B9	D20	M9	READY#	G13	Vss	A2	N.C.	A4
A17	H1	CLK2	F12	D21	N9	RESET	C9	Vss	A6	N.C.	B4
A18	H2	D0	H12	D22	P9	Vcc	A1	Vss	A9	N.C.	B6
A19	H3	D1	H13	D23	N8	Vcc	A5	Vss	B1	N.C.	B12
A20	JI	D2	H14	D24	P7	Vcc	A7	Vss	B5	N.C.	C6
A21	K1	D3	J14	D25	N6	Vcc	A10	Vss	B11	N.C.	C7
A22	K2	D4	K14	D26	P5	Vcc	A14	Vss	B14	N.C.	E13
A23	L1	D5	K13	D27	N5	Vcc	C5	Vss	C11	N.C.	F13

circuitry. For example, using BS16 allows the 80386 to be connected to the 16-bit data bus instead of the 32-bit. The use of NA (next address) provides the option of address pipelining, where the address of the next memory cycle is provided in the last clock cycle of the present memory cycle.

W/R, D/C, and M/IO

These signals provide the bus cycle definitions and the type of the bus cycle according to Table 21-5.

Table 21-5: 80386 Bus Cycle Definition

M/$\overline{IO}$	D/$\overline{C}$	W/$\overline{R}$	Bus Cycle Type
0	0	0	Interrupt acknowledge
0	0	1	Does not occur
0	1	0	Data read (I/O)
0	1	1	Data write (I/O)
1	0	0	Memory code read
1	0	1	Halt (shutdown)
1	1	0	Memory data read
1	1	1	Memory data write

Example 21-9

Indicate which part of the data bus is selected for the following BEs.

(a) BE3 BE2 BE1 BE0 = 0000 (b) BE3 BE2 BE1 BE0 = 0011

(c) BE3 BE2 BE1 BE0 = 1100 (d) BE3 BE2 BE1 BE0 = 1101

Solution:

(a) D31–D0, the entire 32-bit data bus

(b) D31–D16, the upper 16-bit data bus

(c) D15–D0, the lower 16-bit data bus

(d) the 8 bits of D15–D8

Figure 21-5 shows the above data selection graphically. Note that BE is active low.

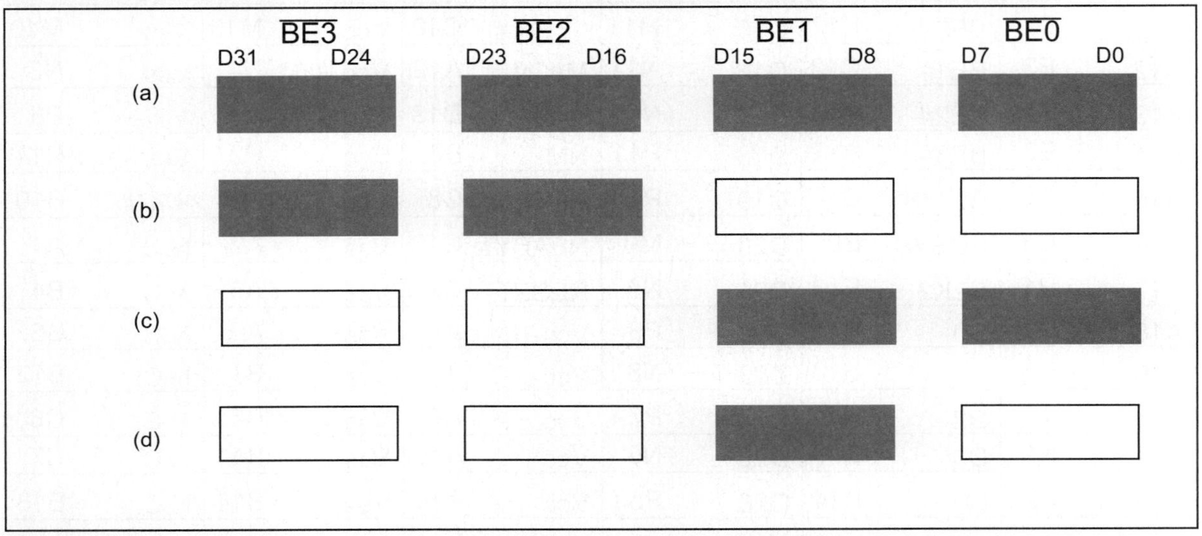

Figure 21-5. Graphical Representation of Example 13-9 (Selected Byte Is Shaded)

Example 21-10

A 80386 system is advertised as 33 MHz. What frequency is connected to CLK2?

Solution:

CLK2 = 66 MHz because the frequency connected to CLK2 is always twice the system frequency.

CLK2

This provides the timing for the 386. The frequency connected to CLK2 is always twice the system frequency. For example, a 16-MHz 386 system requires CLK2 to be 32 MHz.

RESET

This is a level-sensitive input signal into the 80386. When a low-to-high signal is applied to RESET, the 80386 will suspend all operations and the registers are initialized to fixed values. The RESET state of EIP and CS must be noted, along with the state of A31–A2 and BE0–BE3, because this has some major implications as far as where the boot ROM should be located (see Table 21-6). This means that the microprocessor will fetch the first opcode from memory location FFFFFFF0. This is 16 bytes from the 4-gigabyte maximum address range of FFFFFFFFH. At this location, there is either a JMP FAR or a

CALL FAR instruction. Upon executing the JMP or CALL instruction, the 386 makes A31–A20 all zero, thereby forcing it to stay within the 1-megabyte address range of real mode. This is the case for all 386, 486, and Intel Pentium chips. All these processors wake up in real mode, but the address where the first opcode must be found is located in the extended memory space and not in the first megabyte address space of real mode. This means that for 386 and higher PC systems, there are duplicate ROMs in both the 4-gigabyte and 1-megabyte address spaces, as shown in Figure 21-6.

Table 21-6: RESET State

Item	Contents
CS	F000
EIP	0000FFF0
A31 – A2	All high
$\overline{BE3}$ – $\overline{BE0}$	All low

(Reprinted by permission of Intel Corporation, Copyright Intel Corp. 1992)

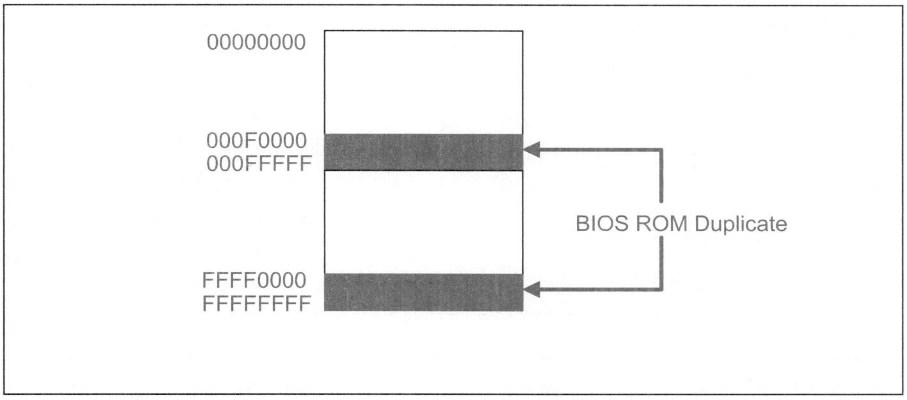

Figure 21-6. BIOS ROM Duplicate for 386/486/Pentium PC

The remaining signals of the 386 are similar to those for the 80286, and readers can refer to Chapter 10 for their meanings.

Bus bandwidth in the 386

With zero wait states, it takes the 386 two clocks to perform the read or write cycle. A 2-clock bus cycle is standard in all high-performance microprocessors, including RISC processors. This leads to a very high bus bandwidth. The two clocks of the 386 memory (or I/O) bus cycle time for zero wait states are shown in Figure 21-7. In the case of pipelined read/write cycle time, the next address is provided in the last T clock of the present cycle, thus providing some extra time for the decoder logic circuitry, path delay, and memory access time. Although in pipelined mode the next address is provided in the

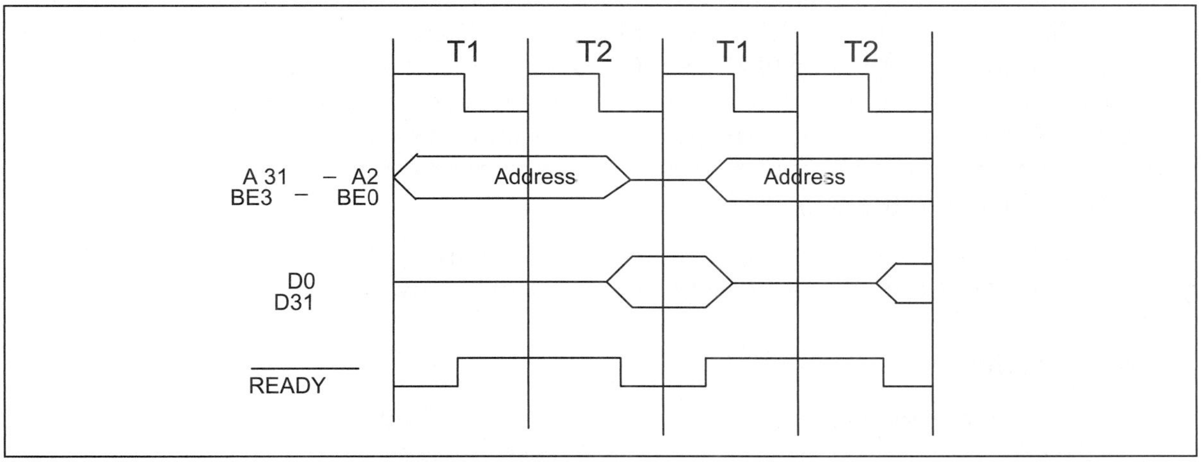

Figure 21-7. 386 Bus Cycle Time (Nonpipelined)

last stage of the present cycle, the read and write cycle time still consists of 2 clocks for the zero-wait-state system.

Example 21-11

Calculate the 386 bus bandwidth of a 33-MHz system with each of the following.
(a) 0 WS (b) 1 WS

Solution:

With the T state of 30 ns (1/33 MHz = 30 ns), we have memory cycle time of 60 ns and 90 ns for (a) and (b), respectively.
(a) The bus bandwidth is (1/60 ns) × 4 = 66.66 megabytes/second.
(b) The bus bandwidth is (1/90 ns) × 4 = 44.44 megabytes/second.

Data misalignment in the 386

The case of misaligned data has a major effect on the 386 bus performance. If the data is aligned, for every memory read cycle the 80386 brings in 4 bytes of data using the D31–D0 data bus. Such data alignment is referred to as doubleword alignment. To make data doubleword aligned, the least significant digits of the hex addresses must be 0, 4, 8, or C (hex). Look at Example 21-12.

Example 21-12

Show the data transfer of the following cases and indicate the memory cycle time if the system frequency is 25 MHz. Assume that EAX = 4598F31EH and the system is in real mode.

(a) MOV [2950],EAX
(b) MOV [299A],EAX

Solution:

The system frequency of 25 MHz makes the cycle time 80 ns (1/25 MHz = 40 ns and each memory cycle is 2 clocks, giving 80 ns).

(a) In this instruction, the 4-byte content of EAX is moved to memory location with starting offset address of 2950H on the 32-bit data bus of D31–D0. This address is doubleword aligned since the least significant digit is 0. Therefore, it takes only one memory cycle or 80 ns to transfer the data.
(b) In the first memory cycle, locations with addresses of 2998H, 2999H, 299AH, and 299BH are accessed, but only 299AH and 299BH are used for storing AL and AH. In the second memory cycle, the address offsets of 299CH, 299DH, 299EH, and 299FH are accessed where only 299CH and 299D are used to store the upper 16 bits of EAX. This means that we have a total of 160 ns. If possible, this must be avoided since nonaligned data slows the data access.

I/O address space in the 386

The 80386 can access a maximum of 65,536 input ports and 65,536 output ports using the IN and OUT instructions. In this regard, the 386 is exactly like the 8088/86/286 microprocessors.

Review Questions

1. BHE and A0 are associated with the ____________ (80386SX, 80386DX) processor.
2. The 80386SX is ____________ (16, 32) bits externally.
3. Exactly how many pins are set aside for the address in the 386?
4. The BE2 pin is associated with which part of the data bus?

5. An 80386 of 20 MHz requires a crystal frequency of __________.
6. Give the first physical address location where the 80386 looks for an opcode upon RESET.
7. With the same frequency, the 80386SX has bus bandwidth ___________ (twice, half) that of the 80386.
8. Find the memory cycle time for an 80386 of 20 MHz.

SECTION 21.3: 80386 PROTECTED MODE

The 80386 protected mode discussion applies equally to 486 and Pentium chips. Due to the complexity associated with 80386 protected mode, many long chapters are needed for this subject, and for this reason, here we simply provide an overview of the 386 in protected mode.

Protection mechanism in the 386

As discussed in Chapter 1, physical addresses in the 8086 are calculated by shifting the segment register left and adding it to the offset. This is also the case for the 80286 and subsequent x86 processors in real mode. In protected mode, however, the physical address of blocks of data or code is held by a look-up table and the segment register is no longer shifted left to calculate the physical address. Instead, it is used as an index into a look-up table in which the physical address of the operand or code is held.

Another important change introduced in the 80386 is the protection mechanism. The lack of protection of the operating system or users' programs is one of the weaknesses of 8088/86-based MS-DOS. This weakness is due to the inability of the 8088/86 to block general instructions from accessing the core (kernel) of the operating system. In the 8088/86, any program can go from any code segment to any code segment, so it is easy to crash the system. In contrast, the 80386 provides resources to the operating system that prevent the user from either accidentally or maliciously taking over the core (kernel) of the operating system and forcing the system to crash. Of course, this idea of protection is nothing new; it is commonly used in mainframes and minicomputers, where it is often referred to as user and supervisor mode. The 386 provides protection by allowing any data or code to be assigned a privilege level. The four privilege levels are 0, 1, 2, and 3, where the privilege level of 0 is the highest and level 3 is the lowest. While operating systems are always assigned the highest privilege level (level 0), the user and applications such as word processors are assigned the lowest privilege level (level 3). Since the user is assigned the lowest privilege level, any attempt by the user to take over the operating system is blocked. Higher privilege levels can access lower levels, but not the other way around. Again, it must be emphasized that the protection mechanism can be used only when the 80386 is switched to protected mode.

Virtual memory

Another major feature of the 80386 is its ability to access virtual memory. A CPU with virtual memory is fooled into thinking that it has access to an unlimited amount of physical (DRAM) memory. DRAM primary memory is also called *main* memory. In this scheme, every time the CPU looks for certain information, the operating system will first search for it in main DRAM memory and if it is not there, it will bring it into RAM from secondary memory (hard disk). What happens if there is no room in RAM? It is the job of the operating system to swap data out of RAM and make room for new data. Which data will be swapped out depends on how the operating system is designed. Some operating systems use the LRU (least recently used) algorithm to swap data in and out of primary memory (DRAM). In the LRU method, the operating system keeps account of which data has been used the least number of times in a certain period, and when there is need for room it will swap out the least recently used data to hard disk to make room for the new data.The total amount of RAM on the computer could be only 2G with a hard disk capacity of 100G bytes, but the CPU is fooled into thinking that it has access to all 100G of memory. Among the operating systems, Microsoft Windows 2000, XP, and Vista, all the variations of Unix, Sun Microsystems' Solaris, and Mac OS X use the capability of the 80386's virtual memory. Since MS-DOS was written for the 8088/86 microprocessors, it does not have virtual memory.

To implement virtual memory, two methods are used: segmentation and paging. In segmentation, the size of the data swapped in and out can vary from 1 byte to a few megabytes (in 80386, 80486, and Pentium, the upper limit can be as high as 4 gigabytes). In paging, the size is a multiple of one page of 4096 (4K) bytes. Paging is used widely since it prevents memory fragmentation, where available memory becomes fragmented into small sections of varied sizes. When this happens, the operating system must continuously move files around to make room for the new files, which could be any size. Paging makes the job of the operating system much easier since all the files will be a multiple of 4K bytes. If the size of a file is not a multiple of 4K bytes (which is the case most of the time), the operating system will leave the unused portion empty and the next file will be placed on a 4K boundary. This is similar to the cluster in floppy and hard disks. As shown in Chapter 19, the disk allocates memory to each file in clusters. For example, if 4 sectors are used for each cluster, each cluster can store 2048 (4 × 512) bytes per sector. If a given file is 12,249 bytes, the operating system will assign a total of 7 clusters or 14,168 (7 × 2024 = 14,168) bytes. All bytes between 14,168 and 12,249 are unused. This results in wasting some memory space on the disk but at the same time makes the design of the disk controller and operating system much easier. This concept applies as well to the paging method of virtual memory as far as the allocation of main memory (DRAM) to data and code is concerned. One can briefly define the segmentation and paging virtual memory mechanisms in the following statement. In segmentation virtual memory, the file can be any byte size, located anywhere it can fit into main memory. In paging virtual memory, the file is always a multiple of 4096 bytes and located on a 4K-byte boundary in main memory.

All high-performance RISC microprocessors use paging virtual memory only and none use the segmentation method. The reason that 386, 486, and Pentium processors support segmentation (in addition to paging) is that they had to stay compatible with the 8086's 64K-byte segment size.

Segmentation and descriptor table

In segmentation virtual memory, the segment registers are used as selectors into the descriptor table, where all the information about a given piece of data and code is kept. The descriptor table uses 8 bytes of space to provide the following information about a given piece of code or data.

1. 4 bytes for the A0–A31 address, where the code (or data) is located in main memory. This allows the 386 to access any memory location within its 4-gigabyte address space. Notice in Figure 21-8 that A23–A0 is provided by bytes 2, 3, and 4, but A31–A24 is provided by byte 7.
2. L0–L19: This 20-bit limit is used for checking the segment size and is limited to 1 megabyte. Notice that bytes 0 and 1 provide L0–L15, and D0–D3 of byte 6 is set aside for L16–L19. This provides the scheme whereby the 1 megabyte limit imposed on data or code is checked. Since the limit for the segment-oriented 8086/286 is 64K bytes (216 = 64K), the upper 4 bits must all be zeros. In the 386, however, the segment limit can be raised to 4 gigabytes. To do that the G (granularity) bit is set to high. If G = 0, L0–L19 is used as a number of bytes for the limit, but if G = 1, L0–L19 is used as a multiple of 4K for the segment limit. This gives 220 × 212 = 232 = 4 gigabytes address range, making it possible for the 386 to have segments as large as 4 gigabytes. This is quite a relief for software writers of database and other application packages since the size of data (e.g., a big array) can go as high as 4 gigabytes and is no longer limited to 64K. In the case of the 286 when the size of the data section of the program was larger than 64K, they had to

do lots of software manipulation to overcome this limitation. This also explains the origins of the memory models of SMALL, MEDIUM, LARGE, and so on, widely used in Assembly and C programs.

3. The access byte allows protection of a given piece of data or code by assigning the privilege levels of 0, 1, 2, and 3 to it, where 0 indicates the highest privilege level and 3 is the lowest privilege level. D0–D7 of the access byte are described next.

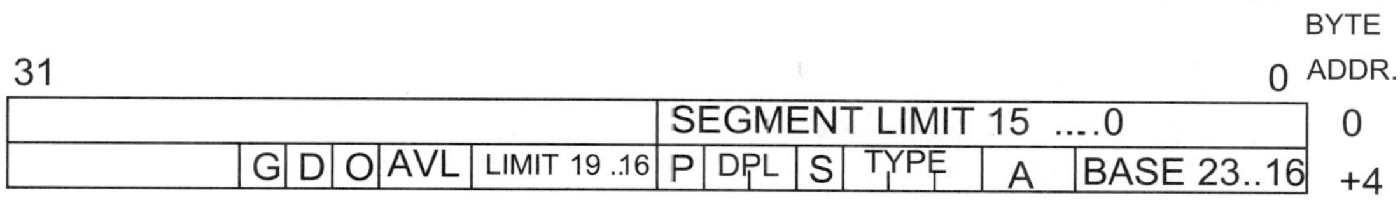

BASE	Base Address of the segment
LIMIT	The length of the segment
p	Present Bit 1 = Present 0 = Not Present
DPL	Descriptor Privilege Level 0–3
S	Segment Descriptor 0 = System Descriptor 1 = Code or Data Segment Descriptor
TYPE	Type of Segment (3 bits: X, E, R/W)
A	Accessed Bit
G	Granularity Bit 1 = Segment length is page granular 0 = Segment length is byte granular
D	
0	Bit must be zero for compatibility with future processors
AVL	Available field for user or OS

Note: In a maximum-size segment (i.e., a segment with G=1 and segment limit 19...0 = FFFFFH), the lowe 12 bits of the segment base should be zero (i.e., segment base 11...000 = 000H).

Figure 21-8. Descriptor Table Entry

A (accessed) bit

If the data or code is accessed (used), A = 1; otherwise, A = 0. This allows the operating system to monitor the A bit periodically to see if the CPU is using this piece of code or data. If a piece of code or data has not been used recently, the next time the operating system needs to make room in main memory for new pieces of code (or data), it can move this code (or data) back to the hard disk. The A bit also allows the operating system to decide if a given piece of information (code or data) needs to be saved. For example, if a piece of data has not been accessed, the operating system can trash it and avoid wasting time saving it on the hard disk. On the other hand, if the data was accessed and it was written into, the operating system must save a copy of it on the hard disk before it abandons it to create room in main memory for some other data or code.

R/W (read/write) bit

This bit allows code or data to be read protected or write protected. For example, the core of the operating system can be write protected, which prevents the user from writing into it and crashing the system. In the case of DOS, any program can use the DEBUG utility and alter the core of the operating system residing in main memory (DRAM), thereby crashing the PC.

X bit

This has a different meaning for the data segment and code segment. In the case of data, it indicates whether the segment should expand downward as the stack segment grows, or upward as the data segment grows. In the case of the code segment, it is used to enforce certain rules of privilege level access.

E bit

This indicates if the information is executable (E = 1), such as code, or nonexecutable (E = 0), such as data and stack. This bit also affects the way the X and R/W bits are interpreted.

S bit

This indicates if the descriptor belongs to the code and data segment (S = 1) or if it is a system segment descriptor (S = 0).

DPL (descriptor privilege level) bits

This allows one of the combinations, 00, 01, 10, or 11, to be assigned to the code or data, indicating the privilege level.

P (present) bit

This indicates if the piece of code or data is present in main memory (DRAM). If it is present (P = 1), the CPU will process it. If it is not present (P = 0), the CPU causes an exception and the exception handler of the operating system will bring the desired piece of code or data into main memory from the hard disk. When the operating system does so, it sets P = 1 to indicate that the information is now present in main memory.

The descriptor table is built by the operating system for every piece of code and data. The descriptor table register (DTR) inside the 386 holds the physical address of where the table is located in the 4-gigabyte address space, which means that the descriptor table register (DTR) is a 32-bit register. When the CPU changes the contents of a segment register (CS, DS, and so on), it uses the segment value as an index into the descriptor table and pulls into the CPU from the descriptor table all 8 bytes belonging to this segment. These 8 bytes are saved in the invisible part of the segment register inside the 386,

Example 21-13

From Figure 21-8 we have the following access byte for code and data.

P	DPL	1	1	...	...	A	(access byte for code segment)
P	DPL	1	0	...	...	A	(access byte for data segment)

Discuss the following access bytes.
(a) 10011011 (b) 10010111 (c) 11110001

Solution:

(a) This is an access byte for code segment, present, accessed, and privilege level of 00 (highest).
(b) This is an access byte for data segment, present, accessed, privilege level of 00 (highest), and both read and write accessible.
(c) This is an access byte for data segment, present, accessed, privilege level of 11 (lowest), and write protected.

which means that every segment register inside the 386 has an 8-byte extension, which is not visible to the programmer. The pulling of an 8-byte table into the CPU for every change of segment register is time consuming but afterward, the CPU has all the information it needs to access a piece of code or data. The addition of two new segment registers, FS and GS, in the 386, plus the presence of CS, DS, SS, and ES, helps the CPU always to have a total of 6 descriptor table entries available inside the CPU. If code or data is not held by one of these 6 descriptor table entries, the CPU must go through the long process (it takes 22 clock cycles) of pulling them into the CPU. As we will show later in this section, this problem is solved in the paging method.

Looking at the 8 bytes of the descriptor table, one might ask why Intel did not assign 32-bit physical addresses of desired code or data in consecutive bytes, instead of using bytes 2, 3, and 4, and then byte 7. The reason is the 80286 CPU. In the 286 protected mode, bytes 2, 3, and 4 are used for the 24-bit address (A0–A23), and bytes 6 and 7 had to be zero. This led Intel to use byte 7 for the A31–A24 part of the physical address of the 386. Byte 6 is used for raising the limit and the G bit, among other things. See Figure 21-8.

Local and global descriptor tables

There are two types of descriptor tables for the 386: the local descriptor table (LDT) and the global descriptor table (GDT). The GDT is used for the system, and individual tasks can have their own LDT. How do the segment registers know which one they are accessing? The third bit (TI) of the segment register (referred to as the selector) always indicates which table should be used. See Figure 21-9.

64 terabytes of virtual memory

As seen in Figure 21-9, the 14 bits of the selector (segment) register can have 16,384 (2^{14}) possible combinations. Each possible value can access a descriptor that can hold addresses of memory chunks as large as 4 gigabytes. Therefore, we have $2^{14} \times 2^{32} =$

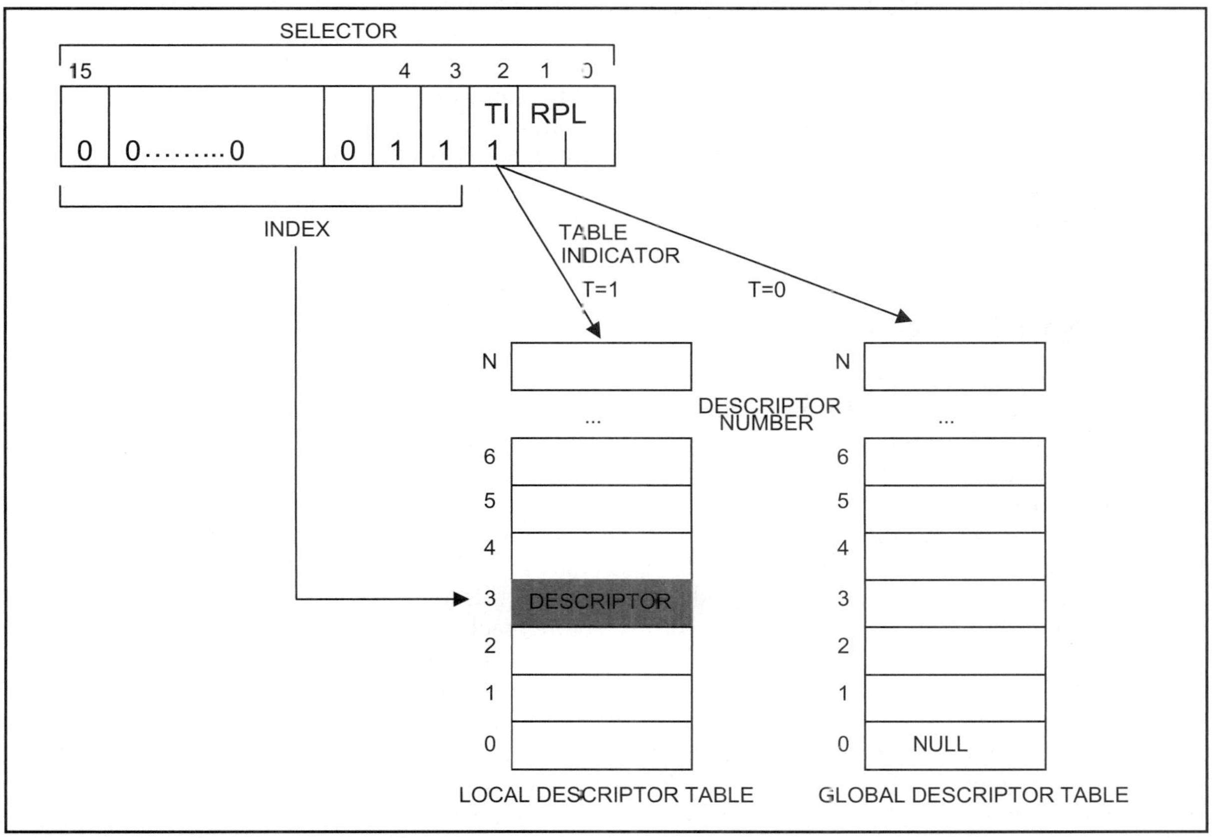

Figure 21-9. LDT and GDT Selection

64 terabytes of virtual memory for the 386 (recall that tera is defined as 2^{40}). To put it another way: The 386 can access 64 terabytes of hard disk (virtual memory) as long as the virtual memory is broken down into 4-gigabyte pieces, since it has only 32 address pins. While the segment limit in the 8086/286 is 64K bytes, the segment limit in the 386 was raised to 4G. One of the drawbacks of 386 segmentation is its variable segment size, which leads to memory fragmentation. Another is the absence of what is called a *dirty bit* in the access byte of the descriptor table. Assume that there is some memory that can be written into. The accessed (A) bit indicates if the data has been accessed but does not indicate if any new data was written into it. Why should the operating system care whether the memory is altered (written into)? If the data is altered, it is the job of the operating system to save it on the disk to make sure that the hard disk always has the latest data. If the dirty bit is zero (D = 0), it means that the data has not been altered and the operating system can abandon it when it needs room for new data (or code) since the original copy is on the hard disk. This will save time for the operating system. If the dirty bit is one (D = 1), the operating system must save the data before it is lost or abandoned. Both problems of variable segment size and lack of a dirty bit in segmentation are fixed in the paging method of virtual memory.

Paging

Paging of virtual memory was a new addition to the 386, but the segmentation method was left over from the 80286. All RISC and Motorola 680x0 processors support paging virtual memory. In paging virtual memory, main memory is divided into fixed 4K-byte chunks instead of variable sizes of 1 byte to 4 gigabytes, as in segmentation. If a given piece of code or data is not present in main memory, the operating system brings it into main memory from the hard disk, 4K at a time.This is a much more manageable size of memory to transfer than, for example, a 64K-byte segment. Since the size of memory is reduced to 4K bytes, the 386 keeps a table for the 32 most recently used pages present in main memory to prevent the CPU from swapping data in and out of main memory unnecessarily. This table is called the *translation lookaside buffer* (*TLB*) and is kept inside the 386. To understand the importance of the TLB, let's look at the way paging works. First, the term linear address in the 386 must be clarified. The 32-bit address of the operand is called the *linear address*. This linear address can be a direct value such as in the instruction "MOV EAX,[50000000]" or may be pointed to by any of registers EDI, ESI, EBX, EDX, and so on, as in the instruction, "MOV EAX,[EBX]". This linear address must be translated into a physical address to be put on the A31–A0 address pins and sent out for the address decoder to find the location in RAM or ROM. In other words, the address 50000000H in instruction "MOV EAX,[50000000]" does not refer to an actual RAM and ROM address 50000000H. See Figure 21-10.

Going from a linear address to a physical address

In paging, the linear address is divided into three parts. The upper 10 bits (A31–A22) are used for an entry into what is called a *page directory*. There is a 32-bit register, CR3, inside the 386 that holds the physical base address of the page directory. Since the upper 10 bits of the linear address point to the entry in the page table directory, there can be 1024 page directories (2^{10} = 1024). Each entry in the page directory is 4 bytes of page table descriptor. Of the 4 bytes of each page table descriptor, the upper 20 bits are used to point to another table, where the physical address of the 4K page frame is held. How is the correct entry in the table located? A21–A11 (10 bits total) of the linear address are used to point to one of the page table entries. Again, each entry in this second table has 4 bytes. The upper 20 bits are for A31–A12 of the physical address of where data is located. The lower 12 bits of the physical address are the lower 12 bits of the linear address. See Figures 21-11 and 21-12. In other words, only the lower 12 bits of the linear address match the lower 12 bits of the physical location in RAM or ROM where data is located, and the upper 20 bits of the linear address must go through two levels of translation tables to get the actual physical address of the beginning page where the data is held. This seems like a very long and inefficient process, and it is. This is the reason for the TLB

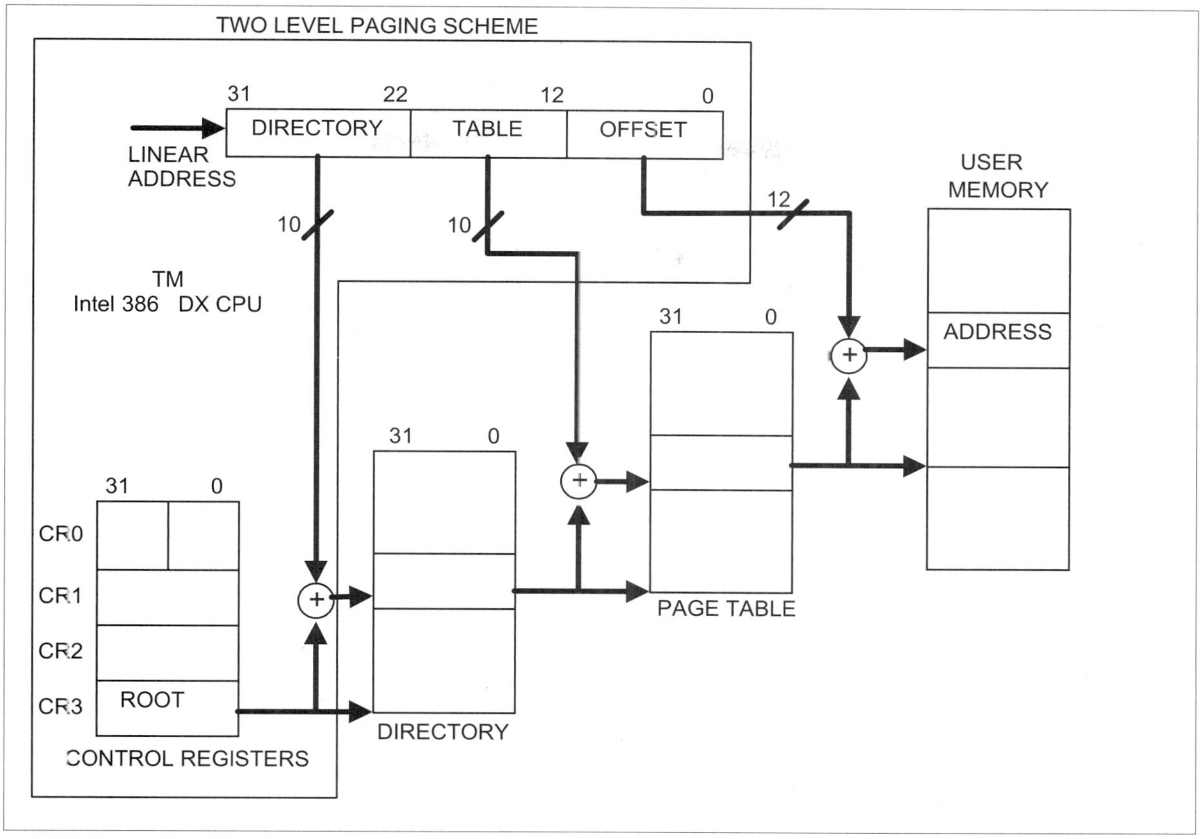

Figure 21-10. Paging Mechanism

31 … 12	11 10 9	8	7	6	5	4	3	2	1	0
PAGE TABLE ADDRESS 31..12	OS RESERVED	0	0	D	A	0	0	U/S	R/W	P

Figure 21-11. Page Directory Entry (Points to Page Table)

31 … 12	11 10 9	8	7	6	5	4	3	2	1	0
PAGE FRAME ADDRESS 31..12	OS RESERVED	0	0	D	A	0	0	U/S	R/W	P

Figure 21-12. Page Table Entry (Points to Page)

(translation lookaside buffer). The TLB inside the 386 holds the list of the most recently (commonly) used physical addresses of the page frames. When the CPU wants to access a piece of information (data or code) by providing the linear address, it first compares the 20-bit upper address with the TLB to see if the table entry for the desired page is already inside the CPU. This results in two possibilities: (1) If it matches, it picks the 20-bit physical address of the page and combines it with the lower 12 bits of the linear address to make a 32-bit physical address to put on the 32 address pins to fetch the data (or code); (2) if it does not match, the CPU must fetch into TLB the page table entry from memory.

Each entry in the page table has 4 bytes. Of these 4 bytes, 20 bits are used to hold the A31–A11 physical address of the page frame. The rest are used for the P (present) bit, D (dirty) bit, R/W (read/write) bit, A (accessed) bit, and finally, U/S (user/supervisor) bit, which indicates the privilege level of given data or code. In the segmentation method there were 2 bits for privilege level, giving rise to four levels of protection of 0, 1, 2, and 3, where level 3 was assigned to the lowest level and level 0 to the highest level. However, in the paging method, there is only 1 bit for privilege level, which is called U/S (user/supervisor). If U/S = 0, it is user privilege level and is equivalent to level 3 in segmentation. If U/S = 1, it is supervisor level, belonging to the operating system and system kernel. The supervisor privilege level is equivalent to levels 0, 1, and 2 in the segmentation method.

The bigger the TLB, the better

Since the TLB in the 386 keeps the list of addresses for the 32 most recently used pages, it allows the CPU to have access to 128K bytes (32 × 4 = 128) of code and data at any time without going through the time-consuming process of converting the linear address to a physical address (two-stage table translation). See Figure 21-13. Therefore, one way to enhance the processor is to increase the number of pages held by the TLB. This is what the Pentium has done, as we will see in Chapter 23. Table 21-7 compares paging and segmentation.

Table 21-7: Paging and Segmentation Comparison

Feature	Paging	Segmentation
Size	4K bytes	Any size
Levels of privilege	2	4
Base address	4K-byte aligned	Any address
Dirty bit	Yes	No
Access bit	Yes	Yes
Present bit	Yes	Yes
Read/write protection	Yes	Yes

Virtual 8086 mode

A major dilemma for designers of the Intel 386 was how to enhance the 386 and still run 8088/86 software based on MS-DOS in protected mode. They solved this dilemma by adding the virtual 8086 mode to the 386. In virtual 8086 mode, the 386 partitions memory into 1-megabyte sections, each assigned to one task. It also runs each task as if it is an 8086 program, not concerned with privilege levels. In other words, the 8086 virtual mode of the 386 microprocessor allows any program written for DOS to be run unchanged under one task, where each task can have its own 1 megabyte of memory. This means that in virtual 8086 mode, the 386 uses the SEG:OFFSET concept used in the 8088/86 microprocessor. Microsoft Windows uses the virtual 8086 mode of the 80386 microprocessor. These operating systems use the 386's virtual 8086 mode to run multiples of programs written for the 8088/86. The difference is that in MS-DOS, only one task can be active at a time and all other tasks are sitting idle (dormant) while one task is being run, but in Windows each task is given a slice of the CPU's time, and many tasks can be active concurrently. For example, a word processor can be used while the modem/FAX is receiving and sending data, a spreadsheet program such as MS Excel is doing some calculations, and a disk is being formatted. Of course, since there is only one microprocessor taking care of all these tasks, it is the job of the multitasking operating system such as Windows to slice the CPU time and assign each task time on a circular rotational basis. If there are

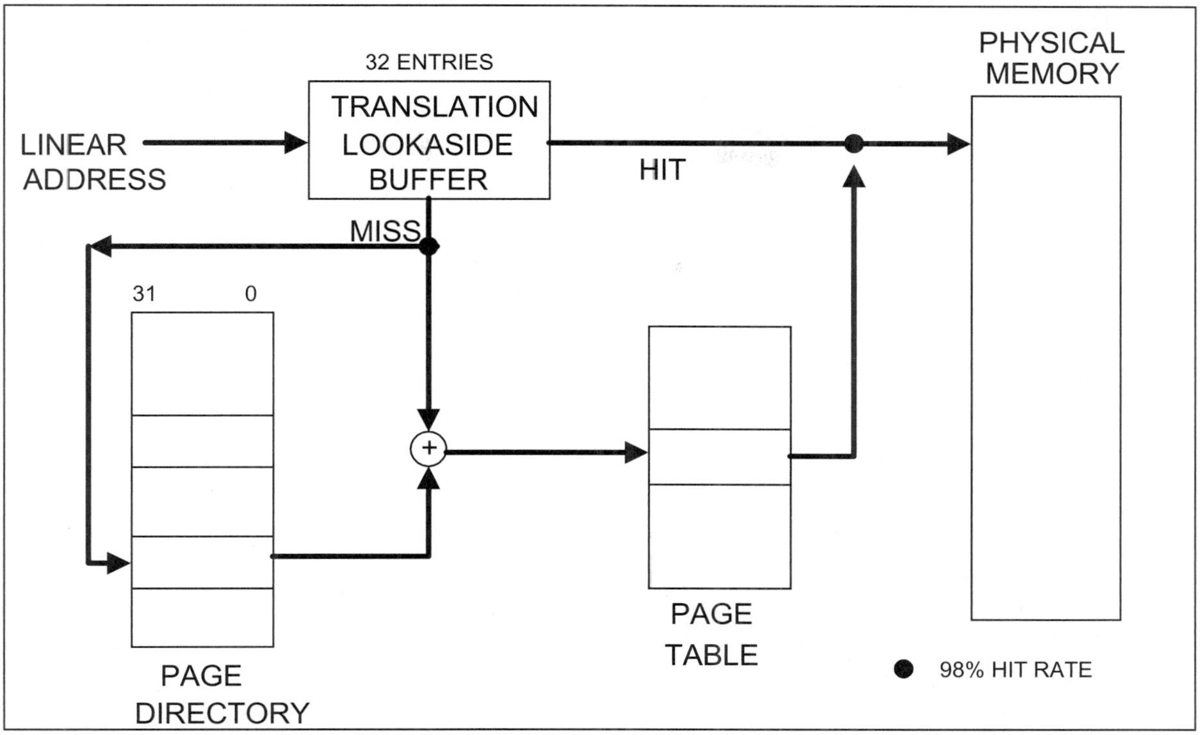

Figure 21-13. Translation Lookaside Buffer

too many tasks and all are active, they all seem to be slow since each task gets less time (attention) from the CPU. Of course, one way to solve this slowness is to use high-performance CPUs with GHz speed. The multitasking operating system can be cooperative or preemptive. In cooperative multitasking, two or more applications cooperate with each other in taking turns to use the CPU alternately. If one application misbehaves, it can cause the whole system to be unstable and crash. In preemptive multitasking, a task can be interrupted preemptively at any point by another program. If a task is interrupted by another task, its present state will be saved by the operating system and it will be serviced after the new task is given a chance to use the CPU.

Review Questions

1. True or false. In protected mode, the 386 physical address is calculated by shifting the segment register value and adding the offset.
2. Virtual memory refers to ___________ (main DRAM, hard disk) memory.
3. How does the operating system decide which code (or data) should be abandoned to make room for new code?
4. In protected mode (segmentation), where is the physical address of the desired code or data located?
5. Of the 8 bytes of the descriptor table entry, which one(s) are used for the physical address? Assume that they are numbered from 0 to 7.
6. When a piece of code is run, which bit of the access byte is modified?
7. In 386 segmentation, level 3 is assigned the _______ (lowest, highest) privilege.
8. In 386 segmentation, level 0 is assigned the _______ (lowest, highest) privilege.
9. How many privilege levels are there in 386 paging?
10. True or false. In 386 paging, the linear and physical addresses are the same.
11. To get the physical address in 386 paging the linear address must go through ___ (1, 2) stage(s) of translation.
12. The virtual 8086 mode was introduced in the _________ (80286, 80386).
13. True or false. In MS-DOS, only one task can be active at a time.
14. Why is Windows (Vista, XP) but not DOS a true multitasking operating system?

PROBLEMS

SECTION 21.1: 80386 IN REAL MODE

1. Which microprocessors support the instructions PUSHA and POPA?
2. Explain the function of PUSHA. It is equivalent to what set of instructions?
3. Explain the function of POPA. It is equivalent to what set of instructions?
4. Which microprocessors support "SHL dest,immediate"?
5. Find the contents of the destination register for each of the following.

```
(a) MOV   AX,43H
    SHL   AX,4
(b) MOV   BX,8000H
    SHR   BX,16
(c) MOV   CX,0AAAH
    ROL   CX,8
(d) MOV   CX,0AAAH
    ROL   CX,12
```

6. True or false. The 80286 was the first x86 to abandon multiplexing of the address and data buses.
7. In which x86 microprocessor was the concept of virtual memory introduced?
8. In which of the x86 microprocessors was the 2-clock memory cycle introduced?
9. Which of the following instructions will cause an error in the 386?
   (a) MOV EBX,AX　　(b) MOV ECX,BX
   (c) ADD ECX,EDX　　(d) ADD EDX,AL
   (e) MOV EBX,SI　　(f) ADD SI,DI
10. Show how data is stored in "MOV [3500],EBX". Assume that EBX = 9834F543H.
11. Show how data is stored in "MOV ES:[1000],ECX" (ECX = 07B324H).
12. Which registers can be used for the scaled index addressing mode?
13. Write a 386 program to add a factor of 100 to an array of 10 DWORD data. Use the scaled index addressing mode.
14. Write a 386 program to add two multibyte data items of 8-byte size and store the result. Use the scaled index addressing mode.
15. Indicate all the registers that can be used for pointers in the 386. Also give their default segments.
16. Find the destination register contents after execution of each of the following.

```
(a) MOV      BX,-12
    MOVSX    EBX,BX
(b) MOV      CL,-8
    MOVSX    EDX,CL
(c) MOV      AH,7
    MOVZX    ECX,AH
(d) MOV      AX,99H
    MOVZX    EBX,AX
```

17. Find the contents of EAX and EBX after execution of the following.

```
    MOV   ECX,307F455H
    BSF   EAX,ECX
    BSR   EBX,ECX
```

18. Find the contents of AX and DX after execution of the following.

```
    MOV   BX,98H
    BSF   AX,BX
    BSR   DX,BX
```

19. What is the purpose of instructions MOVSX and MOVZX?
20. True or false. In the instruction 'MOVSX REG,REG", the source and destination registers must match in size.

SECTION 21.2: 80386: A HARDWARE VIEW

21. BE0–BE3 are active _________ (low, high).
22. True or false. The address and data bus in the 386 are multiplexed.
23. Which part of the data bus is activated if BE0 = 0 and BE1 = 0 (at the same time)?
24. Which part of the data bus is activated if BE2 = 0 and BE3 = 0 (at the same time)?
25. Which part of the data bus is activated if BE0 = 0 and BE3 = 0 (at the same time)?
26. Which part of the data bus is activated if BE1 = 1 and BE2 = 0 (at the same time)?
27. A 25-MHz 386 is connected to CLK2 of _______ MHz.
28. Show the status of CS, IP, A31–A2, and BE3–BE0 in the 386 upon RESET.
29. What are the implications of your answer to Problem 28?
30. For what addresses in the 386 PC is BIOS ROM duplicated, and why?
31. Draw the bus cycle for the nonpipelined 386. Show the address, data, and READY signals.
32. Find the total bus cycle necessary to transfer the operand in the instruction "MOV [2002],ECX".
33. For aligned data, the addresses for DWORD type data in the 386 must have ________ as the lower hex digit.
34. Find the memory cycle time for a 33-MHz 386.

SECTION 21.3: 80386 PROTECTED MODE

35. What is virtual memory?
36. True or false. The CPU requests data from virtual memory before it requests data from main memory.
37. While main memory is made of __________ (DRAM, hard disk), virtual memory is ___________ (DRAM, hard disk).
38. What is the difference between the real and protected modes of the 386 in terms of memory space?
39. To access the entire 4 gigabytes of the 386, the CPU must be in ________ mode.
40. True or false. The 286 supports both segmentation and paging virtual memory.
41. True or false. The 386 supports both segmentation and paging virtual memory.
42. True or false. In the 286, the segment size can be 1 byte to 16 megabytes.
43. True or false. In the 386, the segment size can be 1 byte to 4 gigabytes.
44. For the 386, what is the page size in paging virtual memory?
45. How many bytes does each entry in the descriptor table use?
46. State the difference between real mode and protected mode as far as the physical address of the operand is concerned.
47. How many bits are set aside for the addresses in the descriptor table, and where are they located in the descriptor table?
48. To make the descriptor table of a 386 286-compatible, we must make bytes 7 and 8 all _____ (0s,1s).
49. How many bits are set aside for the segment limits in the descriptor table, and where are they located in the descriptor table?
50. True or false. Every piece of data or code accessed by the 386 in protected mode must have an access byte.

51. 00 is the _________(lowest, highest) privilege level and 11 is the _________ (lowest, highest) one.
52. What is the function of bit A in the access byte of a 386 descriptor table entry?
53. What is the function of bit P in the access byte of a 386 descriptor table entry?
54. State the characteristic of each of the following access bytes. State for each whether it is for code or data.
(a) 10010001 (b) 11110001 (c) 11110011
(d) 11111011 (e) 10011011 (f) 11111011
55. What does TLB stand for, and what is it used for?
56. In the 386, state the difference between the linear and the physical address.
57. To get the address of code or data in paging, the 386 converts from _________ (linear address, physical address) to _________ (linear address, physical address).
58. In the 386, before the address of the data or code is fetched it is checked against the values held by the ___________.
59. What is the number of entries in the 386 TLB?
60. True or false. In virtual 8086, the addresses are calculated by shifting the segment register left and adding it to the offset.
61. State the differences between paging and segmentation virtual memory.
62. How many privilege levels are there in paging?

ANSWERS TO REVIEW QUESTIONS

SECTION 21.1: 80386 IN REAL MODE

1. 16-bit
2. 16-bit
3. 80286
4. 80286
5. c
6. False
7. BL = AA, BH = 77, BX = 77AA, EBX = 99FF77AA
8. Little endian
9. EAX, EBX, ECX, EDX, ESI, and EDI
10. EA = 2000 + 8 × 100 = 2800H
11. b, c, e only
12. EDX = FFFFFFF7H
13. ECX = 000000FBH
14. DX = 3, AX = 000C

SECTION 21.2: 80386: A HARDWARE VIEW

1. 80386SX 2. 16
3. 34 pins since A31–A2 is 30 and 4 pins for the BE0, BE1, BE2, and BE3
4. D23–D16
5. 40 MHz
6. FFFFFFF0H
7. Half
8. 1/20 MHz = 50 ns; therefore, it is 100 ns.

SECTION 21.3: 80386 PROTECTED MODE

1. False
2. Hard disk
3. According to rule of least recently used
4. The descriptor table
5. Bytes 2, 3, 4, and 7
6. A (access) bit
7. Lowest
8. Highest
9. Two: user and supervisor
10. False
11. 2 stages
12. 80386
13. True
14. Because in Windows more than one task can be active, but in DOS only one task is active for a given period

CHAPTER 22

HIGH-SPEED MEMORY DESIGN AND CACHE

OBJECTIVES

Upon completion of this chapter, you will be able to:

- **>> Explain how the introduction of wait states is implemented in the IBM PC to coordinate the memory cycle times of x86 CPUs and high-speed memory**
- **>> Define terms used in memory design, such as memory cycle time and memory access time**
- **>> Describe the various types of DRAM: standard mode, page mode, and static column mode**
- **>> Describe how the interleaving method is implemented to solve the problem of back-to-back DRAM access and the required precharge time**
- **>> Discuss the advantages of using DRAM for main memory and SRAM for cache**
- **>> Diagram the three types of cache organization: fully associative, direct mapped, and set associative**
- **>> Explain the write-back and write-through methods of updating main memory as cache data is altered**
- **>> Describe the cache replacement policies LRU and FIFO**
- **>> Contrast and compare EDO and FPM DRAM**
- **>> Describe the operation and purpose of SDRAM**
- **>> Explain the components and function of Rambus technology**

The potential power of high-performance microprocessors can be exploited only if memory is fast enough to respond to the microprocessor's need to fetch code and data. There is no use in choosing a fast processor and then interfacing it with slow memory. In this chapter we deal with issues of high-speed memory design. In Section 22.1 we look at read and write cycle times of the x86 family. In Section 22.2 we discuss various types of DRAMs, such as page mode and static column mode. In Section 22.3, the cache memory option is discussed, and the way 386, 486, and Pentium processors use cache memory to increase system throughput is examined. Section 22.4 examines the newer and faster DRAMS of EDO, SDRAM, and Rambus technologies.

SECTION 22.1: MEMORY CYCLE TIME OF THE x86

When interfacing a microprocessor to memory, the first issue is how much time is provided by the CPU for one complete read or write cycle. In other words, what is the memory cycle time of the CPU? In the 8088/86 microprocessor, the memory cycle time consists of 4 clocks, which leaves plenty of time to access memory. The slowest 8088/86, with a working frequency of 5 MHz, has an 800-ns memory cycle (4×200 ns = 800, T = 1/5 MHz = 200 ns), and the fastest 8088/86, with 10-MHz speed, will have a 400-ns memory cycle. A memory cycle of 400 ns means that the CPU can access memory every 400 ns, and not faster. This is enough time to access even the slow and inexpensive DRAMs. However, for the 286, 386, 486, and Pentium, memory cycle time consists of only two T clocks. This makes memory design a challenging task, especially when the speed of the CPU goes beyond 20 MHz. Table 22-1 shows the memory cycle times for various speeds of x86 microprocessors. From Table 22-1 it can be seen that as the frequency of the CPU is increased, the maximum amount of time allowed to access memory is decreased, forcing the designer either to use fast and expensive memory or to introduce wait states into the memory cycle.

Introducing wait states into the memory cycle

When the memory timing requirement of the CPU cannot be met, one option that designers have is to introduce wait states. All x86 microprocessors have the READY pin. When the microprocessor initiates the memory cycle, meaning that it puts the addresses on the address bus, the time at which it must have the data at the pins of the data bus is fixed and is shown in Table 22-1. This fixed amount of time can be extended by activating the READY pin. Every time that the READY pin is activated, the CPU adds one extra clock to the memory cycle. For example, the 25-MHz 80386 has a memory cycle of 80 ns ($2 \times 40 = 80$) with zero wait states. If READY is activated only once during the memory cycle, it adds one clock of 40 ns to the memory cycle, thereby giving the memory and decoding circuitry a total of 120 ns to get the information to the data pins of the CPU. This 120 ns is spent on the following parameters: (1) memory decoding logic circuitry and address bus buffers (chipset), (2) access time of memory, and (3) the time it takes for signals to travel from memory data pins to the data pins of the CPU, going through any logic gates on the pathway, such as FPGA chipset. Of these three parameters, memory access time is normally the longest. For more about logic families, see Chapter 26. If the allocated memory cycle time is not enough, more wait states are needed, making the memory cycle time longer.

In a 25-MHz 386 with 1 wait state, there is a 120-ns memory cycle time, meaning that the CPU can perform read and write operations no faster than every 120 ns. What happens if 140 ns is needed? Since the wait state is an integer multiple of the clock cycle (1, 2, 3, and so on), there is no other choice but to have 2 wait states. In other words, there is no such thing as 1.5 wait states. Wait states degrade computer performance, as shown in Example 22-1. It does not make sense to buy a high-frequency CPU, then interface it with slow memory. The next section will look at possible solutions to this problem.

Table 22-1: Memory Cycle Times for x86

Mem. Cycle	CPU	Bus Frequency (MHz)	Clock Cycle (ns)	Memory Cycle (ns)
4 clocks	8088/86	5	200	800
	8088/86	8	125	500
	8086	10	100	400
2 clocks	80286	6	166.6	333.3
	80286	8	125	250
	80286	10	100	200
	80286	16	62.5	125
	80286	20	50	100
2 clocks	80386DX	16	62.5	125
	80386DX	20	50	100
	80386DX	25	40	80
	80386DX	33	30	60
2 clocks*	80486DX	25	40	80
	80486DX	33	30	60
	80486DX	40	25	50
	80486DX	50	20	40
2 clocks*	Pentium	60	16 6	33
	Pentium	100	10	20
	Pentium	150	6.6	13.2

* From external DRAM or the secondary cache .

Note: All memory cycle times are with zero wait states .

Example 22-1

Find the effective memory performance of a 25-MHz 386 CPU with one wait state.

Solution:

Since the 0 WS memory cycle is 80 ns (1/25 MHz = 40 and 2 × 40 = 80 ns), for 1 WS we have a memory cycle time of 120 ns. That means that the memory performance is the same as that of a 16.6-MHz 80386 (120 ns/2 = 60 ns, then 1/60 ns = 16.66 MHz) as far as memory accessing is concerned. This is 67% performance of the 80386 with zero wait states.

Review Questions

1. Find the read/write cycle time of the following systems
   (a) 40-MHz 386 with 0 WS (b) 50-MHz 486 with 1 WS
   (c) 66-MHz Pentium with 1 WS
2. A given CPU has a read/write cycle time of 50 ns. What does this mean?
3. Find the effective working frequency for memory access in each of the following.
   (a) 40-MHz 386 with 1 WS (b) 50-MHz 486 with 1 WS
4. If a given CPU has a read cycle time of 60 ns and 10 ns is used for the decoder and address/data path delay, how much is for memory access time?
5. If a given system is designed with 1 WS and has a 90-ns memory cycle time, find the CPU's frequency if the read/write cycle time of this CPU is 2 clocks.

SECTION 22.2: PAGE AND STATIC COLUMN DRAMS

To understand interfacing memory to high-performance computers, the different types of available RAM must first be understood. Although SRAMs are fast, they are expensive and consume a lot of power due to the use of flip-flops in the design of the memory cell, as we discussed in Chapter 10. At the opposite end of the spectrum is DRAM, which is cheaper but is slow (compared to CPU speed) and needs to be refreshed periodically. The refreshing overhead together with the long access time of DRAM is a major problem in the design of high-performance computers. The problem of the time taken for refreshing DRAM is minimal since it uses only a small percentage of bus time, but the solution to the slowness of DRAM is very involved. One common solution is using a combination of a small amount of SRAM, called *cache* (pronounced "cash"), along with a large amount of DRAM, thereby achieving the goal of near zero wait states. Before we discuss such solutions, we must understand what resources are available to high-performance system designers. To this end, the different types of available DRAM will be discussed, and cache memory is discussed in Section 22.3. First we clarify some widely used terminology such as memory cycle time and memory access time.

Memory access time vs. memory cycle time

Memory access time is defined as the time interval between the moment the addresses are applied to the memory chip address pins and the time the data is available at the memory's data pins. The memory data sheets refer to it as t_{AA} (address access time). Another commonly used time interval is t_{CA} (access time from CS), which is measured from the time the chip select pin of memory is activated to the time the data is available. In some cases, notably EEPROM, t_{OE} is the time interval between the moment OE (READ) is activated to the time the data is available. However, memory access time t_{AA} is the one most often advertised.

Memory cycle time is the time interval between two consecutive accesses to the memory chip. For example, a memory chip of 100 ns cycle time can be accessed no faster than 100 ns, which means that two back-to-back reads can be performed no faster than 200 ns, and 3 back-to-back reads will take 300 ns, and so on. It must be noted that while in SRAM the memory cycle time is equal to memory access time, this is not so in DRAM memory, as discussed next.

Types of DRAM

There are different types of DRAM, which are categorized according to their mode of data access. These modes include standard mode, page mode, static column mode, and nibble mode. Although each mode is discussed separately below, often two of the above modes exist on the same DRAM chip. For example, page mode DRAM has standard mode as well.

DRAM (standard mode)

Standard mode (also called random access) DRAM, which has the longest memory cycle time, requires the row address to be provided first and then the column address for each cell. Each group is latched in by the activation of RAS (row address select) and CAS (column address select) inputs, respectively. The access time is from the time that the row address is provided to the time that the data is available at the output data pin of the DRAM chip. This is the access time that is commonly advertised and is called t_{RAC} (RAS access time, the access time from the moment RAS is provided). This is acceptable if we are accessing a random cell within DRAM. However, since most of the time data and code processed by the CPU are in consecutive memory locations and the CPU does not jump around to random locations (unless there is a JMP or CALL instruction), the DRAM will be accessed with back-to-back read operations. Unfortunately, DRAM cannot provide the code (or data) in the amount of time called tRAC if there is a back-to-back read from the same DRAM chip because DRAM needs a precharge time (t_{RP}) after each RAS has been deactivated to get ready for the next access. This leads us back to the concept of memory cycle time for DRAM memory chips. The memory cycle time for memory chips is the minimum time interval between two back-to-back read/write operations. In SRAM and ROM, the access time and memory cycle time are always equal, but that is not the case for DRAMs. In DRAM, that after RAS makes the transition to the inactive state (going from low to high), it must stay high for a minimum of t_{RP} (RAS precharge) to precharge the internal device circuitry for the next active cycle. Therefore, in DRAM we have the following approximate relationship between the memory access time and memory cycle time.

$t_{RC} = t_{RAC} + t_{RP}$ (This is for standard mode)
read cycle time = RAS access time + RAS precharge time

For example, if DRAM has an access time of 100 ns, the memory cycle time is really about 190 ns (100 ns access time plus 90 ns precharge time). To access a single location in such a DRAM, 100 ns is enough, but to access more than one successively, 190 ns is required for each access due to the precharge time that is needed internally by DRAM to get ready to access the next capacitor cell. Tables 22-2 and 22-3 show DRAM and SRAM memory cycle times, respectively.

Table 22-2: DRAM Access Time vs. Cycle Time (4M × 1)

DRAM	RAS Access (t_{RAC}) (ns)	Read Cycle (t_{RC}) (ns)	RAS Precharge (t_{RP}) (ns)
MCM44100-60	60	110	45
MCM44100-70	70	130	50
MCM44100-80	80	150	60

(Reprinted by permission of Motorola Corporation , Copyright Motorola Corp. 1993)

Table 22-3: SRAM Access Time vs. Cycle Time

SRAM (IDT Product)	Address Access (t_{AA}) (ns)	Read Cycle (t_{RC}) (ns)
IDT71258S25	25	25
IDT71258S35	35	35
IDT71258S45	45	45
IDT71258S25	70	70

(Reprinted by permission of Integrated Device Technology , Copyright IDT, 1993)

The read cycle time not being equal to the access time is one of the major differences between SRAM and DRAM. Although in SRAM the read cycle time is equal to the access time, in DRAM of standard mode the read cycle time is about twice the access time normally advertised (t_{ACC}). This could make a difference in the total time spent by the CPU to access memory. Look at Examples 22-2 and 22-3. From the above discussion and

Example 22-2

Compare the minimum CPU time needed to read 150 random memory locations of a given bank in each of the following.
(a) DRAM with T_{ACC} = 100 ns and T_{RC} = 190 ns
(b) SRAM of T_{ACC} = 100 ns

Solution:

(a) DRAM requires 190 ns to access each location. Therefore, a total of 150 × 190 = 28,500 ns would be spent by the CPU to access all those 150 memory locations.
(b) In the case of SRAM, the CPU spends only 150 × 100 ns = 15,000. This would have been needed since T access = T read cycle ($t_{ACC} = t_{RC}$).

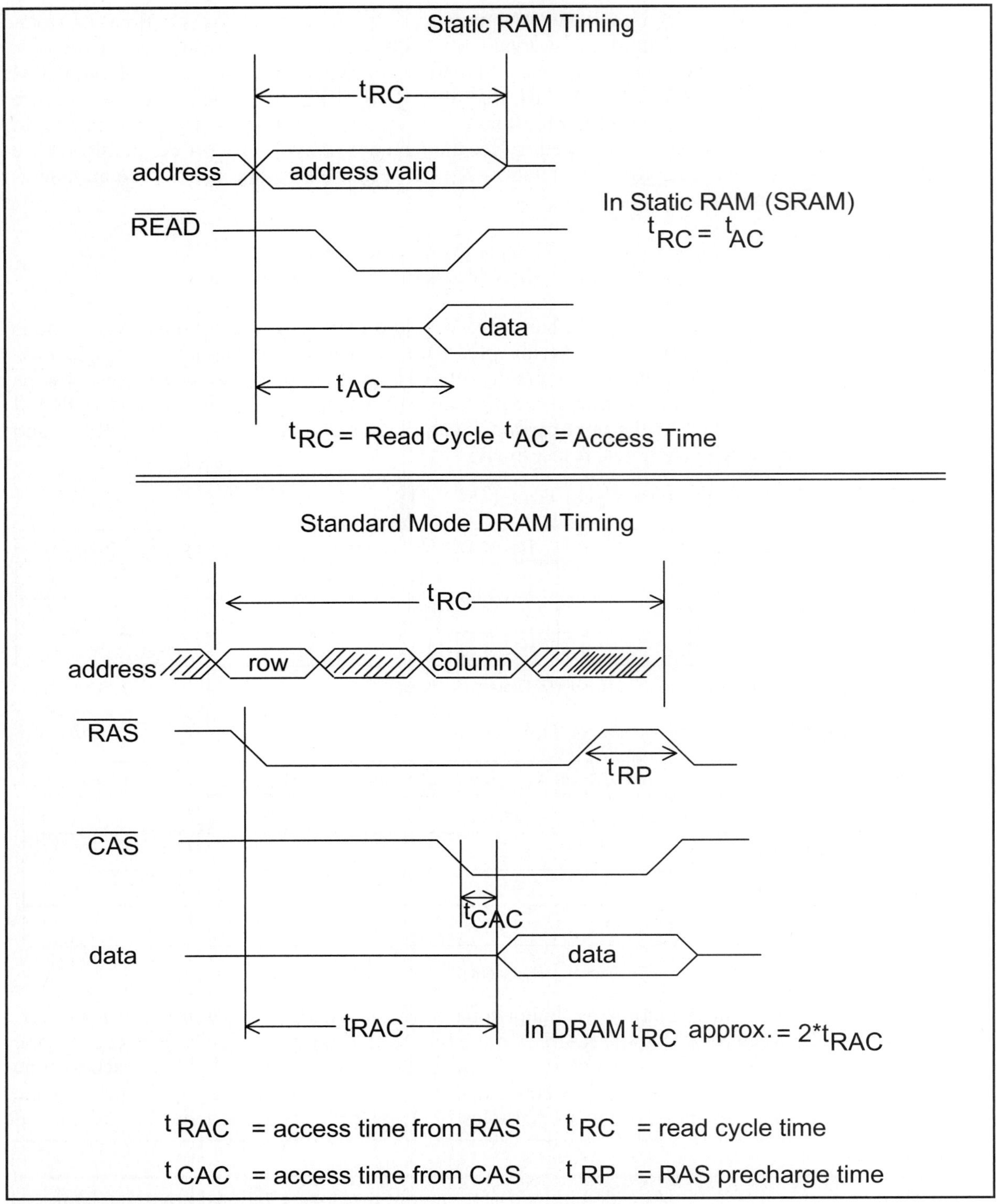

Figure 22-1. DRAM vs. SRAM Timing

Example 22-3
Calculate the time to access 1024 random bits of a 1M × 1 chip if t_{RC} = 85 ns and t_{RAC} = 165 ns. **Solution:** For standard mode (also called random) we have the following for reading 1024 bits: time to read 1024 random bits = 1024 × t_{RC} = 1024 × 165 ns = 168,960 ns

Example 22-2 we can conclude that for successive accesses of random locations inside the DRAM the CPU must spend a minimum of t_{RC} time on each access. See Figure 22-1 for DRAM and SRAM timing.

DRAM interfacing using the interleaving method

One of the methods used to overcome the problem of precharge time in DRAMs is the interleaving method of DRAM interfacing. In this method, two sets of banks are placed next to each other and the CPU accesses each set of banks alternately. In this way the precharge time of one set of banks is hidden behind the access time of the other one. This means that while the CPU is accessing one set of banks, the other set is being precharged. Look at Figure 22-2. Assume that the 80386SX is working on 20 MHz frequency; therefore, the CPU has a memory cycle time of 100 ns. Using DRAM with access time of 70 ns and the precharge of 65 ns gives a DRAM cycle time of 135 ns (70 + 65 = 135). This is much longer than the 100 ns provided by the CPU. Using interleaved memory design can solve this problem. In this case when the 386SX accesses bank set A, it goes on to access bank set B while set A takes care of its precharge time. Similarly, when the CPU accesses set A, the set B banks will have time to precharge.

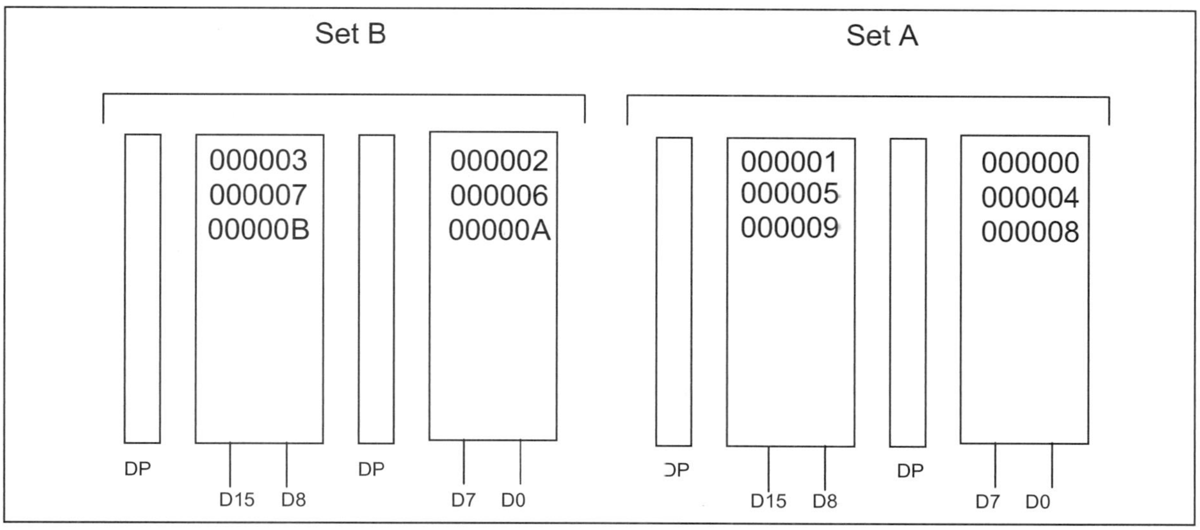

Figure 22-2. Interleaved DRAM Organization

Example 22-4

Show the time needed to access all 1024 memory locations of Example 22-3 if the interleaved method of memory interfacing is used.

Solution:

In the interleaved method, since the precharge time of one bank is hidden behind the access time of the other bank, each memory location is accessed in t_{RAC} as far as the CPU is concerned; therefore, 1024 × 85 = 87,040 ns is the total amount of time spent by the CPU to access 1024 locations.

Interleaved drawback

The major drawback of interleaved memory is memory expansion. In expanding the memory based on the interleaved method, a minimum of two sets of banks must be added every time additional memory is required. Look at Example 22-5. Many inexpensive embedded systems based on 386SX, 386DX, and 486SX of 16–25 MHz frequency use the interleaved memory design method to avoid using expensive cache memory without sacrificing performance.

Example 22-5

Assume that we are using 1M × 1 DRAM organization in Figure 22-2. If each set is 4 megabytes, find the following.
(a) the chip count (b) the minimum memory addition and the chip count

Solution:

(a) Assuming 1M × 9 for each bank where each bank takes care of 8 bits of data, there are 9 chips for every byte. That means a total of 36 DRAM chips for each set, or a total of 72 1M × 1 chips for the first 8 megabytes of interleaved memory.
(b) From then on, any memory addition must be in multiples of 4 megabytes since each set needs 2M; therefore, we need another 36 1M × 1 DRAM chips to raise the total memory of the system to 12M.

Example 22-6

A 386 embedded system has 1M of DRAM installed using the interleaved design method. Show the memory organization and DRAM chip count assuming that only 256K × 1 and 256K × 4 DRAM chips are used.

Solution:
Since the 386SX has a 16-bit data bus, it uses 512K bytes for each set of A and B, or four banks of 256K × 9, where each set consists of two banks of 256K × 9. Therefore, the total chip count is 12 since each bank uses three chips (two 256K × 4 and one 265K × 1 for parity bit). This is shown as follows.

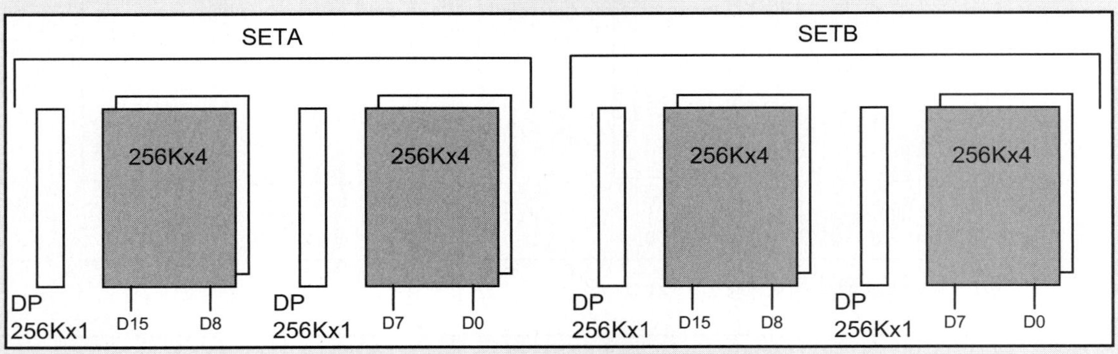

Example 22-7

Show the minimum memory addition and the chip count for Example 22-6. Assume that the available DRAM chips are 256K × 1 and 256K × 4.

Solution:
The minimum memory addition is 1M. Since we have two banks for each set of interleaved memory, we have two 256K × 4 and one 256K × 1 for parity, which means three chips for each bank. Therefore, the minimum memory addition requires 12 chips, eight of which are 256K × 4 and four are 256K × 1 for parity bits, resulting in 1 megabyte.

Page mode DRAM

The storage cells inside DRAM are organized in a matrix of N rows and N columns. In reading a given cell, the address for the row (A1–An) is provided first and RAS is activated; then the address for the column (A1–An) is provided and CAS is activated. In DRAM literature the term *page* refers to a number of column cells in a given row. See Examples 22-8 and 22-9.

Example 22-8

Show how memory storage cells are organized in each of the following DRAM chips.
(a) 256K × 1 (b) 1M × 1 (c) 4M × 1

Solution:

(a) The 256K × 1 has 9 address pins (A0–A8); therefore, cells are organized in a matrix of $2^9 \times 2^9$ = 512 × 512, giving 512 rows, each consisting of 512 columns of cells.
(b) 1024 × 1024
(c) 2048 × 2048

Example 22-9

Assuming that the DRAMs in Example 22-8 are of page mode, show how each chip is organized into pages. Find the number of columns per page for (a), (b), and (c).

Solution:

(a) For 1M × 1 we have 512 pages, where each page has 512 columns of cells.
(b) 1024 pages, where each page has 1024 bits (columns).
(c) 2048 pages each of 2048 bits

The idea behind page mode is that since memory locations are accessed consecutively in most situations, there is no need to provide both the row and column address for each location, as was the case in DRAM with standard timing. Instead, in page mode, first the row address is provided, RAS latches in the row address, and then the column addresses are provided and CAS toggles back and forth, latching in the column addresses until the last column of a given page is accessed. Then the address of the next row (page) is provided and the process is repeated. While the access time of the first cell is the standard access time using both row and column (t_{RAC}), the access time in accessing the second cell on the last cell of the same page (row) is much shorter. This access time is often referred to as t_{CAC} (T of column access). In page mode DRAM when we are in a given page, each successive cell can be accessed no faster than t_{PC} (page cycle time). See Figure 22-3. Table 22-4 gives page mode timing parameters. In DRAM of page mode both the standard mode and page mode are supported.

Table 22-4: Page Mode DRAM Timing Parameters (4M x 1)

Page Mode DRAM	Access Time from RAS, t_{RAC} (ns)	Read Cycle Time, t_{RC} (ns)	Access Time from CAS, t_{CAC} (ns)	Page Cycle Time, t_{PC} (ns)
MCM44100-60	60	110	15	40
MCM44100-70	70	130	20	45
MCM44100-80	80	150	20	50

Static column mode

Static column mode makes accessing all the columns of a given row much simpler by eliminating the need for CAS. In this mode, the first location is accessed with a standard read cycle where the row address is latched by RAS followed by the column address and what is called CS (chip select) clock. From then on, CS is incremented internally. As long as RAS and CS remain low, the contents of successive cells appear at the data output pin of DRAM until the last column of a given row is accessed; then the

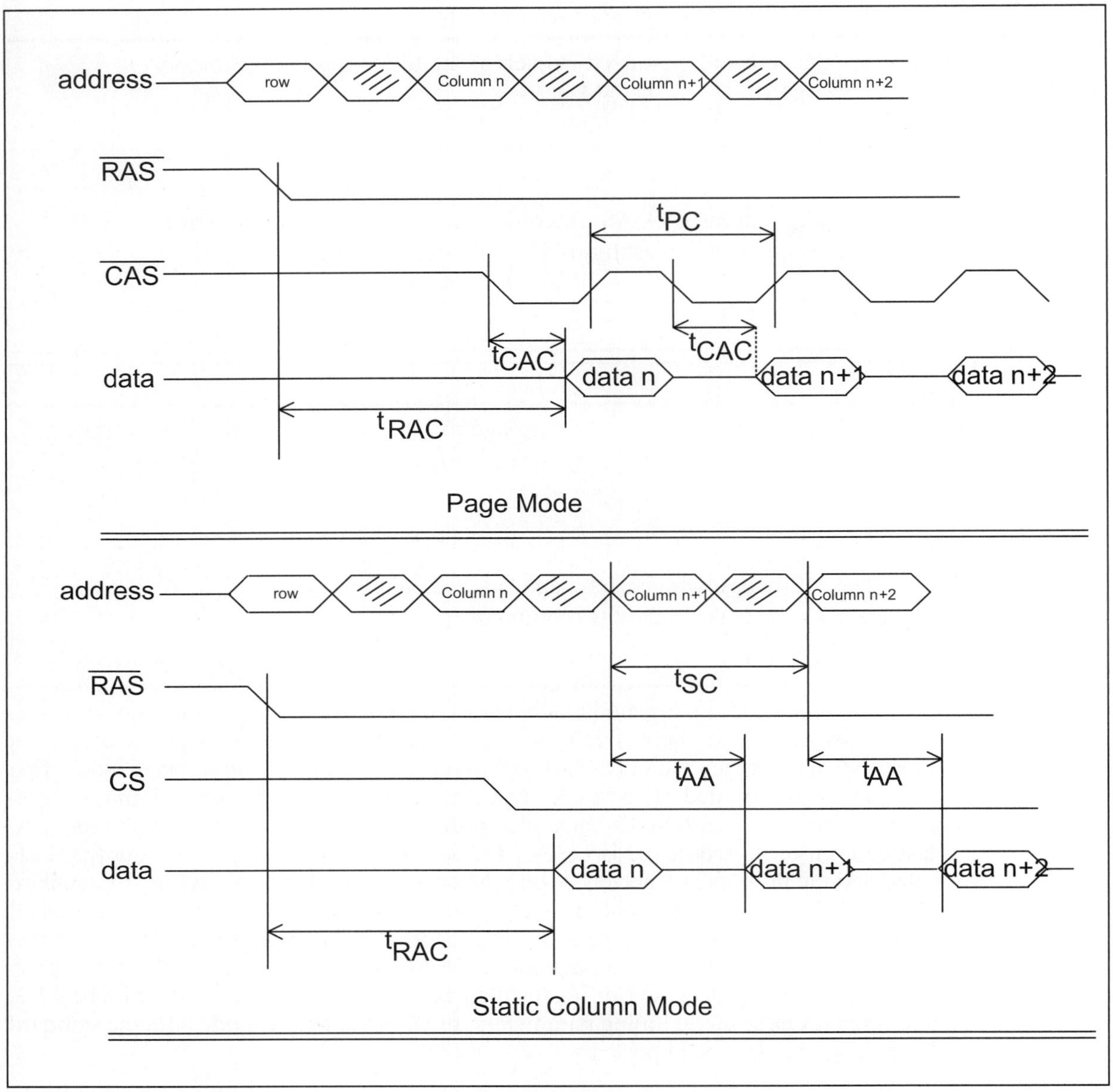

Figure 22-3. DRAM Page and Static Column Modes

process is moved to the next row. This means that the initial access time of the first cell is the standard access time (t_{RAC}), but each subsequent column in that row is accessed in a time called t_{AA} (access time from column address). Due to the fact that there is no setup and hold time for column address select (CAS), the use of static-column-mode DRAM lends itself to memory design of high-frequency systems. A large percentage of 80386 and higher processor computers use static column DRAM for main memory.

In static column mode where the initial standard access time is t_{RAC}, when we are in a given page, any cell can be accessed with the access time of t_{AA}, but all the successive bits can be accessed no faster than t_{SC} (static column cycle time). See Figure 22-3. Table 22-5 gives static column mode timing parameters.

Comparing Examples 22-10 and 22-11, if the time spent by the CPU is the same for both the page mode and static column mode, what is the advantage of static column mode? The answer is that static-column-mode DRAM design is simpler since there is no circuit or timing requirement for the CAS pin. Notice in Figure 22-3 that we need to keep both RAS and CS (chip select) low in order to access successive cells. Here is what Motorola (now Freescale) Application Note AN986 says about the superiority of the static-column-mode DRAM: "This mode is useful in applications that require less noise than

Table 22-5: Static Column DRAM Timing Parameters (4M × 1)

Static Column DRAM	T RAS Access, t_{RAC} (ns)	T Read Cycle, t_{RC} (ns)	T Column Access, t_{AA} (ns)	Cycle Time, t_{SC} (ns)
MCM54102A-60	60	110	30	35
MCM54102A-70	70	130	35	45
35MCM54102A-80	80	150	40	45

(Reprinted by permission of Motorola Corporation , Copyright Motorola Corp., 1993)

Example 22-10

Calculate the total time spent by the CPU to access an entire page of memory if the memory banks are page mode DRAM of 1M × 1 with t_{RC} =165 ns, t_{RAC} = 85 ns, and t_{PC} = 50 ns.

Solution:

For page mode we have the following for reading 1024 bits:

Time to read 1024 bits of the same page $= t_{RAC} + 1023 \times t_{PC}$
$= 85 \text{ ns} + 1023 \times 50 \text{ ns} = 51{,}235 \text{ ns}$

Example 22-11

Calculate the total time spent by the CPU to access the entire page of memory if the memory banks are static-column-mode DRAMs of 1M × 1 with t_{RC} = 165 ns, t_{RAC} = 85 ns, and t_{SC} = 50 ns.

Solution:

For static column mode we have the following for reading 1024 bits:
time to read 1024 bits of the same page $= t_{RAC} + 1023 \times t_{SC}$
$= 85 \text{ ns} + 1023 \times 50 \text{ ns} = 51{,}235 \text{ ns}$

page mode. Output buffers are always on when the device is in this mode and the CS clock is not cycled, resulting in fewer transients and simpler operation.... Static column consists of changing column addresses while holding the RAS and CS clocks active."

Timing comparison of DRAM modes

A summary of DRAM timing is given in Table 22-6. Much of this material is taken from Motorola (Freescale) Application Note AN986.

Table 22-6: Timing for 1M × 1 85 ns DRAM Chip

Access Time (ns)	Standard	Page	Static Column
Access time from row, t_{RAC}	85	85	85
Access time from column, t_{CAC}		25	
Access time from column, t_{AA}			45
Access time from column, t_{NCAC}			
Cycle time			
Read cycle time, t_{RC}	165		
Page mode cycle time, t_{PC}		50	
Static column time, t_{SC}			50

(Reprinted by permission of Motorola Corporation , Copyright Motorola Corp. 1993)

This concludes the discussion of DRAM operation modes. It must be noted that in many systems one of the above modes is implemented in order to eliminate the need for the wait state to access every bit of DRAM. As seen from the above discussion, even the best of any of these modes still cannot eliminate the need for the wait state entirely unless SRAM is used for the entire memory, which is prohibitively expensive. The best solution is to use a combination of SRAM and DRAM, which is discussed next.

Review Questions

1. In which type(s) of memory is the read cycle time equal to the memory access time?
2. A given DRAM is advertised to have an access time of 50 ns. What is the approximate memory cycle time for this DRAM?
3. A given DRAM has a 120-ns memory read cycle time. What is its access time (t_{RAC})?
4. In DRAM, a read cycle consists of ___________ and ____________.
5. Assume an 80386 of interleaved memory with 2M bytes initial DRAM for each of the following.
   (a) Show how the banks are organized.
   (b) What is the minimum memory addition?
6. True or false. In page mode, the initial read takes t_{RAC}.
7. For page mode DRAM, while we are in a given page, we can access successive memory locations no faster than ____________ .
8. Calculate the time the CPU must spend to access 100 locations all within the same page if t_{RAC} = 60 ns and t_{PC} = 30 ns.
9. The higher the system frequency, the less noise can be tolerated in the system. Which is preferable in a 20-MHz system, static column or page mode DRAM?

SECTION 22.3: CACHE MEMORY

The most widely used memory design for high-performance CPUs implements DRAMs for main memory along with a small amount (compared to the size of main memory) of SRAM for cache memory. This takes advantage of the speed of SRAM and the high density and cheapness of DRAM. As mentioned earlier, to implement the entire memory of the computer with SRAM is too expensive and to use all DRAM degrades performance. Cache memory is placed between the CPU and main memory. See Figure 22-4.

When the CPU initiates a memory access, it first asks cache for the information (data or code). If the requested data is there, it is provided to the CPU with zero wait states, but if the data is not in cache, the memory controller circuitry will transfer the data from main memory to the CPU while giving a copy of it to cache memory. In other words, at any given time the cache controller has knowledge of which information (code or data) is kept in cache; therefore, upon request for a given piece of code or data by the CPU the address issued by the CPU is compared with the addresses of data kept by the cache controller. If they match (hit) they are presented to the CPU with zero WS, but if the needed information is not in cache (miss) the cache controller along with the memory controller will fetch the data and present it to the CPU in addition to keeping a copy of it in cache for future reference. The reason a copy of data (or code) fetched from main memory is kept in the cache is to allow any subsequent request for the same information to result in a hit and provide it to the CPU with zero wait states. If the requested data is available in cache memory, it is called a hit; otherwise, if the data must be brought in from main memory, it is a miss.

In most computers with cache, the hit rate is 85% and higher. By combining SRAM and DRAM, cache memory's access time matches the memory cycle of the CPU. In the 80386/486 microprocessor with a frequency of 33 MHz and above, the use of cache

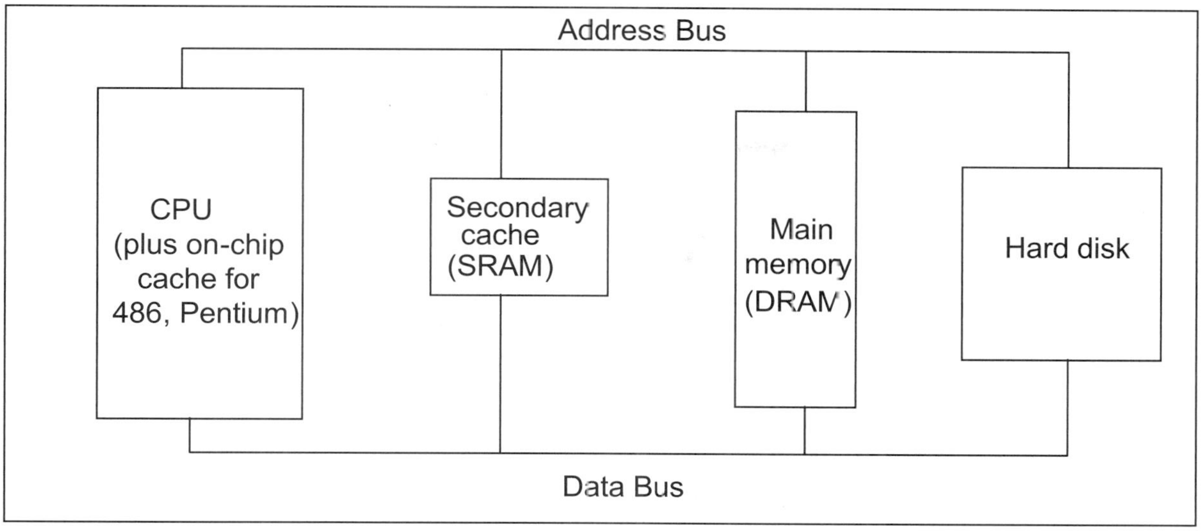

Figure 22-4. CPU and Its Relation to Various Memories

is absolutely essential. For example, in the 33-MHz 80386-based computer with only a 60-ns read cycle time, only static RAM with an access time (cycle time) of 45 ns can provide the needed information to the CPU without inserting wait states. We have assumed that 15 ns (60 – 45 = 15) is used for the delay associated with the address and data path. To implement the entire 16M of main memory of a 33-MHz 386/486 system with 45 ns SRAM is not only too expensive but the power dissipation associated with such a large amount of SRAM would require a complex cooling system used only for expensive mini- and mainframe computers. The problem gets worse if we use a 486 of 50 MHz or a Pentium of 60 MHz.

It must be noted that when the CPU accesses memory, it is most likely to access the information in the vicinity of the same addresses, at least for a time. This is called the *principle of locality of reference*. In other words, even for a short program of 50 bytes, the CPU is accessing those 50 memory locations from cache with zero wait states. If it were not for this principle of locality and the fact that the CPU accesses memory randomly, the idea of cache would not work. This implies that JMP and CALL instructions are bad for the performance of cache-based systems. The hit rate, the number of hits divided by the total number of tries, depends on the size of the cache, how it is organized (cache organization), and the nature of the program.

Cache organization

There are three types of cache organization:

1. fully associative
2. direct mapped
3. set associative

The following is a discussion of each organization with its advantages and disadvantages. For the sake of clarity and simplicity, an 8-bit data bus and a 16-bit address bus are assumed.

Fully associative cache

In fully associative cache, only a limited number of bytes from main memory are held by cache along with their addresses. The SRAMs holding data are called *data cache* and the SRAMs holding addresses of the data are called *tag cache*. This discussion assumes that the microprocessor is sending a 16-bit address to access a memory location that has 8 bits of data and that the cache is holding 128 of the possible 65,536 (216) locations. This means that the width of the tag is 16 bits since it must hold the address, and that the depth is 128. When the CPU sends out the 16-bit address, it is compared with all

128 addresses kept by the tag. If the address of the requested data matches one of the addresses held by the tags, the data is read and is provided to the CPU (a hit). If it is not in the cache (a miss), the requested data must be brought in from main memory to the CPU while a copy of it is given to cache. When the information is brought into cache, the contents of the memory locations and their associated addresses are saved in the cache (tag cache holds the address and data cache holds the data).

In fully associative cache, the more data that is kept, the higher the hit rate. An analogy is that the more books you have on a table, the better the chance of finding the book you want on the table before you look for it on the book shelf. The problem with fully associative is that if the depth is increased to raise the hit rate, the number of comparisons is too time consuming and inefficient. For example, a fully associative cache with a depth of 1024 requires 1024 comparisons, and that is too time consuming even for fast comparators. On the other hand, with a depth of 16 the CPU ends up waiting for data too often. This is because the operating system is swapping information in and out of cache, since its size is too small, and it must save the present data in the cache before it can bring in new data. This replacement policy is discussed later. In the above example of 128 depth, the amount of SRAM for tag is 128 × 16 bits and 128 × 8 for data, that is, 256 bytes for tag and 128 bytes for data cache for a total of 384 bytes. Although the above example used a total of 384 bytes of SRAM, it is said that the system has 128 bytes of cache. In other words, the data cache size is what is advertised. The SRAM inside the cache controller provides the space for storing the tag bits. Tag bits are not included in cache size. In Figure 22-5, DRAM location F992 contains data 85H. The left portion of the figure shows when the data is moved from DRAM to cache.

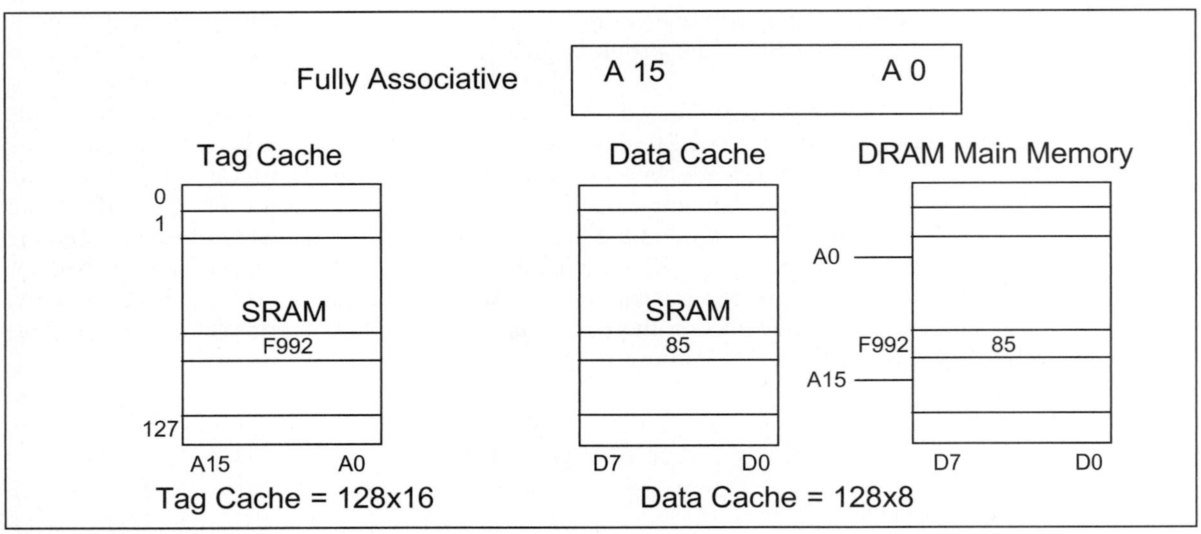

Figure 22-5. Fully Associative Cache

Direct-mapped cache

Direct-mapped cache is the opposite extreme of fully associative. It requires only one comparison. In this cache organization, the address is divided into two parts: the index and the tag. The index is the lower part of the address, which is directly mapped into SRAM, while the upper part of the address is held by the tag SRAM. From the above example, A0 to A10 are the index and A11 to A15 are the tag. Assuming that CPU addresses location F7A9H, the 7A9 goes to the index but the data is not read until the contents of tag location 7A9 is compared with 11110B. If it matches (its content is 11110), the data is read to the CPU; otherwise, the microprocessor must wait until the contents of location F7A9 are brought from main memory DRAM into the CPU while a copy of it is issued to cache for future reference. There is only one unique location with index address of 7A9, but 32 possible tags ($2^5 = 32$). Any of these possibilities, such as C7A9, 27A9, or 57A9, could be in tag cache. In such a case, when the tag of a requested address does not match the tag cache, a cache miss occurs. Although the number of comparisons has been reduced to one, the problem of accessing information from locations with the same index but different tag, such as F7A9 and 27A9, is a drawback. The SRAM requirement for this cache is shown below. While the data cache is 2K bytes, the tag requirement is 2K × 5 = 10K bits or about 1.25K bytes. See Figure 22-6.

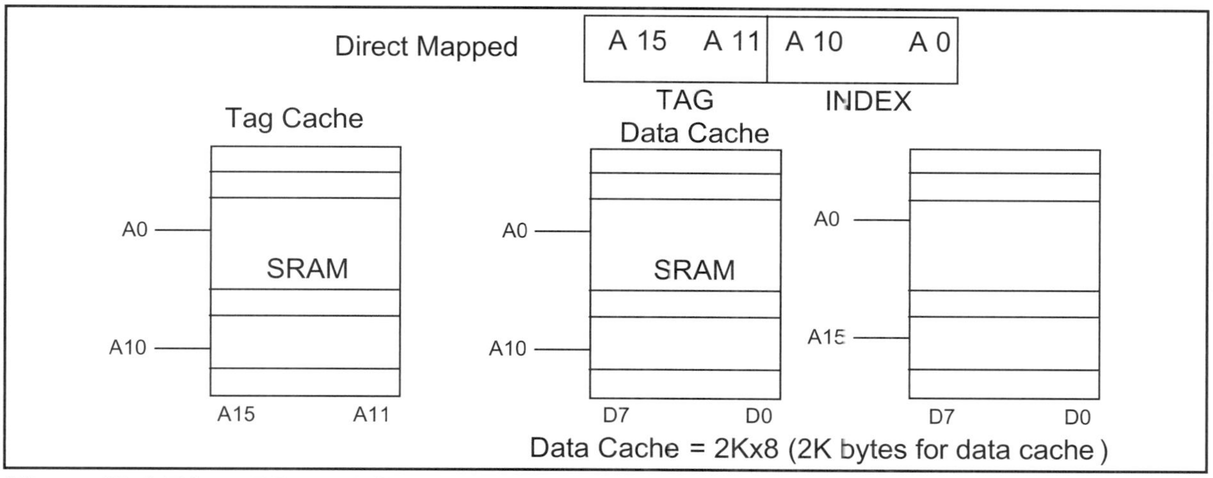

Figure 22-6. Direct-Mapped Cache

Set associative

This cache organization is in between the extremes of fully associative and direct mapped. While in direct mapped there is only one tag for each index, in set associative, the number of tags for each index is increased, thereby increasing the hit rate. In 2-way set associative, there are two tags for each index, and in 4-way there are 4 tags for each index. See Figures 22-7 and 22-8. Comparing direct-mapped and 2-way set associative, one can see that with only a small amount of extra SRAM, a better hit rate can be achieved. In this organization, if the microprocessor is requesting the contents of memory location 41E6H, there are 2 possible tags that could hold it, since cache circuitry will access index 1E6H and compare the contents of both tags with "0100 00". If any of them matches it, the data of index location 1E6 is read to the CPU, and if none of the tags matches "0100 00", the miss will force the cache controller to bring the data from DRAM to cache, while a copy of it is provided to the CPU at the same time. In 4-way set associative, the search for the block of data starting at 41E6 is initiated by comparing the 4 tags with "0100 000", which will increase the chance of having the data in the cache by 50%, compared with 2-way set associative. As seen in the above example, the number of comparisons in set associative depends on the degree of associativity. It is 2 for 2-way set associative, 4 for 4-way set associative, 8 for 8-way, n for n-way set associative, and in the thousands for fully set associative. The higher the set, the better the performance, but the amount of SRAM required for tag cache is also increased, making the 8-way and 16-way associatives' increased costs unjustifiable compared to the small increase in hit rate. The increase in the set also increases the number of tag comparisons. Most cache systems that use this organization are implemented in 4-way set associative (e.g., 80486 on-chip cache).

From a comparison of these two cache organizations, the difference between them in organization and SRAM requirements can be seen. In 2-way, the tag of 1K × 6 and data of 1K × 8 for each set gives a total of 14K bits [2 × (1K × 6 + 1K × 8) = 28K bits]. In 4-way, there is 512 × 7 for the tag and 512 × 8 for data, giving a total of 32K bits [(512 × 7 + 512 × 8) × 4 = 32K bits] of SRAM requirement. Only with an extra 4K bits will the hit rate improve substantially. As the degree of associativity is increased, the size of the index is reduced and added to the tag and this increases the tag cache SRAM requirement, but the size of data cache remains the same for all cases of direct map, 2-way, and 4-way associative. These concepts are clarified further in Examples 22-12, 22-13, and 22-14.

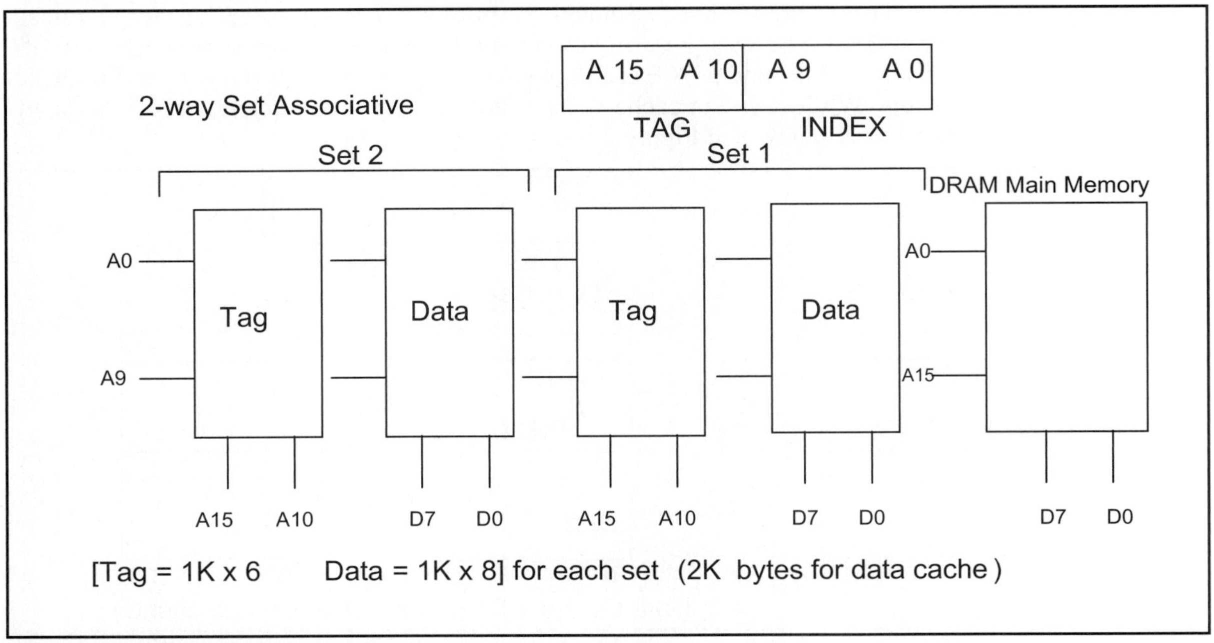

Figure 22-7. Two-way Set Associative

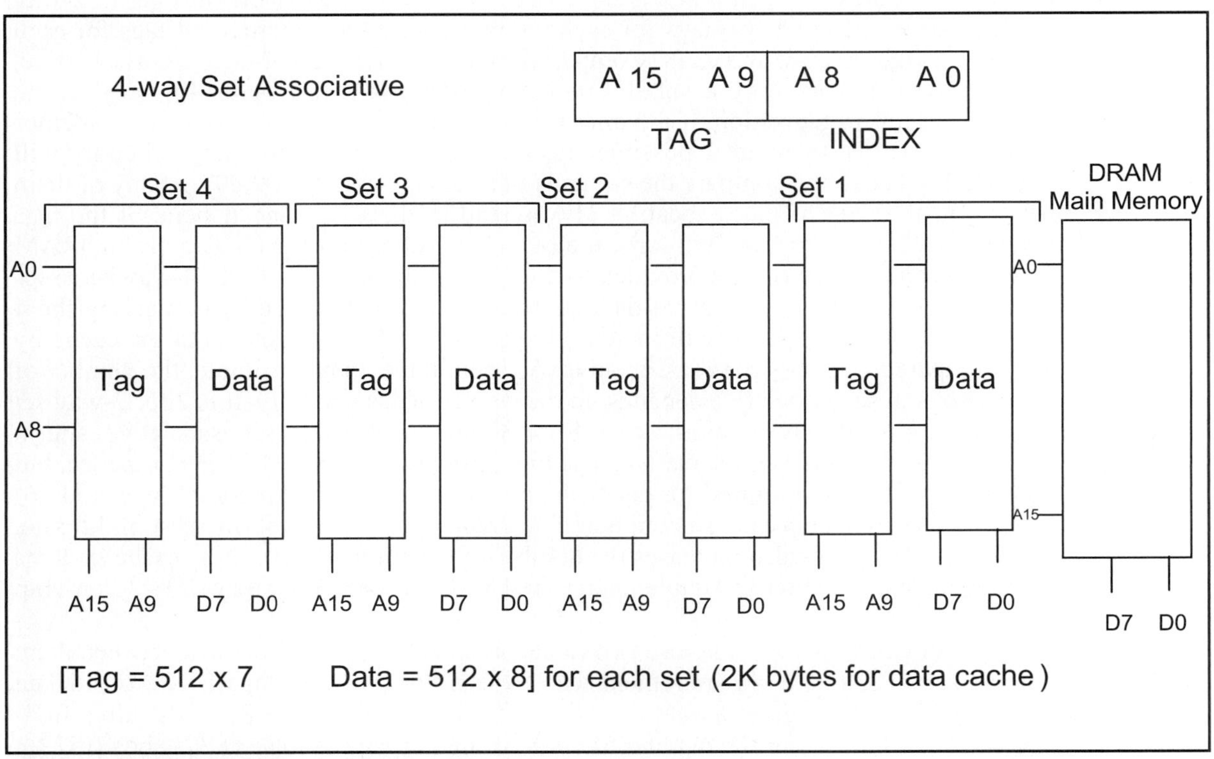

Figure 22-8. Four-way Set Associative

Example 22-12

This example shows directed-mapped cache for 16M main memory.

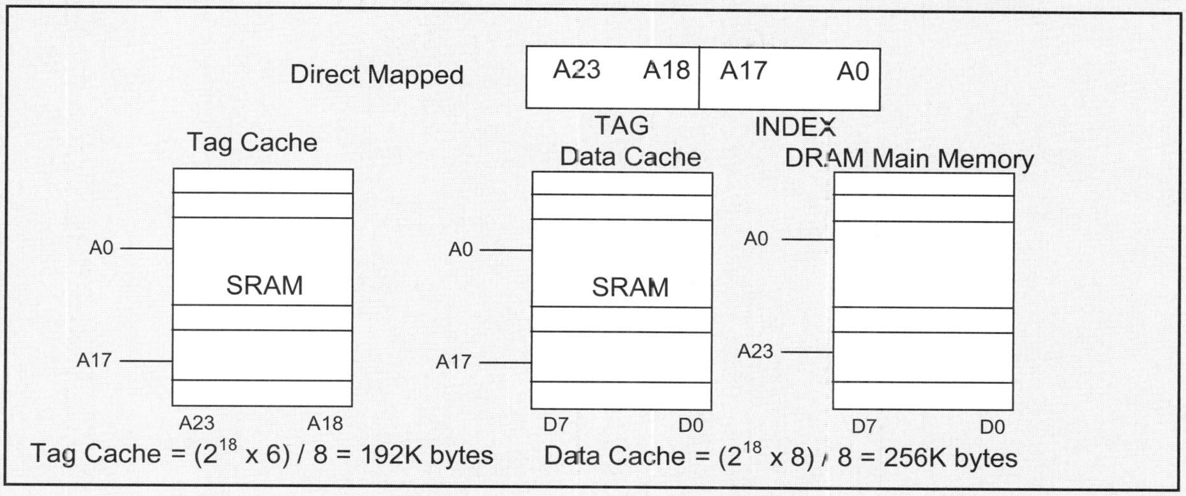

Example 22-13

This example shows 2-way set associative mapped cache for 16M main memory.

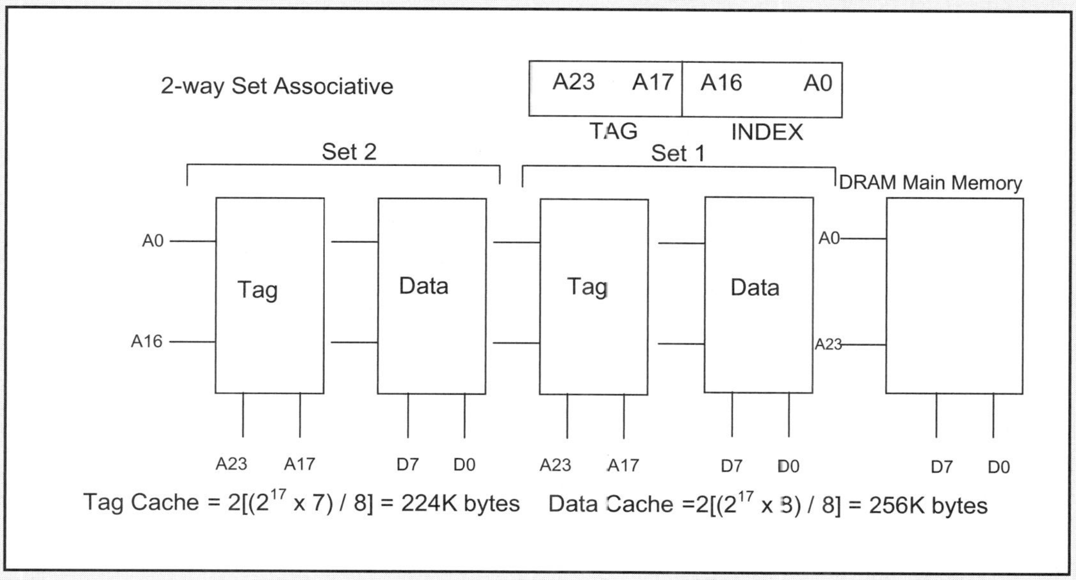

Example 22-14

This example shows 4-way set associative mapped cache for 16M main memory.

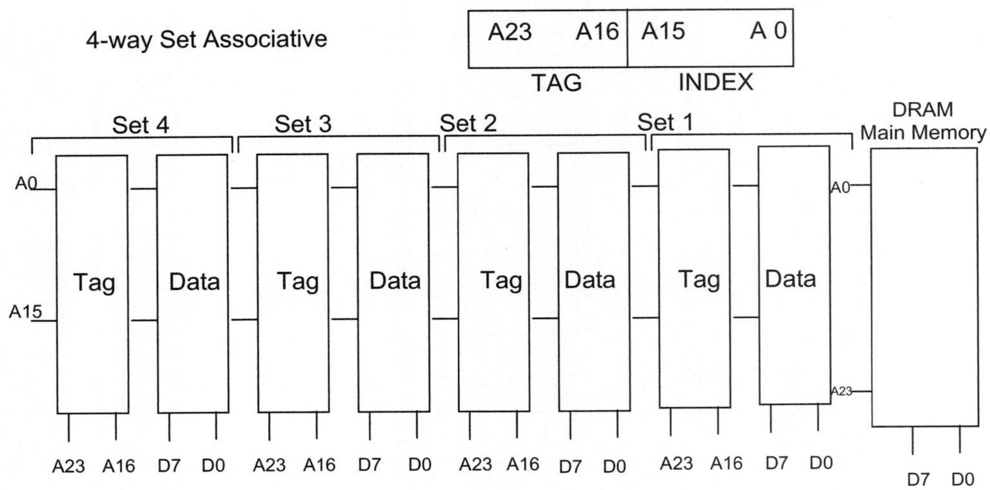

Tag Cache = $4[(2^{16} \times 8) / 8]$ = 256K bytes Data Cache = $4[(2^{16} \times 8) / 8]$ = 256K bytes

Updating main memory

In systems with cache memory, there must be a way to make sure that no data is lost and that no stale data is used by the CPU, since there could be copies of data in two places associated with the same address, one in main memory and one in cache. A sound policy on how to update main memory will ensure that a copy of any new data written into cache will also be written to main memory before it is lost since the cache memory is nothing but a temporary buffer located between the CPU and main memory. To prevent data inconsistency between cache and main memory, there are two major methods of updating the main memory: (1) write-through and (2) write-back. The difference has to do with main memory traffic.

Write-through

In write-through, the data will be written to cache and to main memory at the same time. Therefore, at any given time, main memory has a copy of valid data contained in cache. At the cost of increasing bus traffic to main memory, this policy will make sure that main memory always has valid data, and if the cache is overwritten, the copy of the latest valid data can be accessed from main memory. See Figure 22-9.

Write-back (copy-back)

In the write-back (sometimes called *copy-back*) policy, a copy of the data is written to cache by the processor and not to main memory. The data will be written to main memory by the cache controller only if cache's copy is about to be altered. The cache has an extra bit called the *dirty bit* (also called the *altered bit*). If data is written to cache, the dirty bit is set to 1 to indicate that the cache data is new data that exists only in cache and not in main memory. At a later time, the cache data is written to main memory and the dirty bit is cleared. In other words, when the dirty bit is high it means that the data in cache has changed and is different from the corresponding data in main memory; therefore, the

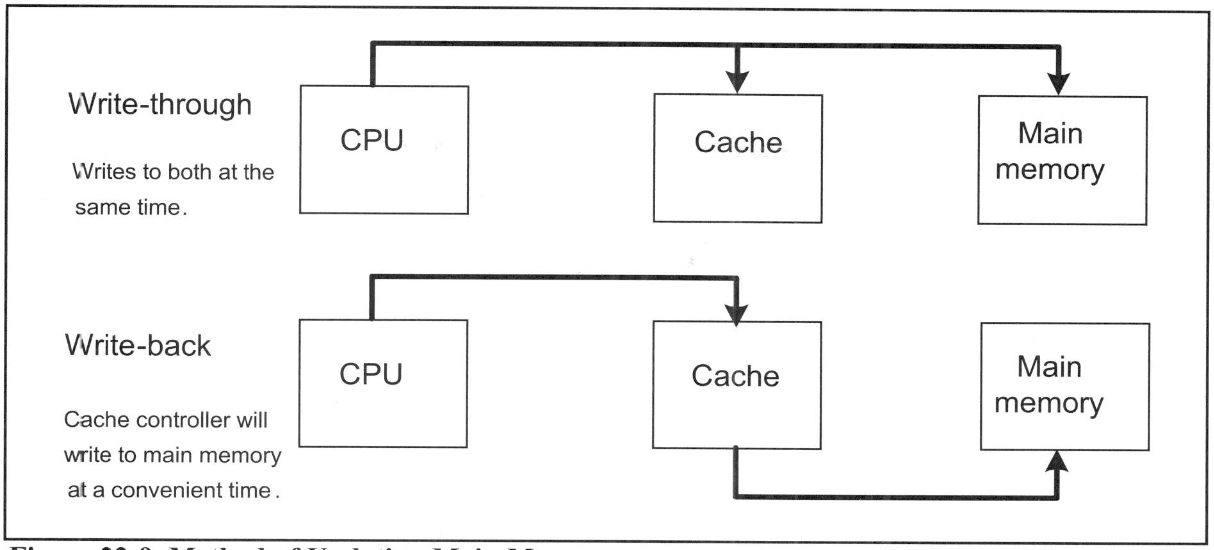

Figure 22-9. Method of Updating Main Memory

cache controller will make sure that before erasing the new data in cache, a copy of it is given to main memory. Getting rid of information in cache is often referred to as *cache flushing*. This updating of the main memory at a convenient time can reduce the traffic to main memory so that main memory buses are used only if cache has been altered. If the cache data has not been altered and is the same as main memory, there is no need to write it again and thereby increase the bus traffic as is the case in the write-through policy. See Figure 22-9.

Before concluding this section, two more cache terminologies that are commonly used in the technical literature will be described: cache coherency and cache replacement policy.

Cache coherency

In systems in which main memory is accessed by more than one processor (DMA or multiprocessors), it must be ensured that cache always has the most recent data and is not in possession of old (or stale) data. In other words, if the data in main memory has been changed by one processor, the cache of that processor will have the copy of the latest data and the stale data in the cache memory is marked as dirty (stale) before the processor uses it. In this way, when the processor tries to use the stale data, it is informed of the situation. In cases where there is more than one processor and all share a common set of data in main memory, there must be a way to ensure that no processor uses stale data. This is called *cache coherency*.

Cache replacement policy

What happens if there is no room for the new data in cache memory and the cache controller needs to make room before it brings data in from main memory? This depends on the cache replacement policy adopted. In the LRU (least recently used) algorithm, the cache controller keeps account of which block of cache has been accessed (used) the least number of times, and when it needs room for the new data, this block will be swapped out to main memory or flushed if a copy of it already exists in main memory. This is similar to the relation between virtual memory and main memory. The other replacement policies are to overwrite the blocks of data in cache sequentially or randomly, or use the FIFO (first in, first out) policy. Depending on the computer's design objective and its intended use, any of these replacement policies can be adopted.

Cache fill block size

If the information asked for by the CPU is not in cache and the cache controller must bring it in from main memory, how many bytes of data are brought in whenever there is a miss? If the block size is too large (let's say 5000 bytes), it will be too slow since the main memory is accessed normally with 1 or 2 WS. At the other extreme, if the block is too small, there will be too many cache misses. There must be a middle-of-the-road approach. The block size transfer from the main memory to CPU (and simultaneous copy to cache) varies in different computers, anywhere between 32 and 512 bytes. If the block size is 32 bytes, then it is called the 8-line cache refill policy, where each line is 4 bytes of the 32-bit data bus.

Level 1, 2, and 3 caches

With advances in IC fabrication we can put hundreds of millions of transistors onto a single chip. This has allowed putting some caches on the CPU chip itself. When the cache is embedded into the CPU die, it is called L1 (level 1) cache. If the cache is on-chip, inside the package but outside of the CPU die, then it is called L2 (level 2) cache, whereas cache outside the CPU residing on the motherboard is called L3 (level 3). See Chapter 24.

Review Questions

1. Cache is made of _________ (DRAM, SRAM).
2. From which does the CPU asks for data first, cache or main memory?
3. Rank the following from fastest to slowest as far as the CPU is concerned.
   (a) main memory (b) register (c) cache memory
4. In fully associative cache of 512 depth, there will be ____ comparisons for each data request.
5. Which cache organization requires the least number of comparisons?
6. A 4-way set associative organization requires _____ comparisons.
7. What does write-through refer to?
8. Which one increases the bus traffic, write-through or write-back?
9. What does LRU stand for, and how is it used?
10. What does cache refill policy of 4 lines refer to?

SECTION 22.4: SDRAM, DDR RAM, AND RAMBUS MEMORIES

In recent years the need for faster memory has led to the introduction of some very high-speed DRAMs. In this section we look at three of them: EDO (extended data-out), SDRAM (synchronous DRAM), and RDRAM (Rambus DRAM). In the mid-1990s, the speed of x86 processors went over 100 MHz and subsequently Intel began talking about 300–400 MHz CPUs. However, a major problem for these high-speed CPUs is the speed of DRAM. After all, cache has to be filled with information residing in main memory DRAM. Before we discuss some high-speed DRAMs, it needs to be noted that "1 GHz" CPU does not mean that the bus speed is also 1 GHz. For microprocessors over 1 GHz, the bus speed is often a fraction of the CPU speed. This is due to the expense and difficulty (e.g., crosstalk, electromagnetic interference) associated with the design of high-speed motherboards and the slowness of memory and logic gates. For example, in many 1 GHz Pentium systems, the bus speed is only 400 MHz.

EDO DRAM: origin and operation

Earlier in this chapter we discussed page mode DRAM. It needs to be noted that page mode DRAM has been modified and now is referred to as fast page mode DRAM. Note that DRAM data books of the mid-1990s refer only to fast page DRAM (FPM DRAM) and not page mode. The following describes the operation and limitations of fast page DRAM and how it led to EDO DRAM.

1. The row address is provided and latched in when RAS falls. This opens the page.
2. The column address is latched in when CAS falls and data shows up after t_{CAC} has elapsed. However, the next column of the same row (page) cannot be accessed faster

than t_{PC} (page cycle time). This means that accessing consecutive columns of opened pages is limited by the t_{PC}. The t_{PC} timing itself is influenced by how long CAS has to stay low before it goes up. Why don't DRAM designers pull up the CAS faster in order to shorten the t_{PC}? This seems like a very logical suggestion. However, there is a problem with this approach in fast page mode: When the CAS goes high, the data output is turned off. So if CAS is pulled high too fast (to shorten the t_{PC}), the CPU is deprived of the data. One solution is to change the internal circuitry of fast page DRAM to allow the data to be available longer (even if CAS goes high). This is exactly what happened. As a result of this change, the name EDO (extended data-out) was given to avoid confusion with fast page mode DRAM. This is the reason that EDO is sometimes called *hyper-page* since it is the hyper version of fast page DRAM. Tables 22-7 and 22-8 show a comparison of FPM and EDO DRAM timing. Notice in both cases that all the parameters are the same except t_{PC}. For the EDO version of page mode, the t_{PC} is 10 ns less than fast page mode.

Table 22-7: 70 ns 4M DRAM Timing

	FPM	EDO
Speed (ns)	70	70
tRAC (ns)	70	70
tRC (ns)	130	130
tPC (ns)	40	30

Note: 256Kx16 DRAM
From Micron Technology

Table 22-8: 60, 50 ns 4M DRAM Timing

	FPM	EDO	EDO
Speed (ns)	60	60	50
tRAC (ns)	60	60	50
tRC (ns)	110	110	100
tPC (ns)	35	25	20

Note: 256Kx16 DRAM
From Micron Technology

In examining t_{PC} timing in Figure 22-10, notice that t_{PC} (page cycle time) consists of two portions: t_{CP} (CAS precharge time) and t_{CAS} (CAS pulse width). The t_{CP} is similar across 70 ns, 60 ns, and 50 ns DRAMs of FPM and EDO (about 10 ns). It is t_{CAS} that varies among these DRAMs. In EDO this portion is made as small as possible. Figure 22-11 compares FPM and EDO timing.

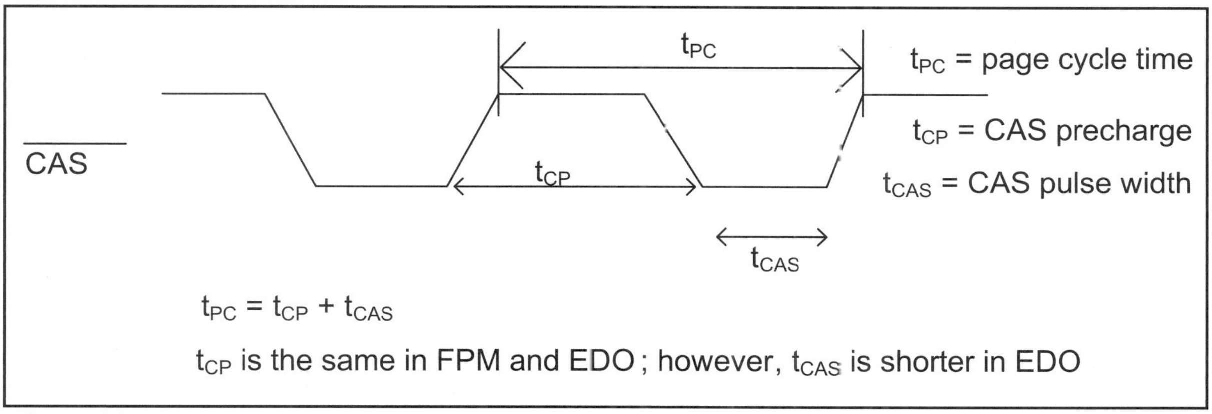

Figure 22-10. t_{PC} Timing in Page Mode DRAM

SDRAM (synchronous DRAM)

When the CPU bus speed goes beyond 75 MHz, even EDO is not fast enough. SDRAM is a memory for such systems. First, let us see why it is called synchronous DRAM. In all the traditional DRAMs (page mode, fast page, and EDO), CPU timing is not synchronized with DRAM timing, meaning that there is no common clock between the CPU and DRAM for reference. In those systems it is said that the DRAM is asynchronous with the microprocessor since the CPU presents the address to DRAM and memory

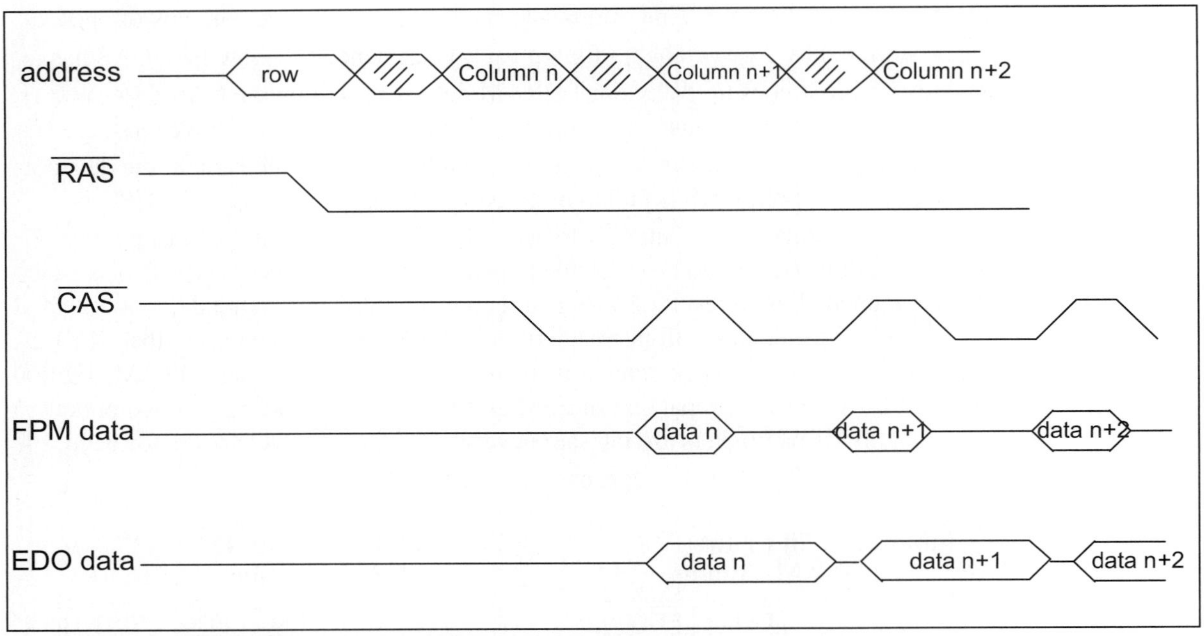

Figure 22-11. Comparison of FPM and EDO Timing

provides the data in the master/slave fashion. If data cannot be provided on time, the CPU is notified with the NOT READY signal. In response to NOT READY, the CPU inserts a wait state into its bus timing and waits until the DRAM is ready. In other words, the CPU bus timing is dependent upon the DRAM speed. This is not the case in synchronous DRAM. In systems with SDRAM, there is a common clock (called the *system clock*) that runs between the microprocessor and SDRAM. All bus activities (address, data, control) between the CPU and DRAM are synchronized with this common clock. That is, the common clock is the point of reference for both the CPU and SDRAM and there is no deviation from it and hence no waiting by the CPU. See Figure 22-12 for SDRAM timing. As shown in Figure 22-12, the system clock is the common clock that the address, data, and control signals are synchronized with. As you examine the timing figures in EDO and page mode, you will not find such a clock.

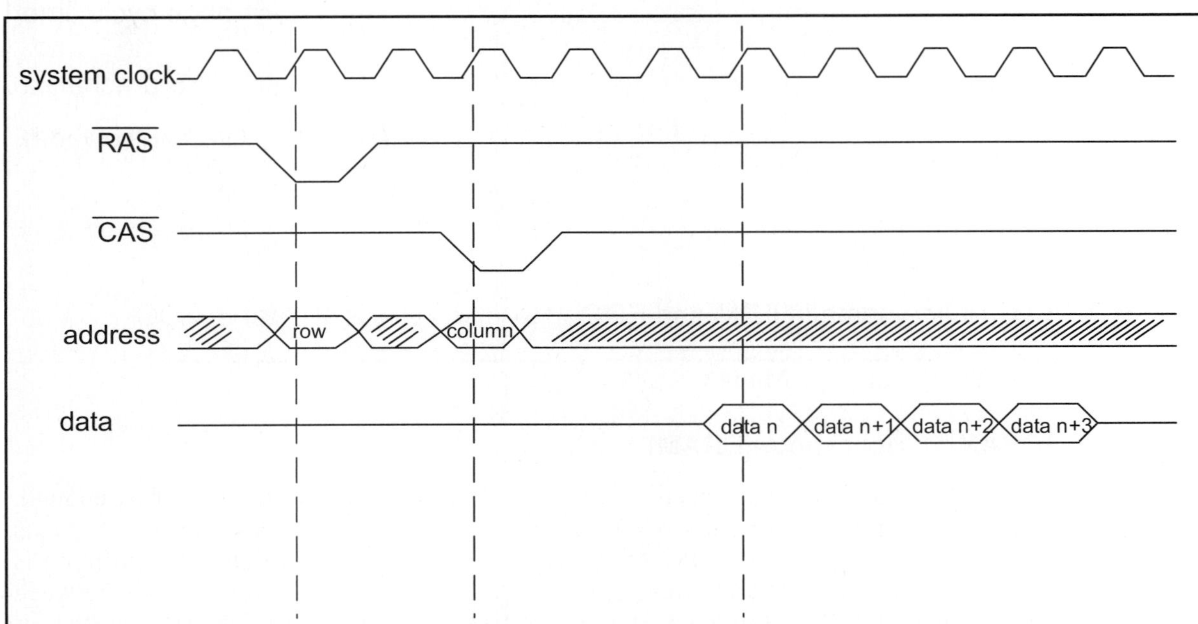

Figure 22-12. SDRAM Timing

SDRAM, DDR RAM, and burst mode

The presence of the common system clock between the CPU and SDRAM lends itself to what is called *burst I/O*. Although burst I/O will do both read and write, we will discuss the read operation for the sake of simplicity. In burst read, the address of the first location is provided as normal. RAS is first, followed by CAS. However, since we read several consecutive locations in the cache fill (depending on whether the cache has 4, 8, 16, or 32 lines), there is no need to provide the full address of each line and pay the timing penalty for address setup and hold time. Why not simply program the burst SDRAM to let it know how many consecutive locations are needed according to the cache design? That is exactly the idea behind many SDRAMs. They are capable of being programmed to output up to 256 consecutive locations inside DRAM. In other words, the number of burst reads can be 1, 2, 4, 8, 16, or 256, and burst SDRAM can be programmed in advance for any number of these reads. The number of burst reads is referred to as *burst length*. In many recent SDRAMs, the burst length can be as high as a whole page. Burst read shortens memory access time substantially. For example, if burst length is programmed for 8, for the first location we need the full address of RAS followed by CAS. However, for the second, third, ..., eighth, we can get the data out of the SDRAM with a minimum delay, limited only by the internal circuitry of DRAM. Starting with the Pentium, processors use the concept of burst read in their bus timing. In DDR (Double Data Rate) RAM the data is provided to the CPU on both the positive and negative edges of the clock.

SDRAM and interleaving

In order to increase performance, SDRAMs use the concept of interleaving discussed in Section 22.2. In traditional interleaved design, the board designer must arrange the DRAM in an interleaved fashion in order to hide the precharge time of one bank behind the access time of the other one. In SDRAM, this interleaving is done internally. In other words inside the SDRAM itself, DRAM cells are organized in such a way that while one bank is being refreshed the other one is being accessed. By incorporating both the burst mode and interleaving concepts into SDRAM, it is predicted that SDRAM memory can be used for a bus frequency as high as 125 MHz but not beyond that.

Figure 22-12 shows SDRAM timing. How many clocks after CAS will the data appear at the data pins? This can be programmed. It is called *read latency* and can be 1, 2, or 3 clocks. In Figure 22-12, the read latency is 3 since the data appears at the data buses 3 clocks after CAS.

It should be noted that SDRAM and EDO standards are both set by industry and every DRAM maker supports them. SDRAM and EDO are not proprietary technologies.

Rambus DRAM

In contrast to EDO and SDRAM, Rambus is proprietary DRAM architecture. DRAM manufacturers license this technology from Rambus Inc. in exchange for royalty payments. DRAMs with Rambus technology are referred to as RDRAM in technical literature.

Overview of Rambus technology

The heart of Rambus technology is a proprietary interface for chip-to-chip bus connection. This high-speed bus technology is composed of three sections: (1) a Rambus interface, (2) a Rambus channel, and (3) Rambus DRAM. The Rambus interface standard must be incorporated into both DRAM and the CPU. While many DRAM makers are introducing DRAM with a Rambus interface (called RDRAM), not every microprocessor is equipped with a Rambus interface. However, Intel has indicated that it will equip future generations of the x86 with a Rambus interface. However, if a given microprocessor is not equipped with a Rambus interface, one can design a memory controller with the Rambus interface and place it between the CPU and RDRAM. Such a controller is referred to as a *Rambus channel master* and the RDRAM is called a *Rambus channel slave*. See Figures 22-13 and 22-14.

In Rambus technology, only the master can generate a request since it contains

Example 22-15

Assume a bus frequency of 100 MHz. Discuss bus timing for (a) EDO of 50 ns speed where t_{CP} = 20, (b) SDRAM of t_{CK} = 10 ns.

Solution:

1/100 MHz = 10 ns is the system clock.

(a) In EDO when the page is opened, the fastest it can provide data is t_{PC}, which is 20 ns. Therefore, we need at least one wait state.
(b) In SDRAM of t_{CK} = 10 ns, the first address is strobed into the DRAM and subsequent data bursts are provided at 10 ns intervals. Therefore, no wait state is needed. Of course, for both of the above cases any bus overhead was ignored.

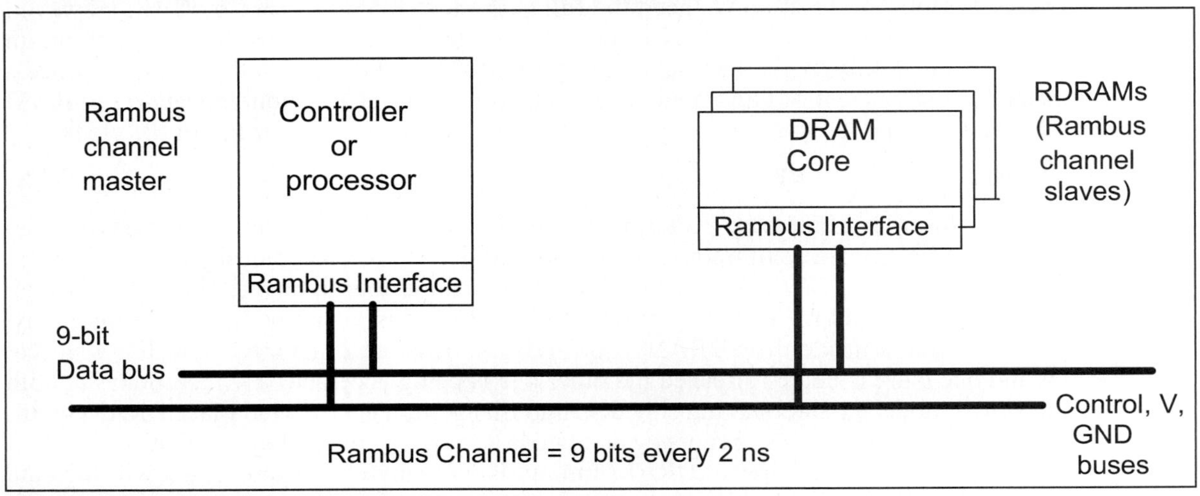

Figure 22-13. A Rambus-Based System

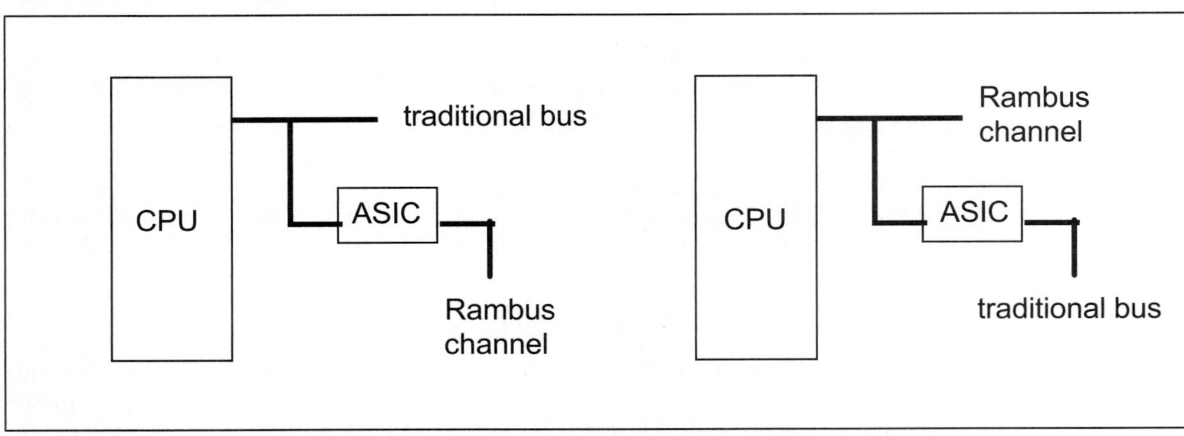

Figure 22-14. CPUs with and without Rambus Channel (Courtesy of Rambus, Inc.)

intelligence. Slave devices such as RDRAM respond to requests by the master. This eliminates any need for addition of intelligence circuitry to the RDRAM, thereby increasing its die size. This also means that data transfers can happen only between master and slave and there is never any direct data transfer between slaves. However, master capability can be added to devices other than the CPU such as peripheral devices, graphic processors, and memory controllers. The following describes additional features of Rambus channel technology.

1. The Rambus channel has only a 9-bit data bus.
2. There are only two DRAM organizations available for RDRAMs: × 8 or × 9. For example, a 16M-bit RDRAM is organized as 2M × 8.This is in contrast to DIMM memory modules where there are 72 pins for data alone. Such a large number of pins without a sufficient number of ground pins limits bus speed for these memory designs to less than 100 MHz. To reduce crosstalk and EMI (electromagnetic interference), we can add ground pins but that in turn makes the DRAM memory module too large (see Chapter 25 for the role of ground in reducing crosstalk).
3. Since the data bus is limited to 9 pins in a Rambus channel, adding a sufficient number of ground pins can push the speed of the bus to 500 MHz. To counter the impact of a limited bus size on bus bandwidth, the Rambus employs the method of block transfer. This is explained next.

Rambus protocol for block transfer

In Rambus the data is transferred in blocks. Such a block-oriented data transfer requires a set of protocols in which the packet types are defined very strictly. There are three types of packets in the Rambus protocol: (a) request, (b) acknowledge, and (c) data. The following steps show how the read operation works according to Rambus protocol. See Figure 22-15.

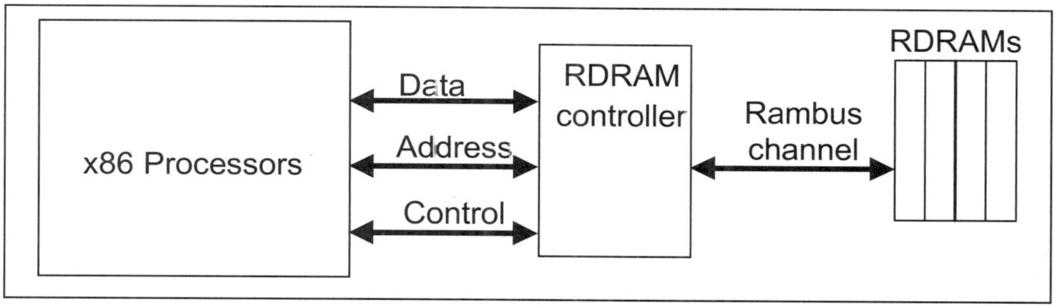

Figure 22-15. x86 System Using Rambus DRAM (Courtesy of Rambus, Inc.)

1. The master issues a request packet specifying the initial starting address of the needed data, plus the number of bytes needed to be transferred (the maximum for byte count is 256 bytes). This is considered one transaction.
2. RDRAM receives the request packet and decodes the addresses and byte count. If it has the requested data, an acknowledge packet is sent back to the master.
3. The acknowledge packet has three possibilities:
   (a) The addressed data does not exist.
   (b) The addressed data does exist but it is too busy to transfer the data. Try again later. This is called *nack.*
   (c) The addressed data does exist and it is ready to transfer them. This is called *okay.*
4. If the acknowledge packet has an okay in it, the RDRAM starts to transfer the data packet immediately.

An interesting aspect of this protocol is that the delay associated with receiving the acknowledge and sending the data packets can be programmed into configuration registers of both master and slave during BIOS system initialization.

At the time of this writing, the Rambus technology has become the standard in x86 PC design.

See the following websites for the latest in memory chip and design:

http://www.micron.com/

http://www.rambus.com

Review Questions

1. A 200-MHz Pentium has a bus frequency of ________.
2. A 100-MHz Pentium has a bus frequency 2/3 of the CPU. What is the read cycle time for this processor?
3. When a page is opened, what limits us in accessing consecutive columns?
4. True or false. In EDO, when CAS goes up the data output is turned off.
5. Which of the following DRAMs has a common synchronous clock with the CPU?
   (a) FPM (b) EDO (c) SDRAM (d) all of the above
6. True or false. SDRAM incorporates interleaved memory internally.
7. Can anyone incorporate the Rambus interface in their device?
8. Who issues the request in a Rambus system?
9. Who issues the acknowledge in a Rambus system?
10. Can normal EDO or FPM DRAMs be used for the Rambus channel?

PROBLEMS

SECTION 22.1: MEMORY CYCLE TIME OF THE x86

1. Calculate the memory cycle time for each of the following systems.
   (a) 386 of 33 MHz, 1 WS (b) 486 of 50 MHz, 1 WS
   (c) 386 of 25 MHz, 2 WS (d) Pentium of 60 MHz, 1 WS
2. If the memory cycle time is 90 ns, a 386 system of 33 MHz needs ___ WS.
3. If the memory cycle time is 80 ns, a 486 system of 50 MHz needs ___ WS.
4. If the memory cycle time is 45 ns, a Pentium system of 66 MHz needs ___ WS.
5. If the memory cycle time is 200 ns, a 386SX system of 20 MHz needs ___ WS.
6. Find the effective memory performance of a 486 system of 50 MHz with 2 WS. Compare the performance degradation with a 0-WS system.
7. Find the effective memory performance of a 386 system of 33 MHz with 1 WS. Compare the performance degradation with a 0-WS system.
8. Find the effective memory performance of a Pentium system of 60 MHz with 1 WS. Compare the performance degradation with a 0-WS system.
9. If a given system with a 2-clock memory cycle time has a memory cycle time of 60 ns and is designed with 1 WS, find the CPU frequency.
10. In a 33-MHz, 0-WS 486 system, a minimum of 20 ns is used for data and address path delay and address decoding. What is the maximum memory cycle time?

SECTION 22.2: PAGE AND STATIC COLUMN DRAMS

11. In which memory are the cycle time and access time equal?
12. What is the difference between the t_{AA} and t_{CA} SRAM data sheet?
13. Define the memory cycle time for a memory chip.
14. Define the memory cycle time for the CPU.

15. What is t_{RC} and t_{RAC} in DRAM? State the difference between them.
16. Show the relation (approximate) between t_{RC}, t_{RAC}, and t_{RP}.
17. A given DRAM has t_{RAC} = 60 ns. What is the t_{RC} (approximate)?
18. A given DRAM has t_{RAC} = 85 ns. What is the t_{RC} (approximate)?
19. A given DRAM has t_{RC} = 110 ns. What is the t_{RAC} (approximate)?
20. A given DRAM has t_{RC} = 90 ns. What is the t_{RAC} (approximate)?
21. Calculate the time needed to access 2048 bits of 60-ns 4M × 1.Use Table 22-2.
22. Calculate the time needed to access 2048 bits of 4M × 1 of 70 ns. Use Table 22-2.
23. Draw a timing diagram for standard mode SRAM and DRAM memory cycle.
24. What is the minimum memory addition to a 386 system with interleaved memory design if each bank of 8 bits is set for 1M × 8? The parity bit is not included.
25. Calculate the chip count for Problem 24 if 1M × 4 chips are used.
26. What is the minimum memory addition to a 486 system with interleaved memory design if each bank of 8 bits is set for 4M × 8? The parity bit is not included.
27. Calculate the chip count for Problem 26 if 4M × 4 chips are used.
28. Show the hex address for 386/486 interleaved memory banks.
29. Calculate the time needed to access 2048 bits of one page for page mode DRAM of 4M × 1 of 70 ns. Use Table 22-4.
30. Calculate the time needed to access 2048 bits of one page for page mode DRAM of 4M × 1 of 60 ns. Use Table 22-4.
31. Calculate the time needed to access 2048 bits of one page for static column mode DRAM of the 4M × 1 of 70 ns. Use Table 22-5.
32. Calculate the time needed to access 2048 bits of one page for static column mode DRAM of 4M × 1 of 60 ns. Use Table 22-5.
33. True or false. In a Pentium of 1 GHz, the bus speed is also 1 GHz.
34. True or false. In recent years, the bus speed has been falling behind the CPU speed.

SECTION 22.3: CACHE MEMORY

35. List the three different cache organizations.
36. What is the principle of locality of reference?
37. What does LRU stand for, and to what does it refer in cache memory?
38. What do write-through and write-back refer to? Define each one and state an advantage and a disadvantage for each.
39. What does a line size of 16 bytes mean?
40. Calculate the tag and data cache sizes needed for each of the following cases if the memory requesting address to main memory is 20 bits (A19–A0). Assume a data bus of 8 bits. Draw a block diagram for each case.
    (a) fully associative of 1024 depth
    (b) direct mapped where A15–A0 is for the index
    (c) 2-way set associative where A14–A0 is for the index
    (d) 4-way set associative (e) 8-way set associative
41. In Problem 40, compare the size of data cache and tag cache parts (b), (c), (d), and (e). What is your conclusion?
42. Calculate the tag and data cache sizes needed for each of the following cases if the memory requesting address to main memory is 24 bits (A23–A0). Assume a data bus of 8 bits. Draw a block diagram for each case.
    (a) fully associative of 1024 depth
    (b) direct mapped where A19–A0 is for the index
    (c) 2-way set associative where A18–A0 is for the index
    (d) 4-way set associative (e) 8-way set associative

43. In Problem 42, compare the size of data cache and tag cache for (b), (c), (d), and (e).What is your conclusion based on this comparison?
44. Give three factors affecting the cache hit.
45. What does the law of diminishing returns mean when applied to cache?
46. Exaplin the difference betwwen the L1 and L2 cache.
47. True or false. In L1 cache speed is the same as CPU die speed.

SECTION 22.4: SDRAM, DDR RAM, AND RAMBUS MEMORIES

48. The CPU speed (in Pentium and higher processors) is often a ________ (multiple, fraction) of the bus speed.
49. Calculate the memory read cycle time of a CPU with a bus speed of 300 MHz. Assume a 2-clock read cycle time.
50. In the above question, discuss the difficulties associated with the design of such a high-speed bus.
51. In Pentium processors with speeds of 100 MHz and higher, the bus speed is _______ the CPU speed.
    (a) the same as (b) a fraction of (c) a multiple of
52. In DRAM technology, EDO stands for ______ and FPM stands for _________.
53. True or false. Both EDO and FPM are page mode DRAMs.
54. What does "opening a page" mean in page mode DRAMs? What is the role of signals RAS and CAS in opening a page?
55. In ______ (EDO, FPM) DRAM, the data is turned off when CAS goes high.
56. In FPM DRAM, what happens if CAS goes high too soon and what is the consequence?
57. When a page is opened, reading consecutive columns is limited by the speed of __________.
58. In the design of DRAM, why is it desired to pull CAS high as soon as possible?
59. What is the t_{PC} for a 50-ns DRAM?
60. In a comparison of EDO and FPM DRAM of 60-ns and 70-ns speed, which timing parameters are the same and which are different?
61. For EDO DRAM, t_{PC} is normally 10 ns _____ (less than, greater than) the t_{PC} of FPM DRAM.
62. The t_{PC} timing is made of two parts. They are _______ and _______. One of them is constant across all DRAMs of 70-ns, 60-ns, and 50-ns speeds. Which one is that?
63. What does SDRAM stand for?
64. What is the most important difference between SDRAM and traditional DRAMs of FPM and EDO?
65. The SDRAM of 75 MHz can provide data every ____ ns after a page has been opened.
66. The SDRAM of 120 MHz can provide data every ____ ns after a page has been opened.
67. It is predicted that SDRAM can be used for bus speeds as high as ____ MHz.
68. What is burst mode memory? Define burst length.
69. In SDRAM, what is the size of burst length?
70. True or false. The x86 processors starting with the 486 support burst mode read.
71. What is the difference between interleaved memory design on board and interleaved in SDRAM?
72. True or false. EDO and SDRAM memories are proprietary technology requiring licenses.

73. What does "RDRAM" stand for, and what is the difference between that and SDRAM and EDO DRAMs?
74. Rambus technology consists of three parts. Name them.
75. True or false. Both master and slave sections of Rambus technology must have a Rambus interface.
76. A Rambus channel has a(n) _____-bit data bus.
77. RDRAM is a ________ (master, slave) in Rambus technology.
78. True or false. In Rambus technology, the data transfer happens between master and slave only.
79. True or false. In Rambus technology the data transfer never happens between slaves.
80. Name and describe the three types of packets for communication protocol in Rambus technology.
81. In the above question, "okay" and "nack" are part of which packet?
82. Explain the role of nack and okay for data transfer in Rambus technology.
83. The bus speed in Rambus can go as high as ______ MHz.
84. What happens if a CPU does not have a Rambus interface?
85. True or false. EDO and SDRAM can be used in place of RDRAM in a Rambus channel.

ANSWERS TO REVIEW QUESTIONS

SECTION 22.1: MEMORY CYCLE TIME OF THE x86

1. (a) 1/40 MHz = 25 ns; therefore, 2 × 25 = 50 ns; (b) 2 × 20 (for 0 WS) + 20 (1 WS) = 60 ns;
   (c) 2 × 15 ns + 15 = 45 ns
2. It means that the CPU cannot access memory faster than every 50 ns.
3. (a) The read cycle time is 75 ns; therefore, the effective working frequency is the same as 26.6 MHz of 0 WS (1/37.5 ns = 26.6 MHz).
   (b) The read cycle time is 60 ns; therefore, the effective working frequency is the same as 33 MHz (1/30 ns = 33 MHz).
4. A total of 50 ns is left for the memory access time.
5. Since 2 + 1 WS = 3 clocks for each read cycle time, 30 ns (90/3 = 30) for the CPU clock duration; therefore, the CPU frequency is 33 MHz (1/30 ns = 33 MHz).

SECTION 22.2: PAGE AND STATIC COLUMN DRAMS

1. SRAM and ROM
2. 100 ns
3. 60 ns
4. t_{RAC} (RAS access time), t_{RP} (RAS precharge time)
5. (a) There are two sets of 1M byte; therefore, each set consists of 4 banks of 256K × 9 memory where each bank belongs to 1 byte of the D31–D0 data bus.
   (b) 2M
6. True
7. t_{PC}
8. Total time = t_{RAC} + 99 × t_{PC} = 60 + 99 × 30 = 3030 ns
9. Static column

SECTION 22.3: CACHE MEMORY

1. SRAM
2. Cache
3. Register, cache, and main memory
4. 512
5. Direct map
6. 4
7. The CPU writes to cache and main memory at the same time when updating main memory.
8. Write-through
9. LRU (least recently used) is a cache replacement policy. When there is a need for room in the cache memory the cache controller flushes the LRU data to make room for new data.
10. When the cache is filled with new data, it is done a minimum of 4 lines ($4 \times 4 = 16$ bytes) at a time.

SECTION 22.4: SDRAM, DDR RAM, AND RAMBUS MEMORIES

1. Often less than 100 MHz; many times it is only 66 MHz.
2. $2/3 \times 100$ MHz = 66 MHz. Now $1/66$ MHz = 15 ns. 2×15 ns = 30 ns read cycle time.
3. The t_{PC} (page cycle time)
4. False
5. SDRAM
6. True
7. Yes, as long as you get a license from Rambus Inc.
8. Master (Rambus controller)
9. RDRAM slave
10. No. It must be RDRAM.

CHAPTER 23

PENTIUM AND RISC PROCESSORS

OBJECTIVES

Upon completion of this chapter, you will be able to:

- >> List the design enhancements of the x86 microprocessors from 80486 to Pentium 4
- >> Discuss the advantages of the 5-stage pipeline over the 2-stage pipeline
- >> Explain how the burst cycle is used to increase memory cycle times for read and write operations
- >> Compare the cache sizes of x86 processors from 486 to Pentium 4
- >> List three ways that designers can increase the processing power of a CPU
- >> List design enhancements of the Pentium over previous-generation x86 microprocessors
- >> Describe the impact on performance of the 64-bit data bus of the Pentium
- >> Describe superscalar architecture and Harvard architecture and their use in the Pentium
- >> List the unique features of RISC architecture compared to CISC and describe the impact on processing speed and program development
- >> Describe the main features introduced or enhanced in the Pentium Pro
- >> Give an overview of how MMX (MultiMedia extension) technology is used in some Intel processors
- >> Describe which aspects of DSP (digital signal processing) were incorporated into MMX technology
- >> Code Assembly language instructions to identify the CPU

The 8088/86 microprocessor is the product of technology of the 1970s. Advances made in integrated circuit technology in the 1980s made ICs with 1 million transistors possible. This led to the design of some very powerful microprocessors. This chapter will look at Intel's 486 and Pentium I microprocessors and will examine the merit of RISC processors and their potential power. In Section 23.1 the 80486 microprocessor is studied. Intel's Pentium I is discussed in Section 23.2. Section 23.3 explores RISC processors, and their performance is compared with that of x86 CISC processors. Section 23.4 discusses the main features of Intel's Pentium Pro processor. Section 23.5 describes MMX (MultiMedia extension) technology used for sound and graphics.

SECTION 23.1: THE 80486 MICROPROCESSOR

The 80486 is the first 1-million-transistor microprocessor (actually, 1.2 million) packaged in 168-pin PGA packaging. It is not only compatible with all previous Intel x86 microprocessors, but is also much faster than the 80386. When Intel went from the 286 to the 386, register widths were increased from 16 bits to 32 bits. In addition, the external data bus size was increased from 16 bits to 32 bits, and the address bus became 32 bits instead of 24 bits as in the 80286. However, the 32-bit core of the 386 microprocessor is preserved in the 486 microprocessor. This is due to the fact that many studies have shown that 32-bit registers can take care of more than 95% of the operands in high-level languages. Like the 386, the 486 has a 32-bit address bus and a 32-bit data bus. The data bus is D0–D31 and the address bus is A2–A31 in addition to BE0–BE3, just as in the 386. In the design of the 486, Intel uses four times as many transistors as used in the 386 to enhance its processing power while keeping a 32-bit microprocessor.

Enhancements of the 486

The following are the ways the 486 is enhanced in comparison to the 386.

Enhancement 1

By heavily pipelining the fetching and execution of instructions, the 486 executes many of its instructions in only 1 clock cycle instead of in 3 clocks as in the 386. By using a large number of transistors, it splits the fetching and execution of each instruction into many stages, all working in parallel. This allows the processing of up to five instructions to be overlapped. Pipelining in the 486 will be discussed further at the end of this section.

Enhancement 2

By putting 8K bytes of cache with the core of the CPU all on a single chip, the 486 eliminates the interchip delay of external cache. In other words, while in the 386 the cache is external, the 486 has 8K bytes of on-chip cache to store both code and data. Although the 486 has 8K bytes of on-chip cache, 128K to 256K bytes of off-chip cache are also present in many systems. Off-chip cache (level two) is commonly referred to as *secondary cache*, while on-chip cache is called *first-level cache*. The 8K on-chip cache of the 486 has 2-way set associative organization and is used for storing both data and code. It uses the write-through policy for updating main memory.

Enhancement 3

Intel used some of 1.2 million transistors to incorporate a math coprocessor on the same chip as the CPU. While in all previous x86 microprocessors the math coprocessor was a separate chip, in the 80486 the math coprocessor is part of a single IC along with the CPU. This reduced the interchip delay associated with multichip systems such as the 386 and 387 but at the same time made the cost of a 80486 high compared to a 386 since the 80486 is in reality two chips in one: the main CPU and math coprocessor. For many people who did not need a math coprocessor this extra price was not justified. Therefore, Intel introduced the 80486SX, which is the main CPU, and a separate math coprocessor named 80487SX.

Enhancement 4

Another major addition to the 486 is the use of 4 pins for data parity (DP), which allows implementation of parity error checking on the system board. The four pins DP0, DP1, DP2, and DP3 are bidirectional, and each is used for 1 byte of the D31–D0 data bus. When the 486 writes data it also provides the even-parity bit for each byte through the DP0–DP3 pins. When it reads the data it expects to receive the even parity bit for each byte on the DP0–DP3 pins. After comparing them internally, if there is a difference between the data written and the data read, it activates the pin PCHK (parity check) to indicate the error. This means that PCHK is an output pin while DP0–DP3 are bidirectional I/O. It must be noted that inconsistency between data written and data read has no effect on the execution of code by the CPU. It is the responsibility of system designers to incorporate error detection by using the PCHK pin in their designs. In the above discussion of parity, the word *data* is meant to refer to both code and data. Figure 23-1 shows the memory organization of the 486.

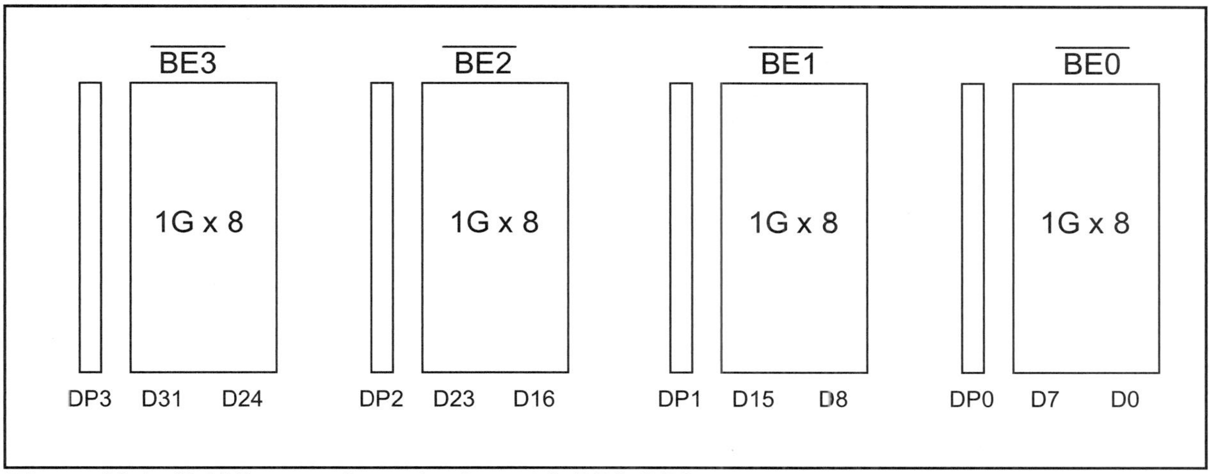

Figure 23-1. 486 Memory Organization with DP0–DP3

Enhancement 5

Another enhancement of the 486 involves the burst cycle. The memory cycle time of the 486 with the normal zero wait states is 2 clocks. In other words, it takes a minimum of 2 clocks to read from or write to external memory or I/O. In this regard, the 486 is like the 386. To increase the bus performance of the 486, Intel provides an additional option of implementing what is called a *burst cycle*. The 486 has two types of memory cycles, nonburst (which is the same as the 386) and burst mode. In the burst cycle, the 486 can perform 4 memory cycles in just 5 clocks. The way the 80486 performs the burst cycle read is as follows. The initial read is performed in a normal 2-clock memory cycle time, but the next three reads are performed each with only one clock. Therefore, four reads are performed in only 5 clocks. This is commonly referred to as 2-1-1-1 read, which means 2 clocks for the first read and 1 clock for each of the following three reads. This is in contrast to 386, which is 2-2-2-2 for reading 4 doublewords of aligned data. Of course, burst cycle reading is most efficient if the data and codes are in 4 doubleword (32-bit) consecutive locations. In other words, the burst cycle can be used to fetch a maximum of 16 bytes of information into the CPU in only 5 clocks, provided that they are aligned on doubleword boundaries. There are two pins, BRDY (burst ready) and BLAST (burst last), used specifically to implement the burst cycle. BRDY is an input into the 486 and BLAST is an output from the 486. See Figure 23-2 and Example 23-1.

Enhancement 6

The 486 supports all 386 instructions in addition to six new ones. They are shown in Table 23-1. Three of the new instructions, INVD, INVLPG, and WBINVD, are added specifically for dealing with the on-chip cache and the TLB entries. The XADD instruc-

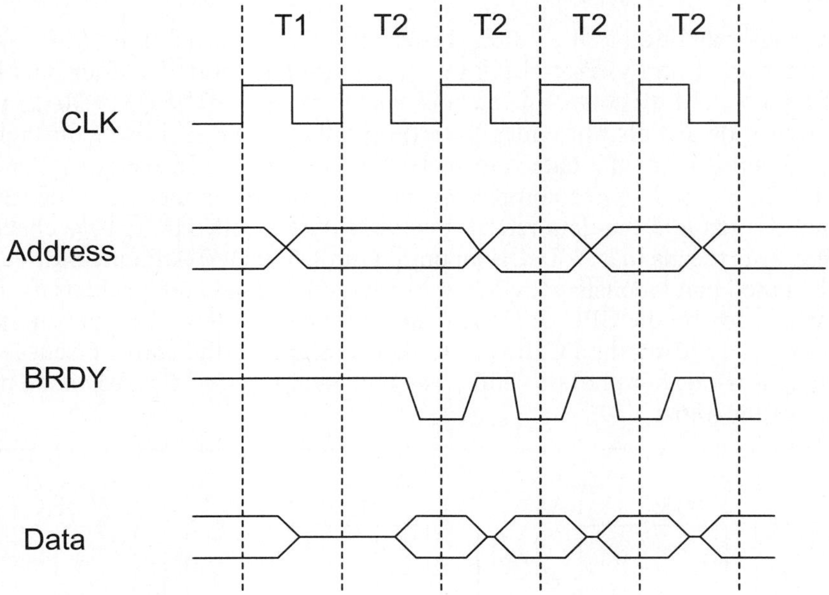

Figure 23-2. Burst Cycle Read in the 486

Example 23-1

Calculate and compare the bus bandwidth of the following systems. Assume that both are working with 33 MHz and that the 386 is 0 WS. Also assume that the data is aligned and is in 4 consecutive doubleword memory locations.
(a) 386 (b) burst mode of the 486

Solution:

(a) In the 386, since each memory cycle time takes 2 clocks we have memory cycle time = 2 (1/33 MHz) = 2 × 30 ns = 60 ns bus bandwidth = (1/60 ns) × 4 bytes = 66 megabytes/second
(b) In burst mode, the 486 performs 4 memory cycles in only 5 clocks; therefore, the average memory cycle time in burst mode is 1.25 (5/4 = 1.25) clocks for each 32-bit (doubleword) of data fetched as long as they are aligned and located in consecutive memory locations. This results in bus bandwidth = [1/(1.25 × 30 ns)] 4 bytes = 106.66 megabytes/second

Table 23-1: New 486 Instructions

Instruction	Meaning
BSWAP	Byte swap
CMPXCHG	Compare and exchange
INVD	Invalidate cache
INVLPG	Invalidate TLB entry
WBINVD	Write back and invalidate cache
XADD	Exchange and add

tion first loads the destination operand into the source and then loads the total sum of both the destination and the original source into the destination. The CMPXCHG instruction compares the accumulator, AL, AX, or EAX, with the destination operand, which could be a register or memory. If they are equal, the ZF = 1 and the source are copied into the destination. If they are not equal, ZF = 0 and the destination are copied into the accumulator. For example, the instruction "CMPXCHG BX,CX" copies CX into BX only if CX = AX; otherwise, it copies BX into accumulator AX.

As mentioned previously, some systems use little endian while others use the big endian convention of storing data. To allow the implementation of either, BSWAP is provided. The BSWAP instruction converts the contents of a 32-bit register from the little endian to big endian, or vice versa. See Example 23-2.

Example 23-2

Find the contents of memory location ES:4000 after running the following program.

```
MOV          EAX,[ 2000]          ;load EAX from memory DS:2000
BSWAP        EAX                  ;change little endian to big endian
MOV          ES:[ 4000] ,EAX      ;save the result at ES:4000
```

Assume that memory locations DS:2000–DS:2003 have the following contents:

DS:2000 = (87)
DS:2001 = (54)
DS:2002 = (F2)
DS:2003 = (99)

Solution:

The first instruction brings in the data in the little endian format where the least significant byte is fetched into the least significant byte of EAX, which is the AL register. BSWAP makes the 87H the most significant byte and puts 99H into the AL register. Therefore, after the execution of the last instruction we have the following:

ES:4000 = (99)
ES:4001 = (F2)
ES:4002 = (54)
ES:4003 = (87)

The addition of the BSWAP instruction makes it much easier for operating system software writers to convert their software from little endian to big endian, or vice versa.

CLK in the 80486

Another difference is the clock frequency provided to the 486. As mentioned in Chapter 22, the CLK input frequency, which provides the fundamental timing for the internal working of the CPU, is twice the system frequency for 386 microprocessors. In the case of the 486, the CLK is the same as the system frequency.

386, 486 performance comparison

As stated earlier, most of the instructions in the 486 are executed with only one clock. See Example 23-3.

Example 23-3

Compare the clock count for the loop part of the following program run on the 386 and 486. This program transfers a block of DWORD data. Assume that the block size is 10.

```
       MOV    CX,10                       ;count=10
       MOV    SI,OFFSET ARRAY1            ;load address of source
       MOV    DI,OFFSET  RESULT                 ;load address of destination
AGAIN:MOV    EAX,DWORD PTR [ SI]          ;get the element
       MOV    [ DI] ,EAX                  ;store it
       ADD    SI,4                        ;point to next element
       ADD    DI,4                        ;point to next element of result
       DEC    CX                          ;decrement the counter
       JNZ    AGAIN                       ;and go back if not zero
```

Solution:

		386	486
AGAIN:	MOV EAX,DWORD PTR [SI]	4	1
	MOV [DI],EAX	2	2
	ADD SI,4	2	1
	ADD DI,4	2	1
	DEC CX	2	1
	JNZ AGAIN	7/3	3/1
Total for one iteration		19	9

Notice the branch penalty for the JNZ instructions. If it goes back, it takes 7 clocks for the 386 and 3 for the 486. If it falls through, it takes only 3 and 1 for the 386 and 486, respectively. Also notice that "MOV [DI],EAX" takes 2 clocks since EAX must be provided first by the previous instruction. This is called a *data dependency*. In the next section we compare 386 and 486 performance with that of the Pentium processor.

More about pipelining

In the 8085 there was no pipelining. At any given moment, it either fetched or it executed. It could not do both at the same time. In the 8085, while the buses were fetching the instructions (opcodes) and data, the CPU was sitting idle, and in the same way, when the CPU was executing instructions, buses were sitting idle. However, in the 8086 the fetch and execute were performed in parallel by two sections inside the CPU called the BIU (bus interface unit) and EU (execution unit). The 8086 has an internal queue where it keeps the opcodes that are prefetched and waiting for the execution unit to process them. In the sequence of instructions, if there is a jump (JMP, JNZ, JNC, and so on) or CALL, the prefetched buffer (queue) is flushed and the bus interface unit of the CPU brings in instructions from the target location while the the execution unit waits for the new instruction. Since the introduction of the 8086 in 1978, microprocessor designers have come to rely more and more on the concept of pipelining to increase the processing power of the CPU. The next development was to expand the concept of a pipeline to the three stages of fetch, decode, and execute. In the 486, the pipeline stage is broken down even further to 5 stages as follows:

1. fetch (prefetch)
2. decode 1
3. decode 2
4. execute
5. register write-back

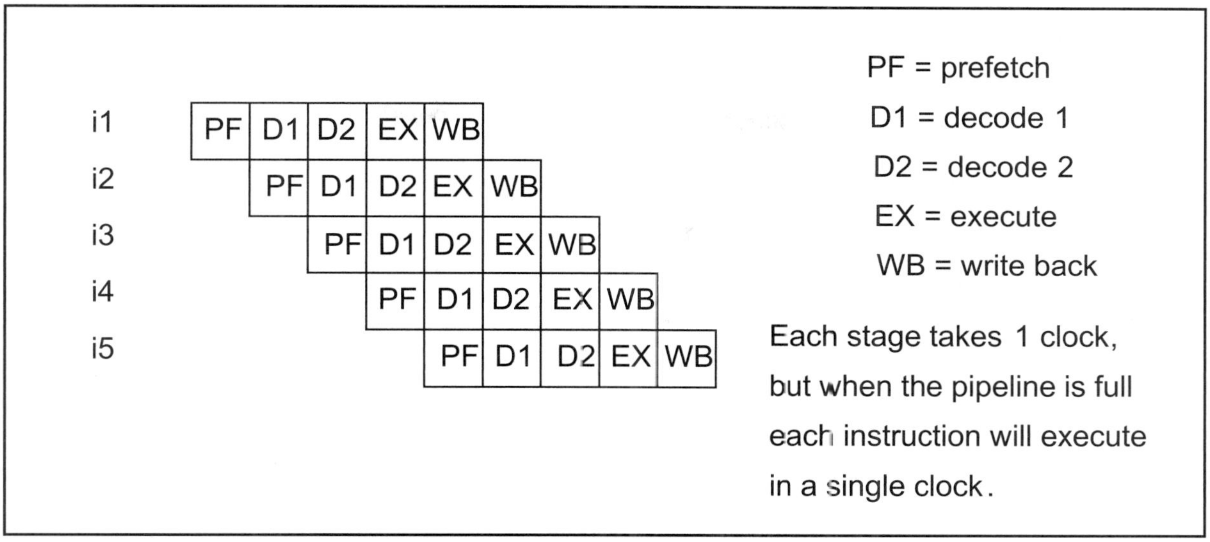

Figure 23-3. 486 Pipeline Stages

Due to such a large number of addressing modes in the x86, a two-stage decoder is used for the calculation and protection check of operand addresses. The register write-back is the stage where the operand is finally delivered to the register. For example, after the instruction "ADD EAX,[EBX+ECX*8+200]" is fetched, the two decoding stages are responsible for calculating the physical address of the source operand, checking for a valid address, and getting it into the CPU. There the operand is added together with EAX during the execution stage, and finally, the addition result is written into EAX, the destination register. Figure 23-3 shows the 486 pipeline.

Review Questions

1. How many pins does the 80486 have, and what kind of packaging is used for it?
2. True or false. The 486 is a 32-bit microprocessor.
3. The 80486 has a(n) _____-bit external and a(n) ________-bit internal data bus.
4. State the difference between the 80486 and the 80486SX.
5. On-chip cache is referred to as _________, while off-chip cache is called _________.
6. State the size of the on-chip cache for the 486 and the cache organization.
7. Calculate the bus bandwidth of a 486 burst read for a 50-MHz system.
8. If the 486 is advertised as 33 MHz, the clock frequency connected to the CLK pin is _________.
9. Pin A20M is an __________ (input, output) signal.
10. A20 (the twentieth address bit) is an _____________ (input, output) signal for the 486.

SECTION 23.2: INTEL'S PENTIUM

Intel put 3.1 million transistors on a single piece of silicon using a 273-pin PGA package to design the next generation of x86. It is called Pentium instead of 80586. The name Pentium was chosen to distinguish it from clones because it is hard to copyright a number such as 80586. There are three ways available to microprocessor designers to increase the processing power of the CPU.

1. Increase the clock frequency of the chip. One drawback of this method is that the higher the frequency, the more the power dissipation and the more difficult and expensive the design of the microprocessor and motherboard.
2. Increase the number of data buses to bring more information (code and data) into the CPU to be processed. While in the case of DIP packaging this option was very expensive and unrealistic, in today's PGA packaging this is no longer a problem.
3. Change the internal architecture of the CPU to overlap the execution of more instructions. This requires a lot of transistors. There are two trends for this option, superpipeline and superscalar. In superpipelining, the process of fetching and executing instructions is split into many small steps and all are done in parallel. In this way the execution of many instructions is overlapped. The number of instructions being processed at a given time depends on the number of pipeline stages, commonly termed the *pipeline depth*. Some designers use as many as 8 stages of pipelining. One limitation of superpipelining is that the speed of the execution is limited to the the slowest stage of the pipeline. Compare this to making pizza. You can split the process of making pizza into many stages, such as flattening the dough, putting on the toppings, and baking, but the process is limited to the slowest stage, baking, no matter how fast the rest of the stages are performed. What happens if we use two or three ovens for baking pizzas to speed up the process? This may work for making pizza but not for executing programs, since in the execution of instructions we must make sure that the sequence of instructions is kept intact and that there is no out-of-step execution. The difficulties associated with a stalled pipeline (a slowdown in one stage of the pipeline, which prevents the remaining stages from advancing) has made CPU designers abandon superpipelining in favor of superscaling. In superscaling, the entire execution unit has been doubled and each unit has 5 pipeline stages. Therefore, in superscalar, there is more than one execution unit and each has many stages, rather than one execution unit with 8 stages as in the case of a superpipelined processor. In some superscalar processors, there are two execution units each with 4 pipeline stages instead of a single execution unit with 8 pipeline stages as superpipelining proponents would have it. In other words, in superscaling we have two (or even three) execution units and as the instructions are fetched they are issued to the various execution units. Using the analogy of pizza, superscalar is like doubling or tripling the entire crew flattening the dough, putting toppings on, and baking. Of course, you will need a lot more people involved in the process and you have to have more ovens, but at the same time you are doubling or tripling the pizza output. In cases of recent microprocessor architecture, a vast majority of designers have chosen superscaling over superpipelining. This requires numerous transistors to duplicate several execution units, just like needing more people in our pizza-making analogy. Fortunately, advances in IC design have allowed designers access to a couple of million transistors to throw around for the implementation of powerful superscaling. There are some problems with superscaling, such as data dependency issues, which can be solved by the compiler, as we will discuss below.

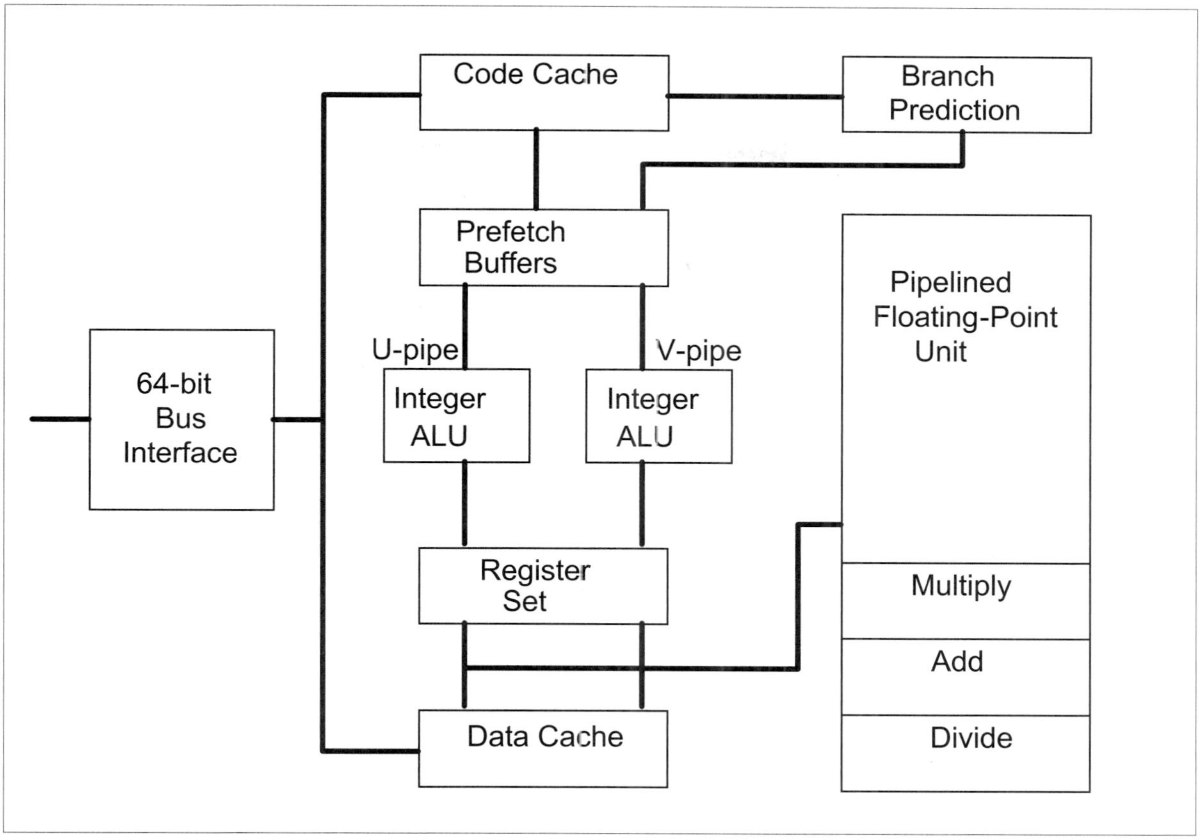

Figure 23-4. Inside the Pentium

Intel used all three methods to increase the processing power of the Pentium I. Intel shipped the 60- and 66-MHz Pentium and eventually reached 3 GHz for Pentium IV, as well. The Pentium has a 64-bit external data bus and is a superscalar processor with two execution units to process integer data. This is in addition to a separate execution unit for floating-point data.

Features of the Pentium

The following are some of the major features of the Pentium processor.

Feature 1

In the Pentium, the external data buses are 64-bit, which will bring twice as much code and data into the CPU as the 486. However, just like the 386 and 486, Pentium registers are 32-bit. Bringing in twice as much as information can work only if there are two execution units inside the processor, and this is exactly what Intel has done. The Pentium uses 64 pins, D0–D63, to access external memory banks, which are 64 bits wide. D0–D7 is the least significant byte, and D56–D63 is the most significant byte. Accessing 8 bytes of external data bus requires 8 BE (byte enable) pins, BE0–BE7, where BE0 is for D0–D7, BE1 for D8–D15, and so on. This is shown in Figure 23-5 and Table 23-2.

While in the 486 there were four DP (data parity) pins, one for each of the 4 bytes of the data bus, in the Pentium there are 8 DP pins to handle the 8 bytes of data pins D0–D63. The Pentium has A31 to A3 for the address buses. This is shown in Figure 23-6. Just like the 486, the Pentium also has the A20M (A20 Mask) input pin for the implementation of HMA (high memory area).

Feature 2

The Pentium has a total of 16K bytes of on-chip cache: 8K is for code and the other 8K is for data. In the 486 there is only 8K of on-chip cache for both code and data. The data cache can be configured as write-back or write-through, but to prevent any acci-

Table 23-2: Pentium Byte Enable Signals

Byte Enable Signal	Associated Data Bus Signals
BE0#	D0–D7 (byte 0, the least significant)
BE1#	D8–D15 (byte 1)
BE2#	D16–D23 (byte 2)
BE3#	D24–D31 (byte 3)
BE4#	D32–D39 (byte 4)
BE5#	D40–D47 (byte 5)
BE6#	D48–D55 (byte 6)
BE7#	D56–D63 (byte 7, the most significant)

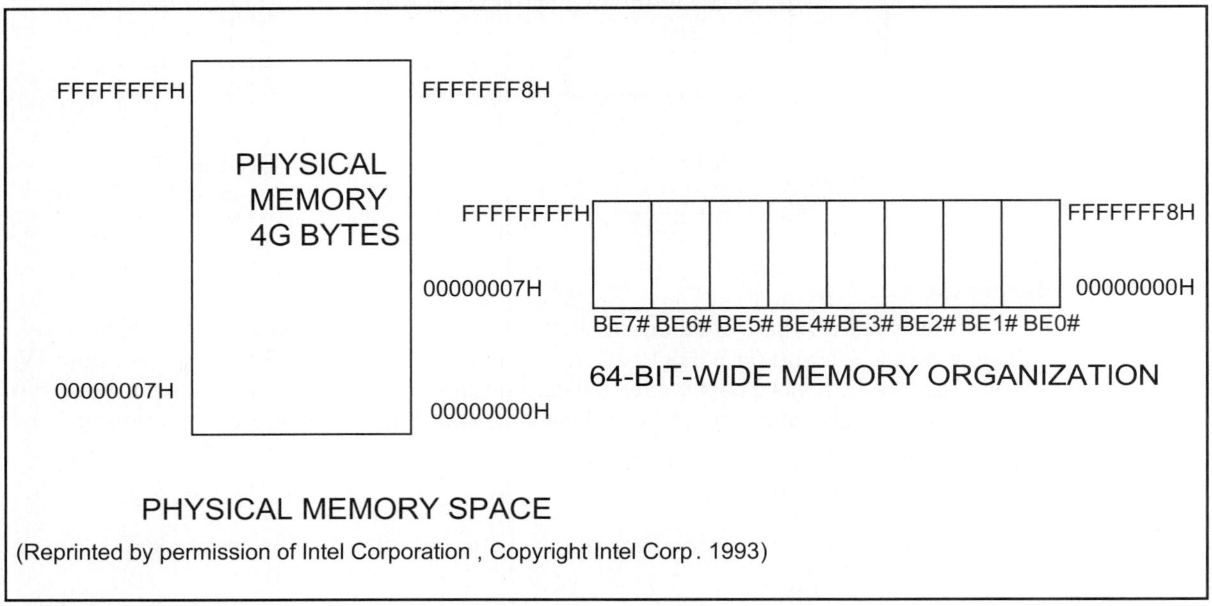

Figure 23-5. Pentium Memory Organization

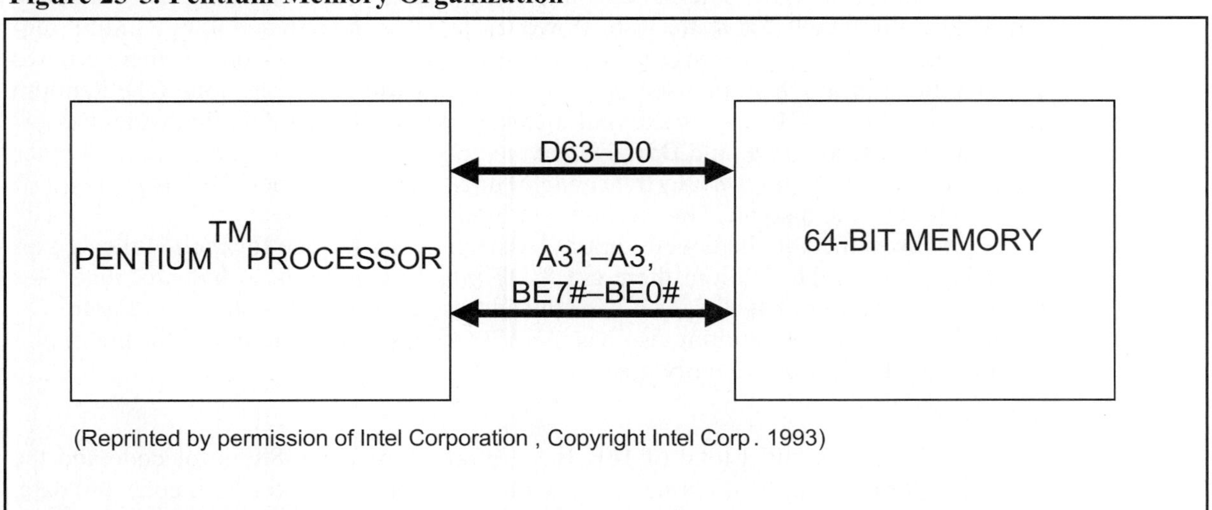

Figure 23-6. Pentium Address Buses

dental writing into code cache, the 8K of code cache is write protected. In other words, while the CPU can read or write into the data cache, the code cache is write protected to prevent any inadvertent corruption. Of course, when there is a cache miss for code cache, the CPU brings code from external memory and stores (writes) it in the cache code, but no instruction executing in the CPU can write anything into the code cache. The replacement policy for both data and code caches is LRU (least recently used).

Both the on-chip data and code caches are accessed internally by the CPU core simultaneously. However, since there is only one set of address buses, the external cache containing both data and code must be accessed one at a time and not simultaneously. Some CPUs, notably RISC processors, use a separate set of address and data pins (buses) for the data and another set of address and data buses for the code section of the program. This is called *Harvard architecture* and will be discussed in the next section. The Pentium accesses the on-chip code and data caches simultaneously using Harvard architecture, but not the secondary (external) off-chip cache and data. The Pentium's cache organization for both the data and code caches is 2-way set associative. Each 8K is organized into 128 sets of 64 bytes, which means $2^7 \times 2^6 = 2^{13} = 8192 = 8K$ bytes. Each set consists of 2 lines of cache, and each line is 32 bytes wide.

Feature 3

The on-chip math coprocessor of the Pentium is many times faster than the one on the 486. It has been redesigned to perform many of the instructions, such as add and multiply, ten times faster than the 486 math coprocessor. In microprocessor terminology, the on-chip math coprocessor is commonly referred to as a *floating point unit* (*FPU*) while the section responsible for the execution of integer-type data is called the *integer unit* (*IU*). The FPU section of the Pentium uses an 8-stage pipeline to process instructions, in contrast to the 5-stage pipeline in the integer unit. See Figure 23-7.

Feature 4

Another unique feature of the Pentium is its superscalar architecture. A large number of transistors were used to put two execution units inside the Pentium. As the instructions are fetched, they are issued to these two execution units. However, issuing two instructions at the same time to different execution units can work only if the execution of one does not depend on the other one, in other words, if there is no data dependency. As an example, look at the following instructions.

```
ADD    EAX,EBX              ;add EBX to EAX
NOT    EAX                  ;take 1's complement EAX
INC    DI                   ;increment the pointer
MOV    [DI],EBX             ;move out EBX
```

In the above code, the ADD and NOT instructions cannot be issued to two execution units since EAX, the destination of the first instruction, is used immediately by the second instruction. This is called *read-after-write dependency* since the NOT instruction wants to read the EAX contents, but it must wait until after the ADD is finished writing it into EAX. The problem is that ADD will not write into EAX until the last stage of the pipeline, and by then it is too late for the pipeline of the NOT instruction. This prevents the NOT instruction from advancing in the pipeline, therefore causing the pipeline to be stalled until the ADD finishes writing and then the NOT instruction can advance through the pipeline. This kind of register dependency raises the clock count from 1 to 2 for the NOT instruction. What if the instructions are rescheduled, as follows?

```
ADD    EAX,EBX              ;add EBX to EAX
INC    DI                   ;increment the pointer
NOT    EAX                  ;take 1's complement of EAX
MOV    [DI],EBX             ;move out EBX
```

If they are rescheduled as shown above, each can be issued to separate execution units, allowing parallel execution of both instructions by two different units of the CPU.

Since the clock count for each instruction is one, just like the 486, having two execution units leads to executing two instructions by pairing them together, thereby using only one clock count for two instructions. In the case of the above program, if it is run on the Pentium it will take only 2 clocks instead of 4 as in the case of the 486 microprocessor, assuming that two instructions are paired together. This reordering of instructions to take advantage of the two internal execution units of the Pentium is the job of the compiler and is called instruction scheduling. Currently, compilers are being equipped to do instruction scheduling to remove dependencies. The role of the compiler to reschedule instructions in order to take advantage of the superscalar capability of the Pentium must be emphasized. The process of issuing two instructions to the two execution units is commonly referred to as *instruction pairing*. The two integer execution units of the Pentium are called "U" and "V" pipes. Each has 5 pipeline stages. While the U pipe can execute any of the instructions in the x86 family, the V pipe executes only simple instructions such as INC, DEC, ADD, SUB, MUL, DIV, NOT, AND, OR, EXOR, and NEG. These simple instructions are executed in one clock as long as the operands are "REG,REG" or "REG,IMM" and have no register dependency. For example, instructions such as "ADD EAX,EBX", "SUB ECX,2000", and "MOV EDX,1500" are simple instructions requiring 1 clock, but not "ADD DWORD PTR [EBX+EDI+500],EAX", which needs 3 clocks.

Feature 5

Branch prediction is another new feature of the Pentium. The penalty for jumping is very high for a high-performance pipelined microprocessor such as the Pentium. For example, in the case of the JNZ instruction, if it jumps, the pipeline must be flushed and refilled with instructions from the target location. This takes time. In contrast, the instruction immediately below the JNZ is already in the pipeline and is advancing without delay. The Pentium processor has the capability to predict and prefetch code from both possible locations and have them advanced through the pipeline without waiting (stalling) for the outcome of the zero flag. The ability to predict branches and avoid the branch penalty combined with the instruction pairing can result in a substantial reduction in the clock count for a given program. See Example 23-4.

Feature 6

As discussed in Chapter 21, the 386/486 has a page size of 4K for page virtual memory. The Pentium provides the option of 4K or 4M for the page size. The 4K page option makes it 386 and 486 compatible, while the 4M page size option allows mapping of a large program without any fragmentation. The 4M page size in the Pentium reduces the frequency of a page miss in virtual memory.

Feature 7

As discussed in Chapter 21, the 386 (and 486) has only 32 entries for the TLB (translation lookaside buffer), which means that the CPU has instant knowledge of the whereabouts of only 128K of code and data. If the desired code or data is not referenced in the TLB, the CPU must go through the long process of converting the linear address to a physical address. The Pentium has two sets of TLB, one for code and one for data. For data, the TLB has 64 entries for 4K pages. This means that the CPU has quick access to 256K ($64 \times 4K = 256K$) of data. The TLB for the code is 32 entries of 4K page size. Therefore, the CPU has quick access to 128K of code at any given time. Combining the TLBs for the code and data, the Pentium has quick access to 384K ($128 + 256$) of code and data before it resorts to updating the TLB for the page miss. Contrast this to 128K for the 486. If the page size of 4M is chosen, the TLB for the data has 8 entries while the TLB for the code has 32 entries.

Feature 8

The Pentium has both burst read and burst write cycles. This is in contrast to the 486, which has only the burst read. This means that in the 486 any write to consecutive doubleword locations must be performed with the normal 2 clock cycles. This is not the case in the Pentium.

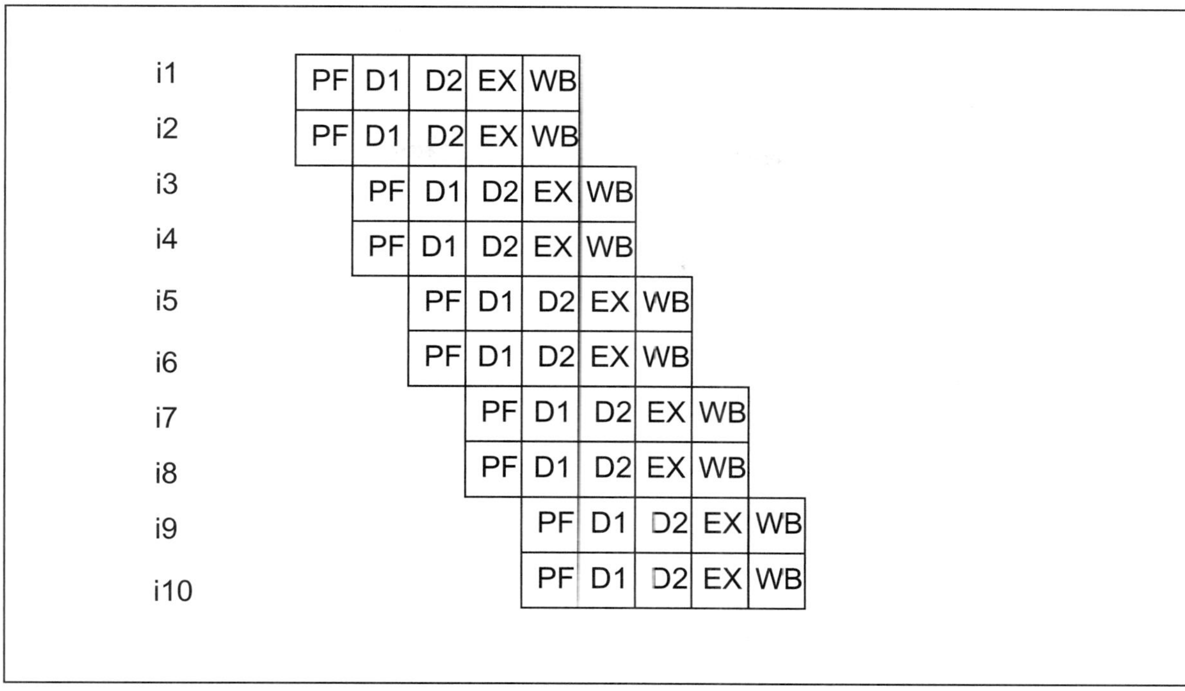

Figure 23-7. Pentium Pipeline

Example 23-4

Compare the clock count for the program in Example 23-3, run on a 486 and a Pentium. Assume that the compiler has done the code scheduling to allow instruction pairing for the Pentium.

Solution:

First compare the following rearranged code with the code in Example 23-3. Here we have rescheduled instructions "ADD SI,4" and "ADD BX,4" to avoid register dependency. This allows pairing of two instructions and issuing them to execution units of the Pentium. For example, the instructions "MOV EAX,DWORD PTR [SI]" and "ADD SI,4" are issued simultaneously, one to each execution unit. This results in the execution of both instructions in only one clock.

```
AGAIN:      MOV    EAX,DWORD PTR [SI]
            ADD    SI,4
            MOV    [DI],EAX
            ADD    DI,4
            DEC    CX
            JNZ    AGAIN
```

Total clock count for one iteration: 3

In the 486, the execution of "JNZ AGAIN" takes 3 clocks every time it jumps to AGAIN, but for the Pentium it takes only 1 clock since the CPU has predicted the branch, fetched it, and the instructions at label AGAIN are in the pipeline advancing. This way, regardless of the outcome of the JNZ, both the instruction below JNZ and the first instruction at label AGAIN are in two separate pipelines, advancing. If ZF = 0, the other pipeline is trashed, and if ZF = 1 (the end of loop), the instruction below the JNZ is executed and the branch prediction pipeline is abandoned. In the above program each iteration takes only 3 clocks on the Pentium compared with 9 clocks in the 486. While branch prediction is performed by the internal hardware of the Pentium, instruction scheduling must be done by the compiler.

The Pentium has features that lend themselves to implementation of multiple microprocessors (multiprocessors) working together. It also has features called *error detection* and *functional redundancy* to preserve and ensure data and code integrity.

Intel's overdrive technology

To increase both the internal and external clock frequency of the CPU requires faster DRAM, high-speed motherboard design, high-speed peripherals, and efficient power management due to a high level of power dissipation. As a result, the system is much more expensive. To solve this problem, Intel came up with what is called *overdrive technology*, also referred to as *clock doubler and tripler*. The idea of a clock doubler or tripler is to increase the internal frequency of the CPU while the external frequency remains the same. In this way, the CPU processes code and data internally faster while the motherboard costs remain the same. For example, the 486DX2-50 uses the internal frequency of 50 MHz but the external frequency by which the CPU communicates with memory and peripherals is only 25 MHz. This allows the instructions stored in the queue of 486 to be executed at twice the speed of fetching them from the system buses. With the advent of 32- and 64-bit external buses, on-chip cache, and burst cycle reading (reading 16 bytes in only 5 clocks), the amount of code and data fetched into the queue of the CPU is sufficient to keep the execution unit of CPU busy even if it is working with twice or three times the speed of external buses. This is the reason that Intel is designing processors with clock triplers. In that case, if the CPU's external buses are working at the speed of 33 MHz, the CPU works at 99 MHz speed. The design of a system board of 33 MHz costs much less than that of a 100-MHz system board. With slower memory and peripherals one can get instruction throughput of three times the bus throughput. As designers move to wider data buses, such as 128-bit-wide buses, the use of clock doublers and triplers is one way of keeping the system board cost down without sacrificing system throughput. The Intel 486DX4 is an example of a clock-tripler CPU. Note that "X4" does not mean that the external frequency is 4 times the internal frequency.

Review Questions

1. The Pentium chip has _______ pins.
2. The Pentium has ______ data pins.
3. True or false. The Pentium is a 32-bit processor.
4. What is the total cache on the Pentium? How much is for data, and how much is for code?
5. Which is write protected, data or code cache?
6. True or false. The on-chip data and code cache are accessed simultaneously.
7. True or false. The branch prediction task is performed by circuitry inside the Pentium.
8. Why is the Pentium called a *superscalar processor*?
9. True or false. Instruction scheduling is done by circuitry inside the Pentium.
10. True or false. The general-purpose registers of the Pentium are the same as those in the 386 and 486.

SECTION 23.3: RISC ARCHITECTURE

In the early 1980s a controversy broke out in the computer design community, but unlike most controversies, it did not go away. Since the 1960s, in all mainframe and minicomputers, designers put as many instructions as they could think of into the microinstructions of the CPU. Some of these instructions performed complex tasks. An example is adjusting the result of decimal addition to get BCD nibble-type data. Naturally, microprocessor designers followed the lead of minicomputer and mainframe designers. Since these microprocessors used such a large number of instructions and many of them performed highly complex activities, they came to be known as CISC (complex instruction set computer). According to several studies in the 1970s, many of these complex instructions etched into the brain of the CPU were never used by programmers and compilers. The huge cost of implementing a large number of instructions (some of them complex) into the microprocessor, plus the fact that more than 60% of the transistors on the chip are used by the instruction decoder, made some designers think of simplifying and reducing the number of instructions. As this concept was developed, it came to be known as RISC (reduced instruction set computer).

Features of RISC

The following are some of the features of RISC. It must be noted that recent CISC processors such as the Pentium have used some of the following features in their design.

Feature 1

RISC processors have a fixed instruction size. In a CISC microprocessor such as the x86, instructions can be 1, 2, or even 6 bytes. For example, look at the following instructions.

```
CLC                     ;a 1-byte instruction
SUB     DX,DX           ;a 2-byte instruction
ADD     EAX,[ SI+8]     ;a 5-byte instruction
JMP     FAR             ;a 5-byte instruction
```

This variable instruction size makes the task of the instruction decoder very difficult since the size of the incoming instruction is never known. In a RISC microprocessor, the size of all instructions is fixed at 4 bytes (32 bits). In cases where instructions do not require all 32 bits, they are filled with zeros. Therefore, the CPU can decode the instructions quickly. This is like a bricklayer working with bricks of the same size as opposed to using bricks of variable sizes. Of course, it is much more efficient to use the bricks of the same size.

Feature 2

RISC uses load/store architecture. In CISC microprocessors, data can be manipulated while it is still in memory. For example, in x86 instructions such as "ADD [BX],AL", the microprocessor must bring the contents of the memory location pointed at by BX into the CPU, add it to AL, then move the result back to the memory location pointed at by BX. In RISC, designers did away with this kind of instruction. In RISC, instructions can only load from memory into registers or store registers into memory locations.

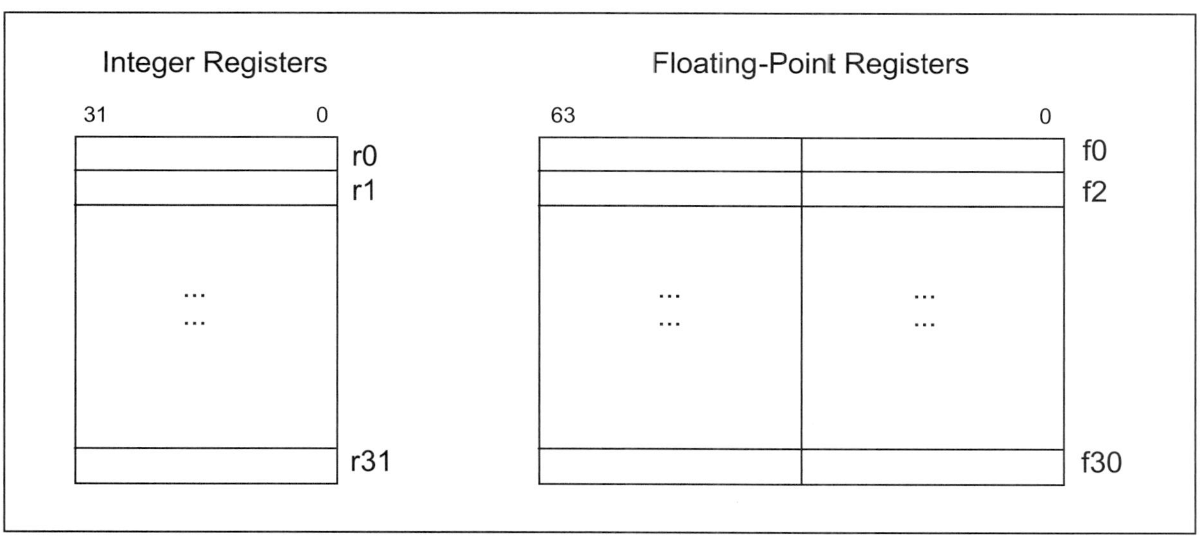

Figure 23-8. RISC Integer and Floating-Point Registers

There is no direct way of doing arithmetic and logic instructions between registers and contents of memory locations. All these instructions must be performed by first bringing both operands into the registers inside the CPU, then performing the arithmetic or logic operation, and then sending the result back to memory. This idea was first implemented by the CRAY 1 supercomputer in 1976 and is commonly referred to as *load/store architecture*.

Feature 3

One of the major characteristics of RISC architecture is a large number of registers. All RISC microprocessors have 32 registers, r0–r31, each 32 bits wide. See Figure 23-8. Of these 32 registers, only a few of them are assigned to a dedicated function. For example, r0 is automatically assigned the value zero and no other value can be written to it. One advantage of a large number of registers is that it avoids the use of the stack to store parameters. Although a stack can be implemented on a RISC processor, it is not as essential as in CISC since there are so many registers available. It must be noted that RISC processors, in addition to 32 general-purpose registers, also have another 32 registers for floating-point operations. The floating-point register can be configured as 64-bit in order to handle double-precision operands.

Feature 4

RISC processors have a small instruction set. RISC processors have only the basic instructions such as ADD, SUB, MUL, DIV, LOAD, STORE, AND, OR, EXOR, SHR, SHL, CALL, and JMP. For example, there are no such instructions as INC, DEC, NOT, NEG, DAA, DAS, and so on. Since RISC has very few instructions, it is the job of the programmer (compiler) to implement those instructions by using available RISC instructions. One example is an immediate load instruction such as "MOV AX,25 ", which does not exist in some RISC processors. Instead, some other instructions, such as the OR instruction, can be used to implement an immediate move as shown in the following example for the given RISC processor.

```
or     25,r0,r8     ;OR 25 with r0 and put result in r8
```

In RISC syntax,the destination register is the last register, r8 in the above example. Since r0 is always zero, ORing any number with it will result in that number. The above example will place 25 in r8. Another example is that there is no INC (increment) command. The ADD instruction is used instead, as in the following example:

```
add    1,r15,r15    ;add 1 to r15 and place result in r15
```

The limited number of instructions is one of the criticisms leveled at the RISC processor since it makes the job of Assembly language programmers much more tedious and difficult compared to CISC Assembly language programming. This is one reason that RISC is used more commonly in high-level language environments such as C rather than Assembly language environments. It is interesting to note that some defenders of CISC have called it "complete instruction set computer" instead of "complex instruction set computer" since it has a complete set of every kind of instruction. How many of them are used is another matter. The limited number of instructions in RISC leads to programs that are large, as Example 23-5 shows. Although this can lead to using more memory, this is not a problem since DRAM memory is so cheap. However, before the advent of semiconductor memory in the 1960s, CISC designers had to pack as much action as possible into a single instruction.

Feature 5

At this point, one might ask, with all the difficulties associated with RISC programming, what is the gain? The most important characteristic of the RISC processor is the fact that more than 99% of its instructions are executed with only 1 clock, in contrast to CISC instructions. Although some instructions in the 80486 microprocessor are executed with 1 clock, with the the use of RISC concepts in designing it, it is still a CISC proces-

Example 23-5

In the x86, the NOT instruction performs the 1's complement operation, but RISC does not have such an instruction. How is the 1's complement operation performed in RISC? Show code to take the 1's complement of 25H on both RISC and x86.

Solution:

RISC has an EXOR instruction. If we EXOR the operand with all 1s, the operand is inverted. In RISC we have

```
or      25H,r0,r8     ;OR 25H with r0 and put result in r8 (r8 = 25H)
or      FFH,r0,r5     ;OR FFH with r0 and put result in r5 (r5 = FFH)
xor     r8,r5,r9      ;XOR r8 with r5 and put result in r9 (r9 = DAH)
```

Since each instruction is 4 bytes (32-bit), the three instructions take a total of 12 bytes of memory. In the x86, a CISC-type processor, we have the following, which takes only 4 bytes of memory (to see, use DEBUG to assemble):

```
MOV    AL,25H
NOT    AL
```

sor for the reasons discussed above. Even the other 1% of the RISC instructions that are executed with 2 clocks can be executed with one clock cycle by juggling instructions around (code scheduling). Code scheduling is the job of the compiler. What did designers do with all those transistors saved using the RISC implementation? In the case of RISC processors, these extra transistors are used to implement the math coprocessor, powerful cache and cache controller, and a very powerful graphics processor all on a single chip. In many computers, such as 386-based systems, all these functions are performed by separate chips.

Feature 6

Since CISC has such a large number of instructions, each with so many different addressing modes, microinstructions (microcode) are used to implement them. The implementation of microinstructions inside the CPU takes more than 60% of transistors in many CISC processors. However, in case of RISC, due to their small set of instructions, they are implemented using the hardwire method. Hardwiring of RISC instructions takes no more than 10% of the transistors. It is interesting to note that in the Pentium, a CISC processor, the V-pipe executes only simple instructions and is hardwired while the U-pipe executes any of the x86 instructions and uses microinstructions.

Harvard and von Neumann architectures

Every microprocessor must have memory space to store program (code) and data. While code provides instructions to the CPU, the data provides the information to be processed. The CPU uses buses (wire traces) to access the code ROM and data RAM memory spaces. The early computers used the same bus for accessing both the code and data. Such an architecture is commonly referred to as *von Neumann* (Princeton) architecture. That means for von Neumann computers, the process of accessing code or data could cause each to get in the other's way and slow down the processing speed of the CPU, because each had to wait for the other to finish fetching. To speed up the process of program execution, some CPUs use what is called *Harvard architecture*. The Harvard architecture has separate buses for the code and data memory. See Figure 23-9. That means that we need four sets of buses: (1) a set of data buses for carrying data into and out of the CPU, (2) a set of address buses for accessing the data, (3) a set of data buses for carrying code into the CPU, and (4) an address bus for accessing the code. See Figure 23-9. This

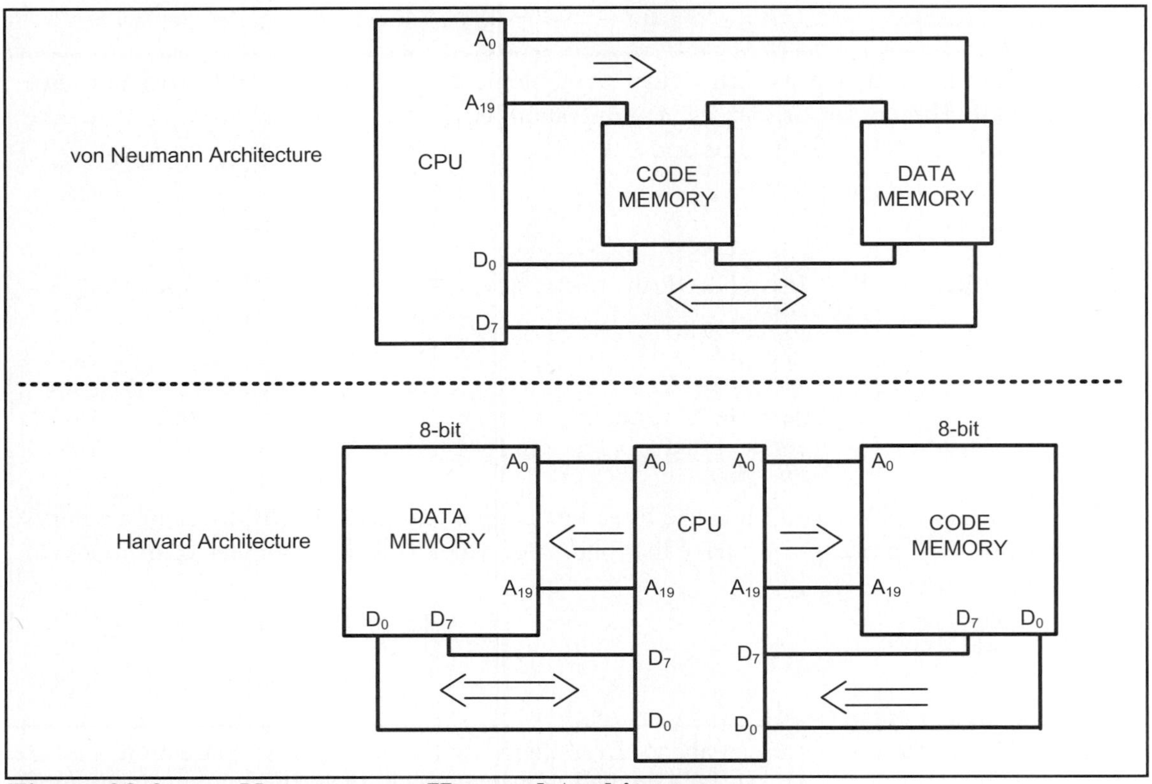

Figure 23-9. von Neumann vs. Harvard Architecture

is easy to implement inside an IC chip such as a Pentium where both code and data are internal (on-chip) and distances are on the micron and millimeter scale. But implementing Harvard architecture for system boards such as x86 PC-type computers is very expensive because the RAM memories that hold code and data are external to the CPU. Separate wire traces for data and code on the motherboard will make the board large and expensive. For example, a Pentium microprocessor with a 64-bit data bus and a 32-bit address bus will need about 100 wire traces on the motherboard if it is von Neumann architecture (96 for address and data, plus a few others for control signals of read and write and so on). But the number of wire traces will double to 200 if we use Harvard architecture. Harvard architecture will also necessitate a large number of pins coming out of the microprocessor itself. For this reason Harvard architecture is not implemented in the world of PC and workstation motherboards. This is also the reason that microprocessors such as Pentium use Harvard architecture internally, but they still use von Neumann architecture when they access external memory. See Chapter 24. The von Neumann architecture was developed at Princeton University, while the Harvard architecture was the work of Harvard University.

Comparison of sample program for RISC and CISC

Since RISC has established itself as a dominant architecture, an example will be given of a program written for Intel's x86 CISC and a RISC processor. Then they will be compared. The next program example will compare total clocks for a program that transfers a block of 32-bit words from some memory location to another memory location. First, several points about the RISC processor must be discussed.

1. In instructions such as "add r3,r5,r2", r3 is added to r5 and placed in r2. This is in contrast to the x86, in which the destination register is the first register.
2. r0 is always equal to zero, regardless of the operation performed on it.
3. The load instruction cannot be followed by the store instruction, which tries to use the value that is just being loaded: in other words, no read after write (RAW).
4. Some instructions, such as branch instructions, are delayed, which means that the

next instruction after the branch will be executed since the pipeline already has fetched it before the branch is taken; therefore, if we cannot put a useful instruction after the branch, a NOP should be used.

5. There are several ways of encoding the NOP instruction. One is "add r0,r0,r0", which adds r0 to r0 and places the result in r0; since r0 is always zero, the instruction does nothing but waste time. Another would be to use shift left, such as "shl r0,r0,r0".
6. Some other instruction must be used to accomplish a MOV.
7. The logic instruction is used to perform the compare job.

Now look at the following RISC program, first written with total disregard of code scheduling and then written with code scheduling. The number of clocks for one round of loop is calculated in each case.

```
;this is a program for a RISC processor to transfer a
;block of 20 dwords (each 32-bit) from memory locations starting at
;the address pointed at by r3 to memory locations pointed at by r4.
;r2 is the counter
;There is no code scheduling in the following example.
                                                             clocks

     or 20,r0,r2       ;load the r2 with 20(count=20)
bak: ld.l 0(r3),r5     ;load r5 from content of mem loc 0+r3       1
     add r0,r0,r0      ;NOP since r5 cannot be used by store       1
     st.l r5,0(r4)     ;store r5 into mem loc of 0+r4              1
     add 4,r3,r3       ;point at the next dword source data        1
     add 4,r4,r4       ;point at the next dword destin data        1
     add -1.r2,r2      ;decrement counter: r2=r2-1                 1
     or r0,r2,r2       ;set condition code to high if r2=0         1
     bnc.t bak         ;go to bak if CC=0. execute next instruction 1
     add r0,r0,r0      ;NOP for delayed branch                     1

                       total clocks for one loop iteration         9
```

In the above program, ld.l and st.l are for 32-bit operands. For byte and 16-bit bit operands, they would be ld.b and st.b, ld.s and st.s, respectively. Next the same program is juggled around and the NOPs are removed for better performance. As mentioned in the last section, this juggling is called *code scheduling*.

```
                                                   number of clocks
     or 20,r0,r2       ;loop starts next
bak: ld.l 0(r3),r5     ;                                           1
     add 4,r3,r3       ;                                           1
     st.l r5,0(r4)     ;                                           1
     add -1,r2,r2      ;                                           1
     or r0,r2,r2       ;                                           1
     bnc.t bak         ;                                           1
     add 4,r4,r4       ;                                           1

                       total clock count for one iteration         7
```

It is assumed that the above program is run on a RISC processor with no superscalar capability. Using a superscalar RISC will cut the clock count to half, or 4 clocks for the same program. This is much better than the 386 and 486 microprocessors shown in Example 23-3. It is comparable with the Pentium as shown in Example 23-4. While the

Pentium uses 3.1 million transistors to achieve such an impressive performance, a RISC processor with the same performance level can be designed using less than 1 million transistors. There is only one problem. It will not run the massive number software packages written for the x86 Windows PC.

IBM/Freescale RISC

IBM and Freescale (formerly Motorola) together have a RISC processor called the Power PC 60x. It uses only 2.8 million transistors, with a power consumption of 8.5 watts versus 16 watts in the Pentium. Apple used the Power PC RISC for the last generations of Macintosh computers. However, recently they switched to x86 processors. No PC maker has used the RISC processors, since RISC computers must emulate to run Windows based software rather than running native. In other words, while MS Windows runs native on the x86, the RISC processor or any non-x86 has no choice but to emulate. In addition to the Power PC 60x RISC processor, there are some other notable RISC processors vying for a share of the embedded market. Among them are ARM and PIC32.

A comparison of the major characteristics of the Pentium and the Power PC 601 is provided in Table 23-3. To take full advantage of the power of RISC, software developers must write applications specifically for RISC, rather than using emulation.

Before concluding this discussion of RISC processors, it is interesting to note that

Table 23-3: Pentium I vs. Power PC 601

Feature	Pentium I	Power PC 601
Number of transistors (million)	3.1	2.8
Power dissipation at 66 MHz (watts)	16	8.5
Die size (mm^2)	262	120
Technology	BICMOS 0.8 µm	CMOS 0.6 µm
Number of general-purpose registers	8	32
Number of floating-point registers	8 (stack base)	32
Cache (K bytes)	16 (code 8K, data 8K)	32K (unified)
Register size	32-bit	32-bit
Running Windows software	Native	Emulation
Number of instructions issued per clock cycle	2	3 (one is FP)
Architecture	Superscalar	Superscalar

RISC technology was explored by the scientists in IBM in the mid-1970s, but it was David Patterson of the University of California at Berkeley who in 1980 brought the merits of RISC concepts to the attention of computer scientists.

Review Questions

1. What do RISC and CISC stand for?
2. True or false. The 386 executes the vast majority of its instructions in 3 clock cycles, while RISC executes them in 1 clock.
3. RISC processors normally have ___ general-purpose registers, each ___ bits.
4. True or false. Instructions such as "ADD AX,[DI]" do not exist in RISC.
5. What is the size of instructions in RISC?
6. True or false. While CISC instructions are variable sizes, RISC instructions are all the same size.
7. Which of the following operations do not exist for the ADD instruction in RISC?
   (a) register to register (b) immediate to register
   (c) memory to register

8. How many floating-point registers do we have in RISC and the 80x87?
9. Why can floating-point registers in RISC be configured as 64-bit?
10. True or false. Harvard architecture uses the same address and data buses to fetch both opcode and data.

SECTION 23.4: PENTIUM PRO PROCESSOR

In this section we discuss the main features of Intel's Pentium Pro processor. Intel's Pentium Pro is the sixth generation of the x86 family of microprocessors. For this reason, early literature about this chip referred to it as P6. Intel officially calls this chip Pentium Pro to emphasize its superiority over the previous Pentium generation. Intel used 5.5 million transistors to make the Pentium Pro. The first Pentium Pro introduced in 1995 had a speed of 150 Mhz and consumed 23 watts of power at that speed. Since then, Intel has introduced Pentium Pro chips with higher speeds and various power consumption ratings.

There are no major surprises in the Pentium Pro in the sense that it runs all the software written for the 8088/86, 286, 386, 486, and Pentium microprocessors and its 32-bit registers are exactly the same as those of the 386. In other words, the register size was not increased to 64 bits as has been done by some RISC processors such as Digital Equipment's Alpha chip.

For the first time, Intel also attached level 2 (L2) cache to the Pentium Pro all on a single package but with two separate dies. This packaging is called *dual cavity* by Intel. The integration of a 256K-byte L2 cache with the processor into a single package reduces interchip delay between the L2 cache and the CPU. While such an integration cut memory access delay, it also made many SRAM makers mad since they lost another chunk of PC business to Intel. Notice that the Pentium Pro CPU has only 16K bytes of L1 cache on the same die, just like the Pentium processor, while 256 KB (or 512 KB) L2 cache is on the separate die. In addition to the 5.5 million transistors used for the Pentium Pro CPU and its 16 KB L1 cache, the L2 cache uses over 10 million transistors depending on the size of L2 cache. See Table 23-4 for further comparison of Pentium and Pentium Pro processors.

Table 23-4: Comparison of Pentium and Pentium Pro

Feature	Pentium	Pentium Pro
Year introduced	1993	1995
Number of transistors	3.3 million	5.5 million
Number of pins	273	387
External data bus	64 bits	64 bits
Address bus	32 bits	36 bits
Physical memory (maximum)	4 GB	64 GB
Virtual memory	64 TB	64 TB
Data types (register sizes)	8, 16, 32 bits	8, 16, 32 bits
Cache (L1)	16K bytes (data 8K, code 8K)	16K bytes (data 8K, code 8K)
Cache (L2)	External	256KB/512KB
Superscalar	2-Way	3-Way
Number of execution units	3	5
Branch prediction	yes	yes
Out-of-order execution	no	yes

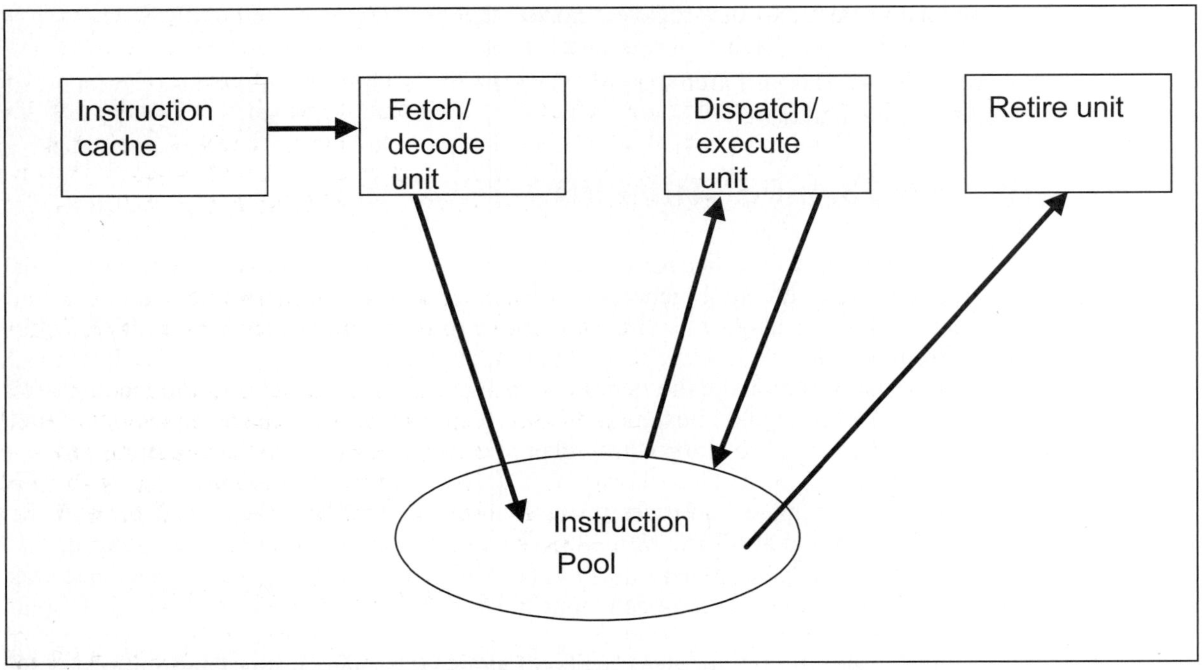

Figure 23-10. Pentium Pro Instruction Execution

Pentium Pro: internal architecture

Intel finally yielded to the rise of RISC concepts in the design of the Pentium Pro. In the Pentium Pro, all x86 instructions brought into the CPU are broken down into one or more small and easy-to-execute instructions. These easily executable instructions are called *micro-operations* (*μops*) by Intel. This is similar to the concept in RISC except that in RISC architecture the instruction set is very simple and easy to execute, and the instructions stored in memory are exactly the same as the ones inside the CPU. In contrast, Intel had to maintain code compatibility for the Pentium Pro with all previous x86 processors, all the way back to 8086. Therefore, Intel had no choice but to convert the x86 instructions produced by the compiler/assembler into micro-operations internally inside the CPU. An interesting aspect of converting x86 instructions into micro-ops internally is that it uses what is called *triadic instruction format*. In triadic instruction format, there are two source registers and one destination register. An example of triadic format is "ADD R1,R5,R8" in which registers R1 and R5 are added together and placed in R8. The contents of source registers R1 and R5 are not altered. Contrast this with "ADD AX,BX" in which there is only one source register (BX). For more examples of the triadic instruction format, see Section 23.3 on the RISC processors.

The use of a triadic instruction set in the Pentium Pro architecture means that a large number of registers inside the Pentium Pro are not accessible or visible to the programmer. In other words, as far as the programmer of the Pentium Pro (or compiler) is concerned, only the traditional register set EAX, EBX, ECX, and so on is available and visible. This ensures that compatibility with previous generations of the x86 is maintained.

Pentium Pro is both superpipelined and superscalar

As mentioned above, in the Pentium Pro all x86 instructions are converted into micro-ops with triadic formats before they are processed. This conversion allows an increase in the pipeline stages with little difficulty. Intel uses a 12-stage pipeline for the Pentium Pro. In contrast to the 5-pipestage Pentium, although each pipestage of the 12-pipestage Pentium Pro performs less work, there are more stages. This means that in the Pentium Pro, more instructions can be worked on and finished at a time. The Pentium Pro with its 12-stage pipeline is referred to as *superpipelined*. Since it also has multiple execution units capable of working in parallel, it is also superscalar. Another advantage of the 12-pipestage concept is that it can achieve a higher clock rate (frequency) with the given transistor technology. This is one reason that the earliest Pentium chips had a frequency of only 60 MHz while the earliest Pentium Pro has a frequency of 150 MHz. Intel also used what is called *out-of-order execution* to increase the performance of the Pentium Pro. This is explained next.

What is out-of-order execution?

In Pentium architecture, when one of the pipeline stages is stalled, the prior stages of fetch and decode are also stalled. In other words, the fetch stage stops fetching instructions if the execution stage is stalled, due, for example, to a delay in memory access. This dependency of fetch and execution has to be resolved in order to increase CPU performance. That is exactly what Intel has done with the Pentium Pro and is called *decoupling* the fetch and execution phases of the instructions. In the Pentium Pro, as x86 instructions are fetched from memory they are decoded (converted) into a series of micro-ops, or RISC-type instructions, and placed into a pool called the *instruction pool.* See Figure 23-9. This fetch/decode of the instructions is done in the same order as the program was coded by the programmer (or compiler). However, when the micro-ops are placed in the instruction pool they can be executed in any order as long as the data needed is available. In other words, if there is no dependency, the instructions are executed out of order, not in the same order as the programmer coded them. In the case of the Pentium Pro, the dispatch/execute unit schedules the execution of micro-ops from the instruction pool subject to the availability of needed resources and stores the results temporarily. Such a speculative execution can go 20–30 instructions deep into the program. It is the job of the retire unit to provide the results to the programmer's (visible) registers (e.g., EAX, EBX) according to the order in which the instructions were coded. Again, it is important to note that the instructions are fetched in the same order that they were coded, but executed out of order if there is no dependency, and ultimately retired in the same order as they were coded. This out-of-order execution can boost performance in many cases. Look at Example 23-6.

Due to the fact that memory fetches (due to cache misses) can take many clock cycles and result in underutilization of the CPU, out-of-order execution is a way of finding something to do for the CPU. Simply put, the idea of out-of-order execution is to look deep into the stream of instructions and find the ones that can be executed ahead of others, providing that resources are available. Again, it is important to note that the Pentium

Example 23-6

For the following code, indicate the instructions that can be executed out of order in the Pentium Pro.

```
(1) LOAD (R2), R4   ;LOAD R4 FROM MEMORY POINTED AT BY R2
(2) ADD  R3,R4,R7   ;R3+R4--->R7
(3) ADD  R6,R8,R10  ;R6+R8--->R10
(4) SUB  R5,R1,R9   ;R5-R1--->R9
(5) ADD  R6,1,R12   ;R6+1---->R12
```

Solution:

Instruction i2 cannot be executed until the data is brought in from memory (either cache or main memory DRAM). Therefore, i2 is dependent on i1 and must wait until the R4 register has the data. However, instructions i3, i4, and i5 can be executed out of order and parallel with each other since there is no dependency among them. After the execution of i2, all the instructions i2, i3, i4, and i5 can be retired instantly since they all have been executed already. This would not be the case if these instructions were executed in the Pentium since its pipeline would be stalled due to the memory access for R4. In that case, instructions i3, i4, and i5 could not even be fetched let alone decoded and executed.

Example 23-7

The following x86 code (a) sets the pointer for three different arrays, and the counter value, (b) gets each element of ARRAY_1, adds a fixed value of 100 to it, and stores the result in ARRAY_2, and (c) complements the element and stores it in ARRAY_3. Analyze the execution of the code in light of the out-of-order execution and branch prediction capabilities of the Pentium Pro.

```
(1)           MOV   EBX,ARRAY_1   ;LOAD POINTER
(2)           MOV   ESI,ARRAY_2   ;LOAD POINTER
(3)           MOV   EDI,ARRAY_3   ;LOAD POINTER
(4)           MOV   ECX,COUNT     ;LOAD COUNTER
(5) AGAIN:    MOV   EAX,[EBX]     ;LOAD THE ELEMENT
(6)           ADD   EAX,100       ;ADD THE FIX VALUE
(7)           ADD   EBX,4         ;UPDATE THE POINTER
(8)           MOV   [ESI],EAX     ;STORE THE RESULT
(9)           ADD   ESI,4         ;UPDATE THE POINTER
(10)          NOT   EAX           ;COMPLEMENT THE RESULT
(11)          MOV   [EDI],EAX     ;AND STORE IT
(12)          ADD   EDI,4         ;UPDATE THE POINTER
(13)          LOOP AGAIN          ;STAY IN THE LOOP
(14)          MOV   AX,4C00H      ;EXIT
(15)          INT 21H
```

Solution:

The fetch/decode unit fetches and converts instructions into micro-ops. Since there is no dependency for instructions i1 through i5, they are dispatched, executed, and retired except for i5. Notice that the pointer values are immediate values; therefore, they are embedded into the instruction when the fetch/decode unit gets them. Now i5 is a memory fetch that can take many clocks, depending on whether the needed data is located in cache or main memory. Meanwhile i6, i8, i10, and i11 must wait until the data is available. However i7, i9, and i12 can be executed out of order knowing that the updated values of pointers EBX, EDI, and ESI are kept internally until the time comes when they will be committed to the visible registers by the retire unit. More importantly, the LOOP instruction is predicted to go to the target address of AGAIN and i5, i6, ... are dispatched once more for the next iteration. This time the memory fetch will take very few clocks since in the previous data fetch, the CPU read at least 32 bytes of data using the Pentium Pro 64-bit (8 bytes) data bus and the burst read mode, transferring four sets of 8-byte data into the CPU. This process will go on until the last round of the LOOP instruction where ECX becomes zero and falls through. At this time, due to misprediction, all the micro-instructions belonging to instructions i5, i6, i7, ... (start of the loop) are removed and the whole pipeline restarts with instructions belonging to i14, i15, and so on.

Pro will not immediately provide the results of out-of-order executions to programmer-visible registers such as EAX, EBX, and so on, since it must maintain the original order of the code. Instead, the results of out-of-order executions are stored in the pool and wait to be retired in the same order as they were coded. Therefore, programmer-visible registers are updated in the same sequence as expected by the programmer.

Branch prediction

The Pentium Pro, like the Pentium before it, has branch prediction, but with greater capability. When the Pentium Pro encounters branch instructions (such as JNZ), it creates a list of them in what is called the branch target buffer (BTB). The BTB predicts the target of the branch and starts executing from there. When the branch is executed, the result is compared with what the prediction section of the CPU said it would do. If they match, the branch is retired. If not, all instructions behind the branch are removed from the pool and the correct branch target address is provided to the BTB. From there the BTB refills the pipeline with instructions from the new target address. See Example 23-7.

Note the following points concerning the reordering of store instructions from Intel documentation: "Stores are never performed speculatively since there is no transparent way to undo them. Stores are also never re-ordered among themselves. A store is dispatched only when both the address and the data are available and there are no older stores awaiting dispatch."

Bus frequency vs. internal frequency in Pentium

Frequently you may see an advertisement for a 1-GHz or 2-GHz Pentium PC. It is important to note that the stated frequency is the internal frequency of the CPU and not the bus frequency. This is due to the fact that designing a 1-GHz motherboard is very difficult and expensive. Such a design requires a very fast logic family and memory in addition to a massive simulation to avoid crosstalk and signal radiation. The bus frequency for such systems is currently less than 1 GHz.

Review Questions

1. Pentium Pro is the official name designated by Intel. What was it called before such a designation?
2. True or false. Both the Pentium and Pentium Pro have 16KB L1 cache.
3. True or false. Both the Pentium and Pentium Pro have L2 cache on the same package.
4. True or false. The x86 instruction set is in triadic form.
5. Which of the x86 processors has out-of-order execution?
6. Which unit inside the Pentium Pro commits the final results of operations to registers EAX, EBX, and so on?
7. True or false. The Pentium Pro is a superpipelined processor.

SECTION 23.5: MMX TECHNOLOGY

In this section, we discuss the MMX (MultiMedia extension) technology used in some of the Intel processors.

DSP and multimedia

To run high-quality multimedia applications with sound and graphics requires very fast and sophisticated mathematical operations. Such complex operations are normally performed by a highly specialized chip called DSP (digital signal processing). DSP chips are the main engines performing tasks such as 2D and 3D graphics, video and audio compression, fax/modem, PC-based telephoning with live pictures, and image processing.

There are three approaches to equip the PC with DSP capability.

1. Use a full-fledged DSP chip on the board along with the main CPU. This is the best and ideal approach since there are some very powerful DSP chips out there. However, the problem is that there is no industry-wide standard to be followed by the PC designers and the lack of such a standard can lead to incompatibility both in hardware and software.
2. Use the x86 and x87 FP (floating-point) instructions to emulate the function of DSP. This is slow and the performance is unacceptable.
3. The third approach is to incorporate some DSP functions into the x86 microprocessor. This approach leaves everyone at the mercy of Intel, yet it brings compatibility and a unified approach to the issue. Although the performance is not as good as with the first approach, it is much better than the second approach.

The third approach is exactly what happened. In early 1997 Intel introduced a series of Pentium and Pentium Pro chips with somewhat limited DSP capability called *MMX technology*. In the case of Intel's MMX technology, software compatibility, both on the BIOS and operating system levels, was the most important goal. It needs to be noted that although MMX does not have the rich set of instructions normally associated with DSP chips such as Texas Instruments' DSP chips, it still performs many of the DSP functions reasonably well.

Register aliasing by MMX

As stated earlier, one of the main goals of MMX technology was to maintain compatibility with other x86 processors with no MMX capability. To assure that, Intel uses the FP (floating-point) register set of the x87 math coprocessor as the working register for MMX instructions instead of introducing a whole new set of registers. This is called *register aliasing*, meaning that the same physical register has different names. While the x87 FP registers are 80 bits wide, the MMX uses only 64 bits of it. The x87 floating-point registers are called ST(0), ST(1), ..., ST(7) when they are used by the x87 instruction set, but the same registers are called MM0, MM1, ..., MM7 when used by the MMX portion of the CPU. See Figure 23-11. Register aliasing by MMX has some major implications:

1. We must not use the registers to store MMX data and FP (floating point) data at the same time since they are the same physical registers.
2. We must not mix MMX instructions with FP instructions. Mixing MMX and FP instructions slows down the application since it takes many clock cycles to switch between MMX and x87 instructions. The best method is to have separate program modules for x87 instructions and MMX instructions with no intermixing.
3. When leaving an MMX program module, make sure that all the MMX registers are cleared before issuing any x87 instructions. The same is true if switching from x87 to MMX. All FP registers must be popped to leave them empty.
4. As shown in Chapter 20, FP registers are accessed by the x87 instructions in the stack format. However, when these same registers are accessed by the MMX instruction set, each one is accessed directly by its name, MM0–MM7. These MMX registers cannot be used to address memory and must be used only to perform calculations on data.

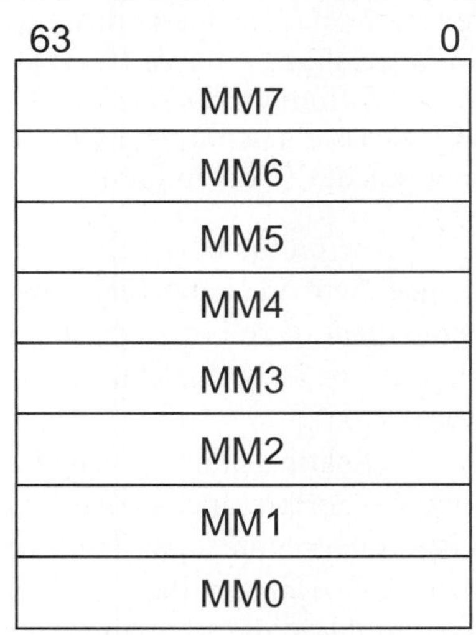

Figure 23-11. MMX Register Set

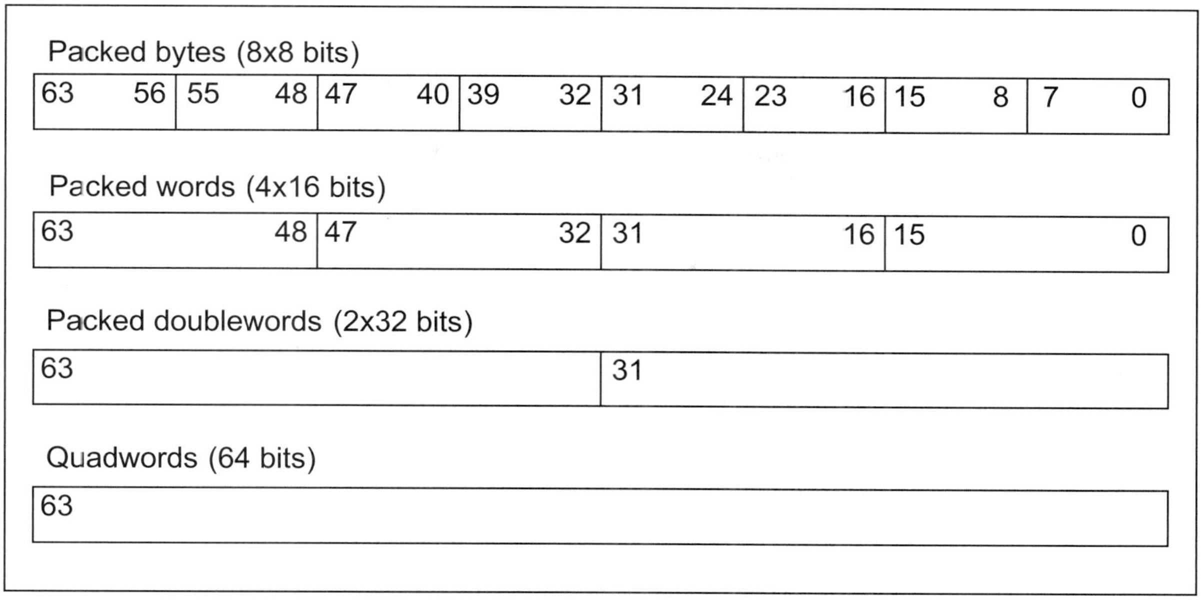

Figure 23-12. MMX Data Types

Data types in MMX

As mentioned earlier, the MMX uses only 64 bits of the 80-bit wide FP registers. Therefore, the largest MMX data size is 64-bit. However, the 64-bit register can be used for four different data types. See Figure 23-12. They are as follows.

1. Quadword (one 64-bit)
2. Packed doubleword (two 32-bit)
3. Packed word (four 16-bit)
4. Packed byte (eight 8-bit)

All four data types of the MMX are integers and are referred to as *packed data*. It must be noted that the contents of the MMX registers can be treated as any of the four different types of eight bytes, four words, two doublewords, or one quadword. It is the job of the MMX instruction to specify the data type. For example, the instruction Packed Add has three different formats depending on the data type. They are as follows:

PADDB (Add Packed Byte) adds two groups of 8 packed bytes
PADDW (Add Packed Word) adds two groups of 4 packed words
PADDD (Add Packed double) adds two groups of 2 packed doublewords

It is interesting to note that Intel has introduced to the x86 instruction set a total of 57 new instructions just for the MMX. Since currently there is no assembler or compiler equipped with the MMX instruction set, we will not give any MMX programming example. To find out if a given Pentium is an MMX chip, we must use a Pentium instruction called CPUID. This is discussed next.

CPU identification for x86

CPU identification is so important to the new generation of operating systems and software packages that starting with the Pentium, Intel has introduced a new instruction to do just that. The problem is how to identify microprocessors prior to the Pentium. According to Intel, to identify the microprocessor by way of software one must examine the bits of the flag register. Notice that in the 8088/86/286 the flag register is a 16-bit register but in the 386/486/Pentium it is a 32-bit register. Table 23-5 shows the status of the flag bits used in identifying the processor type. These bits can be examined at any time and not just at boot-up.

Table 23-5: Flag Bits for CPU Identification

CPU	Flag Bits
8088/86	Bits 12 through 15 are always 1.
80286	Bits 12 through 15 are always 0 (in real mode).
80386	Bit 18 is always 0 (in real and protected mode).
80846	Bit 21 cannot be changed, therefore it is 486; if bit 21 can be changed to 1 and 0, then it must be a Pentium.
Pentium and Pentium Pro	Starting with the Pentium, one can use a new instruction, CPUID, to get information such as family and model of the processor. However, it is the ability to set or reset bit 21 of the flag that indicates whether the CPUID instruction is supported or not. The CPUID instruction can be executed any time in protected mode or real mode. For Intel's Pentium and higher microprocessors, prior to execution of the CPUID instruction we must set EAX = 1. After the execution of CPUID, bits D8–D11 of EAX have the family number. The family number is 5 for the Pentium and is 6 for the Pentium Pro.

It must also be noted that there is no instruction in the x86 family that can exchange the contents of the flag register and a general-purpose register directly. Therefore, to examine the contents of the flag register in an x86, we must use the stack as outlined in the following steps:

1. Push the flag register onto the stack.
2. Get (pop) it into a register such as AX, BX.
3. Manipulate bits d15–d12 (or any other bits).
4. Push it back onto the stack.
5. Pop it back into the flag register from the stack.
6. Push the flag back onto the stack again.
7. Get (pop) the new flag bits back into a register.
8. Examine bits d12–d15 to see if the changes in step 3 took effect.

The following code shows the above steps.

```
Step 1)     PUSHF                ;push the flag into stack
Step 2)     POP    BX            ;and get it into BX
Step 3)     AND    BX,0FFFH      ;mask bits d15-d12
Step 4)     PUSH   BX            ;send it back into stack
Step 5)     POPF                 ;bring it back into flag reg
Step 6)     PUSHF                ;store the flag on stack
Step 7)     POP    BX     ;get it into BX again to examine
Step 8a)    AND    BX,0F000H     ;mask all bits except d12-d15
Step 8b)    CMP    BX,....
```

Notice in the above code that instructions "PUSHF" and "POPF" are used for pushing and popping the 16-bit flag register. However, to access the 32 bits of the flag register in the 386 and Pentium processors, instructions PUSHFD and POPFD must be used. These steps are coded in Program 23-1.

CPUID instruction and MMX technology

Not all Pentium and Pentium Pro microprocessors come with MMX technology. To find out if a microprocessor is equipped with MMX technology, we can use Pentium instruction CPUID. According to Intel, upon return from instruction CPUID, if D23 of EDX is high, the CPU has MMX technology. MMX identification is performed in Program 23-1.

```
;Program 23-1
;after making sure CPUID instruction is supported
      MOV   EAX,1               ;REQUEST FOR FEATURE FLAG
      CPUID                     ;CPUID INSTRUCTION
      TEST  EDX,00800000H       ;BIT 23 OF EDX INDICATES MMX
      JNZ   MMX_YES
      ...                       ;NO MMX
MMX_YES:    ...
```

Note: To assemble the code below you need MASM 6.11 which support the Pentium instructions such as CPUID (the directive .586 is for that purpose). If you are using MASM 5.x then remove both the .586 directive and CPUID instruction and replace them with opcode for CPUID which is (0FA2H) in the following manner

```
            DW    0FA2H         ;opcode for CPUID instruction
```

```
;this routine identifies the PC'S 80X86 microprocessor
;upon return from this subroutine, AX contains microprocessor
code
;where 0=8088/86,1=286,3=386,4=486, 5=Pentium, 6=Pentium Pro

GET_CPUID   PROC
;see if it is 8088/86 by checking bits d12-d15 of flag reg
            PUSHF               ;push the flag into stack
            POP   BX            ;and get it into BX
            AND   BX,0FFFH      ;mask bits d15-d12
            PUSH  BX            ;send it back into stack
            POPF                ;bring it back into flag reg
            PUSHF               ;store the flag back on stack again
            POP   BX            ;and get it back into BX
            AND   BX,0F000H     ;mask all bits except d12-d15
            CMP   BX,0F000H     ;are the d12-d15 all zeros?
            MOV   AX,0          ;make AX=0 code for 8088/86
            JE    OVER          ;if yes then AX=0 code for 8088/86
;see if it is 80286 by checking bits of d12-d15 of flag reg
            OR    BX,0F000H     ;if not try setting d12-d15 to high
            PUSH  BX            ;push it into stack
            POPF                ;make d12-d15 of flag reg all 1s
            PUSHF               ;get the flag back into stack
            POP   BX            ;get it back to examine the bits
            AND   BX,0F000H     ;mask all bits except d12-d15
            CMP   BX,0F000H     ;are d12-d15 all 1s
            MOV   AX,1          ;make AX=1 code for 286
            JE          OVER          ;if yes set AX=1 code for 286
;see if it is 386 by checking bit 18 of flag bit
            .386
            PUSHFD      ;if not it is 386 or higher. push flag
            POP   EBX           ;and get it into EBX
            MOV   EDX,EBX       ;save it
            XOR   EBX,40000H    ;flip bit 18
            PUSH  EBX           ;sent it into stack
            POPFD               ;get it into flag
            PUSHFD              ;get it back into stack
            POP   EBX           ;get the new flag back into EBX
```

```
            MOV    AX,3           ;make AX=3 code for 386
            XOR    EBX,EDX        ;see if bit 18 is toggled
            JE     OVER           ;if yes then AX=3 CODE for the 386
;see if it is 486 or higher. try changing bit 21 of flag reg
            MOV    AX,4;if not it is 486 or higher (AX=4 for 486)
            PUSHFD       ;see if bit 21(ID bit) can be altered
            POP    EBX   ;in order to use the CPUID instruction
            MOV    EDX,EBX        ;save original flag bit in EDX
            XOR    EBX,200000H ;flip bit 21
            PUSH   EBX            ;save it on the stack
            POPFD                 ;get it into flag reg
            PUSHFD                ;get flag back into stack
            POP    EBX            ;get it into EBX to examine bit 21
            XOR    EBX,EDX        ;see if bit 21 changes
            JE     OVER           ;if yes AX=04 code for 486
;see which Pentium (586,or 686) by using CPUID instruction
            MOV    EAX,1          ;set EAX=1 before executing CPUID
            .586                  ;use Pentium instruction
            CPUID
;after execution of CPUID, bits D8-D11 of EAX have family number
            .386                  ;back to 386 instructions
            AND    EAX,0F00 ;mask all bits except the family bit
            SHR    EAX,8 ;move d8-d11 to lower nibble then AX=5
                         ;for Pentium, 6 for Pentium Pro

.8086
OVER: RET                ;return with AX=processor number
GET_CPUID   ENDP
```

For the latest CPU identification program, see http://www.Intel.com.

Review Questions

1. Why do we not use the FP x87 instructions for DSP multimedia?
2. MMX is available for which of the x86 processors?
   (a) 486 (b) Pentium (c) Pentium Pro
3. True or false. MMX aliases the x86 registers.
4. What are the names of the MMX registers?
5. True or false. MMX instructions access registers in stack format.
6. How can a program determine if the processor is a 386?
7. To identify the 486 processor, toggle _____________.
8. CPUID instruction was first introduced with the _______ processor
9. To identify the Pentium, use CPUID with EAX = _____.
10. Can the CPUID instruction help determine if the chip supports MMX technology? If so, how?

PROBLEMS

SECTION 23.1: THE 80486 MICROPROCESSOR

1. The 486 chip uses ________ pins.
2. The 486 is a(n) ______-bit microprocessor.

3. What is the size of on-chip cache in the 486?
4. Off-chip cache is referred to as ________ cache.
5. True or false. On-chip cache in the 486 is used to hold both data and code.
6. State the differences between the 486 and 486SX microprocessors.
7. The 486 has a(n) ______-bit external data bus.
8. How many data parity pins does the 486 have?
9. The 486 can access ________ bytes of memory using the _________ address pins.
10. How many BE pins does the 486 have?
11. Nonburst read and write cycles take _______ clocks.
12. What does "2-2-2-2" cycle mean?
13. What does "2-1-1-1" cycle mean?
14. The 486 fetches _______ bytes of code and data into CPU using a burst cycle. How many clocks does the burst cycle take?
15. Calculate the bus bandwidth of a 486 with 25 MHz in each of the following.
    (a) nonburst cycle (b) burst cycle
16. Calculate the bus bandwidth of a 486 with 33 MHz in each of the following.
    (a) nonburst cycle (b) burst cycle
17. Which 486 instruction converts from the little endian to big endian, or vice versa?
18. Show how the data is placed in memory for the following program before and after the execution of BSWAP.

```
MOV   EAX,23F46512H
MOV   [4000],EAX
BSWAP EAX
MOV   [6000],EAX
```

19. If a 486 is advertised as 25 MHz, what clock frequency is connected to CLK?
20. What is the purpose of the A20M pin in the 486?
21. Assume that CS = FFFFH and IP = 76A0H. Calculate the physical address of the instruction in each of the following states.
    (a) A20M = 0 (b) A20M = 1
22. What is HMA, and how does it relate to the A20M pin in the 486?
23. The 486 uses a pipeline of ________ stages.
24. Give the names of the pipeline stages in the 486.

SECTION 23.2: INTEL'S PENTIUM

25. The number of pipeline stages in a superpipeline system is _________ (less, more) than in a superscalar system.
26. Which has one or more execution units, superpipeline or superscalar?
27. The Pentium uses _______ transistors and has _______ pins.
28. The Pentium has a(n) ______-bit external data bus whose pins are named _______.
29. The Pentium is a(n) _______-bit microprocessor.
30. State how many BE pins the Pentium has and their purpose.
31. BE pins are active ________ (low, high).
32. If BE7–BE0 = 11110000, which part of the data buses is activated?
33. If BE7–BE0 = 00000000, which part of the data buses is activated?
34. How many DP (data parity) bits does the Pentium have?
35. Find and compare the Pentium bus bandwidth for the following 60-MHz systems.
    (a) nonburst mode (b) burst cycle mode
36. What is the size of on-chip cache in the Pentium?
37. In Problem 36, how much cache is for data and how much for code?
38. Which part of on-chip cache in the Pentium is write protected, data or code?

39. True or false. The Pentium has an on-chip math coprocessor.
40 What does instruction pairing mean in the Pentium?
41. The Pentium uses _____________ (superscalar, superpipeline) architecture.
42. What is instruction pairing, and when can it happen?
43. What is data dependency, and how is it avoided?
44. Write a program for the 386/486/Pentium to calculate the total sum of 10 doubleword operands. Use looping.
45. Compare the clock count of the loop in Problem 44 for each of the following. Use branch prediction for (c) and (d). Note that if two instructions are paired and one takes 2 clocks and the other takes only 1 clock, the clock count is 2.
    (a) 386 (b) 486 (c) a Pentium with instruction pairing but no code scheduling
    (d) a Pentium with instruction pairing and code scheduling
46. Calculate the bus bandwidth for a 486DX2-50 and a 486DX4-100. Note that the 486DX2-50 is 25 MHz and the 486DX4-100 is 33 MHz.
47. Draw the pipeline stages for the pairing of instructions in the Pentium.
48. True or false. The Pentium has the A20M pin.

SECTION 23.3: RISC ARCHITECTURE

49. Why is RISC called load/store architecture?
50. In RISC, all instructions are ____-byte.
51. Which of the following instructions do not exist in RISC?
    (a) ADD reg,reg (b) MOV r,immediate (c) OR reg,mem
52. State the steps in a RISC program to add a register to a memory location.
53. What is the advantage of having all the instructions the same size?
54. Why are RISC programs larger than CISC programs?
55. The vast majority of RISC instructions are executed in ______ (1, 2, 3) clocks.
56. What is Harvard architecture? Is it unique to RISC? Can a CISC system use Harvard architecture?
57. Code a RISC program to add 10 operands of 4-byte (doubleword) size and save the result. Do not be concerned with carries.
58. Show the code scheduling and clock count for Problem 57.
59. What is a delayed branch?
60. MS-DOS runs native on which of the following processors?
    (a) IBM/Freescale Power PC RISC (b) Intel x86
    (c) ARM RISC
61. Generally, RISC processors have _____ registers, each 32 bits wide.
62. Which register in RISC always has the value zero in it, no matter what operation is performed on it?
63. Discuss the terms *porting*, *emulating*, and *running native*.

SECTION 23.4: PENTIUM PRO PROCESSOR

64. The Pentium Pro is ____-bit internally and ____-bit externally.
65. The Pentium Pro has _____ address bits.
66. The Pentium Pro is capable of addressing ______ bytes of memory.
67. The Pentium Pro has ______ pins for data bus.
68. A Pentium Pro is advertised as 200 MHz. Is this an internal CPU frequency or a bus frequency?
69. What is the difference between the L2 cache of Pentium and Pentium Pro systems?
70. Do the 5.5 million transistors used for the Pentium Pro include the L2 cache tran-

sistor count?

71. Which of the x86 processors has out-of-order execution?
72. Are the triadic registers of the Pentium Pro visible to the programmer?
73. True or false. Instructions are fetched according to the order in which they were written.
74. True or false. Instructions are executed according to the order in which they were written.
75. True or false. Instructions are retired according to the order in which they were written.
76. The visible registers EAX, EBX, and so on, are updated by which unit of the CPU?
77. True or false. Among the instructions, STOREs are never executed out of order.
78. Which of the x86 processors have branch prediction capability?

SECTION 23.5: MMX TECHNOLOGY

79. True or false. The MMX uses x86 and x87 instructions to emulate DSP functions.
80. A given system has both a general-purpose CPU (such as the x86) and a DSP chip. Discuss the role of each chip.
81. Does the 486 system have MMX technology?
82. Explain the concept of register aliasing.
83. Which registers are aliased in MMX technology?
84. Explain the difference between the way x87 registers are accessed and the way MMX accesses the same registers.
85. Indicate which group of instructions can be intermixed.
    (a) x86, x87 (b) x86, MMX (c) x87, MMX
86. When leaving MMX, what is the last thing that a programmer should do?
87. When leaving x87, what is the last thing that a programmer should do?
88. How many bits of the x87 register are used by MMX?
89. A quadword has _________ bits.
90. Give other data formats besides quadword that can be viewed by MMX instructions.
91. True or false. Every x86 supports the CPUID instruction.
92. Explain how to identify the 386 microprocessor.
93. Explain how to identify the 486 microprocessor.
94. Explain how to identify the Pentium microprocessor.
95. Explain how to identify the Pentium Pro microprocessor.
96. Explain how to identify the Pentium Pro with MMX technology.
97. How can the CPUID instruction be coded into a program if the assembler does not support it?

ANSWERS TO REVIEW QUESTIONS

SECTION 23.1: THE 80486 MICROPROCESSOR

1. 168 pins in PGA
2. True
3. 32, 32
4. The 80486 has the math coprocessor on-chip; the 80486SX has the math coprocessor 80487SX (a separate chip).
5. Primary cache, secondary cache or L1 and L2 cache
6. 8K bytes, 2-way set associative

7. 1/50 MHz = 20 ns is the clock cycle. In burst cycle 5 clocks of 20 ns can access 16 bytes of memory (4 memory cycles × 4 bytes of 32-bit data bus); therefore, (1/100 ns) × 16 bytes = 160 megabytes/second
   Another way would be average clock per 32-bit access is (5 × 20)/4 = 25 ns and (1/25 ns) × 4 bytes = 160 megabytes/second
8. 33 MHz
9. Input
10. Output, like all other address bits in x86 processors

SECTION 23.2: INTEL'S PENTIUM

1. 273
2. 64
3. True
4. 16K bytes: 8K for code and 8K for data
5. Code cache
6. True
7. True
8. Since it has two execution units (pipelines) capable of executing two instructions with one clock
9. False; by the compiler
10. True

SECTION 23.3: RISC ARCHITECTURE

1. Reduced instruction set computer, complex instruction set computer
2. True
3. 32, 32
4. True
5. They are all 4 bytes.
6. True
7. c
8. 32, 8 stack based
9. To take care of double-precision floating-point operands and single-precision operands
10. False. It uses separate buses for data and code.

SECTION 23.4: PENTIUM PRO PROCESSOR

1. P6
2. True
3. False
4. False
5. So far, the Pentium Pro
6. Retire unit
7. True

SECTION 23.5: MMX TECHNOLOGY

1. The x87 instruction set is mainly for math functions such as sine, cosine, log, and so on and does not lend itself to DSP-type operations.

2. (b) & (c)
3. False
4. MA0–MA7
5. False
6. Bit 18 of flag register is always high.
7. Bit 21
8. Pentium
9. EAX=1
10. Yes. After return from CPUID we test bit 23 of the EDX register. If it is high, the chip has MMX technology.

CHAPTER 24

THE EVOLUTION OF x86: FROM 32-BIT TO 64-BIT

OBJECTIVES

Upon completion of this chapter, you will be able to:

- **>> Compare and contrast the variations of Pentium processors**
- **>> Compare and contrast the variations of x86 32-bit architecture**
- **>> Describe the enhancements of 32-bit architecture in x86 processors**
- **>> Describe the role of L1–L3 caches in x86 processors**
- **>> Describe the role of multicore and multithreading feaures of x86**
- **>> Describe 64-bit architecture of x86**

The x86 microprocessor architecture has gone through major changes since the introduction of the first Pentium processor in 1993. This chapter examines some of these changes. In Section 24.1, we examine the evolution of the Pentium family from Pentium II to Pentium IV. The 64-bit architecture of the x86 and Itanium processors and their impact on operating systems such as Vista are discussed in Section 24.2. Although many new features of the x86 are present in both Intel and AMD products, for the sake of convenience, we refer only to the Intel chips.

SECTION 24.1: x86 PENTIUM EVOLUTION

In the last chapter we examined enhancements introduced in the Pentium and Pentium Pro processors. In this section, we continue to explore the evolution of the Pentium from Pentium II to Pentium IV. We will discuss features such as cache, multitreading, and multicore capabilities introduced in later-generation x86 processors.

From Pentium II to Pentium IV: an overview

To increase the processing power of the Pentium series, Intel added new enhancements in each of the successive generations of Pentium II, III, and IV. These enhancements mostly involved the internal architecture and did not change the register size of the original 386, which was 32-bit. Next, we examine the enhancements of the 32-bit architecture for x86 CPUs introduced in recent years.

Level 1 and Level 2 caches

In the Pentium II, the concept of level 2 (L2) cache was introduced. In Pentium Pro, we had 8K bytes of cache for code (instruction) and another 8K bytes of cache for data, feeding code and data to the fetch unit. This is called level 1 (L1) cache. See Figure 24-1.

In the Pentium II the size of L1 cache was increased to 16K bytes each for code and data. It also added 256K bytes of L2 cache. While L1 cache feeds code (instruction) and data into the fetch and execution units and is part of the inner working of the CPU, the Level 2 cache is sitting outside the CPU die but still on the same package as the CPU itself. Since the L1 cache is on the same die as the CPU, it works at the same clock speed as the CPU. For example, if a given Pentium has clock speed of 2 GHz, then the L1 cache feeds the CPU information at that speed. L2 cache works at a fraction of the CPU speed. For example, if a given Pentium has clock speed of 2 GHz, then the L2 cache feeds the CPU information at 1 GHz. When a Pentium with L2 cache brings in code and data from externally located DRAM memory, it places them in L2 cache. Then the memory management unit of the core CPU brings in the information from the L2 cache and separates the code and data, placing each in data or code L1 caches (Harvard architecture). See

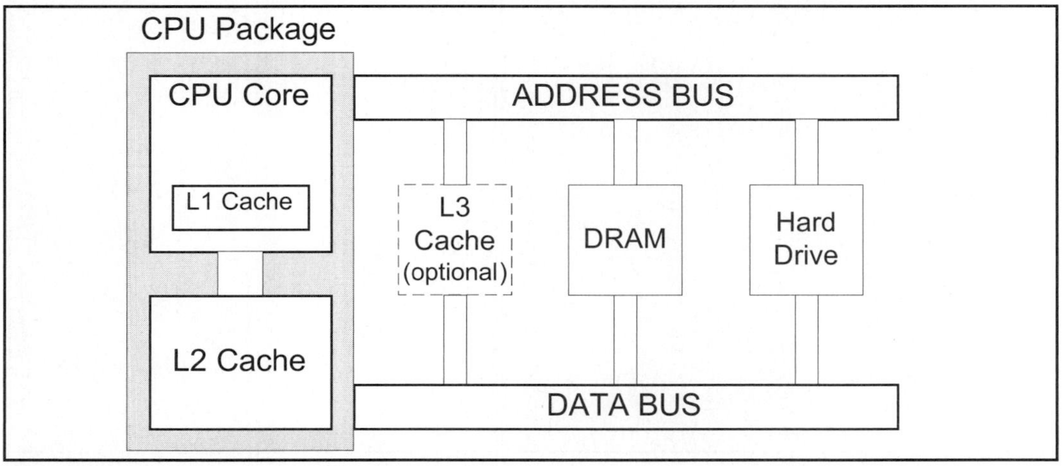

Figure 24-1. L1, L2, and L3 Cache

Figure 24-2. Notice that code and data caches are separate, which is not the case with the L2 cache. L2 cache is *unified cache* meaning that the cache is used for both code and data. The amount allocated to data and code varies dynamically, depending on the nature of the program being run. If the program being run is more data intensive, then more of the L 2 cache is allocated to data. Table 24-1 shows the amount of L1 and L2 cache memories in the x86 processors and their speed at the time of introduction. With CPU speed rising above 1 GHz, the biggest problem is external (that is, external to the CPU chip) memory access time. For that reason, in some high-performance systems the designers place level 3 (L3) cache outside the CPU on the motherboard to speed up the external memory access. This L3 cache is sitting between the CPU chip and DRAM memory module on the motherboard. See Figures 24-1 and 24-2.

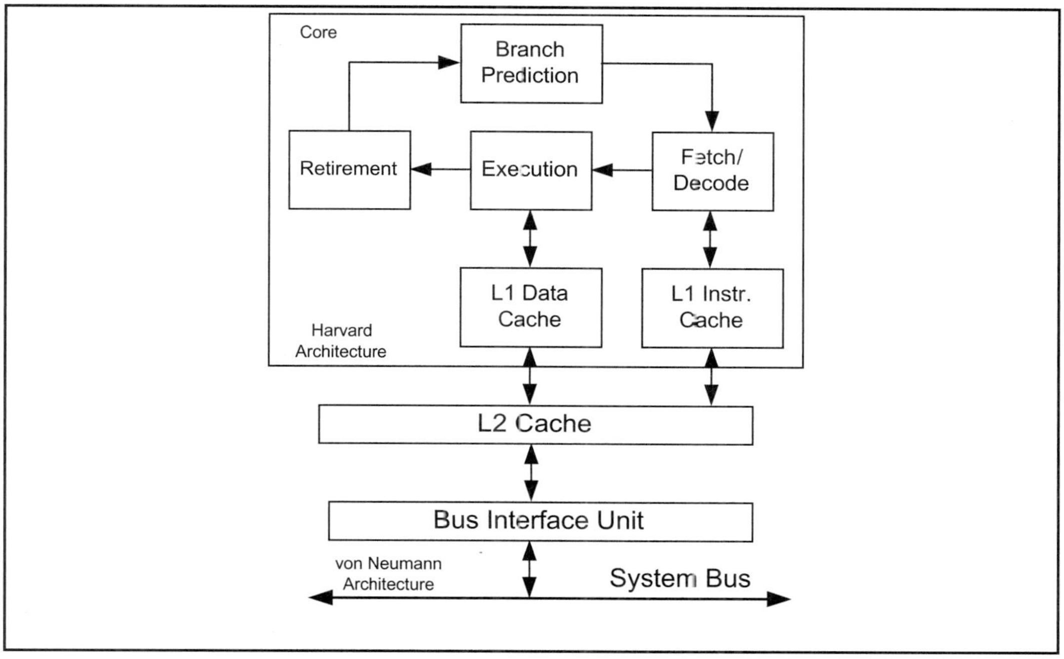

Figure 24-2. L2 Cache Feeding Code and Data to L1 Caches

Table 24-1: Pentium Cache Size and Speed

Product	Year	Transistors	CLK Speed	L1 Cache (Code)	L1 Cache (Data)	L2 Cache
386	1985	275K	20 MHz	0	0	0
486	1989	1.2M	25 MHz	4 KB	4 KB	0
Pentium	1993	3.1M	60 MHz	8 KB	8 KB	0
Pentium Pro	1995	5.5M	200 MHz	8 KB	8 KB	0
Pentium II	1997	7M	266 MHz	16 KB	16 KB	256 KB
Pentium III	1999	8.2M	700 MHz	16 KB	16 KB	512 KB
Pentium IV	2000	42M	1.5 GHz	12 KB	8 KB	256 KB
Pentium M	2004	140M	2 GHz	32 KB	32 KB	2 MB
Pentium IV	2004	125M	3.4 GHz	12 KB	16 KB	1 MB
Pentium Duo	2006	152M	2.16 GHz	32 KB	32 KB	2 MB

Note: The clock speed is the speed at the time of introduction.

In some x86 processors such as the Itanium with multiple cores there is L2 cache on the same package as the CPU, but located outside the CPU die. In such a processor one can summarize the role of L1–L3 caches as follows:

(1) The speed of the L1 cache is a multiple of the CPU speed, since the CPU can retire multiple instructions per clock cycle.

(2) L2 cache works at the same speed as the CPU since it is on the same package as the CPU.

(3) L3 cache works at a fraction of the speed of the CPU since it is located outside the CPU package.

Hyper-Threading Technology (HTT)

In the Pentium IV, the concept of multithreaded execution was introduced. First, the definition of thread: It is a series of parallel programs that can run on different CPUs simultaneously. In the multiprocessor environment, each program is given its own CPU and memory. Intel placed multiple CPUs into a single chip and called it *hyper-threading*. Therefore, hyper-threading in its simplest form is to allow a single CPU to execute two or more threads of code simultaneously. Of course, to do that the CPU must be equipped with internal logic and resources to execute the threads. Many of the early Pentium CPUs were not equipped with hyper-threaded technology, since it requires large amounts of transistors to duplicate many of the resources inside the CPU. As far as the operating system is concerned, the CPU with hyper-threaded capability apears to be multiple logical CPUs inside a single physical CPU. Therefore, to take full advantage of hyper-threading technology, both the operating system and the application must be rewritten (or reconfigured) to make them threaded-aware. For that reason, the latest generation of Microsoft XP and Vista operating systems, along with many widely used applications, are using the hyper-threading features of the x86 processors to speed up program execution. The ideal situation in the multithreaded environment is to write the application programs so that threads can execute independently of each other. However, that is not the case in the real world. Since both logical processors inside the hyper-threaded CPU use the same bus to accesses memory, they can get in each other's way and slow down program execution. Figure 24-3 shows the system bus access for the threaded CPU and multiprocessing. Note that in threaded CPUs, internal logical CPUs must share the system bus access. This is in contrast to using multiple processors in which each CPU has its own access to the system bus. See Figure 24-4.

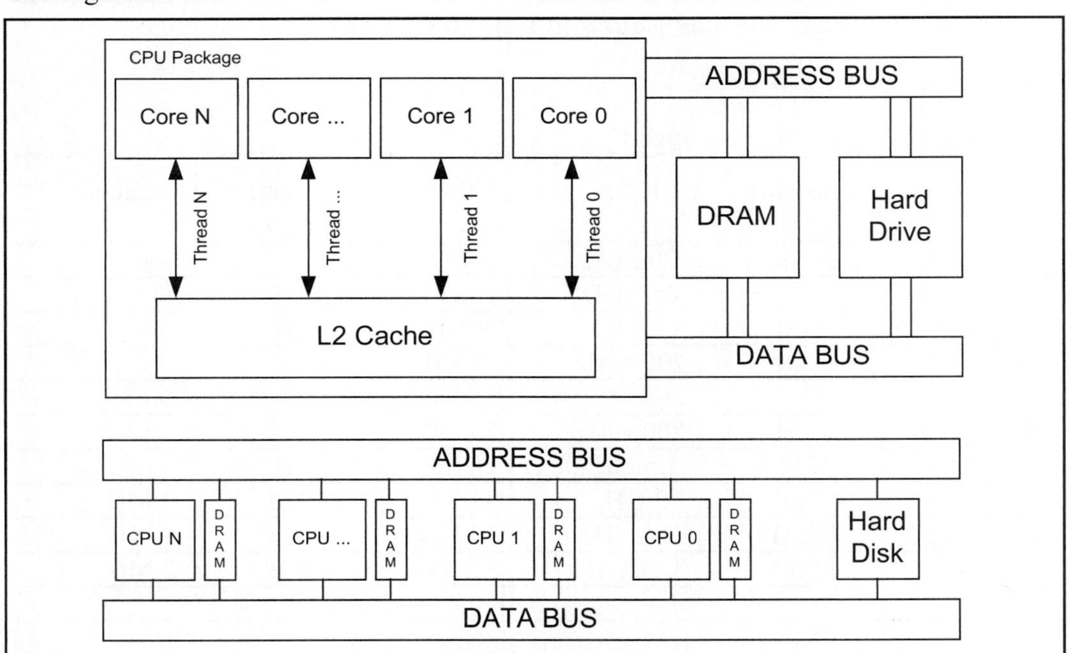

Figure 24-3. Multithreading vs. Multiprocessing

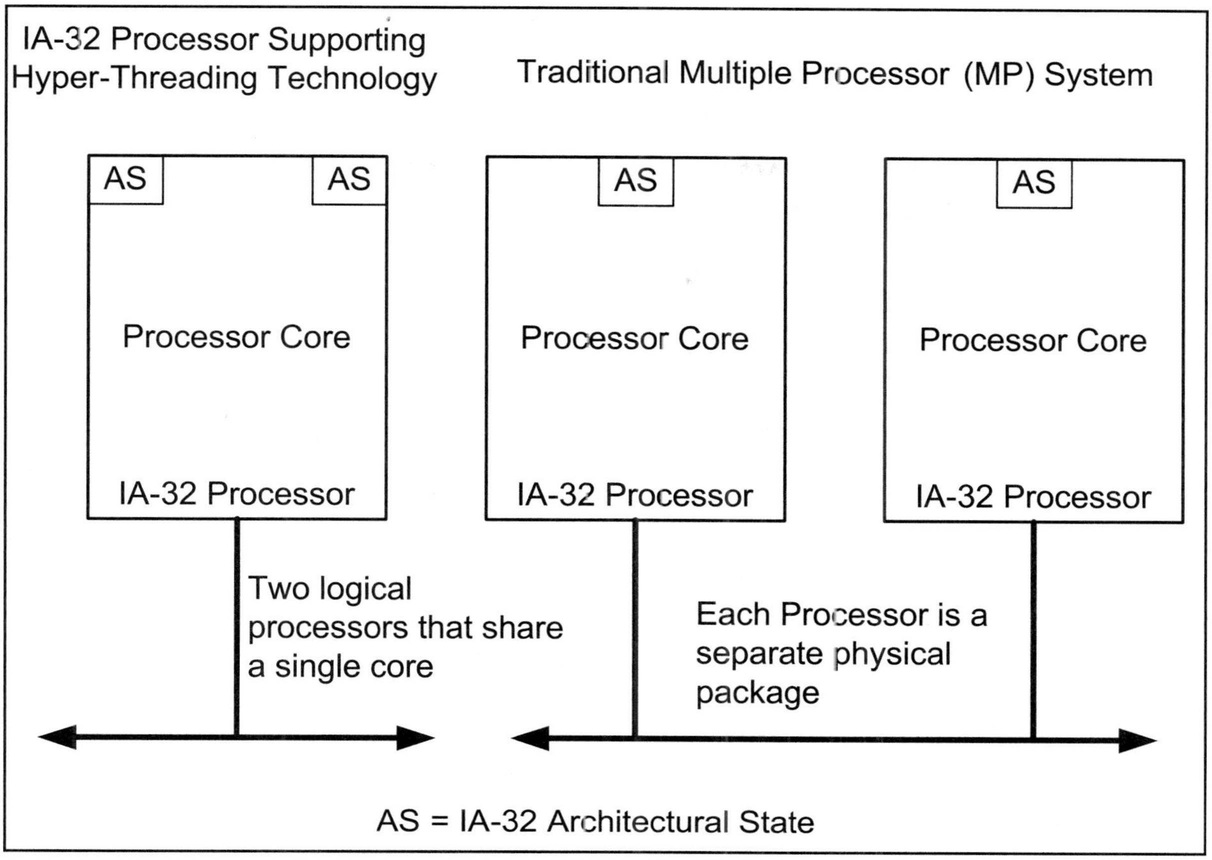

Figure 24-4. Intel Multithreading (Courtesy of Intel Corp.)

In x86 literature the words *threads* and *tasks* are used interchangeably. However, there is a difference between a task and a thread. In multitasking you are running multiple tasks such as playing music, typing into a word processor, and running a virus scan all on a single CPU. In multitasking the CPU switches from one task to another in a round robin (circular) fashion, giving each task a slice of the CPU's time. In contrast, true multithreading attempts to parallelize the execution of a single program in order to speed up the execution of that program. Not all applications lend themselves to parallelization and that is the reason that not all programs benefit equally from multithreaded CPUs. Much of the multithreaded feature of the x86 comes from its multicore capability, which is discussed next. For more discussion of multithreading and multitasking, see the following article:

http://arstechnica.com/articles/paedia/cpu/hyperthreading.ars

Multicore Technology

Many newer-generation Pentiums have what is called *multicore technology*. Multicore packs two or more independent microprocessors (called cores) into a single chip. At this time, many x86 CPUs come with dual-core and quad-core features. Both Intel and AMD are working on processors with 8 cores. In the dual-core CPU, almost everthing is doubled, which is like putting two physical CPUs into a single chip. The difference between multicore and multiprocessor CPUs is that in the multicore CPU there is one pathway to the system memory for the CPU while in the multiprocessor CPUs each processor has its own memory space independent of the others. See Figure 24-5.

See the following for more on multicore CPUs:

http://en.wikipedia.org/wiki/Multicore_(computing)

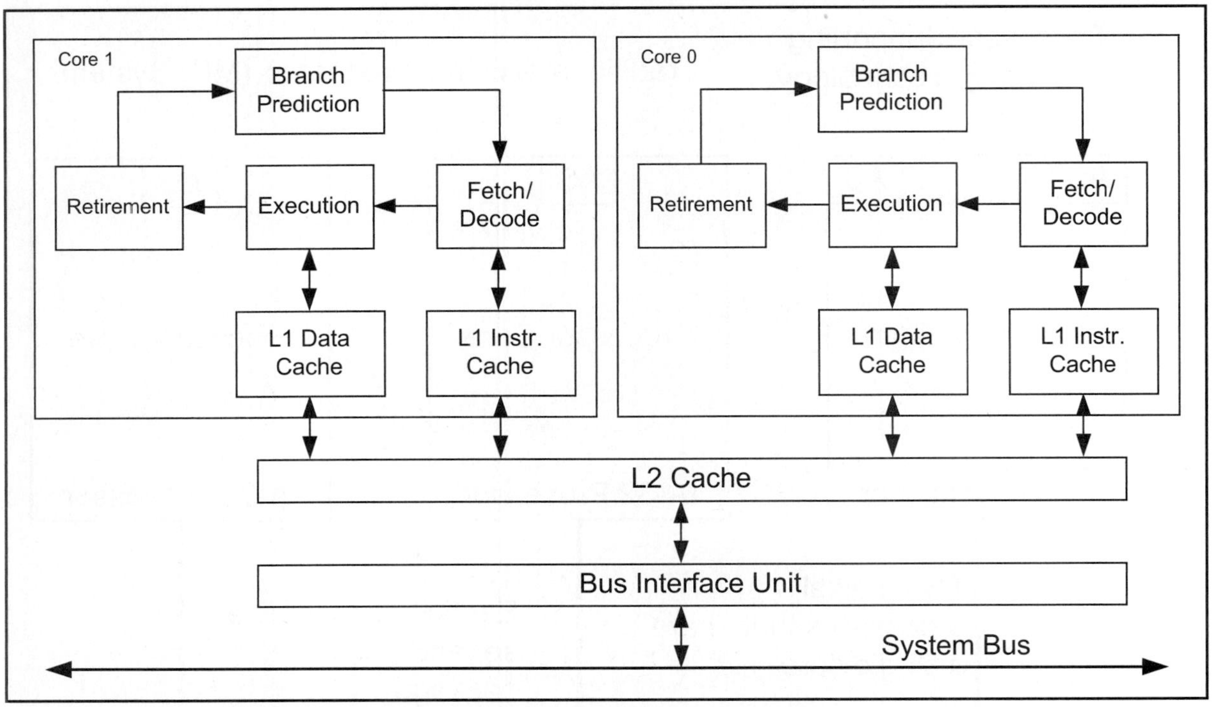

Figure 24-5. Dual-Core Processor from Intel

Streaming SIMD Extension (SSE)

The concept of single instruction multiple data (SIMD) was first incorporated into the Pentium Pro with the introduction of MMX technology, as we saw in the last chapter. The original MMX used SIMD operations on packed bytes of data. Intel expanded the SIMD to include operations on floating-point data and called it *streaming SIMD extension* (*SSE*). The SSE was expanded further to include both single and double double-precision floating points, which are referred to as SSE, SSE2, SSE3, and SSSE3 in the Intel literature. While the original SIMD had one set of MMX registers, the extended SIMD has several sets of MMX registers in addition to a new set of 128-bit registers called XMM. See Figure 24-6.

Other features of Pentium

The Pentium increased the number of pipeline stages to 10 and eventually to 20 to make it a super superpipeline CPU. It also expanded the instruction decoder units to four with a hugely expanded size of instruction queue. These improvements, along with greatly expanded out-of-order executions and branch prediction capability, allowed the CPU to retire up to four instructions per clock cycle. The combination of the above enhancements was still insufficient to keep the high-speed system designers happy. That left x86 CPU companies such as Intel and AMD no choice but to expand the size of general-purpose registers to 64-bit. The change to 64-bit architecture came about only when increase in performance of the 32-bit architecture was no longer possible. For this reason, we see two distinctive architectures for the x86 being discussed in recent years: 32-bit and 64-bit. The performance enhancements of 64-bit architecture for the x86 processors are discussed in the next section.

Review Questions

1. True or false. All Pentiums have 16 KB L1 cache.
2. True or false. All Pentiums have L2 cache on the die.
3. True or false. All Pentiums have the hyper-threading feature.
4. What is the difference between multicore and multiprocessor?
5. True or false. The Pentium IV has a much deeper pipeline than the Pentium I.

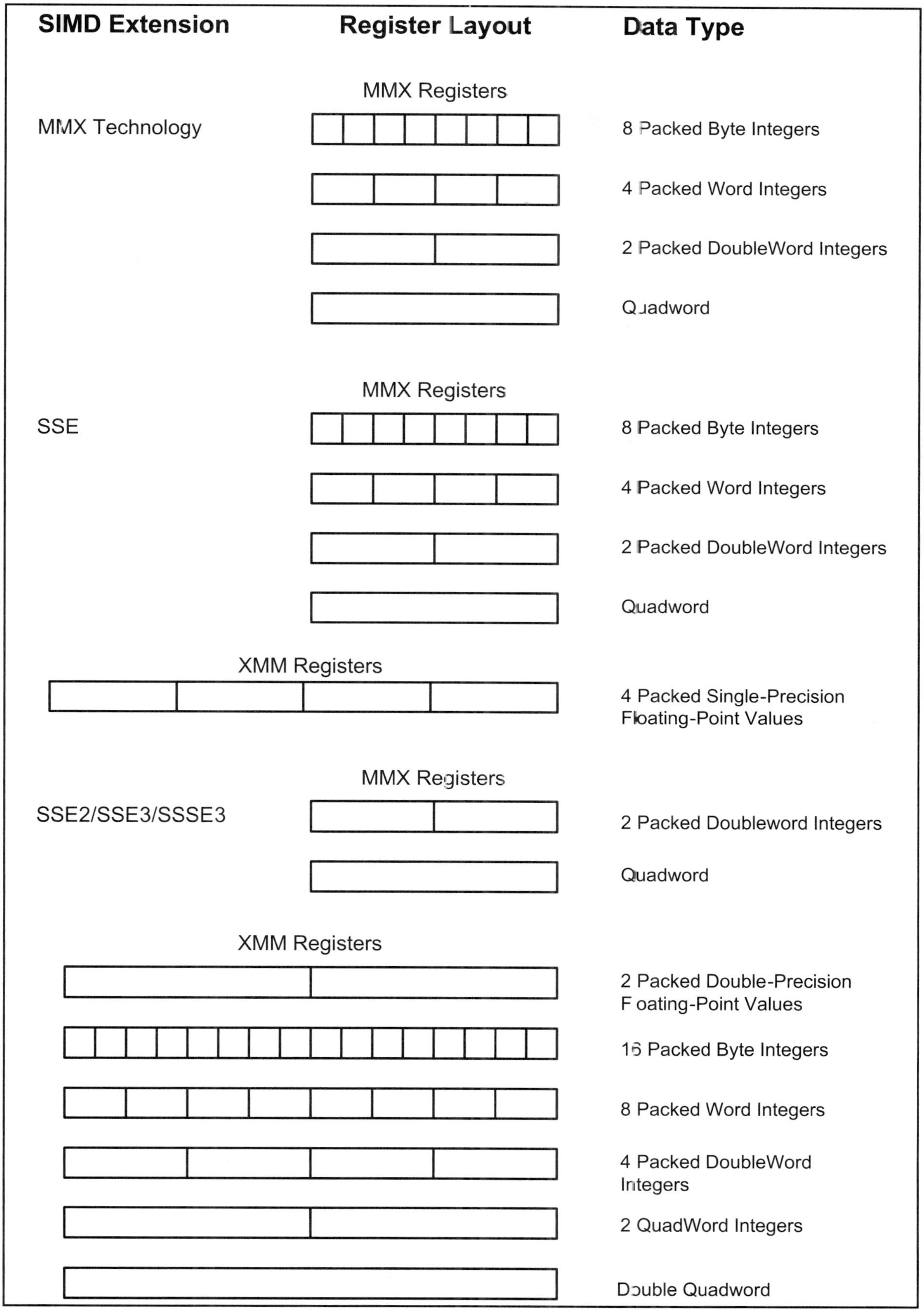

Figure 24-6. SIMD Extension in x86 (Courtesy of Intel Corp.)

SECTION 24.2: 64-BIT PROCESSORS AND VISTA FOR x86

The original 8086/88 was a 16-bit processor. Although the 80286 did improve performance in many aspects, it was still a 16-bit chip. Since IBM used both the 8088 and 80286 in the original PCs, the DOS operating system and early application software were based on 16-bit architecture. The 80386 was the first 32-bit processor in the x86 family introduced in 1989. It was not until the late 1990s that Microsoft took advantage of the 32-bit architecture of the 386 and introduced Windows NT. Subsequent improvements introduced into the original 386 in the form of Pentium I–IV family members allowed Microsoft to bring out more robust operating systems such as Windows 2000 and XP. It took many years and a massive effort to convert 16-bit DOS applications to the Windows NT/2K/XP 32-bit architecture. With the introduction of 64-bit x86 processors, the same efforts are needed again to get applications ready for the 64-bit Microsoft Vista. In this section, we discuss the x86 64-bit processors and their impact on Vista.

64-bit architecture in x86

The design of the 64-bit x86 microprocessor needed to maintain compatibility with the earlier generations of 386 legacy software. This was the same problem that 386 designers faced a decade earlier when they moved away from the 80286. In the case of the 386, the 16-bit registers of AX, BX, and so on from the 8086/286 were extended to 32-bit, and were called EAX, EBX, and so on. In this way, the 8086/286 registers became a subset of 386 registers. See Figure 24-7. To make the transition from 32-bit to 64-bit data types easier, the 64-bit architecture can work in two modes: (1) compatibility mode, and (2) 64-bit mode. In compatibility mode, most 16-bit and 32-bit legacy application software can be run without recompilation under 64-bit operating systems such as Microsoft Vista. Compatibility mode uses the 8-bit, 16-bit, and 32-bit data sizes and limits the memory space to 4G bytes, just like 386 architecture. In 64-bit mode, the 386 registers AEX, EBX, and so on are expanded to 64-bit, and are designated as RAX, RBX, and so on. It also added another 8 new 64-bit general purpose registers called R8–R15. That gives us a total of 16 64-bit registers. The new R8–R15 registers can also be accessed as 8-bit, 16-bit, and 32-bit registers. See Figure 24-8.

External memory space of 64-bit architecture

The 32-bit architecture of the Pentium limited the external memory space to 4G bytes. It also used segment registers to convert the linear addresses to physical addresses, as we saw in previous chapters. The 64-bit architecture stopped using segment registers for the most part, and introduced a flat memory space of 2^{64} bytes. The EIP (instruction pointer) of the 386 has been expanded from 32-bit to 64-bit, and is called *RIP*. It allows access of up to 2^{64} bytes of code and data stored in external memory using address A0–A63 pins. There is also a new 64-bit data pointer register for accessing externally located data memory using the A0–A63 address pins. However, to save pins not all family members have all the 64 address pins. In many cases, the 64-bit x86 processors have only 36 pins (A0–A35), which allow access of up to 64 G bytes of external memory. In some products, we also see 40 pins (A0–A39) for external memory connections, giving us a maximum of 2^{40} bytes of memory space. The expansion of the instruction pointer from 32-bit to 64-bit will assure sufficient memory space for the forseeable future. This especially will help in the implementation of multiprocessor systems using a large number of processors, giving each one its own memory space of 4 to 64 gigabytes. In 32-bit architectures, register sizes match the 32-bit external data bus. In CPUs such as the Pentium Pro, the external data bus was expanded to 64-bit, even though the registers were left at 32-bit. The reason for having more external data pins than the register size is to increase the bus bandwidth and therefore bring in more code and data to keep the CPU busy. In the case of 64-bit registers, the external data bus is also 64-bit. To increase the bus bandwidth of the 64-bit x86 processors, it is very likely we will see 128 pins for the data bus in the near future.

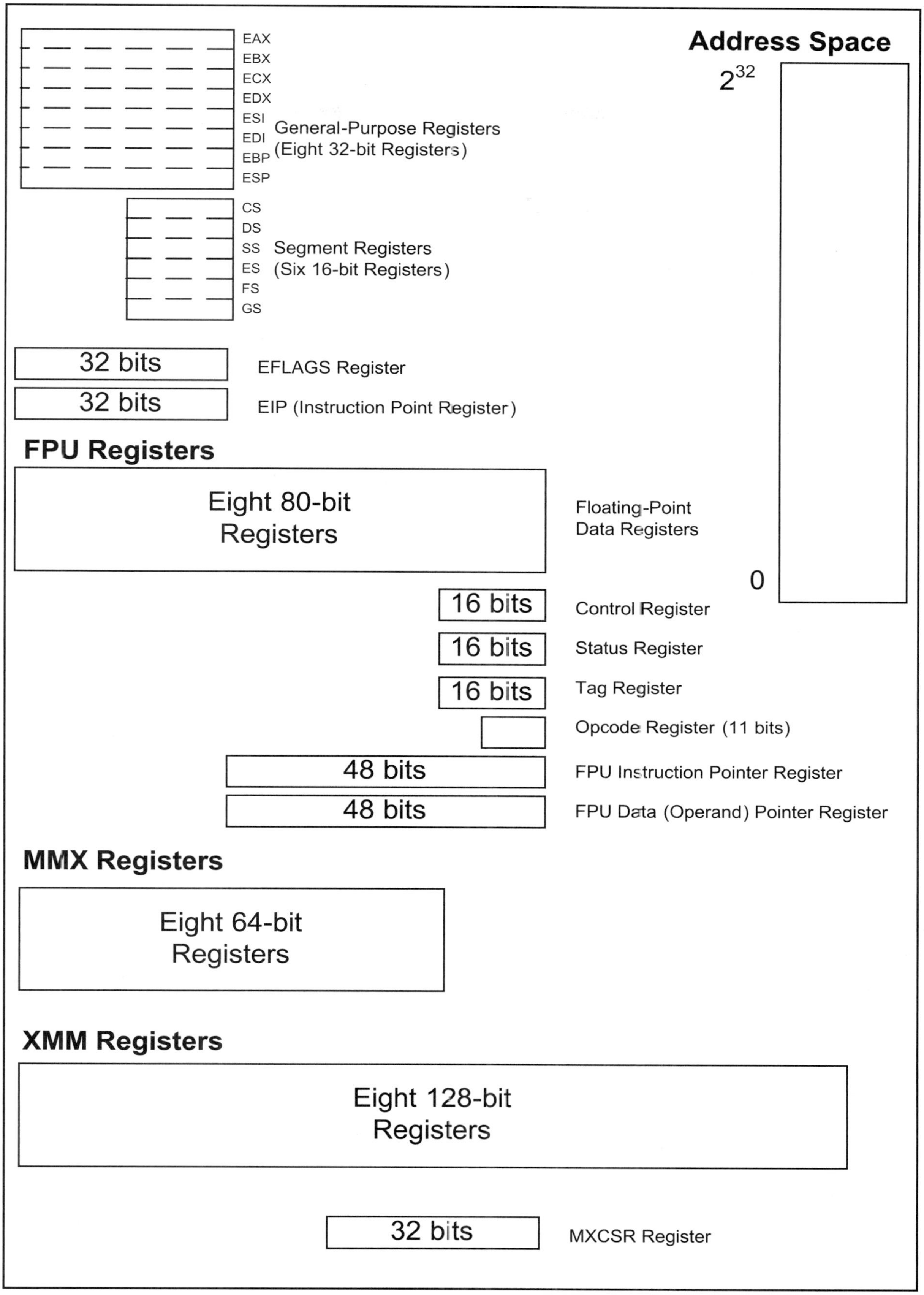

Figure 24-7. Intel Architecture - 32 (IA-32) Registers

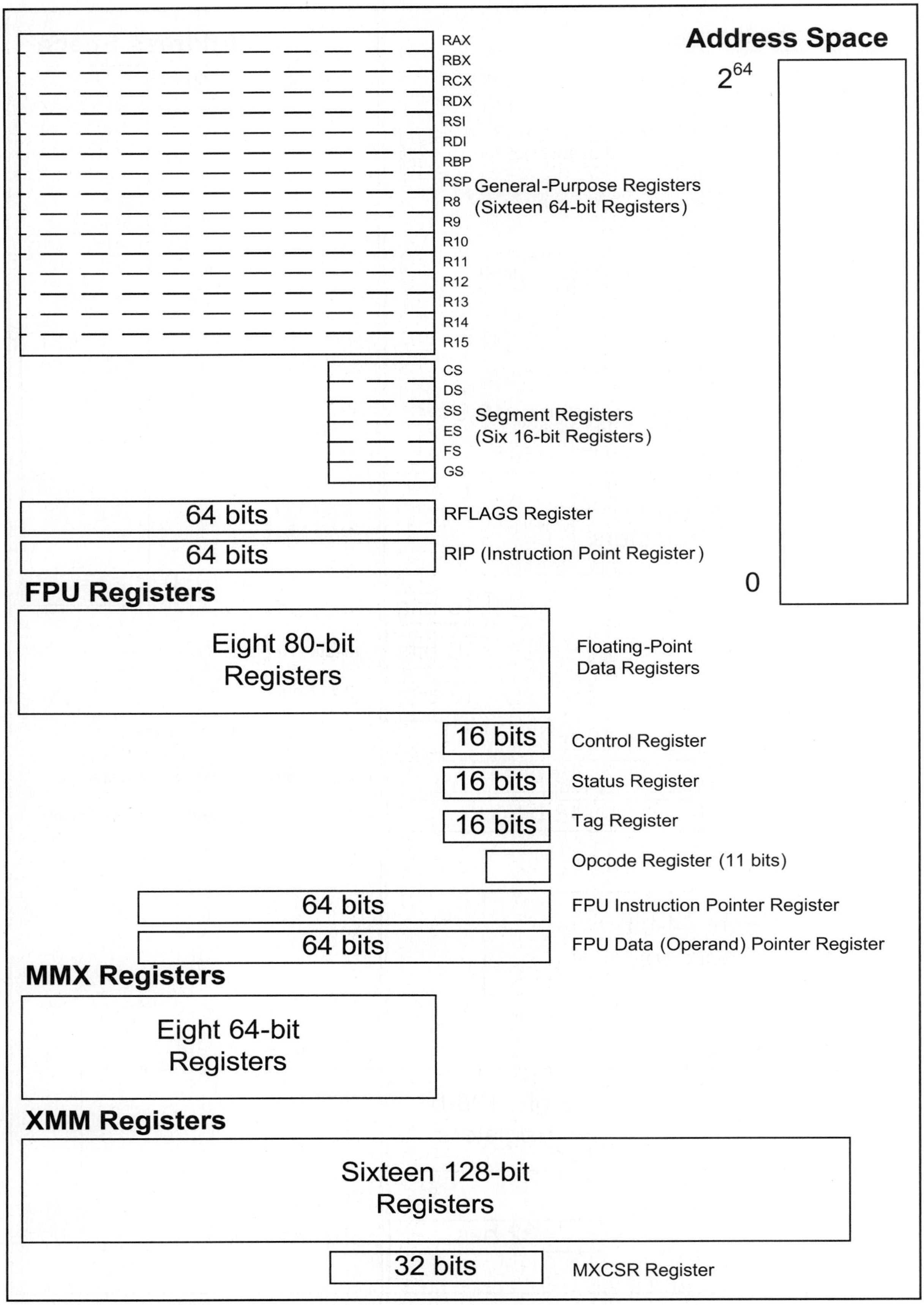

Figure 24-8. 64-Bit Registers in 64-Bit x86 Processors

x86 64-bit processors and Vista

Both Intel and AMD have been producing 64-bit x86 processors since the early 2000s. Microsoft used some aspects of the 64-bit architecture in Windows XP and called it XP x64. It was only after the introduction of Vista that Microsoft took full advantage of the 64-bit architecture. But due to legacy software and the need for the transition to the new 64-bit systems, there are two versions of Vista: Vista and Vista x64. While Vista will run on any x86 32-bit architecture, Vista x64 requires a 64-bit x86 processor and at least 1G byte of DRAM on the motherboard. In Vista-32 and XP, the operating system divided 4GB of physical memory space into two sections, and allocated one half for user applications and the other half for the kernel, drivers, video, and other needs of the OS. Vista x64 raised this memory from 4G bytes to 16 terabytes (TB) (2^{44}), and uses 8 terabytes for itself and the other 8 terabytes for user applications. Of course, it is not practical to have 8 TB of DRAM on the motherboard at this time. For that reason, Vista x64 limits the memory to 128 GB DRAM for a single processor.

Moore's Law

In the mid 1960s, Intel cofounder Gordon Moore made the following astounding prediction: "The number of transistors that would be incorporated on a silicon die would double every 18 months for the next several years." Examining Tables 24-1 and 24-2 shows how this prediction has come true.

Table 24-2: Some 64-Bit Processors in x86 Family

Product	Year	Transistors	Maximum Ext. Memory
Xeon	2004	125 M	64 GB
Xeon Processor MP	2005	675 M	1024 GB (1 TB)
Pentium 4	2005	164 M	64 GB
Pentium 840	2005	230 M	64 GB
Dual-Core Xeon	2005	321 M	64 GB
Pentium 4 672	2005	164 M	64 GB
Pentium 955	2006	376 M	64 GB
Core 2 X6800	2006	291 M	64 GB
Xeon 5160	2006	291 M	64 GB
Xeon 7140	2006	1.3 B	64 GB
Core 2 QX6700	2006	582 M	64 GB
Quad-Core Xeon 5355	2006	582 M	64 GB

Note: For the registers associated with these 64-bit processors see Figure 24-6.

Review Questions

1. Name the two modes of operation in 64-bit x86 processors.
2. True or false. All 64-bit x86 processors have L3 cache.
3. True or false. Both the compatibility mode and 64-bit mode can use 64-bit registers.
4. The size of external memory in compatibility mode is limited to______ bytes.
5. How many 64-bit regsiters do we have in 64-bit x86 processors?

PROBLEMS

SECTION 24.1: x86 PENTIUM EVOLUTION

1. True or false. L1 caches for code and data are separate.
2. True or false. Code and data use the same L2 cache.
3. What is the size of L1 cache in the Pentium III?
4. What is the size of L2 cache in the Pentium III?
5. What is the size of L1 cache in the Pentium IV (2006 version)?
6. What is the size of L2 cache in the Pentium IV (2006 version)?

7. What is the size of L1 cache in the Pentium Duo (2006 version)?
8. What is the size of L2 cache in the Pentium Duo (2006 version)?
9. What are differences between L1 and L2 caches?
10. What is L3 cache?
11. What is the size of L3 cache in the Pentium IV (2006 version)?
12. Compare the speed of the L1–L3 caches in terms of the CPU speed.
13. True or false. All Pentiums have the hyper-threading feature.
14. True or false. All Pentiums have the multicore feature.
15. Explain the differences between multithreading and multitasking.
16. Explain the differences between multicore and multiprocessor.
17. Explain the difference between MMX and XMM.
18. True or false. The latest MMX supports both single- and double-precision floating-point data.
19. True or false. The latest x86 Pentium uses out-of-order execution.
20. True or false. The latest x86 Pentium uses branch prediction.

SECTION 24.2: 64-BIT PROCESSORS AND VISTA FOR x86

21. True or false. The 32-bit applications written for the 32-bit 386 architecture cannot be run on 64-bit x86 processors.
22. Name all the 64-bit general-purpose registers of x86 64-bit processors.
23. Name all the 32-bit general-purpose registers of x86 64-bit processors.
24. Name all the 16-bit general-purpose registers of x86 64-bit processors.
25. Name all the 8-bit general-purpose registers of x86 64-bit processors.
26. What are the main modes of operation for 64-bit x86 processors?
27. What is the maximum memory accessible by the compatibility mode of the 64-bit x86?
28. What is the maximum memory accessible by the 64-bit mode of the x86?
29. What is the maximum memory accessible by the XP for 32-bit architecture?
30. What is the maximum memory accessible by the Vista x64? How much of that is implemented now?
31. Can we access the 64-bit registers in all modes?
32. The Vista x64 limits maximum memory to _______GB.

ANSWERS TO REVIEW QUESTIONS

SECTION 24.1: x86 PENTIUM EVOLUTION

1. False
2. False
3. False
4. In multicore, we have several CPUs inside a single chip accessing the same main memory space, while in multiprocessor, each CPU has its own memory space.
5. True

SECTION 24.2: 64-BIT PROCESSORS AND VISTA FOR x86

1. Compatibility mode and 64-bit mode
2. False
3. False
4. 4G
5. 16

CHAPTER 25

SYSTEM DESIGN ISSUES AND FAILURE ANALYSIS

OBJECTIVES

Upon completion of this chapter, you will be able to:

- **>> Contrast and compare MOS and bipolar transistors**
- **>> Evaluate logic families according to speed, power dissipation, noise immunity, input/output interface compatibility, and cost**
- **>> Trace the evolution of Intel x86 microprocessors in terms of IC technology**
- **>> Define IC fan-out and describe why connecting an output to too many inputs can cause false logic**
- **>> Describe capacitance derating and its effect on system design and the use of buffers to decrease its effect**
- **>> Discuss power in system design, including static and dynamic currents, power dissipation, and sleep mode for peripherals**
- **>> Define ground bounce and VCC bounce and describe methods that designers use to avoid the false signal generation that they may cause**
- **>> Define crosstalk and describe ways to avoid crosstalk in system design**
- **>> Define transmission line ringing and discuss ways to reduce its effect**
- **>> Define measures of system reliability: FIT and MTBF**

The invention of the transistor and the subsequent advent of integrated circuit (IC) technology is believed by many to be the start of the second industrial revolution. In this chapter we provide an overview of IC technology and interfacing. In addition, we look at the computer system as a whole and examine some general considerations in system design. In Section 25.1 we provide an overview of IC technology. IC interfacing and system design considerations are examined in Section 25.2. In Section 25.2 we also discuss failure analysis in systems.

SECTION 25.1: OVERVIEW OF IC TECHNOLOGY

In this section we provide an overview of IC technology and discuss recent developments in advanced logic families.

The transistor was invented in 1947 by three scientists at Bell Laboratory. In the 1950s, transistors replaced vacuum tubes in many electronics systems, including computers. It was not until in 1959 that the first integrated circuit was successfully fabricated and tested by Jack Kilby of Texas Instruments. Prior to the invention of the IC, the use of transistors, along with other discrete components such as capacitors and resistors, was common in computer design. Early transistors were made of germanium, which was later abandoned in favor of silicon. This was due to the fact that the slightest rise in temperature resulted in massive current flows in germanium-based transistors. In semiconductor terms, it is because the band gap of germanium is much smaller than that of silicon, resulting in a massive flow of electrons from the valence band to the conduction band when the temperature rises even slightly. By the late 1960s and early 1970s, the use of the silicon-based IC was widespread in mainframes and minicomputers. Transistors and ICs were based on P-type materials. Due to the fact that the speed of electrons is much higher (about two and a half times) than the speed of the holes, N-type devices replaced P-type devices. By the mid-1970s, NPN and NMOS transistors had replaced the slower PNP and PMOS transistors in every sector of the electronics industry, including in the design of microprocessors and computers. Since the early 1980s, CMOS (complementary MOS) has become the dominant method of IC design. Next we provide an overview of differences between MOS and bipolar transistors.

MOS vs. bipolar transistors

There are two type of transistors: bipolar and MOS (metal-oxide semiconductor). Both have three leads. In bipolar transistors, the three leads are referred to as the *emitter*, *base*, and *collector*, while in MOS transistors they are named *source*, *gate*, and *drain*. In bipolar, the carrier flows from the emitter to the collector and the base is used as a flow controller. In MOS, the carrier flows from the source to the drain and the gate is used as a flow controller. In NPN-type bipolar transistors, the electron carrier leaving the emitter must overcome two voltage barriers before it reaches the collector (see Figure 25-1). One is the N-P junction of the emitter-base and the other is the P-N junction of the base-collector. The voltage barrier of the base-collector is the most difficult one for the electrons to overcome (since it is reversed biased) and it causes the most power dissipation. This led to the design of the unipolar type transistor called *MOS*. In N-channel MOS transistors, the electrons leave the source reaching the drain without going through any voltage barrier. The absence of any voltage barrier in the path of the carrier is one reason why MOS dissipates much less power than bipolar transistors. The low power dissipation of MOS allows putting millions of transistors on a single IC chip. In today's million-transistor microprocessors and DRAM memory chips, the use of MOS technology is indispensable. Without the MOS transistor, the advent of desktop personal computers would not have been possible, at least not so soon. The use of bipolar transistors in both the mainframe and minicomputer of the 1960s and 1970s required expensive cooling systems and large rooms due to their bulkiness. MOS transistors do have one major drawback: They are slower than bipolar transistors. This is due partly to the gate capacitance of the MOS transistor. For MOS to be turned on, the input capacitor of the gate takes time to charge up to the turn-on (threshold) voltage, leading to a longer propagation delay.

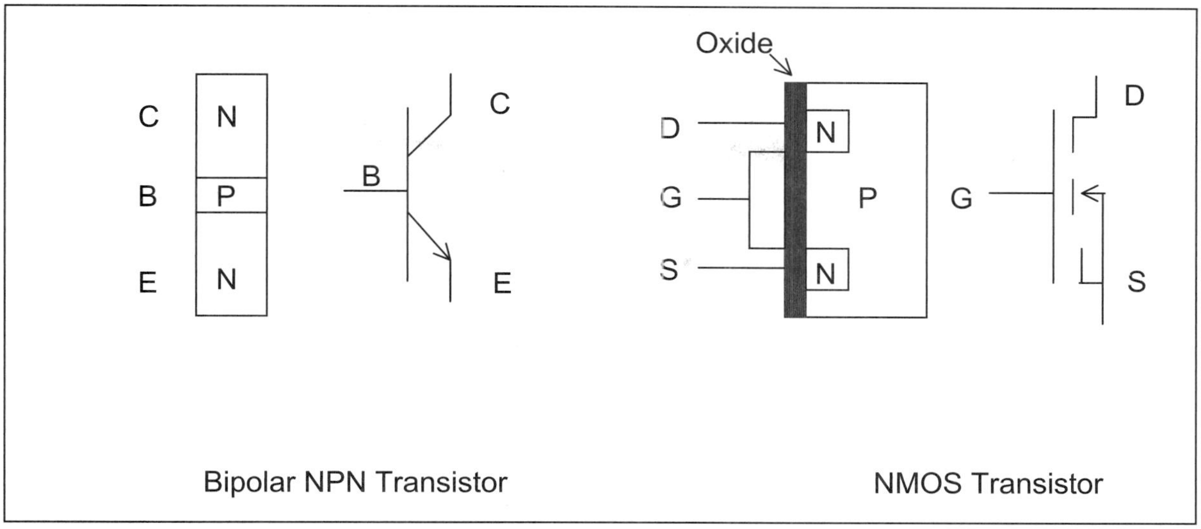

Figure 25-1. Bipolar vs. MOS Transistors

Overview of logic families

Logic families are judged according to (1) speed, (2) power dissipation, (3) noise immunity, (4) input/output interface compatibility, and (5) cost. Desirable qualities are high speed, low power dissipation, and high noise immunity (since it prevents the occurrence of false logic signals during switching transition). In interfacing logic families, the more inputs that can be driven by a single output, the better. This means that high-driving-capability outputs are desired. This plus the fact that the input and output voltage levels of MOS and bipolar transistors are not compatible means that one must be concerned with the ability of one logic family driving the other one. In terms of the cost of a given logic family, it is high during the early years of its introduction and prices decline as production and use rise.

The case of inverters

As an example of logic gates, we look at a simple inverter. In a one-transistor inverter, while the transistor plays the role of a switch, R is the pull-up resistor. See Figure 25-2. However, for this inverter to work effectively in digital circuits, the R value must be high when the transistor is "on" to limit the current flow from VCC to ground in order to have low power dissipation (P = VI, where V = 5 V). In other words, the lower the I, the lower the power dissipation. On the other hand, when the transistor is "off", R must be a small value to limit the voltage drop across R, thereby making sure that VOUT is close to VCC. This is a contradictory demand on R. This is one reason that logic gate designers use active components (transistors) instead of passive components (resistors) to implement the pull-up resistor R.

The case of a TTL inverter with totem pole output is shown in Figure 25-3. In Figure 25-3, Q3 plays the role of a pull-up resistor.

CMOS inverter

In the case of CMOS-based logic gates, PMOS and NMOS are used to construct a CMOS (complementary MOS) inverter as shown in Figure 25-4. In CMOS inverters, when the PMOS transistor is off, it provides a very high impedance path, making leakage current almost zero (about 10 nA); when the PMOS is on, it provides a low resistance on the path of VDD to load. Since the speed of the hole is slower than that of the electron, the PMOS transistor is wider to compensate for this disparity; therefore, PMOS transistors take more space than NMOS.

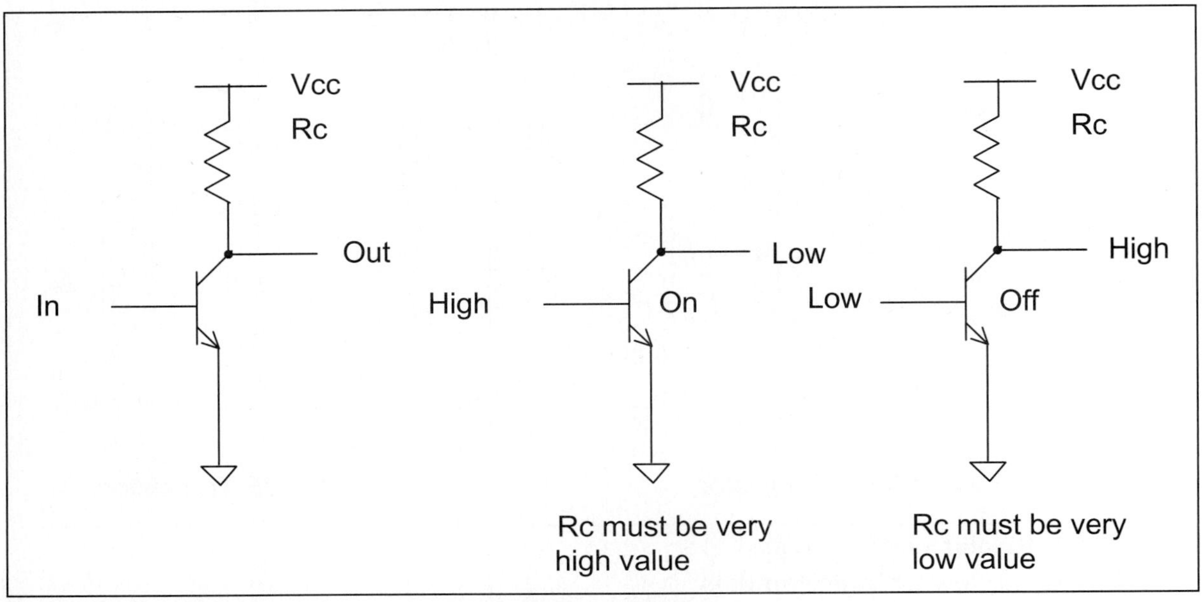

Figure 25-2. One-Transistor Inverter with Pull-up Resistor

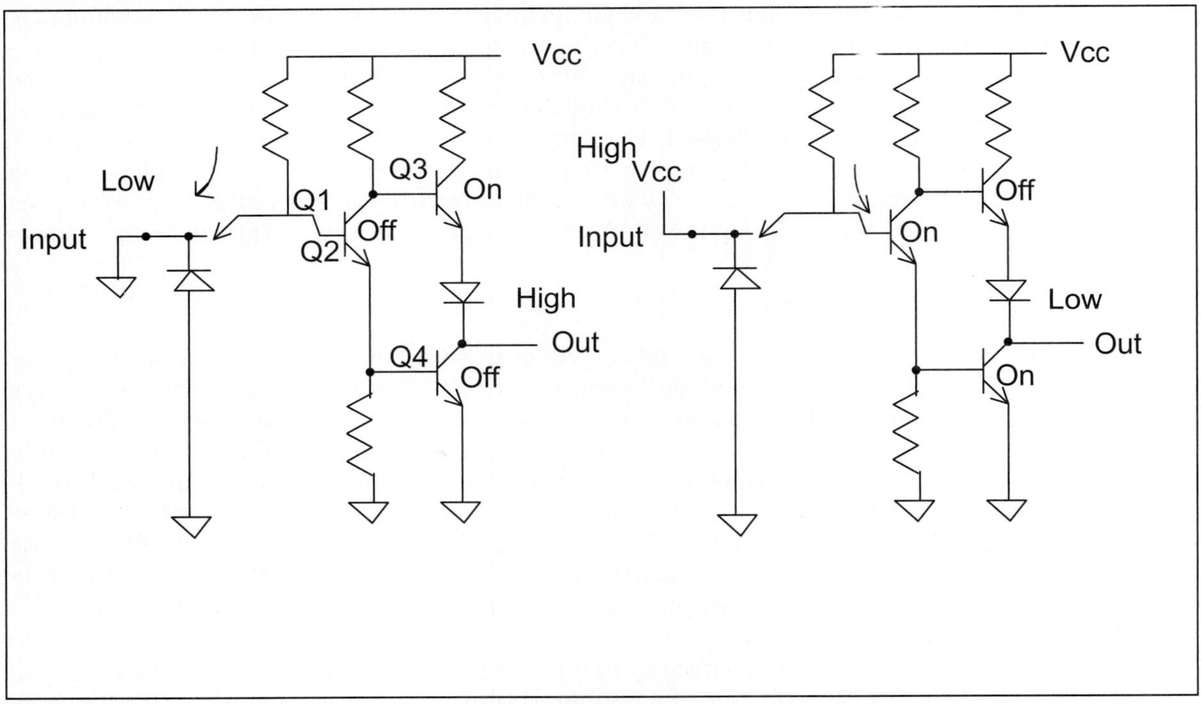

Figure 25-3. TTL Inverter with Totem-Pole Output

Input, output characteristics of some logic families

In 1968 the first logic family made of bipolar transistors was marketed. It was commonly referred to as the standard TTL (transistor-transistor logic) family. The first MOS-based logic family, the CD4000/74C series, was marketed in 1970. The addition of the Schottky diode to the base-collector of bipolar transistors in the early 1970s gave rise to the S family. The Schottky diode shortens the propagation delay of the TTL family by preventing the collector from going into what is called *deep saturation*. Table 25-1 lists major characteristics of some logic families. In Table 25-1, note that as the CMOS circuit's operating frequency rises, the power dissipation also increases. This is not the case for bipolar-based TTL.

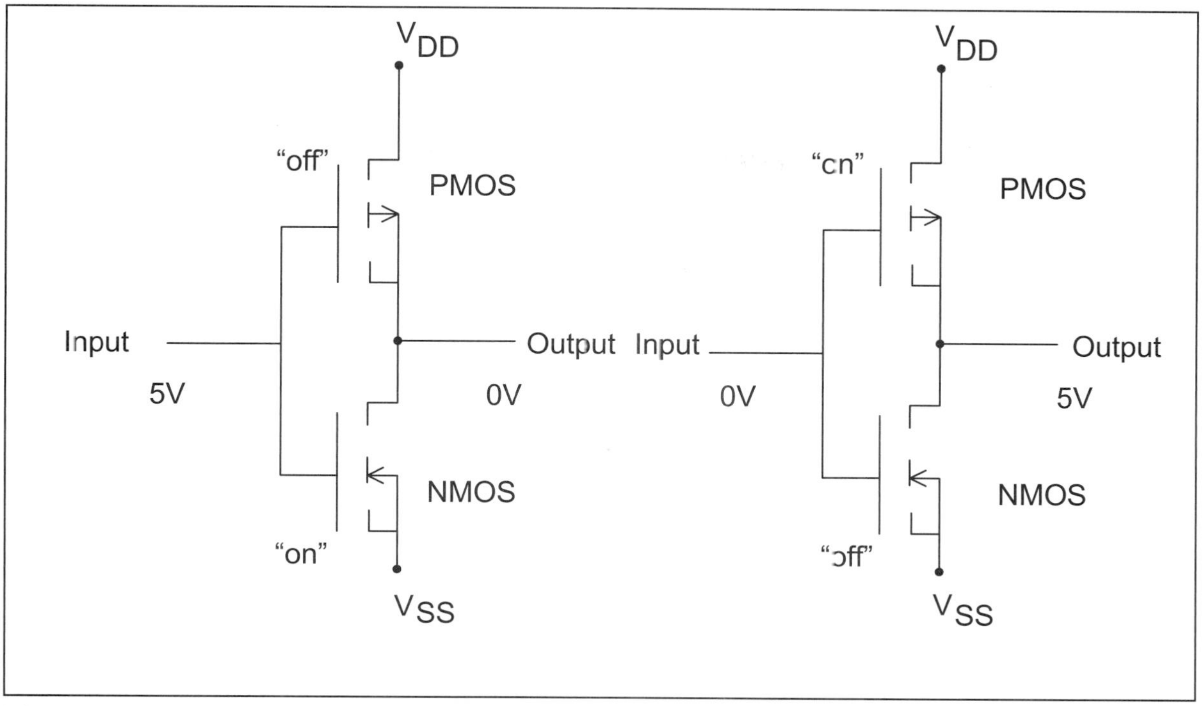

Figure 25-4. CMOS Inverter

Table 25-1: Characteristics of Some Logic Families

Characteristic	STD TTL	LSTTL	ALSTTL	HCMOS
V_{CC}	5 V	5 V	5 V	5 V
V_{IH}	2.0 V	2.0 V	2.0 V	3.15 V
V_{IL}	0.8 V	0.8 V	0.8 V	1.1 V
V_{OH}	2.4 V	2.7 V	2.7 V	3.7 V
V_{OL}	0.4 V	0.5 V	0.4 V	0.4 V
I_{IL}	–1.6 mA	–0.36 mA	–0.2 mA	–1 μA
I_{IH}	40 μA	20 μA	20 μA	1 μA
I_{OL}	16 mA	8 mA	4 mA	4 mA
I_{OH}	–400 μA	–400 μA	–400 μA	4 mA
Propagation delay	10 ns	9.5 ns	4 ns	9 ns
Static power dissipation (f = 0)	10 mW	2 mW	1 mW	0.0025 nW
Dynamic power dissipation at f = 100 kHz	10 mW	2 mW	1 mW	0.17 mW

History of logic families

Early logic families and microprocessors required both positive and negative power voltages. In the mid-1970s, 5V VCC became standard. For example, Intel's 4004, 8008, and 8080 all used negative and positive voltages for the power supply. In the late 1970s, advances in IC technology allowed combining the speed and drive of the S family with the lower power of LS to form a new logic family called FAST (Fairchild Advanced

Table 25-2: Logic Family Overview

Product	Year Introduced	Speed (ns)	Static Supply Current (mA)	High/Low Family Drive (mA)
Std TTL	1968	40	30	–2/32
CD4K/74C	1970	70	0.3	–0.48/6.4
LS/S	1971	18	54	–15/24
HC/HCT	1977	25	0.08	–6/–6
FAST	1978	6.5	90	–15/64
AS	1980	6.2	90	–15/64
ALS	1980	10	27	–15/64
AC/ACT	1985	10	0.08	–24/24
FCT	1986	6.5	1.5	–15/64

Reprinted by permission of Electronic Design Magazine, c. 1991.

Schottky TTL). In 1985, AC/ACT (Advanced CMOS Technology), a much higher speed version of HCMOS, was introduced. With the introduction of FCT (Fast CMOS Technology) in 1986, at last the speed gap between CMOS and TTL was closed. Since FCT is the CMOS version of FAST, it has the low power consumption of CMOS but the speed is comparable with TTL. Table 25-2 provides an overview of logic families up to FCT.

Recent advances in logic families

As the speed of high-performance microprocessors such as the 386 and 486 reached 25 MHz, it shortened the CPU's cycle time, leaving less time for the path delay. Designers normally allocate no more than 25% of a CPU's cycle time budget to path delay. Following this rule means that there must be a corresponding decline in the propagation delay of logic families used in the address and data path as the system frequency is increased. In recent years, many semiconductor manufacturers have responded to this need by providing logic families that have high speed, low noise, and high drive. Table 25-3 provides the characteristics of high-performance logic families introduced in recent years. ACQ/ACTQ are the second-generation advanced CMOS (ACMOS) with much lower noise. While ACQ has the CMOS input level, ACQT is equipped with TTL-level input. The FCTx and FCTx-T are second-generation FCT with much higher speed. The x in the FCTx and FCTx-T refers to various speed grades, such as A, B, and C, where the A designation means low speed and C means high speed. For designers who are well versed in using the FAST logic family, the use of FASTr is an ideal choice since it is faster than FAST, has higher driving capability (I_{OL}, I_{OH}), and produces much lower noise than FAST. At the time of this writing, next to ECL and gallium arsenide logic gates, FASTr is the fastest logic family in the market (with the 5V VCC), but the power consumption is high relative to other logic families, as shown in Table 25-3. In recent years, a 3.3V VCC with higher speed and lower power consumption has become standard. The combining of

Table 25-3: Advanced Logic General Characteristics

Family	Year	Number Suppliers	Tech Base	I/O Level	Speed (ns)	Static Current	I_{OH}/I_{OL}
ACQ	1989	2	CMOS	CMOS/CMOS	6.0	80 μA	–24/24 mA
ACTQ	1989	2	CMOS	TTL/CMOS	7.5	80 μA	–24/24 mA
FCTx	1987	3	CMOS	TTL/CMOS	4.1–4.8	1.5 mA	–15/64 mA
FCTxT	1990	2	CMOS	TTL/TTL	4.1–4.8	1.5 mA	–15/64 mA
FASTr	1990	1	Bipolar	TTL/TTL	3.9	50 mA	–15/64 mA
BCT	1987	2	BICMOS	TTL/TTL	5.5	10 mA	–15/64 mA

Reprinted by permission of Electronic Design Magazine, c. 1991.

high-speed bipolar TTL and the low power consumption of CMOS has given birth to what is called *BICMOS*. Although BICMOS seems to be the future trend in IC design, at this time it is expensive due to the extra steps required in BICMOS IC fabrication, but in some cases there is no other choice. For example, Intel's Pentium microprocessor, a BICMOS product, had to use high-speed bipolar transistors to speed up some of the internal functions in order to keep up with RISC processor performance Table 25-3 provides advanced logic characteristics. Table 25-4 shows logic families used in systems with different speeds. The x is for the different speeds where A, B, and C are used for designation. A is the slowest one while C is the fastest one. The above data is for the 'LS244 buffer.

Table 25-4: Importance of Speed

System Clock Speed (MHz)	Clock Period (ns)	Predominant Logic for Path
2–10	100–500	HC, LS
10–30	33–100	ALS, AS, FAST, FACT
30–66	15–33	FASTr, BCT, FCTA

Evolution of IC technology in Intel's x86 microprocessors

Since 1971, when Intel introduced the first microprocessor, the 4004, until the introduction of the Pentium microprocessor, IC technology has gone through some massive changes. The early processors (4004 and 8008) used PMOS. The 8080, 8085, 8088, 8086, and 80286 all used NMOS when first introduced. In recent years, CMOS versions of the 8088, 8086, and 286 have been introduced for power-efficient systems. Currently, CMOS is the universal technology in the design of microprocessors. Only BICMOS could allow designers to put over 3 million transistors on a single chip, make it work at 1 GHz, and consume around 10 watts of power. There has been a steady decline in the transistor's dimensions throughout the 1980s, 1990s, and 2000s. The design rule, the thickness of the lines inside the IC, has come down from a few microns to a fraction of a micron during this time. See Table 25-5.

Table 25-5: Intel Microprocessor Evolution

Microprocessor	Year	IC Tech	Line Thickness (μm)	Power Supply (V)	Number of Transistors
8086	1978	NMOS	3.0	5	29,000
80286	1982	NMOS	1.5	5	134,000
80386	1985	CMOS	1.5	5	275,000
80486	1989	CMOS	1.0	5	1.2 million
Pentium I	1993	BICMOS	0.8	5	3.1 million
Pentium II	1997	BICMOS	0.25	3.3	7.5 million
Pentium III	1999	BICMOS	0.18	3.3	9.5 million
Pentium IV	2000	BICMOS	0.18	3.3	42 million

The early microprocessors used power supplies with negative (–) and positive (+) voltages. For example, the 4004 used –10 and +5 V. The 8008 used –9 and +5 V, and the 8080 used –5, +5, +12 V. Since the introduction of the 8085, the use of a +5 V power supply has become standard in all microprocessors. To reduce power consumption, 3.3V VCC is being embraced by many designers. The lowering of VCC to 3.3 V has two major advantages: (1) it lowers the power consumption, resulting in prolonged battery life in systems such as a laptop PC or hand-held personal digital assistant, and (2) it allows a further reduction of line size (design rule) to submicron dimensions. This reduction results in putting more transistors in a given die size. The decline in the line size has reached 0.1 micron, while the transistor density per chip is over 300 million.

Review Questions

1. State the main advantages of MOS and bipolar transistors.
2. True or false. In logic families, the higher the noise margin, the better.
3. True or false. Generally, high-speed logic consumes more power.
4. Power dissipation increases linearly with the increase in frequency in ___________ (CMOS, TTL).
5. In a CMOS inverter, indicate which transistor is on when the input is high.
6. For system frequencies of 10–30 MHz, which logic families are used for the address and data path?

SECTION 25.2: IC INTERFACING AND SYSTEM DESIGN ISSUES

There are several issues to be considered in designing a microprocessor-based system. They are IC fan-out, capacitance derating, ground bounce, VCC bounce, crosstalk, transmission lines, power dissipation, and chip failure analysis. This section provides an overview of these design issues in order to provide a sampling of what is involved in high-performance system design.

IC fan-out

In IC interfacing, fan-out/fan-in is a major issue. How many inputs can an output signal drive? This question must be addressed for both logic "0" and logic "1" outputs. Fan-out for low and fan-out for high are as follows:

$$\text{fan-out (of low)} = \frac{I_{OL}}{I_{IL}} \qquad \text{fan-out (of high)} = \frac{I_{OH}}{I_{IH}}$$

Of the above two values the lower number is used to ensure the proper noise margin. Figure 25-5 shows the sinking and sourcing of current when ICs are connected.

In Figure 25-5, as the number of inputs connected to the output increases, I_{OL} rises, which causes V_{OL} to rise. If this continues, the rise of V_{OL} makes the noise margin smaller, and this results in the occurrence of false logic due to the slightest noise.

In designing the system, very often an output is connected to various kinds of inputs. See Examples 25-1 and 25-2.

The total I_{IL} and I_{IH} requirement of all the loads on a given output must be less than the driver's maximum I_{OL} and I_{OH}. This is shown in Example 25-3.

In cases such as Example 25-3 where the receiver current requirements exceed the drivers' capability, we must use a buffer (booster), such as the 74xx245 and 74xx244. The 74xx245 is used for bidirectional and the 74xx244 for unidirectional signals. See current 74LS244 and 74LS245 characteristics in Table 25-6.

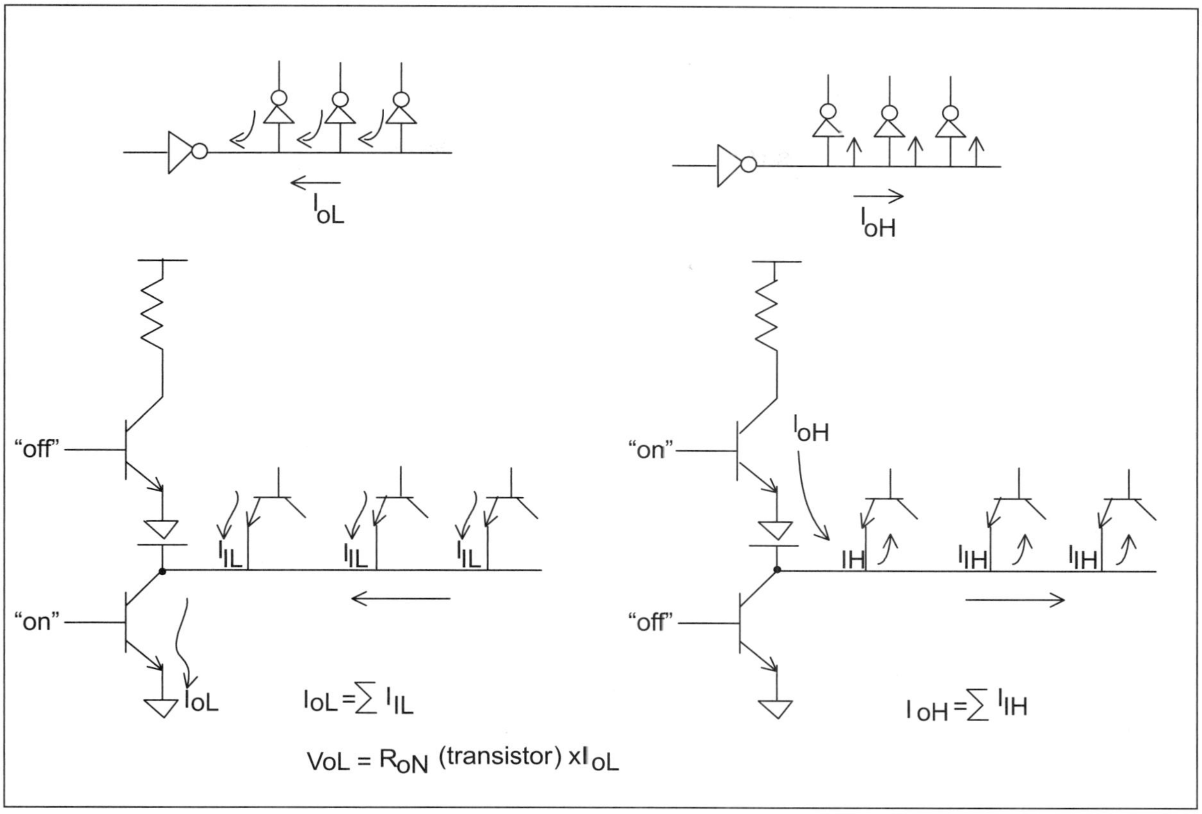

Figure 25-5. Current Sinking and Sourcing in TTL

Example 25-1

Find how many unit loads (UL) can be driven by the output of the LS logic family.

Solution:

The unit load is defined as I_{IL} = 1.6 mA and I_{IH} = 40 A. Table 25-1 shows I_{OH} = 400 A and I_{OL} = 8 mA for the LS family. Therefore, we have

fan-out (low) = I_{OL}/I_{IL} = 8 mA / 1.6 mA = 5

fan-out (high) = I_{OH}/I_{IH} = 400 μA / 40 μA = 10

This means that the fan-out is 5. In other words, the LS output must not be connected to more than 5 inputs with unit load characteristics.

Table 25-6: Electrical Specifications for Buffers

Buffer	I_{OH} (mA)	I_{OL} (mA)	I_{IH} (μA)	I_{IL} (mA)
74LS244	3	12	20	0.2
74LS254	3	12	20	0.2

Note: V_{OL} = 0.4 V and V_{OH} = 2.4 V are assumed.

Example 25-2

An address pin needs to drive 5 standard TTL loads in addition to 10 CMOS inputs of DRAM chips. Calculate the minimum current to drive these inputs for both logic "0" and "1".

Solution:

The standard load for TTL is I_{IH} = 40 A and I_{IL} = 1.6 mA, and for CMOS, $I_{IL} = I_{IH}$ = 10 A.
minimum current for "0" = total of all I_{IL} = 5 × 1.6 mA + 10 × 10 A = 8.1 mA
minimum current for "1" = total of all I_{IH} = 5 × 40 A + 10 × 10 A = 300 A

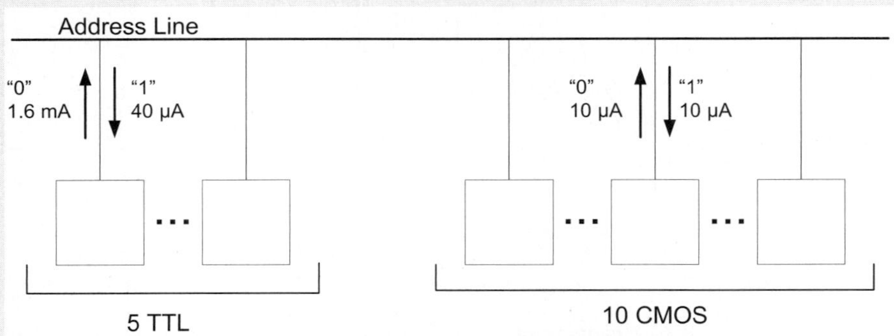

Example 25-3

Assume that the microprocessor address pin in Example 25-2 has specifications I_{OH} = 400 A and I_{OL} = 2 mA. Do the input and output current needs match?

Solution:

For a high output state, there is no problem since $I_{OH} > I_{IH}$. However, the number of inputs exceeds the limit for I_{OL} since an I_{IL} of 8.1 mA is much larger than the maximum I_{OL} allowed by the microprocessor.

Capacitance derating

Next we study what is called *capacitance derating* and its impact in system design. A pin of an IC has an input capacitance of 5 to 7 pF.This means that a single output that drives many inputs sees a large capacitance load since the inputs are in parallel and therefore added together. Look at the following equations.

$$Q = CV \tag{25-1}$$

$$\frac{Q}{T} = \frac{CV}{T} \tag{25-2}$$

$$F = \frac{1}{T} \tag{25-3}$$

$$I = CVF \tag{25-4}$$

In Equation (25-4), I is the driving capability of the output pin, C is C_{IN} as seen by the output, and V is the voltage. The equation indicates that as the number of C_{IN} loads goes up, there must be a corresponding increase in IO, the driving capability of the output. In other words, outputs with high values of I_{OL} and I_{OH} are desirable. Although recently there have been some logic families with I_{OL} = 64 mA and I_{OH} = 15 mA, their power consumption is high. Equation (25-4) indicates that if I = constant, as C goes up, F must come down, resulting in lower speed. The most widely accepted solution is the use of a large number of drivers to reduce the load capacitance seen by a given output. Assume that we have a single address bus line driving 16 banks of 32-bit-wide memory. Each bank has 4 chips of 64K × 8 organization, which results in 16 × 4 = 64 memory chips, or 16 × 64K × 32 = 32M bytes of SRAM. Depending on how many 244s are used to drive the memory addresses, the delay due to the address path varies substantially. To understand this we examine four cases.

Case 1: Two 244 drivers

This option uses two 244 drivers, one for A0–A7 and one for A8–A15. An output of the 244 drives 16 banks of memory, each with 4 inputs. Assuming that each memory input has 5 pF capacitance, this results in a total of 4 ×16 × 5 = 320 pF capacitance load seen by the 244 output. However, the 244 output can handle no more than 50 pF. As a result, the delay due to this extra capacitance must be added to the address path delay. For each 50 to 100 pF of capacitance, an extra 3 ns delay is added to the address path delay. In our calculation, we use 3 ns for each 100 pF of capacitance. Figure 25-6 shows driving memory inputs by two 244 chips. See Example 25-4.

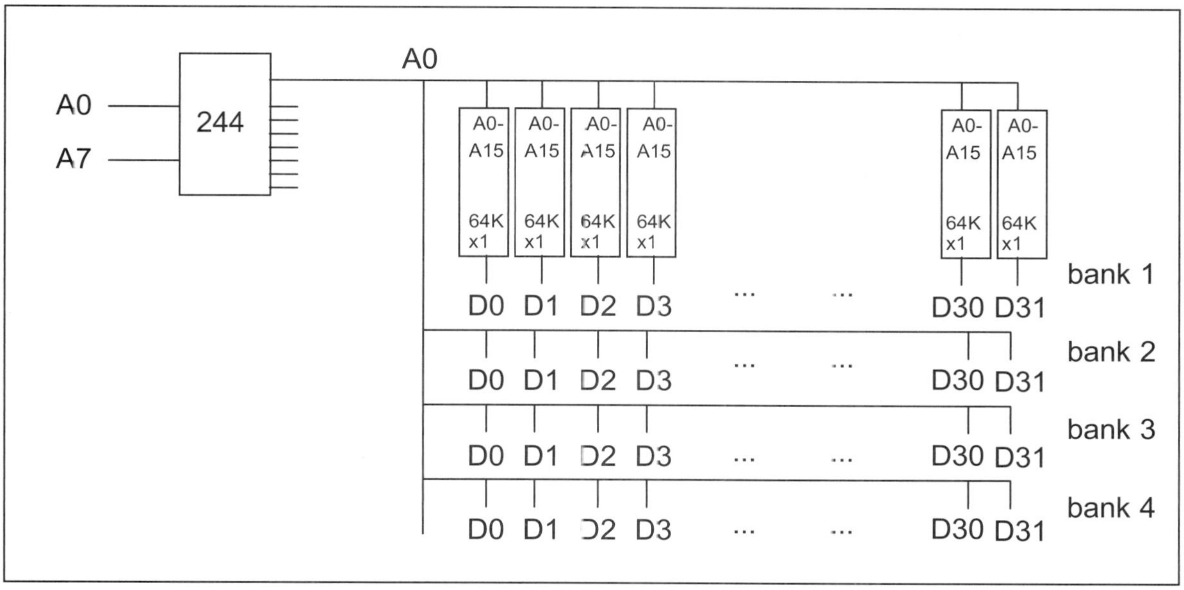

Figure 25-6. Case 1: Two 244 Address Drivers (the second 244 for A8–A15 is not shown)

Example 25-4

Calculate the following for Figure 25-6, assuming a memory access time of 25 ns and a propagation delay of 10 ns for the 244.
(a) delay due to capacitance derating on the address path
(b) the total address path delay for case 1

Solution:

(a) Of the 320 pF capacitance seen by the 244, only 50 pF is taken care of; the rest, which is 270 (320 – 50 = 270), causes a delay. Since there are 3 ns for each extra 100 pF, we have the following delay due to capacitance derating, (270/100) × 3 ns = 8.1 ns.
(b) Address path delay = 244 buffer propagation delay + capacitance derating delay + memory access time = 10 ns + 8.1 ns + 25 ns = 43.1 ns.

Case 2: Doubling the number of 244 buffers

Doubling the number of 244 buffers will reduce the address path delay. A single 244 drives only 8 banks, or a total of 32 inputs, since there are 4 inputs in each bank. As a result, a 244 output will see a capacitance load of $32 \times 5 = 160$ pF. In this case, we use only four 244 buffer chips, as shown in Figure 25-7 and Example 25-5.

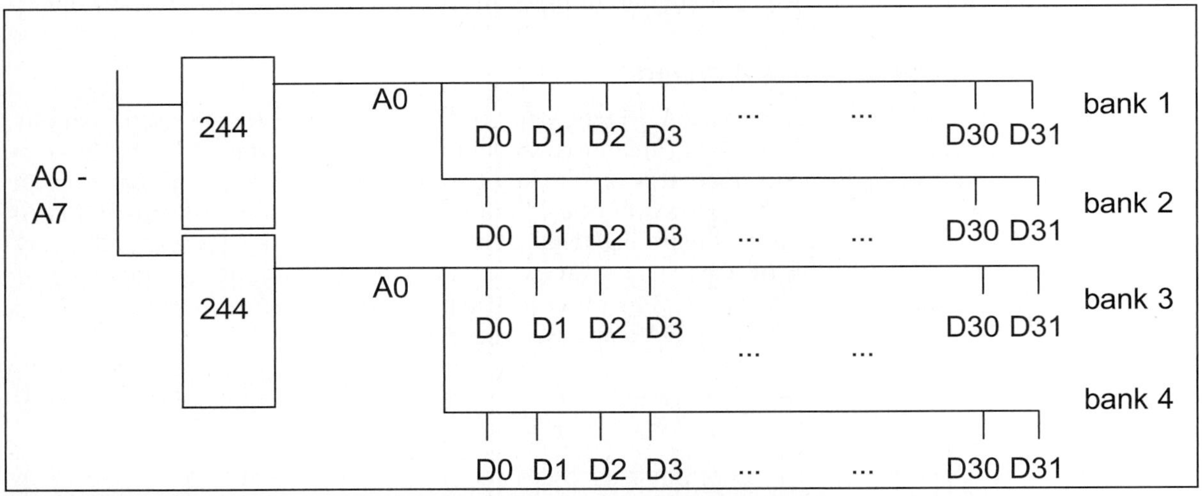

Figure 25-7. Case 2: Four 244 Address Drivers (the two 244s for A8–A15 are not shown)

Example 25-5

Calculate (a) delay due to capacitance derating on the address path, and (b) total address path delay for case 2. Assume a memory access time of 25 ns and a propagation delay of 10 ns for the 244.

Solution:

(a) Of the 160 pF capacitance seen by the 244, only 50 pF is taken care of; the rest, which is 110 pF, causes a delay. Since there are 3 ns for each extra 100 pF, we have $(110/100) \times 3$ ns = 3.1 ns delay due to capacitance derating.
(b) The address path delay = 244 buffer propagation delay + capacitance derating delay + memory access time = 10 ns + 3.1 ns + 25 ns = 28.1 ns.

Case 3: Doubling again

In this case, we double the number of 244 buffers again, so that an output of the 244 drives four banks, each with 4 inputs. This results in a total capacitance load of $4 \times 4 \times 5 = 80$ pF. Only 50 pF of it is taken care of by the 244, leaving 30 pF, causing a delay. See Figure 25-8.

Examining cases 1 through 3 shows that for high-speed system design we must accept a higher cost due to extra parts and higher power consumption.

Power dissipation considerations

Power dissipation of a system is a major concern of system designers, especially for laptop and hand-held systems. Although power dissipation is a function of the total current consumption of all components of a system, the impact of VCC is much more pronounced, as shown next. Earlier we showed in Equation (26-4) that I = CFV. Substituting this in equation P = VI yields the following:

$$F = VI = CFV^2 \qquad (25\text{-}5)$$

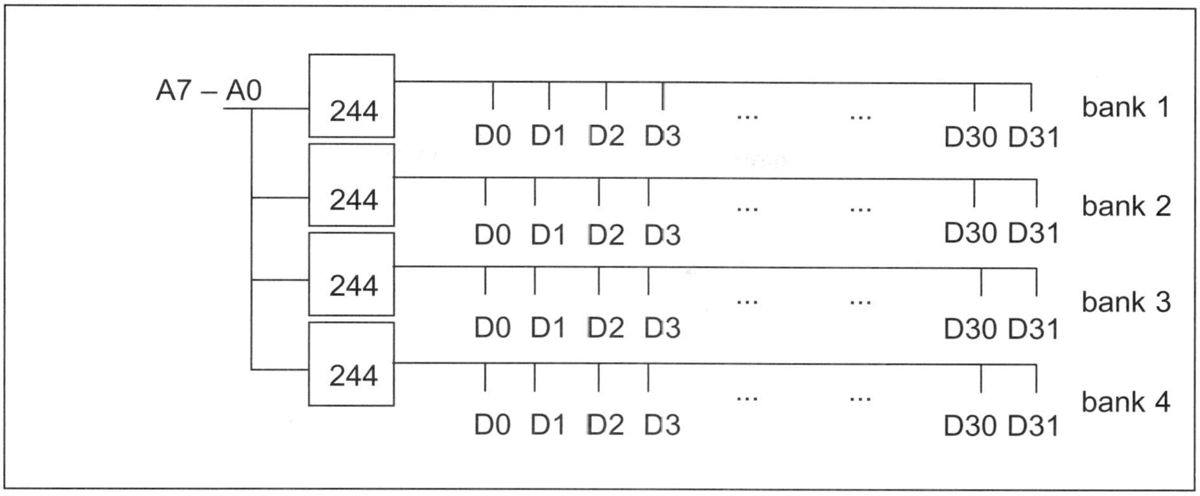

Figure 25-8. Case 3: A Single 244 Address Driver for Each Bank (A8–A15 are not shown)

Example 25-6

Prove that a 3.3 V system consumes 56% less power than a system with a 5 V power supply.

Solution:

Since P = VI, by substituting I = V/R, we have $P = V^2/R$. Assuming that R = 1, we have $P\ (3.3)^2 = 10.89$ W and $P = (5)^2 = 25$ W. This results in using 14.11 W less (25 – 10.89 = 14.11), which means a 56% power saving (14.11 W/25 W × 100 = 56%).

In Equation (25-5), the effects of frequency and VCC voltage should be noted. While the power dissipation goes up linearly with frequency, the impact of the power supply voltage is much more pronounced (squared). See Example 25-6.

Dynamic and static currents

There are two major types of currents flowing through an IC: dynamic and static. A dynamic current is a function of the frequency under which the component is working, as seen in Equation (25-4). This means that as the frequency goes up, the dynamic current and power dissipation go up. The static current, also called *dc*, is the current consumption of the component when it is inactive (not selected).

Power-down option

The popularity of laptop PCs has led microprocessor designers to make an all-out effort to conserve battery power. Today processors have what is called *system management mode* (*SMM*), which reduces energy consumption by turning off peripherals or the entire system when not in use. According to Intel, SMM can put the entire system, including the monitor, into sleep mode during periods of inactivity, thereby reducing "power from 250 watts to less than 30 watts." The effects on the 3.3 V power supply alone translate into a power savings of up to 56% over systems with a 5 V power supply, as was shown in Example 25-6.

Ground bounce

One of the major issues that designers of high-frequency systems must grapple with is ground bounce. Before we define ground bounce, we will discuss lead inductance of IC pins. There is a certain amount of capacitance, resistance, and inductance associated with each pin of the IC. The size of these elements varies depending on many factors such as length, area, and so on. Figure 25-9 shows the lead inductance and capacitance of the 24 pins of a DIP IC.

Pin	Self-inductance	Capacitance
1	15.10 nH	1.86 pF
2	12.20 nH	1.70 pF
3	9.54 nH	1.29 pF
4	7.44 nH	0.95 pF
5	5.31 nH	0.61 pF
6	3.73 nH	0.43 pF
7	3.41 nH	0.43 pF
8	4.66 nH	0.61 pF
9	6.95 nH	0.95 pF
10	8.96 nH	1.29 pF
11	11.70 nH	1.70 pF
12	14.50 nH	1.86 pF
13	14.50 nH	1.86 pF
14	11.70 nH	1.70 pF
15	8.96 nH	1.29 pF
16	6.95 nH	0.95 pF
17	4.66 nH	0.61 pF
18	3.41 nH	0.43 pF
19	3.73 nH	0.43 pF
20	5.31 nH	0.61 pF
21	7.44 nH	0.95 pF
22	9.54 nH	1.29 pF
23	12.20 nH	1.70 pF
24	15.10 nH	1.86 pF

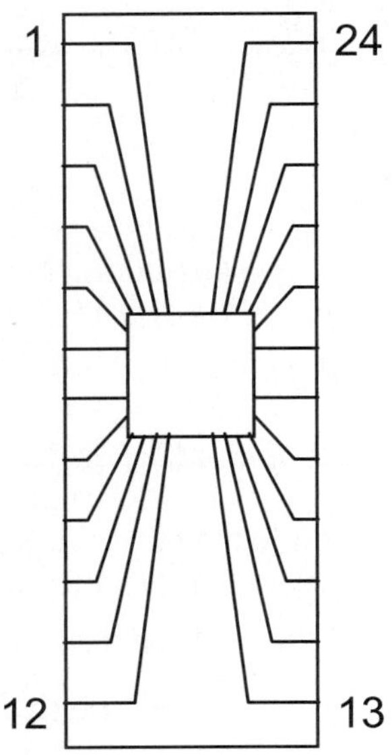

Figure 25-9. Inductance and Capacitance of 24-pin DIP

Reprinted by Permission of Electronic Design Magazine, c. 1992.

The inductance of the pins is commonly referred to as *self-inductance* since there is also what is called *mutual inductance*, as we will show below. Of the three components of capacitance, resistance, and inductance, self-inductance is the one that causes the most problems in high-frequency system design since it can result in ground bounce. Ground bounce is caused when a massive amount of current flows through the ground pin when multiple outputs change from high to low all at the same time. The voltage relation to the inductance of the ground lead follows:

$$V = L \frac{d_i}{d_t} \tag{25-6}$$

As we increase the system frequency, the rate of dynamic current, di/dt, is also increased, resulting in an increase in the inductance voltage L (di/dt) of the ground pin. Since the low state (ground) has a small noise margin, any extra voltage due to the inductance voltage can cause a false signal. To reduce the effect of ground bounce, the following steps must be taken where possible.

1. The VCC and ground pins of the chip must be located in the middle rather than at the opposite ends of the IC chip (the 14-pin TTL logic IC uses pins 14 and 7 for ground and VCC). This is exactly what we see in high-performance logic gates such as Texas Instrument's advanced logic AC11000 and ACT11000 families. For example, the ACT11013 is a 14-pin DIP chip where pins 4 and 11 are used for the ground and VCC instead of 7 and 14 as in the TTL. We can also use the SOIC packages instead of DIP. The self-inductance of the leads is shown in Table 25-7.
2. Use logics with a minimum number of outputs. For example, a 4-output is preferable to an 8-output. This explains why many designers of high-performance systems avoid using memory chips or the drivers and buffers of 16- or 32-bit-wide outputs since all the outputs switching at the same time will cause a massive flow of current in the ground pin, and hence cause ground bounce (see Figure 25-10).
3. Use as many pins for the ground and VCC as possible to reduce the lead length, since the self-inductance of a wire with length l and a cross section of B × C is:

Table 25-7: 20-Pin DIP and SOIC Lead Inductance

Pins	DIP (nH)	SOIC (nH)
1,10,11,20	13.7	4.2
2,9,12,19	11.1	3.8
3,8,13,18	8.6	3.3
4,7,14,17	6.0	2.9
5,6,15,16	3.4	2.4

Courtesy of Texas Instruments

$$L = 0.002 \ln \left(\frac{2l}{B + C} + \frac{l}{2} \right) \tag{25-7}$$

As seen in Equation (25-7), the wire length, l, contributes more to self-inductance than does the cross section. This explains why all high-performance microprocessors and logic families use several pins for the VCC and ground. For example, in the case of Intel's Pentium processor there are over 50 pins for the ground and another 50 pins for the VCC.

The discussion of ground bounce is also applicable to VCC when a large number of outputs changes from the low to high state and is referred to as VCC bounce. However, the effect of VCC bounce is not as severe as ground bounce since the high ("1") state has wider noise margin than the low ("0") state.

Filtering the transient currents using decoupling capacitors

In the TTL family, the change of the output from low to high can cause what is called *transient current*. In totem-pole output, when the output is low, Q4 is on and saturated, whereas Q3 is off. By changing the output from the low to high state, Q3 becomes on and Q4 becomes off. This means that there is a time that both transistors are on and drawing currents from the VCC. The amount of current depends on the RON values of the two transistors, and that, in turn, depends on the internal parameters of the transistors. However, the net effect of this is a large amount of current in the form of a spike for the output current, as shown in Figure 25-10. To filter the transient current, a 0.01 F or 0.1 F ceramic disk capacitor can be placed between the VCC and ground for each TTL IC.

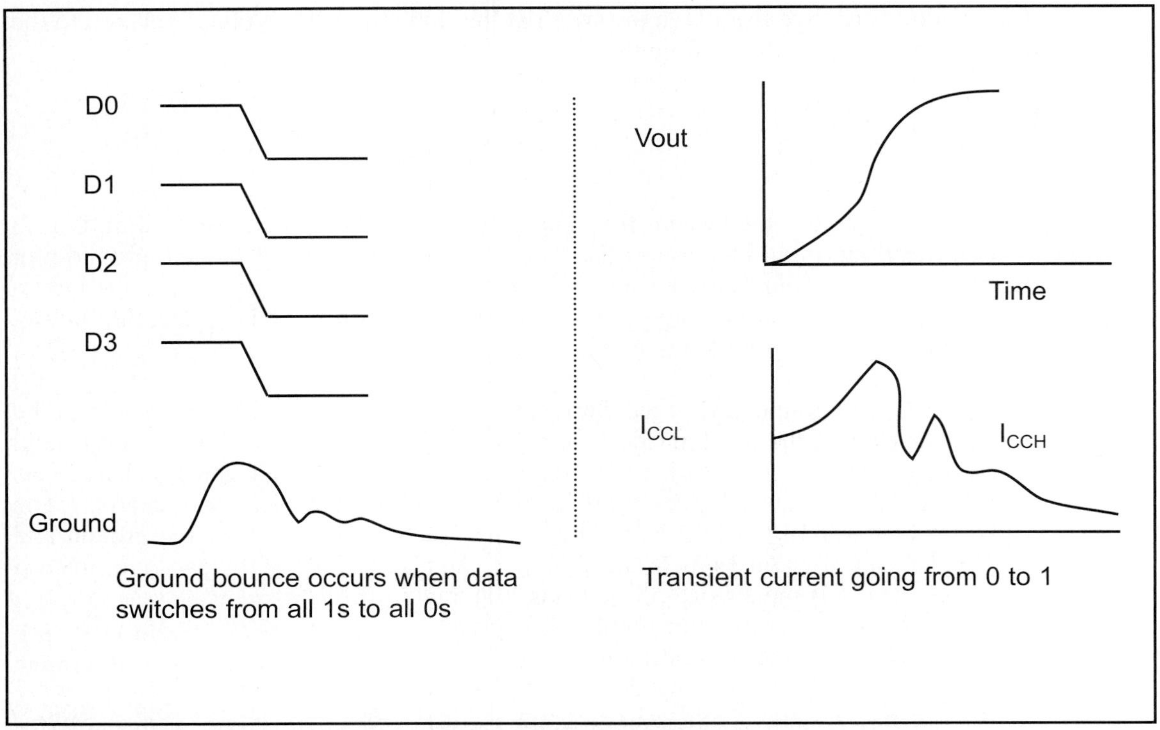

Figure 25-10. (a) Ground Bounce (b) Transient Current

However, the lead for this capacitor should be as small as possible since a long lead results in a large self-inductance and that results in a spike on the VCC line [V = L (di/dt)]. This is also called *VCC bounce*. The ceramic capacitor for each IC is referred to as a *decoupling capacitor*. There is also a bulk decoupling capacitor, as described next.

Bulk decoupling capacitor

As many IC chips change state at the same time, the combined currents drawn from the board's VCC power supply can be massive and cause a fluctuation of VCC on the board where all the ICs are mounted. To eliminate this, a relatively large (relative to an IC decoupling capacitor) tantalum capacitor is placed between the VCC and ground lines. The size and location of this tantalum capacitor vary depending on the number of ICs on the board and the amount of current drawn by each IC, but it is common to have a single 22 F to 47 F capacitor for each of the 16 devices, placed between the VCC and ground lines. See Technical Notes TN0006 and TN4602 from Micron Technology.

http://download.micron.com/pdf/technotes/TN0006.pdf
http://download.micron.com/pdf/technotes/DDR/TN4602.pdf

Crosstalk

Crosstalk is due to mutual inductance. See Figure 25-11. Previously, we discussed self-inductance, which is inherent in a piece of conductor. Mutual inductance is caused by two electric lines running parallel to each other. It is calculated as follows:

$$M = 0.002\, l \ln\frac{2l}{d} - \ln\left(\frac{2l}{B+C} + \frac{l}{2}\right) \qquad (25\text{-}8)$$

where l is the length of two conductors running in parallel, and d is the distance between them, and the medium material placed in between affects K. Equation (25-8) indicates that the effect of crosstalk can be reduced by increasing the distance between the parallel or adjacent lines (in printed circuit boards, these will be traces). In many cases, such as printer and disk drive cables, there is a dedicated ground for each signal. Placing ground lines (traces) between signal lines reduces the effect of crosstalk. This method is used even in some ACT logic families where VCC and GND pins are next to each other. Crosstalk is also called *EMI* (*electromagnetic interference*). This is in contrast to ESI (electrostatic interference), which is caused by capacitive coupling between two adjacent conductors.

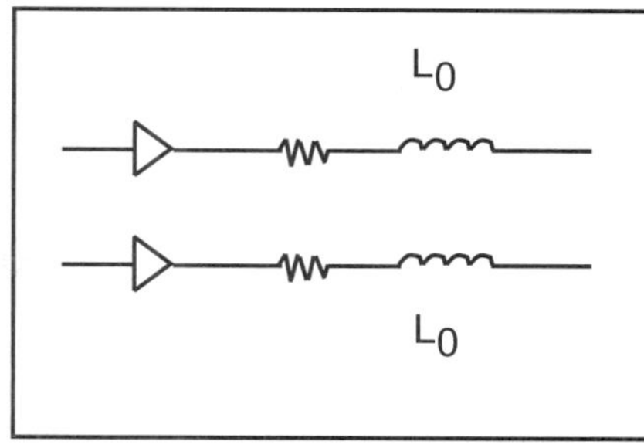

Figure 25-11. Crosstalk (EMI)

Transmission line ringing

The square wave used in digital circuits is in reality made of a single fundamental pulse and many harmonics of various amplitudes. When this signal travels on the line, not all the harmonics respond the same way to the capacitance, inductance, and resistance of the line. This causes what is called *ringing*, which depends on the thickness and the length of the line driver, among other factors. To reduce the effect of ringing, the line drivers are terminated by putting a resistor at the end of the line. See Figure 25-12. There are three major methods of line driver termination: parallel, serial, and Thevenin. In many systems resistors of 30–50 ohms are used to terminate the line. The parallel and Thevenin methods are used in cases where there is a need to match the impedance of the line with the load impedance. This requires a detailed analysis of the signal traces and load impedance, which is beyond the scope of this volume. In high-frequency systems, wire traces on the printed circuit board (PCB) behave like transmission lines, causing ringing. The severity of this ringing depends on the speed and the logic family used. Table 25-8 provides the length of the traces, beyond which the traces must be looked at as transmission lines.

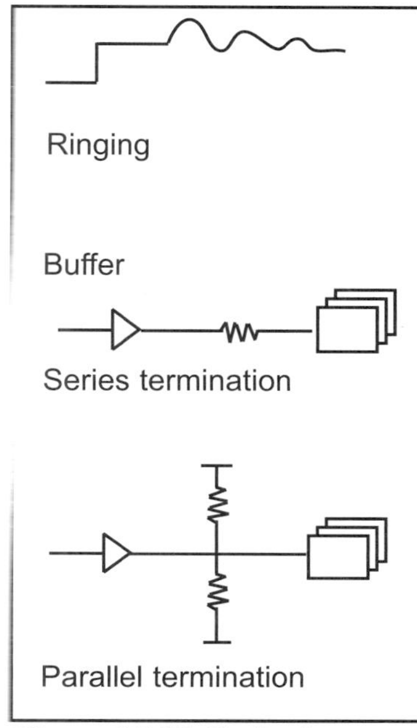

Figure 25-12. Reducing Transmission Line Ringing

FIT and failure analysis

Chip manufacturers provide a parameter called *FIT* (*failure in time*) to measure the reliability for a single chip. The FIT of a single chip is the number of expected failures in a billion (10^9) hours of operation. If a chip has FIT of 300, then there will be 300 failures per billion device hours of operation. To reduce the number of device failures, manufacturers use burn-in to eliminate the early failures before the product is shipped to the customer. This is commonly referred to as infant mortality since the failure rate starts high and eventually levels off to a constant level. See Figure 25-13. Although we can eliminate the early failures using burn-in, we can never reduce the failure rate to zero due to wearout and other factors such as soft error. This is discussed next.

Table 25-8: Line Length Beyond Which Traces Behave Like Transmission Lines

Logic Family	Line Length (in.)
LS	25
S, AS	11
F, ACT	8
AS, ECL	6
FCT, FCTA	5

(Reprinted by permission of Integrated Device Technology, copyright IDT 1991)

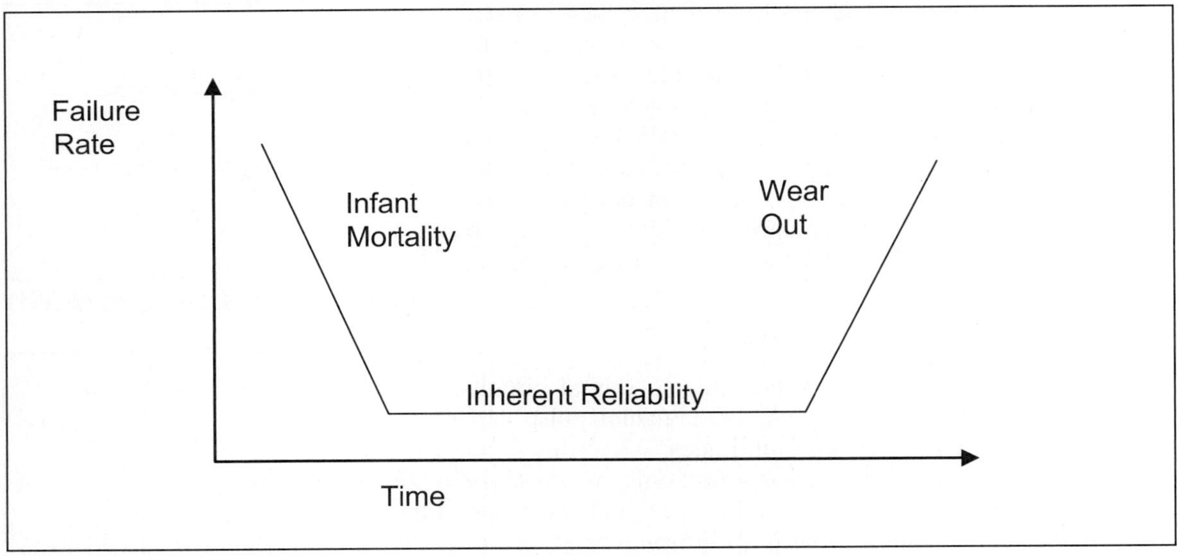

Figure 25-13. Bathtub Failure Rate

Soft error and hard error

In memory there are two kinds of errors that can cause a bit to change: *soft* error and *hard* error. If the cell bit gets stuck permanently in a "high" or "low" state, this is referred to as a hard error. Hard error is due to deterioration of the cell caused by wear-out (see Figure 25-13). There is no remedy for hard error except to replace the defective RAM chip since the damage is permanent. The other kind of error, a soft error, alters the cell bit from 1 to 0 or from 0 to 1, even though the cell is perfectly fine (no hard error). Soft error is caused by alpha particle radiation and power surges. The sources of the alpha particles are the radiation in the air or the materials in the plastic package enclosing the RAM die. The occurrence of a soft error as a result of alpha particles ionizing the charges in a RAM cell is a greater source of concern since it is 5 times more likely to happen than a hard error. As the density of RAM chips increases and the size of the RAM cell goes down, the probability of a soft error for a given cell goes up, but the relation is not linear.

Mean time between failures (MTBF) for system

Reliability of system depends directly on two factors: a) the FIT (failures in time) value of a single part, and b) the number of parts in the system. We use these two factors to calculate what is called *MTBF* (*mean time between failures*). The MTBF predicts the average time before the first failure happens. The MTBF for a single chip is calculated using the FIT as follows:

$$\text{MTBF} = \frac{1{,}000{,}000{,}000 \text{ hours}}{\text{FIT}} \qquad (25\text{-}9)$$

To get the MTBF rate for the system, we must divide the single-chip MTBF by the number of chips in the system.

$$\text{MTBF of system} = \frac{\text{MTBF of one chip}}{\text{number of chips in system}} \qquad (25\text{-}10)$$

See Examples 25-7 and 25-8.

See Technical Notes TN-00-14 and TN-00-18 on the www.micron.com website.

http://download.micron.com/pdf/technotes/TN0014.pdf

http://download.micron.com/pdf/technotes/TN0018.pdf

There is a paper called "Testing RAM for Embedded Systems" by Jack Ganssle and available from the folllowing website:

http://www.ganssle.com/testingram.pdf

Also see the article "Thirteen feet of concrete won't shield your RAM from the perils of cosmic rays. What's the solution?" by Jack Ganssle in Dr. Dobb's Journal. It is available from the following website:

http://www.ddj.com/dept/debug/196800160

Example 25-7

Assuming that the FIT for a single chip is 252, calculate the MTBF for:
(a) a single chip
(b) a system with 512 chips

Solution:

(a) The MTBF for a single chip is as follows: MTBF for 1 chip = 1,000,000,000 hr / 252 = 3,968,254 hr = 453 years
(b) The MTBF for 512 chips is = 453 years / 512 chips = 0.884 year = 323 days

Example 25-8

Calculate the system MTBF for the system in Example 25-7 if FIT = 745.

Solution:

MTBF for a single chip = 10^9 / 745 hrs. = 153 years. For the system it is 153 years / 512 = 109 days.

ECL and gallium arsenide (GaAs) chips

The use of secondary cache (L2: level 2 cache, as many call it) and EDC in systems with speeds of 66 MHz and higher is adding to the data and address path delay. This is forcing designers to resort to using ECL and GaAs chips. Due to the fact that ECL chips have a very high power dissipation, they are not used in low-cost x86 design. However, GaAs chips are showing up in high-speed x86 and RISC-based computers. This is especially the case for the GaAs EDC and cache controller chips. The mass of electrons in GaAs is lighter than in silicon, due to its quantum mechanics structure. As a result, the electrons in GaAs have a much higher speed. This means that GaAs chips can achieve a much higher speed than silicon. The power dissipation of the GaAs transistor is comparable to the silicon-based MOS transistor. Therefore, GaAs technology might appear to provide the ideal chip since it has the speed of ECL (it is even faster than ECL) and the power dissipation of CMOS. However, it has the following disadvantages.

1. Unlike silicon, of which there is a plentiful supply in nature in the form of sand, GaAs is a rare commodity, and therefore more expensive.
2. GaAs is a compound made of two elements, Ga and As, and therefore is unstable at

high temperatures.
3. It is very brittle, making it impossible to have large wafers. As a consequence, at this time no more than 100,000 transistors can be placed on a single chip. Contrast this to the millions of transistors for silicon-based chips.
4. The GaAs yields are much lower than for silicon, making the cost per chip much more expensive than for silicon chips.

These problems make the building of an entire computer based on GaAs a visionary project, if not an impossible one. This was the case for the CRAY III supercomputer, which was based on GaAs, and the buses ran at speeds of multiple GHz; but the project was also several years behind and millions of dollars over budget, so it was eventually abandoned and the company went out of business.

Review Questions

1. What is the fan-out of the "0" state?
2. If the fan-out of "low" and "high" are 10 and 15, respectively, what is the fan-out?
3. If I_{OL} = 12 mA, I_{OH} = 3 mA for the driver, and I_{IL} = 1.6 mA, I_{IH} = 40 A for the load, find the fan-out.
4. Why do I_{IL} and I_{OH} have negative signs in many TTL books?
5. What are the 74xx244 and 74xx245 used for?
6. What is capacitive derating?
7. Ground bounce happens when the output makes a transition from _______ to _______.
8. Give one way to reduce ground bounce.
9. Transient current is due to transition of output from _______ to _______.
10. Why do high-speed logic gates using DIP packaging put the VCC and ground pins in the middle instead of the corners?
11. True or false. Soft error is permanent.
12. True or false. Hard error is permanent.
13. Alpha particle radiation causes ____________ (soft, hard) errors.
14. FIT is in ________ (hours, months, years) of device operation.
15. What is the MTBF for 512 megabytes of memory if DRAM chips used are 16M × 8 with FIT = 252?
16. What is the MTBF for 512 megabytes of memory if DRAM chips used are 16M × 8 with FIT = 1000?

PROBLEMS

SECTION 25.1: OVERVIEW OF IC TECHNOLOGY

1. Why do bipolar transistors dissipate more power?
2. Why is the MOS transistor slower than the bipolar?
3. Why has the use of NMOS replaced PMOS?
4. For a TTL inverter indicate which transistors are "on" and "off" for the following. (a) input = high (b) input = low
5. Repeat Problem 4 for CMOS.
6. Why in CMOS does the current dissipation rise as the frequency goes up?
7. What is the purpose of the Schottky diode in the 74LSxx family?
8. What is the noise margin for "0" and "1" in the LS family?
9. What is the noise margin for "0" and "1" in the HCMOS family?
10. Which one uses more static current, LS or HCMOS?
11. Which one is faster, LS or ALS?
12. Which one is more power efficient, LS or ALS?
13. Which one is faster, AC/ACT or FCT? Which one is more power efficient?
14. What is the FCT logic family?
15. What is the advantage of FASTr over FAST logic?

16. What is the BCT logic family?
17. True or false. The LS family is used for system frequency of less than 10 MHz.
18. True or false. The BCT family is used for system frequency of less than 15 MHz.
19. Pentium uses a line size of ______ micron.
20. Why is CMOS the technology of choice in microprocessor design?

SECTION 25.2: IC INTERFACING AND SYSTEM DESIGN ISSUES

21. Calculate the fan-out if LS drives ALS.
22. Calculate the fan-out for LS driving unit loads.
23. Calculate the fan-out for ALS driving unit loads.
24. Calculate the number of LS that the 74LS244 can drive,
25. Find I_{OL} and I_{OH} needed to drive 10 LS and 20 CMOS input loads.
26. True or false. Capacitance derating is a function of frequency.
27. To minimize capacitance derating, use _____ (high, low) drive capability logics.
28. Calculate the path delay if one 244 is driving 2 banks each with 32 inputs and Cin for each input is 7 pF. Assume that the 244 delay is 8 ns and memory access time = 15 ns.
29. Repeat Problem 28 where the number of 244s is doubled.
30. Repeat Problem 29 where the number of 244s is doubled again.
31. Which current is a function of frequency, dynamic or static?
32. Give the advantages of lower VCC.
33. If VCC = 3.7 V, compare the power dissipation in comparison with VCC = 5 V.
34. If VCC = 1.8 V, compare the power dissipation in comparison with VCC = 5 V.
35. Repeat Problem 34 using VCC = 3.3 V.
36. In many IC chips VCC and GND are located at the middle of the DIP package instead of in the corners. Discuss the effect of VCC and GND pin locations in such cases.
37. Discuss the causes and cures for ground bounce.
38. Why is the effect of VCC bounce less severe than ground bounce?
39. Discuss the cause of transient current and ways to reduce its effects.
40. Discuss why many 245 drivers are used for the system data bus.
41. Discuss the cause of crosstalk and methods to reduce it.
42. What is the cause of ringing?
43. PCB traces behave like transmission lines most in the ______ logic family and least in the _______ logic family.
44. What is the purpose of 30–50 ohms resistance at the end of lines driving the DRAM arrays?
45. Discuss the difference between soft error and hard error.
46. Give the main causes of soft error and hard error.
47. Calculate the MTBF for 16M of memory using 1M × 1 and FIT = 252.
48. Calculate the MTBF for Problem 47 with EDC if the word size is 32-bit.

ANSWERS TO REVIEW QUESTIONS

SECTION 25.1: OVERVIEW OF IC TECHNOLOGY

1. MOS is more power efficient, while bipolar is faster.
2. True
3. True
4. CMOS
5. NMOS
6. In the lower end, ALS, and in the higher end, FAST

SECTION 25.2: IC INTERFACING AND SYSTEM DESIGN ISSUES

1. It is the number of loads that the driver can support and it is calculated by I_{OL}/I_{IL}.
2. 10
3. I_{OL}/I_{IL} = 12 mA/1.6 mA = 7 and I_{OH}/I_{IH} = 3 mA/40 μA = 75. Fan-out is 7, a lower number.
4. The negative sign indicates that these currents are flowing out of the IC (conventional current flow).
5. They are used for the line driver: the 74xx244 for unidirectional and 74xx245 for bidirectional lines.
6. It is signal delay caused by excessive load capacitance.
7. High, low
8. Make the ground pin length as small and short as possible.
9. Low, high
10. To make the self-inductance of pins VCC and GND small in order to reduce the ground and VCC bounce
11. False
12. True
13. Soft
14. Hours
15. 453/32 = 14.1 years since we have 512M × 8/16M × 8 = 32 chips
16. 3.56 years (114.15 years for one DRAM divided by 32 chips)

CHAPTER 26

ISA, PC104, AND PCI BUSES

OBJECTIVES

Upon completion of this chapter, you will be able to:

>> Describe the ISA bus signals for I/O interfacing
>> Describe the ISA bus signals for memory interfacing
>> Calculate I/O cycle time and bus bandwidth for the ISA bus
>> Define the meaning of the terms master, slave, bus arbitration, bus protocol, and bus bandwidth and describe their importance in PC design
>> Describe the evolution of bus architecture from ISA to PCI
>> List the limitations of the ISA bus
>> List the major characteristics of PCI architecture
>> List the enhancements of the PCI bus over the ISA bus
>> Contrast and compare ISA and PCI in terms of bus bandwidth
>> Define the term local bus and describe its merits
>> List the major characteristics of the PCI local bus

This chapter explores the ISA and PCI expansion slot buses. In Section 26.1 we present ISA bus memory signals. The I/O signals and specifications for the ISA bus are discussed in Section 26.2. The PCI bus is covered in Section 26.3.

SECTION 26.1: ISA BUS MEMORY SIGNALS

In Chapter 9 we covered the basics of ISA bus signals. In this section we provide more details of the ISA bus for memory interfacing including the memory read/write cycle time. In PCs with x86 microprocessors, the signals for the ISA expansion slots are provided by the chipset. The chipset makes sure that the signals for the ISA slot conform with the ISA bus standard regardless of the CPU's speed and data width. The ISA bus specifications and timing for memory are precise and must be understood if we want to design an ISA plug-in card with on-board memory. Next, we review once more the signals related to memory in the ISA expansion slot.

Address bus signals

SA0–SA19 (system address)

The system address bus provides the address signals for the desired memory (or I/O) location. The chipset latches these signals and holds them valid throughout the bus cycle. See Figure 26-1.

LA17–LA23 (latchable address)

These signals, along with SA0–SA19, allow access to 16M bytes of memory space from the ISA expansion slot. The chipset does not latch these signals. They must be latched by the board designed for the expansion slot.

SBHE (system byte high enable)

Because it is an active low signal, when it is low it indicates that data is being transferred on the upper byte (D8–D15) of the data bus.

SD0–SD15 (system data bus)

The system data bus (SD0–SD15) is used to transfer data between the CPU, memory, and I/O devices.

Memory control signals

MEMW (memory write)

An active-low control signal is used to write data into the memory chip. This signal is connected to the WE (write enable) pin of the memory chip. This signal can be used to access the entire 16M allowed by the ISA bus.

MEMR (memory read)

This active-low control signal is used to read data from the memory chip. It is connected to the OE (output enable) pin of the memory chip. This signal can be used to access the entire 16M allowed by the ISA bus. MEMW and MEMR are used for 16M memory. However, if the 1M memory 00000–FFFFFH is chosen, the following control signals must be used.

SMEMW (system memory write)

An active-low control signal used to write data into a memory chip. This signal is connected to the WE pin of the memory chip. This signal goes low when accessing addresses between 0 and FFFFFH (0 and 1M bytes).

SMEMW (system memory write)

An active-low control signal used to read data from the memory chip. This signal is connected to the OE pin of the memory chip. This signal goes low when accessing addresses between 0 and FFFFFH (0 and 1M bytes).

Although the ISA bus has a 16-bit data bus (D0–D15), either the 8-bit section (D0–D7) or the entire 16 bits (D0–D15) can be used. This is decided by the input pin MEMCS16, as explained next.

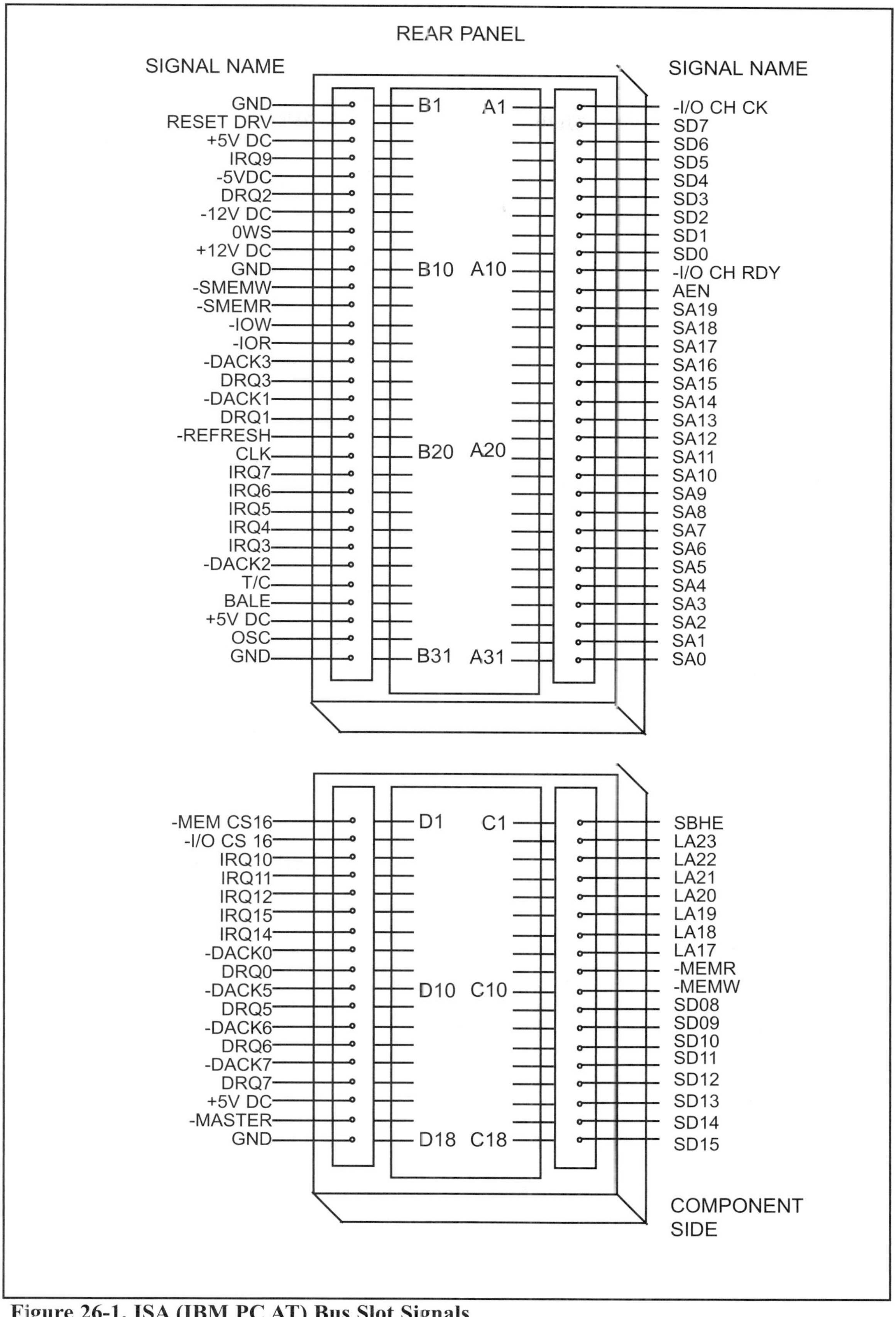

Figure 26-1. ISA (IBM PC AT) Bus Slot Signals
(Reprinted by permission from "IBM Technical Reference" c. 1985 by International Business Machines Corporation)

MEMCS16 (memory chip select 16)

This is an input signal and is active low. When not asserted, it indicates to the chipset that only the D0–D7 portion of the data bus is being used. Notice that the 8-bit portion is the default mode and it is achieved by doing nothing to this pin. In contrast, when this signal is asserted low, both the low byte and high byte of the data bus (D0–D15) will be used for data transfer. Therefore, to use the entire 16-bit data bus, this pin must be low.

The ISA bus allows the interfacing of slow memories by inserting wait states into the memory cycle time. This prolonging of memory cycle time is available for both 8-bit and 16-bit data transfers. The standard 8-bit data transfer has 4 WS in the memory read cycle time. As a result, the default memory cycle time is 6 clocks. The standard 16-bit data transfer uses 1 WS in the memory cycle time. That results in 3 clocks for the read/write cycle time. To shorten the memory cycle time we use the ZEROWS pin as explained next.

ZEROWS (zero wait state)

This is an input signal and is active low. The standard 16-bit ISA bus cycle time contains one WS unless ZEROWS is activated. By activating this pin (making it low), we are telling the CPU that the present memory cycle can be completed without a wait state. That results in performing the bus cycle time in 2 clocks. The standard 8-bit ISA bus cycle time contains 4 WS unless ZEROWS is activated. That means that when both MEMCS16 and ZEROWS are high (without asserting them low), the 8-bit data bus (D0–D7) is being used and the data transfer is completed in 6 clocks. This default 8-bit read/write cycle time with its 4 WS is sufficient for interfacing even slow ROMs to the ISA bus. If MEMCS16 = 1 and ZEROWS = 0, the 8-bit memory cycle time has 1 WS instead of 4.

SYSCLK (system clock)

This is is an output clock providing the standard 8 MHz ISA bus clock. The 8 MHz clock results in a 125 ns (1/8 MHz = 125 ns) period and all memory and I/O ISA bus timing is based on this. Therefore, a zero WS read cycle time for a 16-bit bus takes 2×125 ns = 250 ns. The standard 8-bit bus with its 4 WS will be $(2 + 4) \times 125$ ns = 750 ns. The 8-bit bus with ZEROWS asserted low has 1 WS, making its memory cycle time $(2 + 1) \times 125$ ns = 375 ns. The SYSCLK pin is located on the B side of the 62-pin section of the ISA bus.

IOCHRDY (IO Channel Ready)

This is an input signal into the ISA bus and is active low. By driving it low, we are asking the system to extend the standard ISA bus cycle time. In response to the asserting of this signal, the system will insert wait states into the memory (or I/O) cycle time until it is deasserted. This is rarely used for memory interfacing since the standard memory cycle time of the ISA bus provides plenty of time.

ISA bus timing for memory

Suppose that we are designing a data acquisition board for an ISA expansion slot, and the board requires ROM. What would be the best approach? We can use either the 8-bit or 16-bit data section of the ISA bus. There are some major differences between them that must be noted. Next we look at each separately.

8-bit memory timing for ISA bus

In the case of using the 8-bit data bus, we use D0–D7 and A0–A19 of the 62-pin section of the bus. More importantly, the ISA bus provides plenty of time for slow memories by inserting wait states into the read cycle time. First, we must remember that the memory cycle time is only 2 clocks (with 0 WS) and the maximum speed is 8 MHz. Since 1/8 MHz = 125 ns, the bus cycle time is $2 \times 125 = 250$ ns. In the case of the standard 8-bit read/write cycle time, the chipset inserts 4 WS clocks into the read cycle time as shown in Figure 26-2. In the case where ZEROWS is asserted low and MEMCS16 = 1, the read/write cycle time has 1 WS for the 8-bit bus as shown in Figure 26-3.

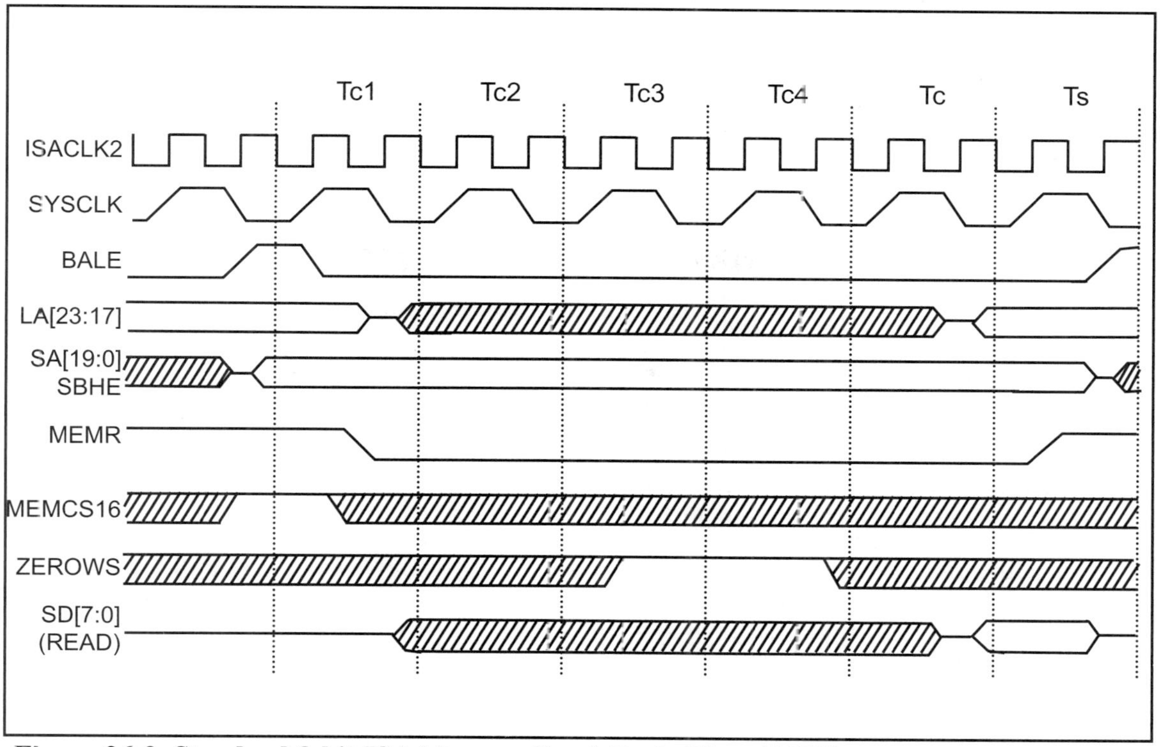

Figure 26-2. Standard 8-bit ISA Memory Read Cycle Time (4 WS)

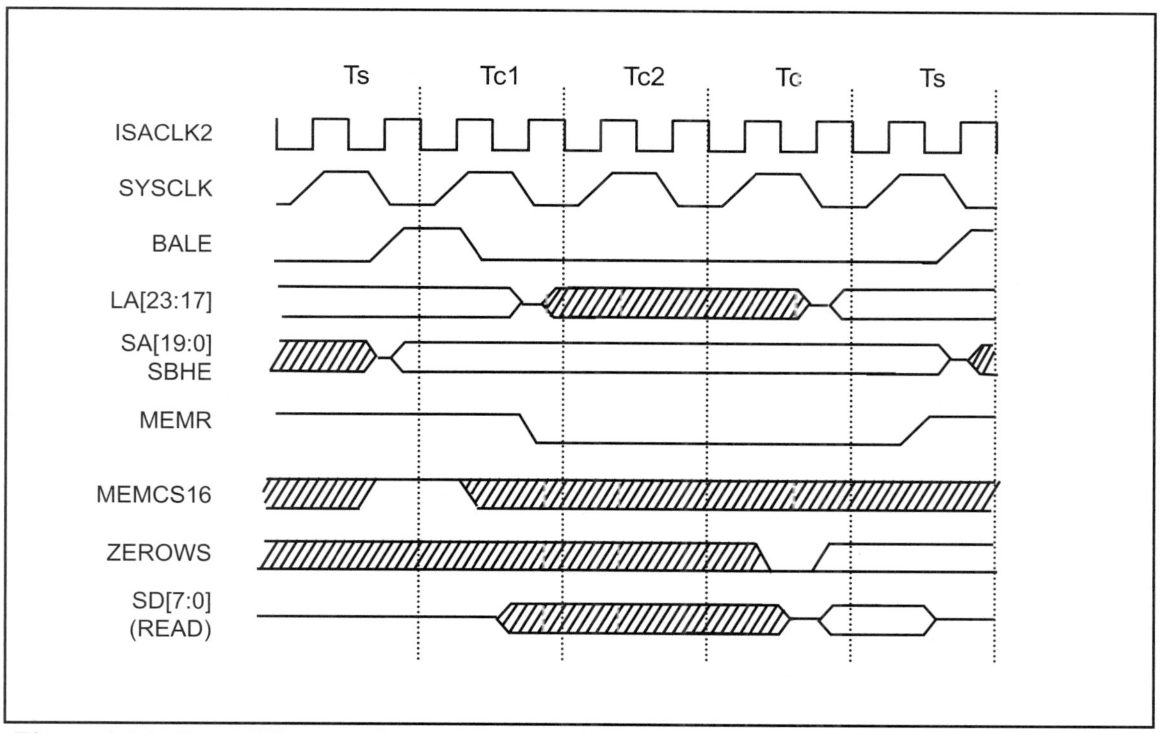

Figure 26-3. Zero WS 8-bit ISA Memory Read Cycle Time (1 WS)

Figure 26-4 shows the 8-bit memory interfacing for the ISA bus.

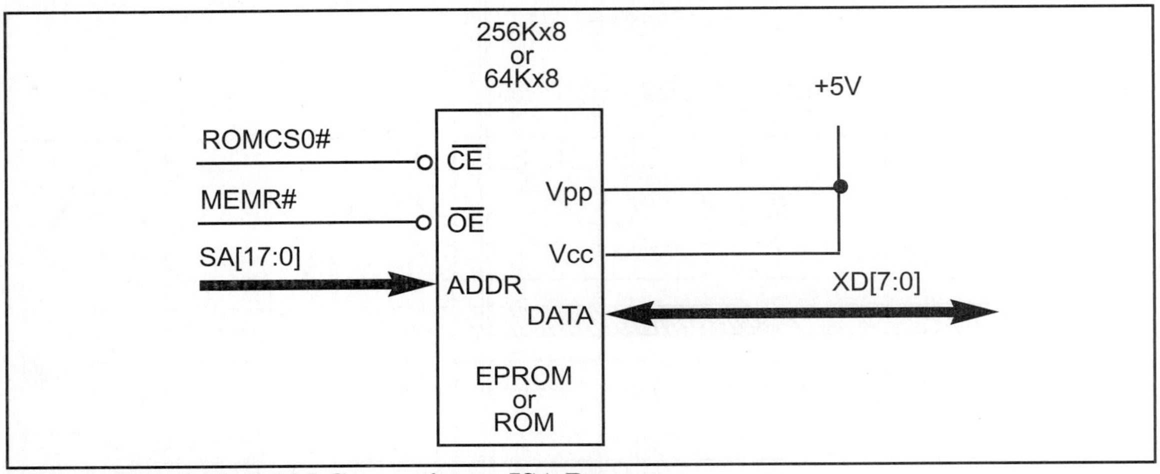

Figure 26-4. 8-bit ROM Connection to ISA Bus

16-bit memory timing for ISA bus

As mentioned earlier, the 8-bit data transfer is the default mode for the ISA bus expansion slots. In order to perform the 16-bit data transfer using lines D0–D15, we must assert the MEMCS16 pin low. A 16-bit data bus transfer is twice as fast as an 8-bit data bus transfer. However, it also requires twice the board space in addition to having a higher power consumption. The 16-bit data read cycle time for the ISA bus with zero WS is shown in Figure 26-6. Notice that in all the 2-clock bus cycle CPUs and systems, the first clock is set aside for the addresses and the second clock is for data. In order to get a zero wait state bus activity, the ZEROWS pin (active low) must be activated. This is because the standard ISA bus cycle contains one wait state. The standard 16-bit ISA bus cycle timing with 1 WS is shown in Figure 26-5, and Figure 26-6 shows the ISA cycle with 0 WS. Combining the effect of MEMCS16 and ZEROWS pins gives the information in Table 26-1. See Example 26-1.

Table 26-1: ISA Bus Memory Read/Write Cycle Time

MEMCS16	ZEROWS	Data Bus	Read Cycle Time
0	0	D0–D15	250 ns
0	1	D0–D15	375 ns
1	0	D0–D7	375 ns
1	1	D0–D7	750 ns

Example 26-1

Calculate the bus bandwidth of the ISA bus for (a) 0 WS, and (b) 1 WS. Assume that all the data transfers are 16-bit (D0–D15).

Solution:

Since the ISA bus speed is 8 MHz, we have 1/8 MHz = 125 ns as a clock period. The bus cycle time for zero wait state is 2 clocks; therefore, we have:

(a) ISA bus cycle time with 0 WS is 2×125 ns = 250 ns.
Bus bandwidth = 1/250 ns $\times$ 2 bytes = 8 megabytes/second.
(b) ISA bus cycle time with 1 WS is 250 ns + 125 ns = 375 ns.
Bus bandwidth = 1/375 ns $\times$ 2 bytes = 5.33 megabytes/second.

Another major issue in 16-bit ISA bus interfacing is the problem of odd and even banks. For example, assume that we are interfacing two ROM chips to D0–D15 of the ISA bus, one connected to D0–D7 and the other to D8–D15. We must divide our information (code or data) into two parts and burn each part into one of the ROMs. In many ROM burners, there is an option for splitting the data into odd- and even-addressed bytes to sup-

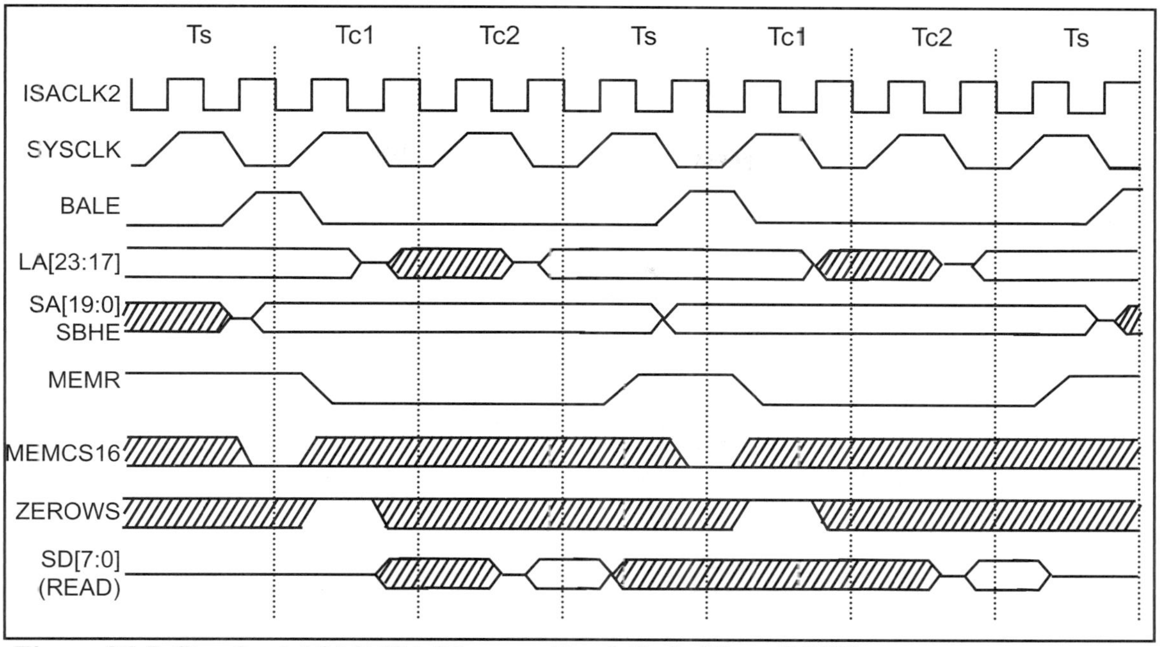

Figure 26-5. Standard 16-bit ISA Memory Read Cycle Time (1 WS)

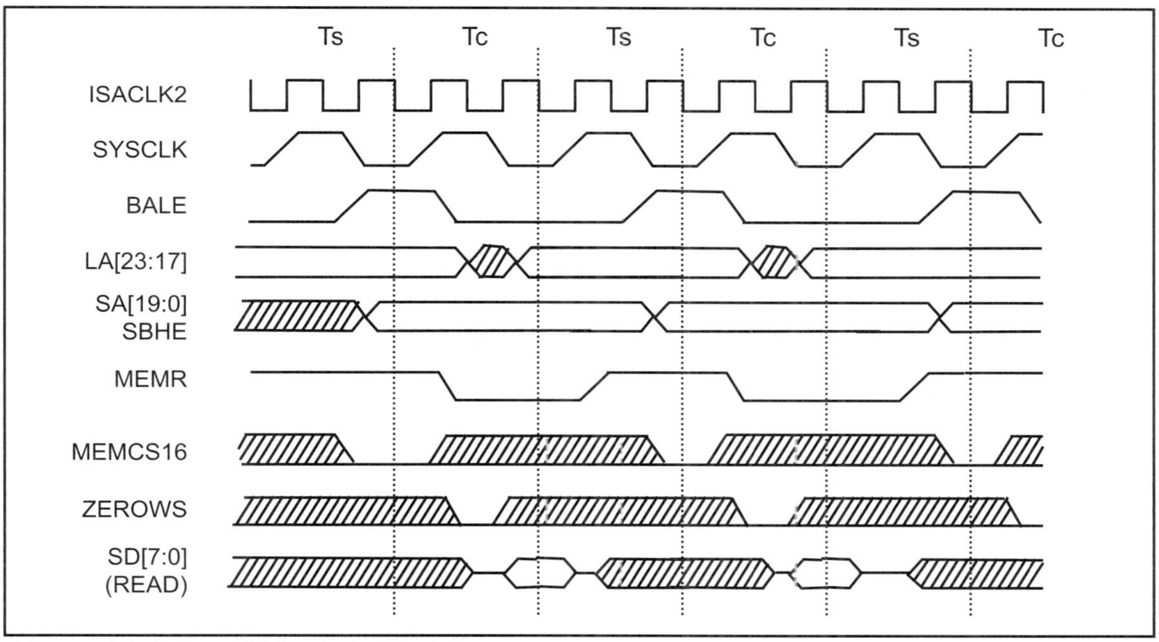

Figure 26-6. Zero WS 16-bit ISA Memory Read Cycle Time (0 WS)

port 16-bit data systems. The 16-bit data connection to two ROMs for the ISA slot is shown in Figure 26-7.

DIMM and SIMM memory modules

In the 1980s, PCs had sockets on the motherboard for DRAM chips. To expand the memory of a PC, you had to buy memory chips and plug them into the sockets. With the introduction of the 16-bit ISA bus, memory expansion boards became common. The problem with the memory expansion cards was that you were limited to the bus speed of the ISA expansion slot no matter how fast the memory chip. This led to the idea of a memory module as a way to expand memory for x86 PCs (386 and higher). The connectors for memory modules are much smaller in size and accommodate much faster memory than

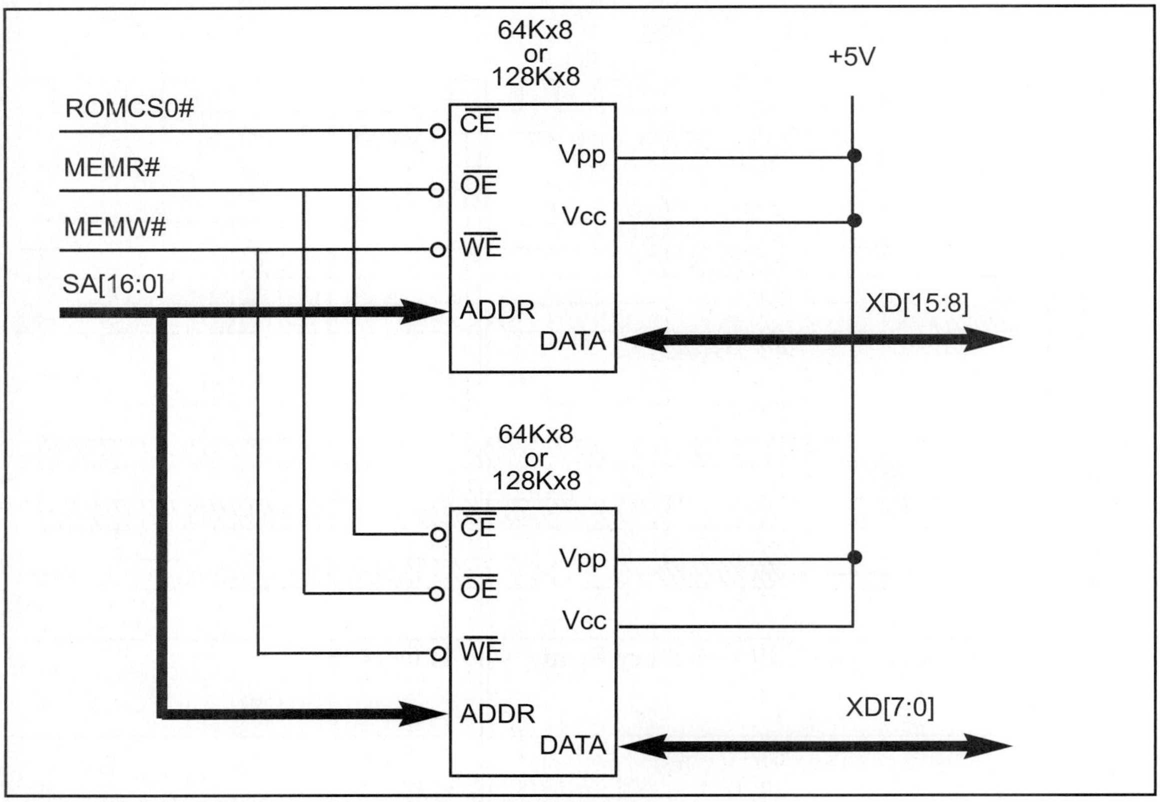

Figure 26-7. 16-bit ISA Bus Connection to ROM

do ISA connectors. It is also much easier to insert them into the motherboard. The only problem was the lack of a standard connector. This was solved with the introduction of SIP (single in-line package). Later, the SIMM (single in-line memory module) and DIMM (dual in-line memory module) were introduced. Currently, SIP is no longer in use, and SIMM and DIMM are the dominant memory modules. It is important to notice that the use of memory modules frees the motherboard designer from the agonizing choice of which organization and speed of DRAM to use. All that is required is to incorporate various organizations and speeds into the design of the motherboard and let the user select options via the CMOS set-up process.

ROM duplicate and x86 PC memory map

The memory map for the 1-megabyte memory range 00000 to FFFFFH is the same for all x86-based PCs. See Figure 26-8. As was shown in Chapter 9, when 286 microprocessors are powered up, the CPU is in real mode and fetches the first opcode from physical memory location FFFFF0H. This is because CS = F000H, IP = FFF0H, and A20–A23 are all high. This leads to a physical address of FFFFF0H, which is 16 bytes below the top of FFFFFFH, the 16-megabyte maximum memory range of the 286. After execution of the first opcode, address pins A20–A23 all become 0s. Address pins A20–A23 will not be activated again unless the 286 mode of operation is changed to protected mode. In other words, when the CPU wakes up in real mode at address FFFF0H, the first opcode is fetched from FFFFF0H because A20–A23 are all high. This is one reason that there is an exact duplicate of ROM at addresses 0F0000–0FFFFF and FF0000–FFFFFF. This duplication allows access of BIOS ROM in both real and protected modes. This concept applies to the 386/486 and Pentium PCs and is shown in Figure 26-9a. In these processors, the 32-bit address bus provides the memory space of 00000000 to FFFFFFFFH. We can verify this by using the system tools software that comes with Windows. See Figure 26-9b. You can experiment with this by going to *Accessories*, clicking on *System Tools*, and then clicking on *System Information*. Click on *Hardware Resources* and then click on *Memory*.

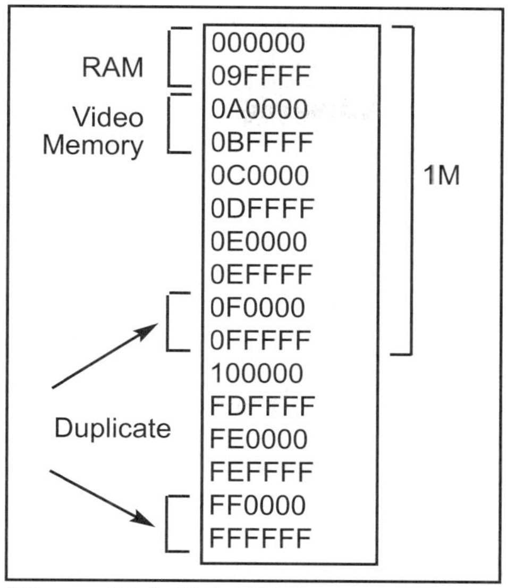

Figure 26-8. PC AT Memory Map

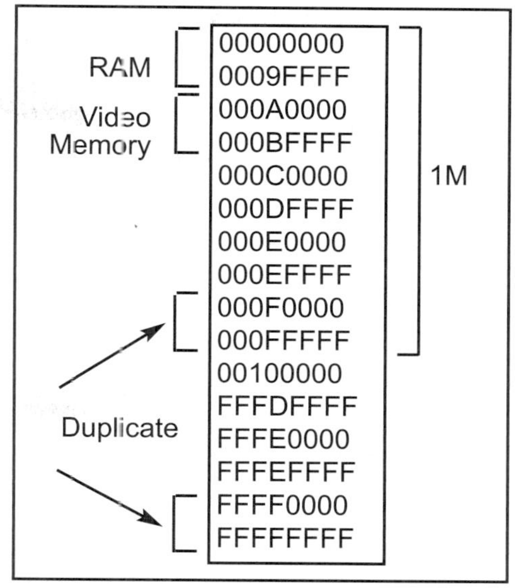

Figure 26-9a. x86 PC Memory Map

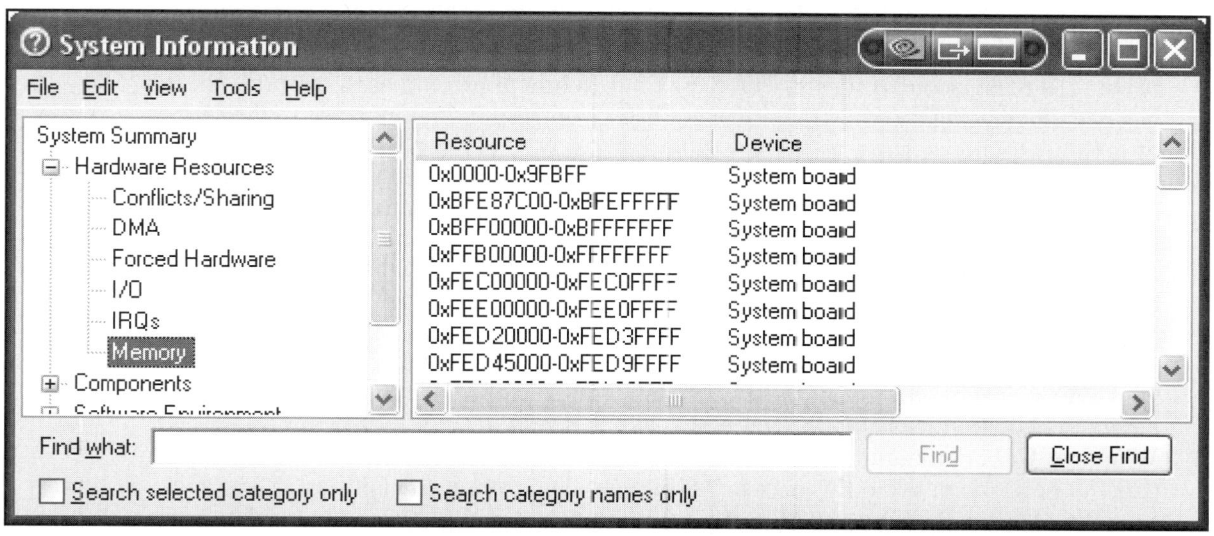

Figure 26-9b. x86 PC Memory Map from System Information Tool in Windows

Shadow RAM

By using *System Tools* to explore the system memory of Pentium PCs, we can see the duplicates of ROM in the RAM memory space. The reason for that is the fact that the ROM access time is too slow for the 100 MHz bus speed. To speed up the ROM access time, its contents are copied into RAM and write protected. This is called *shadow RAM* and provides the ROM's contents to CPU at a much faster speed than ROM. From this point forward, every time the CPU needs to access the ROM's contents, it will get the information from RAM at a very high speed. As long as the PC is on, the DRAM containing the ROM information is write protected and will not be corrupted. It must be noted that the process of creating shadow RAM is done when the system is booted, and when the PC is turned off, shadow RAM's contents are lost.

Review Questions

1. The MEMCS16 pin of the ISA bus is an active- ______ (low, high) signal.
2. If MEMCS16 = high, which portion of the data bus is used?
3. The ZEROWS pin of the ISA bus is an active- ______ (low, high) signal.
4. If ZEROWS = high, the 8-bit memory cycle time takes ______.
5. The ISA bus has a maximum frequency of 8 MHz. Find its bus bandwidth for 0 WS.

SECTION 26.2: I/O BUS TIMING IN ISA BUS

As we have seen, interfacing 8-bit devices such as the 8255 to the x86 PC is a straightforward process because the 8-bit data pins match D0–D7 of the ISA bus. Just as in 16-bit memory interfacing, there is a problem when we want to use the 16-bit (D0–D15) data bus of the x86 for I/O operations. In this section we look at the timing and design of 16-bit data I/O and compare it with 8-bit I/O operations. First, we examine a few issues concerning the ISA bus.

8-bit and 16-bit I/O in ISA bus

The term *ISA computers* encompasses IBM PC AT, PS, and any x86 PCs with AT-type expansion slots, as explored in Chapter 9. These computers could use the 286, 386, 486, Intel Pentium, or any x86 microprocessor from AMD as their CPU, but they have ISA-type expansion slots. The following points must be noted about these types of computers:

1. In communications between the x86 CPU and I/O ports, typically I/O devices are slow and cannot respond to the CPU's normal speed. In such situations, wait states must be inserted into the I/O cycle. The 80286 and all higher microprocessors (386, 486, Pentium, etc.) have two clocks for the I/O cycle time when they are designed with 0 WS. In this regard, it is the same as memory cycle time for such processors. For example, a 100-MHz Pentium processor provides a total of 20 ns (2×10 ns since 1/200 MHz = 10 nanoseconds) for the I/O cycle time. While in recent years, memory speed has been increasing steadily, there has not been a corresponding increase in the speed of I/O components such as ADCs (analog-to-digital converters). In general, I/O devices are much slower than memory since the I/O is interfaced with nature whereas memory is a semiconductor device. For example, a temperature sensor converts temperatures to voltage levels and the voltages are converted to binary numbers using ADCs before they are provided to the CPU. The delays associated with each stage of conversion add to the I/O response time, causing a severe bottleneck. This is only one of the reasons why the ISA expansion slot speed is limited to 8 MHz, in spite of the fact that the CPU bus speed is 100 MHz in the Pentium. Therefore, to interface with slow I/O devices, one must insert wait states into the I/O cycle time to match the device speed.
2. When the CPU communicates with an ISA expansion slot, be it memory or a peripheral I/O port, it can use only an 8- or 16-bit data bus, even if the CPU is 32-bit, such as a 386, 486, or Pentium. In 386 and 486 PCs where the CPU has a 32-bit data bus, memory (or even a peripheral) on the motherboard uses a 32-bit data bus, but when it goes to an ISA expansion slot it must use a 16-bit bus. When designing a plug-in card for the ISA expansion slot of a motherboard with a 32-or 64-bit CPU, we use the 8- or 16-bit data bus but not the 32/64 bit data buses. To access the x86's entire 32- or 64-bit data bus through the expansion slot, we must use a PCI slot. This is discussed in Section 26.3.
3. The ISA bus speed is limited to 8 MHz. It does not matter that the x86 CPU works on a frequency of 10 MHz or 100 MHz or even 1 GHz: When it communicates with devices (memory or I/O ports) through the ISA expansion slots it must slow down to 8 MHz. That means inserting many wait states in order to access the boards connected to ISA expansion slots. The good news is that the chipset on the motherboard will do all the above tasks.

These three limitations are commonly referred to as *I/O bottleneck* since they slow down the flow of information to/from the CPU when using I/O devices.

I/O signals of the ISA bus

Just as in memory, the ISA bus supports both 8- and 16-bit I/O operations. Furthermore, the problems associated with 8- and 16-bit data memory interfacing discussed in Chapter 10 also exist in I/O interfacing of the ISA bus. The following ISA slot pins are associated with I/O interfacing and must be understood.

SA0–SA9 (system address)

The system address A0–A9 bus provides the I/O addresses. This limits the I/O addresses supported by the ISA slot to 1024 ports.

SD0–SD7 (system data bus)

The system D0–D7 8-bit data bus or D0–D15 16-bit data bus provides the data path between the CPU and the I/O device.

IOR and IOW

The IOR and IOW control signals are both active low and are connected to the read and write pins of I/O devices.

IOCS16 (I/O chip select 16)

This is an input into the ISA bus and is an active-low signal. It informs the system that the I/O operation uses the entire 16-bit data bus D0–D15. If this input is not driven low, the ISA bus uses the 8-bit data bus D0–D7. The IOCS16 input pin is used to tell the motherboard circuitry that the present I/O cycle is a 16-bit data transfer. When the data transfer is a 16-bit transfer to an 8-bit peripheral, if the IOCS16 pin is not pulled low the data transfer is performed in two consecutive I/O cycles, each transferring one byte at a time, which requires more time. Add-in cards with a 16-bit data path use the IOCS16 pin to instruct the motherboard not to convert a word transfer into a byte transfer. This pin must be driven with an open collector or tri-state driver capable of sinking 20 mA. Bus data transfers are summarized as shown in Table 26-2.

Table 26-2: ISA Bus Data Transfer Summary

Data Transfer Type (from Source to Destination)	Number of WS	Number of Clocks per Cycle
From 16-bit to 16-bit	1	3
From 8-bit to 8-bit	4	6
From 16-bit to 8-bit	10	12

Reminder: In 286/386/486/Pentium machines, the memory (or I/O) cycle consists of 2 clocks when it is a zero-wait-state cycle.

ZEROWS (zero wait state)

ZEROWS is an input pin into the ISA bus and is active low. If this signal is driven low, it tells the system that I/O and memory operations can be completed without any WS. We will see how this applies to I/O operation in this section.

IOCHRDY (I/O channel ready)

This is an input signal into the ISA bus and is active low. By driving it low, we are asking the system to extend the standard ISA bus cycle. In response to asserting this pin, the system will insert wait states into the I/O or memory cycle until it is deasserted. The function of this pin is the opposite of ZEROWS. In other words, by asserting this pin we are extending the I/O (or memory) read and write cycle time to allow the interfacing of slow devices to the ISA bus.

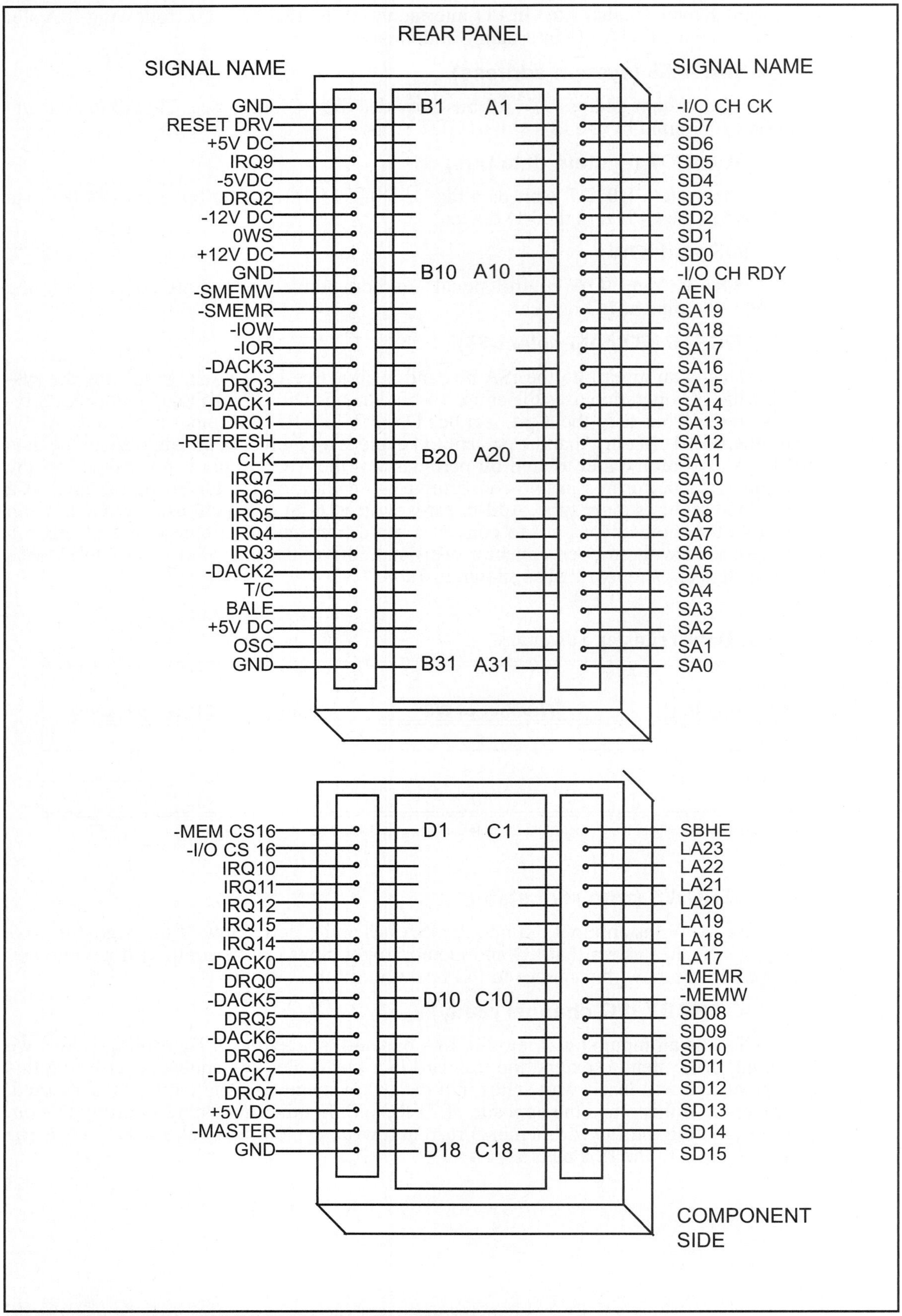

Figure 26-10. ISA (IBM PC AT) Bus Slot Signals

(Reprinted by permission from "IBM Technical Reference" c. 1985 by International Business Machines Corporation)

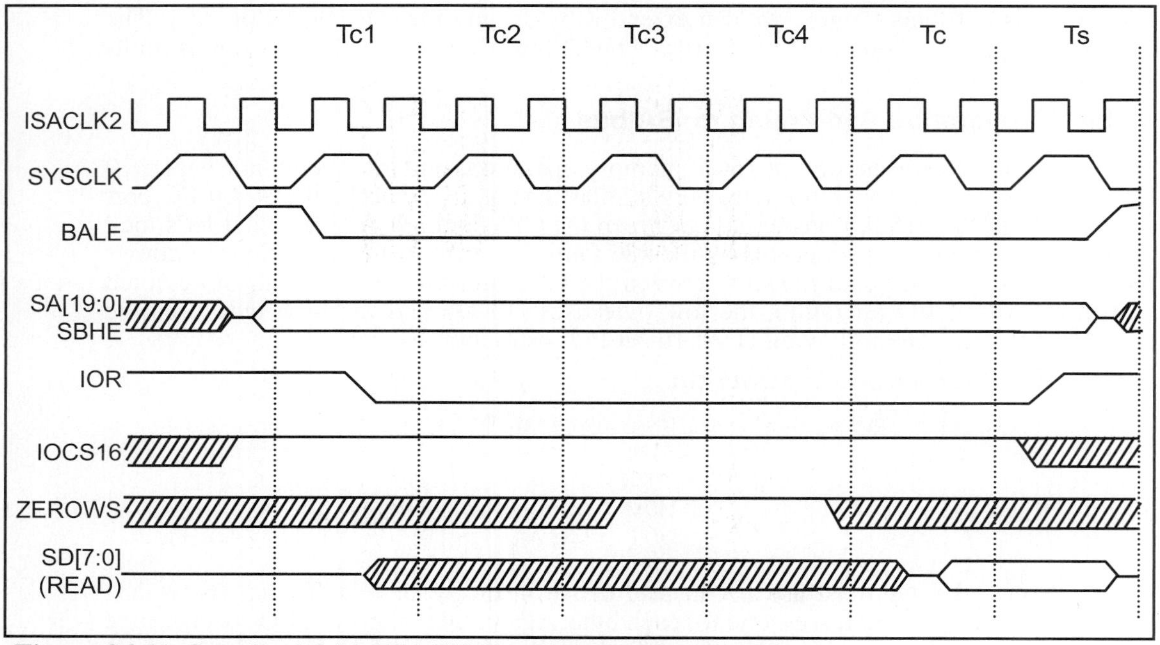

Figure 26-11. Standard 8-bit ISA I/O Read Cycle Time (4 WS)

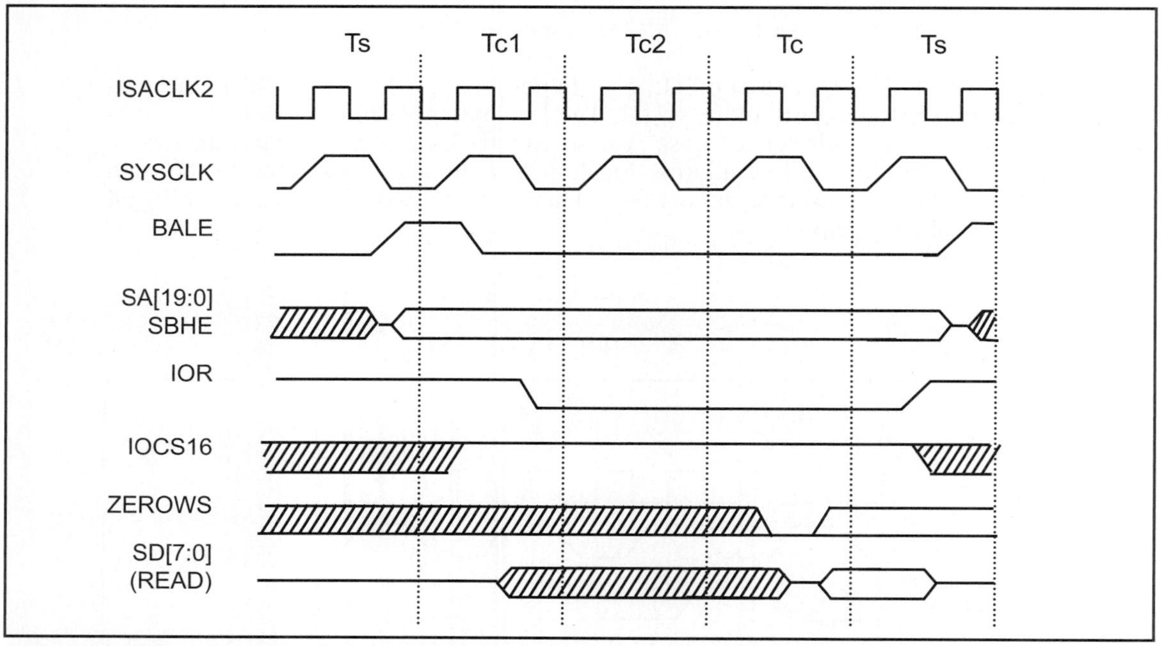

Figure 26-12. Zero WS 8-bit ISA I/O Read Cycle Time (1 WS)

8-bit I/O timing and operation in ISA bus

It would be helpful to review the discussion of 8-bit memory operation and interfacing in Chapter 10 since memory and I/O operations are very similar. The I/O operation of the ISA bus defaults to 8-bit and uses the D0–D7 data bus to transfer data between the I/O device and the CPU. It completes the read (or write) cycle in 6 (2 + 4 WS) clocks if ZEROWS is not asserted low. See Figure 26-11. In other words, the default I/O operation for ISA bus has 4 WS and uses the 8-bit data bus. This is an 8-bit standard I/O operation and is shown in Figure 26-11. With a maximum of 8 MHz for the ISA bus clock, I/O takes a total of 125 ns × 6 = 750 ns. Just like memory cycle time, we can shorten the I/O cycle time by asserting the ZEROWS pin low. This will cause the I/O operation to be completed in 3 (2 + 1 WS) clocks instead of 6 (2 + 4 WS). Notice that the default is 4 WS unless ZEROWS is asserted. This is shown in Figure 26-12. Now if the default 4 WS I/O

cycle time is not long enough, we can extend it by driving the IOCHRDY pin low. The extension happens as long as the IOCHRDY pin is low. The maximum extension is limited to 10 WS.

16-bit I/O operation and timing in ISA bus

Review the discussion of 16-bit memory operations and interfacing in Chapter 10 since memory and I/O operations are very similar. Just as in memory, the 16-bit I/O port uses data bus D0–D15 to transfer data between the CPU and I/O devices. First let's look at 16-bit I/O instructions supported by the x86 family. Similar to memory, the x86 allows the use of D0–D15 for data transfer between the CPU and I/O devices. It must be noted that in the 16-bit I/O operation, the low byte uses D0–D7 and the high byte uses the D8–D15 data bus. The following is the 16-bit I/O instruction format.

16-bit data ports instruction

```
        Inputting Data             Outputting Data
(1)     IN    AX,port#             OUT   port#,AX

(2)     MOV   DX,port#             MOV   DX,port#
        IN    AX,DX                OUT   DX,AX
```

Notice that we must use AX instead of AL in the 16-bit I/O. To use 16-bit data I/O, we need two port addresses, one for each byte. Again, this is because I/O space is byte addressable, just like memory space. Look at the following:

```
MOV    AX,98F6H
OUT    40H,AX    ;send out AX to port 308H & 309H
```

In this case F6H, the content of AL, goes to port address 40H while 98H, the content of AH, is transferred to port address 41H. The low byte goes to the low port address and the high byte to the high port address. This is exactly like memory data transfers in that the low byte goes to the low address location and the high byte goes to the high address location (the little endian convention). This principle works the same for 16-bit port addresses as shown below:

```
MOV    DX,310H
MOV    AX,98F6H
OUT    DX,AX    ;send out AX to port 310H & 311H
```

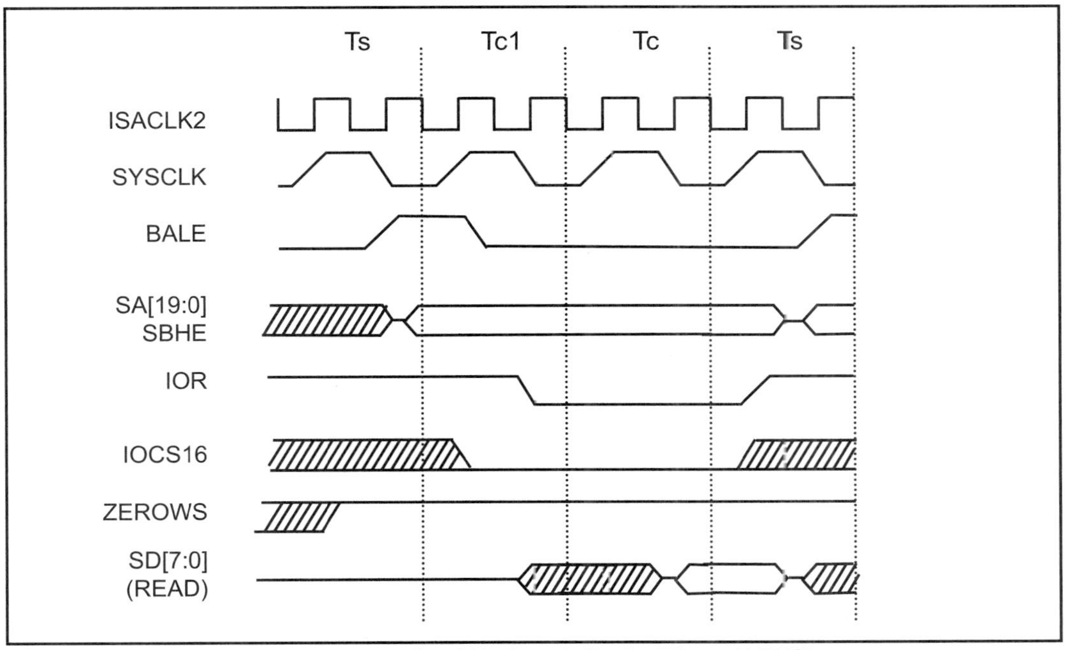

Figure 26-13. Standard 16-bit ISA I/O Read Cycle Time (1 WS)

Again the F6H is sent to port address 310H using data path D0–D7 and 98H goes to port address 311H using the D8–D15 data path. Next we will contrast 8-bit and 16-bit I/O via the ISA bus.

16-bit I/O timing and operation via ISA bus

As mentioned earlier, 8-bit data transfer is the default mode for the ISA bus expansion slots. To perform 16-bit data transfers using D0–D15, we must assert the IOCS16 pin low. The 16-bit bus transfers data twice as fast as an 8-bit data bus. However, it requires twice the board space in addition to an increase in power consumption. The 16-bit data read cycle time for the ISA bus with one WS is shown in Figure 26-13. This is the standard read cycle time in the ISA bus. In other words, unlike memory, you cannot get zero wait state bus activity for 16-bit I/O operations. Therefore, ZEROWS has no effect on 16-bit I/O operations, and the standard 16-bit ISA bus cycle timing is completed in 3 clocks. While you cannot shorten the 16-bit I/O bus cycle time, you can extend it by asserting the IOCHRDY pin.

I/O bus bandwidth for ISA

The I/O bus bandwidth is the rate of data transfer between the CPU and I/O devices and is dictated by the bus speed and the data bus width used. Example 26-2 shows the calculation of bus bandwidth for the ISA bus.

Example 26-2

Find the ISA bus bandwidth for (a) 8-bit standard, (b) 8-bit with ZEROWS asserted, and (c) 16-bit standard.

Solution:

Since the ISA bus speed is 8 MHz, we have 1/8 MHz = 125 nanoseconds for the bus clock.

(a) The standard 8-bit I/O bus cycle for ISA uses 4 WS. Therefore, the bus cycle time is 6 clocks. Now cycle time = 6 × 125 ns = 750 and the bus bandwidth is 1/750 ns × 1 bytes = 1.33 megabytes/second.

(b) If ZEROWS is asserted in 8-bit I/O, we have 3 clocks for the I/O cycle time. Therefore, we have a cycle time of 3 × 125 = 375 ns and 1/375 ns × 1 byte = 2.66 megabytes/second for bus bandwidth.

(c) For 16-bit I/O data transfers we must assert the IOCS16 pin. The I/O cycle time is 3 clocks. Therefore, we have a bus bandwidth of 1 / (3 × 125) × 2 = 5.33 megabytes/second. For 16-bit I/O, we cannot assert ZEROWS to shorten the cycle time.

Interfacing 8-bit peripherals to a 16-bit data bus

As mentioned in the last chapter, microprocessors with a 16-bit data bus use odd and even byte spaces. This is done with the help of the A0 and BHE pins as shown below:

BHE	A0	
0	0	Even-addressed words (uses D0–D15)
0	1	Odd-addressed byte (uses D8–D15)
1	0	Even-addressed byte (uses D0–D7)

In interfacing 16-bit I/O, the main issue is how to deal with odd- and even-address ports. The fact that data for even-addressed ports is carried on data bus D0–D7 and data for odd-addressed ports is carried on D8–D15 makes port design a challenging issue. There are two solutions to this problem.

1. Simply use two separate PPI devices, such as the 8255. One is used for odd address-

es and the other for even addresses. For example, in a design using this method, if port 74H is assigned to port A of the 8255, then port B has the address 76H, port C the address 78H, and so on. Another problem is outputting the contents of register AX in an instruction such as "OUT 76H,AX". In this case, AL is carried to the 8255 with odd port addresses on D0–D7 and AH is carried to the other 8255 with even port addresses on D8–D15. This is extremely awkward and confusing for the programmer. Figure 26-14 shows the 8255 with odd and even port addresses.

2. The second solution is to connect all 8-bit peripheral ports to data bus D0–D7. This is exactly what IBM PC/AT designers, and indeed all makers of the x86 ISA bus, have done. In such a design, one problem must be solved. What happens when instructions such as "OUT 75H,AL" are executed? This is the odd-addressed port and the data is provided by the CPU on D8–D15, but the port is connected to D0–D7. To solve this problem, one must use a latch to grab the data from bus D8–D15 and provide it to D0–D7, where the port is connected. The latch responsible for this is called the *Hi/Lo byte copier* in ISA bus literature. In order for the Hi/Lo byte copier to work properly, it needs some logic circuitry. It is the function of the bus control logic circuitry to detect the following cases and activate the Hi/Lo byte copier.

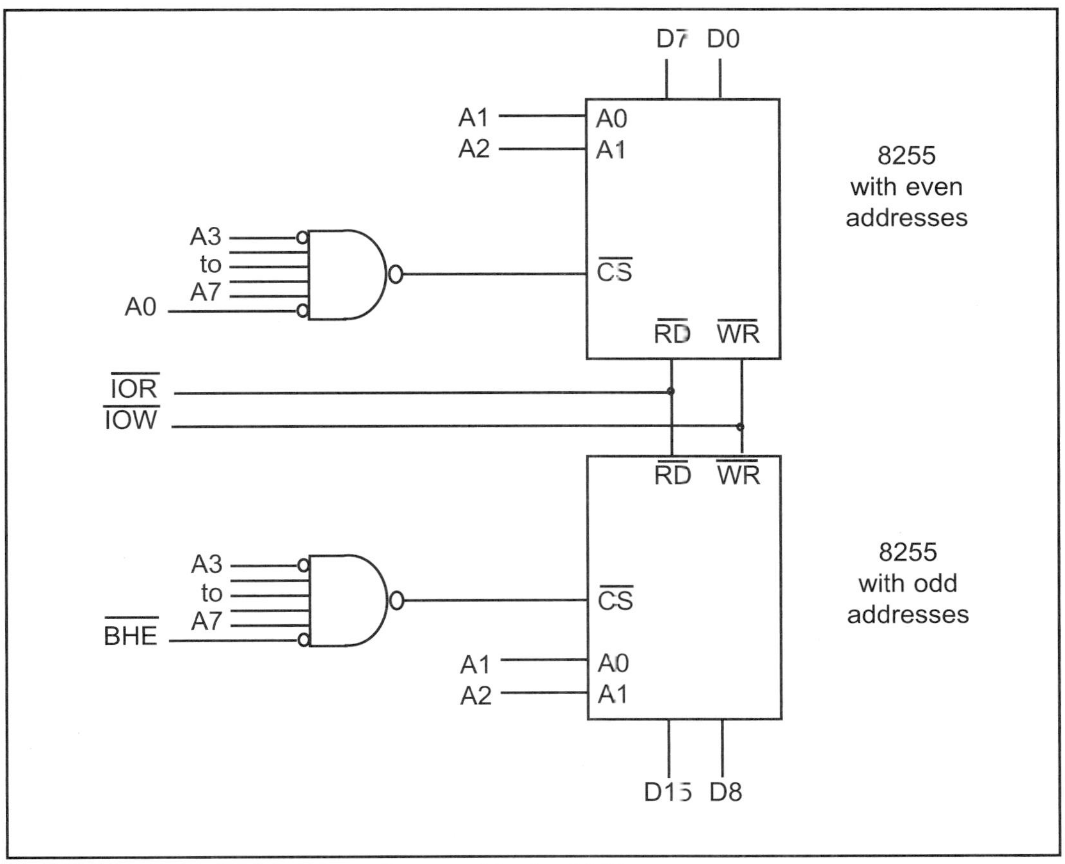

Figure 26-14. Odd and Even Ports with the 8255

Case 1: Outputting a byte to odd-addressed ports

To write a byte to an odd-addressed port, the CPU provides the data on its upper data bus (D8–D15) and makes A0 = 1 and $\overline{BHE}$ = 0 since the port address is an odd address. For example, in the instruction "OUT 41H,AL", the contents of AL are provided to D8–D15 while $\overline{BHE}$ = 0 and A7–A0 = 0100 0001. The bus control logic circuitry senses that the CPU is trying to send 8-bit data to an odd address through its D8–D15 data bus. It activates the Hi/Lo byte copier, which copies the data from D8–D15 to D0–D7. The data is presented to the 8-bit peripheral device, which is connected to the lower data bus, D0–D7.

Case 2: Inputting a byte from odd-addressed ports

To read a byte from an odd-addressed port, the CPU expects to receive the data on its upper data bus (D8–D15) and makes A0 = 1 and BHE = 0. For example, in the instruction "IN AL,43", A7–A0 = 0100 0011 and BHE = 0. The CPU expects the data to come in through D8–D15. The input port device is connected to D0–D7. The bus control logic circuitry senses that the CPU is trying to get 8 bits of data from a peripheral device through its D8–D15 data pins. The port is connected to D0–D7. It activates the Hi/Lo byte copier and copies the data from D0–D7 to D8–D15 and the data is presented to the CPU. The details of bus control logic circuits are quite involved and in today's PCs are buried in the chipsets of x86 PCs.

Limitations of the ISA bus

In 1984 IBM extended the life of the PC/XT bus by adding an extra 36 pins. Although this made the AT bus a 16-bit bus, it did not solve some other problems associated with the AT bus. In 1985 with the introduction of the 386 chip, a microprocessor with a 32-bit data bus, it was obvious that something had to be done about the limitations of the AT bus. The limitations of the ISA are as follows.

1. The data path is limited to 16 bits; therefore, it is unable to accommodate the 32-bit data bus of the 386/486/Pentium microprocessors.
2. The 24-bit address bus limits the maximum memory accessible through the expansion slot to 16M. Therefore, it is unable to accommodate the 32-bit address bus (4-gigabyte address space) of the 386/Pentium.
3. In the ISA motherboard, there could be up to 8 ISA expansion slots. The expansion slot is bulky and has a large surface contact, resulting in a massive amount of capacitance and inductance load on each signal. The accumulated capacitance and inductance associated with all the slots, plus the problem of crosstalk, limits the working frequency of the expansion slot of the ISA bus to 8 MHz. That means that the CPU can be 20, or 33, or even 500 MHz, but when it is communicating with the expansion slot it must slow down to 8 MHz. The absence of extra ground pins to reduce the effects of crosstalk and radio-frequency emissions makes the ISA bus irredeemable for good.
4. Since the interrupts (IRQs) are edge triggered, each can be assigned only to a single device and there cannot be any sharing of the interrupt between two or more devices. In high-frequency systems, the edge-triggered interrupt can also result in false activation of the interrupt due to a spike or noise on the IRQ input.
5. The PC/XT had three 8-bit channels (channels 1–3) for DMA as was shown in Chapter 15. Channel 0 was used for DRAM refreshing. The ISA released channel 0 from the task of refreshing the DRAM and added three more DMA channels, all with 16-bit data transfer capability. This made the ISA bus capable of handling a total of 7 DMA channels, four 8-bit channels and three 16-bit channels. Another major problem of DMA channels in ISA is the 16M address space limitation, due to the availability of the A0–A23 address bus. This means that 386/486/Pentium machines, with their 4-gigabyte memory space, cannot be used for DMA bus activity to transfer data to memory space located beyond 16M.

The combined effects of the above limitations means that the performance of a system with a powerful and fast microprocessor such as the 386/486/Pentium is limited by its expansion slot and system design. This fact led IBM and other PC makers to search for new solutions. While IBM decided to design a whole new bus standard, radically different from the ISA bus, called IBM Micro Channel, other PC makers decided to go for a local bus or extending and improving the ISA bus, which they called the EISA bus. Micro Channel was not made an open architecture by IBM. The industry developed other, more powerful buses; consequently, Micro Channel never became popular and it was eventually discontinued by IBM. The EISA bus was the PC-compatible makers' enhancement to the ISA bus. Although many of the limitations of the ISA bus discussed earlier were removed from the EISA bus, there remained one major one in that the EISA bus was limited to 8 MHz, just like the ISA. This is due to the fact that the PC industry wanted to keep the EISA bus ISA-bus compatible, down to the smallest detail. The EISA bus was in reality an upgrade of ISA and consequently carried many of its limitations, but it had the A23–A31 address lines and D16–D31 data lines to accommodate the 386/486/Pentium.

The low speed and other limitations of the EISA bus led the x86 PC makers to come up with a whole new bus. It is called PCI and we discuss it in the next section.

PC104 bus and embedded PC

The ISA bus has found a new life in the form of the PC104 bus. The name PC104 comes from the fact that 6 additional pins are added to the ISA making it a 104-pin bus. Since ISA has 62 pins for the A & B sides and 36 pins for the C & D sides we have a total of 98 pins for the ISA. Of the 6 additional pins, 5 of them are ground and 1 is the key. See Table 26-3. Examining Table 26-3, we see that all the ISA signals are the same as for PC104. The only difference is the extra ground pins and the key pin. It must be noted that the voltage and current characteristics of PC104 are the same as for ISA signals. The PC104 form-factor is much smaller than the original ISA form-factor and is widely used in embedded systems due to lower power consumption.

For more information on PC104 see http://www.pc104.org.

Review Questions

1. In the ISA bus, we use address ________ to locate the I/O device.
2. What is the maximum number of I/O devices the ISA bus supports?
3. What is role of the ZEROWS pin for I/O cycle time?
4. The IOCS16 pin is an ___________ (input, output) signal.
5. To use the 16-bit I/O capability of the ISA bus we must assert pin _________.
6. What is the minimum I/O cycle time for the 8-bit ISA bus?
7. What is the maximum bus bandwidth for the ISA bus?

SECTION 26.3: PCI BUS

In this section we provide an overview of the PCI bus. But before we do that we need to explain some widely used terminology.

Master and slave

Many devices connected together communicate with each other through address, data, and control buses. When one device wishes to communicate with another, it sends an address to distinguish it from others since each device is assigned a unique address. It also sends a read or write signal to indicate its intention. The master device is the one that initiates and controls the communication while the responding device is called the *slave*. In x86-based PCs, the CPU is an example of a master and memory is an example of a slave.

Bus arbitration

There is only one set of global address, data, and control buses available in a given system. This means that requests by more than one master to use the buses must be arbitrated in an orderly fashion, since no bus can serve two masters at the same time. For a master to access the buses, it must ask permission from the central bus arbitrator and wait for a response before it proceeds. Depending on the system design, the central arbitrator can assign access to each master according to a priority scheme or on a first-come-first-served basis.

Bus protocol

To coordinate activity among various parts of the system, buses must follow a strict set of timing and signal specifications. The term *bus protocol* refers to these specifications for a given bus. The two major bus protocols are synchronous and asynchronous. In synchronous protocol, bus activity is synchronized according to a central frequency, the

Table 26-3: PC104 Signals

J2/P2			J1/P1		
Pin	Row D	Row C	Pin	Row B	Row A
0	**GND**	**GND**	1	GND	IOCHK*
1	MEMCS16*	SBHE*	2	RESET	SD7
2	IOCS16*	LA23	3	+5V	SD6
3	IRQ10	LA22	4	IRQ9	SD5
4	IRQ11	LA21	5	-5V	SD4
5	IRQ12	LA20	6	DRQ2	SD3
6	IRQ15	LA19	7	-12V	SD2
7	IRQ14	LA18	8	SRDY*	SD1
8	DACK0*	LA17	9	+12V	SD0
9	DRQ0	MEMR*	10	KEY	IOCHRDY
10	DACK5*	MEMW*	11	SMEMW*	AEN
11	DRQ5	SD8	12	SMEMR*	SA19
12	DACK6*	SD9	13	IOW*	SA18
13	DRQ6	SD10	14	IOR*	SA17
14	DACK7*	SD11	15	DACK3*	SA16
15	DRQ7	SD12	16	DRQ3	SA15
16	+5V	SD13	17	DACK1*	SA14
17	MASTER*	SD14	18	DRQ1	SA13
18	GND	SD15	19	REFRESH*	SA12
19	**GND**	**KEY**	20	BCLK	SA11
			21	IRQ7	SA10
			22	IRQ6	SA9
			23	IRQ5	SA8
			24	IRQ4	SA7
			25	IRQ3	SA6
			26	DACK2*	SA5
			27	TC	SA4
			28	BALE	SA3
			29	+5V	SA2
			30	OSC	SA1
			31	GND	SA0
			32	**GND**	**GND**

system frequency. In the x86 PC, the CPU accesses memory using synchronous protocol since memory cannot deviate from the timing specifications of the central clock oscillator. Asynchronous protocol obeys its own timing in that it decides when it is ready and does not operate according to the central clock frequency. Printer interfacing in the x86 PC is an example of asynchronous bus protocol. As discussed in Chapter 18, if the CPU is to send data to the printer, it must continuously monitor the printer's busy signal; only when the printer is not busy (ready) can it issue data to the printer's data bus. The CPU must also signal the availability of the data to the printer by the strobe signal and wait for its acknowledgment. In the asynchronous method of CPU-printer communication, the CPU is the master and the printer is the slave. The slave (printer) obeys its own timing for

the acknowledge signal, independent of the system frequency. However, in CPU-memory communication, memory timing specifications are according to the system frequency and the CPU does not poll memory to see if it is ready to accept data. Asynchronous protocol is used when there is a mismatch between the bus timing of the master and slave. Normally, the slave is slower than the master and has self-timing, whereas in synchronous protocol, the timing of the master and slave match. Synchronous protocol generally has a higher rate of data transfer than asynchronous protocol.

Definition and merits of local bus

Just as a high-performance car needs high-performance roads (no bumps, no speed limit) to explore its full potential, high-performance CPUs also require high-performance buses. While microprocessor performance is rapidly rising, buses are not keeping up. Many high-performance supercomputers and mainframes use their own proprietary buses, but their limited use makes them nonstandard and consequently expensive. When 286 microprocessors of 10 to 16 MHz were used, many manufacturers resorted to proprietary buses to overcome the 8-MHz limitation associated with the ISA bus. This was especially the case where memory was concerned. In 80286/386 systems with 16- or 20-MHz speed, memory boards plugged into expansion slots could be accessed no faster than 8 MHz. This fact led manufacturers such as Compaq to have their own memory expansion modules. In such systems, while ISA expansion slots are used for peripheral boards such as video, hard disk, or network cards, memory expansion was done by a specially designed slot on the motherboard used only for memory modules. These memory modules work at the same speed as the CPU, or close to it. These systems were often advertised as dual-bus systems. One bus was for the ISA cards and the other one was for the memory modules. In the late 1990s with the widespread adaptation of SIP (single in-line pin), SIMM (single in-line memory module), and DIMM (dual in-line memory module), this problem was resolved. However, the lack of a bus standard for video and other adapter cards such as disk controllers forced PC board designers to come up with what is called a *local bus*. The idea of a local bus is to access the system buses at the same speed as the microprocessor, or close to it. In a 33-MHz microprocessor system with both ISA and local buses, the speed of the ISA bus signals is limited to 8 MHz, but the local bus signals are accessed at the same speed as the CPU, 33 MHz. In PC/XT systems of 4.7 MHz, the XT buses were accessed at the same speed as the 8088 microprocessor. The gap between CPU speed and expansion slot speed started to develop when the 80286 speed exceeded 8 MHz. In those days, there were not many devices that needed speed beyond 8 MHz. This changed with the introduction of graphical user interface (GUI) software such as Microsoft Windows. In ISA bus systems, even the 16-bit video card plugged into the ISA expansion slot was not fast enough to keep up with the demand of the graphics software. This led some PC manufacturers to embed the video card into the motherboard and bypass the use of an ISA expansion slot for the video board. The problem with this option is that if the video section of the motherboard goes bad, one must either discard the motherboard or connect a video card to the expansion slot, depending on how the system board is designed. To solve the problem of slow video speed in ISA systems, some video board makers used a graphics processor to relieve the main CPU, the 386/486, from the burden of manipulation of graphic data stored in video RAM. In the absence of a graphic processor on the video board, the main CPU is responsible for graphic data manipulation, which means that it must go through the slow ISA bus to access the data since the x86 CPU is connected to the video RAM through the ISA bus. The use of a specially designed processor called a *graphic processing unit* (*GPU*) on the video board with the sole responsibility of taking care of the calculation-intensive work of graphics provided a major improvement in the video system of the PC. However, it has one limitation. If the graphic data needs to be transferred from the disk to video RAM (or vice versa), it must still go through the slow ISA bus. Table 26-4 shows the bus bandwidth requirements for graphics and real-time video. Table 26-4 and Example 26-3 explain why the push for a high-speed bus was led by Intel: It wanted an improvement in bus performance lest their processors be buried under slow buses. The resulting bus standard is called *PCI* (*peripheral component interconnect*).

Table 26-4: Bus Bandwidth Requirements for Graphics and Real-Time Video

Graphics			
Resolution (pixels)	**Colors (bits/pixels)**	**Redraw Rate (updates/s)**	**Bandwidth (bytes/s)**
640 × 480	8	10	2.9M
1024 × 768	16	10	15M
1028 × 1024	24	10	37.5M
Real-Time Video			
Frame Size (Pixels)	**Colors (bits/s)**	**Frame Rate (frames/s)**	**Bandwidth (bytes/s)**
160 × 120	8	15	288K
320 × 240	24	15	3.5M
640 × 480	24	30	26.3M
1024 × 768	24	30	67.5M

Example 26-3

Verify the bus bandwidth requirement for each of the following.
(a) 1024 × 768 resolution, 16 colors, 10 redraw rate
(b) 640 × 480 resolution, 24 colors, 30 frames per second

Solution:

(a) Bus bandwidth = 1024 × 768 × 16 × 10 = 125,829,120 bits/second = 15 megabytes/second
(b) Bus bandwidth = 640 × 480 × 24 × 30 = 221,184,000 bits/second = 26.3 megabytes/second

PCI bus

High-performance microprocessors such as the Pentium require a high bus bandwidth to take advantage of their full potential. Therefore, it is not surprising that Intel became involved in defining a new bus standard. Although Intel came up with the specifications of the PCI local bus, this standard has become available free of charge to all PC and add-in board manufacturers. PCI was conceived as a specification standard for peripheral connections for Intel's high-performance microprocessors such as the 80486 and Pentium. Later, with encouragement and input from the PC industry, it became a local bus standard with the pin-out for expansion slot connections. It has incorporated the following major characteristics: (a) burst mode data transfer, (b) level-triggered interrupts, (c) bus mastering, (d) automatic configuration, and (e) high bus bandwidth. More important, it has a bridge, which allows any kind of add-in card based on ISA to be plugged into the PCI local bus. PCI local bus characteristics are listed next.

PCI local bus characteristics

1. It has a maximum speed of 33 MHz.
2. It has 32- and 64-bit data paths.
3. It supports burst mode data transfer of 2-1-1-1 used by microprocessors such as the Pentium.
4. It supports bus mastering, allowing the implementation of multiprocessors where any number of microprocessors can become master and take control of the buses.
5. It is compatible with ISA. With implementation of a bus bridge, it supports the slow ISA bus as shown in Figure 26-15. Buffers in the bridge allow the microprocessor to write into the buffer and go about its own business, leaving the task of handling the slow ISA to the bridge.
6. The PCI local bus is processor independent. It can be used with any microprocessor, not just Intel x86. For this reason, all companies have also supported the PCI and

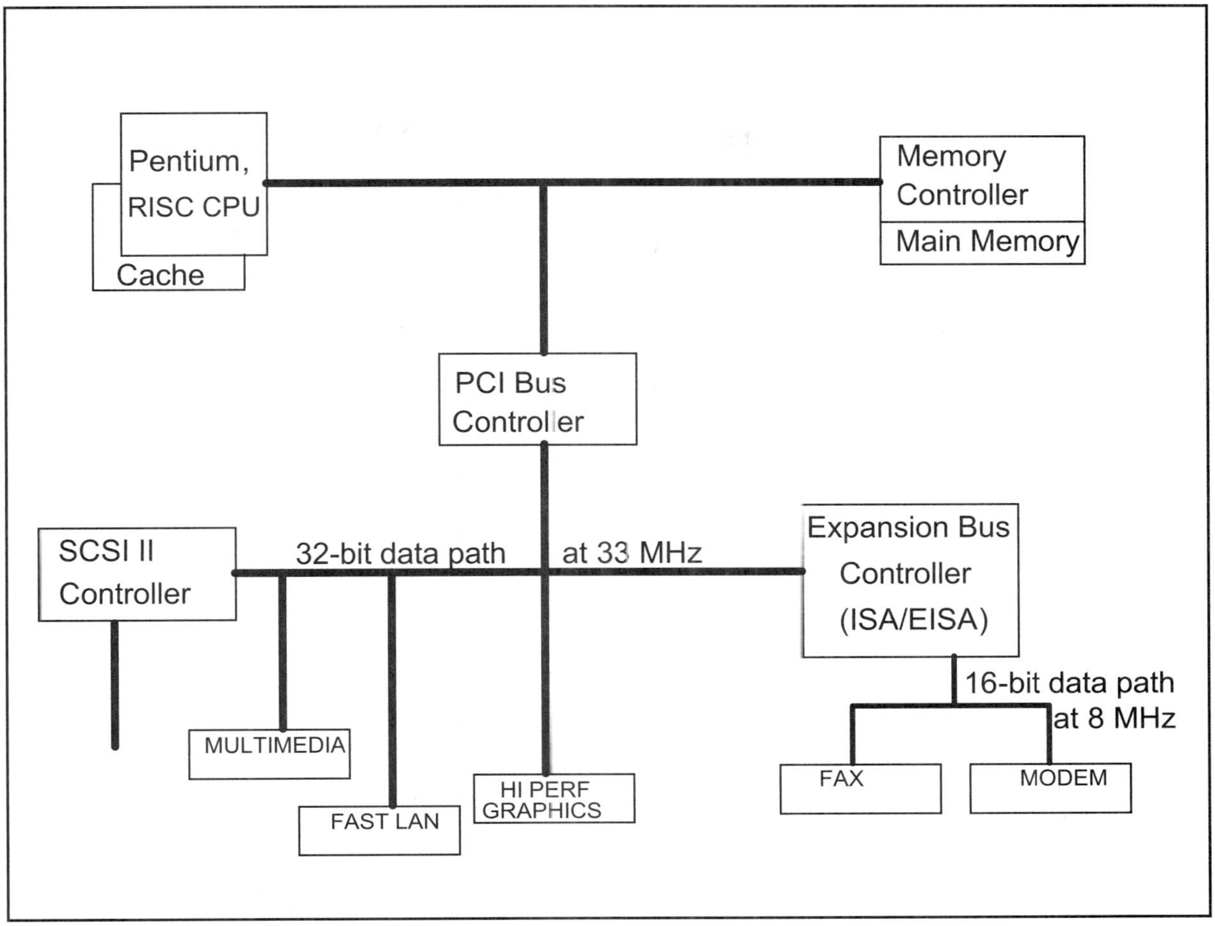

Figure 26-15. PCI Local Bus Architecture
(Reprinted by permission of Intel Corporation, Copyright Intel, 1993)

used it with their non-x86 microprocessors. See Figure 26-16.

7. It supports both 5- and 3.3-V expansion cards, allowing smooth transition from 5- to 3.3-V systems. The placing of small cutouts (keys) prevents users from plugging a card with one voltage into a motherboard with a different voltage.
8. It provides autoconfiguration capability, where a user can install a new add-in card without setting DIP switches, jumpers, and selecting the interrupt. Configuration software automatically selects an unused address and interrupt to resolve conflicts.
9. It has a ground or VCC pin between every two signals to reduce crosstalk and radio-frequency emissions. See Figure 26-17.
10. It implements level-triggered interrupts, which support interrupt sharing.
11. It supports up to 10 peripherals. Some of the peripherals must be embedded into the motherboard.
12. The maximum number of expansion slots working at 33 MHz varies, depending on the 5-V or the 3.3-V implementation. The increase in the number of expansion slots beyond 5 means a speed lower than 33 MHz. The use of a highly refined connector with a small area of contact makes the PCI bus a high-frequency bus.

Plug-and-play feature

The PCI is equipped with the autoconfiguration feature but at the same time it has a slot for the ISA bus, in which autoconfiguration is not supported. How can this work? This lack of autoconfiguration is a major headache for computer users and network managers. This led Microsoft and Intel to work together to equip the ISA bus with the autoconfiguration feature. This feature is often referred to as *plug and play*. The PCI autoconfiguration feature can work completely only after the ISA cards and BIOS are equipped with autoconfiguration (plug and play). Plug and play falls into the following three categories.

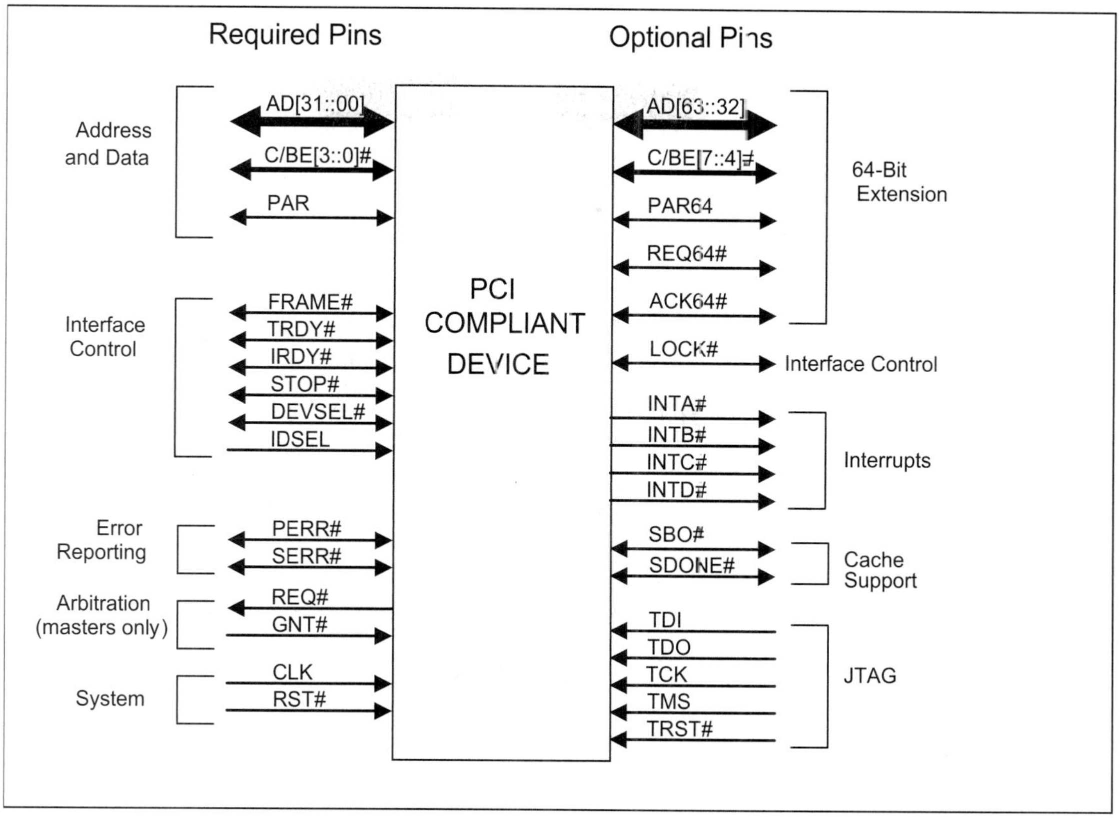

Figure 26-16. PCI Pin List
(Reprinted by permission of PCI Special Interest Group, Copyright 1992, 1993)

1. Neither the motherboard BIOS nor the add-in card is equipped with the plug-and-play feature. This is sometimes called "plug and pray." You may get it to work by trial and error.
2. The motherboard BIOS is equipped with plug and play, but the add-in card is not. In this case, setup software will help you to assign the I/O addresses, IRQs, and DMA channels.
3. Both the motherboard BIOS and the add-in card are equipped for plug and play. In this case, autoconfiguration will take care of everything. It will assign I/O addresses, IRQs, and DMA channels without any user involvement.

PCI connector

A few points must be noted about the PCI connector. First, notice in Figure 26-17 that very few PCI signals match the signals of x86 microprocessors. The reason is that PCI is a mezzanine bus, meaning that the PCI controller sits between the CPU and the external bus connection. In this way, any CPU can be used with the PCI bus. Standardizing the bus connection frees the CPU buses from any restriction. This is in contrast to some other buses, in which signals come directly from the x86 pins and have the same name. When new signals were added to the Pentium, the bus had to be upgraded. PCI solves this problem by being microprocessor independent.

Another point to be noted is the multiplexing of address and data on the PCI bus, since the same pins are used for address and data. In the first clock, the address is provided, and in the second clock, the data is provided. Therefore, the PCI bus has a cycle time of 2 clocks in nonburst mode, just like the Pentium. For burst mode, in the first clock the address is provided and in each subsequent clock a word (32-bit) of data is provided.

Another point to be noted is the 64-bit extension for the PCI bus. The PCI bus can be implemented for a 32-bit data bus or a 64-bit data bus. The 32-bit sections end at pin

pin	5V Environment		3.3V Environment	
pin	Side B	Side A	Side B	Side A
1	-12V	TRST#	-12V	TRST#
2	TCK	+12V	TCK	+12V
3	Ground	TMS	Ground	TMS
4	TDO	TDI	TDO	TDI
5	+5V	+5V	+5V	+5V
6	+5V	INTA#	+5V	INTA#
7	INTB#	INTC#	INTB#	INTC#
8	INTD#	+5V	INTD#	+5V
9	PRSNT1#	Reserved	PRSNT1#	Reserved
10	Reserved	+5V(I/O)	Reserved	+3.3V(I/O)
11	PRSNT2#	Reserved	PRSNT2#	Reserved
12	Ground	Ground	CONNECTOR KEY	
13	Ground	Ground	CONNECTOR KEY	
14	Reserved	Reserved	Reserved	Reserved
15	Ground	RST#	Ground	RST#
16	CLK	+5V(I/O)	CLK	+3.3V(I/O)
17	Ground	GNT#	Ground	GNT#
18	REQ#	Ground	REQ#	Ground
19	+5V(I/O)	Reserved	+3.3V(I/O)	Reserved
20	AD[31]	AD[30]	AD[31]	AD[30]
21	AD[29]	+3.3V	AD[29]	+3.3V
22	Ground	AD[28]	Ground	AD[28]
23	AD[27]	AD[26]	AD[27]	AD[26]
24	AD[25]	Ground	AD[25]	Ground
25	+3.3V	AD[24]	+3.3V	AD[24]
26	C/BE[3]#	IDSEL	C/BE[3]#	IDSEL
27	AD[23]	+3.3V	AD[23]	+3.3V
28	Ground	AD[22]	Ground	AD[22]
29	AD[21]	AD[20]	AD[21]	AD[20]
30	AD[19]	Ground	AD[19]	Ground
31	+3.3V	AD[18]	+3.3V	AD[18]
32	AD[17]	AD[16]	AD[17]	AD[16]
33	C/BE[2]#	+3.3V	C/BE[2]#	+3.3V
34	Ground	FRAME#	Ground	FRAME#
35	IRDY#	Ground	IRDY#	Ground
36	+3.3V	TRDY#	+3.3V	TRDY#
37	DEVSEL#	Ground	DEVSEL#	Ground
38	Ground	STOP#	Ground	STOP#
39	LOCK#	+3.3V	LOCK#	+3.3V
40	PERR#	SDONE	PERR#	SDONE
41	+3.3V	SBO#	+3.3V	SBO#
42	SERR#	Ground	SERR#	Ground
43	+3.3V	PAR	+3.3V	PAR
44	C/BE[1]#	AD[15]	C/BE[1]#	AD[15]
45	AD[14]	+3.3V	AD[14]	+3.3V
46	Ground	AD[13]	Ground	AD[13]
47	AD[12]	AD[11]	AD[12]	AD[11]
48	AD[10]	Ground	AD[10]	Ground
49	Ground	AD[09]	Ground	AD[09]

pin	5V Environment		3.3V Environment	
pin	Side B	Side A	Side B	Side A
50	CONNECTOR KEY		Ground	Ground
51	CONNECTOR KEY		Ground	Ground
52	AD[08]	C/BE[0]#	AD[08]	C/BE[0]#
53	AD[07]	+3.3V	AD[07]	+3.3V
54	+3.3V	AD[06]	+3.3V	AD[06]
55	AD[05]	AD[04]	AD[05]	AD[04]
56	AD[03]	Ground	AD[03]	Ground
57	Ground	AD[02]	Ground	AD[02]
58	AD[01]	AD[00]	AD[01]	AD[00]
59	+5V(I/O)	+5V(I/O)	+3.3V(I/O)	+3.3V(I/O)
60	ACK64#	REQ64#	ACK64#	REQ64#
61	+5V	+5V	+5V	+5V
62	+5V	+5V	+5V	+5V
	CONNECTOR KEY		CONNECTOR KEY	
	CONNECTOR KEY		CONNECTOR KEY	
63	Reserved	Ground	Reserved	Ground
64	Ground	C/BE[7]#	Ground	C/BE[7]#
65	C/BE[6]#	C/BE[5]#	C/BE[6]#	C/BE[5]#
66	C/BE[4]#	+5V(I/O)	C/BE[4]#	+3.3V(I/O)
67	Ground	PAR64	Ground	PAR64
68	AD[63]	AD[62]	AD[63]	AD[62]
69	AD[61]	Ground	AD[61]	Ground
70	+5V(I/O)	AD[60]	+3.3V(I/O)	AD[60]
71	AD[59]	AD[58]	AD[59]	AD[58]
72	AD[57]	Ground	AD[57]	Ground
73	Ground	AD[56]	Ground	AD[56]
74	AD[55]	AD[54]	AD[55]	AD[54]
75	AD[53]	+5V(I/O)	AD[53]	+3.3V(I/O)
76	Ground	AD[52]	Ground	AD[52]
77	AD[51]	AD[50]	AD[51]	AD[50]
78	AD[49]	Ground	AD[49]	Ground
79	+5V(I/O)	AD[48]	+3.3V(I/O)	AD[48]
80	AD[47]	AD[46]	AD[47]	AD[46]
81	AD[45]	Ground	AD[45]	Ground
82	Ground	AD[44]	Ground	AD[44]
83	AD[43]	AD[42]	AD[43]	AD[42]
84	AD[41]	+5V(I/O)	AD[41]	+3.3V(I/O)
85	Ground	AD[40]	Ground	AD[40]
86	AD[39]	AD[38]	AD[39]	AD[38]
87	AD[37]	Ground	AD[37]	Ground
88	+5V(I/O)	AD[36]	+3.3V(I/O)	AD[36]
89	AD[35]	AD[34]	AD[35]	AD[34]
90	AD[33]	Ground	AD[33]	Ground
91	Ground	AD[32]	Ground	AD[32]
92	Reserved	Reserved	Reserved	Reserved
93	Reserved	Ground	Reserved	Ground
94	Ground	Reserved	Ground	Reserved

Figure 26-17. Pinout of the PCI Connector

(Reprinted by permission of PCI Special Interest Group, Copyright 1992, 1993)

62. The pinout is shown in Figure 26-17. Pins 63 through 94 are used for 64-bit data/address extension only.

Notice also in Figure 26-17 that every third pin is dedicated to ground or VCC. This eliminates the crosstalk problem and allows the bus to be used for frequencies up to 33 MHz.

Figure 26-18 illustrates the PCI board connectors.

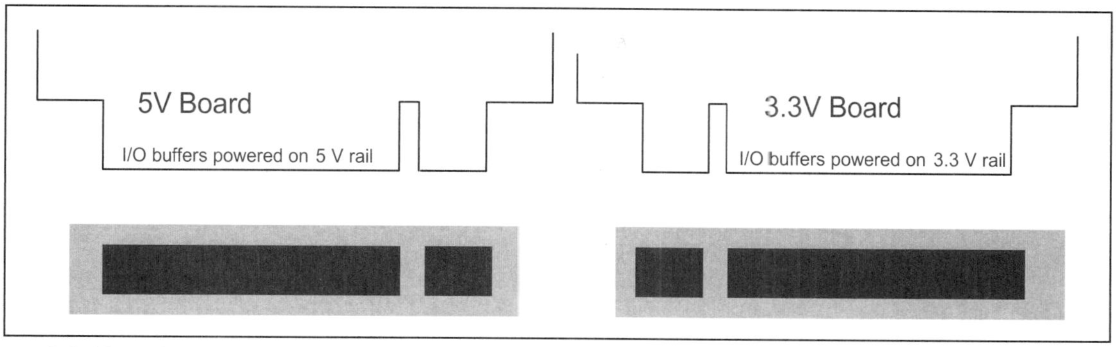

Figure 26-18. PCI Board Connectors
(Reprinted by permission of PCI Special Interest Group, Copyright 1992, 1993)

PCI performance

The PCI local bus supports both single memory cycle and burst mode. In the single cycle, it takes 2 clocks to read or write a word of data. In the first clock, the address is provided and in each subsequent clock, the data is accessed. This makes it 2-1-1-1-1-1.... Example 26-4 calculates the bus bandwidth for the PCI.

Table 26-5 provides the performance comparison of all the buses for non-burst mode data transfer. Table 26-6 shows the performance of buses and ports.

Example 26-4

Calculate the bus bandwidth of PCI for (a) single and (b) burst transfer, both on a 32-bit data path.

Solution:

PCI can work up to a maximum of 33 MHz. The clock period is 30 ns.
(a) For the single transfer, each transfer takes 2 clocks or a total of 60 ns to transfer 4 bytes (32 bits) of data. Therefore, bus bandwidth = (1/60 ns) × 4 bytes = 66.6 megabytes/second
(b) In burst mode, ignoring the overhead of the first clock for the address, it takes 1 clock or 30 ns to transfer 32-bit data. Therefore, bus bandwidth = (1/30 ns) × 4 bytes = 133 megabytes/second.

Table 26-5: ISA, EISA, and PCI Local Bus Bandwidth Comparison

	ISA	EISA	PCI	PCI
Data path (bits)	16	32	32	64
Bus speed (MHz)	8	8.3	33	33
Bandwidth (megabytes/s)	8	16	66	133

Note: In the bus bandwidth calculation, 2 clocks per memory cycle are assumed.

Table 26-6: Data Transfer Rate for Buses and Ports

Bus/Port	Data Path bit	Maximum Bus Bandwidth
ISA (8 MHz)	16	8M bytes/second
PCI (33 MHz)	32	66M bytes/second
PCI (33 MHz)	64	133M bytes/second
PCI (66 MHz)	64	266M bytes/second
USB 2.0	1	480M bits/second
LPT	8	500K bytes/second
COM port	1	56K bits/second

Note: In the bus bandwidth calculation , 2 clocks per memory cycle are assumed .

Review Questions

1. What is the local bus?
2. The memory expansion connections to 16-MHz CPUs are through the _______ (ISA bus, local bus).
3. What is a dual bus system?
4. Which needs a local bus, the modem or the hard disk controller?
5. True or false. PCI is a 32- and a 64-bit bus.
6. How has PCI reduced the effects of crosstalk (EMI) for high-frequency systems?

> PCI Express (PCIe) is becoming the standard for 64-bit systems.
> Use Google to research the PCIe bus.

PROBLEMS

SECTION 26.1: ISA BUS MEMORY SIGNALS

1. Which of the control signals are used for ISA memory interfacing if the address of memory is in the range of F0000–FFFFFH?
2. The ISA bus can access a maximum of _____ bytes. Why?
3. The MEMCS16 is an active- _____ signal. Is this an input signal?
4. Explain the use of the MEMCS16 pin.
5. The ZEROWS is an active- ____ signal. Is this an input signal?
6. Explain the use of the ZEROWS pin.
7. If the MEMCS16 pin is high, what portion of the data bus is being used?
8. If the MEMCS16 pin is low, what portion of the data bus is being used?
9. If the ZEROWS pin is high, give the memory cycle time for an 8-bit data transfer.
10. If the ZEROWS pin is high, give the memory cycle time for a 16-bit data transfer.
11. If the ZEROWS pin is asserted low, give the memory cycle time for a 16-bit data transfer.
12. If the ZEROWS pin is asserted low, give the memory cycle time for an 8-bit data transfer.
13. To achieve the best data transfer rate for ISA bus memory interfacing, what should be the status of the MEMCS16 and ZEROWS pins?

14. Fill the blanks for following cases.

MEMCS16	ZEROWS	Data bus used	Read Cycle time	Bus Bandwidth
0	0	______	_______	_______
0	1	______	_______	_______
1	0	______	_______	_______
1	1	______	_______	_______

15. Why do we use DIMM and SIMM sockets for memory expansion instead of the ISA bus slot?

SECTION 26.2: I/O BUS TIMING IN ISA BUS

16. Explain the role of ZEROWS in I/O timing.
17. Explain the role of IOCS16 in 16-bit I/O.
18. In the ISA bus, the default mode for I/O operation is ________ (8-bit, 16-bit).
19. What is the clock speed for the ISA bus?
20. How many WS are used in the 8-bit standard I/O cycle?
21. In Problem 20, how much does it take to complete one I/O read cycle?
22. How many WS are used in the 8-bit I/O cycle when ZEROWS is asserted?
23. In Problem 21, how much does it take to complete one I/O read cycle?
24. How many WS are used in the 16-bit standard I/O cycle?
25. In Problem 24, how much does it take to complete one I/O read cycle?
26. The IOCS16 pin is an ______ (input, output) and an active- ______ (low, high) signal.
27. The ZEROWS pin is an ________ (input, output) and an active- ______ (low, high) signal.
28. Calculate the bus bandwidth for 8-bit standard I/O of the ISA bus.
29. Calculate the bus bandwidth for 16-bit standard I/O of the ISA bus.
30. What is the function of the CHANRDY pin?
31. Explain how we can extend the I/O cycle time of the ISA bus.

SECTION 26.3: PCI BUS

32. The PCI has a maximum speed of ____ MHz.
33. True or false. The PCI bus can accommodate 64-bit data buses of the Pentium.
34. True or false. The PCI bus supports autoconfiguration.
35. True or false. Interrupt sharing is not allowed in PCI.
36. Calculate and compare the maximum bus bandwidth for the 32-bit PCI. Assume that it is non-burst mode.
37. Calculate the maximum bus bandwidth for the following. Assume burst mode.
    (a) 32-bit PCI (b) 64-bit PCI

ANSWERS TO REVIEW QUESTIONS

SECTION 26.1: ISA BUS MEMORY SIGNALS

1. Low
2. D0–D7
3. Low
4. 6 clocks since 2 + 4 WS = 6
5. 8 megabytes /sec.

SECTION 26.2: I/O BUS TIMING IN ISA BUS

1. A0–A9
2. With ten address lines, A0–A9, we get 1024 I/O devices.

3. By asserting it, we tell the system board to shorten the I/O bus cycle time.
4. Input
5. IOCS16
6. 3 clock cycles (only 1 WS) if we assert ZEROWS. That gives us 3 × 125 ns = 375 ns.
7. The maximum bus bandwidth is achieved with the 16-bit data bus and it is 1/(3 × 125 ns) × 2 = 5.33 megabytes per second.

SECTION 26.3: PCI BUS

1. It is the bus that is closely attached to the CPU and works at the same frequency as the CPU (or close to it).
2. Local bus
3. A system that has both a PCI bus and a DIMM memory bus.
4. Hard disk controller
5. True
6. By placing a ground or Vcc pin between every two signal lines

CHAPTER 27

USB PORT PROGRAMMING

OBJECTIVES

Upon completion of this chapter, you will be able to:

- **>>** **Compare USB performance with PCI**
- **>>** **Compare USB performance with LPT and serial COM ports**
- **>>** **Understand the difference between the host and peripheral devices**
- **>>** **Define the terms upstream and downstream**
- **>>** **Describe the role of hubs in USB port expansion**
- **>>** **Understand the difference between A-type and B-type connectors**
- **>>** **Describe the USB cable signals**
- **>>** **Describe the differences between bus and self-powered hubs in USB**
- **>>** **Understand the current needs of peripheral devices and hosts**
- **>>** **Program USB devices in the Windows operating system using C#**

This chapter deals with the basics of USB ports and how we program them in the x86 PC using the C language. In Section 27.1, we examine some general features of the USB port and compare its performance with the PCI bus and COM port. Section 27.2 deals with USB port expansion, power management, and cables. USB programming using Microsoft's C# is discussed in Section 27.3.

SECTION 27.1: USB PORTS: AN OVERVIEW

USB stands for *universal serial bus*. Next to PCI, it is one of the most important additions to the PC system in recent years. In this section we provide an overview of the USB bus. To see the need for the USB bus, we first review the limitations and benefits of ISA and PCI buses, as well as serial and parallel ports. ISA and PCI buses provide a high rate of data transfer between the CPU and the outside world. This is because they have a wide data path, high-frequency bus speed, and communicate directly with the CPU. As mentioned in previous chapters, the ISA bus is a 16-bit bus and has a speed of 8 MHz. For the PCI bus, the data path is 64 bits wide and has a maximum speed of 64 MHz. The following are some of the major limitations of the ISA and PCI buses:

1. Both ISA and PCI are inside the PC; therefore, to access them you need to open the PC's case and plug the card into an expansion slot.
2. The PCI and ISA expansion slots take too much physical space on the motherboard. This limits the number of expansion slots that are available on a given motherboard.
3. Both ISA and PCI buses require too much power. For every ISA expansion slot, an extra 25 watts must be incorporated into the PC's power supply. Every PCI slot requires an extra 10 watts. As a result, a motherboard with 2 ISA and 3 PCI expansion slots has burdened the PC power supply with an additional 80 watts of power ($2 \times 25 + 3 \times 10 = 80$). This can make a significant difference in handheld and laptop systems, and that is the reason that neither of these devices has any expansion slots on its boards.

The most important disadvantage of serial and parallel ports is the limit of 4of each in a given motherboard. The PC BIOS limits the number of serial (COM) and parallel (LPT) ports to 4. Of course, this is a theoretical limitation imposed by the BIOS. Practically speaking, there are a limited number of IRQs to assign to all of these LPT and COM ports. This is the reason that there are no more than 2 COM ports and 1 LPT port on motherboards. There are other limitations associated with the COM and LPT ports that come to light only when compared with the major features of the USB port.

Major features of USB

Here are some of the most important features of the USB. They demonstrate why this is one of the most important additions to PC architecture in recent years.

1. A single USB port can accommodate up to 127 devices such as a mouse, scanner, printer, modem, and so on. The devices are daisy chained together with the help of external hubs. The devices are recognized automatically by the PC. Many devices such as printers and monitors can be equipped with a hub, thereby saving the additional expense of buying a separate hub. More importantly, daisy chaining the devices via hubs requires no opening of the PC case when connecting additional devices to a PC.
2. The data transfer rates for USB 1.1 are 1.5 and 12 megabits per second (Mb/s). The USB 2.0 raised the maximum limit to 480 Mbs.
3. USB is hot-pluggable. This means that new devices can be connected to the USB port without first turning off the PC. Remember that this was not the case with ISA and PCI. In all these devices, the PC had to be turned off prior to installing the device and configuring the system. The hot-pluggability of USB is one of its most important features.
4. USB does not have to burden the system's power supply. Unlike ISA and PCI, connecting additional USB devices to a PC does not require an exorbitant amount of

power from the PC power supply. Each new device requires no more than a maximum of 500 mW from the PC power supply. More importantly, the USB hubs sitting outside the PC can have their own power supply, thereby relieving the motherboard's power supply of the burden of providing power to every device. For example, a printer or monitor that is equipped with a USB hub can provide power to all external USB devices. The USB is also equipped with power managing capability. This allows a device that is not being used for a period of time to be powered down into sleep mode. Starting with Windows 98, USB device management is part of the operating system; the same is true for Windows NT/2000, XP, and Vista.

Bus comparison

Table 27-1 shows the comparison of ISA, PCI, COM, LPT, and USB. Notice in the calculation of the data transfer rate (bus bandwidth) for ISA and PCI buses, that a 2-clock read and write cycle is assumed. For the LPT port, the 2-microsecond timing for the parallel port is assumed. Both Microsoft and Intel have worked hard to eliminate ISA bus, LPT, and COM ports from PC motherboards. In many of today's desktop PCs we have PCI and USB ports only.

Table 27-1: USB Bus Performance Comparison

Bus/Port	Data Path (bits)	Maximum Bus Bandwidth
ISA (8 MHz)	16	8M bytes/second
PCI (33 MHz)	32	66M bytes/second
PCI (33 MHz)	64	133M bytes/second
PCI (66 MHz)	64	266M bytes/second
USB 1.0	1	1.5M bits/second
USB 1.1	1	12M bits/second
USB 2.0	1	480M bits/second
LPT	8	500K bytes/second
COM	1	56K bits/second

Note: In the bus bandwidth calculation, 2 clocks per memory cycle are assumed.

Review Questions

1. True or false. Both PCI and USB are hot-pluggable.
2. What advantage do USB devices have over other devices in terms of system power requirements?
3. True or false. Each USB device must have its own USB connection to the system via an expansion slot.

SECTION 27.2: USB PORT EXPANSION AND POWER MANAGEMENT

Host, peripherals, and hubs

In discussing USB, we clarify some widely used terminology. In the USB, the host is the *master* and the peripheral I/O device is the *slave,* meaning that in any given system there is only one host (master), which controls many peripheral I/O devices (slaves). The data going toward the host is called *upstream* and data flowing from the host toward peripherals is referred to as *downstream*. Many x86 IBM PC motherboards come with a single USB host that can be connected to peripheral I/O devices. Since a motherboard has only one host and it needs to connect to many peripheral I/O devices, we need what is called a *hub*. In motherboards with two or more USB ports, the hub is incorporated into the host. This hub on the x86 motherboard's host is commonly referred to as the *root hub*. See Figure 27-1. The root hub on an x86 motherboard has anywhere from two to six ports, where each port can be connected to a peripheral I/O device or another hub.

If we need to attach more peripheral I/O devices, then we use an external hub to expand the number of USB ports. A USB hub has one upstream port and several downstream ports and we must use a special USB cable to connect the external hubs together. The special cable used for connecting the hubs together has A-type and B-type connectors. The A-type connector is connected to one of the USB ports while the B-type connector is connected to the upstream port of the hub. Examining the USB cable with both A-type and B-type connectors, you will notice that they are shaped differently. See Figure 27-2. This should prevent any mixing of upstream and downstream connections. The vast majority of the USB cables you see in everyday use are USB extension cables. The USB extension cable has A-type connectors on both ends. It is used for connecting a peripheral I/O device such as a Flash memory stick to the USB port of the x86 PC. Such a cable has a male (plug) A-type connector on one end and a female (receptacle) A-type connector on the other end. See Figure 27-2.

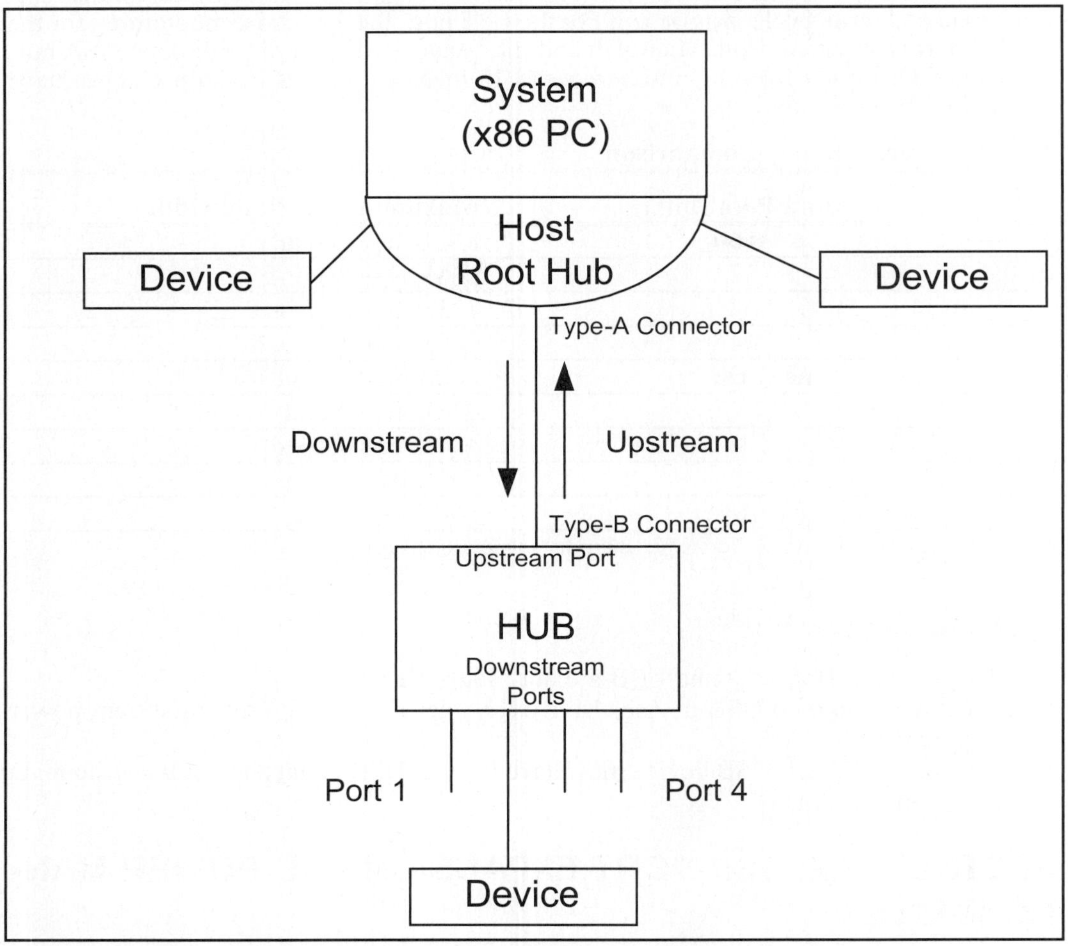

Figure 27-1. Host and Root Hub in x86 PC

Expanding USB ports with external hubs

As we discussed earlier, we can use an external hub to connect many peripheral I/O devices to a single host. There are two types of external hubs: bus-powered and self-powered. In the bus-powered hub, the power source comes from the root hub of the host PC via a USB cable. The self-powered hub has its own power source, and that is the reason it is also referred to as as an externally powered hub.

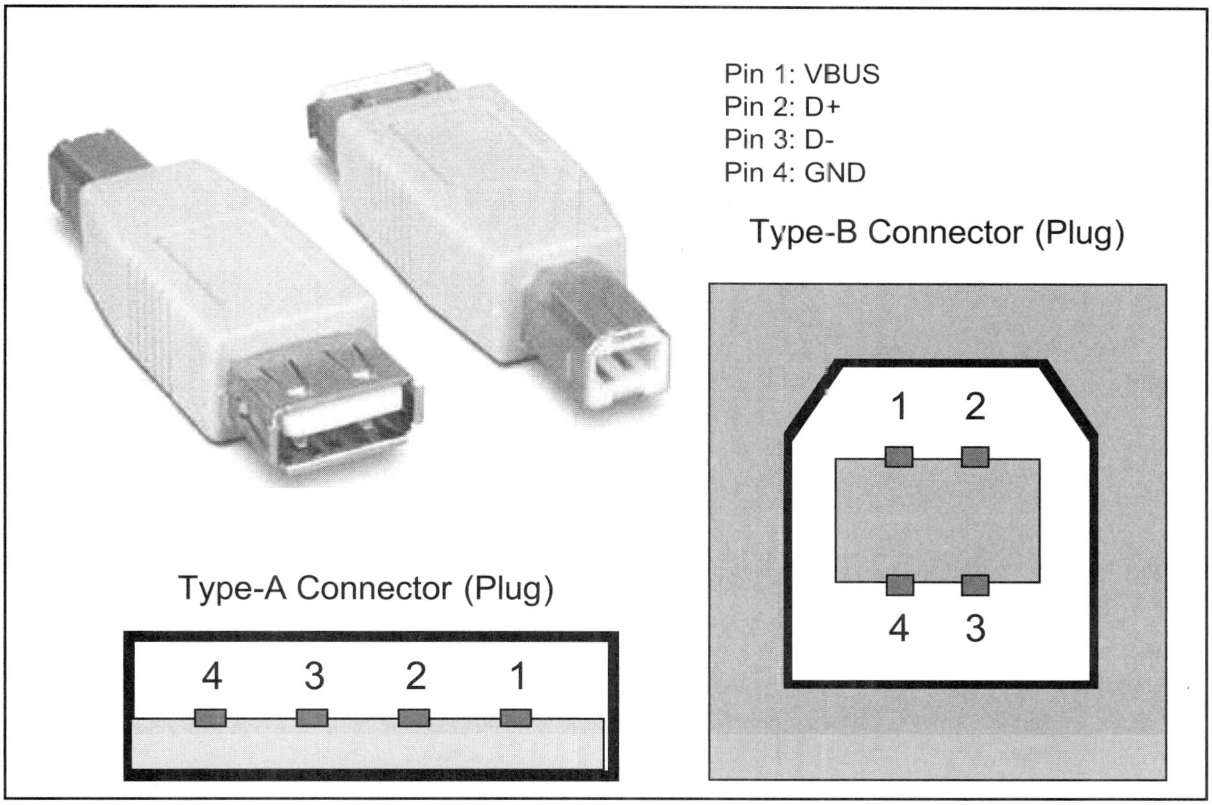

Figure 27-2. USB Connectors

Bus-powered hubs

The vast majority of peripheral devices that we use in everyday life, such as Flash memory sticks, are bus-powered, which means that the USB cable provides power. In the x86 PC, the root hub provides power to all USB peripheral devices connected to many of its USB ports. These self-powered devices use USB cables to get power from the x86 PC's power supply. Since the maximum current usage by a downstream peripheral device is 100 mA, a root hub with 6 USB ports uses no more than 600 (6 × 100 mA = 600 mA) current from the PC's power supply. By connecting the bus-powered external hubs, we can increase the number of USB ports. In that case, the host can provide a maximum of 500 mA of current (and 5V power source) per A-connector. Of the 500 mA provided to the bus-powered external hub, it uses 100 mA for its own internal circuitry. That leaves only 400 mA to be given to the downstream peripheral ports. Because each downstream peripheral port must be given 100 mA, the number of downstream peripheral ports in the bus-powered hub is limited to 4 (4 × 100 mA + 100 mA = 500 mA).

Self-powered hubs

In the self-powered external hub, we need a power source other than the motherboard. This generally comes from the wall outlet. Notice that the self-powered hub is also referred to as an externally powered hub. In the case of the self-powered external hub, each peripheral port is provided 500 mA. That means that an external self-powered hub with four downstream ports uses at least 2000 mA (500 mA × 4 = 2000 mA) in addition to the power it needs for its own internal circuitry. There are some hubs with more than 4 USB connectors. They are self-powered with their own power source, since they need more than 2000 mA current. Printers with a USB port often have a self-powered hub, drawing from the printer's power supply. The devices with I/O functions and the hub are often called *compound devices* in the USB literature. See Figure 27-3.

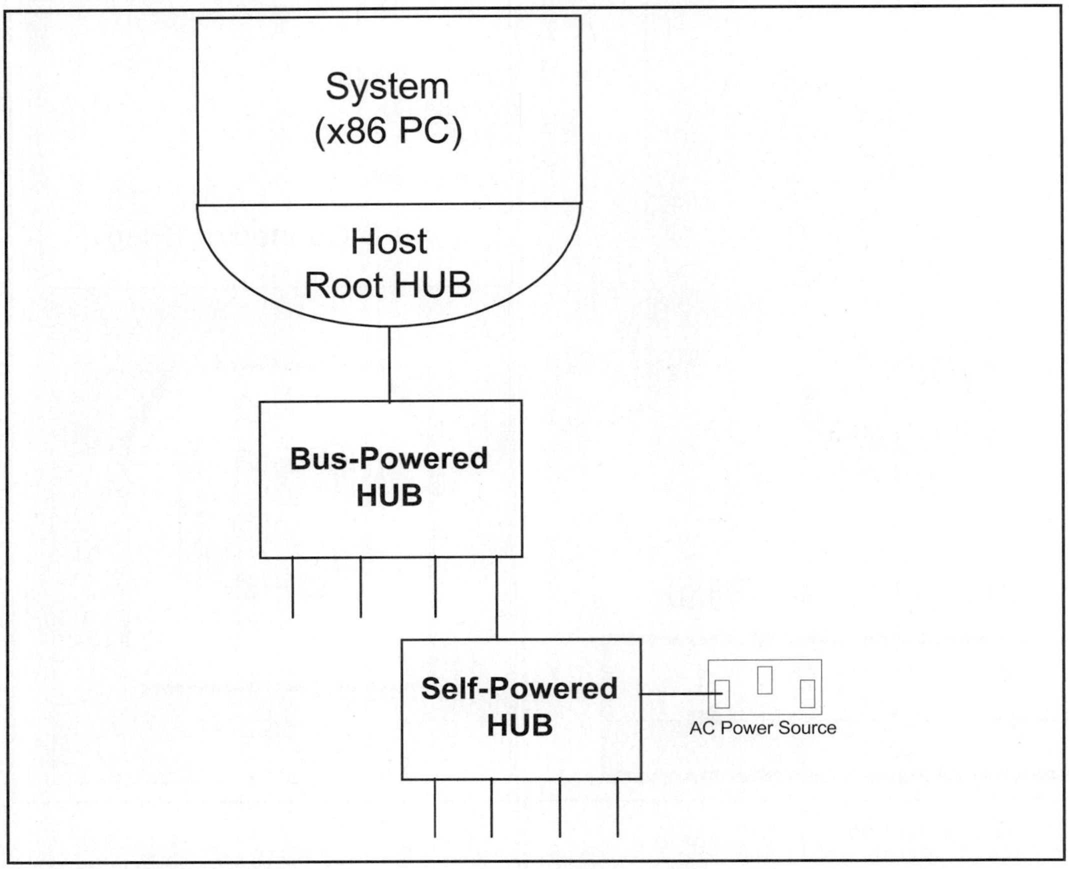

Figure 27-3. Bus-Powered and Self-Powered Hubs

Daisy chaining the hubs

The USB bus allows up to five levels of external hubs to be connected to the root hub. Of course, due to the limited power provided by the root hub, many of these hubs must be self-powered. Notice that USB documentation terminology uses the word *tier* instead of level. It also states that the total number of tiers (levels) cannot be more than seven. This means the number of levels from the farthest peripheral device to the host (root hub) must be no more than seven. The host (root hub) is counted as the first tier (level), and the number of external hubs that can be cascaded together is limited to five. This makes the peripheral device connected to the last hub tier number seven. See Figure 27-4.

USB cable signals

The USB cable has four wires inside it. They are Vcc, Gnd, D+, and D–. While Vcc and Gnd provide the power source, D+ and D– are for the data path. See Figure 27-2. The D+ and D– wires provide the data path between the host and the peripheral device using the half-duplex method of data transfer. That means that at any given time the data either is going from the device to the host or coming from the host to the device, but never at the same time. This is in contrast to the RS232 serial port, which has full-duplex capability since it has both TxD (transmit data) and RxD (receive data) pins. Notice that just like the RS232, the USB cable does not have any wire for clock to synchronize data transfer. That is the reason that USB is called an *asynchronous* bus. The USB port uses the D+ and D– pins to implement the asynchronous method of data transfer. It uses NRZI (none return to zero inverted) encoding. The maximum cable length for the USB is 5 meters (15 feet). In high-speed data transfers, the maximum length needs to be less than 5 meters.

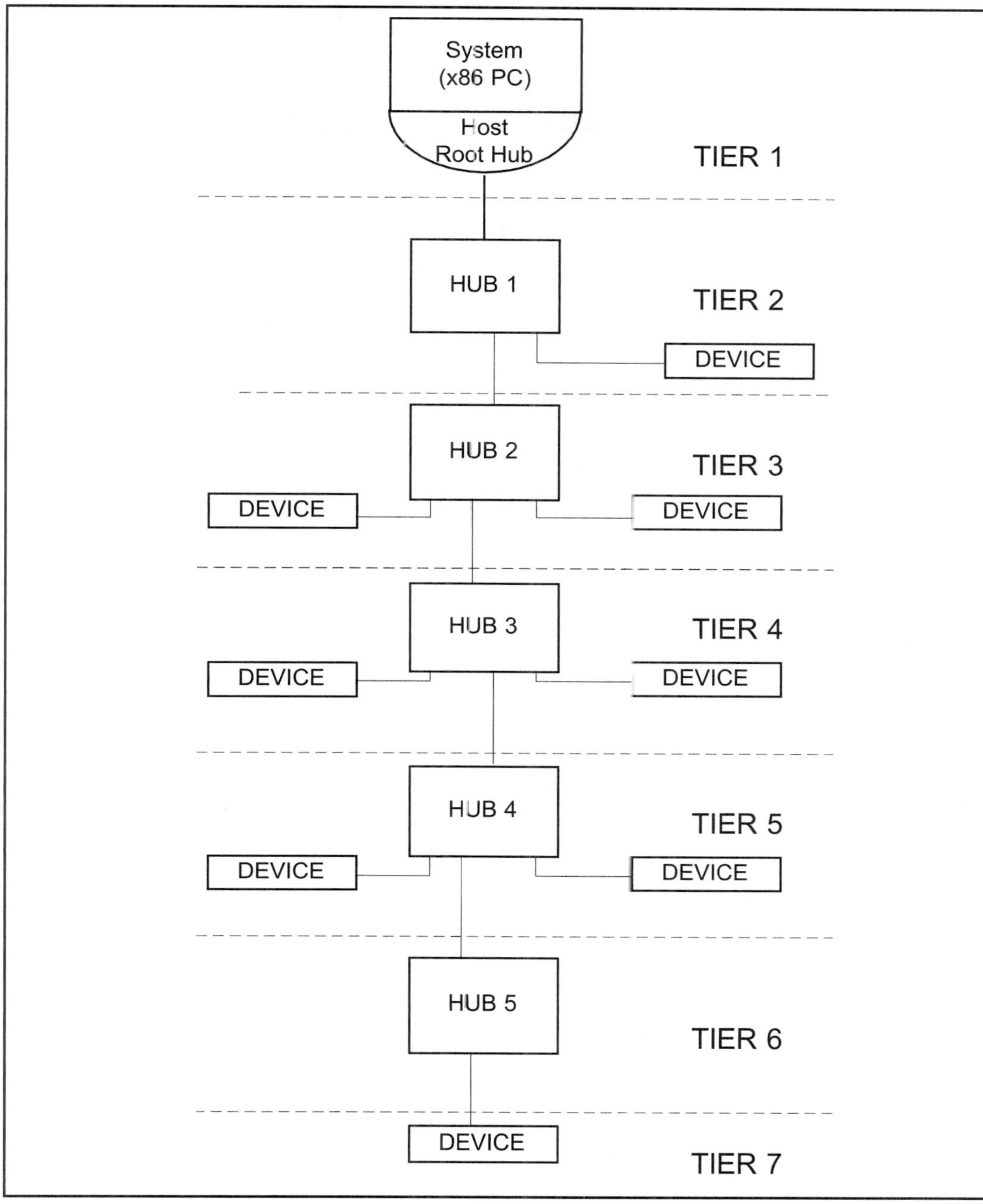

Figure 27-4. Daisy Chaining Hubs

USB enumeration

The USB devices are hot-pluggable, meaning that the moment we attach a USB peripheral device to the USB port of the x86 PC, it is recognized by the host. That means that there is no need to turn off the PC and then turn it back on in order for the device to be recognized. When we connect a peripheral device to a USB port, the host senses the voltage change on the data line via the cable. The host enquires about this USB device and after receiving a satisfactory answer, the peripheral device is assigned an address. The address value has seven bits, which can take values from 0 to 127 (0–7F in hex, or 0000000–1111111 in binary). However, the host assigns addresses between 1 and 127, since the address 0 is the default address of any peripheral device before it is recognized by the host. This process of recognizing a device and assigning it an address is called *enumeration* in the USB literature. As long as this peripheral device is connected to the host and the host is powered, this unique address value is not changed. If we disconnect the peripheral device from the host, the host will take back the address and keep it in the pool of 127 addresses available for future assignments.

Review Questions

1. True or false. The USB is hot-pluggable
2. True or false. Each USB peripheral device must have its own power source.
3. True or false. Data going from the host to the peripheral is called upstream.
4. What is the maximum current provided by the host to each peripheral?
5. What is the difference between the A side and the B side of a USB cable?

SECTION 27.3: USB PORT PROGRAMMING

In this section we discuss how to program USB devices in the Windows operating system. We use C# .NET software, which is available for free from Microsoft's web site.

In programming the peripherals connected to the x86 PC USB host, the following points must be remembered:

1. A single USB host can connect with up to 127 peripheral devices.
2. Although the x86 PC can have multiple USB hosts, we are assuming that there is only one host in the x86 PC.
3. In the Windows OS, opening a connection to a peripheral device is like opening a file.
4. The API provided with the driver handles where the data will go to or come from.

Communicating via USB-Bluetooth

As an example of programming a USB port using C#, we can use a USB-to-Bluetooth device to communicate between two PCs. In such cases, each PC will have a USB-Bluetooth device connected to its USB port. See Figure 27-5. Before we show the programming of the USB port, we can see the communications between two PCs via Bluetooth using HyperTerminal.

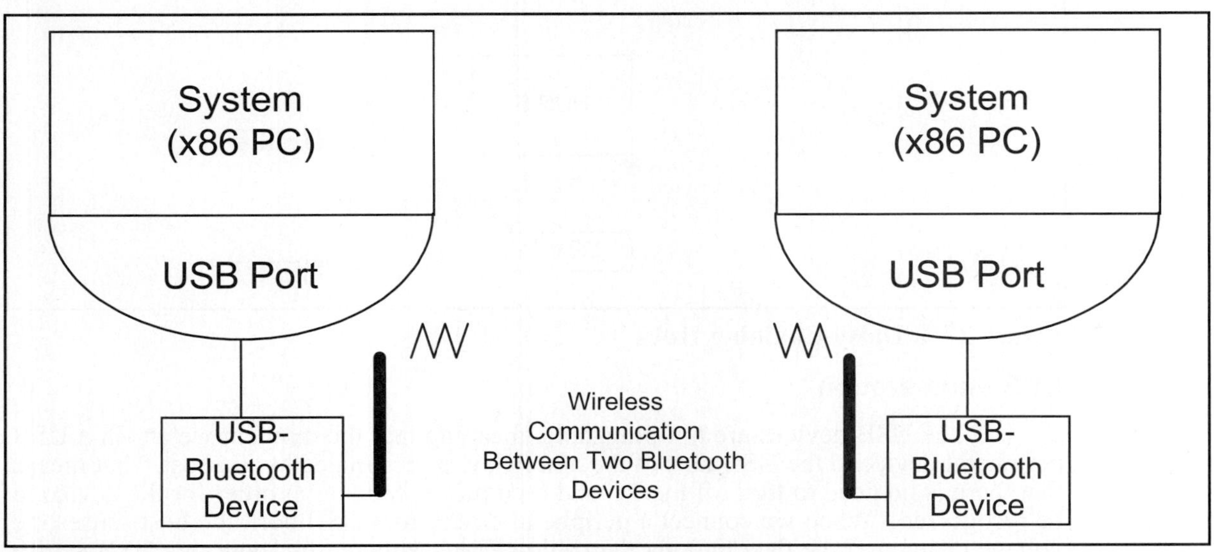

Figure 27-5. PC Communications via USB-Bluetooth Devices

Assuming each PC has a USB-Bluetooth device connected to its USB port, the following steps show how to use HyperTerminal to establish communication between them:

1. Use the Bluetooth utility provided with the Bluetooth device to open a virtual COM port on the PC.

2. Configure HyperTerminal to communicate on the virtual COM port assigned by the Bluetooth utility.
3. Perform the same tasks on the second PC.
4. Send messages between the two PCs to verify communications.

In Program 27-1 we write the message "Hello World" to the USB-Bluetooth device. In Program 27-2 we receive the message from the USB-Bluetooth device.

```
//Program 27-1 : Send data through a USB COM port.
//Must be compiled in Visual C# 2005 Express, which is
//availabe for free from Microsoft's website.

using System;
using System.IO.Ports;

namespace SerialComm {
 class SerialOut {
  static void Main ()
  {
  // The following line will set the COM port parameters in C#.
   SerialPort com1 = new SerialPort( "COM12", 9600, Parity.None,
                      8, StopBits.One );
   com1.Open (); // Open virtual COM port 12.
   do
   {
    // Send the data through the COM port.
    com1.WriteLine ( "Hello World!!!" );
   }
   while (!Console.KeyAvailable);
   com1.Close (); // Close virtual COM port 12.
  }
 }
}

//Program 27-2 : Getting data through a USB COM port.

using System;
using System.IO.Ports;

namespace SerialComm {
 class SerialIn {
  static void Main ()
  {
  // The following line will set the COM port parameters in C#.
   SerialPort com1 = new SerialPort( "COM12", 9600, Parity.None,
                      8, StopBits.One );
   com1.Open (); // Open virtual COM port 12.
   do
   {
    // Send the data through the COM port.
    Console.WriteLine(com1.ReadLine());
   }
   while (!Console.KeyAvailable);
   com1.Close (); // Close virtual COM port 12.
  }
 }
}
```

By running the above two programs, the message is transferred via the USB port to Bluetooth, and from Bluetooth wirelessly to another PC. Programs 27-1 and 27-2 are similar to the programs we wrote for serial communication in Chapter 17.

Comparing Figure 27-6 with the null modem connection discussed in Chapter 17, one might wonder, is there any way to communicate between two PCs via a null USB cable? The answer is absolutely not, since two USB hosts cannot talk to each other directly. In other words, the USB host can only talk to a USB device and that is the reason we are using two USB devices to establish communication between two hosts.

Review Questions

1. True or false. A USB host can talk to another USB host.
2. True or false. A USB device can talk to another USB device on the same USB bus.
3. True or false. A USB device can talk to a USB host.
4. True or false. In the Windows OS, opening a connection to a USB device is like opening a file.
5. Generally, how many hosts exist in a single x86 PC?

PROBLEMS

SECTION 27.1: USB PORTS: AN OVERVIEW

1. Give the maximum number of USB, LPT, and COM ports that the BIOS of the x86 PC can support.
2. Give the power allocation from the motherboard for PCI, ISA, and USB.
3. Give the maximum data transfer rate for the USB specifications 1.0, 1.1, and 2.0.
4. What does it mean if a device is hot pluggable?
5. Of the PCI, ISA, LPT, COM, and USB, indicate which one is hot pluggable.

SECTION 27.2: USB PORT EXPANSION AND POWER MANAGEMENT

6. True or false. The x86 PC with a USB port comes with the host capability.
7. True or false. The x86 PC with multiple USB ports comes with a root hub.
8. True or false. The x86 PC with a root hub provides sufficient current to all the peripheral devices connected to it.
9. True or false. The x86 PC with a USB port needs an external power supply for any peripheral device connected to it.
10. True or false. Every hub must have one upstream port.
11. Define downstream and upstream as used in USB terminology.
12. Define host, peripheral device, and hub as used in USB terminology.
13. How much current does a host provide to a peripheral device?
14. How much current does a hub provide to a peripheral device?
15. In daisy-chaining USB, what is the maximum number of layers (tiers) we can have?
16. What is the difference between self-powered and bus-powered hubs?
17. Where do we use the USB cable with an A-type connector on both ends?
18. Where do we use the USB cable with an A-type connector on one end and a B-type connector on the other end?
19. How many pins does the USB cable have?
20. Give the function of each pin of the USB cable.
21. What is the maximum length of a USB cable?
22. What method does USB use to represent 0s and 1s?
23. True or false. The USB is half-duplex.
24. What is enumeration in USB?
25. Who performs the enumeration?
26. What is the maximum current provided by the host to a bus-powered hub?
27. What is the difference between the A-side and B-side of a USB cable?
28. What is the maximum number of USB peripheral devices that a single host can enumerate?

29. What is the range of numbers used for USB enumeration?
30. True or false. The USB host changes the enumeration number periodically to make sure the peripheral device is alive and responding.
31. True or false. Data going from a peripheral device to the host is called upstream.
32. What is the maximum current usage by a downstream peripheral device?

SECTION 27.3: USB PORT PROGRAMMING

33. Combine Programs 27-1 and 27-2 into a single program and test it.

ANSWERS TO REVIEW QUESTIONS

SECTION 27.1: USB PORTS: AN OVERVIEW

1. False
2. Devices connected together via a USB hub can share a power supply.
3. False

SECTION 27.2: USB PORT EXPANSION AND POWER MANAGEMENT

1. True
2. False
3. False
4. 100 mA
5. The B side is connected to the upstream port on a hub while the A side is for downstream ports. They are also shaped differently to prevent confusion.

SECTION 27.3: USB PORT PROGRAMMING

1. False
2. False
3. True
4. True
5. 1

APPENDIX A

DEBUG PROGRAMMING

OVERVIEW

DEBUG is a program included in the MS-DOS operating systems that allows the programmer to monitor a program's execution closely for debugging purposes. Specifically, it can be used to examine and alter the contents of memory, to enter and run programs, and to stop programs at certain points in order to check or even change data. This appendix provides a tutorial introduction to the DEBUG program. You will learn how to enter and exit DEBUG, how to enter, run, and debug programs, how to examine and alter the contents of registers and memory, plus some additional features of DEBUG that prove useful in program development. Numerous examples of Assembly language programming in DEBUG are given throughout and the appendix closes with a quick reference summary of the DEBUG commands.

First, a word should be said about the examples in this appendix. Within examples, what you should type in will be represented in plain text caps:

PLAIN TEXT REPRESENTS WHAT THE USER TYPES IN

and the response of the DEBUG program will be in bold caps:

BOLD CAPS REPRESENT THE COMPUTER'S RESPONSE

The examples in this appendix assume that the DEBUG program is in drive A and that your programs are on drive B. If your system is set up differently, you will need to keep this in mind when typing in drive specifications (such as "B:"). It is strongly suggested that you type in the examples in DEBUG and try them for yourself. The best way to learn is by doing!

SECTION A.1: ENTERING AND EXITING DEBUG

To enter the DEBUG program, simply type its name at the command prompt:

```
C:\>DEBUG <return>
-
```

"DEBUG" may be typed in either uppercase or lowercase. Again let us note that this example assumes that the DEBUG program is on the disk in drive A. After "DEBUG" and the carriage return (or enter key) is typed in, the DEBUG prompt "-" will appear on the following line. DEBUG is now waiting for you to type in a command.

Now that you know how to enter DEBUG, you are ready to learn the DEBUG commands. The first command to learn is the quit command, to exit DEBUG.

The quit command, Q, may be typed in either uppercase or lowercase. This is true for all DEBUG commands. After the Q and the carriage return have been entered, DEBUG will return you to the command prompt. This is shown in Example A-1.

Example A-1 : Entering and Exiting DEBUG

```
C:\>DEBUG <return>
-Q  <return>
C:\>
```

SECTION A.2: EXAMINING AND ALTERING REGISTERS

The register command allows you to examine and/or alter the contents of the internal registers of the CPU. The R command has the following syntax:

```
R  <register name >
```

The R command will display all registers unless the optional <register name> field is entered, in which case only the register named will be displayed.

Example A-2 : Using the R Command to Display All Registers

```
C:\>DEBUG <return>
-R
AX=0000 BX=0000 CX=0000 DX=0000 SP=FFEE BP=0000 SI=0000 DI=0000
DS=15EF ES=15EF SS=15EF CS=15EF IP=0100 NV UP EI PL NZ NA PO NC
15EF:0100 0AE4          OR      AH,AH
```

After the R and the carriage return are typed in, DEBUG responds with three lines of information. The first line displays the general-purpose, pointer, and index registers' contents. The second line displays the segment registers' contents, the instruction pointer's current value, and the flag register bits. The codes at the end of line two, "NV UP DI ... NC", indicate the status of eight of the bits of the flag register. The flag register and its representation in DEBUG are discussed in Section A.6. The third line shows some information useful when you are programming in DEBUG. It shows the instruction pointed at by CS:IP. The third line on your system will vary from what is shown above. For the purpose at hand, concentrate on the first two lines. The explanation of the third line will be postponed until later in this appendix.

When you enter DEBUG initially, the general-purpose registers are set to zero and the flag bits are all reset. The contents of the segment registers will vary depending on the system you are using, but all segment registers will have the same value, which is decided by the operating system. For instance, notice in Example A-2 above that all segment registers contain 0C44H. It is strongly recommended not to change the contents of the segment registers since these values have been set by the operating system. Note: In a later section of this appendix we show how to load an Assembly language program into DEBUG. In that case the segment registers are set according to the program parameters and registers BX and CX will contain the size of the program in bytes.

If the optional register name field is specified in the R command, DEBUG will display the contents of that register and give you an opportunity to change its value. This is seen next in Example A-3.

Example A-3 : Using the R Command to Display/Modify Registers

```
(a) Modifying the contents of a register
    -R CX
    CX   0000
    :FFFF
    -R CX
    CX   FFFF
    :

(b) DEBUG pads values on the left with zero
    -R AX
    AX   0000
    :1
    -R AX
    AX   0001
    :21
    -R AX
    AX   0021
    :321
    -R AX
    AX   0321
    :4321
    -R AX
    AX   4321
    :54321
         ^ Error
(c) Entering data into the upper byte
    -R DH
    BR Error
    -R DX
    DX   0000
    :4C00
    -
```

Part (a) of Example A-3 first showed the R command followed by register name CX. DEBUG then displayed the contents of CX, which were 0000, and then displayed a colon ":". At this point a new value was typed in, and DEBUG prompted for another command with the "-" prompt. The next command verified that CX was indeed altered as requested. This time a carriage return was entered at the ":" prompt so that the value of CX was not changed.

Part (b) of Example A-3 showed that if fewer than four digits are typed in, DEBUG will pad on the left with zeros. Part (c) showed that you cannot access the upper and lower bytes separately with the R command. If you type in any digit other than 0 through F (such as in "2F0G"), DEBUG will display an error message and the register value will remain unchanged.

See Section A.6 for a discussion of how to use the R command to change the contents of the flag register.

SECTION A.3: CODING AND RUNNING PROGRAMS IN DEBUG

In the next few topics we explore how to enter simple Assembly language instructions, and assemble and run them. The purpose of this section is to familiarize the reader with using DEBUG, not to explain the Assembly language instructions found in the examples.

A, the assemble command

The assemble command is used to enter Assembly language instructions into memory.

```
A  <starting address>
```

The starting address may be given as an offset number, in which case it is assumed to be an offset into the code segment, or the segment register can be specified explicitly. In other words, "A 100" and "A CS:100" will achieve the same results. When this command is entered at the command prompt "-", DEBUG will begin prompting you to enter Assembly language instructions. After an instruction is typed in and followed by <return>, DEBUG will prompt for the next instruction. This process is repeated until you type a <return> at the address prompt, at which time DEBUG will return you to the command prompt level. This is shown in part (a) of Example A-4.

Before you type in the commands of Example A-4, be aware that one important difference between DEBUG programming and Assembly language programming is that DEBUG assumes that all numbers are in hex, whereas most assemblers assume that numbers are in decimal unless they are followed by "H". Therefore, the Assembly language instruction examples in this section do not have "H" after the numbers as they would if an assembler were to be used. For example, you might enter an instruction such as "MOV AL,3F". In an Assembly language program written for MASM, for example, this would have been typed as "MOV AL,3FH".

As you type the instructions, DEBUG converts them to machine code. If you type an instruction incorrectly such that DEBUG cannot assemble it, DEBUG will give you an error message and prompt you to try again. Again, keep in mind that the value for the code segment may be different on your machine when you run Example A-4. Notice that each time DEBUG prompts for the next instruction, the offset has been updated to the next available location. For example, after you typed the first instruction at offset 0100, DEBUG converted this to machine language, stored it in bytes 0100 to 0102, and prompted you for the next instruction, which will be stored at offset 0103. Note: Do not assemble beginning at an offset lower than 100. The first 100H (256) bytes are reserved by the operating system and should not be used by your programs. This is the reason that examples in this book use "A 100" to start assembling instructions after the first 100H bytes.

U, the unassemble command: looking at machine code

The unassemble command displays the machine code in memory along with its equivalent Assembly language instructions. The command can be given in either format

Example A-4 : Assemble, Unassemble, and Go Commands

```
(a) Assemble command
    -A 100
    103D:0100 MOV AX,1
    103D:0103 MOV BX,2
    103D:0106 MOV CX,3
    103D:0109 ADD AX,BX
    103D:010B ADD AX,CX
    103D:010D INT 3
    103D:010E
    -
(b) Unassemble command
    -U 100 10D
    103D:0100 B80100        MOV     AX,0001
    103D:0103 BB0200        MOV     BX,0002
    103D:0106 B90300        MOV     CX,0003
    103D:0109 01D8          ADD     AX,BX
    103D:010B 01C8          ADD     AX,CX
    103D:010D CC            INT     3
    -
(c) Go command
    -R
AX=0000 BX=0000 CX=0000 DX=0000 SP=CFDE BP=0000 SI=0000 DI=0000
DS=103D ES=103D SS=103D CS=103D IP=0100 NV UP DI PL NZ NA PO NC
103D:0100 B80100          MOV    AX,0001
    -G
AX=0006 BX=0002 CX=0003 DX=0000 SP=CFDE BP=0000 SI=0000 DI=0000
DS=103D ES=103D SS=103D CS=103D IP=010D NV UP DI PL NZ NA PE NC
103D:010D CC              INT    3
    -
```

shown below.

```
U <starting address >  <ending address>
U <starting address >  < L number of bytes>
```

Whereas the assemble instruction takes Assembly language instructions from the keyboard and converts them to machine code, which it stores in memory, the unassemble instruction does the opposite. Unassemble takes machine code stored in memory and converts it back to Assembly language instructions to be displayed on the monitor. Look at part (b) of Example A-4. The unassemble command was used to unassemble the code that was entered in part (a) with the assemble command. Notice that both the machine code and Assembly instructions are displayed. The command can be entered either with starting and ending addresses, as was shown in Example A-4: "U 100 10D", or with a starting address and a number of bytes in hex. The same command in the second format would be "U 100 LD", which tells DEBUG to start unassembling at CS:100 for D bytes. If the U command is entered with no addresses after it: "U <return>", then DEBUG will display 32 bytes beginning at CS:IP. Successively entering "U <return>" commands will cause DEBUG to display consecutive bytes of the program, 32 bytes at a time. This is an easy way to look through a large program.

G, the go command

The go command instructs DEBUG to execute the instructions found between the two given addresses. Its format is

```
G < = starting address>  <stop address(es)>
```

If no addresses are given, DEBUG begins executing instructions at CS:IP until a breakpoint is reached. This was done in part (c) of Example A-4. Before the instructions were executed, the R command was used to check the values of the registers. Since CS:IP pointed to the first instruction, the G command was entered, which caused execution of instructions up until "INT 3", which terminated execution. After a breakpoint is reached, DEBUG displays the register contents and returns you to the command prompt "-". Up to 10 stop addresses can be entered. DEBUG will stop execution at the first of these breakpoints that it reaches. This can be useful for programs that could take several different paths.

At this point the third line of the register dump has become useful. The purpose of the third line is to show the location, machine code, and Assembly code of the next instruction to be executed. In Example A-5, look at the last line in the register dump given after the G command. Notice, at the leftmost part of line three, the value CS:IP. The values for CS and IP match those given in lines one and two. After CS:IP is the machine code, and after the machine code is the Assembly language instruction.

Part (a) of Example A-5 is the same as part (c) of Example A-4. The go command started at CS:IP and executed instructions until it reached instruction "INT 3". Part (b) gave a starting address but no ending address; therefore, DEBUG executed instructions from offset 100 until "INT 3" was reached. This could also have been typed in as "G =CS:100". Part (c) gave both starting and ending addresses. We can see from the register

Example A-5 : Various Forms of the Go Command

```
The program is first assembled:
    -A 100
    103D:0100 MOV AX,1
    103D:0103 MOV BX,2
    103D:0106 MOV CX,3
    103D:0109 ADD AX,BX
    103D:010B ADD AX,CX
    103D:010D INT 3
    103D:010E
(a) Go command in form  "G"
    -G
AX=0006 BX=0002 CX=0003 DX=0000 SP=CFDE BP=0000 SI=0000 DI=0000
DS=103D ES=103D SS=103D CS=103D IP=010D NV UP DI PL NZ NA PE NC
103D:010D CC            INT     3
    -
(b) Go command in form "G  = start address"
    -G =100
AX=0006 BX=0002 CX=0003 DX=0000 SP=CFDE BP=0000 SI=0000 DI=0000
DS=103D ES=103D SS=103D CS=103D IP=010D NV UP DI PL NZ NA PE NC
103D:010D CC            INT     3
    -
(c) Go command form "G  = start address  endingaddress"
    -G =100 109
AX=0001 BX=0002 CX=0003 DX=0000 SP=CFDE BP=0000 SI=0000 DI=0000
DS=103D ES=103D SS=103D CS=103D IP=0109 NV UP DI PL NZ NA PE NC
103D:0109 01D8          ADD     AX,BX
    -
(d) Go command format "G   address"
    -R IP
    IP 0109
    :0100
    -G 109
AX=0001 BX=0002 CX=0003 DX=0000 SP=CFDE BP=0000 SI=0000 DI=0000
DS=103D ES=103D SS=103D CS=103D IP=0109 NV UP DI PL NZ NA PE NC
103D:0109 01D8          ADD     AX,BX
    -
```

results that it did execute from offset 100 to 109. Part (d) gave only the ending address. When the start address is not given explicitly, DEBUG uses the value in register IP. Be sure to check that value with the register command before issuing the go command without a start address.

T, the trace command: a powerful debugging tool

The trace command allows you to trace through the execution of your programs one or more instructions at a time to verify the effect of the programs on registers and/or data.

```
T <= starting address>   <number of instructions>
```

This tells DEBUG to begin executing instructions at the starting address. DEBUG will execute however many instructions have been requested in the second field. The default value is 1 if no second field is given. The trace command functions similarly to the go command in that if no starting address is specified, it starts at CS:IP. The difference between this command and the go command is that trace will display the register contents after each instruction, whereas the go command does not display them until after termination of the program. Another difference is that the last field of the go command is the stop address, whereas the last field of the trace command is the number of instructions to execute.

Example A-6 shows a trace of the instructions entered in part (a) of Example A-4. Notice the way that register IP is updated after each instruction to point to the next instruction. The third line of the register display shows the instruction pointed at by IP, that is, the next instruction to be executed. Tracing through a program allows you to examine what is happening in each instruction of the program. Notice the value of AX after each instruction in Example A-6.

The same trace as shown in Example A-6 could have been achieved with the command "-T 5", assuming that IP = 0100. Experiment with the various forms of the trace command. "T" with no starting or count fields will execute one instruction starting at CS:IP. If no first field is given, CS:IP is assumed. If no second field is given, 1 is assumed.

If you trace a large number of instructions, they may scroll upward off the screen faster than you can read them. <Ctrl-num lock > can be used to stop the scrolling temporarily. To resume the scrolling, enter any key. This works not only on the trace command, but for any command that displays information to the screen.

Example A-6 : Trace Command

```
-T=100 5
AX=0001 BX=0000 CX=0000 DX=0000 SP=CFDE BP=0000 SI=0000 DI=0000
DS=103D ES=103D SS=103D CS=103D IP=0103 NV UP DI PL NZ NA PO NC
103D:0103 BB0200          MOV     BX,0002

AX=0001 BX=0002 CX=0000 DX=0000 SP=CFDE BP=0000 SI=0000 DI=0000
DS=103D ES=103D SS=103D CS=103D IP=0106 NV UP DI PL NZ NA PO NC
103D:0106 B90200          MOV     CX,0003

AX=0001 BX=0002 CX=0003 DX=0000 SP=CFDE BP=0000 SI=0000 DI=0000
DS=103D ES=103D SS=103D CS=103D IP=0109 NV UP DI PL NZ NA PO NC
103D:0109 01D8            ADD     AX,BX

AX=0003 BX=0002 CX=0003 DX=0000 SP=CFDE BP=0000 SI=0000 DI=0000
DS=103D ES=103D SS=103D CS=103D IP=010B NV UP DI PL NZ NA PE NC
103D:010B 01C8            ADD     AX,CX

AX=0006 BX=0002 CX=0003 DX=0000 SP=CFDE BP=0000 SI=0000 DI=0000
DS=103D ES=103D SS=103D CS=103D IP=010D NV UP DI PL NZ NA PE NC
103D:010D CC              INT     3
-
```

Loading the 8-bit and 16-bit registers is shown in Example A-7.

Example A-7 : Moving Data into 8- and 16-bit Registers

```
C:>DEBUG
-R
AX=0000 BX=0000 CX=0000 DX=0000 SP=CFDE BP=0000 SI=0000 DI=0000
DS=103D ES=103D SS=103D CS=103D IP=0100 NV UP DI PL NZ NA PO NC
103D:0100 B664          MOV     DH,64
-A 100
103D:0100 MOV AL,3F
103D:0102 MOV BH,04
103D:0104 MOV CX,FFFF
103D:0107 MOV CL,BH
103D:0109 MOV CX,1
103D:010C INT 3
103D:010D
-T =100 5
AX=003F BX=0000 CX=0000 DX=0000 SP=CFDE BP=0000 SI=0000 DI=0000
DS=103D ES=103D SS=103D CS=103D IP=0102 NV UP DI PL NZ NA PO NC
103D:0102 B704          MOV     BH,04

AX=003F BX=0400 CX=0000 DX=0000 SP=CFDE BP=0000 SI=0000 DI=0000
DS=103D ES=103D SS=103D CS=103D IP=0104 NV UP DI PL NZ NA PO NC
103D:0104 B9FFFF        MOV     CX,FFFF

AX=003F BX=0400 CX=FFFF DX=0000 SP=CFDE BP=0000 SI=0000 DI=0000
DS=103D ES=103D SS=103D CS=103D IP=0107 NV UP DI PL NZ NA PO NC
103D:0107 88F9          MOV     CL,BH

AX=003F BX=0400 CX=FF04 DX=0000 SP=CFDE BP=0000 SI=0000 DI=0000
DS=103D ES=103D SS=103D CS=103D IP=0109 NV UP DI PL NZ NA PO NC
103D:0109 B90100        MOV     CX,0001

AX=003F BX=0400 CX=0001 DX=0000 SP=CFDE BP=0000 SI=0000 DI=0000
DS=103D ES=103D SS=103D CS=103D IP=010C NV UP DI PL NZ NA PO NC
103D:010C CC            INT     3
 -
```

Example A-8 is stored starting at CS:IP of 1132:0100. This logical address corresponds to physical address 11420 (11320 + 0100).

SECTION A.4: DATA MANIPULATION IN DEBUG

Next are described three DEBUG commands that are used to examine or alter the contents of memory.

```
F     the fill command fills a block of memory with data
D     the dump command displays contents of memory to the screen
E     the enter command examines/alters the contents of memory
```

F, the fill command: filling memory with data

The fill command is used to fill an area of memory with a data item. The syntax of the F command is as follows:

```
F <starting address >  <ending address>  <data>
F <starting address > <L number of bytes> <data>
```

Example A-8 : Assembling and Unassembling a Program

```
C:>DEBUG
-R
AX=0000 BX=0000 CX=0000 DX=0000 SP=CFDE BP=0000 SI=0000 DI=0000
DS=1132 ES=1132 SS=1132 CS=1132 IP=0100 NV UP DI PL NZ NA PO NC
1132:0100 BED548          MOV     SI,48D5
-A 100
1132:0100 MOV AL,57
1132:0102 MOV DH,86
1132:0104 MOV DL,72
1132:0106 MOV CX,DX
1132:0108 MOV BH,AL
1132:010A MOV BL,9F
1132:010C MOV AH,20
1132:010E ADD AX,DX
1132:0110 ADD CX,BX
1132:0112 ADD AX,1F35
1132:0115
-U 100 112
1132:0100 B057      MOV   AL,57
1132:0102 B686      MOV   DH,86
1132:0104 B272      MOV   DL,72
1132:0106 89D1      MOV   CX,DX
1132:0108 88C7      MOV   BH,AL
1132:010A B39F      MOV   BL,9F
1132:010C B420      MOV   AH,20
1132:010E 01D0      ADD   AX,DX
1132:0110 01D9      ADD   CX,BX
1132:0112 05351F    ADD   AX,1F35
-
```

This command is useful for filling a block of memory with data, for example, to initialize an area of memory with zeros. Normally, you will want to use this command to fill areas of the data segment, in which case the starting and ending addresses would be offset addresses into the data segment. To fill another segment, the register should precede the offset. For example, the first command below would fill 16 bytes, from DS:100 to DS:10F with FF. The second command would fill a 256-byte block of the code segment, from CS:100 to CS:1FF with ASCII 20 (space).

```
F 100 10F FF
F CS:100 1FF 20
```

Example A-9 demonstrates the use of the F command. The data can be a series of items, in which case DEBUG will fill the area of memory with that pattern of data, repeating the pattern over and over. For example:

```
F 100 L20 00 FF
```

The command above would cause 20 hex bytes (32 decimal) starting at DS:100 to be filled alternately with 00 and FF.

D, the dump command: examining the contents of memory

The dump command is used to examine the contents of memory. The syntax of the D command is as follows:

```
D <start address >  <end  address>
D <start address >  <L number of bytes>
```

Example A-9 : Filling and Dumping a Block of Memory

```
(a) Fill and dump commands
  C:>DEBUG
  -F 100 14F 20
  -F 150 19F 00
  -D 100 19F
103D:0100   20 20 20 20 20 20 20 20-20 20 20 20 20 20 20 20
103D:0110   20 20 20 20 20 20 20 20-20 20 20 20 20 20 20 20
103D:0120   20 20 20 20 20 20 20 20-20 20 20 20 20 20 20 20
103D:0130   20 20 20 20 20 20 20 20-20 20 20 20 20 20 20 20
103D:0140   20 20 20 20 20 20 20 20-20 20 20 20 20 20 20 20
103D:0150   00 00 00 00 00 00 00 00-00 00 00 00 00 00 00 00
103D:0160   00 00 00 00 00 00 00 00-00 00 00 00 00 00 00 00
103D:0170   00 00 00 00 00 00 00 00-00 00 00 00 00 00 00 00
103D:0180   00 00 00 00 00 00 00 00-00 00 00 00 00 00 00 00
103D:0190   00 00 00 00 00 00 00 00-00 00 00 00 00 00 00 00
-
(b) Filling and dumping selected memory locations
  -F 104 10A FF
  -D 104 10A
103D:0104   FF FF FF FF-FF FF FF
  -D 100 10F
103D:0100   20 20 20 20 FF FF FF FF-FF FF FF 20 20 20 20 20
-
(c) Filling and dumping code segment memory
  -F CS:100 12F 20
  -D CS:100 12F
103D:0100   20 20 20 20 20 20 20 20-20 20 20 20 20 20 20 20
103D:0110   20 20 20 20 20 20 20 20-20 20 20 20 20 20 20 20
103D:0120   20 20 20 20 20 20 20 20-20 20 20 20 20 20 20 20
  -
```

The D command can be entered with a starting and ending address, in which case it will display all the bytes between those locations. It can also be entered with a starting address and a number of bytes (in hex), in which case it will display from the starting address for that number of bytes. If the address is an offset, DS is assumed. The D command can also be entered by itself, in which case DEBUG will display 128 consecutive bytes beginning at DS:100. The next time "D" is entered by itself, DEBUG will display 128 bytes beginning at wherever the last display command left off. In this way, one can easily look through a large area of memory, 128 bytes at a time.

Example A-10 demonstrates the use of the fill and dump commands. Part (a) shows two fill commands to fill areas of the data segment, which are then dumped. Part (b) was included to show that small areas of memory can be filled and dumped. Part (c) shows how to fill and dump to memory from other segments. Keep in mind that the values for DS and CS may be different on your machine.

E, the enter command: entering data into memory

The fill command was used to fill a block with the same data item. The enter command can be used to enter a list of data into a certain portion of memory. The syntax of the E command is as follows:

```
E <address>   <data list>
E <address>
```

Part (a) of Example A-10 showed the simplest use of the E command, entering the starting address, followed by the data. That example showed how to enter ASCII data, which can be enclosed in either single or double quotes. The E command has another powerful feature: the ability to examine and alter memory byte by byte. If the E com-

Example A-10 : Using the E Command to Enter Data into Memory

```
(a) Entering data with the E command
   -E 100 'John Snith'
   -D 100 10F
103D:0100 4A 6F 68 6E 20 53 6E 69-74 68 20 20 20 20 20 20   John Snith

(b) Altering data with the E command
   -E 106
   103D:0106   6E.6D
   -D 100 10F
103D:0100 4A 6F 68 6E 20 53 6D 69-74 68 20 20 20 20 20 20   John Smith

(c) Another way to alter data with the E command, hitting the
     space bar to go through the data a byte at a time
   -E 100
103D:0100 4A. 6F. 68. 6E. 20. 53. 6E.6D
   -D 100 10F
103D:0100 4A 6F 68 6E 20 53 6D 69-74 68 20 20 20 20 20 20   John Smith
   -

(d)  Another way to alter data with the E command
   -E 107
   103D:0107 69.-
   103D:0106 6E.6D
   -
```

mand is entered with a specific address and no data list, DEBUG assumes that you wish to examine that byte of memory and possibly alter it. After that byte is displayed, you have four options:

1. You can enter a new data item for that byte. DEBUG will replace the old contents with the new value you typed in.
2. You can press <return>, which indicates that you do not wish to change the value.
3. You can press the space bar, which will leave the displayed byte unchanged but will display the next byte and give you a chance to change that if you wish.
4. You can enter a minus sign, "-", which will leave the displayed byte unchanged but will display the previous byte and give you a chance to change it.

Look at part (b) in Example A-10. The user wants to change "Snith" to "Smith". After the user typed in "E 106", DEBUG responded with the contents of that byte, 6E, which is ASCII for n, and prompted with a ".". Then the user typed in the ASCII code for "m", 6D, entered a carriage return, and then dumped the data to see if the correction was made. Part (c) of Example A-10 showed another way to make the same correction. The user started at memory offset 100 and pressed the space bar continuously until the desired location was reached. Then he made the correction and pressed carriage return.

Finally, part (d) showed a third way the same correction could have been made. In this example, the user accidentally entered the wrong address. The address was one byte past the one that needed correction. The user entered a minus sign, which caused DEBUG to display the previous byte on the next line. Then the correction was made to that byte. Try these examples yourself since the E command will prove very useful in debugging your future programs.

The E command can be used to enter numerical data as well:

```
E 100 23 B4 02 4F
```

Example A-11 gives an example of entering code with the assemble command, entering data with the enter command, and running the program.

Example A-11 : Entering Data and Code and Running a Program

```
C:>DEBUG
-A 100
103D:0100 MOV AL,00
103D:0102 ADD AL,[ 0200]
103D:0106 ADD AL,[ 0201]
103D:010A ADD AL,[ 0202]
103D:010E ADD AL,[ 0203]
103D:0112 ADD AL,[ 0204]
103D:0116 INT 3
103D:0117
-E DS:0200 25 12 15 1F 2B
-D DS:0200 020F
103D:0200 25 12 15 1F 2B 02 00 E8-51 FF C3 E8 1E F6 74 03 %...+..hQ.Ch.vt.
-G =100 116
AX=0096 BX=0000 CX=0000 DX=0000 SP=CFDE BP=0000 SI=0000 DI=0000
CS=103D ES=103D SS=103D CS=103D IP=0116 OV UP DI NG NZ AC PE NC
103D:0116 CC            INT     3
-
```

SECTION A.5: EXAMINING/ALTERING THE FLAG REGISTER IN DEBUG

The discussion of how to use the R command to examine/alter the contents of the flag register was postponed until this section, so that program examples that affect the flag bits could be included. Table A-1, on the following page, gives the codes for 8 bits of the flag register, which are displayed whenever a G, T, or R DEBUG command is given.

Table A-1: Codes for the Flag Register

Flag	Code When Set (= 1)	Code When Reset (= 0)
OF overflow flag	OV (overflow)	NV (no overflow)
DF direction flag	DN (down)	UP (up)
IF interrupt flag	EI (enable interrupt)	DI (disable interrupt)
SF sign flag	NG (negative)	PL (plus, or positive)
ZF zero flag	ZR (zero)	NZ (not zero)
AF auxiliary carry flag	AC (auxiliary carry)	NA (no auxiliary carry)
PF parity flag	PE (parity even)	PO (parity odd)
CF carry flag	CY (carry)	NC (no carry)

If all the bits of the flag register were reset to zero, as is the case when DEBUG is first entered, the following would be displayed for the flag register:

NV UP DI PL NZ NA PO NC
Similarly, if all the flag bits were set to 1, the following would be seen:
OV DN EI NG ZR AC PE CY

Example A-12 shows how to use the R command to change the setting of the flag register.

Example A-12 showed how the flag register can be examined, or examined and then altered. When the R command is followed by "F", this tells DEBUG to display the contents of the flag register. After DEBUG displays the flag register codes, it prompts with another "-" at the end of the line of register codes. At this point, flag register codes may be typed in to alter the flag register, or a simple carriage return may be typed in if no changes are needed. The register codes may be typed in any order.

Example A-12 : Changing the Flag Register Contents

```
-R F
NV UP DI PL NZ NA PO NC -DN OV NG
-R F
OV DN DI NG NZ NA PO NC -
-
```

Impact of instructions on the flag bits

Example A-13, on the following page, shows the effect of ADD instructions on the flag register. The ADD in part (a) involves byte addition. Adding 9C and 64 results in 00 with a carry out. The flag bits indicate that this was the result. Notice the zero flag is now ZR, indicating that the result is zero. In addition, the carry flag was set, indicating the carry out. The ADD in part (b) involves word addition. Notice that the sign flag was set to NG after the ADD instruction was executed. This is because the result, CAE0, in its binary form will have a 1 in bit 15, the sign bit. Since we are dealing with unsigned addition, we interpret this number to be positive CAE0H, not a negative number. This points out the fact that the microprocessor treats all data the same. It is up to the programmer to interpret the meaning of the data. Finally, look at the ADD in part (c). Adding AAAAH and 5556H gives 10000H, which results in BX = 0000 with a carry out. The zero flag indicates the zero result (BX = 0000), while the carry flag indicates that a carry out occurred.

Examples A-14 and A-15 show how to code a simple program in DEBUG, set up the desired data, and execute the program. This program includes a conditional jump that will decide whether to jump based on the value of the zero flag. This example also points out some important differences between coding a program in DEBUG and coding a program for an assembler such as MASM. First notice the JNZ instruction. If this were an Assembly language program, the instruction might be "JNZ LOOP_ADD", where the label LOOP_ADD refers to a line of code. In DEBUG we simply JNZ to the address. Another important difference is that an Assembly language program would have separate data and code segments. In Example A-14, the test data was entered at offset 0200, and consequently, BX was set to 0200 since it is being used as a pointer to the data. In an Assembly language program, the data would have been set up in the data segment and the instruction might have been "MOV BX,OFFSET DATA1" where DATA1 is the label associated with the data directive that stored the data.

Example A-13 : Observing Changes in the Flag Register

```
(a)
    C:>DEBUG
    -A 100
    103D:0100 MOV AL,9C
    103D:0102 MOV DH,64
    103D:0104 ADD AL,DH
    103D:0106 INT 3
    103D:0107
    -T 3
AX=009C BX=0000 CX=0000 DX=0000 SP=CFDE BP=0000 SI=0000 DI=0000
DS=103D ES=103D SS=103D CS=103D IP=0102 NV UP DI PL NZ NA PO NC
103D:0102 B664          MOV      DH,64

AX=009C BX=0000 CX=0000 DX=6400 SP=CFDE BP=0000 SI=0000 DI=0000
DS=103D ES=103D SS=103D CS=103D IP=0104 NV UP DI PL NZ NA PO NC
103D:0104 00F0          ADD      AL,DH

AX=0000 BX=0000 CX=0000 DX=6400 SP=CFDE BP=0000 SI=0000 DI=0000
DS=103D ES=103D SS=103D CS=103D IP=0106 NV UP DI PL ZR AC PE CY
103D:0106 CC            INT      3
    -

(b)
    -A 100
    103D:0100 MOV AX,34F5
    103D:0103 ADD AX,95EB
    103D:0106 INT 3
    103D:0107
    -T =100 2
AX=34F5 BX=0000 CX=0000 DX=6400 SP=CFDE BP=0000 SI=0000 DI=0000
DS=103D ES=103D SS=103D CS=103D IP=0103 NV UP DI PL NZ NA PO NC
103D:0103 05EB95        ADD      AX,95EB

AX=CAE0 BX=0000 CX=0000 DX=6400 SP=CFDE BP=0000 SI=0000 DI=0000
DS=103D ES=103D SS=103D CS=103D IP=0106 NV UP DI NG NZ AC PO NC
103D:0106 CC              INT      3
    -
(c)
    -A 100
    103D:0100 MOV BX,AAAA
    103D:0103 ADD BX,5556
    103D:0107 INT 3
    103D:0108
    -G =100 107
AX=34F5 BX=0000 CX=0000 DX=6400 SP=CFDE BP=0000 SI=0000 DI=0000
DS=103D ES=103D SS=103D CS=103D IP=0107 NV UP DI PL ZR AC PE CY
103D:0107 CC             INT      3
    -
```

Example A-14 : Tracing through a Program to Add 5 Bytes

```
C:>DEBUG
-A 100
103D:0100 MOV CX,05
103D:0103 MOV BX,0200
103D:0106 MOV AL,0
103D:0108 ADD AL,[ BX]
103D:010A INC BX
103D:010B DEC CX
103D:010C JNZ 0108
103D:010E MOV [ 0205] ,AL
103D:0111 INT 3
103D:0112
-E 0200 25 12 15 1F 2B
-D 0200 020F
103D:0200  25 12 15 1F 2B 9A DE CE-1E F3 20 20 20 20 20 20  %...+.^n.
-G =100 111
AX=0096 BX=0205 CX=0000 DX=0000 SP=CFDE BP=0000 SI=0000 DI=0000
DS=103D ES=103D SS=103D CS=103D IP=0111 NV UP DI PL ZR NA PE NC
103D:0111 CC            INT      3
-D 0200 020F
103D:0200  25 12 15 1F 2B 96 DE CE-1E F3 20 20 20 20 20 20  %...+.^n.
```

Example A-15 : Data Transfer Program in DEBUG

```
C:>DEBUG
-A 100
103D:0100 MOV SI,0210
103D:0103 MOV DI,0228
103D:0106 MOV CX,6
103D:0109 MOV AL,[ SI]
103D:010B MOV [ DI] ,AL
103D:010D INC SI
103D:010E INC DI
103D:010F DEC CX
103D:0110 JNZ 0109
103D:0112 INT 3
103D:0113
-E 0210 25 4F 85 1F 2B C4
-D 0210 022F
103D:0210 25 4F 85 1F 2B C4 43 0C-01 01 01 00 02 FF FF FF %O..+DC.....
103D:0220 FF FF FF FF FF FF FF FF-FF FF FF FF 45 0D CA 2A ........E.J*
-G =100
AX=00C4 BX-0000 CX=0000 DX=0000 SP=CFDE BP=0000 SI=0216 DI=022E
DS=103D ES=103D SS=103D CS=103D IP=0112 NV UP DI PL ZR NA PE NC
103D:0112 CC            INT      3
-D 0210 022F
103D:0210 25 4F 85 1F 2B C4 43 0C-01 01 01 00 02 FF FF FF %O..+DC.....
103D:0220 FF FF FF FF FF FF FF FF-25 4F 85 1F 2B C4 CA 2A ....%O..+DJ*
-
```

Table A-2: Summary of DEBUG Commands

Function	Command Options
Assemble	A <starting address>
Compare	C <start address> <end address> <compare address> C <start address> <L number of bytes> <compare address>
Dump	D <start address> <end address> D <start address> <L number of bytes>
Enter	E < address> <data list> E < address>
Fill	F <start address> <end address> <data> F <start address> <L number of bytes> <data>
Go	G < = start address> <end address(es)>
Hexarith	H <number 1> <number 2>
Load	L <start address> <drive> <start sector> <sectors>
Move	M <start address> <end address> <destination> M <start address> <L number of bytes> <destination>
Name	N <filename>
Procedure	P < = start address> <number of instruction>
Register	R <register name>
Search	S <start address> <end address> <data> S <start address> <L number of bytes> <data>
Trace	T < = start address> <number of instruction>
Unassemble	U <start address> <end address> U <start address> <L number of bytes>
Write	W <start address> <drive> <start sector> <sectors>

Notes:
1. All addresses and numbers are given in hex.
2. Commands may be entered in lowercase or uppercase, or a combination.
3. Ctrl-c will stop any command.
4. Ctrl-Num Lock will stop scrolling of command output. To resume scrolling, enter any key.

APPENDIX B

x86 INSTRUCTIONS DESCRIPTION

OVERVIEW

In this appendix, we list the instructions of the 8086, give their format and expected operands, and describe the function of each instruction. Where pertinent, programming examples have been given. These instructions will operate on any x86 processor. There are additional instructions for the 80186 processor and above (80286, 80386, 80486, and Pentium); however, these instructions are not given in this list and can be found in their datasheets.

SECTION B.1: THE 8086 INSTRUCTION SET

AAA ASCII Adjust after Addition

```
Flags:      Affected: AF and CF. Unpredictable: OF, SF, ZF, PF.
Format:     AAA
```

Function: This instruction is used after an ADD instruction has added two digits in ASCII code. This makes it possible to add ASCII numbers without masking off the upper nibble "3". The result will be unpacked BCD in AL with the carry flag set if needed. This instruction adjusts only on the AL register. AH is incremented if the carry flag is set.

```
Example 1:
        MOV     AL,31H      ;AL=31 THE ASCII CODE FOR 1
        ADD     AL,37H      ;ADD 37 (ASCII FOR 7) TO AL;   AL=68H
        AAA                 ;AL=08 AND CF=0
```

In the example above, ASCII 1 (31H) is added to ASCII 7 (37H). After the AAA instruction, AL will contain 8 in BCD and CF = 0. The following example shows another ASCII addition and then the adjustment:

```
Example 2:
        MOV    AL,'9'       ;AL=39 ASCII FOR 9
        ADD    AL,'5'       ;ADD 35 (ASCII FOR 5) TO AL THEN AL=6EH
        AAA                 ;NOW AL=04 CF=1
        OR     AL,30H       ;converts result to ASCII
```

AAD ASCII Adjust before Division

```
Flags:      Affected: SF, ZF, PF.  Unpredictable: OF, AF, CF.
Format:     AAD
```

Function: Used before the DIV instruction to convert two unpacked BCD digits in AL and AH to binary. A better name for this would be BCD to binary conversion before division. This allows division of ASCII numbers. Before the AAD instruction is executed, the ASCII tag of 3 must be masked from the upper nibble of AH and AL.

```
Example:
        MOV    AX,3435H     ;AX=3435 THE ASCII FOR 45
        AND    AX,0F0FH     ;AX=0405H UNPACKED BCD FOR 45
        AAD                 ;AX=002DH HEX FOR 45
        MOV    DL,07        ;DL=07
        DIV    DL           ;2DH DIV BY 07 GIVES AL=06,AH=03
        OR     AX,3030H     ;AL=36=QUOTIENT AND AH=33=REMAINDER
```

AAM ASCII Adjust after Multiplication

```
Flags:      Affected: AF, CF.  Unpredictable: OF, SF, ZF, PF.
Format:     AAM
```

Function: Again, a better name would have been BCD adjust after multiplication. It is used after the MUL instruction has multiplied two unpacked BCD numbers. It converts AX from binary to unpacked BCD. AAM adjusts only AL, and any digits greater than 9 are stored in AH.

```
Example:
        MOV   AL,'5'        ;AL=35
        AND   AL,0FH        ;AL=05 UNPACKED BCD FOR 5
        MOV   BL,'4'        ;BL=34
        AND   BL,0FH        ;BL=04 UNPACKED BCD FOR 4
        MUL   BL            ;AX=0014H=20 DECIMAL
        AAM                 ;AX=0200
        OR    AX,3030H      ;AX=3230 ASCII FOR 20
```

AAS ASCII Adjust after Subtraction

```
Flags:      Affected: AF, CF.  Unpredictable: OF, SF, ZF, PF.
Format:     AAS
```

Function: After the subtraction of two ASCII digits, this instruction is used to convert the result in AL to packed BCD. Only AL is adjusted; the value in AH will be decremented if the carry flag is set.

```
Example:
        MOV   AL,32H        ;AL=32 ASCII FOR 2
        MOV   DH,37H        ;DH=37 ASCII FOR 7
        SUB   AL,DH         ;AL-DH=32-37=FBH WHICH IS -5 IN 2'S COMP
                            ;CF=1 INDICATING A BORROW
        AAS                 ;NOW AL=05 AND CF=1
```

ADC Add with Carry

```
Flags:      Affected: OF, SF, ZF, AF, PF, CF.
Format:     ADC   dest,source         ;dest = dest + source + CF
```

Function: If CF = 1 prior to this instruction, then after execution of this instruction, source is added to destination plus 1. If CF = 0, source is added to destination plus 0. Used widely in multibyte and multiword additions.

ADD Signed or Unsigned ADD

```
Flags:      Affected: OF, SF, ZF, AF, PF, CF.
Format:     ADD   dest,source         ;dest = dest + source
```

Function: Adds source operand to destination operand and places the result in destination. Both source and destination operands must match (e.g., both byte size or word size) and only one of them can be in memory.

Unsigned addition:

In addition of unsigned numbers, the status of CF, ZF, SF, AF, and PF may change, but only CF, ZF, and AF are of any use to programmers. The most important of these flags is CF. It becomes 1 when there is carry from D7 out in 8-bit (D0–D7) operations, or a carry from D15 out in 16-bit (D0–D15) operations.

```
Example 1:
        MOV   BH,45H        ;BH=45H
        ADD   BH,4FH        ;BH=94H (45H+4FH=94H)
                            ;CF=0,ZF=0,SF=1,AF=1,and PF=0
Example 2:
        MOV   AL,FEH        ;AL=FEH
        MOV   DL,75H        ;DL=75H
        ADD   AL,DL         ;AL=FE+75=73H
                            ;CF=1,ZF=0,AF=0,SF=0,PF=0
```

```
Example 3:
        MOV     DX,126FH        ;DX=126FH
        ADD     DX,3465H        ;DX=46D4H (126F=3465=46D4H)
                                ;CF=0,ZF=0,AF=1,SF=0,PF=1
        MOV     BX,0FFFFH
        ADD     BX,1                ;BX=0000 (FFFFH+1=0000)
                                ;AND CF=1,ZF=1,AF=1,SF=0,PF=1
```

Signed addition:

In addition of signed numbers, the status of OF, ZF, and SF must be noted. Special attention should be given to the overflow flag (OF) since this indicates if there is an error in the result of the addition. There are two rules for setting OF in signed number operation. The overflow flag is set to 1:

1. If there is a carry from D6 to D7 and no carry from D7 out in an 8-bit operation or a carry from D14 to D15 and no carry from D15 out in a 16-bit operation
2. If there is a carry from D7 out and no carry from D6 to D7 in an 8-bit operation or a carry from D15 out but no carry from D14 to D15 in a 16-bit operation

Notice that if there is a carry both from D7 out and from D6 to D7, then OF = 0 in 8-bit operations. In 16-bit operations, OF = 0 if there is both a carry out from D15 and a carry from D14 to D15.

```
Example 4:
        MOV     BL,+8           ;BL=0000 1000
        MOV     DH,+4           ;DH=0000 0100
        ADD     BL,DH           ;BL=0000 1100 SF=0,ZF=0,OF=0,CF=0
```

Notice SF = D7 = 0 since the result is positive and OF = 0 since there is neither a carry from D6 to D7 nor any carry beyond D7. Since OF = 0, the result is correct [(+8) + (+4) = (+12)].

```
Example 5:
        MOV     AL,+66          ;AL=0100 0010
        MOV     CL,+69          ;CL=0100 0101
        ADD     CL,AL           ;CL=1000 0111 = -121 (INCORRECT)
                                ;CF=0,SF=1,ZF=0, AND OF=1
```

In Example 5, the correct result is +135 [(+66) + (+69) = (+135)], but the result was –121. The OF = 1 is an indication of this error. Notice that SF = D7 = 1 since the result is negative; OF = 1 since there is a carry from D6 to D7 and CF = 0.

```
Example 6:
        MOV     AL,-12          ;AL=1111 0100
        MOV     BL,+18          ;BL=0001 0010
        ADD     BL,AL           ;BL=0000 0110 (WHICH IS +6 )
                                ;SF=0,ZF=0,OF=0, AND CF=1
```

Notice above that OF = 0 since there is a carry from D6 to D7 and a carry from D7 out.

```
Example 7:
        MOV     AH,-30          ;AH=1110 0010
        MOV     DL,+14          ;DL=0000 1110
        ADD     DL,AH           ;DL=1111 0000 (WHICH IS -16 AND CORRECT)
                                ;AND SF=1,ZF=0,OF=0, AND CF=0
```

OF = 0 since there is no carry from D7 out nor any carry from D6 to D7.

```
Example 8:
        MOV   AL,-126        ;AL=1000 0010
        MOV   BH,-127        ;BH=1000 0001
        ADD   AL,BH          ;AL=0000 0011 (WHICH IS +3AND WRONG)
                             ;AND SF=0,ZF=0 AND OF=1
```

OF = 1 since there is carry from D7 out but no carry from D6 to D7.

AND Logical AND

```
Flags:      Affected: CF = 0, OF = 0, SF, ZF, PF.
            Unpredictable: AF.
Format:     AND  dest,source
```

Function: Performs logical AND on the operands, bit by bit, storing the result in the destination.

```
Example:
        MOV   BL,39H         ;BL=39
        AND   BL,09H         ;BL=09
        ;39   0011 1001
        ;09   0000 1001
        ;--   ---------
        ;09   0000 1001
```

A	B	A AND B
0	0	0
0	1	0
1	0	0
1	1	1

CALL Call a Procedure

```
Flags:      Unchanged.
Format:     CALL proc    ;transfer control to procedure
```

Function: Transfers control to a procedure. RET is used to return control to the instruction after the call. There are two types of CALLs: NEAR and FAR. If the target address is within the same code segment, it is a NEAR call. If the target address is outside the current code segment, it is a FAR CALL. Each is described below.

NEAR CALL: If calling a near procedure (the procedure is in the same code segment as the CALL instruction) then the content of the IP register (which is the address of the instruction after the CALL) is pushed onto the stack and SP is decremented by 2. Then IP is loaded with the new value, which is the offset of the procedure. At the end of the procedure when the RET is executed, IP is popped off the stack, which returns control to the instruction after the CALL. There are three ways to code the address of the called NEAR procedure:

1. Direct:

```
      CALL  proc1
      ...
proc1 PROC NEAR
      ...
      RET
proc1 ENDP
```

2. Register indirect:

```
      CALL  [SI]   ;transfer control to address in SI
```

3. Memory indirect:

```
      CALL  WORD  PTR [DI]     ;DI points to the address that

                         ;contains IP address of proc
```

FAR CALL: When calling a far procedure (the procedure is in a different segment from the CALL instruction), the SP is decremented by 4 after CS:IP of the instruction following the CALL is pushed onto the stack. CS:IP is then loaded with the segment and offset address of the called procedure. In pushing CS:IP onto the stack, CS is pushed first and then IP. When the RETF is executed, CS and IP are restored from the stack and execution continues with the instruction following the CALL. The following addressing modes are supported:

1. Direct (but outside the present segment):

```
      CALL  procl
      ...
procl PROC  FAR
      ...
      RETF
procl ENDP
```

2. Memory indirect:

```
      CALL  DWORD PTR [ DI]      ;transfer control to CS:IP where
                                 ;DI and DI+1 point to location of
                                 ;CS and DI+2 and DI+4 point to
                                 ;location of IP
```

CBW **Convert Byte to Word**

```
Flags:      Unchanged.
Format:     CBW
```

Function: Copies D7 (the sign flag) to all bits of AH. Used widely to convert a signed byte in AL to a signed word to avoid the overflow problem in signed number arithmetic.

```
Example:
      MOV   AX,0
      MOV   AL,-5          ;AL=(-5)=FB in 2's complement
                           ;AX = 0000 0000 1111 1011
      CBW                  ;now AX=FFFB
                           ;AX = 1111 1111 1111 1011
```

CLC **Clear Carry Flag**

```
Flags:      Affected: CF.
Format:     CLC
```

Function: Resets CF to zero (CF = 0).

CLD **Clear Direction Flag**

```
Flags:      Affected: DF.
Format:     CLD
```

Function: Resets DF to zero (DF = 0). In string instructions if DF = 0, the pointers are incremented with each execution of the instruction. If DF = 1, the pointers are decremented. Therefore, CLD is used before string instructions to make the pointers increment.

CLI Clear Interrupt Flag

```
Flags:      Affected: IF.
Format:     CLI
```

Function: Resets IF to zero, thereby masking external interrupts received on INTR input. Interrupts received on NMI input are not blocked by this instruction.

CMC Complement Carry Flag

```
Flags:      Affected: CF.
Format:     CMC
```

Function: Changes CF from 0 to 1 or from 1 to 0.

CMP Compare Operands

```
Flags:      Affected: OF, SF, ZF, AF, PF, CF.
Format:     CMP dest,source    ;sets flags as if "SUB dest,source"
```

Function: Compares two operands of the same size. The source and destination operands are not altered. Performs comparison by subtracting the source operand from the destination and sets flags as if SUB were performed. The relevant flags are as follows:

	CF	ZF	SF	OF
dest > source	0	0	0	SF
dest = source	0	1	0	SF
dest < source	1	0	1	inverse of SF

CMPS/CMPSB/CMPSW Compare Byte or Word String

```
Flags:      Affected: OF, SF, ZF, AF, PF, CF.
Format:     CMPSx
```

Function: Compares strings a byte or word at a time. DS:SI is used to address the first operand; ES:DI is used to address the second. If DF = 0, it increments the pointers SI and DI. If DF = 1, it decrements the pointers. It can be used with prefix REPE or REPNE to compare strings of any length. The comparison is done by subtracting the source operand from the destination and sets flags as if SUB were performed.

CWD Convert Word to Doubleword

```
Flags:      Unchanged.
Format:     CWD
```

Function: Converts a signed word in AX into a signed doubleword by copying the sign bit of AX into all the bits of DX. Often used to avoid the overflow problem in signed number arithmetic.

```
Example:
      MOV   DX,0
      MOV   AX,-5           ;AX=(-5)=FFFB in 2's complement
      ;DX = 0000H
      CWD
      ;DX = FFFFH
```

DAA Decimal Adjust after Addition

```
Flags:          Affected: SF, ZF, AF, PF, CF, OF.
Format:         DAA
```

Function: This instruction is used after addition of BCD numbers to convert the result back to BCD. It adds 6 to the lower 4 bits of AL if it is greater than 9 or if AF = 1. Then it adds 6 to the upper 4 bits of AL if it is greater than 9 or if CF = 1.

```
Example 1:
        MOV    AL,47H          ;AL=0100 0111
        ADD    AL,38H          ;AL=47H+38H=7FH.    invalid BCD
        DAA                    ;NOW AL=1000 0101 (85H IS VALID BCD)
```

In this example, since the lower nibble was larger than 9, DAA added 6 to AL. If the lower nibble is smaller than 9 but AF = 1, it also adds 6 to the lower nibble.

```
Example 2:
        MOV    AL,29H          ;AL=0010 1001
        ADD    AL,18H          ;AL=0100 0001 INCORRECT RESULT
        DAA                    ;AL=0100 0111 A VALID BCD FOR 47H.
```

The same thing can happen for the upper nibble.

```
Example 3:
        MOV    AL,52H          ;AL=0101 0010
        ADD    AL,91H          ;AL=1110 0011 AN INVALID BCD
        DAA                    ;AL=0100 0011 AND CF=1
```

Again the upper nibble can be smaller than 9 but because CF = 1, it must be corrected.

```
Example 4:
        MOV    AL,94H          ;AL=1001 0100
        ADD    AL,91H          ;AL=0010 0101 INCORRECT RESULT
        DAA                    ;AL=1000 0101 A VALID BCD FOR 85 AND CF=1
```

It is entirely possible that 6 is added to both the high and low nibbles.

```
Example 5:
        MOV    AL,54H          ;AL=0101 0100
        ADD    AL,87H          ;AL=1101 1011 INVALID BCD
        DAA                    ;AL=0100 0001 AND CF=1  (141 IN BCD)
```

DAS Decimal Adjust after Subtraction

```
Flags:          Affected: SF, ZF, AF, PF, CF.  Unpredictable: OF.
Format:         DAS
```

Function: This instruction is used after subtraction of BCD numbers to convert the result to BCD. If the lower 4 bits of AL represent a number greater than 9 or if AF = 1, then 6 is subtracted from the lower nibble. If the upper 4 bits of AL are now greater than 9 or if CF = 1, 6 is subtracted from the upper nibble.

```
Example:
        MOV    AL,45H          ;AL=0100 0101 BCD for 45
        SUB    AL,17H          ;AL=0010 1110 AN INVALID BCD
        DAS                    ;AL=0010 1000 BCD FOR 28(45-17=28)
```

For more examples of problems associated with BCD arithmetic, see DAA.

DEC Decrement

```
Flags:      Affected: OF, SF, ZF, AF, PF.  Unchanged: CF.
Format:     DEC dest    ;dest = dest-1
```

Function: Subtracts 1 from the destination operand. Note that CF (carry/ borrow) is unchanged even if a value 0000 is decremented and becomes FFFF.

DIV Unsigned Division

```
Flags:      Unpredictable: OF, SF, ZF, AF, PF, CF.
Format:     DIV source          ;divide AX or DX:AX by source
```

Function: Divides either an unsigned word (AX) by a byte or an unsigned doubleword (DX:AX) by a word. If dividing a word by a byte, the quotient will be in AL and the remainder in AH. If dividing a doubleword by a word, the quotient will be in AX and the remainder in DX. Divide by zero causes interrupt type 0.

ESC Escape

```
Flags:      Unchanged.
Format:     ESC
```

Function: This instruction facilitates the use of math coprocessors (such as the 8087), which share data and address buses with the microprocessor. ESC is used to pass an instruction to a coprocessor and is usually treated as NOP (no operation) by the main processor.

HLT Halt

```
Flags:      Unchanged.
Format:     HLT
```

Function: Causes the microprocessor to halt execution of instructions. To get out of the halt state, activate an interrupt (NMI or INTR) or RESET.

IDIV Signed Number Division

```
Flags:      Unpredictable: OF, SF, ZF, AF, PF, CF.
Format:     IDIV  source       ;divide AX or DX:AX by source
```

Function: This division function divides either a signed word (AX) by a byte or a signed doubleword (DX:AX) by a word. If dividing a word by a byte, the signed quotient will be in AL and the signed remainder in AH. If dividing a doubleword by a word, the signed quotient will be in AX and the signed remainder in DX. Divide by zero causes interrupt type 0.

IMUL Signed Number Multiplication

```
Flags:      Affected: OF, CF.  Unpredictable: SF, ZF, AF, PF.
Format:     IMUL source    ;AX =source x AL or DX:AX =source x AX
```

Function: Multiplies a signed byte or word source operand by a signed byte or word in AL or AX with the result placed in AX or DX:AX.

IN Input Data from Port

```
Flags:      Unchanged.
Format:     IN  accumulator,port     ;input byte into AL or
                                     ;word into AX
```

Function: Transfers a byte or word to AL or AX from an input port specified by the second operand. The port address can be direct or register indirect:

1. Direct: the port address is specified directly and cannot be larger than FFH.

```
Example 1:
        IN      AL,99H          ;BRING A BYTE INTO AL FROM PORT 99H

Example 2:
        IN      AX,78H          ;BRING A WORD FROM PORT ADDRESSES 78H
                                ;AND 79H.  THE BYTE FROM PORT 78 GOES
                                ;TO AL AND BYTE FROM PORT 79H TO AH.
```

2. Register indirect: the port address is kept by the DX register. Therefore, it can be as high as FFFFH.

```
Example 3:
        MOV     DX,481H         ;DX=481H
        IN              AL,DX           ;BRING THE BYTE TO AL FROM THE
PORT
                                        ;WHOSE ADDRESS IS POINTED BY DX

Example 4:
        IN      AX,DX           ;BRING A WORD FROM PORT ADDRESS POINTED
                                ;TO BY DX. THE BYTE FROM PORT AT
                                ;DX GOES TO AL AND BYTE FROM PORT AT
                                ;DX+1 TO AH.
```

INC Increment

```
Flags:       Affected: OF, SF, ZF, AF, PF.  Unchanged: CF.
Format:      INC destination    ;dest = dest + 1
```

Function: Adds 1 to the register or memory location specified by the operand. Note that CF is not affected even if a value FFFF is incremented to 0000.

INT Interrupt

```
Flags:       Affected:  IF, TF.
Format:      INT   type   ;transfer control to INT type
```

Function: Transfers execution to one of the 256 interrupts. The vector address is specified by the type number, which cannot be greater than FFH (0 to FF = 256 interrupts).

The following steps are performed for the interrupt:

1. SP is decremented by 2 and the flags are pushed onto the stack.
2. SP is decremented by 2 and CS is pushed onto the stack.
3. SP is decremented by 2 and the IP of the next instruction after the interrupt is pushed onto the stack.
4. Multiplies the type number by 4 to get the address of the vector table. Starting at this address, the first 2 bytes are the value of IP and the next 2 bytes are the value for CS of the interrupt handler (interrupt handler is also called interrupt service routine).
5. Resets IF and TF.

Interrupts are used to get the attention of the microprocessor. In the 8088/86 there are a total of 256 interrupts: INT 00, INT 01, INT 02, ... , INT FF. As mentioned above, the address that an interrupt jumps to is always four times the value of the interrupt number. For example, INT 03 will jump to memory address 0000CH ($4 \times 03 = 12 = 0CH$). Table B-1 is a partial list of the interrupts, commonly referred to as the interrupt vector table.

Every interrupt has a program associated with it called the interrupt service routine (ISR). When an interrupt is invoked, the CS:IP address of its ISR is retrieved from the vector table (shown above). The lowest 1024 bytes ($256 \times 4 = 1024$) of RAM are set aside for the interrupt vector table and must not be used for any other function.

Table B-1: Interrupt Vector Table

INT # (hex)	Physical Address	Logical Address
INT 00	00000	0000:0000
INT 01	00004	0000:0004
INT 02	00008	0000:0008
INT 03	0000C	0000:000C
INT 04	00010	0000:0010
INT 05	00014	0000:0014
...	...	...
INT FF	003FC	0000:03FC

Example: Find the physical and logical addresses of the vector table associated with (a) INT 14H and (b) INT 38H.

Solution:

(a) The physical address for INT 14H is 00050H–00053H (4 × 14H = 50H). That gives the logical address of 0000:0050H–0000:0053H.

(b) The physical address for INT 38H is 000E0H–000E3H, making the physical address 0000:00E0H–0000:00E3H.

The difference between INTerrupt and CALL instructions

The following are some of the differences between the INT and CALL FAR instructions:

1. While a CALL can jump to any location within the 1-megabyte address range (00000–FFFFF) of the 8088/86 CPU, "INT nn" jumps to a fixed location in the vector table as discussed earlier.
2. While the CALL is used by the programmer at a predetermined point in a program, a hardware interrupt can come in at any time.
3. A CALL cannot be masked (disabled), but "INT nn" can be masked.
4. While a "CALL FAR" automatically saves on the stack only the CS:IP of the next instruction, "INT nn" saves the FR (flag register) in addition to the CS:IP.
5. While at the end of the procedure that has been CALLed the RETF (return FAR) is used, for "INT nn" the instruction IRET (interrupt return) is used.

The 256 interrupts can be categorized into two different groups: hardware and software interrupts.

Hardware interrupts

The 8088/86 microprocessors have two pins set aside for inputting hardware interrupts. They are INTR (interrupt request) and NMI (nonmaskable interrupt). Although INTR can be ignored through the use of software masking, NMI cannot be masked using software. These interrupts are activated externally by putting 5 volts on the hardware pins of NMI or INTR. Intel has assigned INT 02 to NMI. When it is activated it will jump to memory location 00008 to get the address (CS:IP) of the interrupt service routine (ISR). Memory locations 00008, 00009, 0000A, and 0000B contain the 4-byte CS:IP. There is no specific location in the vector table assigned to INTR because INTR is used to expand the number of hardware interrupts and should be allowed to use any "INT nn" instruction that has not been assigned previously. In the IBM PC, one Intel 8259 PIC (programmable interrupt controller) chip is connected to INTR to add a total of eight hardware interrupts to the microprocessor. IBM PC AT, PS/2 80286, and x86 computers use two 8259 chips to allow up to 15 hardware interrupts.

Software interrupts

These interrupts are called software interrupts since they are invoked as a result of the execution of an instruction and no external hardware is involved. In other words, these interrupts are invoked by executing an "INT nn" instruction such as the DOS function call "INT 21H" or video interrupt "INT 10H". These interrupts can be invoked by a program at any time, the same as any other instruction. Many of the interrupts in this category are used by the DOS operating system and IBM BIOS to perform the essential tasks that every computer must provide to the system and the user. Also within this group of interrupts are predefined functions associated with some of the interrupts. They are "INT 00" (divide error), "INT 01" (single step), "INT 03" (breakpoint), and "INT 04" (signed number overflow). Each one is described below. These interrupts are shown in Table B-2. Looking at Table B-2, one can say that aside from "INT 00" to "INT 04", which have predefined functions, the rest of the interrupts, from "INT 05" to "INT FF", can be used to implement either software or hardware interrupts.

Table B-2: IBM PC Interrupt System

Interrupt	Logical Address	Physical Address	Purpose
0	00E3:3072	03EA2	Divide error
1	0600:08ED	068ED	Single step (trace command in DEBUG)
2	F000:E2C3	FE2C3	Nonmaskable interrupt
3	0600:08E6	068E6	Breakpoint
4	0700:0147	00847	Signed number arithmetic overflow
5	F000:FF54	FFF54	Print screen (BIOS)
10	F000:F065	FF065	Video I/O (BIOS)
...	...	...	...
21	relocatable	---	DOS function calls
...	...	...	...

Functions associated with "INT 00" to "INT 03"

As mentioned earlier, interrupts "INT 00" to "INT 03" have predefined functions and cannot be used in any other way. The function of each is described next.

INT 00 (divide error)

This interrupt, sometimes referred to as a conditional or exception interrupt, is invoked by the microprocessor whenever there is a condition that it cannot take care of, such as an attempt to divide a number by zero. "INT 00" is invoked by the microprocessor whenever there is an attempt to divide a number by zero. In the IBM PC and compatibles, the service subroutine for this interrupt is responsible for displaying the message "DIVIDE ERROR" on the screen if a program such as the following is executed:

```
MOV   AL,25        ; put 25 into AL
MOV   BL,00        ; put 00 into BL
DIV   BL           ; divide 25 by 00
```

This interrupt is also invoked if the quotient is too large to fit into the assigned register when executing a DIV instruction.

INT 01 (single step)

There is often a need to execute a given program one instruction at a time and then inspect the registers (possibly memory as well) to see what is happening inside the CPU. This is commonly referred to as single-stepping. IBM and Microsoft call it TRACE in the DEBUG program. To allow the implementation of single-stepping, Intel has set aside "INT 01" specifically for that purpose. For the Trace command in DEBUG after execution of each instruction, the CPU jumps automatically to physical location 00004 to fetch the 4 bytes for CS:IP of the interrupt service routine. One of the functions of this ISR is to dump the contents of the registers onto the screen.

INT 02 (nonmaskable interrupt)

This interrupt is used in the PC to indicate memory errors, among other problems.

INT 03 (breakpoint)

While in single-step mode, one can inspect the CPU and system memory after execution of each instruction. A breakpoint allows one to do the same thing, after execution of a group of instructions rather than after each instruction. Breakpoints are put in at certain points of a program to monitor the flow of the program and to inspect the results after certain instructions. The CPU executes the program to the breakpoint and stops. One can proceed from breakpoint to breakpoint until the program is complete. With the help of single-step and breakpoints, programs can be debugged and tested more easily. The Intel 8088/86 CPUs have set aside "INT 03" for the sole purpose of implementing breakpoints. When the instruction "INT 03" is placed in a program the CPU will execute the program until it encounters "INT 03", and then it stops. One interesting point about this interrupt is that it is a one-byte instruction, in contrast to all other interrupt instructions, "INT nn", which are two-byte instructions. This allows the user to insert 1 byte of code and remove it to proceed with the execution of the program. The opcode for INT 03 is "CC".

IBM PC and DOS assignment of interrupts

When the IBM PC was being developed, the designers at IBM had to coordinate the assignment of the 256 available interrupts for the 8088/86 with Microsoft, the developer of the DOS operating system, lest a conflict occur between the BIOS and DOS interrupt designations. The result of cooperation in assigning interrupts to IBM BIOS subroutines and DOS function calls is shown in Table B-2. The table gives a partial listing of interrupt numbers from 00 to FF, the logical address of the service subroutine for each interrupt, their physical addresses, and the purpose of each interrupt. It must be mentioned that depending on the computer and the DOS version, some of the logical addresses could be different from Table B-2.

How to get the vector table of any PC

One can get the vector table of any x86 IBM-compatible computer and inspect the logical address assigned to each interrupt. To do that use DEBUG's DUMP command "-D 0000:0000", as shown next.

```
A>debug
-D 0000:0000
0000:0000  E8 56 2B 02 56 07 70 00-C3 E2 00 F0 56 07 70 00   ...........
0000:0010  56 07 70 00 54 FF 00 F0-47 FF 00 F0 47 FF 00 F0   ...........
```

Note: The contents of the memory locations could be different, depending on the DOS version.

Example:

From the dump above, find the CS:IP of the service routine associated with INT 5.

Solution:

To get the address of "INT 5", calculate the physical address of 00014H ($5 \times 4 =$ 00014H). The contents of these locations are 00014 = 54, 00015 = FF, 00016 = 00, and 00016 = F0. This gives CS = F000 and IP = FF54.

INTO Interrupt on Overflow

```
Flags:     Affected:  IF, TF.
Format:    INTO
```

Function: Transfers execution to an interrupt handler written for overflow if OF (overflow flag) has been set. Intel has set aside INT 4 for this purpose. Therefore, if OF = 1 when INTO is executed, the CPU jumps to memory location 00010H ($4 \times 4 = 16 = 10H$). The contents of memory locations 10H, 11H, 12H, and 13H are used as IP and CS of the interrupt handler procedure. This instruction is widely used to detect overflow in signed number addition. In signed number operations, OF becomes 1 in two cases:

1. Whenever there is a carry from D6 to D7 in 8-bit operations and no carry from D7 out (or in 16-bit operations when there is carry from D14 to D15 and CF = 0)
2. When there is carry from D7 out and no carry from D6 to D7 (or in the case of 16-bit operations when there is a carry from D15 out and no carry from D14 to D15)

IRET Interrupt Return

```
Flags:     Affected: OF, DF, IF, TF, SF, ZF, AF, PF, CF.
Format:    IRET
```

Function: Used at the end of an interrupt service routine (interrupt handler), this instruction restores all flags, CS, and IP to the values they had before the interrupt so that execution may continue at the next instruction following the INT instruction. While the RET instruction is used at the end of the subroutine associated with the CALL instruction, IRET must be used for the subroutine associated with the "INT XX" instruction or the hardware interrupt handler.

JUMP Instructions

The following instructions are associated with jumps (both conditional and unconditional). They are categorized according to their usage rather than alphabetically.

```
J condition

Flags:     Unchanged.
Format:    Jxx   target       ;jump to target upon condition
```

Function: Used to jump to a target address if certain conditions are met. The target address cannot be more than –128 to +127 bytes away. The conditions are indicated by the flag register. The conditions that determine whether the jump takes place can be categorized into three groups:

1. flag values,
2. the comparison of unsigned numbers, and
3. the comparison of signed numbers.

Each is explained next.

1. "J condition" where the condition refers to flag values. The status of each bit of the flag register has been decided by execution of instructions prior to the jump. The following "J condition" instructions check if a certain flag bit is raised or not.

```
JC    Jump Carry           jump if CF=1
JNC   Jump No Carry        jump if CF=0
JP    Jump Parity          jump if PF=1
JNP   Jump No Parity       jump if PF=0
JZ    Jump Zero            jump if ZF=1
JNZ   Jump No Zero         jump if ZF=0
JS    Jump Sign            jump if SF=1
JNS   Jump No Sign         jump if SF=0
JO    Jump Overflow        jump if OF=1
JNO   Jump No Overflow     jump if OF=0
```

Notice that there is no "J condition" instruction for AF.

2. "J condition" where the condition refers to the comparison of unsigned numbers. After a compare (CMP dest,source) instruction is executed, CF and ZF indicate the result of the comparison, as follows:

	CF	ZF
destination > source	0	0
destination = source	0	1
destination < source	1	0

Since the operands compared are viewed as unsigned numbers, the following "J condition" instructions are used.

```
JA     Jump Above                 jump if CF=0 and ZF=0
JAE    Jump Above or Equal        jump if CF=0
JB     Jump Below                 jump if CF=1
JBE    Jump Below or Equal        jump if CF=1 or ZF=1
JE     Jump Equal                 jump if ZF=1
JNE    Jump Not Equal             jump if ZF=0
```

3. "J condition" where the condition refers to the comparison of signed numbers. In the case of the signed number comparison, although the same instruction, "CMP destination,source", is used, the flags used to check the result are as follows:

destination > source	OF=SF or ZF=0
destination = source	ZF=1
destination < source	OF inverse of SF

Consequently, the "J condition" instructions used are different. They are as follows:

```
JG     Jump Greater               jump if ZF=0 or OF=SF
JGE    Jump Greater or Equal      jump if OF=SF
JL     Jump Less                  jump if OFSF
JLE    Jump Less or Equal         jump if ZF=1 or OFSF
JE     Jump if Equal              jump if ZF = 1
```

There is one more "J condition" instruction:

```
JCXZ          ;Jump if CX is Zero.  ZF is ignored.
```

All "J condition" instructions are short jumps, meaning that the target address cannot be more than –128 bytes backward or +127 bytes forward from the IP of the instruction following the jump. What happens if a programmer needs to use a "J condition" to go to a target address beyond the –128 to +127 range? The solution is to use the "J condition" along with the unconditional JMP instruction, as shown next.

```
            ADD    BX,[ SI]
            JNC    NEXT
            JMP    TARGET1
NEXT:       ....
            ...
TARGET1:    ADD    DI,10
            ...
```

JMP Unconditional Jump

```
Flags:      Unchanged.
Format:     JMP [ directives]  target        ;jump to target address
```

Function: This instruction is used to transfer control unconditionally to a new address. The difference between JMP and CALL is that the CALL instruction will return and continue execution with the instruction following the CALL, whereas JMP will not return. The target address could be within the current code segment, which is called a near jump, or outside the current code segment, which is called a far jump. Within each category there are many ways to code the target address, as shown next.

1. Near jump

(a) direct short jump: In this jump the target address must be within –128 to +127 bytes of the IP of the instruction after the JMP. This is a 2-byte instruction. The first byte is the opcode EBH and the second byte is the signed number displacement, which is added to the IP of the instruction following the JMP to get the target address. The directive SHORT must be coded, as shown next:

```
        JMP   SHORT OVER
        ...
OVER:   ...
```

If the target address is beyond the –128 to +127 byte range and the SHORT directive is coded, the assembler gives an error.

(b) Direct jump: This is a 3-byte instruction. The first byte is the opcode E9H and the next two bytes are the signed number displacement value.The displacement is added to the IP of the instruction following the JMP to get the target address. The displacement can be in the range –32,768 to +32,767. In the absence of the SHORT directive, the assembler in its first pass always uses this kind of JMP, and then in the second pass if the target address is within the –128 and +127 byte range, it uses the NOP opcode 90H for the third byte. This is the reason to code the directive SHORT if it is known that the target address of the JMP is within the short range.

(c) Register indirect jump: In this jump the target address is in a register as shown next:

```
        JMP    DI   ;jump to the address found in DI
```

Any nonsegment register can be used for this purpose.

(d) Memory indirect jump: In this jump the target address is in a memory location whose address is pointed at by a register:

```
        JMP    WORD PTR [SI]  ;jump to the address found at the address in SI
```

The directive WORD PTR must be coded to indicate this is a near jump.

2. Far jump

In a far JMP, the target address is outside the present code segment; therefore, not only the offset value but also the segment value of the target address must be given. A far jump is a 5-byte instruction: the opcode EAH and 4 bytes for the offset and segment of the target address. The following shows the two methods of coding the far jump.

(a) Direct far jump: This requires that both CS and IP be updated. One way to do that is to use the LABEL directive:

```
               JMP    TARGET2
               ...
TARGET2        LABEL FAR
ENTRY:         ...
```

This is exactly what IBM did in the BIOS of the IBM PC/XT when the computer is booted. When the power to the PC is turned on, the 8088/86 CPU begins to execute at address FFFF:0000H. IBM uses a FAR jump to make it go to location F000:E05BH, as shown next:

```
                        ;CS=FFFF and IP=0000
```

```
0000  EA5BE000F0    JMP    RESET
                    ;CS=F000
                    ORG    0E05BH
E05B  RESET         LABEL    FAR
E05B  START:
E05B                CLI
E05C                ...
```

The EXTRN and PUBLIC directives also can be used for the same purpose.

(b) Memory indirect far jump: The target address (both CS:IP) is in a memory location pointed to by the register:

```
JMP   DWORD   PTR   [ BX]
```

The DWORD and PTR directives must be used to indicate that it is a far jump.

LAHF Load AH from Flags

```
Flags:     Unchanged.
Format:    LAHF
```

Function: Loads the lower 8 bits of the flag register into AH.

LDS Load Data Segment Register

```
Flags:     Unchanged.
Format:    LDS  dest,source;load dest and DS starting at source
```

Function: Loads into destination (which is a register) the contents of two memory locations indicated by source and loads DS with the contents of the two succeeding memory locations. This is useful for accessing a new data segment and its offset.

Example:

Assume the following memory locations with the contents:

```
;DS:1200=(46)
;DS:1201=(10)
;DS:1202=(38)
;DS:1203=(82)
LDS   DI,[ 1200]    ;now DI=1046 and DS=8238.
```

LEA Load Effective Address

```
Flags:     Unchanged.
Format:    LEA dest,source   ;dest = OFFSET source
```

Function: Loads into the destination (a 16-bit register) the effective address of a direct memory operand.

Example 1:

```
      ORG    0100H
DATA  DB     34,56,87,90,76,54,13,29
      ...
                          ;to access the sixth element:
      LEA    SI,DATA+5    ;SI=100H+5=105 THE EFFECTIVE ADDRESS
      MOV    AL,[ SI]     ;GET THE SIXTH ELEMENT
```

Example 2:

```
                          ;if BX=2000H and SI=3500H
      LEA    DX,[ BX][ SI] +100H
                          ;DX=effective address=2000+3500+100=5600H
```

The following two instructions show two different ways to accomplish the same thing:

```
MOV   SI,OFFSET DATA    ;advantage: executes faster
LEA   SI,DATA
```

LES Load Extra Segment Register

```
Flags:      Unchanged.
Format:     LES  dest,source;load dest and ES starting at source
```

Function: Loads into destination (a register) the contents of two memory locations indicated by the source and loads ES with the contents of the two succeeding memory locations. Useful for accessing a new extra segment and its offset. This instruction is similar to LDS except that the ES and its offset are being loaded.

LOCK Lock System Bus Prefix

```
Flags:      Unchanged.
Format:     LOCK        ;used as a prefix before instructions
```

Function: Used in microcomputer systems with more than one processor to prevent another processor from gaining control over the system bus during execution of an instruction.

LODS/LODSB/LODSW Load Byte or Word String

```
Flags:      Unchanged.
Format:     LODSx
```

Function: Loads AL or AX with a byte or word from the memory location pointed to by DS:SI. If DF = 0, SI will be incremented to point to the next location. If DF = 1, SI will be decremented to point to the next location. SI is incremented/decremented by 1 or 2, depending on whether it is a byte or word string.

LOOP Loop until CX = 0

```
Flags:      Unchanged.
Format:     LOOP  target;DEC CX, then jump to target if CX not 0
```

Function: Decrements CX by 1, then jumps to the offset indicated by the operand if CX is not zero; otherwise continues with the next instruction below the LOOP. This instruction is equivalent to

```
DEC   CX
JNZ   target
```

LOOPE/LOOPZ LOOP if Equal / Loop if Zero

```
Flags:  Unchanged.
Format: LOOPx target   ;DEC CX, jump to target if CX  0 and ZF=1
```

Function: Decrements CX by 1, then jumps to location indicated by the operand if CX is not zero and ZF is 1; otherwise continues with the next instruction after the LOOP. In other words, it gets out of the loop only when CX becomes zero or when ZF = 0.

```
Example: Assume that 200H memory locations from offset 1680H
should contain 55H.  LOOPE can be used to see if any of these
locations does not contain 55H:

      MOV   CX,200       ;SET UP THE COUNTER
      MOV   SI,1680H     ;SET UP THE POINTER
```

```
BACK: CMP   [SI],55H      ;COMPARE THE 55H WITH MEM LOCATION
                          ;POINTED AT BY SI
      INC   SI            ;INCREMENT THE POINTER
      LOOPE BACK          ;CONTINUE THE PROCESS UNTIL CX=0 OR
                          ;ZF=0.  IN OTHER WORDS EXIT IF ONE
                          ;LOCATION DOES NOT HAVE 55H
```

LOOPNE/LOOPNZ LOOP While CF Not Zero and ZF Equal Zero

```
Flags:  Unchanged.
Format: LOOPxx target    ;DEC CX, then jump if CX and ZF not zero
```

Function: Decrements CX by 1, then jumps to location indicated by the operand if CX and ZF are not zero; otherwise continues with the next instruction below the LOOP. In other words it will exit the loop if CX becomes 0 or ZF = 1.

Example:

Assume that the daily temperatures for the last 30 days have been stored starting at memory location with offset 1200H. LOOPE can be used to find the first day that had a 90-degree temperature.

```
      MOV   CX,30         ;SET UP THE COUNTER
      MOV   DI,1200H      ;SET UP THE POINTER
AGAIN:CMP   [DI],90
      INC   DI
      LOOPNE AGAIN
```

MOV Move

```
Flags:      Unchanged.
Format:     MOV   dest,source         ;copy source to dest
```

Function: Copies a word or byte from a register, memory location, or immediate number to a register or memory location. Source and destination must be of the same size and cannot both be memory locations.

MOVS/MOVSB/MOVSW Move Byte or Word String

```
Flags:      Unchanged.
Format:     MOVSx
```

Function: Moves byte or word from memory location pointed to by DS:SI to memory location pointed to by ES:DI. If DF = 0, both pointers are incremented; otherwise, they are decremented. SI and DI are incremented/decremented by 1 or 2 depending on whether it is a byte or word string. When used with the REP prefix, CX is decremented each time until CX is zero.

MUL Unsigned Multiplication

```
Flags:  Affected: OF, CF.  Unpredictable:  SF, ZF, AF, PF.
Format: MUL source      ;AX = source × AL or DX:AX = source × AX
```

Function: Multiplies an unsigned byte or word indicated by the operand by a unsigned byte or word in AL or AX with the result placed in AX or DX:AX.

NEG Negate

```
Flags:      Affected: OF, SF, ZF, AF, PF, CF.
Format:     NEG   dest   ;negates operand
```

Function: Performs 2's complement of operand. Effectively reverses the sign bit of the operand. This instruction should only be used on signed numbers.

NOP **No Operation**

```
Flags:      Unchanged.
Format:     NOP
```

Function: Performs no operation. Sometimes used for timing delays to waste clock cycles. Updates IP to point to next instruction following NOP.

NOT **Logical NOT**

```
Flags:      Unchanged.
Format:     NOT dest      ;dest = 1's complement of dest
```

Function: Replaces the operand with its negation (the 1's complement). Each bit is inverted.

OR **Logical OR**

```
Flags:      Affected: CF=0, OF=0, SF, ZF, PF. Unpredictable: AF.
Format:     OR dest,source    ;dest= dest OR source
```

Function: Performs logical OR on the bits of two operands, replacing the destination operand with the result. Often used to turn a bit on.

A	B	A OR B
0	0	0
0	1	1
1	0	1
1	1	1

OUT **Output Byte or Word**

```
Flags:      Unchanged.
Format:     OUT   dest,acc     ;transfer acc to port dest
```

Function: Transfers a byte or word from AL or AX to an output port specified by the first operand. Port address can be direct or register indirect as shown next:

1. Direct: port address is specified directly and cannot be larger than FFH.

```
Example 1:
        OUT    68H,AL        ;SEND OUT A BYTE FROM AL TO PORT 68H

               or

        OUT    34H,AX        ;SEND OUT A WORD FROM AX TO PORT
                             ;ADDRESSES 34H AND 35H. THE BYTE
                             ;FROM AL GOES TO PORT 34H AND
                             ;THE BYTE FROM AH GOES TO PORT 35H
```

2. Register indirect: port address is kept by the DX register. Therefore, it can be as high as FFFFH.

```
Example 2:
        MOV    DX,64B1H      ;DX=64B1H
        OUT    DX,AL         ;SENT OUT THE BYTE IN AL TO THE PORT
                             ;WHOSE ADDRESS IS POINTED TO BY DX
               or

        OUT    DX,AX         ;SEND OUT A WORD FROM AX TO PORT
                             ;ADDRESS POINTED TO DX.   THE BYTE
                             ;FROM AL GOES TO PORT DX AND AND BYTE
                             ;FROM AH GOES TO PORT DX+1.
```

POP POP Word

```
Flags:      Unchanged.
Format:     POP dest    ;dest = word off top of stack
```

Function: Copies the word pointed to by the stack pointer to the register or memory location indicated by the operand and increments the SP by 2.

POPF POP Flags off Stack

```
Flags:      OF, DF, IF, TF, SF, ZF, AF, PF, CF.
Format:     POPF
```

Function: Copies bits previously pushed onto the stack with the PUSHF instruction into the flag register. The stack pointer is then incremented by 2.

PUSH PUSH Word

```
Flags:      Unchanged.
Format:     PUSH  source      ;PUSH source onto stack
```

Function: Copies the source word to the stack and decrements SP by 2.

PUSHF PUSH Flags onto stack

```
Flags:      Unchanged.
Format:     PUSHF
```

Function: Decrements SP by 2 and copies the contents of the flag register to the stack.

RCL/RCR Rotate Left through Carry and Rotate Right through Carry

```
Flags:  Affected: OF, CF.
Format: RCx dest,n;dest = dest rotate right/left n bit positions
```

Function: Rotates the bits of the operand right or left. The bits rotated out of the operand are rotated into the CF and the CF is rotated into the opposite end of the word or byte. Note: "n" must be 1 or CL.

RET Return from a Procedure

```
Flags:      Unchanged.
Format:     RET  [n]    ;return from procedure
```

Function: Used to return from a procedure previously entered by a CALL instruction. The IP is restored from the stack and the SP is incremented by 2. If the procedure was FAR, then RETF (return FAR) is used, and in addition to restoring the IP, the CS is restored from the stack and SP is again incremented by 2. The RET instruction may be followed by a number that will be added to the SP after the SP has been incremented. This is done to skip over any parameters being passed back to the calling program segment.

ROL/ROR Rotate Left and Rotate Right

```
Flags:  Affected: OF, CF.
Format: ROx dest,n      ;rotate dest right/left n bit positions
```

Function: Rotates the bits of a word or byte indicated by the second operand right or left. The bits rotated out of the word or byte are rotated back into the word or byte at the opposite end. Note: "n" must be 1 or CL.

SAHF Store AH in Flag Register

```
Flags:      Affected: SF, ZF, AF, PF, CF.
Format:     SAHF
```

Function: Copies AH to the lower 8 bits of the flag register.

SAL/SAR Shift Arithmetic Left/ Shift Arithmetic Right

```
Flags:   Affected: OF, SF, ZF, PF, CF.  Unpredictable: AF.
Format: SAx dest,n ;shift signed dest left/right n bit positions
```

Function: Shifts a word or byte left /right. SAR/ SAL arithmetic shifts are used for signed number shifting. In SAL, as the operand is shifted left bit by bit, the LSB is filled with 0s and the MSB is copied to CF. In SAR, as each bit is shifted right, the LSB is copied to CF and the empty bits filled with the sign bit (the MSB). SAL/SAR essentially multiplies/divides destination by a power of 2 for each bit shift. Note: "n" must be 1 or CL.

SBB Subtract with Borrow

```
Flags:      Affected: OF, SF, ZF, AF, PF, CF.
Format:     SBB  dest,source        ;dest = dest - CF - source
```

Function: Subtracts the source operand from the destination, replacing the destination. If CF = 1, it subtracts 1 from the result; otherwise, it executes like SUB.

SCAS/SCASB/SCASW Scan Byte or Word String

```
Flags:      Affected: OF, SF, ZF, AF, PF, CF.
Format:     SCASx
```

Function: Scans a string of data pointed by ES:DI for a value that is in AL or AX. Often used with the REPE/REPNE prefix. If DF is zero, the address is incremented; otherwise, it is decremented.

SHL/SHR Shift Left/Shift Right

```
Flags:   Affected: OF, SF, ZF, PF, CF.  Unpredictable: AF.
Format: SHx dest,n ;shift unsigned dest left/right n bit positions
```

Function: These are logical shifts used for unsigned numbers, meaning that the sign bit is treated as data. In SHR, as the operand is shifted right bit by bit and copied into CF, the empty bits are filled with 0s instead of the sign bit as is the case for SAR. In the case of SHL, as the bits are shifted left, the MSB is copied to CF and empty bits are filled with 0, which is exactly the same as SAL. In reality, SAL and SHL are two different mnemonics for the same opcode. SHL/SHR essentially multiplies/divides the destination by a power of 2 for each bit position shifted. Note: "n" must be 1 or CL.

STC Set Carry Flag

```
Flags:      Affected: CF.
Format:     STC
```

Function: Sets CF to 1.

STD Set Direction Flag

```
Flags:      Affected: DF.
Format:     STD
```

Function: Sets DF to 1. Used widely with string instructions. As explained in the string instructions, if DF = 1, the pointers are decremented.

STI Set Interrupt Flag

```
Flags:      Affected: IF.
Format:     STI
```

Function: Sets IF to 1, allowing the hardware interrupt to be recognized through the INTR pin of the CPU.

STOS/STOSB/STOSW Store Byte or Word String

```
Flags:       Unchanged.
Format:      STOSx
```

Function: Copies a byte or word from AX or AL to a location pointed to by ES:DI and updates DI to point to the next string element. The pointer DI is incremented if DF is zero; otherwise, it is decremented.

SUB Subtract

```
Flags:       Affected: OF, SF, ZF, AF, PF, CF.
Format:      SUB   dest,source         ;dest = dest - source
```

Function: Subtracts the source from the destination and puts the result in the destination. Sets the carry and zero flags according to the following:

	CF	ZF	
dest > source	0	0	the result is positive
dest = source	0	1	the result is 0
dest < source	1	0	the result is negative in 2's comp

The steps for subtraction performed by the internal hardware of the CPU are as follows:

1. Takes the 2's complement of the source
2. Adds this to the destination
3. Inverts the carry and changes the flags accordingly

The source operand remains unchanged by this instruction.

TEST Test Bits

```
Flags:       Affected: OF, SF, ZF, PF, CF.  Unpredictable: AF.
Format:      TEST  dest,source         ;performs dest AND source
```

Function: Performs a logical AND on two operands, setting flags but leaving the contents of both source and destination unchanged. While the AND instruction changes the contents of the destination and the flag bits, the TEST instruction changes only the flag bits.

```
Example: Assume that D0 and D1 of port 27 indicate conditions A
and B, respectively, if they are high and only one of them can
be high at a given time.  The TEST instruction can be used as
follows:

        IN      AL,PORT_27
        TEST AL,0000 0001B        ;CHECK THE CONDITION A
        JNZ    CASE_A             ;JUMP TO INDICATE CONDITION A
        TEST AL,0000 0010B        ;CHECK FOR CONDITION B
        JNZ    CASE_B             ;JUMP TO INDICATE CONDITION B
        ....                      ;THERE IS AN ERROR SINCE NEITHER
        ....                      ;   A NOR B HAS OCCURRED.
CASE_A:    ....
        ....
CASE_B:    ....
```

WAIT Puts Processor in WAIT State

```
Flags:       Unchanged.
Format:      WAIT
```

Function: Causes the microprocessor to enter an idle state until an external interrupt occurs. This is often done to synchronize it with another processor or with an external device.

XCHG Exchange

```
Flags:       Unchanged.
Format:      XCHG   dest,source        ;swaps dest and source
```

Function: Exchanges the contents of two registers or a register and a memory location.

XLAT Translate

```
Flags:       Unchanged.
Format:      XLAT
```

Function: Replaces contents of AL with the contents of a look-up table whose address is specified by AL. BX must be loaded with the start address of the look-up table and the element to be translated must be in AL prior to the execution of this instruction. AL is used as an offset within the conversion table. Often used to translate data from one format to another, such as ASCII to EBCDIC.

XOR Exclusive OR

```
Flags:       Affected: CF = 0, OF = 0, SF, ZF, PF.
             Unpredictable: AF.
Format:      XOR   dest,source
```

Function: Performs a logical exclusive OR on the bits of two operands and puts the result in the destination. "XOR AX,AX" can be used to clear AX.

A	B	A XOR B
0	0	0
0	1	1
1	0	1
1	1	0

APPENDIX C

ASSEMBLER DIRECTIVES AND NAMING RULES

OVERVIEW

This appendix consists of two sections. The first section describes some of the most widely used directives in 80x86 Assembly language programming. In the second section Assembly language rules and restrictions for names and labels are discussed and a list of reserved words is provided.

SECTION C.1: x86 ASSEMBLER DIRECTIVES

Directives, or as they are sometimes called, pseudo-ops or pseudo-instructions, are used by the assembler to translate Assembly language programs into machine language. Unlike the microprocessor's instructions, directives do not generate any opcode; therefore, no memory locations are occupied by directives in the final ready-to-run (exe) version of the assembly program. To summarize, directives give directions to the assembler program to tell it how to generate the machine code; instructions are assembled into machine code to give directions to the CPU at execution time. The following are descriptions of the some of the most widely used directives for the 80x86 assembler. They are given in alphabetical order for ease of reference.

ASSUME

The ASSUME directive is used by the assembler to associate a given segment's name with a segment register. This is needed for instructions that must compute an address by combining an offset with a segment register. One ASSUME directive can be used to associate all the segment registers. For example:

```
ASSUME      CS:name1,DS:name2,SS:name3,ES:name4
```

where name1, name2, and so on, are the names of the segments. The same result can be achieved by having one ASSUME for each register:

```
ASSUME      CS:name1
ASSUME      DS:name2
ASSUME      SS:name3
ASSUME      ES:nothing
ASSUME      nothing
```

The key word "nothing" can be used to cancel a previous ASSUME directive.

DB (Define Byte)

The DB directive is used to allocate memory in byte-sized increments. Look at the following examples:

```
DATA1       DB          23
DATA2       DB          45,97H,10000011B
DATA3       DB          'The planet Earth'
```

In DATA1 a single byte is defined with initial value 23. DATA2 consists of several values in decimal (45), hex (97H), and binary (10000011B). Finally, in DATA3, the DB directive is used to define ASCII characters. The DB directive is normally used to define ASCII data. In all the examples above, the address location for each value is assigned by the assembler. We can assign a specific offset address by the use of the ORG directive.

DD (Define Doubleword)

To allocate memory in 4-byte (32-bit) increments, the DD directive is used. Since word-sized operands are 16 bits wide (2 bytes) in 80x86 assemblers, a doubleword is 4 bytes.

```
VALUE1      DD          4563F57H
RESULT      DD          ?               ;RESERVE   4-BYTE LOCATION
DAT4        DD          25000000
```

It must be noted that the values defined using the DD directive are placed in memory by the assembler in low byte to low address and high byte to high address order. This convention is referred to as little endian. For example, assuming that offset address 0020 is assigned to VALUE1 in the example above, each byte will reside in memory as follows:

```
DS:20=(57)
DS:21=(3F)
DS:22=(56)
DS:23=(04)
```

DQ (Define Quadword)

To allocate memory in 8-byte increments, the DQ directive is used. In the 80x86 a word is defined as 2 bytes; therefore, a quadword is 8 bytes.

```
DAT_64B       DQ          5677DD4EE4FF45AH
DAT8          DQ          10000000000000
```

DT (Define Tenbytes)

To allocate packed BCD data, 10 bytes at a time, the DT directive is used. This is widely used for memory allocation associated with BCD numbers.

```
DATA          DT          399977653419974
```

Notice there is no H for the hexadecimal identifier following the number. This is a characteristic particular to the DT directive. In the case of other directives (DB, DW, DD, DQ), a number with no H at the end, is assumed to be in decimal and will be converted to hex by the assembler. Remember that the little endian convention is used to place the bytes in memory, with the least significant byte going to the low address and the most significant byte to the high address. DT can also be used to allocate decimal data if "d" is placed after the number:

```
DATA          DT          65535d      ;stores hex FFFF in a 10-byte
location
```

DUP (Duplicate)

The DUP directive can be used to duplicate a set of data a certain number of times instead of having to write it over and over.

```
DATA1 DB    20 DUP (99)        ;DUPLICATE 99 20 TIMES
DATA2 DW    6 DUP (5555H)      ;DUPLICATE 5555H 6 TIMES
DATA3 DB    10 DUP (?)         ;RESERVE 10 BYTES
DATA4 DB    5 DUP (5 DUP (0))  ;25 BYTES  INITIALIZED TO ZERO
DATA5 DB    10 DUP (00,FFH)    ;20 BYTES ALTERNATE 00, FF
```

DW (Define Word)

To allocate memory in 2-byte (16-bit) increments, the DW directive is used. In the 80x86 family, a word is defined as 16 bits.

```
DATAW_1       DW          5000
DATAW_2       DW          7F6BH
```

Again, in terms of placing the bytes in memory the little endian convention is used with the least significant byte going to the low address and the most significant byte going to the high address.

END

Every program must have an entry point. To identify that entry point the assembler relies on the END directive. The labels for the entry and end point must match.

```
HERE:         MOV    AX,DATASEG  ;ENTRY POINT OF THE PROGRAM
              ...
              ...
              END    HERE        :EXIT POINT OF THE PROGRAM
```

If there are several modules, only one of them can have the entry point, and the name of that entry point must be the same as the name for the END directive as shown below:

```
;from the main program:
                   EXTRN   PROG1:NEAR
                   ...
MAIN_PRO:    MOV   AX,DATASG          ;THE ENTRY POINT
                   MOV   DS,AX
                   ...
                   CALL  PROG1
                   ...
                   END   MAIN_PRO           ;THE EXIT POINT

;from the module PROG1:
                   PUBLIC PROG1
PROG1              PROC
                   ...
                   RET                      ;RETURN TO THE MAIN
MODULE
PROG1              ENDP
                   END                      ;NO LABEL IS GIVEN
```

Notice the following points about the above code:

1. The entry point must be identified by a name. In the example above the entry point is identified by the name MAIN_PRO.
2. The exit point must be identified by the same name given to the entry point, MAIN_PRO.
3. Since a given program can have only one entry point and one exit point, all modules called (either from main or from the submodules) must have directive END with nothing after it.

ENDP (see the PROC directive)

ENDS (see the SEGMENT and STRUCT directives)

EQU (Equate)

To assign a fixed value to a name, one uses the EQU directive. The assembler will replace each occurrence of the name with the value assigned to it.

```
FIX_VALU     EQU   1200
PORT_A       EQU   60H
COUNT        EQU   100
MASK_1       EQU   00001111B
```

Unlike data directives such as DB, DW, and so on, EQU does not assign any memory storage; therefore, it can be defined at any time and at any place, and can even be used within the code segment.

EVEN

The EVEN directive forces memory allocation to start at an even address. This is useful due to the fact that in 8086 and 286 microprocessors, accessing a 2-byte operand located at an odd address takes extra time. The use of the EVEN directive directs the assembler to assign an even address to the variable.

```
            ORG     0020H
DATA_1              DB          34H
            EVEN
DATA_2              DW          7F5BH
```

The following shows the contents of memory locations:

```
DS:0020 = (34)
DS:0021 = (? )
DS:0022 = (5B)
DS:0023 = (7F)
```

Notice that the EVEN directive caused memory location DS:0021 to be bypassed, and the value for DATA_2 is placed in memory starting with an even address.

EXTRN (External)

The EXTRN directive is used to indicate that certain variables and names used in a module are defined by another module. In the absence of the EXTRN directive, the assembler would search for the definition and give an error when it couldn't find it. The format of this directive is

```
            EXTRN   name1:typea [ ,name2:typeb]
```

where type will be NEAR or FAR if name refers to a procedure, or will be BYTE, WORD, DWORD, QWORD, TBYTE if name refers to a data variable.

```
;from the main program:
            EXTRN   PROG1:NEAR
            PUBLIC DATA1
            ...
MAIN_PRO    MOV     AX,DATASG           ;THE ENTRY POINT
            MOV     DS,AX
            ...
            CALL    PROG1
            ...
            END     MAIN_PRO            ;THE EXIT POINT

;PROG1 is located in a different file:
            EXTRN DATA1:WORD
            PUBLIC PROG1
PROG1       PROC
            ...
            MOV     BX,DATA1
            ...
            RET                         ;RETURN TO THE MAIN MODULE
PROG1       ENDP
            END
```

Notice that the EXTRN directive is used in the main procedure to identify PROG1 as a NEAR procedure. This is needed because PROG1 is not defined in that module. Correspondingly, PROG1 is defined as PUBLIC in the module where it is defined. EXTRN is used in the PROG1 module to declare that operand DATA1, of size WORD, has been defined in another module. Correspondingly, DATA1 is declared as PUBLIC in the calling module.

GROUP

The GROUP directive causes the named segments to be linked into the same 64K-byte segment. All segments listed in the GROUP directive must fit into 64K bytes. This can be used to combine segments of the same type, or different classes of segments. An example follows:

```
SMALL_SYS   GROUP       DTSEG,STSEG,CDSEG
```

The ASSUME directive must be changed to make the segment registers point to the group:

```
ASSUME      CS:SMALL_SYS,DS:SMALL_SYS,SS:SMALL_SYS
```

The group will be listed in the list file, as shown below:

```
Segments and Groups:

Name                        Length      Align       Combine Class

SMALL_SYS . . . . . . .             GROUP
  STSEG . . . . . .         0040        PARA        NONE
  DTSEG . . . . . .         0024        PARA        NONE
  CDSEG . . . . . .         005A        PARA        NONE
```

INCLUDE

When there is a group of macros written and saved in a separate file, the INCLUDE directive can be used to bring them into another file. In the program listing (.lst file), these macros will be identified by the symbol "C" (or "+" in some versions of MASM) before each instruction to indicate that they are copied to the present file by the INCLUDE directive.

LABEL

The LABEL directive allows a given variable or name to be referred to by multiple names. This is often used for multiple definitions of the same variable or name. The format of the LABEL directive is

```
name  LABEL type
```

where type may be BYTE, WORD, DWORD, or QWORD. For example, a variable of name DATA1 is defined as a word and also needs to be accessed as 2 bytes, as shown in the following:

```
DATA_B      LABEL BYTE
DATA1       DW          25F6H

            MOV   AX,DATA1          ;AX=25F6H
            MOV   BL,DATA_B         ;BL=F6H
            MOV   BH,DATA_B +1            ;BH=25H
```

The following shows the LABEL directive being used to allow accessing a 32-bit data item in 16-bit portions.

```
DATA_16     LABEL WORD
DATDD_4     DD          4387983FH
            ...
            MOV   AX,DATA_16        ;AX=983FH
            MOV   DX,DATA_16 + 2    ;DX=4387H
```

The following shows its use in a JMP instruction to go to a different code segment.

```
            ....
            JMP   PROG_A
            ....
PROG_A      LABEL FAR
```

PUBLIC

To inform the assembler that a name or symbol will be referenced by other modules, it is marked by the PUBLIC directive. If a module is referencing a variable outside itself, that variable must be declared as EXTRN. Correspondingly, in the module where the variable is defined, that variable must be declared as PUBLIC in order to allow it to be referenced by other modules. See the EXTRN directive for examples of the use of both EXTRN and PUBLIC.

SEG (Segment Address)

The SEG operator is used to access the address of the segment where the name has been defined.

```
DATA1        DW          2341H
             ...
             MOV   AX,SEG DATA1       ;AX=SEGMENT ADDRESS OF DATA1
```

This is in contrast to the OFFSET directive, which accesses the offset address instead of the segment.

SEGMENT and ENDS

In full segment definition these two directives are used to indicate the beginning and the end of the segment. They must have the same name for a given segment definition. See the following example:

```
DATSEG       SEGMENT
DATA1        DB          2FH
DATA2        DW          1200
DATA3        DD          99999999H
DATSEG       ENDS
```

There are several options associated with the SEGMENT directive, as follows:

```
name1 SEGMENT  [ align] [ combine] [ class]

name1 ENDS
```

ALIGNMENT: When several assembled modules are linked together, this indicates where the segment is to begin. There are many options, including PARA (paragraph = 16 bytes), WORD, and BYTE. If PARA is chosen, the segment starts at a hex address divisible by 10H. PARA is the default alignment. In this alignment, if a segment for a module finished at 00024H, the next segment will start at address 00030H, leaving from 00025 to 0002F unused. If WORD is chosen, the segment is forced to start at a word boundary. In BYTE alignment, the segment starts at the next byte and no memory is wasted. There is also the PAGE option, which aligns segments along the 100H (256) byte boundary. While all these options are supported by many assemblers, such as MASM and TASM, there is another option supported only by assemblers that allow system development. This option is AT. The AT option allows the program to assign a physical address. For example, to burn a program into ROM starting at physical address F0000, code

```
ROM_CODE     SEGMENT     AT F000H
```

Due to the fact that option AT allows the programmer to specify a physical address that conflicts with DOS's memory management responsibility, many assemblers such as MASM will not allow option AT.

COMBINE TYPE: This option is used to merge together all the similar segments to create one large segment. Among the options widely used are PUBLIC and STACK. PUBLIC is widely used in code segment definitions when linking more than one module. This will consolidate all the code segments of the various modules into one large code segment. If there is only one data segment and that belongs to the main module, there is no need to define it as PUBLIC since no other module has any data segment to combine with. However, if other modules have their own data segments, it is recommended that they be made PUBLIC to create a single data segment when they are linked. In the absence of that, the linker would assume that each segment is private and they would not be combined with other similar segments (codes with codes and data with data). Since there is only one stack segment, which belongs to the main module, there is no need to define it as PUBLIC. The STACK option is used only with the stack segment definition and indicates to the linker that it should combine the user's defined stack with the system stack to create a single stack for the entire program. This is the stack that is used at run time (when the CPU is actually executing the program).

CLASS NAME: Indicates to the linker that all segments of the same class should be placed next to each other by the LINKER. Four class names commonly used are 'CODE', 'DATA', 'STACK', and 'EXTRA'. When this attribute is used in the segment definition, it must be enclosed in single quotes in order to be recognized by the linker.

SHORT

In a direct jump such as "JMP POINT_A", the assembler has to choose either the 2-byte or 3-byte format. In the 2-byte format, one byte is the opcode and the second byte is the signed number displacement value added to the IP of the instruction immediately following the JMP. This displacement can be anywhere between –128 and +127. A negative number indicates a backward JMP and a positive number a forward JMP. In the 3-byte format the first byte is the opcode and the next two bytes are for the signed number displacement value, which can range from –32,768 to 32,767. When assembling a program, the assembler makes two passes through the program. Certain tasks are done in the first pass and others are left to the second pass to complete. In the first pass the assembler chooses the 3-byte code for the JMP. After the first pass is complete, it will know the target address and fill it in during the second pass. If the target address indicates a short jump (less than 128) bytes away, it fills the last byte with NOP. To inform the assembler that the target address is no more than 128 bytes away, the SHORT directive can be used. Using the SHORT directive makes sure that the JMP is a 2-byte instruction and not 3-byte with 1 byte as NOP code. The 2-byte JMP requires 1 byte less memory and is executed faster.

SIZE

The size operator returns the total number of bytes occupied by a name. The three directives LENGTH, SIZE, and TYPE are somewhat related. Below is a description of each one using the following set of data defined in a data segment:

```
DATA1 DQ            ?
DATA2 DW            ?
DATA3 DB            20 DUP (?)
DATA4 DW            100 DUP (?)
DATA5 DD            10 DUP (?)
```

TYPE allows one to know the storage allocation directive for a given variable by providing the number of bytes according to the following table:

```
bytes
1           DB
2           DW
4           DD
8           DQ
10          DT
```

For example:

```
      MOV   BX,TYPE DATA2      ;BX=2
      MOV   DX,TYPE DATA1      ;DX=8
      MOV   AX,TYPE DATA3      ;AX=1
      MOV   CX,TYPE DATA5      ;CX=4
```

When a DUP is used to define the number of entries for a given variable, the LENGTH directive can be used to get that number.

```
MOV   CX,LENGTH DATA4    ;CX=64H      (100 DECIMAL)
MOV   AX,LENGTH DATA3    ;AX=14H      (20 DECIMAL)
MOV   DX,LENGTH DATA5    ;DX=0A       (10 DECIMAL)
```

If the defined variable does not have any DUP in it, the LENGTH is assumed to be 1.

```
MOV   BX,LENGTH DATA1    ;BX=1
```

SIZE is used to determine the total number of bytes allocated for a variable that has been defined with the DUP directive. In reality the SIZE directive basically provides the product of the TYPE times the LENGTH.

```
MOV   DX, SIZE DATA4     ;DX=C8H=200 (100 x 2=200)
MOV   CX, SIZE DATA5     ;CX=28H=40 (4 x 10=40)
```

STRUC (Structure)

The STRUC directive indicates the beginning of a structure definition. It ends with an ENDS directive, whose label matches the STRUC label. Although the same mnemonic ENDS is used for end of segment and end of structure, the assembler knows which is meant by the context. A structure is a collection of data types that can be accessed either collectively by the structure name or individually by the labels of the data types within the structure. A structure type must first be defined and then variables in the data segment may be allocated as that structure type. Looking at the following example, the data directives between STRUC and ENDS declare what structure ASC_AREA looks like. No memory is allocated for such a structure definition. Immediately below the structure definition is the label ASC_INPUT, which is declared to be of type ASC_AREA. Memory is allocated for the variable ASC_INPUT. Notice in the code segment that ASC_INPUT can be accessed either in its entirety or by its component parts. It is accessed as a whole unit in "MOV DX,OFFSET ASC_INPUT". Its component parts are accessed by the variable name followed by a period, then the component's name. For example, "MOV BL,ASC_INPUT.ACT_LEN" accesses the actual length field of ASC_INPUT.

```
;from the data segment:
ASC_AREA          STRUC                   ;defines struc for string
input
MAX_LEN           DB     6        ;  maximum length of input string
ACT_LEN           DB     ?        ;  actual length of input string
ASC_NUM           DB     6 DUP (?)    ;  input string
ASC_AREA          ENDS            ;end struc definition
ASC_INPUT   ASC_AREA     <>       ;allocates memory for struc

;from the code segment:
                  ...
GET_ASC:          MOV    AH,0AH
                  MOV    DX,OFFSET ASC_INPUT
                  INT         21H
                  ...
      MOV   SI,OFFSET ASC_INPUT.ASC_NUM   ;SI points to ASCII num
      MOV   BL,ASC_INPUT.ACT_LEN          ;BL holds string length
      ...
```

TITLE

The TITLE directive instructs the assembler to print the title of the program on top of each page of the ".lst" file. What comes after the TITLE pseudo-instruction is up to the programmer, but it is common practice to put the name of the program as stored on the disk right after the TITLE pseudo-instruction and then a brief description of the function of the program. Whatever is placed after the TITLE pseudo-instruction cannot be more than 60 ASCII characters (letters, numbers, spaces, punctuation).

TYPE

The TYPE operator returns the number of bytes reserved for the named data object. See the SIZE directive for examples of its use.

SECTION C.2: RULES FOR LABELS AND RESERVED NAMES

Labels in 80x86 Assembly language for MASM 5.1 and higher must follow these rules:

1. Names can be composed of:
   alphabetic characters: A–Z and a–z
   digits: 0–9
   special characters: "?" "." "@" "_" "$"

2. Names must begin with an alphabetic or special character. Names cannot begin with a digit.

3. Names can be up to 31 characters long.

4. The special character "." can only be used as the first character.

5. Uppercase and lowercase are treated the same. "NAME1" is treated the same as "Name1" and "name1".

Assembly language programs have five types of labels or names:

1. Code labels, which give symbolic names to instructions so that other instructions (such as jumps) may refer to them

2. Procedure labels, which assign a name to a procedure

3. Segment labels, which assign a name to a segment

4. Data labels, which give names to data items

5. Labels created with the LABEL directive

Code labels

These labels will be followed by a colon and have the type NEAR. This enables other instructions within the code segment to refer to the instruction. The labels can be on the same line as the instruction:

```
        ...
ADD_LP: ADD   AL,[ BX]    ;label is on same line as the instruction
        ...
        ...
        LOOP ADD_LP
```

or on a line by themselves:

```
        ...
ADD_LP:                   ;label is on a line by itself
```

```
ADD   AL,[BX]      ;ADD_LP refers to this instruction
...
...
LOOP ADD_LP
```

Procedure labels

These labels assign a symbolic name to a procedure. The label can be NEAR or FAR. When using full segment definition, the default type is NEAR. When using simplified segment definition, the type will be NEAR for compact or small models but will be FAR for medium, large, and huge models. For more information on procedures, see PROC in Section C.1.

Segment labels

These labels give symbolic names to segments. The name must be the same in the SEGMENT and ENDS directives. See SEGMENT in Section C.1 for more information. Example:

```
DAT_SG      SEGMENT
SUM         DW          ?
DAT_SG      ENDS
```

Data labels

These labels give symbolic names to data items. This allows them to be accessed by instructions. Directives DB, DW, DD, DQ, and DT are used to allocate data. Examples:

```
DATA1       DB          43H
DATA2       DB          F2H
SUM         DW          ?
```

Labels defined with the LABEL directive

The LABEL directive can be used to redefine a label. See LABEL in Section C.1 for more information.

Reserved Names

The following is a list of reserved words in 80x86 Assembly language programming. These words cannot be used as user-defined labels or variable names.

Register Names:

```
AH   AL   AX   BH   BL   BP   BX   CH   CL   CS   CX   DH
DI   DL   DS   DX   ES   SI   SP   SS
```

Instructions:

```
AAA      AAD      AAM      AAS      ADC      ADD
AND      CALL     CBW      CLC      CLD      CLI
CMC      CMP      CMPS     CWD      DAA      DAS
DEC      DIV      ESC      HLT      IDIV     IMUL
IN       INC      INT      INTO     IRET     JA
JAE      JB       JBE      JCXZ     JE       JG
JGE      JL       JLE      JMP      JNA      JNAE
JNB      JNBE     JNE      JNG      JNGE     JNL
JNLE     JNO      JNP      JNS      JNZ      JO
JP       JPE      JPO      JS       JZ       LAHF
LDS      LEA      LES      LOCK     LODS     LOOP
LOOPE    LOOPNE   LOOPNZ   LOOPZ    MOV      MOVS
MUL      NEG      NIL      NOP      NOT      OR
OUT      POP      POPF     PUSH     PUSHF    RCL
RCR      REP      REPE     REPNE    REPNZ    REPZ
```

RET	ROL	ROR	SAHF	SAL	SAR
SBB	SCAS	SHL	SHR	STC	STD
STI	STOS	SUB	TEST	WAIT	XCHG
XLAT	XOR				

Assembler operators and directives:

$ * + - . / = ? [ ]

ALIGN	ASSUME	BYTE	COMM	COMMENT	DB
DD	DF	DOSSEG	DQ	DS	DT
DW	DWORD	DUP	ELSE	END	ENDIF
ENDM	ENDS	EQ	EQU	EVEN	EXITM
EXTRN	FAR	FWORD	GE	GROUP	GT
HIGH	IF	IFB	IFDEF	IFDIF	IFE
IFIDN	IFNB	IFNDEF	IF1	IF2	INCLUDE
INCLUDELIB	IRP	IRPC	LABEL	LE	LENGTH
LINE	LOCAL	LOW	LT	MACRO	MASK
MOD	NAME	NE	NEAR	NOTHING	OFFSET
ORG	PAGE	PROC	PTR	PUBLIC	PURGE
QWORD	RECORD	REPT	REPTRD	SEG	SEGMENT
SHORT	SIZE	STACK	STRUC	SUBTTL	TBYTE
THIS	TITLE	TYPE	WIDTH	WORD	.186
.286	.286P	.287	.386	.386P	.387
.8086	.8087	.ALPHA	.CODE	.CONST	.CREF
.DATA	.DATA?	.ERR	.ERR1	.ERR2	.ERRB
.ERRDEF	.ERRDIF	.ERRE	.ERRIDN	.ERRNB	.ERRNDEF
.ERRNZ	.FARDATA	.FARDATA?	.LALL	.LFCOND	.LIST
.MODEL	%OUT	.RADIX	.SALL	.SEQ	.SFCOND
.STACK	.TFCOND	.TYPE	.XALL	.XCREF	.XLIST

APPENDIX D

INTERRUPT CALLS AND LEGACY SOFTWARE

OVERVIEW

This appendix lists many of the interrupt calls for INT 21H, INT 10H, and so on, which are used primarily for input, output, and file and memory management.

SECTION D.1: 21H INTERRUPTS

AH Function of INT 21H

00 Terminate the program

Additional Call Registers	Result Registers
CS = segment address of PSP (program segment prefix)	None

Note: Files should be closed previously or data may be lost.

01 Keyboard input with echo

Additional Call Registers	Result Registers
None	AL = input character

Note: Checks for Ctrl-Break.

02 Output character to monitor

Additional Call Registers	Result Registers
DL = character to be displayed	None

03 Asynchronous input from auxiliary device (serial device)

Additional Call Registers	Result Registers
None	AL = input character

04 Asynchronous character output

Additional Call Registers	Result Registers
DL = character to be output	None

05 Output character to printer

Additional Call Registers	Result Registers
DL = character to be printed	None

06 Console I/O

Additional Call Registers	Result Registers
DL = OFFH if input or character to be displayed, if output	AL = 0H if no character available = character that was input, if input successful

Note: If input, ZF is cleared and AL will have the character. ZF is set if input and no character was available.

AH Function

07 Keyboard input without echo

Additional Call Registers	Result Registers
None	AL = input character

Note: Does not check for Ctrl-Break.

08 Keyboard input without echo

Additional Call Registers	Result Registers
None	AL = input character

Note: Checks for Ctrl-Break.

09 String output

Additional Call Registers	Result Registers
DS:DX = string address	None

Note: Displays characters beginning at address until a '$' (ASCII 36) is encountered.

0A String input

Additional Call Registers	Result Registers
DS:DX = address at which to store string	None

Note: Specify the maximum size of the string in byte 1 of the buffer. DOS will place the actual size of the string in byte 2. The string begins in byte 3.

0B Get keyboard status

Additional Call Registers	Result Registers
None	AL = 00 if no character waiting = 0FFH if character waiting

Note: Checks for Ctrl-Break.

0C Reset input buffer and call keyboard input function

Additional Call Registers	Result Registers
AL = keyboard function number 01H, 06H, 07H, 08H or 0AH	None

Note: This function waits until a character is typed in.

AH Function

0D Reset disk

Additional Call Registers	Result Registers
None	None

Note: Flushes DOS file buffers but does not close files.

0E Set default drive

Additional Call Registers	Result Registers
DL = code for drive (0 = A, 1 = B, 2 = C, etc.)	AL = number of logical drives in system

0F Open file

Additional Call Registers	Result Registers
DS:DX = address of FCB	AL = 00 if successful = 0FFH if file not found

Note: Searches current directory for file. If found, FCB is filled.

SECTION D.2: MOUSE INTERRUPTS 33H

AX Function of INT 33H

00 Initialize the mouse

Additional Call Registers	Result Registers
None	AX = 0H if mouse not available = FFFFH if mouse available BX = number of mouse buttons

Note: This function is called only once to initialize the mouse. If mouse support is present, AX = FFFFH, and the mouse driver is initialized, the mouse pointer is set to the center of the screen and concealed.

01 Display mouse pointer

Additional Call Registers	Result Registers
None	None

Note: This function displays the mouse pointer and cancels any exclusion area.

02 Conceal mouse pointer

Additional Call Registers	Result Registers
None	None

Note: This function hides the mouse pointer but the mouse driver monitors its position. Most programs issue this command before they terminate.

AX Function

03 Get mouse location and button status

Additional Call Registers	Result Registers
None	BX = mouse button status bit 0 -- left button bit 1 -- right button bit 2 -- center button = 0 if up; = 1 if down CX = horizontal position DX = vertical position

Note: The horizontal and vertical coordinates are returned in pixels.

04 Set mouse pointer location

Additional Call Registers	Result Registers
CX = horizontal position DX = vertical position	None

Note: The horizontal and vertical coordinates are in pixels. Will display the mouse pointer only within set limits; will not display in exclusion areas.

05 Get button press information

Additional Call Registers	Result Registers
BX = button: 0 for left; 1 for right; 2 for center	AX = button status bit 0 -- left button bit 1 -- right button bit 2 -- center button = 0 if up; = 1 if down BX = button press count CX = horizontal position DX = vertical position

Note: This returns the status of all buttons as well as the number of presses for the button indicated in BX when called. The position of the mouse pointer is given in pixels and represents the position at the last button press.

AX Function

06 Get button release information

Additional Call Registers	Result Registers
BX = button: 0 for left; 1 for right; 2 for center	AX = button status bit 0 -- left button bit 1 -- right button bit 2 -- center button = 0 if up; = 1 if down BX = button release count CX = horizontal position DX = vertical position

Note: This returns the status of all buttons as well as the number of releases for the button indicated in BX when called. The position of the mouse pointer is given in pixels and represents the position at the last button release.

07 Set horizontal limits for mouse pointer

Additional Call Registers	Result Registers
CX = minimum horizontal position DX = maximum horizontal position	None

Note: This sets the horizontal limits (in pixels) for the mouse pointer. After this call, the mouse will be displayed within these limits.

08 Set vertical limits for mouse pointer

Additional Call Registers	Result Registers
CX = minimum vertical position DX = maximum vertical position	None

Note: This sets the vertical limits (in pixels) for the mouse pointer. After this call, the mouse will be displayed within these limits.

10 Set mouse pointer exclusion area

Additional Call Registers	Result Registers
CX = upper left horizontal coordinate DX = upper left vertical coordinate SI = lower right horizontal coordinate DI = lower right vertical coordinate	None

Note: This defines an area in which the mouse pointer will not display. An exclusion area can be cancelled by calling functions 00 or 01.

AX Function

24 Get mouse information

Additional Call Registers	Result Registers
None	BH = major version BL = minor version CH = mouse type CL = IRQ number

Note: This returns the version number (e.g., version 7.5: BH = 7, BL = 5). Mouse type: 1 for bus; 2 for serial; 3 for InPort; 4 for PS/2; 5 for HP; IRQ = 0 for PS/2; otherwise = 2, 3, 4, 5 or 7.

SECTION D.3: INT 10H

AH Function

00 Set video mode

Additional Call Registers	Result Registers
AL = video mode	None

See Table D-2 for a list of available video modes and their definitions.

01 Set cursor type

Additional Call Registers	Result Registers
CH = beginning line of cursor (bits 0–4) CL = ending line of cursor (bits 0–4)	None

Note: All other bits should be set to zero. The blinking of the cursor is hardware controlled.

02 Set cursor position

Additional Call Registers	Result Registers
BH = page number DH = row DL = column	None

Note: When using graphics modes, BH must be set to zero. Text coordinates of the upper left-hand corner will be (0,0).

AH	Function

03 Read cursor position and size

Additional Call Registers	Result Registers
BH = page number	CH = beginning line of cursor
	CL = ending line of cursor
	DH = row
	DL = column

Note: When using graphics modes, BH must be set to zero.

04 Read light pen position

Additional Call Registers	Result Registers
None	AH = 0 if light pen not triggered
	= 1 if light pen triggered
	BX = pixel column
	CH = pixel row (modes 04H–06H)
	CX = pixel row (modes 0DH–13H)
	DH = character row
	DL = character column

05 Select active display page

Additional Call Registers	Result Registers
AL = page number (see Table D-1 below)	None

Table D-1: Display Pages for Different Modes and Adapters

Mode G	Pages	Adapters
00H	0–7	VGA
01H	0–7	VGA
02H	0–3	CGA
	0–7	VGA
03H	0–3	CGA
	0–7	VGA
07H	0–7	VGA
0DH	0–7	VGA
0EH	0–3	VGA
0FH	0–1	VGA
10H	0–1	VGA

All other mode-adapter combinations support only one page.

AH Function

06 Scroll window up

Additional Call Registers	Result Registers
AL = number of lines to scroll	None
BH = display attribute	
CH = *y* coordinate of top left	
CL = *x* coordinate of top left	
DH = *y* coordinate of lower right	
DL = *x* coordinate of lower right	

Note: If AL = 0, the entire window is blank. Otherwise, the screen will be scrolled upward by the number of lines in AL. Lines scrolling off the top of the screen are lost, and blank lines are scrolled in at the bottom according to the attribute in BH.

07 Scroll window down

Additional Call Registers	Result Registers
AL = number of lines to scroll	None
BH = display attribute	
CH = *y* coordinate of top left	
CL = *x* coordinate of top left	
DH = *y* coordinate of lower right	
DL = *x* coordinate of lower right	

Note: If AL = 0, the entire window is blank. Otherwise, the screen will be scrolled down by the number of lines in AL. Lines scrolling off the bottom of the screen are lost, and blank lines are scrolled in at the top according to the attribute in BH.

08 Read character and attribute at cursor position

Additional Call Registers	Result Registers
BH = display page	
AH = attribute byte	
AL = ASCII character code	

09 Write character and attribute at cursor position

Additional Call Registers	Result Registers
AL = ASCII character code	None
BH = display page	
BL = attribute	
CX = number of characters to write	

Note: Does not update cursor position. Use interrupt 10 Function 2 to set cursor position.

AH Function

0A Write character at cursor position

Additional Call Registers	Result Registers
AL = ASCII character code BH = display page BL = graphic color CX = number of characters to write	None

Note: Writes character(s) using existing video attribute. Does not update cursor position. Use interrupt 10 Function 2 to set cursor position.

0B Set color palette

Additional Call Registers	Result Registers
BH = 00H to set border or background colors = 01H to set palette BL = palette/color	None

Note: If BH = 00H and in text mode, this function will set the border color only. If BH = 00H and in graphics mode, this function will set background and border colors. If BH = 01H, this function will select the palette. In 320 × 200 four-color graphics, palettes 0 and 1 are available:

Pixel Colors for Palettes 0 and 1

Pixel	Palette 0	Palette 1
0	background	background
1	green	cyan
2	red	magenta
3	brown/yellow	white

0C Write pixel

Additional Call Registers	Result Registers
AL = pixel value CX = pixel column DX = pixel row BH = page	None

Note: Coordinates and pixel value depend on the current video mode. Setting bit 7 of AL causes the pixel value in AL to be XORed with the current value of the pixel.

0D Read pixel

Additional Call Registers	Result Registers
CX = pixel column DX = pixel row BH = page	AL = pixel value

0E TTY character output

Additional Call Registers	Result Registers
AL = character BH = page BL = foreground color	None

Note: Writes a character to the display and updates cursor position. TTY mode indicates minimal character processing. ASCII codes for bell, backspace, linefeed, and carriage return are translated into the appropriate actions.

AH AL Function

0F XX Get video mode

Additional Call Registers	Result Registers
None	AH = width of screen in characters AL = video mode BH = active display page

Note: See Table D-2 for a list of possible video modes.

10 00 Subfunction 00H: set palette register to color correspondence

Additional Call Registers	Result Registers
AL = 00H BH = color CL = palette register (00H to 0FH)	None

10 01 Subfunction 01H: set border color

Additional Call Registers	Result Registers
AL = 01H BH = border color	None

10 02 Subfunction 02H: set palette and border

Additional Call Registers	Result Registers
AL = 02H ES:DX = address of color list	None

13 Write string

Additional Call Registers	Result Registers
AL = write mode = 00H, attribute in BL, cursor not moved = 01H, attribute in BL, cursor moved = 02H, attributes follow char. cursor not moved = 03H, attributes follow char. cursor moved ES:BP = address of string CX = character count DH = initial row position DL = initial column position BH = page	None

Note: For AL = 00 and 01, the string consists of characters only, which will all be displayed with the attribute in BL. For AL = 02 and 03, the data is stored with the attributes (char, attrib, char, attrib, and so on).

Table D-2: Video Modes and Their Definition

AL	Pixels	Characters	Char box	Text/ graph	Colors	Adapter	Max pages	Buffer start
00H	320 × 200	40 × 25	8 × 8	text	16 *	CGA	8	B8000h
	320 × 350	40 × 25	8 × 14	text	16 *	EGA	8	B8000h
	360 × 400	40 × 25	9 × 16	text	16 *	VGA	8	B8000h
	320 × 400	40 × 25	8 × 16	text	16 *	MCGA	8	B8000h
01H	320 × 200	40 × 25	8 × 8	text	16	CGA	8	B8000h
	320 × 350	40 × 25	8 × 14	text	16	EGA	8	B8000h
	360 × 400	40 × 25	9 × 16	text	6	VGA	8	B8000h
	320 × 400	40 × 25	8 × 16	text	6	MCGA	8	B8000h
02H	640 × 200	80 × 25	8 × 8	text	16 *	CGA	8	B8000h
	640 × 350	80 × 25	8 × 14	text	16 *	EGA	8	B8000h
	720 × 400	80 × 25	9 × 16	text	16 *	VGA	8	B8000h
	640 × 400	80 × 25	8 × 16	text	16 *	MCGA	8	B8000h
03H	640 × 200	80 × 25	8 × 8	text	16	CGA	8	B8000h
	640 × 350	80 × 25	8 × 14	text	16	EGA	8	B8000h
	720 × 400	80 × 25	9 × 16	text	16	VGA	8	B8000h
	640 × 400	80 × 25	8 × 16	text	16	MCGA	8	B8000h
04H	320 × 200	40 × 25	8 × 8	graph	4	CGA	1	B8000h
	320 × 200	40 × 25	8 × 8	graph	4	EGA	1	B8000h
	320 × 200	40 × 25	8 × 8	graph	4	VGA	1	B8000h
	320 × 200	40 × 25	8 × 8	graph	4	MCGA	1	B8000h
05H	320 × 200	40 × 25	8 × 8	graph	4 *	CGA	1	B8000h
	320 × 200	40 × 25	8 × 8	graph	4 *	EGA	1	B8000h
	320 × 200	40 × 25	8 × 8	graph	4 *	VGA	1	B8000h
	320 × 200	40 × 25	8 × 8	graph	4 *	MCGA	1	B8000h
06H	640 × 200	80 × 25	8 × 8	graph	2	CGA	1	B8000h
	640 × 200	80 × 25	8 × 8	graph	2	EGA	1	B8000h
	640 × 200	80 × 25	8 × 8	graph	2	VGA	1	B8000h
	640 × 200	80 × 25	8 × 8	graph	2	MCGA	1	B8000h
07H	720 × 350	80 × 25	9 × 14	text	mono	MDA	8	B0000h
	720 × 350	80 × 25	9 × 14	text	mono	EGA	4	B0000h
	720 × 400	80 × 25	9 × 16	text	mono	VGA	8	B0000h
08H	reserved							
09H	reserved							
0AH	reserved							
0BH	reserved							
0CH	reserved							
0DH	320 × 200	40 × 25	8 × 8	graph	16	EGA	2/4	A0000h
	320 × 200	40 × 25	8 × 8	graph	16	VGA	8	A0000h
0EH	640 × 200	80 × 25	8 × 8	graph	16	EGA	1/2	A0000h
	640 × 200	80 × 25	8 × 8	graph	16	VGA	4	A0000h
0FH	640 × 350	80 × 25	9 × 14	graph	mono	EGA	1	A0000h
	640 × 350	80 × 25	8 × 14	graph	mono	VGA	2	A0000h
10H	640 × 350	80 × 25	8 × 14	graph	4	EGA	1/2	A0000h
	640 × 350	80 × 25	8 × 14	graph	16	VGA	2	A0000h
11H	640 × 480	80 × 30	8 × 16	graph	2	VGA	1	A0000h
	640 × 480	80 × 30	8 × 16	graph	2	MCGA	1	A0000h
12H	640 × 480	80 × 30	8 × 16	graph	16	VGA	1	A0000h
13H	320 × 200	40 × 25	8 × 8	graph	256	VGA	1	A0000h
	320 × 200	40 × 25	8 × 8	graph	256	MCGA	1	A0000h

* color burst off

SECTION D.4: INT 12H

Get conventional memory size

Call Registers	Result Registers
None	AX = memory size (KB)

Note: Returns amount of conventional memory.

SECTION D.5: INT 14H

AH Function

00 Initialize COM port

Additional Call Registers	Result Registers
AL = parameter (see below)	AH = port status (see below)
DX = port number (0 if COM1, 1 if COM2, etc.)	AL = modem status (see below)

Note 1: The parameter byte in AL is defined as follows:

```
7 6 5 4 3 2 1 0   Indicates
x x x             Baud rate (000=110, 001=150,
                  010=300, 011=600, 100=1200,
                  101=2400, 110=4800, 111=9600)
      x x         Parity (01=odd, 11=even, x0=none)
          x       Stop bits (0 = 1, 1 = 2)
            x x   Word length (10=7 bits, 11=8 bits)
```

Note 2: The port status returned in AH is defined as follows:

```
7 6 5 4 3 2 1 0   Indicates
1                 Timed-out
  1               Transmit shift register empty
    1             Transmit holding register empty
      1           Break detected
        1         Framing error detected
          1       Parity error detected
            1     Overrun error detected
              1   Received data ready
```

Note 3: The modem status returned in AL is defined as follows:

```
7 6 5 4 3 2 1 0   Indicates
1                 Received line signal detect
  1               Ring indicator
    1             DSR (data set ready)
      1           CTS (clear to send)
        1         Change in receive line signal detect
          1       Trailing edge ring indicator
            1     Change in DSR status
              1   Change in CTS status
```

AH	Function

01 Write character to COM port

Additional Call Registers	Result Registers
AL = character DX = port number (0 if COM1, 1 if COM2, etc.)	AH bit 7 = 0 if successful, 1 if not AH bits 0–6 = status if successful AL = character

Note: The status byte in AH, bits 0–6, after the call is as follows:

```
6 5 4 3 2 1 0     Indicates
1                 Transmit shift register empty
  1               Transmit holding register empty
    1             Break detected
      1           Framing error detected
        1         Parity error detected
          1       Overrun error detected
            1     Receive data ready
```

02 Read character from COM port

Additional Call Registers	Result Registers
DX = port number (0 if COM1, 1 if COM2, etc.)	AH bit 7 = 0 if successful, 1 if not AH bits 0–6 = status if successful AL = character read

Note: The status byte in AH, bits 1–4, after the call is as follows:

```
4 3 2 1     Indicates
1           Break detected
  1         Framing error detected
    1       Parity error detected
      1     Overrun error detected
```

03 Read COM port status

Additional Call Registers	Result Registers
DX = port number (0 if COM1, 1 if COM2, etc.)	AH = port status AL = modem status

Note: The port status and modem status returned in AH and AL are the same format as INT 14H function 00H, described above.

04 Extended initialize COM port

Additional Call Registers	Result Registers
AL = 00H (break), 01H (no break) DX = port number (0 if COM1, 1 if COM2, etc.) BH = parity = 00H none = 01H odd = 02H even	AH = port status (see function AH = 0) AL = modem status (see function AH = 0)

= 03H stick parity odd
= 04H stick parity even
BL = stop bits = 00H (one stop bit)
= 01H (1.5 bits for 5-bit word)
= 01H (2 bits for > 5-bit word)
CH = word length = 00H 5-bit
= 01H 6-bit
= 02H 7-bit
= 03H 8-bit
CL = baud rate = 00H 110 baud
= 01H 150 baud
= 02H 300 baud
= 03H 600 baud
= 04H 1200 baud
= 05H 2400 baud
= 06H 4800 baud
= 07H 9600 baud
= 08H 19200 baud

05 Extended COM port control

Additional Call Registers	Result Registers
AL = 00H (read control register),	If read subfunction,
= 01H (write to control register)	BL = modem control register
DX = port number (0 if COM1, 1 if COM2, etc.)	If write subfunction,
	AL = modem status
BL = Modem control register	(see Figure 9-16)
(see Figure 9-14)	AH = line status
bits 7–5: reserved	(see Figure 9-15)
bit 4: loop	
bit 3: out2	
bit 2: out1	
bit 1: RTS	
bit 0: DTR	

Note: Subfunction AL = 00H returns the modem control register contents in BL. Subfunction AL = 01H writes the contents of BL into the modem control register and returns modem and line status register contents in AL and AH.

SECTION D.6: INT 16H -- KEYBOARD

AH Function

00H Keyboard read

Additional Call Registers	Result Registers
None	AH = key scan code
	AL = ASCII char

Note: Reads one character from the keyboard buffer and updates the head pointer.

AH Function

01H Get keyboard status

Additional Call Registers	Result Registers
None	If no key waiting, ZF = 1. If key waiting, ZF = 0, AH = key scan code, AL = ASCII char.

Note: If a key is waiting, the scan code and character are returned in AH and AL, but the head pointer of the keyboard buffer is not updated.

02H Get shift status

Additional Call Registers	Result Registers
None	AL = status byte bit 7: Insert pressed bit 6: Caps Lock pressed bit 5: Num Lock pressed bit 4: Scroll Lock pressed bit 3: Alt pressed bit 2: Ctrl pressed bit 1: Left Shift pressed bit 0: Right Shift pressed

Note: The keyboard status byte returned in AL indicates whether certain keys have been pressed. If the bit = 1, the key has been pressed.

03H Set typematic rate

Additional Call Registers	Result Registers
AL = 05H BH = repeat delay (see below) BL = repeat rate (see below)	None

Note: Sets the rate at which repeated keystrokes are accepted.

The delay value in BH can be 00H (for 250), 01H (for 500), 02H (for 750), or 03H (for 1000). All values are in milliseconds. The repeat rate in BL represents the number of characters per second. Options are:

```
00H: 30.0   0BH: 10.9   16H: 4.3
01H: 26.7   0CH: 10.0   17H: 4.0
02H: 24.0   0DH: 9.2    18H: 3.7
03H: 21.8   0EH: 8.6    19H: 3.3
04H: 20.0   0FH: 8.0    1AH: 3.0
05H: 18.5   10H: 7.5    1BH: 2.7
06H: 17.1   11H: 6.7    1CH: 2.5
07H: 16.0   12H: 6.0    1DH: 2.3
08H: 15.0   13H: 5.5    1EH: 2.1
09H: 13.3   14H: 5.0    1FH: 2.0
0AH: 12.0   15H: 4.6    20H to FFH - reserved
```

AH Function

10H Extended keyboard read

Additional Call Registers	Result Registers
None	AH = key scan code AL = ASCII char

Note: Used in place of INT 16H function 00H to allow program to detect F11, F12, and other keys of the extended keyboard. After the read, the head pointer of the keyboard buffer is updated.

11H Extended keyboard status

Additional Call Registers	Result Registers
None	If no key waiting, ZF = 1. If key waiting, ZF = 0, AH = key scan code, AL = ASCII char.

Note: This function is used instead of INT 16H function 01H so that programs can detect keys of the extended keyboard such as F11 and F12. If a key is waiting, the scan code and character are returned in AH and AL, but the head pointer of the keyboard buffer is not updated.

12H Extended shift status

Additional Call Registers	Result Registers
None	AL = shift status bit 7: Insert locked bit 6: Caps Lock locked bit 5: Num Lock locked bit 4: Scroll Lock locked bit 3: Alt pressed bit 2: Ctrl pressed bit 1: Left Shift pressed bit 0: Right Shift pressed AH = extended shift status bit 7: SysRq pressed bit 6: Caps Lock pressed bit 5: Num Lock pressed bit 4: Scroll Lock pressed bit 3: Right Alt pressed bit 2: Right Ctrl pressed bit 1: Left Alt pressed bit 0: Left Ctrl pressed

Note: The keyboard status bytes returned in AL and AH indicate whether certain keys have been pressed. If the bit = 1, the key has been pressed.

SECTION D.7: INT 1AH

AH Function

00H Read system-timer time counter

Additional Call Registers	Result Registers
None	CX = high portion of count DX = low portion of count AL = 0 if 24 hours have not passed since last read > 0 if 24 hours have passed since last read

Note: This function returns the number of ticks since midnight. A second is about 18.2 ticks. When the number of ticks indicates that 24 hours have passed, AL is incremented and the tick count is reset to zero. Calling this function resets AL so that whether 24 hours have passed can only be determined once a day.

01H Set system-timer time counter

Additional Call Registers	Result Registers
CX = high portion of tick count DX = low portion of tick count	None

Note: Calling this function will cause the timer overflow flag to be reset.

02H Read real-time clock time

Additional Call Registers	Result Registers
None	CH = hours CL = minutes DH = seconds DL = 01 for daylight savings option = 00 for no option CF = 0 if clock operating, otherwise = 1

Note: Hours, minutes, and seconds are returned in BCD format. This function is used to get the time in the CMOS time/date chip.

03H Set real-time clock time

Additional Call Registers	Result Registers
CH = hours CL = minutes DH = seconds DL = 01 for daylight savings option = 00 for no option	None

Note: Hours, minutes, and seconds are in BCD format. This function is used to set the time in the CMOS time/date chip.

AH Function

04H Read real-time clock date

Additional Call Registers	Result Registers
None	CH = century (19 or 20)
	CL = year
	DH = month
	DL = day
	CF = 0 if clock operating, otherwise = 1

Note: Century, year, month, and day are in BCD format. This function is used to get the date in the CMOS time/date chip.

05H Set real-time clock date

Additional Call Registers	Result Registers
CH = century (19 or 20)	None
CL = year	
DH = month	
DL = day	

Note: Century, year, month, and day are in BCD format. This function is used to set the date in the CMOS time/date chip.

APPENDIX E

I/O ADDRESS MAPS

SECTION E.1: ORIGINAL 80286 IBM PC I/O ADDRESS MAP

Hex Range	Device
000–01F	DMA controller 1, 8237A-5
020–03F	Interrupt controller 1, 8259A, Master
040–05F	Timer, 8254-2
060–06F	8042 (keyboard)
070–07F	Real-time clock, NMI (nonmaskable interrupt) mask
080–09F	DMA page register, 74LS612
0A0–0BF	Interrupt controller 2, 8237A-5
0C0–0DF	DMA controller 2, 8237A-5
0F0	Clear math coprocessor busy
0F1	Reset math coprocessor
0F8–0FF	Math coprocessor
1F0–1F8	Fixed disk
200–207	Game I/O
20C–20D	Reserved
21F	Reserved
278–27F	Parallel printer port 2
2B0–2DF	Alternate enhanced graphics adapter
2E1	GPIB (adapter 0)
2E2 & 2E3	Data acquisition (adapter 0)
2F8–2FF	Serial port 2
300–31F	Prototype card
360–363	PC network (low address)
364–367	Reserved
368–36B	PC network (high address)
36C–36F	Reserved
378–37F	Parallel printer port 1
380–38F	SDLC, bisynchronous 2
390–393	Cluster
3A0–3AF	Bisynchronous 1
3B0–3BF	Monochrome display and printer adapter
3C0–3CF	Enhanced graphics adapter
3D0–3DF	Color/graphics monitor adapter
3F0–3F7	Disk controller
3F8–3FF	Serial port 1
6E2 & 6E3	Data acquisition (adapter 1)
790–793	Cluster (adapter 1)

AE2 & AE3	Data acquisition (adapter 2)
B90–B93	Cluster (adapter 2)
EE2 & EE3	Data acquisition (adapter 3)
1390–1393	Cluster (adapter 3)
22E1	GPIB (adapter 1)
2390–2393	Cluster (adapter 4)
42E1	GPIB (adapter 2)
62E1	GPIB (adapter 3)
82E1	GPIB (adapter 4)
A2E1	GPIB (adapter 5)
C2E1	GPIB (adapter 6)
E2E1	GPIB (adapter 7)

SECTION E.2: Dell x86 PC I/O ADDRESS MAP

Use the System Information utility in Windows to get the I/O address map for your x86 PC.

Hex Range	Device
0x00000000–0x00000CF7	PCI bus
0x00000000–0x00000CF7	Direct memory access controller
0x00000010–0x0000001F	Direct memory access controller
0x00000020–0x00000021	System board
0x00000024–0x00000025	Programmable interrupt controller
0x00000028–0x00000029	Programmable interrupt controller
0x0000002C–0x0000002D	Programmable interrupt controller
0x0000002E–0x0000002F	System board
0x00000030–0x00000031	Programmable interrupt controller
0x00000034–0x00000035	Programmable interrupt controller
0x00000038–0x00000039	Programmable interrupt controller
0x0000003C–0x0000003D	Programmable interrupt controller
0x00000040–0x00000043	System timer
0x0000004E–0x0000004F	System board
0x00000050–0x00000053	System timer
0x00000060–0x00000060	Standard 101/102-Key PS/2 Keyboard
0x00000061–0x00000061	System speaker
0x00000062–0x00000062	Standard 101/102-Key PS/2 Keyboard
0x00000063–0x00000063	System speaker
0x00000064–0x00000064	Standard 101/102-Key PS/2 Keyboard
0x00000065–0x00000065	System speaker
0x00000066–0x00000066	Standard 101/102-Key PS/2 Keyboard
0x00000067–0x00000067	System speaker
0x00000070–0x00000071	System CMOS/real-time clock
0x00000072–0x00000077	System CMOS/real-time clock
0x00000080–0x00000085	Direct memory access controller
0x00000086–0x00000086	System board
0x00000087–0x0000008F	Direct memory access controller
0x00000090–0x00000091	Direct memory access controller
0x00000092–0x00000092	System board
0x00000093–0x0000009F	Direct memory access controller
0x000000A0–0x000000A1	System board
0x000000A4–0x000000A5	Programmable interrupt controller
0x000000A8–0x000000A9	Programmable interrupt controller

0x000000AC–0x000000AD	Programmable interrupt controller
0x000000B0–0x000000B1	Programmable interrupt controller
0x000000B2–0x000000B2	System board
0x000000B3–0x000000B3	System board
0x000000B4–0x000000B5	Programmable interrupt controller
0x000000B8–0x000000B9	Programmable interrupt controller
0x000000BC–0x000000BD	Programmable interrupt controller
0x000000C0–0x000000DF	Direct memory access controller
0x000000F0–0x000000FF	Numeric data processor
0x00000170–0x00000177	Secondary IDE Channel
0x000001F0–0x000001F7	Primary IDE Channel
0x00000274–0x00000277	ISAPNP Read Data Port
0x00000279–0x00000279	ISAPNP Read Data Port
0x00000376–0x00000376	Secondary IDE Channel
0x000003B0–0x000003BB	Mobile Intel(R) 955XM/945GM/PM/GMS/940GML Express PCI Express Root Port - 27A1
0x000003B0–0x000003BB	NVIDIA Quadro FX 1500M
0x000003C0–0x000003DF	Mobile Intel(R) 955XM/945GM/PM/GMS/940GML Express PCI Express Root Port - 27A1
0x000003C0–0x000003DF	NVIDIA Quadro FX 1500M
0x000003F6–0x000003F6	Primary IDE Channel
0x000004D0–0x000004D1	System board
0x00000809–0x00000809	System board
0x00000910–0x0000091F	System board
0x00000920–0x0000092F	System board
0x00000930–0x0000097F	System board
0x00000A79–0x00000A79	ISAPNP Read Data Port
0x00000C80–0x00000CAF	System board
0x00000CB0–0x00000CBB	System board
0x00000CBC–0x00000CBF	System board
0x00000CC0–0x00000CFF	System board
0x00000D00–0x0000FFFF	PCI bus
0x00001000–0x00001005	System board
0x00001006–0x00001007	System board
0x00001008–0x0000100F	System board
0x0000100A–0x00001059	System board
0x00001010–0x0000102F	System board
0x00001060–0x0000107F	System board
0x00001080–0x000010BF	System board
0x000010C0–0x000010DF	Intel(R) 82801G (ICH7 Family) SMBus Controller - 27DA
0x000010C0–0x000010DF	System board
0x0000BF20–0x0000BF3F	Intel(R) 82801G (ICH7 Family) USB Universal Host Controller - 27CB
0x0000BF40–0x0000BF5F	Intel(R) 82801G (ICH7 Family) USB Universal Host Controller - 27CA
0x0000BF60–0x0000BF7F	Intel(R) 82801G (ICH7 Family) USB Universal Host Controller - 27C9
0x0000BF80–0x0000BF9F	Intel(R) 82801G (ICH7 Family) USB Universal Host Controller - 27C8
0x0000BFA0–0x0000BFAF	Intel(R) 82801GBM/GHM (ICH7-M Family) Serial

	ATA Storage Controller - 27C4
0x0000D000–0x0000DFFF	Intel(R) 82801G (ICH7 Family) PCI Express Root Port - 27D6
0x0000E000–0x0000EFFF	Mobile Intel(R) 955XM/945GM/PM/GMS/940GML Express PCI Express Root Port - 27A1
0x0000EF00–0x0000EF7F	NVIDIA Quadro FX 1500M
0x0000F400–0x0000F4FE	System board

APPENDIX F

ASCII CODES

Ctrl	Dec	Hex	Ch	Code
^@	0	00		NUL
^A	1	01	☺	SOH
^B	2	02	☻	STX
^C	3	03	♥	ETX
^D	4	04	♦	EOT
^E	5	05	♣	ENQ
^F	6	06	♠	ACK
^G	7	07	•	BEL
^H	8	08	◘	BS
^I	9	09	○	HT
^J	10	0A	◙	LF
^K	11	0B	♂	VT
^L	12	0C	♀	FF
^M	13	0D	♪	CR
^N	14	0E	♫	SO
^O	15	0F	☼	SI
^P	16	10	►	DLE
^Q	17	11	◄	DC1
^R	18	12	↕	DC2
^S	19	13	‼	DC3
^T	20	14	¶	DC4
^U	21	15	§	NAK
^V	22	16	▬	SYN
^W	23	17	↨	ETB
^X	24	18	↑	CAN
^Y	25	19	↓	EM
^Z	26	1A	→	SUB
^[	27	1B	←	ESC
^\	28	1C	∟	FS
^]	29	1D	↔	GS
^^	30	1E	▲	RS
^_	31	1F	▼	US

Dec	Hex	Ch
32	20	
33	21	!
34	22	"
35	23	#
36	24	$
37	25	%
38	26	&
39	27	'
40	28	(
41	29	)
42	2A	*
43	2B	+
44	2C	,
45	2D	-
46	2E	.
47	2F	/
48	30	0
49	31	1
50	32	2
51	33	3
52	34	4
53	35	5
54	36	6
55	37	7
56	38	8
57	39	9
58	3A	:
59	3B	;
60	3C	<
61	3D	=
62	3E	>
63	3F	?

Dec	Hex	Ch
64	40	@
65	41	A
66	42	B
67	43	C
68	44	D
69	45	E
70	46	F
71	47	G
72	48	H
73	49	I
74	4A	J
75	4B	K
76	4C	L
77	4D	M
78	4E	N
79	4F	O
80	50	P
81	51	Q
82	52	R
83	53	S
84	54	T
85	55	U
86	56	V
87	57	W
88	58	X
89	59	Y
90	5A	Z
91	5B	[
92	5C	\
93	5D	]
94	5E	^
95	5F	_

Dec	Hex	Ch
96	60	`
97	61	a
98	62	b
99	63	c
100	64	d
101	65	e
102	66	f
103	67	g
104	68	h
105	69	i
106	6A	j
107	6B	k
108	6C	l
109	6D	m
110	6E	n
111	6F	o
112	70	p
113	71	q
114	72	r
115	73	s
116	74	t
117	75	u
118	76	v
119	77	w
120	78	x
121	79	y
122	7A	z
123	7B	{
124	7C	¦
125	7D	}
126	7E	~
127	7F	⌂

Dec	Hex	Ch
128	80	Ç
129	81	ü
130	82	é
131	83	â
132	84	ä
133	85	à
134	86	å
135	87	ç
136	88	ê
137	89	ë
138	8A	è
139	8B	ï
140	8C	î
141	8D	ì
142	8E	Ä
143	8F	Å
144	90	É
145	91	æ
146	92	Æ
147	93	ô
148	94	ö
149	95	ò
150	96	û
151	97	ù
152	98	ÿ
153	99	Ö
154	9A	Ü
155	9B	¢
156	9C	£
157	9D	¥
158	9E	₧
159	9F	ƒ

Dec	Hex	Ch
160	A0	á
161	A1	í
162	A2	ó
163	A3	ú
164	A4	ñ
165	A5	Ñ
166	A6	ª
167	A7	º
168	A8	¿
169	A9	⌐
170	AA	¬
171	AB	½
172	AC	¼
173	AD	¡
174	AE	«
175	AF	»
176	B0	░
177	B1	▒
178	B2	▓
179	B3	│
180	B4	┤
181	B5	╡
182	B6	╢
183	B7	╖
184	B8	╕
185	B9	╣
186	BA	║
187	BB	╗
188	BC	╝
189	BD	╜
190	BE	╛
191	BF	┐

Dec	Hex	Ch
192	C0	└
193	C1	┴
194	C2	┬
195	C3	├
196	C4	─
197	C5	┼
198	C6	╞
199	C7	╟
200	C8	╚
201	C9	╔
202	CA	╩
203	CB	╦
204	CC	╠
205	CD	═
206	CE	╬
207	CF	╧
208	D0	╨
209	D1	╤
210	D2	╥
211	D3	╙
212	D4	?
213	D5	╒
214	D6	╓
215	D7	╫
216	D8	╪
217	D9	┘
218	DA	┌
219	DB	█
220	DC	▄
221	DD	▌
222	DE	▐
223	DF	▀

Dec	Hex	Ch
224	E0	α
225	E1	β
226	E2	Γ
227	E3	π
228	E4	Σ
229	E5	σ
230	E6	µ
231	E7	τ
232	E8	Φ
233	E9	Θ
234	EA	Ω
235	EB	δ
236	EC	∞
237	ED	ø
238	EE	∈
239	EF	∩
240	F0	≡
241	F1	±
242	F2	≥
243	F3	≤
244	F4	⌠
245	F5	⌡
246	F6	÷
247	F7	≈
248	F8	≈
249	F9	·
250	FA	·
251	FB	√
252	FC	^{n}
253	FD	^{2}
254	FE	■
255	FF	

INDEX

!

A

B

C

D

E

F

H

I

K

L

M

N

P

R

S

T

U

V

X